THE FLORA

OF

WEST YORKSHIRE.

THE FLORA

OF

WEST YORKSHIRE

WITH A

SKETCH OF THE CLIMATOLOGY AND LITHOLOGY IN CONNECTION THEREWITH

BY

FREDERIC ARNOLD LEES

M.R.C.S.Eng.; L.R.C.P.Lond.;

RECORDER FOR THE BOTANICAL RECORD CLUB, AND PRESIDENT OF THE BOTANICAL SECTION OF THE YORKSHIRE NATURALISTS' UNION.

WITH A NEW FOREWORD
AND BIBLIOGRAPHY OF THE AUTHOR

BY

M. R. D. SEAWARD

WHOSE END, BOTH AT THE FIRST AND NOW, WAS AND IS, TO HOLD, AS 'TWERE, THE MIRROR UP TO NATURE.'—*Hamlet*, Act iii, Sc. 2.

EP Publishing Limited
1978

This is a facsimile of the first edition of 1888 originally published by Lovell, Reeve & Co., London, as the second volume of the Botanical Series of the Transactions of the Yorkshire Naturalists' Union.

The map in the original edition was coloured and has been replaced by a black and white version.

Republished 1978 by
EP Publishing Limited
East Ardsley, Wakefield
West Yorkshire, England

ISBN 0 7158 1339 0

British Library Cataloguing in Publication Data
Lees, Frederic Arnold
The flora of West Yorkshire.
1. Botany – England – West Riding
I. Title
581.9'428'1 QK306
ISBN 0-7158-1339-0

Please address all enquiries to EP Publishing Limited
(address as above)

Printed in Great Britain by
The Scolar Press Limited
Ilkley, West Yorkshire

Plate 1. Frederic Arnold Lees (1847–1921)

FOREWORD

'. . . *monumentum aere perennius*'

Horace, *Odes*, III, xxx, i.

Frederic Arnold Lees was born at Burmantofts Hall, near Leeds, on 20 January 1847. Frederic (often misspelt Frederick) was educated at Leeds Grammar School and at Durham University, and in 1871 qualified as a Member of the Royal College of Surgeons and as Licentiate of the Royal College of Physicians. His practices took him to many parts of England, including Thorner, Hartlepool, Middleton-in-Teesdale, Dalton-in-Furness, Market Rasen, Birstall, Warrington, Wetherby (when for a time he was daily attending St. Paul's Dispensary, Leeds), Kidderminster, Reading, Hawes, Heckmondwike, Harrogate and finally Leeds. (His travels outside England are not documented, beyond a trip he made to Italy and France in 1873.) Could he resist the powerful temptation to see a new area and make new plant records? According to James Britten (1846–1924) in his obituary of Lees (*J.Bot.*, 1922, **60**: 99), "the conflicting claims of botany and his medical practice were sometimes decided, to his temporal disadvantage, in favour of the former, if we may accept the legend which tells how, crossing a moor which promised interesting botanical results, Lees went off in search of them and returned home without visiting the patient whom he started out to see". Certainly the labels of his considerable herbarium bear witness to this side of his life, for it is as a botanist, rather than as a practitioner, that he is remembered.

Lees noted in an autobiographical sketch (*Description of the Lees Herbarium and Library*, 1910, Bradford) that he owed his first instruction in botany to Rev. F. G. Fleay, an under-master at Leeds Grammar School. Fleay had a "personal flair for the naturalistic, imaginative side of things" and instituted a course of natural history teaching through which Lees developed a love of fieldwork and collecting. Lees "chose medicine as his *métier*, that career early seeming to him to be one which conferred most *power* on the individual and allowed congenial pursuance of chemic and botanic study with least dislocation of the demands of making a livelihood."

Other than the collection of plant material for the Leeds Medical School herbarium, Lees had little time for fieldwork between his leaving school, when he was first apprenticed to a Headingley surgeon, and his qualifying as a medical practitioner. The presentation of Babington's *Manual of British Botany* in 1866 for his service to the Leeds Medical School stimulated him further towards the formation of a herbarium, but not until 1872 did he "study field-botany analytically as well as lovingly". In the latter year Lees was elected a Fellow of the Linnean Society, being sponsored by John G. Baker (1834–1920) and Rev. William W. Newbould (1819–1886), the foremost Yorkshire botanists of that period. It was Newbould who on 21 June 1872 inducted Lees "as at the conferring of a degree, with mingled solemnity and formality into the art and craft of botanical observation". Newbould's notes, particularly those on the published work of earlier botanists in Yorkshire, were used extensively by Lees in the preparation of *The Flora of West Yorkshire*; the notes were incorporated in Lees' Library which is now in the possession of the Bradford Public Library. Lees' Library was acquired by the Bradford Libraries Committee in 1906; there are 535 items, mainly botanical works, which are listed in a *Catalogue of the Lees Botanical Collection in the Reference Library* published by the Bradford Public Libraries in 1909.

In 1872 Lees began to assemble a herbarium which in his words "for volume and character should be a testament to the comital distribution of plants in Britain and Ireland . . . each sheet a floral missal . . . pregnant with the data of time, place and circumstance, and which when the whole has been arranged, mounted, and (as far as possible) fully labelled, . . . should be and continue to be an evergreen although 'dried-garden' for the mind". When Lees sold his herbarium to the Bradford Corporation in 1906, it contained approximately 24,000 specimens (20,000 vascular plants and 4,000 non-vascular cryptogams; see Hartley, M. M., *Naturalist*

1977: 25–30), and he claimed that almost every British flowering plant was represented with the exception of the Ghost Orchid. The Lees herbarium was arranged in 1935 according to the 11th edition of *The London Catalogue of British Plants* (1925), but this system proved difficult to operate. In 1969 all non-Lees British phanerogamic material at Bradford was incorporated into the collection (now housed in the Cartwright Memorial Hall, Bradford), the method of storage was rationalised and the contents arranged according to Dandy's *List of British Plants* (1958); Lees' material is clearly indicated. Storage of the bryophytes and lichens too has been reorganised.The lichens have been re-determined and lists published (Seaward, M. R. D., *Naturalist* 1968: 133–135; 1970: 125–129; 1973: 35–36), and similar work on the bryophytes is progressing slowly. Lees continued to accumulate specimens, many of them supplied by correspondents and friends, after disposing of his herbarium, and these formed a second collection which was of considerable size at the time of his death. This second collection went to the British Museum (Natural History). An important collection of his bryophytes is now housed in the Lincoln Museum, and numerous other specimens are to be found in herbaria scattered throughout the British Isles, mainly as a result of distribution through the Botanical Locality Record Club.

From 1870 onwards a steady flow of publications resulted from his labours. He took over the Recordership of the Botanical Locality Record Club one year after its formation in 1873. As a highly methodical worker Lees was ideally suited for this position, which involved a detailed collation of British distributional data, not only to supplement the vice-county recording of higher plants as formulated in *Topographical Botany*, compiled by Hewett Cotterell Watson (1804–1881) and published in 1873–1874, but also to give similar treatment to other plant groups (mainly bryophytes). Until the Club ceased in 1886, Lees also edited the *Botanical Locality Record Club Reports* (reviewed in *J.Bot.*, 1883, **21**: 58–59 and 1888, **26**: 91–92), which are a model of clarity and provide a wealth of distributional data and a considerable amount of systematic information. It is a pity that this succinctness of expression was lost, for Lees' publications after about 1890 suffer, in the words of Dr W. A. Sledge, from "florid phraseology and eccentricity of style".

His main botanical interest lay in Yorkshire. His first major contribution, *West Yorkshire: An Account of its Geology, Physical Geography, Climatology and Botany*, was written in association with James W. Davis and published in 1878 (and reprinted in 1880); the work reflects Lees' interdisciplinary approach to his subject – a forerunner of the science of ecology which only achieved academic recognition this century. This publication coincided with his practice in Lincolnshire and appears to have been highly successful in terms of botanical discovery as can be gauged from his contributions to *The Naturalist* and the *Botanical Locality Record Club Reports*. His discovery of the lichen *Cetraria islandica* at Linwood Warren, Lincolnshire (where it still flourishes) and his description of the first British record of *Selinum carvifolia* (in *J.Bot.*, 1882, **20**: 129–133), discovered by his friend Rev. William Fowler (1835–1912) at Broughton in the same county (last seen there in 1931), are but two of the many notable discoveries with which he was associated at this time.

Two years of work in Lincolnshire provided the foundation for his *Botany and Outline Flora of Lincolnshire* (reviewed in *J.Bot.*, 1893, **31**: 123–124), published in White's *Directory of Lincolnshire* for 1892 and also issued as separates in the same year. Lees' meticulous recording is exemplified in his notebooks, now in the possession of the Botany School, Cambridge, the Leeds Public Library and Dr W. A. Sledge of Leeds, and in a leather-bound and interleaved copy of his *Outline Flora* (dated 1893), annotated in his neat handwriting (cf. plate 2) and containing a useful *pot-pourri* of cuttings, correspondence, photographs and herbarium specimens. An unfinished manuscript, 'A year's contribution to Lincolnshire botany, 1878', and a valuable collection of correspondence from Rev. E. A. Woodruffe-Peacock (1858–1922), the foremost Lincolnshire botanist of that period (see Seaward, M. R. D., 1971, *Lincs. Hist. & Archaeol.* **6**: 113–124), to Lees (dated 1892–1921) are also housed in the Botany School, Cambridge. The annotation of published plant lists is also a feature of Lees' recording (see Simpson, N. D., 1960, *A Bibliographical Index of the British Flora*, p.183; Bradford Public Libraries, 1909, *Catalogue of the Lees Botanical Collection in the Reference Library*, p.5, 7, 9, 11–12, 15–17, 20–23), and a copy of *The London Catalogue of British Plants* (now in the possession of Dr W. A. Sledge, see plate 3) has been suitably marked in respect of 'The Wetherby District Flora'. Additional manuscripts, annotated books (including an interleaved copy of Baines' *Flora of Yorkshire*, 1840), check-lists, catalogues and notebooks (including one of newspaper

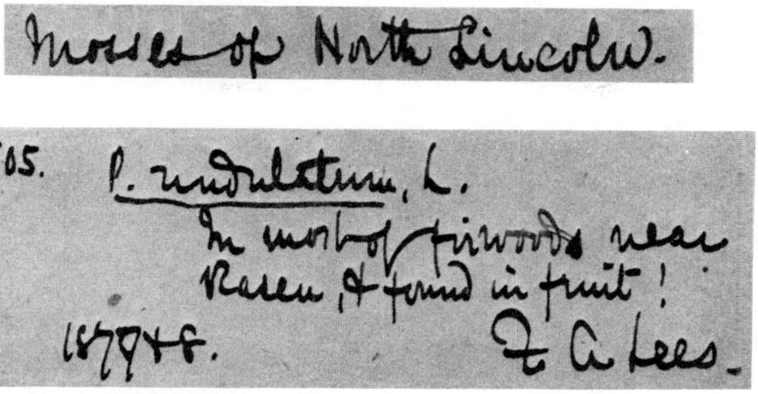

Plate 2. Lees' handwriting – herbarium label for the bryophyte *Plagiothecium undulatum* (Herb. M. R .D. Seaward)

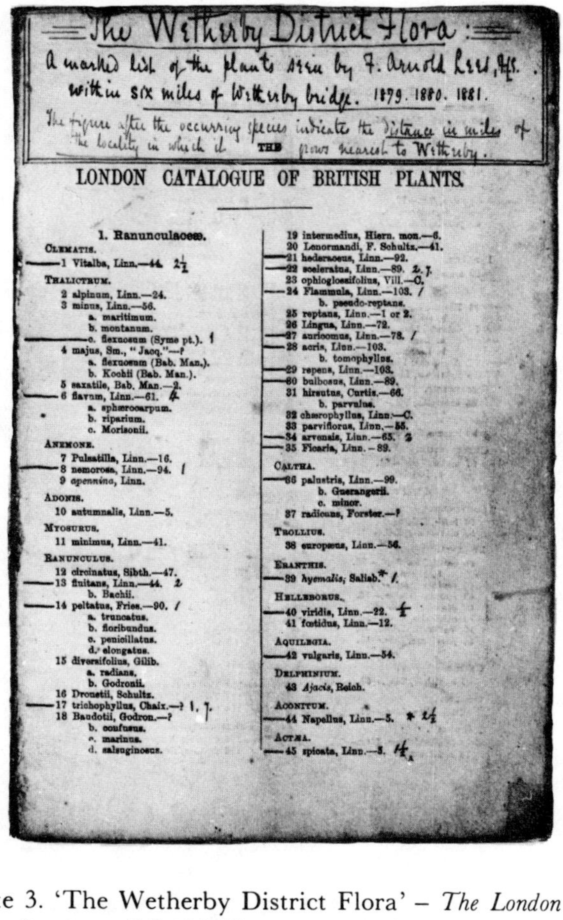
The Wetherby District Flora:
A marked list of the plants seen by F. Arnold Lees, F.L.S.
within six miles of Wetherby bridge. 1879. 1880. 1881.
The figures after the occurring species indicate the distance in miles of the locality in which it grows nearest to Wetherby.

THE

LONDON CATALOGUE OF BRITISH PLANTS.

1. Ranunculaceæ.

CLEMATIS.
1 Vitalba, Linn.—44. 2½

THALICTRUM.
2 alpinum, Linn.—24.
3 minus, Linn.—56.
a. maritimum.
b. montanum.
c. flexuosum (Syme pt.). 1
4 majus, Sm., "Jacq."—?
a. flexuosum (Bab. Man.).
b. Kochii (Bab. Man.).
5 saxatile, Bab. Man.—2.
6 flavum, Linn.—61. 4
a. sphærocarpum.
b. riparium.
c. Morisonii.

ANEMONE.
7 Pulsatilla, Linn.—16.
8 nemorosa, Linn.—94. 1
9 *apennina*, Linn.

ADONIS.
10 autumnalis, Linn.—5.

MYOSURUS.
11 minimus, Linn.—41.

RANUNCULUS.
12 circinatus, Sibth.—47.
13 fluitans, Linn.—44. 2
b. Bachii.
14 peltatus, Fries.—90. 1
a. truncatus.
b. floribundus.
c. penicillatus.
d. elongatus.
15 diversifolius, Gilib.
a. radians.
b. Godronii.
16 Drouetii, Schultz.
17 trichophyllus, Chaix.—? 1, 7.
18 Baudotii, Godron.—?
b. confusus.
c. marinus.
d. salsuginosus.
19 intermedius, Hiern. mon.—6.
20 Lenormandi, F. Schultz.—41.
21 hederaceus, Linn.—92.
22 sceleratus, Linn.—89. 2. 7.
23 ophioglossifolius, Vill.—C.
24 Flammula, Linn.—103. 1
b. pseudo-reptans.
25 reptans, Linn.—1 or 2.
26 Lingua, Linn.—72.
27 auricomus, Linn.—78. 1
28 acris, Linn.—103.
b. tomophyllus.
29 repens, Linn.—103.
30 bulbosus, Linn.—89.
31 hirsutus, Curtis.—66.
b. parvulus.
32 chærophyllus, Linn.—C.
33 parviflorus, Linn.—55.
34 arvensis, Linn.—65. 2
35 Ficaria, Linn.—89.

CALTHA.
36 palustris, Linn.—99.
b. Guerangerii.
c. minor.
37 radicans, Forster.—?

TROLLIUS.
38 europæus, Linn.—56.

ERANTHIS.
39 *hyemalis*, Salisb.* 1.

HELLEBORUS.
40 viridis, Linn.—22. ½
41 fœtidus, Linn.—12.

AQUILEGIA.
42 vulgaris, Linn.—54.

DELPHINIUM.
43 *Ajacis*, Reich.

ACONITUM.
44 Napellus, Linn.—5. * 2½

ACTÆA.
45 spicata, Linn.—3. 1½

Plate 3. 'The Wetherby District Flora' – *The London Catalogue of British Plants* annotated by Lees (Coll. W. A. Sledge)

and manuscript excerpts on the botany of Cornwall, Lincolnshire, Yorkshire, Somerset, Stirlingshire and Derbyshire for the period 1877–1902) are included in the Lees Library housed at the Bradford Public Library.

Lees' greatest botanical achievements were, however, connected with his native county. A period of intense activity in the field and poring over books and manuscripts culminated in the publication of *The Flora of West Yorkshire* in 1888; this bulky work formed the second volume of the Botanical Series of the *Transactions of the Yorkshire Naturalists' Union*. It was the most comprehensive county flora of the nineteenth century: its scope is broad in terms of the plant groups studied (viz. 997 spermatophytes, 47 pteridophytes, 456 bryophytes, 258 lichens, 1009 fungi and 394 algae), the extent (*c.* 2760 square miles) and ecological variability of the area surveyed (viz. the whole of vice-counties 63 and 64, and the south-western corner of vice-county 65), and the extensive literature search involved. Some useful features of the *Flora* are: the inclusion of local plant-names (Lees had a considerable knowledge of dialects, rare and obsolete words, etc.), distribution data based on ten drainage districts, detailed introductory matter (82 pages) on climatology and lithology and the relationship of the flora to them, and a bibliography containing 209 entries. The book received a mixed reception at the hands of the reviewers. Many praised the work, but some were less favourable, e.g. the *Grevillea* reviewer criticised the cryptogamic component, especially the out-of-date classification. The only major criticisms of one reviewer (J. Britten in *J.Bot.*, 1888, **26**: 219–222) appear to be the use of "certain eccentricities of diction" and indexes of common, rustic, family and generic, but not specific, names. Reviews are to be found in many journals and newspapers, including *Bulletin of the Torrey Botanical Club, Field, Gardener's Chronicle, Grevillea, Journal of Botany, Midland Naturalist, Nature* and *Wesley Naturalist*. Lees effectively forestalled some criticism by stating in his foreword that "in mitigation of adverse judgment, it should also be remembered, as an undoubted psychological fact, that the continuous mental strain of application necessary to a correct appreciation and orderly reference of many thousands of facts (over 40,000 localities alone, referring, variously repeated, to over 3,000 species) has a great tendency to blunt perception of relative values . . . &c".

In 1914 the *Journal of Botany* (**52**: 22–23) announced that "an entirely new work [*The Vegetation of Yorkshire* by F. A. Lees] bringing the vegetational history of the county quite up-to-date" was to be published by Brown & Sons of Hull and would be available to subscribers at 12s.6d. net. The work would embody "an analysis of the wild vegetation of England's largest and most vari-surface county" and "no source of science delving into the facts of the past" had been neglected. An entry in Lees' *Note-Book No. 9* (now housed in Leeds Public Library) for 15 April 1914 reads "finished and sent off to Hull, 'Vegetation' MSS. for 2nd time". According to Dr W. A. Sledge (*Naturalist* 1961: 148), a considerable part of the voluminous manuscript "was devoted to speculation on the 'assurgent' and 'declinent' status of each species and the ecological connotation of the title was misleading. The work was much inferior to his earlier Flora. . . . There was never any real hope of its being published in the form in which its author cast it and it became a source of embarrassment to the [Yorkshire Naturalists'] Union during his lifetime". Some important manuscripts relating to this work are now housed in the Leeds Public Library. Eventually the valuable records were extracted by C. A. Cheetham and published in *The Naturalist* (1937: 121–128, 177–184, 233–240, 289–296; 1938: 50–52, 101–108, 157–164, 213–220, 293–300, 323–324; 1939: 59–64, 113–120, 145–152, 225–232, 277–280, 321–328; 1940: 81–88, 185–192, 259–264). In 1941 this series of papers was embodied in *A Supplement to the Yorkshire Floras* edited by C. A. Cheetham and W. A. Sledge; the original title, retained in the serial instalments, had been changed to convey more accurately the scope of the work.

Autobiographical notes are to be found in *Description of the Lees Herbarium and Library* published in 1910, and further biographical and bibliographical information is to be found in: Desmond, R., 1977, *Dictionary of British and Irish Botanists and Horticulturists*, London; Gibbons, E. J., 1975, *The Flora of Lincolnshire*, Lincoln; Hawksworth, D. L. and Seaward, M. R. D., 1977, *Lichenology in the British Isles 1568–1975*, Richmond; Seaward, M. R. D., *Naturalist* 1967: 77–80, 1968: 133–135, 1970: 125–129; Simpson, N. D., 1960, *A Bibliographical Index of the British Flora*, Bournemouth; Sledge, W. A., *Naturalist* 1961: 145–150; *Yorkshire Who's Who*, 1912, London. Obituaries are to be found in the *Botanical Society Exchange Club of the British Isles Report* (1922: 358–363), *Journal of Botany* (1922, **60**: 97–100), *The Naturalist* (1921: 372–373) and the *Yorkshire Post* (19/9/1921, p.5).

Lees died at Meanwood, Leeds, on 17 September 1921. He received few honours during his lifetime – he was, however, elected an honorary member of the Yorkshire Naturalists' Union, the society to which he devoted so much of his energy and expertise. *Carex pilulifera* var. *leesii* Ridley (see *J.Bot.*, 1883, **21**: 58–59), the plant named after him, has unfortunately been sunk in taxonomic revision. Lees' greatest memorial is *The Flora of West Yorkshire* – its reprinting, ninety years after its first publication, will provide a wider audience with a wealth of botanical information, assembled through a lifetime's diligence and dedication.

SELECTED BIBLIOGRAPHY

Only Lees' major botanical works are itemised below. The list is not exhaustive: some important contributions are hard to track down owing to continual misquotation in reference lists, to reprinting of articles (often from newspapers) without due citation of source, and to material, often in the form of chapters within books of topographical interest, bearing no clear signature. A considerable number of Lees' minor published works, which include articles of zoological as well as botanical interest, obituaries, poetry, and book reviews, are to be found mainly in *The Naturalist*, *Journal of Botany* and *Hardwicke's Science-Gossip*; some of these contributions bear the abbreviated signature 'F.A.L.'. Lees' indirect but important role in compiling and editing journals, reports, catalogues, etc. and in submitting textual matter, records, etc. for others, needs further attention beyond the scope of this listing.

1872

Notes on West Riding botany. *Yorks. Nat. Recorder* no. 3: 37–39; no. 6: 93–96.

1873

The peculiarities of plant-distribution in the Leeds district. *J.Bot., Lond.* **11**: 67–72.

Notes on West Riding botany. *Yorks. Nat. Recorder* no.7: 111–112; no.10: 141–144; no.11: 160–163; no.12: 179–182; no.13: 195–196; no.14: 208–211.

1874

On the flora of the Leeds and Bradford district. *J.Bot., Lond.* **12**: 78–83.

On the flora of the Yorkshire coalfield. *J.Bot., Lond.* **12**: 144–146.

1875

Report of the Recorder for 1874. *Bot. Loc. Record Club Rep.* 1875: 33–88.

1876

Report of the Recorder for 1875. *Bot. Loc. Record Club Rep.* 1876: 89–152.

1877

Two unrecorded Lincolnshire plants. *Naturalist, Hull* **3**: 4–6.

Recent moss discoveries in West Yorkshire. *Naturalist, Hull* **3**: 18–21.

On the proposed new map of Yorkshire for the Yorkshire Naturalists' Union. *Naturalist, Hull* **3**: 51–57.

1878

(with J. W. Davis) *West Yorkshire: An Account of its Geology, Physical Geography, Climatology and Botany.* London. [Reprinted in 1880.]

A gossip about two new Lincolnshire cryptogams: (*Hypnum salebrosum* and *Cetraria islandica*). *Naturalist, Hull* **3**: 97–102.

Observations concerning three new West Yorkshire mosses. *Naturalist, Hull* **4**: 85–87.

Report of the Recorder for 1877. *Bot. Loc. Record Club Rep.* 1878: 193–258.

Quinquennial appendix: A summary of county distribution additional to 'Topographical Botany'. *Bot. Loc. Record Club Rep.* 1878: 259–307.

Summary of Comital Plant-Distribution, additional to that detailed in (Watson's) Topographical Botany. [Reprint, compiled from *Bot. Loc. Record Club Reports.*]

1879

Note upon *Hypnum salebrosum*, Hoffm., and its British distribution. *J.Bot., Lond.* **17**: 359–360.

Report of the Recorder for 1878. *Bot. Loc. Record Club Rep.* 1879: 1–41.

The autumn flora of Whernside: An account of an excursion in search of mosses. *Naturalist, Hull* **4**: 136–137.

1880

Phanerogamic report for the year 1879 and cryptogamic report for 1879 and 1880. *Bot. Loc. Record Club Rep.* 1880: 43–118.

1881

(ed.) *The London Catalogue of British Mosses and Hepatics.* 2nd ed. London.

(with W. West) A local ramble and fungus hunt. *Naturalist, Hull* **7**: 61–63.

1882

Report of Yorkshire botany for 1880, prepared for the Botanical Section of the Yorkshire Naturalists' Union. *Trans. Yorks. Nat. Un., ser. Bot.* **1**: 112–122.

Report for the year 1880. *Bot. Loc. Record Club Rep.* 1882: 119–175.

On a new British Umbellifer. *J.Bot., Lond.* **20**: 129–133.

1883

Phanerogamic report for the years 1881 and 1882. *Bot. Loc. Record Club Rep.* 1883: 177–255.

1884

Lastraea cristata near Thorne: Its discoverer and its history. *Naturalist, Hull* **9**: 164–166.

Report for the year 1883. *Bot. Loc. Record Club Rep.* 1884: 1–78.

1885

Additions to the Wensleydale flora: Notes supplementary to Baker's 'North Yorkshire'. *Naturalist, Hull* 1885: 134–138. [Reprinted in different format as *Botany and Natural History of Upper Wensleydale*, Hawes].

More additions to the Wensleydale flora. *Naturalist, Hull* 1885: 306.

Botany of the Ilkley district. In: *Ancient and Modern Ilkley* (R. Collyer and J. H. Turner), Part II, pp. xxix–lxi. Otley.

1887

Botanical notes from north-east Yorkshire. *Naturalist, Hull* 1887: 273–274.

Report for the years 1884, 1885, 1886. *Bot. Record Club Report* 1887: 79–156.

1888

The Flora of West Yorkshire. London. [As *Trans. Yorks. Nat. Un., ser. Bot.* **2**: i–xiv, 1–843].

Notes on the list of Ingleton plants. *Naturalist, Hull* 1888: 160.

A new West Yorkshire plant (*Pyrola rotundifolia*). *Naturalist, Hull* 1888: 267.

Notes on the West Yorkshire flora. *Naturalist, Hull* 1888: 303–306.

1892

Botany and outline flora of Lincolnshire. In: *History, Gazetteer and Directory of the County of Lincolnshire* (W. White, ed.), pp.37–63. Sheffield.

1894

The botany of Nidderdale and the Garden of the Nidd. In: *Nidderdale and the Garden of the Nidd: A Yorkshire Rhineland* (H. Speight), pp.17–63. London.

(with J. Farrah and A. Millward) Additions and corrections to the Nidderdale flora. *Naturalist, Hull* 1894: 55–60.

1895

A lowland sand-loving Gentian. *Naturalist, Hull* 1895: 225–228.

Further remarks anent the new London Catalogue of Plants. *Naturalist, Hull* 1895: 239–241.

1899

The florula of Bare, west Lancaster. *Naturalist, Hull* 1899: 299–303..

The Cambridge and Lincoln Selinum. *J.Bot., Lond.* **37**: 326–327.

1900

West Lancaster indigenes. *Naturalist, Hull* 1900: 3–4.

An old Leeds herbary. *Naturalist, Hull* 1900: 51–61.

Iris spuria, British? *Naturalist, Hull* 1900: 92.

In defence of James Bolton, the fungologist. *Naturalist, Hull* 1900: 225–226.

The volteface of flora: A rejoinder. *Naturalist, Hull* 1900: 229–236.

Spring's pageant in Westmorland and Lancashire. *Naturalist, Hull* 1900: 277–284.

1901

Danewort in mid-Wharfedale, Yorkshire. *Naturalist, Hull* 1901: 231.

Wetherby re-visited by the Yorkshire naturalists. *Naturalist, Hull* 1901: 325–332.

1902

The wild flowers of Elmete. In: *The Old Kingdom of Elmet: York and the Ainsty District* (E. Bogg), pp.257–262. York. [2nd ed. in 1904.]

Some flower features of the southern Ainsty. *Naturalist, Hull* 1902: 63–64.

Lathyrus ochrus: A new Yorkshire colonist. *Naturalist, Hull* 1902: 315–316.

1904

The wild garden of Grassington. In: *Picturesque Fell Land Round About Grassington, Skipton, Burnsall, and Kettlewell* (E. Bogg), pp.45–51. Leeds.

The flower chronicle of Wharfe. In: *Lower Wharfedale* (E. Bogg), pp.405–414. York.
The flower-land of higher Wharfe. In: *Higher Wharfedale: The Dale of Romance, from Ormscliffe to Cam Fell* (E. Bogg), pp.296–304. York.

1906
The wild flower gardens of Swale and Yore. In: *Richmondshire and the Vale of Mowbray* (E. Bogg), Vol.2, pp.257–268. Leeds.
The tale of Wensley's wild flowers. In: *Richmondshire and the Vale of Mowbray* (E. Bogg), Vol.2, pp.690–696. Leeds.

1908
A Tale of Wensley's Wild Flowers. Leeds.
A trio of British and alien plant-lists. *Naturalist, Hull* 1908: 311–319, 337–344.
Note on *Juncus acutus* L. *Naturalist, Hull* 1908: 423.

1909
The wild flower gardens of Swale and Yore. In: *Regal Richmond, and the Land of the Swale* (E. Bogg), pp. 257–268. London. [Cf. Vol. 2 of 1906 ed. of *Richmondshire and the Vale of Mowbray*.]
Note on *Carex sylvatica* var. *capillariformis*. *Naturalist, Hull* 1909: 350–351.

1910
Description of the Lees Herbarium and Library. Bradford. [Includes autobiographical notes.]
Note on *Helleborine atro-rubens* var. *crowtheri*. *Naturalist, Hull* 1910: 129–131.

1914
A new British and Yorkshire St. John's Wort. *Naturalist, Hull* 1914: 10.

1917
Hedge Bedstraw among stone walls. *Naturalist, Hull* 1917: 328–329.
The colonist-alien Heron-bills of Yorkshire. *Naturalist, Hull* 1917: 379–386.

1918
(ed.) A floral film of 1831 (by E. G. Bayford). *Naturalist, Hull* 1918: 89–92.
The floral sanctuary of a Meanwoodside garden. *Naturalist, Hull* 1918: 373–376.

1919
The nightmare of names in the bed of roses. *Naturalist, Hull* 1919: 211–213.
The distribution of *Gentiana verna*. *Naturalist, Hull* 1919: 390–391.

1939
(ed. A. A. Dallman.) *The Vegetation of Craven in Wharfedale, with its Adjacencies in Aire and Ribble: An Analysis of its Flora in Advent and Decline*. Arbroath. [Reprint of a series of papers first appearing in *North Western Naturalist*.]

1941
(ed. C. A. Cheetham and W. A. Sledge.) *A Supplement to the Yorkshire Floras*. London. [Compiled mainly from a series of papers first appearing in *The Naturalist*.]

I wish to express my sincere thanks to Dr W. A. Sledge for his valuable criticism of the draft manuscript, and to Mrs J. Hillman and Mr A. Henderson for their help in the preparation of the plates and text respectively.

M. R. D. SEAWARD

University of Bradford, 1977

The West Riding of Yorkshire showing botanical districts

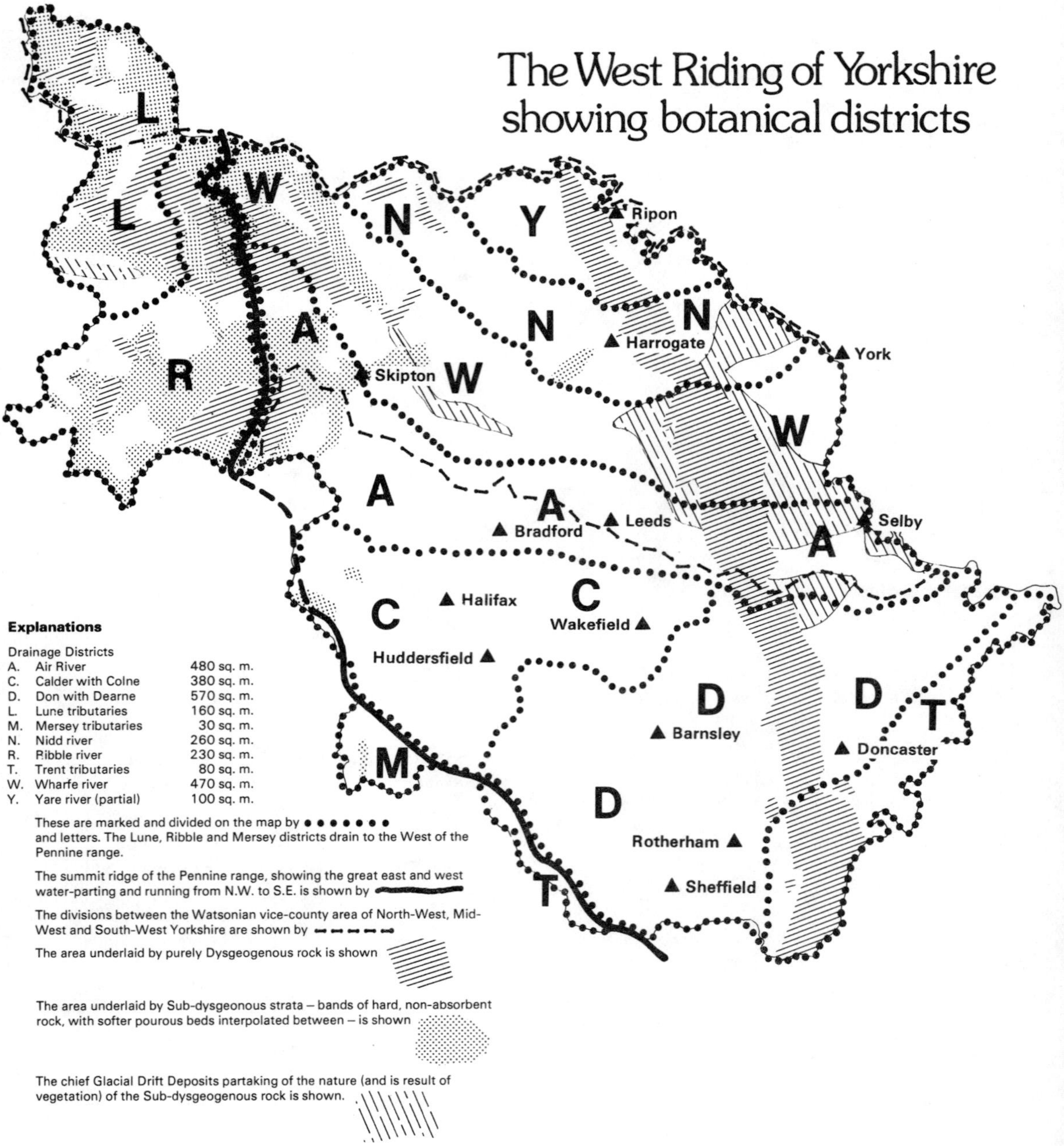

TO THE MEMORY

OF

W. WILLIAMSON NEWBOULD, M.A., F.L.S.,

THAN WHOM NO ONE DID MORE FOR

YORKSHIRE BOTANY

IN THE WIDEST AND MOST ERUDITE WAY,

THIS WORK IS INSCRIBED,

IN GRATEFUL ACKNOWLEDGMENT OF AID RECEIVED,

BY

THE AUTHOR.

'TREES FOR THE BUILDER'S YARD, FLOWERS FOR THE BRIDE'S CHAMBER, GRAIN FOR THE GRANARY, MOSS FOR THE GRAVE.'—*Ruskin.*

FOREWORD.

AFTER a long delay—due to expansion of the scheme originally projected by the inclusion of the collateral subjects, Climatology and Lithology—the Author is at last in the happy position of being able to invite the verdict of the botanical public upon a completed Work that has (truly) been to him a many years' labour of love.

The object of this work is a triple one:—(1) to sketch the climate of the West Riding of York, inclusive of the Ainsty, and show the limits imposed by its various factors upon those *floral* integers that make up the sum of its vegetation; (2) to connect the facts concerning Soils and Rocks, in respect to their behaviour under disintegration, which are in hidden relation as cause and effect, and indicate the influence exerted by their conditions upon horizontal species-distribution; and (3) to furnish botanical collectors or students with a 'Flora'—a list of the species found within the Riding, that shall be not merely a guide to the localities where they grow, but a history (with dates), so far as the area is concerned, of each one as well: briefly, to summarise the Climate, Strata, Scenery, and Vegetation, pointing out the extent of their interdependence, and indicating the rôle of each set of environments in determining the floral features of the surface.

Those who are acquainted with the rare book, 'NORTH YORKSHIRE' (by J. G. Baker, F.R.S.), still of its kind an unsurpassed work, albeit a pioneer dating back a quarter of a century, will have no difficulty in recognising the model upon which the present volume has been moulded. It is hoped that this acknowledged adoption of an admirable method of inquiry and statement, will not be regarded as too servile.

'Most can raise the flower now,
For all have got the seed'

as Tennyson says, but certainly no one can regard the praise implied in the imitation as mere flattery if they are at all acquainted with the original.

Independent of Botany technically, the Author must, first of all, state that to the secretarial staff of the Yorkshire Naturalists' Union he lies under a profound obligation. It is to the energetic initiative of Messrs. W. Eagle Clarke and W. Denison Roebuck that the Flora (laid by for some years in an incomplete state) sees the light at all when it does. During its construction there was necessarily a heavy draught on the help of Friends, too numerous for the credit due to each one to be apportioned individually, but *all* are here thanked most sincerely for their readily-proffered aid, in the shape of lists, specimen-vouchers, etc. All sought assistance will be found duly acknowledged in the section to which it relates, e.g., Mr. W. H. Pearson's collaboration in the chapter on Hepatics, Mr. H. T. Soppitt's in that on Fungi, Mr. W. West's in those of Musci, Algæ, and Cryptogams generally. In addition, grateful recognition of the services rendered by Mr. W. Denison Roebuck (to whose particular share fell the editorial supervision, which one or other of the Secretaries of the Yorkshire Naturalists' Union exercises over all its publications) is especially due: since covering the whole work, connexion of those services with any particular part would be inadequate and underfair. To his unwearied critical outlook, and the friend-like breadth of his unselfish aid, through all the delays and dangers of press-

production, the completed product owes not a little of such freedom from typographical blemishes and clerical gaucheries as it may possess. The Rev. G. P. Harris gave scholarly help on various points relating to the origin and etymology of many of the rustic, local flower-names. Whilst going through the press voluntary proof-readers have been found in Mr. J. G. Baker, Mr. John Emmet, Mr. Chas. P. Hobkirk, Dr. H. F. Parsons, Mr. P. Fox Lee, Mr. Geo. Roberts, and Mr. Wm. Whitwell, whose eye-scrutiny, in part or throughout all the sheets, has proved of great benefit as regards detection of 'slips' of all sorts. And to Mr. Whitwell's patient toil is due the preparation of the Indexes—most thankless of all literary tasks.

The Coloured Map is, perhaps, an attempt at expressing in a small space too great a variety of information, at the expense of clearness; but considerations of cost precluded alike a larger size of sheet or a duplication thereof.

In conclusion, the Author would add a few words in anticipation of such flaws as critical acumen may certainly discover in his book. No one who knows the peculiar danger of literary labour which involves at once compilation and generalisation, will expect an absolute absence of discrepancies: statements in the earlier pages may find modified expression, or even correction, in the latter parts of the book. The stational records, especially, have been brought up to date concurrently with the process of proof-revision; and some books and papers, turning up too late to be inserted in the main Bibliographical list, have been quoted onwards from the point at which they became available. In mitigation of adverse judgment, it should also be remembered, as an undoubted psychological fact, that the continuous mental strain of application necessary to a correct appreciation and orderly reference of many thousands of facts (over 40,000 localities alone, referring, variously repeated, to over 3,000 species) has a great tendency to blunt perception of relative values, leading in a certain degree to a loss of a sense of due proportion; making it inevitable that some mistakes will be

fallen into, some fallacious data allowed to pass unchallenged. Every care has been taken, however; every record of a plant name, or locality even, has been mentally scrutinised—put on its trial, so to say, as to its *possible* unveracity, by the light not merely of general, but of that special knowledge, which is conferred by an intimate acquaintance with the local features of every district, to one accustomed to the somewhat complicated problems met with, on every hand, by the student of Botanology in the side-walk of plant-distribution.

LEEDS, *December 31st, 1887.*

CONTENTS.

'He that hath a good Harvest may be content with some Thistles.'

ERRATA.

Page 4, lines 6 and 7, *for* 'Green-which' *read* Greenwich.
,, 7, line 11, *for* 'encompassed' *read* embraced.
,, 11, line 14, *after* 'themselves' *insert* or *before* 'in.'
,, 15, line 18, *for* 'little' *read* like.
,, 38, line 37, *for* 'after' *read* often.
,, 91, lines 11 and 12, *for* '1812' (as regards Hailstone's List in Whitaker's Craven) *read* 1805. The list in the 2nd ed. of Whitaker (1812) was precisely that of the 1st ed.
,, 92, line 19, *after* 'The British Flora' *add* Ed. ii. (Ed. i, 1830; ed. iii, 1835.)
,, 100, lines 15 and 19 (twice), *for* 'Wandsworth' *read* Balham.
,, 106, line 5, *insert* 'A.D.' *before* 1670, to remove the ambiguity.
,, 107, line 16, *for* 'Berkenhont' *read* Berkenhout.
,, 126, line 18, *for* 'G. E. Smith' *read* H. Ecroyd Smith.
,, 171, line 2 from bottom, *for* 'M' *read* C. Marsden is in the Colne river-basin.
,, 183, line 13 from bottom, *for* '*ciconium* Willd. *read* *cygnorum* Nees.
,, 185, last line but one, *for* 'Sherburn' *read* Ledsham.
,, 244, top line. The Poppleton station for *Sedum Telephium* is in the Nidd river-basin, not that of Wharfe.
,, 259, line 10. The Steeton locality is in W. river-basin.
,, 266, line 21. The Steeton Hall locality is in the Wharfe river-basin. This applies to all Mr. Whitwell's stations. Steeton Hall lies east of Tadcaster.
,, 314, last line, *for* 'to' *read* 'for.'
,, 332, line 19. The Sherburn churchyard locality is in A. (Aire) river-basin, including as it does the Selby-Dam dike.
,, 342, lines 28 and 29. The Carlton Ing-fields station ranged under W. is near Bingley, and in the A. river-basin.
,, 446, line 15. The Tadcaster road locality for *Allium vineale* is in W. river-basin.
,, 485, line 14. The Knavesmire station is within W. river-basin, not that of Nidd.
,, 561, line 7. The Poppleton locality (not the Acomb one) is in district N.
,, 564, line 6 from foot. Nether Poppleton station is in N. river-basin.
,, 730, line 17 *for* **ELVELLACEI** *read* **HELVELLACEI.**
,, 748, line 1, *for* **Raphidium** *read* **Rhaphidium.**
,, 749, lines 28, 30, and 33, *for* **Sphœrozosma** *read* **Sphærozosma.**
,, 765, line 15, *for* **LYNGBYÆ** *read* **LYNGBYEÆ.**
,, 782, lines 31 et sequ., *read*:

Casually, on waste ground where manure had lain, in field by canal near Calverley! Doubtful how brought, perhaps *Lepidium Iberis*, or an allied prostrate-growing species with cordate-ovate silicles, pinnatifid lowest leaves, entire upper leaves, and very small densely racemed flowers with lilac-white petals, but 'in too young a state to name with safety' (J. G. B.).]

,, 783, line 26, for *Sebtaria* read *Setaria*.

CLIMATOLOGY.

CLIMATOLOGY.

CLIMATE has been defined as 'the sum of certain complex atmospheric conditions, the factors in which are Heat and Cold, Light and Darkness, Dryness and Humidity'; or, in other words, it is the term for the way in which these phenomena of Heat, Cold, &c., express themselves from inter-action over any given area, varying as they do in degree and duration, and so forming a great number of combinations. To the factors named above, Wind—that is, Air in motion—may be added; for that, by carrying suspended water in particular directions, will often greatly modify conditions that would otherwise obtain over a tract. The comparative abundance of trees, or the treeless character of an area, also indirectly exerts a minor influence upon the climate of that area, by affecting its average humidity.

The climate of any tract depends upon a variety of circumstances. The broader features are determined by the accidents of its environment, its position with regard to an ocean-tract and the gulf-stream, its distance from the poles or equator, and the regular or irregular character of its surface. If there be a distinct mountain chain, much will depend upon the direction it runs in, and the average height of its summit ridge.

The mean temperature of Britain is mainly fixed by its insular position and its relation to the gulf-stream, these materially modifying the temperature that without their influence our latitude would determine. The position of our County and our West Riding, with its hill-chain in the western fourth running from N.W. to S.E., and the Wold hills of the East Riding in comparative proximity to the N.E., further modifies our climate by protecting us to some extent from the full force and effect of N.W. and N.E. winds. Proximity to the sea always conduces to equability of climate, lowering the æstival maximum of heat, reducing the hibernal maximum of cold. Great Britain is in the tract of that current of heated water which originates in the equatorial Carribean Sea, and gives up its last heat as it finally washes the western shores of Europe, and to this fact is due our comparatively warm climate. The result (in which our area partakes) is that our mean temperature is raised high above that of other tracts in the same latitude. Labrador, for instance, being unwashed by the gulf-stream, has a climate as arctic as Iceland, although no farther from the equator than Edinburgh; and the average annual temperature of Middlesex (49° Fahr.) is as high as that of the western seaboard of North America or the centre of Tartary, at a latitude many degrees nearer the equator. The influence of our sea-girt position is seen in a comparison of our hiberno-æstival differences of temperature with those of inland tracts on the same latitudinal parallel. Paris has an annual mean of 51° Fahr.; Central France, 52° Fahr., 3° higher than London, but London is subject to smaller extremes of temperature—the difference between summer and winter is much less. The mean winter temperature at Leeds and Greenwhich is about 39° 2′ Fahr., in West Cornwall it rises to 45° Fahr., at Edinburgh about 38° Fahr.; whilst in central Russia (same latitude as Edinburgh) it is 14° Fahr., in central France 32° to 35°, in Poland and central Germany about 31°, and at Vienna 32°.

But when we compare the summer averages, the converse occurs. All the places named, and even some further north (for instance, Stockholm and other places in Scandinavia) have a much higher summer mean than we have. The summer mean of S.W. Cornwall and Greenwich is very nearly 60°, that of Leeds about 58°, whilst at Stockholm it reaches 62°, in central Germany 65°, Poland 68°, central Russia 66°, Vienna 68°, and in central France 66°.

The difference in annual mean temperature, produced mainly by our insular position and the relation of the gulf-stream to our shores, can be seen in yet another way. In central Europe, coincident with each degree of latitude northwards, there is a lowering in the annual mean of one degree Fahrenheit. In Great Britain this lessening for latitude is not nearly so great. Our Riding is from 2 to 3 degrees north of London, and yet, on the lower levels, the average annual temperature is only about one degree less!

The humidity and the rainfall of our Riding are determined mainly by the physical peculiarities of the surface, and the relation of the mountain range to sea winds. The flora of our tract depends quite as much upon the degree of moisture prevalent as upon temperature. Every plant has no doubt a certain range in regard to temperature, within which it flourishes best, and beyond which it either dies of cold or withers from heat; but in this matter we have not merely to consider the powers of resistance possessed by plants as regards heat and cold. It has to be remembered that plants have periods of vegetative activity and quiescence, and this new factor makes all the difference in our calculations. On the time when the cold comes, and when the heat comes, well-nigh everything depends. Plants are dependent very much more upon the maximum summer temperature, and the degree and duration of this and of the winter minimum, than upon the mean amount received during the year. In our British flora, as a whole, this absence of extremes in temperature has an important influence on the distribution of integral species; but through a small area like West Yorkshire it is not so easy to trace the result further, and see how far the equability of our Riding has determined what of those British species we should have, and what we should not have. The differences relating to temperature alone, to be dealt with here, become almost too small to be appreciable. The southern species, that run out just before

reaching West Yorkshire, yet reach Derbyshire and Notts (such as *Pyrus torminalis* or *Campanula Trachelium*), probably stop short not because of a difference in temperature, but as a result of some other factor in the sum of distribution at which we can only guess vaguely. Before dealing, however, with effects upon plant-distribution, we will briefly consider the climate of the Riding, first as a whole, in regard to position, and then more in detail as to temperature, rainfall, and winds.

I. *Position as affecting Climate.* We find appreciable climatic differences within our area between the surfaces on the east and west of its rocky backbone—the Pennine chain of hills. On the west side, the effect of proximity to the Irish Sea (warmer than the German Ocean) renders the western slopes warmer in late autumn and winter than the eastern declivities. The surface on the eastern side, a long slope from the bed of the Ouse, is open to the cold easterly winds coming from the Baltic and Germanic tracts of Europe, and from this results a somewhat colder winter. On the other hand, the eastern levels, and the greater part of the gradual slope upwards to the west until the higher hills are reached, are hotter and drier in summer than the corresponding slopes on the west. Much of this is due to the fact of the summit ridge of the Pennine chain protecting them from the humid Atlantic breezes, causing these to deposit much of their moisture in their passage eastward. Comparing the lower levels of the West Riding with those portions of the East Riding east of the Yorkshire Wolds, we find that they are not so humid, since the hills named, to the east of our area, though not lofty, play the same part in cutting off sea-fog and moisture-laden east wind as the Pennine chain does in cutting off western mist. These are the advantages the West Riding gains from its position between two hill ranges. In other respects its climate is that of the eastern half of England generally.

II. *Aerial Temperature in the Riding.* The climate differs considerably in different parts, if we take altitude and humidity into consideration; but speaking broadly of it as a whole, it is mild, and its salient feature is its equability. The mean average temperature of the year for the lowest levels is about 48·5° Fahr., this being slightly above the average for the whole east coast of England from Greenwich to Berwick, and also a little over the average annual mean at York. At Leeds it has been computed as 48·5; for Bradford, 48·3 is given in Abbey's returns; for Sheffield, 48·7; for Halifax, 48·3; for Goole, 48·5; and Sedbergh, 48·1: as against 48·2 at York, 49·0 for Greenwich, 50·6 for Exeter, and 52 for Helston.

There is here, we see, no very great variation; and it is really in the unequal distribution of temperature in the different seasons that we find the peculiarities of local climate expressed. Judging from mean alone, one might gather that Halifax had a better climate than York, and Sheffield than either, whereas it is not so. The mean of the West Riding, which lies within 53° 20′ and 54° 25′ N. Lat., naturally within such narrow latitude varies very little in the lower lands, and the difference longitudin-

ally, between the Ouse valley on the east and the Lune valley on the west, is not above a fifth of a degree; for if one side is warmer in winter and autumn, it is cooler in spring and summer, so that the balance is maintained. In this balance the indigenous vegetation is so intimately concerned, that, as is stated on good authority, were the *mean* temperature to vary as much as five degrees, every one of our species would die; and this, although many of the species upon the higher peaks (plants of the arctic-alpine type) are capable of resisting very intense cold if it only come during the period of vegetable quiescence, and even considerable temporary variations during that 'renewal of life' which is possible to them in spring. The variation in mean annual temperature is rarely beyond a degree either way from the average mean, according as the seasons happen to combine a mild winter with a hot summer, or a cold winter with a wet, cool summer. Luckily for agriculture, this does not occur often.

While the foregoing data may be taken as substantially correct, it must not be forgotten that the temperature of any place is often regulated or modified considerably by the accident of its position with regard to adjacent elevations—by exposure or shelter—independent of mere altitude. Hill spurs running from east to west cut off the bitterness of northern blasts for dwellings on their southern side; and the mean temperature of a farmhouse in a sheltered valley at, say, 600 feet elevation, will be greater than that of another, situated also at an altitude of 600 feet, on an open tableland considerably distant from the true hill country.

Of the daily range of temperature in exposure as contrasted with that in shadow, for West Yorkshire, I have little information. The difference between them must be considerable, and, judging from returns for places outside the Riding, is apparently as much as 20°. The average daily range—the variation between maxima and minima—is about 12° to 13° in the shade, and 32° to 34° under exposure to the direct rays of the sun. The greatest difference is, of course, found in the summer season; and at this time, from the moment the sun sets, the lowest temperature of the ground is always below the lowest temperature in the air. It is in elevated stations amongst the hills that the power of the sun's rays show the greatest result, since they there act through more rarified atmospheric strata; and it is there also that the air cools quickest, because of the greater extent of rock surface presented to the atmosphere, and the comparatively greater readiness with which it absorbs the heat. We have here an effect the exact opposite of that previously indicated as resulting from exposure as regards cold air-currents; but the compensation is not perfect, probably because the power the sun has, in most months of the year, of imparting heat during the time it shines, is a transient power, and irregular or uncertain in its operation, since liable to be interfered with by cloudy, opaque conditions of the aerial strata; whilst the influence of exposure, in lowering the annual mean, is an influence constantly in operation.

The effect of direct sun-rays upon the growth of plants is well known to be considerable; but what is the influence, if any, upon distribution of species is, I believe, not ascertained with any certainty—at any rate, I have no evidence that shows it. Exposure to the sun accelerates the growth and blooming, and so concludes earlier the seed-ripening process in most plants, especially annuals; as may easily be proved by sowing seeds of the same species, some in shade, and others in an open situation; but, on the other hand, many species, of which the Killarney Fern and the moss *Hookeria lucens* are examples, flourish best in deep shade, and die if planted in exposed situations, even though their requirements in other respects (moisture, &c.) are encompassed. Here it is the preference of a species for a certain condition, that of light, which governs distribution. All plants need light and warmth in some degree, but nearly all would probably live without the direct rays of the sun falling on them. The direct rays, then, only influence distribution very broadly, in so far as they affect the general daily temperature during a certain season of the year; and within such narrow limits as our area, or our whole island, where the length of day is practically the same, can have no effect sufficiently distinctive for us to appreciate it.

Something must be said with respect to seasonal differences of temperature, and the relative monthly averages. [The difference between the means of summer and winter in the Riding I am not in a position to give exactly from any observations of my own, and almost every thermometric table differs in some degree from others for the same districts; but the following averages will give a notion sufficiently accurate for our purpose.] At Bradford the mean temperature has been stated to be, upon the average, 60·7 for summer (June, July, and August), and 36·3 for the winter (December, January, February), making the difference 24·4. At Leeds it has been stated as 60·3 for summer, and 39·5 for winter, giving a difference of only 20·8; for York, 59·8 for summer, 36·4 for winter, yielding a difference of 23·4; for Goole, in the extreme east, under the influence, evidently to some extent, of the tidal Humber, 40·8 for the winter season, and 58·4 for the summer, showing a variation of 17·6 only. This would probably, however, be less than the mean variation, if an extended series of observations had been made. In the year 1875-6, beginning with Dec. '75, it was 22·4, but that was an exceptionally warm summer.

For the whole of the Riding, if we calculate the hiberno-æstival difference at 22°, we shall probably not be far from the truth. With regard to other parts of England it may be said that our winters are not colder than those of Middlesex, very little, if at all, warmer than those of Newcastle, and are in fact about similar to most other towns in the eastern counties between the two, not actually on the sea-coast. On the sea-coast, from the influence exerted by the sea, the difference between summer and winter sinks about five degrees, it being given as 17·7 for Scarborough, and 16·1 for Shields.

On the other hand, our spring season is colder than that of Middlesex, but not so cold as that at the seaside. Amongst the Pennine peaks it is likewise colder, and for the same reason—viz., that both sea-water and mountain masses having towards the close of winter cooled down to their lowest degree, greedily absorb the heat as the season wears on. During autumn the hill masses and the sea have been parting with the heat stored up in summer, and therefore, in both cases, the spring vegetation that coincides with the warm weather is delayed beyond what is found to be the case on the inland level tracts.

In summer the inland levels are still the warmest; a seaside locality is cooler because of the equalising influence the sea, still absorbing heat, exerts; and amongst the hills it is cooler because of the elevation, the exposure, and also because the rock masses are likewise absorbing heat.

As autumn wears on, the relations are reversed. Then the seaside and hill stations have a temperature comparatively higher than the level inland ones. Now the sea is giving back to the air the heat acquired during the summer—now the hill masses are doing the same thing precisely. The latter 'postpone cold,' as Professor Phillips has pointed out, by doing in autumn what has just been indicated; for when autumn has gone, and they have parted with their heat, they must re-absorb it again in spring and early summer, and thus make the greatest air-warmth come on later amongst them than at lower levels. Hence it is that at all stations in our Riding (all of them inland ones, unless Goole may be considered as falling partially under the influence of the sea) below an elevation of 900 feet above sea-level, January is the coldest month in the year, February the next coldest, December the next, with March almost on a par with it, whilst November is generally a little warmer than March. April is the next in point of temperature, but a great rise has occurred. October is a trifle warmer than April. The hottest month is July, with August often running it very close, and in the northern parts of the Riding occasionally exceeding it, whilst September appears to be constantly a trifle warmer than June. May generally has a temperature exceeding that of October, but in very late seasons is sometimes actually colder. This refers to the lower levels. Amongst the mountains of the N.W., February is generally the coldest month, August the hottest, and September often equal in temperature to that of July, whilst March is little, if anything, warmer than January; and in vegetation the seasons from November to June a fortnight to three weeks behind the plains.

The hibernal variations, according as winter is 'severe' or 'open,' are not nearly so extreme in England as inland in Europe, and would appear to be much the same for the whole of the eastern half of England. Latitude in winter, with us, makes little if any difference. The accessions of cold, indeed, seem rather to be distributed from north to south than from east to west. The only appreciable differences with us occur in regard to the dividing line of the Pennine chain running from N.W. to S.E. Just as for all England the western side is more equable, so in the West Riding the hibernal range of temperature is in favour of that fifth of our area west of the dividing line alluded to, which comes markedly within the influence of Morecambe Bay. The variation in winter means between one year and another may be stated approximately at between 42 and 32 for the east side, as against 44 and 34 for the west slope of the Riding; so that as regards equability in winter, the lower areas of the Lune and Ribble valleys take the place which the seaboard does in the maritime eastern counties. What of the winter amongst the hills? A very common notion is that it is more severe than on the plains of our area (of course equal elevations pre-supposed), but Settle and Arncliffe observations do not bear this out, and for winter the word 'spring' should be substituted. The winter in the deep valleys of the hills about Whernside, Penyghent, and Ingleborough occupies, as regards severity, a position about midway between that of the plain of the Ouse on the east, and that of the open shallow Wenning valley on the west. How the sea cuts off extremes both in summer and winter has been indicated, and *à priori* we might expect huge masses of rock to enact the same *rôle*, remembering Professor Phillips' dictum as to the hills 'postponing cold' by absorbing heat in summer, and giving it back in autumn. No doubt the giving up goes on far into winter —as long, indeed, as the rocks descending deeply into the earth are warmer interiorly than the air; and this tends to give the villages in the sheltered hill valleys an advantage over those of the exposed plains.

In this connection a word must be said as to the hibernal *minima*; for these occasional extreme degrees of frost, transient though they are with us, and less frequent than on the Continent, have a visible effect on vegetation, in importance far beyond that of hibernal range. The most striking fact in the distribution of these winter *minima* seems to be that they are lower in the plains, and most frequent there, on the east side the Pennine chain. The sea, western position and aspect, and hill masses reduce them. Amongst the hills they do *not* grow greater as elevation increases, as one might perhaps expect. The opposite is the case, according to a comparison of average minima for Leeds, York, Doncaster, Wakefield, Bradford, Settle, Sedbergh, &c. On the nights of past years, when the thermometer sank below zero at the five places first named, the average was *minus* 5. On the same nights, at Harrogate, Ilkley, Settle, the average did not fall below *plus* 5. As to the effect upon vegetation, a few general remarks must suffice. Whenever we have an unusually severe frost, the effect is sure to be seen most strikingly amongst the ornamental aliens of our shrubberies; but native trees and shrubs suffer too. After the foregoing outline of the distribution of cold *minima*, the qualification must, however, be made—that such is the case provided the elevation is small and the position exposed. The gardens in the plains suffer most and oftenest: the night that kills Laurels and Araucarias at Wakefield will only partially kill them at Settle. If only nipped, and not cut to the ground in the plains, they most probably escape all injury in the park shrubberies of higher Wharfedale, Airedale, and Ribblesdale. Remembering this, however, it may be said that in West Yorkshire a fall of the thermometer to zero nips or kills Holly, Ivy, Whin, Broom, Privet, *Rosa arvensis*, *Acer campestre*, and *Populus nigra*, of indigenous plants; and of shrubbery or orchard trees and shrubs has a like deadly effect on the False Acacia, Strawberry-tree, Plane-tree, Tamarisk, *Hippophäe* (though this stands seaside winters), Laburnum, Walnut, Pear-tree, Medlar, Rhododendron, *Garrya elliptica*, Deodar Cedar, Live Oak, Syringa, Lilac, False Tea-tree, Box, all the ornamental Laurels, the Bay-tree, Aucuba 'Laurel,' Laurustinus, Araucaria, Tulip-tree, Mulberry, white-flowered Jasmine, and Rosemary. The smaller shrubs and herbaceous plants that deep snow can cover up generally of course escape injury.

Yet the *Garrya elliptica*, it was stated to the writer, stood the severe winters of 1860, 1865, and 1874, without protection, on the Castle bank in the grounds at Tickhill; and a magnificent *Araucaria imbricata* in a garden at Middleton, at an elevation of 750 feet amongst the hills, had escaped injury during a like time. *Cydonia japonica* in another garden, at the same place, had likewise suffered nothing.

Lastly, such herbaceous plants as the Dahlia, the Ice-plant, and the Tropœolums of our garden-beds, that droop at the first nip of frost, are often cut down on the lower levels about Leeds a fortnight or more before oncoming winter affects them in gardens amongst the hills, showing how much later the onset of frost usually is there, and how, when it does come, it approaches from the plains.

The temperatures hitherto given have all referred to the lower levels of the Riding, and the standard annual mean has been stated at 48·5 degrees: nothing has been said about the constant ratio of decrease in connection with altitude.

In its relation to plant distribution, however, the effect of elevation above sea-level upon mean temperature out-weighs all others in importance, as will be shown hereafter, when the vertical range of vegetation is divided into zones of altitude, corresponding with imaginary zones of temperature closely, although not exactly, being complicated by the factor of moisture, increasing as we ascend.

It is altitudinally only that within our limits we get any constant variation in mean temperature, capable of precise exposition in figures, but by arranging the thermometric data in regard to this it is easy to formulate certain zones or belts on our hillsides, divided by isothermic contour lines; and by these we can indicate climate, heat diminishing and humidity increasing in inverse ratio as we ascend.

The variation in mean annual temperature in our Riding, due to latitude on the levels below a hundred feet, has already been stated to be so small (not above half a degree), that in adopting a standard from which to reckon in constructing a ladder of temperatures for fixed altitudes, we might almost dismiss it altogether from our calculations; and the more so since it is probable that at each rung of elevation counterbalancing conditions come into play, by which the small initial difference is further reduced. As the safest basis, however, for estimating zonal ranges of temperature, we will adopt that of 48°, dropping the decimal fraction of 5 in the annual mean, as stated on page 5, by way of allowance for such latitudinal differences as may exist at the most distant parts.

The lowering of the temperature of the air coincident with vertical rise above sea-level was first, many years back, declared by Dalton to be 1° Fahr. for each successive ascent of a hundred yards: a decrease over twelve hundred times as rapid in its rate as that resulting from latitude. Dr. H. C. Sorby, in a paper on the temperature of springs, to which I shall refer presently, seems to have arrived at a conclusion—whether from a comparison of spring-water temperatures with elevation by themselves, in connection with air-temperatures, does not appear—somewhat at variance with Dalton's: he gives the diminution at ·9° Fahr. for each hundred feet of elevation. This, probably, is not far from correct as between sea-level and the highest mountain peaks, dividing the differences between the extremes by the number of hundred feet leaps, as if the diminution was *regular*; but it is not so in fact. It is not uniform as between the base and the summit of the more lofty hills; for the higher we rise, the more rapid does the rate of diminution become; although it is quite true, as Dr. Sorby points out, that aspect, exposition, and other more transient conditions interfere to vary any estimate. That of Dalton was given as applying to hill ascents in the North of England, and so far as concerns the air in proximity to rock masses of limited height and large superficies, is doubtless nearest absolute accuracy; but for lofty isolated peaks (such as Teneriffe), where the gross surface of rock (possessing powers of storing up and yielding heat) is comparatively small, 1° Fahr. for each hundred yards is too slow a rate of diminution. It is certainly short of the lowering always experienced by aeronauts in their cloudland flights, where are no solid points of radiation at all. The average of results obtained in seven balloon ascents is 1° for 270 feet; and a rule-of-three calculation of the difference furnished by the means of thermometric records (for long periods), in stations of considerable elevation, with others at low levels, gives a rate of decrease still more rapid: 1° for every 250 feet. Dalton's estimate, however, is the most convenient one for calculations that must be understood as approximate, not absolute; and since it has been accepted by authorities on botanical geography, such as Henfrey and J. G. Baker (of whose writings I may here, as elsewhere, acknowledge availing myself largely) as sufficiently accurate for the indication of temperature range as associated with the vertical range of plants, it is adopted here. Taking 48°, then, as the average mean of the lower levels, we may construct a scale for the different heights as follows:—

TABLE OF TEMPERATURE CORRELATED TO THE ELEVATION FOR THE WEST RIDING OF YORK.

Elevation.	Corr.Temp.Mean.	Corr.Clim.Zone.	Zonal Range of Temp.
0.*	48°		
At 300 ft.	47°		
At 600 ft.	46°	Zone 2.	48° to 45°
At 900 ft.	45°		
900 ft.	45°		
At 1200 ft.	44°		
At 1500 ft.	43°	Zone 3.	45° to 42°
At 1800 ft.	42°		
1800 ft.	42°		
At 2100 ft.	41°	Zone 4.	42° to 40°
At 2400 ft.†	40°		

* The lowest levels in the Riding are about 25 *feet above* sea-level.

† Whernside, the loftiest hill, is 2,414 feet in height.

We thus see that within the compass of West Yorkshire, although it extends over only 1° of latitude, by reason of its highest summits we have a difference of above 8° in mean annual temperature; and it becomes possible to see in our vegetation an expression of that difference, equivalent though it is to over 8° of latitude. Nay, it is more than this, really; for in the *zones* given in the table—to be explained presently—the lowest one has been excluded, having no direct connection with altitude. The West Riding, as a parliamentary area, includes a district geographically part of Nottinghamshire, certain elements in the vegetation of which proclaim it as belonging to a first zone. Its streams fall into the Trent, its mean temperature is certainly not less than 48·7, and it has no eminences of any account. The difference, then, between such places as Roche Abbey, Anston, Bawtry (and perhaps Goole, almost on sea-level)—about 48·7, and the peaks of Whernside, Ingleborough, and Penyghent (about 39·9), is much nearer nine degrees than eight; and to arrive at a point near sea-level where the mean annual temperature is as low as that on those hill summits, and the vegetation of the same alpine-arctic character, we should have to go as far north as Drontheim in Norway, or the south coast of Iceland! If we remember what was said as to the effect a variation of 5° in the mean annual temperature would have on our vegetation, we shall see, in the fact of there being so great a difference as nearly 9° between Anston and the peak of Whernside, the reason why the vegetation at the two places is so absolutely different—not one of the species found at the former place occurring at the latter; instead of them a few species looking very stunted and insignificant, but nevertheless wonderfully hardy, constituted to bud and flower, and the seeds germinate

with very little warmth, and endure for ages an annual temperature that, if it came for one year into exceptional operation over the lowland, would leave everything dead.

III. With regard to *springs* and their temperature, little need be said. As we might expect, there is a certain correspondence between the mean annual air temperatures of our various districts, be they lowland or upland, and the water temperatures of the springs. This mean annual temperature must always be somewhere about that of the great bulk of the strata from which they collect, and on the whole equal to, and if of very deep origin somewhat higher than, that of the air. We may, by carefully testing springs twice each season after such weather extremes as they afford, or by calculating from the known temperatures of springs (certain very deep and very shallow kinds rejected), ascertain the mean air temperatures of the places where such springs occur, and use them as checks against errors in, or as supplementary aids to, imperfect thermometric records. On the other hand, if we know the mean annual temperature of the air, and something as to the nature of the spring, whether it is deep or shallow, intermittent or constant in its supply, we may make a very good guess at what its temperature ought to be if our knowledge of its character is correct. Again, in temperatures of springs in different seasons we learn something of their origin, and in the varying figures of the shallower ones, determined by the heat or coldness of the soil they rise in and run through, we get by correspondence a corroboration of the statements already made as to heat absorption and seasonal retardation by mountain masses, etc. We know this by observations so repeatedly verified, that we may safely take them as undoubted facts. Near the surface the temperature varies with the seasons just as in the air, but is always somewhat below it in heat maximum, above it in cold minimum, and behind it in its range of variation. At a yard's depth, instead of there being the 22° named as the average hiberno-æstival difference of the air, it falls about 5°, to an average of 17°, the times of greatest heat and cold falling about July and August, and January and February, just as in the air. At two yards' depth there is another fall in range almost as great in extent, making the seasonal variation only about 12°; and further, at this depth the period of greatest heat falls later in summer (August and September), and the time of greatest cold (March) rather in early spring than winter. At thirty feet or so below ground the range has been reduced to not more than 3°, and the postponement carried so far that now the maximum temperature is when the surface is coldest. And so on, the deeper we penetrate the earth, the less does the difference become, although after a while the mean temperature begins to augment, until at one hundred feet below it is absolutely alike at all seasons, and always higher than the air at the surface.

The temperature, then, of a spring depends very much upon the depth whence it wells up, its run to the surface resulting probably in very little variation either by gain or loss.

A popular notion is that the deepest springs are the coldest, but this is not so in fact. They are spoken of as 'ice-cold' in summer, but the term is relative only; they are never colder in summer than in winter, and if from a moderate depth rarely more than three to four degrees colder in winter than summer. The *warmest* ones are probably the *deepest*, and these vary on the average about one degree. The summer and winter range of the sulphur and chalybeate springs is not much over, and often much under, one degree, showing amongst other things the depth at which they originate. As a rule, the springs that gush out at the foot of the steep *limestone* scars of Craven *vary the least*. Some are said not to vary at all, many not above half a degree. These are not the warmest, but they have a much higher mean nevertheless than the shallower springs upon the gritstone. The nature of the limestone hill country favours deep springs in this way: huge masses of porous strata lie superimposed with little intermixture of retentive beds, allowing the water shed as rain to percolate downwards to great depths before checked and turned upwards by an impervious stratum. The temperature of these little varying springs is nearly always between 46° and 48° Fahr., which is about the mean annual temperature of the districts (according to the elevation) in which they are chiefly found. The deeper springs on the various sandstones have usually a greater range of seasonal variation, but with a mean temperature but little different (45° to 49°, according to elevation) from those on the limestone. The mean temperature of all our springs (except the warm mineral ones) diminishes with elevation, but not always in the same ratio as that of the air. The springs that are least uniform are judged to be the shallower, because their varying temperatures follow the thermometric means of the seasons in just the same way as it has been explained that the ground does down to twelve or fifteen feet, knowledge of geological conditions as to the character of the stratified beds helping to the conclusion. In such the variation between their lowest and their highest temperatures is frequently over twelve, and not seldom as much as sixteen degrees. Where this is the case, the water is coldest in February or March (being then 42° to 45°), and warmest in August or September (58° to 55°), the deeper the spring the more the extremes being cut off. Two ordinary springs of this character (accounted cold ones), one at 'the well-house,' Meanwood (slightly chalybeate), and the other by the cottage on Adel Blackmoor, both rising through millstone grit at about 250 feet elevation, varied ten and twelve degrees respectively, the figures being 45° and 55° for the first, and 44° and 56° for the second. I only made two observations of each, however, during February and August, and so do not know the monthly variation or annual mean of either. The average mean annual temperature of ordinary shallow springs is about 48·8° for the lower levels. Dr. H. C. Sorby has given the means and range of difference for four springs near Sheffield, all above 1,000 feet in elevation, stating them as follows: 1,050 feet, mean $45\frac{3}{4}$°; 1,075 feet, mean $45\frac{1}{2}$°; 1,225 feet, mean $45\frac{1}{8}$°; 1,275 feet, mean $45\frac{1}{8}$°; but these,

although ordinary springs, had deep-seated sources, for the *range of variation* between summer and winter was in the first case only $\frac{1}{2}$, in the second 3, in the third $3\frac{3}{4}$, and in the fourth $4\frac{1}{4}$ degrees. The one at highest elevation is evidently the shallowest, though of course in springs gushing out at elevations above 1,000 feet, in an area where the summit ridge of the hills is only 1,700 feet or thereabouts, the strata to percolate through are not extensive vertically, and the springs must be comparatively shallow. Leaving range out of consideration, however, we have in Dr. Sorby's figures, contrasted with that of a spring at Upper Heeley, at 350 feet, with a mean of $48\frac{1}{2}$°, which is also given by him, a difference in mean annual temperature of $2\frac{3}{4}$° for 700 feet rise in elevation; and $3\frac{3}{8}$° for 925 feet rise, to contrast with a probable fall *in the air* of $2\frac{1}{3}$° for the 700, and $3\frac{1}{10}$ Fahr. for the 925 feet rise. The agreement is not very close, but near enough if we bear in mind that it is almost certain conditions of site and aspect—shelter from sun under cliffs in a deep glen, or exposure on a long bare hill-slope, for example—influence the mean annual temperature of springs, complicating the sum by additional factors, in precisely the same way (if not, perhaps, to a little extent) as they do the temperature of the air.

IV. *Atmospheric Humidity and Rainfall*, in their distribution, and in their relative value as factors in the sum of vegetable climatology, next engage our attention. To begin with, however, we must dismiss from our mind any idea as to there being any necessary connection between them, any fixed relation of the nature of cause and effect; for the one in no way hangs upon the other. It is well to bracket them in their indirectly great influence upon species-limitation as jointly determining the compound climatic condition of moisture; but it is to be understood that we cannot formulate their influence separately with any exactitude. We know their combined effect to be great, but we can only trace even that very broadly in the peculiarities of the distribution of plants through our area. The rainfall of a district to some extent determines the average moisture of the soils of a district, but air-humidity has relations much nearer to altitude or position with regard to evaporating surfaces such as those of seas or large lakes; and the prevailing degree of soil-moisture depends very greatly upon the character of the rocks below the soil, in reference to their capacity for retaining the water they do get. A porous, calcareous stratum in a district where the rainfall is annually sixty inches (say) may yet have a drier soil, and no greater atmospheric humidity, than obtains in another district where the rainfall is only half as much, if the stratified beds are such as to ensure the retention of the lesser gross amount of water shed upon them as rain. The fact is, that in every part of the Riding the salient climatic feature is the equality of its comparatively great humidity, a condition of things already given as characterising Britain as a whole, compared with the inland continental tracts of Europe. The annual mean in the Riding varies but little, the extremes being about 84 and 90 (100 representing saturation, and the mean estimated by the wet and dry bulb thermometer), as compared with 82 at Greenwich, and 80 in the vicinity of Salisbury Plain. Thus the mean does not range, in the parts most fully in contrast, through anything like the degrees that the rainfall does. Goole, for example, in the low country, uninfluenced by hills, has a mean humidity, according to observations of Dr. H. F. Parsons, about the same as that of Hull (88),* being open to the influence of the broad sea arm of the Humber; whilst Sedbergh, amongst the hills in the north-west, west of the summit ridge, has a mean of not more than 90. Yet the rainfall at the latter place is double that of the former, the humidity being due to the altitude of the mountains in its vicinity arresting and retaining as impalpable particles the water in the sea mists rolling up from the Irish Sea. Observations hitherto made are sufficient to warrant a belief that such narrow differences as do exist are found in the warm deep wooded glens of the higher river valleys and on the level central plain occupied by the Permian and Triassic formations overlaid with drift. The mean humidity at York is given as 86; at Ripon, Pontefract, and the Roche Abbey district, 85 and 84 respectively; at Halifax, 86; and at Sedbergh, about 90. In the drier mountain-limestone districts, such as Settle, it is not so much as 90; and the maximum humidity is probably reached in the deep-sheltered glens amongst the gritstone and Yoredale hill districts, above such places as Penistone, Dunford Bridge, Todmorden, Slaidburn, &c., where the moisture-loving mosses and ferns have their head-quarters. Wooded regions are well known to be damper than treeless plains. The planting of trees extensively often as wholly alters the character of a district, raising its humidity, as the drainage and cultivation of a moorland waste does its climate in a contrary sense. But the influence of moisture in its collective sense on the distribution of our plants will be examined more fully in the next chapter, when treating of the specially damp-loving species. Upon the whole, the humidity is greater by a fifth over the tracts to the west of the hills, whilst the rainfall for the western half of the Riding is double that of the eastern half. With regard to the seasonal variation in humidity, there seems to be a gradual increase (with slight monthly recessions occasionally in dry autumns) from July to December, and a gradual decrease from January to June.

The rain in the atmosphere at any one time has been computed to constitute less than one-hundredth part of its bulk. The average annual rainfall for the whole of Yorkshire has been given at about thirty inches; the proportion in the different Ridings being about twenty-three for the East, thirty-two for the North, and thirty-five for the West parliamentary divisions; but in its distribution it is not only unequal, but at any and every spot irregular as to the time when it falls. There is a certain constancy and agreement, however, as to the seasons in which the greatest and least amounts fall. The least quantity appears, upon the average of

* In 1876 it was 89·9 (a wet year), with the rainfall 25·69; whilst in 1877 it was only 88·3, with the rainfall 26·00.

many years, to fall in early spring and early summer; the greatest amount in late summer and early winter. February and May appear to be the driest months, July and August, September and January, the wettest. The following table gives the monthly and annual means for four places chosen out of a much larger number because they may be supposed to represent very fairly the extremes upon the lower levels; viz., York, in the central plain; Goole, at the south-east corner, almost at sea-level; Settle in the Craven limestone district; with Huddersfield occupying an intermediate position, midway between the plains and the hill country. It will be seen that at York we have the least rainfall, with February the driest and August the wettest month; at Goole, May the driest and August again the wettest; at Huddersfield, also, May the driest, with September and October the wettest, whilst at Settle, May is again the least, and October the most wet month.

TABLE OF MONTHLY AND ANNUAL MEAN RAINFALL AT FOUR SELECTED LOCALITIES IN THE WEST RIDING.

MONTHS.	York. 10 years' average. T. Ford.	Settle. 18 years' average T. Tatham.	Goole. 3 years' average. H. F. Parsons.	Huddersfield. 10 years' average J. W. Robson.
January	1·77	4·23	1·92	3·33
February	1·05	3·00	1·55	2·59
March	1·27	2·77	1·69	2·13
April	1·56	2·23	1·48	2·70
May	1·62	2·17	0·95	1·91
June	2·37	3·63	1·93	2·15
July	2·74	4·22	2·54	2·88
August	2·95	3·29	3·26	2·44
September	2·21	3·36	2·73	3·48
October.	2·12	4·72	2·35	3·48
November	1·13	4·25	2·36	2·90
December	1·22	3·72	2·24	3·25
Yearly Total	22·71	41·59	25·00	32·60

The difference in the quantity of rainfall in different parts presents an even still wider range than the temperature in regard to altitude. As throughout Britain generally, the greatest rainfall is in the western half, so in the West Riding the annual average, *cæteris paribus*, is greater in the western half than in the eastern, and greatest of all to the west of the Pennine summit of drainage, because the higher mountain masses are grouped thickest there. A moment's glance at the following short table, giving the mean annual rainfall (determined, unfortunately, from records varying in length of period, and therefore in value) for a number of localities, will show that the average annual amount steadily increases from east to west.

ANNUAL MEAN RAINFALL AT TWELVE WEST RIDING STATIONS.

York	22·71	Halifax	35·00
Goole.	25·00	Sheffield	36·00
Wakefield. . . .	23·40	Skipton	34·00
Leeds.	24·00	Settle	41·59
Huddersfield . . .	32·60	Bolton-by-Bolland . .	47·00
Barnsley	32·28	Sedbergh	49·50

An 'inch' in rainfall is equivalent to a fall of nearly 100 tons upon an acre; or, estimated in another way, 2¼ gallons of water distributed over a square yard of surface.

The rainfall also regularly increases with altitude, as the rivers are followed to their sources at the dale-heads, and up the mountain sides themselves, *up to a certain point*. On the loftiest peaks, such as Whernside, Ingleborough, Penyghent, Baugh Fell, White Fell, Cam Fell, and Great Whernside, all above 2,000 feet, less rain appears to fall. The summit is oftener cloud-capped, and the saturation of the atmosphere in excess of what it is on the hill-slopes a thousand feet lower; and when it does rain, the higher the altitude, the less in size are the raindrops: necessarily so, since they have less distance to fall, less space and time for augmentation by aggregation of watery particles, than in their fall upon lower levels. The greatest actual rainfall, then, is found to occur in the deep and high glens and cloughs riven between the peaks by centuries of weathering. These, in tempestuous weather, act as basins for the reception of dense masses of mist and cloud, filling to overflow, and then, because unable to eddy upwards and float away farther on account of the presence of more cloud cumuli above them, the entrapped clouds condense and deposit all their water as fine rain. The rainfall of such mountain basins, amongst the peaks named, although not systematically ascertained for obvious reasons, is doubtless often as much as eighty, ninety, or even a hundred inches annually. This may seem to some too great an estimate; but I believe it rather under than over the mark. It is nothing near the maximum quantity of rain that can be measured as falling. In the year 1877, according to Symons' Meteorological Magazine, a Mr. Birkett measured 180·4 inches at Seathwaite, in Borrowdale, 4·78 inches falling on September 12th within one day. At Arncliffe, in our own Riding, under the lee of Fountain's Fell, 1876, 3·25 inches fell in one day, and 5·35 inches in the three days, 14th, 15th, and 16th of July. The explanation of the greater rainfall in mountain districts has been often given before, and is well known to most persons; still, in a work of this character, which pretends to be a *resumé* of previous facts rather than an original treatise, it will be, perhaps, no plagiarism to repeat it. Warm air has a capacity for taking up and holding a much greater amount of moisture than cold air. The nearer to the earth do air-currents circulate, the

warmer they are, and therefore are capable of being more highly charged with moisture. Elevation of ground—the abrupt cliff or steep slope of a hill-side—causes a deflection in such currents, and forces them upward into regions that, as has been explained, are cooler by one degree for every hundred yards. What happens then? They expand by diminution of gravity, are cooled by the lower temperature, are unable to retain all the moisture they held as vapour at lower levels, and so some of it separates (condenses) into mist that floats about the summits, or into the larger particles of raindrops that fall again to the earth.

The rainfall of the Pennine chain generally is about fifty-five to sixty inches annually, or about nearly double the average rainfall of England. In the valleys of the Lune, Greta, Ribble, Upper Nidderdale, Airedale, and Dondale, it is about forty to forty-five, the annual average upon the gritstone moorlands, at elevations of from 1,400 to 1,000 feet, and in such places as Keighley, Halifax, Marsden, and Bradfield Moors, being somewhere between the means. Proceeding eastward down the valleys towards the lowland plain of the Ouse, the average annual rainfall is stated to diminish an inch for every twelve miles. At Pateley Bridge, Skipton, and Huddersfield it has fallen to between thirty-seven and thirty-two inches. Still further east, about Leeds, it is twenty-four; and upon the eastern border of the Riding, upon the open Triassic plain, it is only twenty-two or twenty-three inches. The difference between the extreme is thus as much as thirty-one inches! The annual rainfall of Yorkshire, as a whole, is not much over the average for all England; but that of the West Riding taken by itself is some four inches in excess of it. This is in part due to the position of the West Riding, but the comparatively greater extent of the surface at a high elevation is the great factor, another result of which is the plentiful streams of the district carrying off the water from the hills. West Yorkshire thus owes not a little of its industrial prosperity to its rainfall, through its rivers; for before steam took precedence of water-power, they were the initial integers in its grinding, weaving, dyeing, spinning, and other manufactures.

V. *Winds*.—It is not necessary to dwell long upon these. They modify climate as one of the minor elements, but their effect upon plant distribution is, in great part, masked or out-balanced by the major conditions of temperature and moisture, with the exception, however, of their one inherent quality—that of force. In understanding the significance of their extraneous qualities—their non-essential, as to temperature and humidity—it must be remembered that those are regulated very much by the point of the compass whence they come, and the situation with relation to hill-ranges of the area acted upon.

In West Yorkshire the *coldest* winds are those blowing from the hills in the track of the longest line overland; the *warmest*, those reaching us by the shortest overland route. The N.W. and N.N.W. are quite the coldest with us, blowing direct as they do from the group of Cumbrian summits to the west of the northern third of the Pennine chain; the next coldest, that coming from due north; the next, that from the north-east. The warmest winds are, in order of highest temperature, the south-west, south, and west, coming as they do from latitudes that, whatever the season, are warmer than those in other directions. These estimates are based on observations from which yearly averages have been struck. Seasons make a difference in the relative warmth of some of them. Taking winter alone, from December to March, the east wind is the most bitter, but upon the whole year the north-west is the coldest, and the east wind somewhat less cold than the south-east. The force of the north-east wind is, for us, somewhat broken by the York wold-range.

As regards humidity, the south-east and south are the *dampest* winds, from having traversed the German Ocean just before reaching us, and having encountered no intervening hill-range to eliminate their contained water. With us the west comes next, the north-east having discharged its moisture, usually, upon the Yorkshire Wolds and on this side of them. Our *driest* is apparently the north-east, although the south-west is often dry, and in summer brings fine weather, because it has deposited its water upon the summits, ridge, and high moorlands of the Pennine chain.

With respect to prevalence, the south-west and west winds hold the highest position, blowing twice as often as the north and north-east upon the average of the whole year, that due east being least frequent of all; but in the spring season the north-east is with us a very prevalent wind, and being also dry, is one of the causes of our small rainfall in May.

The position of our great included hill-range, however, changes altogether the character and significance of many of the winds in the tracts east and west of it, from its power of reducing the temperature and density of the air-currents deflected upwards and over its serrated ridge and between its summits. On the east side the Pennine chain the west and south-west winds oftenest bring fine dry weather. If they are cooled somewhat, they also lose much of their moisture before reaching the lower parts of the dales. These same winds, however, are wettest on the west side of the chain, when north-east and south-east and east winds usually are dry, and bear a significance just the opposite of what they do upon the eastern plains. Amongst the hills themselves, any wind containing moisture loses some of it as rain, and those which are on the average driest ensure finest weather. Such are the north-east, the south-east, and the east within our area.

But, as already stated, winds have their most important bearing upon plant-distribution in connection with their quality of force; their average as to this throughout the year, and at the several seasons in any locality, determines in no small degree the climate of that locality—the climate in its turn the vegetation. The humidity and repressive effect upon vegetation, seen in the lesser variety and dwarfed stature of herbs and shrubs in situations exposed constantly to powerful blasts, is evident to every one;

but the grand *rôle* of wind is doubtless the conveyance of seeds from one place to another. The species of certain orders and genera are much more prone to dispersion in this way than others, and such families are generally found to have a wide distribution. We see the most notable examples in certain classes of the Cryptogams, Ferns, Mosses, &c., having spores of microscopic size—therefore light, and easily carried long distances. Other instances, hardly less frequent, occur in the orders of Compositæ, Gramina, &c. The pollen of flowering plants is likewise transmissible with the greatest facility, and to its promiscuous spread hybridation is probably—in part, at least—due; although oftenest brought about by the visits of insects. We can hardly take the flora of such a small area as ours, however, and point out the species spread over it by wind; only in a few instances can we make probable guesses even. In the West Riding the rivers in the spates so frequent in autumn, when seeds are mostly ripe, and when the snow melts, oftenest afford clear proof of their power to colonise species in spots foreign to them; and the *rôle* of birds (devouring seeds with fruit in one spot, and voiding them unhurt in others far distant, where they not seldom germinate) is the one next oftenest seen. But in a small residuum of cases we have to fall back upon a blown-by-wind theory. One of such is the occurrence of the Atlantic type Fern, *Asplenium lanceolatum*, in the moorland district at the head of Bradfield dale, north-west of Sheffield, where, *if indigenous*, or rather, I should say, if not purposely introduced by human hands, the only probable hypothesis that will explain its presence there, in a place considerably north of, and a hundred miles outside the area in Britain in which it grows undoubtedly native, seems to be that spores have been blown by a high south-west wind from mid-Wales, and, catching in crevices of stone looking west, have found a suitable nidus for perpetuation of life there. But however far out of its previously known range, and whether indigenous or alien, the fern not merely lives but thrives there, which proves the locality to be within the bounds of a climate possible for it. It is a tender fern that severe cold nips; but if buried in drifted snow in winter, as it would be from its situation in the recesses of rock near the bottom of a moorland clough, it is quite conceivable it would be protected from frost. Lest this be thought an almost too improbable supposition, I may mention a fact coming twice under the notice of James Backhouse, of York, viz., that during high south-west winds windows outside York facing west have 'on two occasions been completely encrusted with spray from the western sea!' And if particles of salt and sand, why not the infinitesimal spores of Ferns? Bearing upon the subject of spore transmission by wind, I may also quote two remarks made by the same authority on ferns when communicating the fact just mentioned: 'How comes it,' he asks, 'that the moisture-loving *Hymenophyllum tunbridgense*, whose head-quarters are south-west Ireland, should be the *only Hymenophyllum* of eastern and central Yorkshire? Speaking of 'range,' I found some years ago grand examples of

Ajuga pyramidalis on one of the mountains north-east of Ambleside—a long way from Sutherland; and also a solitary fine patch of *Dryas octopetala* on Helvellyn, and two casual plants of *Potentilla fruticosa!*' The Fern has a small spore, the Dryas a feathery-tailed seed easily carriable by wind. I would add the queries, How came the Killarney Fern first of all into Yorkshire? How explain the exceptional station of *Asplenium viride* (a limestone fern usually, a grit-rock one only at very high altitudes) in the Wessenden valley above Marsden? But to do so is almost useless, when in the present state of our knowledge it is easier to frame a hundred suggestive hypotheses than bring forward one plain answering fact. It is a curious fact, though, that although winds may and do of course transport seeds and spores in *any* direction, successful colonisation through the medium of air-currents generally does take place from east to west or vice versâ, and only exceptionally along parallels of longitude; the seeds borne in the latter direction (over great distances, at any rate) being much less likely to find, where finally deposited, the climatic conditions of temperature and moisture necessary to their existence, than along tracks more likely to be isothermally analogous. A more fitting place, however, for exposition upon such matters as these will offer when we come to consider the genesis of the West Riding Flora.

VI. *Temperature in relation to Vegetation.*—The relations borne by heat to plant-distribution are intimate, but so much involved in the other factors of climate, that it is difficult to define them separately. Nowhere can we see temperature exerting its influence alone, uninterfered with by the accident of exposure or condition of humidity. Certain general lines of action may, however, be outlined, not as elucidating the subject in detail, but merely by way of suggesting the direction in which inquiry will be best repaid.

The temperatures governing plant-range are the sums of summer heat and winter cold. The annual means are of little account, since there is no essential agreement between them and the hiberno-æstival differences. If we suppose two regions with a uniformity as to annual mean, the one may have a very hot summer and a very cold winter, whilst the other may have neither. A region may have a mean of 55°, and yet the summer never hot enough to ripen wheat; another may have a mean of 33°, and yet, because the winter extreme falls far below zero, the æstival heat may rise quite high enough, and last sufficiently long, for a rapid process of growth, ending in seed-maturation, to be consummated. We see this requirement as to gross amount of heat operate most remarkably in governing the cultivation of cereals in different parts of the world. At the equator, maize cannot be grown on the Andean slopes, where the annual mean is 55°, because the temperature is the same nearly all the year round. The opposite is the case in Scandinavia, for there barley can be grown up to the seventieth parallel of latitude, where the annual mean has fallen to about 32°; for there the summer, if short, owing to the sun being so long above the

horizon, is very hot, and grain sown promptly, as soon as the snow melts, gets its sum of heat compressed into six or seven weeks: on the equatorial slope it never gets it at all! Our own county offers an example midway between the two extremes cited. From our insular climate, we can plant either in autumn or spring, and yet can only grow cereals at an altitude corresponding to an annual mean of 45° or 44° Fahrenheit. At a height on our hill-slopes, 1,200 feet in the West Riding (see table on page 12), with a mean of 44°, the sum of summer heat is never great enough to ripen even barley. Wheat is very little cultivated above 750 feet, where the annual mean is about 45·5, and the summer mean 56·5, a figure corresponding very closely with the limit of possible cultivation in Scotland, reached, according to Dr. Stark, on the line of a summer temperature of 56° on the lowest levels. Oats and barley can be grown in the Riding, however, up to 1,100 feet, although the area in cultivation above 900 is very small. And 900 feet, or thereabouts, is the height at which a zone called the super-agrarian (to be explained presently) commences. It may be added here, *en passant*, that the influence already alluded to more than once as exercised by the Pennine range of hills is seen, in agricultural matters, in the fact that cereals are a better and more certain crop on the eastern side of the hills than on the western. On the latter side grass, clover, and tap-rooted vegetables succeed better than on the former, liking more moisture, and needing a smaller total of summer heat to exhibit them in perfection.

The varying climatic conditions under which grain can be grown, give, however, a second fact in connection with temperature and plant-life. There is a certain latitude as to the length of the time over which the sum of heat that any plant requires may be spread: so that the gross amount is obtained, the period within which it gets it may be long or short (within certain limits), without entailing failure in the result. If two batches of seed (annuals) be planted at the same time, one in an open sunny spot, the other in the shade of a tree or wood, the ones in exposure grow, flower, and ripen seed much sooner than those in shadow; the deficiency of sunlight retards the reproductive processes, but eventually the end is attained equally. The only difference is that a longer time is required for the appropriation of the necessary amount of heat, since a less quantity is forthcoming each day under shelter than in exposure. Everyone knows how a cold spring keeps back vegetation, but how soon the backwardness is made up when the warmth does come. A late spring season does not inevitably entail a late harvest; an early spring with a cloudy, sunless midsummer-tide always lengthens the time needed for ripening: the winter-sown and spring-sown wheat yellow together very nearly. The range in mean annual temperature which can be endured by plants is not, however, nearly so wide as one might perhaps expect. It is stated on the best authority, that a variation of 5° Fahrenheit would kill all or nearly all the species existing in any region. The totals of summer heat needed by various annuals and perennials to perfect themselves has been calculated by the great De Candolle in his *Géographie Botanique*, the holly being said to need 2,200°, the beech 2,500°, etc.; but such figures can at best only be approximations to the truth, and the gross amount needed by our native wild-flowers can only be vaguely guessed at.

The power of resisting cold varies in plants much more than in animals; and the extremes in annual and winter means limit range markedly. Cold when it reaches a certain intensity, kills some species and not others; but long before this point is reached growth is arrested at 32° Fahrenheit. Everyone knows this; but it is not equally palpable, although equally true, that the temperature at which certain species begin to renew their life varies greatly. Some seeds begin to germinate, some plants to leaf and bloom at a very low temperature; for example, the Chickweed, Groundsel, *Draba verna*, Snowdrop, Pilewort, Spring Gentian, *Lamium purpureum*, etc.; whilst others respond in no way to the awakening influence of returning warmth, until the vernal temperature reaches its maximum, as in the case of the later spring flowers, the Ash-tree, Walnut-tree, etc. There is, then, a certain degree of cold below which all life is arrested; and another point, varying with the constitution of each plant, at which sap begins to circulate, and development becomes possible. Looking upon a plant as a vital thermometer (not a new illustration, but no less happy for being borrowed from De Candolle), this growing-point may be said to be its zero; and upon this zero the range of each depends, no less than upon the gross sum of heat needed to perfect its seed. Probably no two species are exactly alike in both respects, although of course, in proportion as the zero is low, so will be the fixed heat requirement; the growing-point running parallel with amount of heat ultimately needed, although the relation is not a fixed one. If a plant has a high zero, say 66° Fahrenheit, like the Palms, it will flourish in no zone save in the torrid, or on its confines; if the zero be intermediate, like that of the Holly (44·6° Fahrenheit), it will flourish in the lowlands of the temperate zone, but its polar limit will soon be reached; whilst if its zero be a very low one, like that of most of the arctic Saxifrages, Cruciferæ, or Ranunculi (33° to 35°), it will be found running far to the north, and upon the peaks of the temperate zone. Its zero does not altogether, however, determine its polar limit, unless it be a perennial; for if an annual like barley, it may have time to complete its vegetative cycle in the summer season, many seeds being able to endure severe frost without destruction.

As regards our own area, on the hill-slopes and summits we have many alpine species of plants, which, since they live year after year in an annual temperature of 40° Fahrenheit, must have a zero at least below that; and judging from their survival on still higher peaks, probably have a zero not much over 35°. Chickweed, Groundsel, etc., are said to have a zero not much over freezing-point. Our montane species generally have a zero of about 38° to 40°, and our lowland ones of 42° to 45° or 46°.

Clematis vitalba has a zero of about 46·4°, Beech 41°, Barley 43°, Holly 44·6°. The figures have been ascertained for but very few of our native species, but an approximation might be got at by ascertaining the elevation of the highest point in Britain where they grow and seed, and the monthly means at that height, taking these in conjunction with the time when they first show signs of life-renewal in spring. A comparison of the table on page 12 with that giving the upward limit of our montane species a few pages beyond this, will give the figures above ventured on. In elucidation of this matter Mr. J. G. Baker gives the common barley (zero 43°) as an example. Planted in October, its seeds germinate and grow on through November, the average daily temperatures in the sun of the two months being 59° and 49° respectively. Its growth then stops, and during December (45° in sun), January (43°), and February (43·5°), it advances nothing, but in the coldest periods is usually covered up in snow, and so protected from the extremity of frost. The heat does not begin to influence it appreciably until March is well on or over; but in April (52° in shade, 69° in sun) it makes a great leap, and thereafter progresses rapidly.

The extremes of cold that kill plants, and determine their polar limit, vary in different species, and are rendered inoperative in many ways. Perennials have to endure through many winters, but may continue to live by their roots in a temperature not high enough to allow of seed ripening. Annuals that only recur when they get heat enough to perfect their seed, can only be affected by extremes of frost during the time they are growing. Most of the annuals of our cornfields are Germanic or eastern in type, and come from countries that have greater cold extremes than ours; with some perennials, such as Holly and Furze, such is not the case. We find these reaching their northern limit in Europe with us, and thus trace in their presence the influence of gulf-stream and insular position in cutting off extremes of temperature. The Plane-tree of south-eastern Europe, the Sweet Chestnut, the Holly, the two Gorses, endure our winters well in general, but the Gorses and Holly are often injured when the thermometer falls to zero, and are restricted to the south-western half of Europe, the latter just reaching south Scandinavia. Here it is not the sum of summer heat that gives our island the advantage, but our comparatively high winter temperature. For the same reason many other species having a comparatively high zero are found within our area.

To summarise the ways in which temperature influences plant distribution, we may consider it under three heads: first, as regards the gross amount of heat required to complete the vegetative cycle from seed to seed again; second, as regards the time when the heat comes, a certain amount being required in a given period; and third, as regards the growing point or zero of each species, that in England, and in our still smaller area, limiting its range altitudinally.

VII. *Climatic and Botanic Zones.*—We come at last to a correlative application of the foregoing climatological data to the distribution of plants.

The student of phyto-geography, as formulated for Great Britain by Hewett Cottrell Watson in *Cybele Britannica*, its *Compendium*, and *Topographical Botany*, will find much of what immediately follows concerning climatic zones of altitude and vegetation to be for the most part an abstract or paraphrase of premisses laid down in the introductory chapters of the works named; but, nevertheless, it may be as well to outline such premisses here for two reasons: first, because those works are scarce, and probably not familiar to above half the readers of these pages; and second, because Mr. Watson's methods of expressing the relations of plants to climate are to be applied in detail to the species of which the West Yorkshire flora is composed.

Every botanist who has had experience of any large area knows that the flora is not everywhere alike when its floral integers come to be reckoned up, but that, on the contrary, the species characteristic of one part are replaced by different ones in another part; and it would also be seen that distance vertically, rising from the plains to the summits, intensifies the change in the flora much more rapidly and thoroughly than distance horizontally. Over the horizontal surface, be it elevated or near sea-level, so long as the variation in level is not above 100 yards, it is latitude that determines the temperature: and since this diminishes but very little for each degree compared to what it does on the Continent, the flora is indirectly impressed by heat, only very gradually. On the other hand, altitude acts quickly and impresses the flora most unmistakably, since it produces the same differences in the temperature over a small expanse of surface that latitude does over the larger one: it also, as we have seen, augments the rainfall, and so the soil moisture, as to which certain genera and species of our flora require a comparative repletion.

The way in which cold (increasing coincidently with elevation) acts most directly upon the flora is in limiting the amount, and perhaps also lowering the intensity, of the competition for growing space amongst the integral forms. The number of species whose organisms permit of their enduring and thriving under lack of heat below a temperature of 45°, become fewer and fewer as the cold becomes greater. Above 45°, up to 70°, the higher the heat the larger the number of species whose structures are suited thereby, and the closer the contest amongst them for supremacy over the surface. As a consequence of this fierce competition, we find that those species which characterise the colder tracts are mostly unable to hold their own in the warmer ones, and the examples of such species amongst the heat-lovers and cold-fearers become few and far between. On the colder tracts, the more elevated or northerly, the few (which are *not* cold-fearers) have the ground to themselves: invaders ascending from the lower levels, whether as seeds carried by wind, dropped by birds, or what not, either soon perish or fail to establish themselves at all, although something of this is due, doubtless, to the restriction of suitable situations that follows from the restricted superficies on the higher levels, and not to causes connected

with inimical temperatures. Invaders descending into tracts with a higher temperature than that they may be said to normally occupy are more common. Seeds may be carried downwards by wind and by birds, and if they find a bare corner in the larger area, are more likely to germinate and establish themselves, since they do not at all fear the extra warmth, as some at least of the ascending trespassers do the increased cold; but in the hill districts it is generally the streams, in spate in autumn and winter when seeds are ripe, that carry, with the down-falling water, species to warmer and lower tracts. If such once get a footing, they often make a brave fight of it, growing more luxuriantly than in their high-land homes, being often what are termed 'strong species,' i.e., of sturdy constitution, strongly individualised, as shown by their tendency to sport into numerous varieties. The Saxifrages, *umbrosa*, *Geum*, and *hypnoides*, are examples of this character. On the mountain slopes themselves the streams are also said to 'bring down the cold' of the higher levels to lower ones, and make the temperature of the rill-sides and faces of rocks, where we generally find the alpine species descend lowest, approximate to those of rocks higher up; but I do not think this operates very vigorously in determining the downward course of mountain species, since it is *not* cold that such love, or warmth that they cannot brook, so much as competition that they fear. If the rill-sides are kept colder by the drip of water about them to any appreciable extent, the cold, as cold, rather limits extension upwards than favours extension in a downward direction.

But in thus writing of limits, upward and downward, as imposed upon plants by cold and moisture, it must not be supposed that such limits are, or can be, fixed absolutely at a given temperature or a given height. The climate natural to any elevation may be varied by local peculiarities, such as southern exposition, shadow, atmospheric humidity, etc.; and a plant may therefore, under some such condition, be found to ascend higher, or descend lower, on one hill-range than it does on another. This, which is true of Great Britain as a whole (the absolute height at which the same species may be found thriving varying by many hundred feet between, say, the Welsh hills and the Grampians), is equally true within the narrower limits of West Yorkshire; on account of accidents of soil and exposure, not merely the absolute but the relative elevation attained by different species is altered. On Malham Moor I have found *Limosella aquatica*, *Hippuris vulgaris*, and *Potamogeton densus*, at about 1,300 to 1,500 feet; and the sub-maritime, lowland Brookweed (*Samolus valerandi*) grows in a moorland hollow near Ringinglow, above Sheffield, at over 1,000 feet above the sea, whilst *Draba incana* descends to 150 feet, *Gnaphalium dioicum* to 100 feet, *Polypodium calcareum* to 250 feet, *Actæa*, *Myrrhis*, *Sagina nodosa*, *Empetrum*, and *Nephrodium oreopteris*, all descend below 100, and some to as low as 25 feet above sea-level, within our area.

But in spite of these reversals, a careful comparison of the facts as to vertical range on one chain of hills with those on another enabled H. C.

Watson to evolve a certain order out of the apparent chaos in which the disturbing and qualifying influences directly and indirectly affecting plant-distribution had steeped British geographical botany; and lay down the general rule that *there is a tendency amongst our species to cease at different altitudes*, whether as ascending or descending forms, and *to have upper and lower limits; to separate themselves into groups that exhibit a certain parallelism in respect of the span and reach of those limits.* These limits we have seen depend, directly or indirectly, upon the fact of species having different zeros and different climatic penchants, and upon altitudes having different temperatures and degrees of humidity; and these, taken in conjunction, made it possible to divide the surface from coast-line to hill-peak into two great divisions—an Agrarian region, within which cultivation is practicable, and an Arctic region, within the bounds of which it is impracticable. Both these regions were again sub-divided into three equal portions by two lines, each one representing the upward limit of certain common and conspicuous species that cease at a considerable distance from each other, this distance approximating tolerably well to 900 feet, the spaces between these ladder-like lines parcelling out the hill slopes into what are styled, by H. C. Watson, Ascending or Climatic Zones. It follows from what has been said as to the causes of species-limitation, that these floral belts or zones represent not merely altitude and latitude, with their coincident degrees of mean temperature, but refer also to the compound factor of climate (determined by many conditions more local than general), and indicate broadly the relative distribution of species due to it. It was inevitable that these zones and their limiting lines should be arbitrary and artificial, and so far untrue to nature, for of course the species on any slope seldom if ever stop abruptly at the given line. The change in vegetation is very gradual, species becoming less abundant as individuals, and isolated in their dispersions, before they totally cease, so that draw the line where one may, it unavoidably severs that which is nearly alike, as H. C. Watson clearly expresses it, the vegetation having a very close similarity immediately on each side of any dividing line, and a wide dissimilarity on the opposite sides of the belt of surface between two dividing lines. Nevertheless, these 'Regions' and 'Zones' do broadly correspond with actual botanical contrasts, the florula as a whole, not less than the specific integers, of each zone showing clearly wide divergencies from the zones next in order above or below.

The six zones of *Cybele Britannica* and the *Compendium* thereof 'traced upwards, from south to north, from lower grounds to higher,' are as follows:—(1) Infer-Agrarian, the tract of Clematis and Rubia; (2) Mid-Agrarian, characterised by Rhamnus catharticus and Cornus sanguinea without Clematis and Rubia; (3) Super-Agrarian, limit of Pteris and Cultivation without the Rhamnus and Cornus; (4) Infer-Arctic, with Bell Heath present, but no Pteris; (5) Mid-Arctic, Ling being found, but no Heath; and (6) Super-Arctic without even Ling.

These six zones were differentiated with the whole surface of Britain under consideration; and, for such a wide extent, are considered by Mr. Watson to be quite as many as can be satisfactorily distinguished botanically, being framed not merely with reference to altitude, but so as to embrace the effect of the eleven degrees of latitude through which Britain stretches from the Lizard Point at 41° 50′ to North at 61°. But with a smaller area, such as the county of York, exclusively in view, more might be pretty safely ventured on. Were there any great need or practical value in such further sub-division, it is, for example, quite feasible to divide the area of West Yorkshire into eight upward leaps or belts of 300 feet, and eliminate certain small groups (ten species in each) characteristic of them in so far as ascending or descending only to such zones. Such a list will presently be framed, but unfortunately it would almost certainly fail to hold good for any other area of equal extent—such, for example, as the group of counties forming Mid-Wales, the Lake District, or the Lowlands of the Scottish border. Or, again, the lowest limits of a few boreal species that happen to be pretty fixed ones, such as that of the Cloudberry, might, *where such species occur*, be used to trace altitudinal lines intermediate between those bounding the Watsonian zones, but they could only have a very local application, and consequently no very considerable value.

One other point in connection with Mr. Watson's zones must be touched upon, as having an application to our area, before the zones exemplified in West Yorkshire are considered seriatim. A reference to the table on p. 12 will show that the range of temperature of each zone is about 3°, so that small as is the latitudinal diminution in temperature at sea-level in Britain, compared with what it is in central Europe, it is yet, between the extreme latitudes of 50° and 61°, much more than equivalent to one zone in altitude. The result of this is necessarily a *successive overlapping of the zones as we proceed from south to north*, the zones at sea-level changing from Infer-Agrarian at latitude 50° to Mid-Agrarian at latitude 53½°, and to Super-Agrarian at about 57°; or, to express it in other words, we may say that the altitudinal limit of the lowest zone declines from 900 feet above sea at 50°, to sea-level at 53° 30′ whilst the upper limit of the Mid-Agrarian zone declines from 900 feet above sea at 53° 30′ to sea-level at 57°, whilst the limit of possible cultivation marking the Agrarian region (upper line of Super-Agrarian zone), which is about 1,800 feet in West Yorkshire, declines to about 1,200 feet in the Scottish Highlands, and actually touches sea-level in the Shetland Islands at latitude 61°. The following little woodcut (the idea of which is borrowed from one somewhat less elaborate given upon the map accompanying *Cybele Britannica*), will, however, convey diagrammatically a more correct impression of the way in which the zones run out northwards than any words.

None of the British counties include all the six zones, unless it be Caernarvon that, with its Snowdon (3,751 feet), and its sea-level about 53° lat., may be considered as doing so, because the loftiest hills all happen to occur

DIAGRAM OF THE ASCENDING OR CLIMATIC ZONES, ILLUSTRATING THEIR DECLINATION AND OVERLAP FOR LATITUDE.

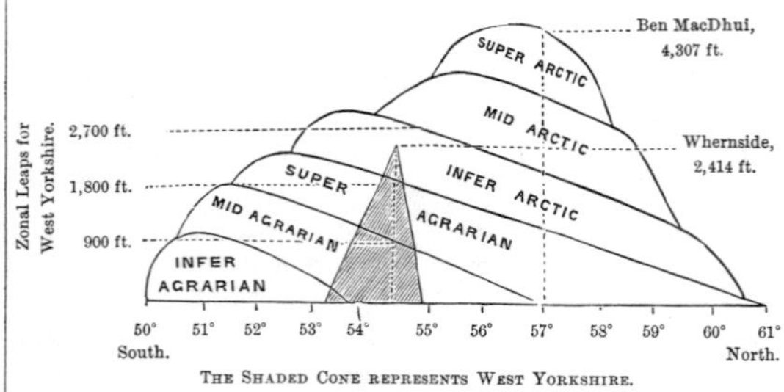

THE SHADED CONE REPRESENTS WEST YORKSHIRE.

in the north of England or Scotland. In most of the southern counties of England only one of these zones, the lowest, can be found expressed; in a few others, two; in several others, three, either the first, second, and third, or the second, third, and fourth; in most of the hillier Scottish counties—two, three, four, and five, or three, four, five, and six, while in but one, I believe, can five be counted. The West Riding of Yorkshire, then, having *four* out of the six within its limits, may be taken as favourably situated for illustrating zonal distribution, seeing that, although for the most part in too high a latitude to afford a full exemplification of its botany, it yet has a portion of its surface clearly within the Infer-Agrarian zone. We get in West Yorkshire neither of the two highest Arctic zones, since the three highest summits only just touch upon 2,400 feet.

With regard to the rest of its surface, the limits of its several zones may be taken as reached in leaps of 300 yards; its lowland surface generally, like that of North Yorkshire, up to 900 feet, forms the Mid-Agrarian zone; its hill-slopes between 900 and 1,800 feet as lying within the Super-Agrarian zone; and those portions of its hill-ridges and peaks above 1,800 feet, as reaching more or less distinctly into that lower third of the Arctic region styled Infer-Arctic. In the Flora which will form the latter portion of this work, these zones will, for brevity, be designated the first, second, third, and fourth, as in the following table, where they are given co-ordinated with the position, altitudes, average temperatures, and botanical limitations they express.

TABLE SHOWING THE ASCENDING CLIMATIC ZONES OF WEST YORKSHIRE, WITH THE ALTITUDES, MEAN TEMPERATURE, SIZE OF AREA, AND BOTANICAL CHARACTERISTICS.

Zones.		Situation.	Altitudinal Range.	Mean Temperature.	Approximate Area.	Botanical Characteristics.
I. Infer-Agrarian	Agrarian Region.	Surface drained by the tributaries of the Trent. South of Lat. 53° 30′.	50 ft. to 350 ft.	49° to 48° Fahr.	50 sq. miles.	Clematis, Viburnum Lantana, and Carduus acaulis as natives.
II. Mid-Agrarian		Lowland area of all the river-basins north of Lat. 53° 30′.	25 ft. to 900 ft.	48° to 45° Fahr.	2,000 sq. miles.	No Clematis or Viburnum as natives. Wheat cultivation. Rhamnus catharticus, Cornus sanguinea and Acer campestre as natives.
III. Super-Agrarian		Moorlands and hill-slopes generally, west of a line drawn from Ripon to Sheffield.	900 ft. to 1,800 ft.	45° to 42° Fahr.	600 sq. miles.	No Cornus or Rhamnus as natives. Oat and Barley cultivation, up to 1,600 feet, but no Wheat. Hawthorn, Hazel, and Cloudberry.*
IV. Infer-Arctic........		The summit ridges of the highest hill-spurs, and the actual peaks north and west of a line drawn from Ripon to Colne.	1,800 ft. to 2,414 ft.	42° to about 40° Fahr.	30 sq. miles.	No cultivation possible. No Hawthorn or Oak. Rowan-tree, Birch, Juniper, still present; with Cloudberry, Sedum Rhodiola, Salix herbacea, Carex rigida, Calluna, Juncus squarrosus and Nardus.

* NOTE.—Cloudberry occurs as low as 1,600 feet, i.e., within the 200 uppermost feet of this zone; but, strictly, its presence shows that that upper portion *at our latitude* is within the Infer-Arctic zone, or at least to the extent of 200 feet overlapped by it.

We must now amplify somewhat the bare indications given in the second, third, fourth, and fifth columns of our table, before a detailed examination of the distribution of species is attempted.

VIII. *Limits and Areas of the four West Yorkshire Zones.* Mr. Watson, in *Cybele Britannica*, defines his Infer-Agrarian zone as embracing all the *lowland* surface of England 'southward from the Dee and Humber (continued into the Trent).' This, if I understand it rightly, would divide the water-shed area of the latter river into two portions: that on the north of the main stream in the Mid-Agrarian, and that on the south in the Infer-Agrarian zone; and would, moreover, make the temperature coinciding with those zones overlap, and be irrespective of latitude at equal elevations: it would place North Lincolnshire in the lower zone, and such towns as Derby and Nottingham, with most of the county of Notts, all of them half a degree in latitude further south, in the colder zone. The mean temperature compatible with the Infer-Agrarian zone in Britain, I take at 51° to 48°, that of the Mid-Agrarian at 48° to 45°, and applying this and the botanical test afforded by the presence of *Clematis* and *Viburnum lantana* as natives, I have, in this work, considered one portion of West Yorkshire on the extreme south-east as belonging properly to the Infer-Agrarian zone. It is only a very small area, not above fifty square miles in extent, but I have been constrained to take this course, seeing that its average warmth is distinctly higher, and its floral and agricultural features distinctly richer, than any other part of the lowland area in West Yorkshire. I thus demarcate by the hydrographical method a first, or lowest, climatic zone.

With regard to the rest of the surface, I follow a plan adopted by J. G. Baker, when treating of the North Riding, in using the contour lines of 900 and 1,800 feet (indicated on the six-inch Ordnance Survey Maps) as separating the surface into other three zones, taking the limits of each as reached in upward leaps of 300 yards. The lowland area in all the river-basins north of 53°30′, below 900 feet in elevation, is in this way taken as constituting the second or Mid-Agrarian zone; the hill-slopes, moors, and other parts of the surface, ranging in height between 900 and 1,800 feet, embracing the third or Super-Agrarian zone; and the rest, a now very restricted area, confined to the higher moorland plateaux, summit-ridges, and peaks, over 1,800 feet in elevation, as making a fourth or Infer-Arctic zone. If there be any error in these altitudinal limitations, it is, I believe, an error in a safe direction—making the zones rather wider than need be, the two higher Agrarian ones especially. As a matter of truth, with regard to the West Riding alone (owing to accidents of position, etc., which it is not necessary to name here), I believe 800 feet leaps would more nearly express the present *facts* with regard to the growth of wheat, the occurrence of *Rhamnus catharticus*, the cultivation of oats and barley, and the limit of the production of garden vegetables; but what the *possibilities*, under a more thorough system of drainage, etc., on the higher levels might be, I am not prepared to decide. Formerly, in the North-West Yorkshire dales,

'forests' of 'scrub' and thickets covered elevated surfaces at 1,600 feet to 1,800 feet, where are *now* only wastes of peat haggs, etc.

Reclaiming land always improves its climate in some degree, and on the higher gritstone levels does so very considerably; for the climatic conditions in such places arise mainly from an excess of evaporation consequent upon a surcharge of water in the soil.

As respects the relative extent of the three upper zones in West Yorkshire, it is in an inverse ratio to altitude. The closest estimate I have been able to make, from an examination of the area included within the several contour lines as viewed on the flat upon the map, is that given in the foregoing table; viz., 30 square miles for the Arctic zone, a little over 1 per cent. of the whole surface of the West Riding (2,760 square miles); 600 square miles for the Super-Agrarian zone, or about 22 per cent. of the whole area; the remaining portion, nearly 2,000 square miles, three-quarters of the whole, coming undoubtedly within the Mid-Agrarian zone.

IX. *Physical and Botanical Characteristics of the West Yorkshire Zones. The Infer-Agrarian Zone.*—We see this only within the river-basin of the Trent's tributaries. The surface is highly tilled, and agriculture brought to a state of perfection rivalled only in a few other parts of the rich warm-soiled Permian limestone tracts. Rocky cliffs are not rare, but limited in extent, and somewhat miniature as to size, occurring mainly where the streams have cut channels for themselves through the limestones: they are, for the most part, well clothed with trees or an indigenous undergrowth of Hazel, Ash, Yew, Privet, etc. Lindrick Common is the only barren rocky area of any extent; and, so far as I know, this zone is wholly wanting in those patches of lowland heath or bog, such as still exist in many parts of the Mid-Agrarian zone. The highest point of the pleasantly diversified surface is Moses' Seat (330 feet), in the vicinity of Lindrick Common. The Maltby Wood-Lea weather-worn rocks are about 250 feet above the sea. There are more elevated points on the western escarpment of the limestone about Harthill and Wickersly, whence brooks drain westward into the Rother, but they do not properly come within the first zone.

The mean temperature has an annual average of 49° to 48·5° according to elevation, and nowhere seems to fall below 48°, with a hiberno-æstival difference of 22° to 23°. The effect of the comparatively warm summers may be seen in such agri- and horti-cultural data as I have been able to gather from residents. Wheat is the principal grain-crop, and the best; but barley and oats succeed nearly equally well, and the yield is above the average of the rest of the Riding. The gardens are rich in their vegetable produce, and in their shrubbery, evergreens, etc., show their climatic advantages in their luxuriance. *Garrya elliptica* and *Arbutus Unedo* appear to thrive well, and at Tickhill the first-named has borne uninjured such severe frosts as have visited that place. The apricot ripens its fruit well on the sunny walls of the village cottages, and the walnut ripens its

fruit almost every year. The common black grape will yield passable fruit without protection, most years. The finer garden fruits, such as the peach, in the high-walled park gardens do well. Of the wild berry-bearing shrubs, the common sloe generally produces its pretty bloomy fruit more abundantly and regularly than it does in most other parts of the Riding; whilst the Purging Buckthorn, *Cornus sanguinea*, Privet, Yew, and Spindle-tree—in characteristic profusion everywhere—ripen their berries super-abundantly.

In the total number of its indigenous plants the area under consideration is too small for successful vying with the Mid-Agrarian zone; but, for its size, is richer in some classes of the specially lowland species. The *Clematis*, *Berberis*, *Viburnum Lantana*, *Onobrychis sativa*, *Bupleurum rotundifolium*, *Humulus* (which fruits), *Carduus acaulis*, and *Antirrhinum Orontium*, occur as true natives or as constant colonists; whilst the district also yields *Dipsacus pilosus*, *Centunculus*, *Nepeta Cataria*, and *Calamintha officinalis*, some of which do not occur elsewhere in West Yorkshire, and the others are incomparably more frequent than in the limestone tracts of the Mid-Agrarian zone.

The Mid-Agrarian Zone, in regard to its botanical characteristics, needs no more detailed description than that already given; as we have seen, it composes three-fourths of the whole surface. Rocks and cliffs are not frequent in the lower two-thirds of any of the river-valleys, save such as stand out at Hackfall, Knaresborough, Thorp Arch, Ledstone Beacon (278 feet), Fairburn, Kirk-Smeaton, Clifton Beacon, near Rotherham (417 feet), and Conisborough, where the rivers break through the belt of Magnesian limestone. The exceptions to pleasant agricultural levels in the eastern plain are found in the solitary eminences of Brayton Barf and Hambleton Haugh. If a line be drawn from Ripon southwards to Sheffield, through Leeds, Wakefield, and Barnsley, nine-tenths of the uncultivated land in this second zone will lie upon the west side of the line, where the 'dales' may be said to begin, as distinguished from the plain-like valleys of the river-basins on the east of such a line. The exceptions as to rocky 'edges' of gritstone to the east of it occur at Spofforth and Plumpton, Etchell Crags, Scarcroft, Woolley Edge, and Monk Bretton. There are, however, yet unreclaimed, considerable stretches of lowland heath and boggy moor in the vicinity of Thorne and Goole, with smaller bits near Rawcliffe, Gateforth, Askham, Marston Moor, Alwoodley, and Staveley. Throughout this eastern two-thirds, well-tilled arable land, mowing-grass fields and pastures are the rule, and woods the exception. Some few of the districts have an abundance of wood in the form of non-indigenous 'plantations' around parks, etc.; but at the present day at least three-fourths of the forest-trees are planted ones. Bishop Wood, near Selby, is the largest of these, and pretty extensive areas of wooded ground may be found at Hackfall, in Studley-Royalty, at Bramham, Kidhall, Micklefield, Ledsham, Aberford, Campsall,

Wentworth, Tinsley, &c.; but little of it is really indigenous. The oaks at Studley, Aberford, and Wharncliffe are fine, and the beeches at Thorp Arch and Ledsham. The native wood, however, is mainly confined to the aboriginal scrub and brushwood on the confines of the uncultivated lowland heaths already mentioned.

In the area to the west of the imaginary line indicated, the case is different. From Sheffield, Barnsley, Leeds, Harewood and Ripley, where the 'western dale' portions of the rivers flowing east may be said to commence, moory uncultivated ground, low rocky banks, or wooded slopes begin to occur more frequently, and for long distances up the dales, until they branch off into smaller glens and cloughs, the surface in the immediate vicinity of the rivers continues to fall within this second zone. For example, in the far north-west corner, Rawthey Bridge, at the head of Rawthey dale, is only 700 feet above the sea; St. Mark's Chapel, Rawthey dale, is 550 feet; the confluence of Cautley Beck with the Rawthey, 600 feet; Sedbergh town, 400 feet; the Crook of Lune, amidst frowning precipitous heights, only 590 feet; Howgill Chapel, 600 feet. Again, in Ribblesdale, grand as are the mountain masses at its head, two-thirds of its area within our Riding lies below 450 feet in elevation; Horton Bridge is 750 feet; Settle Bridge, 440 feet. In Nidderdale, Ripley lies at 240 feet; Lofthouse Bridge is only 540 feet, and Harlow Hill, Harrogate, 596 feet; and yet these places are all for the most part, surrounded by a mountain flora that, as we shall see presently, when discussing the descent of 'montane' species, follows the river-banks even down to lower levels, until such as *Thalictrum flexuosum*, *Myrrhis odorata*, *Stellaria nemorum*, *Rumex aquaticus*, and *Allium complanatum* mingle with characteristically lowland forms in the eastern plain. It is the same in Wharfedale and Airedale: in the former, Buckden Bridge is only 732 feet above the sea; Kettlewell Bridge, 669 feet; the Inn at Kilnsey, 618 feet; Bolton Bridge, 358 feet; Otley Bridge, 220 feet; and Harewood Bridge, 100 feet. In the latter dale, Malham Cove foot, amidst magnificent scenery, is only 680 feet; Gargrave Bridge, 320 feet; Shipley Bridge, 200 feet; and the Aire at Kirkstall Abbey, 150 feet.

Still, much of the wood in the western dales, within this zone, off the river-banks themselves, cannot safely be considered indigenous; and up the dales, to 750 feet at least, Larches or Sycamores interspersed amongst Ash, Hazel, Rowan-tree, and Alder, will reveal the artificial origin of most, if not all, of the wood.

Very nearly all the arable land in the Riding falls within this second zone, and two-thirds of it within the area east of the line drawn from Ripon to Sheffield, the other third being under grass. On the west side of the line the proportions are just reversed; the two-thirds is that verdant limestone grazing pasture for which the 'Craven' district is renowned, and the arable land, chiefly used for green crops and oats, is the third part.

The Riding contains 1,768,000 acres. Of these one-third (568,000 acres) is under cultivation, another third (600,000 acres) is in the form of permanent grass pasture, and the remaining third (600,000 acres) is heathery moor and unreclaimable mountain surface. Proportionately to the amount in the Super-Agrarian zone, the unreclaimable surface in this, below 900 feet, is very small indeed, but I am not able to give it in figures. All the uncultivated land in the eastern part of this zone could be, and in time no doubt will be, reclaimed; but some part of that in the western half, although below 900 feet, must no doubt, from its bare, rocky, soilless character, remain ever as now. The fields in the eastern half are generally partitioned off by hedges; in the western half, by stone walls.

The mean temperature of this second zone ranges from about 48° to 45°, there being a lowering of one degree for every 300 feet. Wheat is principally cultivated up to 600 feet, and may occasionally be seen up to 750 feet. I have not, however, seen any to my knowledge above 700 feet; and I very much doubt if it is ever grown up to within 100 feet of the upward limit of this zone in this Riding, owing more, of course, to the nature of the surface at that height than to the temperature. Oats and barley are grown successfully quite up to the limit, of course; but again, owing to the nature of the surface in the narrow western dales, the area of arable land is very much restricted. Fields of Sainfoin, Lucerne, Flax, and Comfrey are not unfrequent over the belt of Permian strata; the Teazle is a common crop about Milford, Sherburn, etc.; and both Mustard and Chicory may be seen on the Trias.

The hiberno-æstival difference is found to vary from 22° to 20°, and the widest variation, with the hottest summers, is found in the Ouse-bounded plains to the east of Ripon, Ripley, Otley, Leeds, Wakefield, and Conisborough. Within the cordon so drawn those fruit-trees more sensitive to cold, and which need more warmth to ripen their crop than the Apple, Cherry, and hardier Pears and Plums (all of which are grown against house-sides up to 650 or 700 feet in the western half), succeed in warm seasons quite as well as within the Infer-Agrarian zone at Firbeck and Maltby, but are found to fail somewhat oftener than they do there. The Apricot, Greengage, Nectarine, and Peach, ripen fairly well, most years. The Mulberry at Wighill, Riccall, near Cawood, and other places, ripens its fruit tolerably, if not fully, save in exceptional seasons; and the same can be said of the Walnut at Meanwood and Collingham. The common black Grape and Spanish Chestnut do not appear to ripen, so far as my inquiries have gone; but Mr. Baker, in his 'North Yorkshire' (p. 52), remarks that at Thirsk the former does do so, and that the nuts of the latter are quite eatable after being kept awhile.

The Super-Agrarian Zone. The line of demarcation, at 900 feet, between the Mid- and Super-Agrarian zones, runs with great tortuosity round the lower fourth of the bases of the greater mountain masses, and about half-way up round the gritstone spurs that stretch south and east

between the great river-basins. These lines, and the included area, are best seen on the six-inch Ordnance Map—pages of description would be necessary to make them clear to the mind without. This third zone all lies to the west of a line drawn from Masham on the north, through Poole, Bramley, Ossett, and Wortley, to Eccleshall, on the S.W. of Sheffield. The greater part of the surface consists of rolling heather-clad moors given over to Red Grouse. Rocky edges of weathered Millstone Grit and Flag Rock are a characteristic of this zone in its lower parts, and in the districts of Carboniferous Sandstones may be seen everywhere, bare or clothed with a partly indigenous growth of Oak, Birch, Rowan-tree, etc., at the edges of the undulating peat moors, where they overhang the dales; the surface generally between the levels of 1,200 and 900 feet declining rapidly, in steps of rock that appear (and often truly) as though they had slipped from higher levels, or had been dropped by icy agencies half-way down the débris-clad declivities. Other like masses of rock, weathered into a vast number of weird forms, often stand out amongst the rolling sea-like swells and stretches of heatherland. Examples of both these may be seen in the Cow-and-Calf Rocks at Ilkley (900 feet), those below the 'edge' of Otley Chevin (925 feet), Hardcastle Crags (1,225 feet), Brimham Rocks (990 feet), the white-pebbled grit-boulders on Bowland Knotts (1,389 feet), the "rocking-stones" on most of the higher grit moors, etc.

This Super-Agrarian zone is very different in character, however, where Limestone is at the surface instead of Sandstone, as is the case over the greater extent of surface in the Craven district, about Ingleton, Settle, Skipton, the plateaux of Malham Moor, Bcrdley Mastiles, Coniston Moor, etc. No rolling ridges of heath or peat here: the surface, where tolerably level, consists of 'pavements' of tabular blocks of limestone, with a peculiar growth of plants in the innumerable fissures; and the vertical range between 900 and 1,800 feet is either attained by sheer or terraced leaps of wall-like cliffs or rounded slopes of pasture, beautifully green with fine herbage.

The mountain 'passes' between one dale and another, within the Riding, as likewise those between our various dale-heads and the adjacent parts of the North Riding, Lancashire, Cheshire, and Derbyshire, with two exceptions only, all come within this third zone. The valleys of the main rivers cease at about 900 to 1,000 feet, and the elevations reached by the bridle paths (carriage roads some of them) over the ridges into other watersheds are as follows:—From Sheffield, by Ringinglow to Hathersage, 1,300 feet; and by the Rivelin to Derwent Chapel, 1,180 feet. From Penistone, by Salterscroft, to Woodhead, 1,050 feet. From Holmfirth, by Holme Moss, to Woodhead, 1,750 feet; and across Dean Head Moss to Greenfield, 1,650 feet. From Marsden, over Standedge, to Delph, 1,450 feet. From Stainland, over Blacker Edge, to Denshaw, 1,400 feet. From Ripponden to Littleborough, over Blackstone Edge, 1,350 feet. From Hebden Bridge, by Widdop Cross, to Colne, 1,500 feet; and across

Oxenhope Moor to Haworth, 1,400 feet. From Haworth to Colne, by Combe Hill Cross, 1,200 feet. From Slaidburn, over Whin's Brow, to Lancaster, 1,450 feet; over Botton Head Fell to Wray, 1,200 feet; and over Bowland Knotts to Clapham, 1,350 feet. From Kingsdale into the Deepdale of Dentdale, 1,650 feet. From Ingleton, by Wethercote Cave (900 feet), into Upper Ribblesdale, 1,059 feet. From Barton into Dentdale, 1,025 feet. From Dentdale Head into Widdale, between Wold Fell and Widdale Pike, 1,700 feet. From Upper Ribblesdale into Langstrothdale 1,350 feet. From the Gearstones Inn, Ribblesdale (1,050 feet), by Newby Head Inn (1,400 feet), into Widdale, 1,421 feet. From Buckden to Kidstones, in Bishopdale, 1,400 feet. From Kettlewell to Coverdale Head, over 1,600 feet. From Lofthouse, in Upper Nidderdale, to West Scrafton in Uredale, the road under Great Haw attains 1,590 feet. From Littondale by Heselden Gill, between the peaks of Fountains Fell and Penyghent, the road attains 1,390 feet; that between Kilnsey and Malham, 1,284 feet; whilst that between Malham Tarn House and Stainforth is 1,336 feet. The good road between Garsdale and Upper Uredale, at the summit of the pass west of the Moor Cock Inn (1,020 feet), is 1,050 feet. The pass from Sedbergh, up Rawtheydale to Ravenstonedale, is lower; it attains only 950 feet, and where it leaves the Riding at Rawthey Bridge is only 750 feet.

In this third zone the truly indigenous woods of the Riding are for the most part found. On the rockier limestone or gritstone slopes the smaller trees and brush-wood are chiefly made up of Oak, Wych-Elm, Ash, Hazel, Rowan-tree, *Pyrus rupicola*, Holly, Elder, Yew, Birch, and Juniper. The Hawthorn is quite rare above 900 feet. Some of the trees and shrubs mentioned are almost confined to the limestone, aboriginal 'scrub,' but the contrasts due to the kind of rock at the surface will be brought out more fully in the chapter on Lithology; and the principal species of plants that are characteristic of, or found in greatest profusion in this zone will be found indicated in tabulated lists a few pages further on.

The mean temperature of this zone ranges according to elevation from 45° to 42°; and its upper line is supposed to mark the limit of possible cultivation. But within our area, at any rate, vegetables, fruit-trees, and field-crops, associated with the dale-head farms, cease to exist practically at 1,100 feet; the surface capable of conversion into arable land being in this zone so very small, probably not above one per cent. of the total acreage under the plough. Fields of Oats and Barley (after remaining upon the steeps unripened in November) may be noticed now and then up to 1,000 feet, and small walled enclosures of potatoes, with a few gooseberry bushes, in the shelter of farm buildings or planted coppices of Larch, occur somewhat higher on the Malham plateau, as at Middle House Farm; and certain of the summer vegetables, together with rhubarb, are grown up to a like elevation.

The Infer-Arctic Zone.—The only continuous tracts of any size that lie within this zone occur in connection with the undulating peaty plateaux of

Howgill Fells, Baugh Fell, Ingleborough, Fountains Fells, the ridge bordering Langstrothdale on the north from Cam Fell to Buckden Pike, and the area lying between the several peaks of Great Whernside. The highest summit of all, Whernside (2,414 feet), although rising 600 feet above the lower contour line of this zone, is merely a long ridge; although here, and upon Cam and Widdale Fells, the Scar Limestone reaches its highest elevation in the form of a limited tabular and mural terrace. There are peaks rising into this fourth or uppermost zone; but the extent of their united areas does not exceed, at the very most, thirty square miles. It may, perhaps, be worth while to mention the respective heights of these peaks and their characteristics, since they, much more than lower eminences, have a bearing upon the altitudinal range of certain species in the flora of the Riding. It is nearly always *when the limestone is thrown up highest* on such peaks that the hardier plants of the lower levels are found to attain their loftiest ascensions; finding there, in the sheltered fissures of the drier rock, local conditions that to some extent balance or nullify the extreme degrees of cold and moisture that properly belong to the altitude. Beginning at the extreme N.W. corner of the Riding, overlooking Lowgill Station in the Lune Valley, are Fell Head (2,045 feet) and Wind Scarth (2,006 feet), with Black Force (near which grow *Sedum anglicum* and *Epilobium alsinifolium*) on their northern slope. Two miles to the south-east the White Fell plateau culminates in the Calf (2,220 feet), Arant Haw (1,989), and Great Dummack Pike (2,150), a mile or two apart from one another, with the stupendous crags of Cautley and the 'Spout' Fall (1,800 feet), where *Pyrola secunda* grows, for the eastern declivity. All these summits are of Silurian Slate, and the plateau between them is boggy and heathy in part, and lowland plants conspicuously absent. On the south slope the Rowan-tree attains its highest limit (1,950 feet) in Yorkshire. Five miles to the east the grand mass of Baugh Fell has at least four square miles of surface in this upper zone; but it is, for the most part, dreary peat bog and swamp: its summits are West Baugh Fell (with a tarn), 2,070 feet; East Baugh Fell, 2,216 feet; Tarn Hill (above four eastern tarns), 1,900 feet. All the higher surface here is gritstone, with a florula of the type so often mentioned. North of Baugh Fell, across the Rawthey-gill ravine, is the Swarth Fell ridge, rising from 2,050 feet to 2,235 feet. To the south of Baugh Fell the western peak of Rysell attains 1,890 feet. East of this, Widdale Fell stands out boldly, with its pike, 2,203 feet, its tarn, 2,000 feet, and its southern buttress of Wold Fell, 1,829 feet above sea-level; and on its south and east declivity there are terraces of limestone that crop out as high as 1,900 feet. South-east of Wold Fell, across Newby Head Moss, rises the lithologically similar eminence of Cam Fell (1,831 and 1,877 feet), also with a limestone escarpment, nearer the summit, but not really much, if at all, above 1,800 feet in elevation. West and south of Cam Fell are the Arctic zone ridges and peaks of Whernside (1,886 feet at south end, 2,414 feet in the centre, and

1,920 at the north end of the ridge), with a bleak rocky summit, and limestone escarpments at the base of its last slope; Gragreth or Dent Crag, 2,250 feet; Barbon or Middleton Calf top, 1,999 feet; Ingleborough, 2,373 feet; Simon's Fell, 2,088 feet; and Gallbec Park Fell, 1,836 feet,—all with bleak grassy or rocky summits. The peaks of the ridge on the north of Langstrothdale, most of it heathery or turfy moor, rising into this Arctic zone, are Kidhow (1,890 feet), Fleet Moss Peak (1,850 feet), Outershaw Cairn (1,937 feet), Yockenthwaite Pike (2,109 feet), Buckden Pike (2,302 feet), Starbotton Mere top (2,100 feet). In this vicinity are the only two passes that rise into the Infer-Arctic zone: that from Buckden over Stake Fell to Semmer-water (1,818 feet at the summit), and that from Outershaw over by Greenside to Hawes, which is 1,850 feet. South of Cam Fell, between Langstrothdale and Littondale, the dividing ridge culminates in Firth Fell (1,850 feet), Birk Fell (2,001 feet), Horsehead Pike (1,985 feet), and Cosh or Cush Knott (1,959 feet). Further to the south are the grand masses of Penyghent and Fountains Fell, considerable areas of grassy and turfy surface within this zone lying between their highest points: Foxup Pike, the N.E. extremity of Penyghent, being 2,231 feet, the south end 2,273 feet; whilst Fountains Fell is 1,944 feet at its southern rim of Knowe Fell, rising to 2,164 feet above Thorogill Beck, 2,191 feet at the summit above a lonely tarn, and 2,048 feet at its Scoska Moor or Darnbrook Fell extremity. That buttress of Great Whernside overlooking Coverdale is 2,187 feet, and Little Whernside to the east is 1,984 feet. South of these points there is a considerable area of turfy swamp above the contour line: the south peak of Great Whernside is 2,245 feet, the highest point 2,310 feet, and a little below Stonehead Pike (2,000 feet) is a Well called Crooka—the highest in the Riding, I believe. From here the hills south and east decline, and only a very few summits surpass 1,800 feet, of which Meugher Dike is 1,883 feet. On the slopes of this group of Fells the limestone is not thrown up so high as it is to the westward. In the Aire basin, Grisedales Great Scar (1,815 feet), two miles due west of Malham Cove top, is the only part touching this zone; but hereabout the limestone rises nearly to it, if not quite, and *Pyrus rupicola* attains its highest limit. Of the moorland ridges of the South Yorkshire moors, where the Colne and Calder tributaries take their rise, the only one touching this zone is on Holme Moss, where an elevation of 1,859 feet is said to be attained.

The botanical characteristics of this zone can best be indicated when analysing the zonal and altitudinal range of all the native species in the West Riding flora. The Brake-fern and brushwood of the slopes from 1,200 to 1,700 feet is completely absent above 1,800 feet. The only wood occurs in the shape of solitary weather-beaten shrubs and stunted trees of Rowan, Birch, Juniper, and Hazel, growing from the crevices of the slate or limestone rocks in two or three sheltered 'combes' and 'gills' with a south or west aspect; and nearly all the alpine Sedums, Saxifrages,

Epilobiums, Ferns, Mosses, and Lichens, that constitute the interesting species, amongst a monotonous iteration of individuals of commoner type, descend into the Super-Agrarian zone. Only three species appear to be absolutely confined to the Infer-Arctic zone.

X. *Zonal Range of West Yorkshire Flora.*—We have now to deal with the vertical distribution of all our indigenous species; but, in referring them to such and such zones, let it be noted that there is no absolutely fixed line where any two species meet, and cease, abruptly. Individual elasticity in power of adaptation to climatic and stational conditions, brings about an inter-mingling on the confines of each zone or belt.

The most striking effect of a change of level is seen in passing from the well-cultivated eastern plain upwards through the slopes of the dales to the western mountain summits, or vice versâ. The species whose presence on the lowest levels is connected with the warmer summer temperature obtaining there, are observed to become scarcer; then, as the slopes are climbed, yet other species, whose constitutions are not so sensitive to cold, but which begin to find the kind of stations they affect decrease, begin to thin out. This diminution does not proceed evenly and without any compensation, however. An infusion of new forms gradually takes the place of those familiar to the eye on the lower levels, and they become as individuals relatively the most abundant, but in number of species they never counterbalance the absentees.

Throughout all Britain the two lowest zones have incomparably the greatest number of species, the diminution proceeding in accelerated proportion as the Super-Agrarian and Infer-Arctic zones are successively reached. This is not wholly due, however, to the lowering in temperature and increase in humidity, but to the greatly smaller area of surface (meaning restricted variety of station) in the higher zones.

But before we can contrast the vertical range of species in other parts of the country, it will be necessary to show what that vertical range is with us. This is accordingly done in the following table:—

ANALYSIS OF THE ZONAL RANGE OF WEST YORKSHIRE FLORA, TREATING ONLY OF THE 904 INDIGENOUS SPECIES.

CLIMATIC ZONES.	Area of Zones in Square Miles.	Percentage of Total Area.	Number of Species occurring.	Percentage to Total Number.
I.—Infer-Agrarian (Trent District)	50	$1\frac{1}{2}$	523	58 per cent.
II.—Mid-Agrarian (0—900 ft.) ..	2,080	76	873	97 ,,
III.—Super-Agrarian (900 to 1,800 ft.)	600	22	450	50 ,,
IV.—Infer-Arctic (1,800 to 2,400 ft.)	30	$1\frac{1}{9}$	132	13 ,,

The percentages in the above table are, I believe, in no wise vitiated by the exclusion of the Denizens and Colonists in our flora: if anything they should be rendered more accurate, seeing that accident regulates

their dispersion much more than it does that of indigenous aboriginal species. Very few of the cornfield Colonists ascend into the middle zone, and still fewer of the viatical species that accompany human habitations. A few, no doubt, ascend as high as 1,200 feet in oatfields and potato patches (I once saw *Galeopsis versicolor* at 1,500 feet!), but my notes do not furnish above a dozen altogether; and corn is not grown in the West Riding up to 1,200 feet to anything like the extent it is said to be in North Yorkshire (upon the Hambleton plateau) by Mr. J. G. Baker. He, however, includes both Denizens—of which very few indeed occur above 900 feet in West Yorkshire—and Colonists in his estimate of zonal range for North Yorkshire and Northumberland and Durham. Mr. Watson does the same in his estimate of the zonal range of the 1,425 species treated as the total number of distinct British forms in his *Cybele*, but that their inclusion makes very little difference the following table will show. In it the species found in the respective zones of three widely different areas will be found contrasted with the figures just given for our own.

TABLE SHOWING THE PERCENTAGES OF SPECIES OUT OF THE TOTAL FLORAS FOUND IN THE DIFFERENT ZONES, FOR ALL BRITAIN, FOR NORTHUMBERLAND WITH DURHAM, AND FOR NORTH AND WEST YORKSHIRE.

AREAS.	Square Miles.	Total No. of Species used in Estimate.	PERCENTAGES OF TOTAL FLORA IN			
			Infer-Agrarian Zone.	Mid-Agrarian Zone.	Super-Agrarian Zone.	Infer-Arctic Zone.
All Britain .. .. (H. C. Watson.)	88,781	1,425	86	75	54	20
Northumberland with Durham .. (J. G. Baker.)	2,905	935	—†	98	45	12
West Yorkshire .. .. .. (F. A. Lees.)	2,760	904*	58	97	50 } 52	13 } 16
North Yorkshire .. .. .. (J. G. Baker.)	2,112	992	—†	95	42 }	13 }

* Colonists and Denizens excluded in making zonal estimate.

† No part of surface within Infer-Agrarian Zone in this area.

Before seeking into the reasons for the differences and agreements foreshadowed in this table, it may be as well to add an analysis of a somewhat more elaborate kind, supplementary to it, and calculated to bring out into still greater relief certain striking facts. In it the rate of diminution in number of species, coincident with altitude, in West Yorkshire, is given for successive elevations, with 300 feet instead of 900 feet intervals; and the figures for corresponding heights in Durham and North Yorkshire, given by Mr. Baker in his floras for those areas, are placed by the side of those taken from my own notes, as in the last table.

TABLE SHOWING THE NUMBER OF SPECIES, AND THEIR PERCENTAGES TO THE TOTAL FLORA, OCCURRING AT ALTITUDES INTERMEDIATE BETWEEN THOSE MARKING THE ZONES, IN WEST YORKSHIRE, NORTH YORKSHIRE, AND DURHAM.

ALTITUDES.	WEST YORKSHIRE. (Lees.) 904 Species.	NORTH YORKSHIRE. (Baker.) 992 Species.	NORTHUMBERLAND WITH DURHAM. (Baker.) 935 Species.
Below 300 feet..	818 sp. = 91 per cent.	854 sp. = 86 per cent.	882 sp. = 97 per cent.
At 600 ,, ..	612 sp. = 68 ,,	636 sp. = 64 ,,	541 sp. = 60 ,,
,, 900 ,, ..	471 sp. = 52 ,,	496 sp. = 50 ,,	450 sp. = 49 ,,
,, 1,200 ,, ..	374 sp. = 41 ,,	365 sp. = 37 ,,	375 sp. = 40 ,,
,, 1,500 ,, ..	270 sp. = 30 ,,	268 sp. = 27 ,,	300 sp. = 32 ,,
,, 1,800 ,, ..	162 sp. = 18 ,,	167 sp. = 17 ,,	135 sp. = 15 ,,
,, 2,100 ,, ..	64 sp. = 7 ,,	108 sp. = 11 ,,	66 sp. = 7 ,,
,, 2,400 ,, ..	27 sp. = 3 ,,	71 sp. = 7 ,,	34 sp. = 3 ,,

The names of the species stopping short at each successive height will be given hereafter; the bulk of them do not concern us in seeking an explanation of the discrepancies in altitudinal range shown in the last table especially.

The first thing to be noticed in the earlier table on page 42, is the reduced percentage of species in the Infer-Agrarian zone in West Yorkshire, as compared with Britain generally. The reason lies, of course, in the very limited Infer-Agrarian surface in our area; but when we reflect that it covers barely fifty square miles, and is pretty uniform in character of its soil and sub-jacent rock, so far from the proportion of 58 to 86 seeming small, the wonder is that it is so large. It shows that, however dissimilar the floral units may be, the number of species occurring in different districts of equal area is very much alike throughout the Infer-Agrarian zone. The variety of forms to be met with over fifty square miles of New Forest woodland, Dorset heath or Cambridge fen, is very much the same in proportion, though very different in kind, as that of South Yorkshire Permian or Nottingham Triassic tracts.

Conversely, the greater percentage of species, out of the total flora, found in the Mid-Agrarian zone in West and North Yorkshire and Durham, than in that zone for all Britain, is explicable in the fact that the Mid-Agrarian surface bears a much greater proportion to the whole area in the three small tracts to what it does in the united larger one. The surface coming within the Mid-Agrarian zone in Britain is not a half of the whole, whilst in West Yorkshire it forms three-fourths of it.

Still keeping to the first table, we observe how much nearer to the percentage, for the Super-Agrarian zone of all Britain, the percentages for the same zone in North and West Yorkshire are becoming. Fifty per cent. for West Yorkshire, with its comparatively small area in the Super-Agrarian zone, is much more than equivalent really to the fifty-four for all

Britain, and points to a more than average variety of soil and situation *in that zone with us*, as we shall see further on. It even exceeds the percentage for Durham, and by a good deal that for North Yorkshire; but this difference is more strikingly brought out in the second table, in the percentages for 1,200 and 1,500 feet.

Looking at the column for the Infer-Arctic zone, as we should expect from the very limited area reaching into it in West Yorkshire, the percentage sinks considerably below that for Britain, and seems to be the same or thereabouts as the percentages for North Yorkshire and Northumberland; but so far as the latter go, although it is probably correct for this zone *from 1,800 to 2,400 feet*, the figures are fallacious and mask a difference that is very real, for elevations intermediate between the contour lines of the zone. To clearly grasp the fact to be learnt, the second table on page 43 must now be turned to.

Why the percentage of species found below 300 feet should be 5 per cent. greater in West Yorkshire than North Yorkshire I do not know, unless it be that the variety of situation is with us somewhat greater. I am not aware that it is so, however: both areas have their tracts of lowland, water-logged, sandy, and peaty heaths, and some extent of brackish estuarine marsh. Still, it must be recollected that in the extreme south of the Riding we have a small tract which, looking to its mean temperature, and its enrichment by an infusion of Lower Zone plants (e.g., *Clematis Vitalba, Viburnum Lantana, Pyrus torminalis, Antirrhinum Orontium*, etc.), does not represent only the Mid-Agrarian zone, but, essentially, that zone *plus* the Infer-Agrarian. But Mr. Baker gives the percentage in Northumberland with Durham as greater still; and if the possession of a long coast-line with its concomitant littoral species explains it, as I assume it does in part at least, then North York, which has a fair stretch of cliffy coast too, ought to have at least as large a proportion of species below 300 feet as West Yorkshire.

But to continue, it will be observed that the percentage of species remains greater in West than in North Yorkshire at elevations up to 1,800 feet, and is notably in excess at 1,200 and 1,500 feet. At and above 1,800 feet a disproportionate fall occurs. The first table clearly showed us that the Super-Agrarian zone in West Yorkshire is richer in species than in North Yorkshire; and this second table intensifies the variation, and, further, localises the point of greatest divergence at the middle third of the zone. Why is this? In the explanation we find a confirmation of what has been already stated with regard to the causes of the lessened number of species on medium levels, that it *is* situation which plays the most powerful part in restricting ascension of species. Here a knowledge of what species they are that are found ascending into the Super-Agrarian zone in West Yorkshire, and not found to do so in North Yorkshire, gives us the clue. In our area between 900 and 1,800 feet the range of situation is undoubtedly greater than in North York, whatever it may be below 300 feet. There

happen to be not merely large areas of Limestone 'pavements' and scars, on a much more extensive scale than any in the sister Riding, but also a fine sheet of water (Malham Tarn) at an altitude of 1,250 feet, and a considerable tract of Silurian Slate fells—the two latter not occurring in North Yorkshire at all. In the augmented variety of species we see the result. In and around Malham Tarn we find *Hippuris vulgaris, Limosella aquatica, Carex limosa*, several Potamogetons and many xerophilous plants, and on the Slate fells *Alchemilla alpina, Sedum Anglicum*, etc., either not found in North Yorkshire at all or not at such a high level. They simply abstain, or rather they are prevented from ascending, because the stational conditions their habits demand are wanting. Such local peculiarities as these it is, then, which heighten the percentage of Super-Agrarian zone plants for West Yorkshire.

Passing on to the contrasted percentages at our highest levels, the figures and the conditions are exactly reversed. For 2,100 and 2,400 feet, 108 and 71 species respectively are given by Mr. Baker as occurring in North Yorkshire, making the percentages eleven and seven; whereas within the West Riding at the same heights only 64 and 27 species are known to me as growing, yielding percentages of seven and three only—exactly the proportion at the same elevations in Northumberland with Durham. In this case I know for certain that the variety of surface is greatest in North Yorkshire, and therefore that fact alone may be held sufficient to account for the greater variety of plant form: limestone scars (not on a large scale, truly) crop up and terrace the very loftiest hill slopes at 2,400, and 2,100, and 1,950 feet, the highest of such in West Yorkshire being under 1,900 feet, and the area of limestone at that height only very limited indeed. On this dry, fissured, and honey-combed rock in North Yorkshire between 20 and 30 species of xerophilous type are present: its absence with us shows us why we also miss seeing the floral accompaniment at equal elevations. True, with regard to some of the species whose highest limits I have set down at 1,800 and 1,850 feet, I have no doubt that others a trifle (perhaps as much as 100 feet) higher might be discovered by carefu exhaustive work instituted specially for the purpose, but I hardly think they could alter the percentage very materially. If they exist I have not seen them, and (such as they are) the figures given stand as the result of actual observation and not of approximate guess-work.

XI. *Regional Types seen in Vertical Range.*—The correlation of horizontal distribution outside our area, with the altitudinal range of species within it, now calls for elucidation.

By way of indicating the headquarters beyond Britain of the species making up the British flora, and at the same time suggesting the centres of dispersion from whence Britain was clothed with its present vegetation, Mr. H. C. Watson, in his *Cybele Britannica*, defined several 'types of distribution'—English, Germanic, Atlantic, British, Intermediate, Scottish, and Highland—and referred each distinct species to one or other of them; but

in the vertical range of our West Yorkshire species over our very much smaller area, we are able only to discern *three*, each of them inclusive of two or more of Mr. Watson's Type. These three are—1st, a Northern group, answering generally to the Highland, Scottish, and Intermediate types; 2nd, a Southern group, corresponding with the Germanic and (in the main) with the English type; and 3rd, an Indifferent group, intermediate between the two as regards thermal flexibility of constitution, almost equivalent to Mr. Watson's British type. These three indicate roughly the regions to which the West Riding flora owes its integers.

In horizontal distribution of its indigenous species throughout Great Britain, 37 per cent. are found generally and abundantly between the extremities, whilst 46 per cent. show a tendency to favour the southern and western half, and the remaining 17 per cent. show still more clearly a boreal restriction, being most abundant in the northern half of the island, and appearing further south only at some elevation in the more mountainous districts. The intimate association of warmer temperatures with the Southern group is one of cause and effect; and the same may be premised of the 17 per cent. of the Boreal group, and the colder temperatures of the region they especially affect. The 37 per cent., of so-called British type, are clearly comparatively indifferent as to the thermal means; or rather, perhaps, local conditions connected with the nature of the soil and subjacent rock, with them go much farther towards counterbalancing increased asperity of temperature than like conditions do the southern group. Still, from this we must *not* too hastily conclude that the vertical and horizontal range of species depends at all inconsiderably upon the capacity of individual organism to endure cold. We know that over large areas the facts all point the other way.

The matter, then, comes to this. From the peculiarities in horizontal distribution alluded to, and the fact of the diminution of species upwards affecting certain forms differently, it is possible to discriminate three distinct groups in respect of their climatic and altitudinal limitations. These groups correspond with the three types last distinguished, answering to the 37, the 17, and the 46 per cent. respectively; but having regard to our area alone, I shall designate them as—

1st, General or Ascending Species.
2nd, Montane or Descending Species.
3rd, Lowland or Stationary Species.

The second is a compound group, as I shall define it, made up of species agreeing in the essential character of being montane within our area, but differing somewhat in their horizontal distribution outside our boundaries, some of them being rather western than northern in their range.

Each of these classes in turn calls for an examination in detail.

XII. *The Ascending Species of the West Riding Flora* are those which are much the commonest on the lower levels, lessening in variety and in

abundance as individuals on the slopes, but which do, nevertheless, ascend to heights more or less pronounced, ninety of them even into the Infer-Arctic zone. Mostly belonging to the British type of Watson, they are horizontally distributed throughout the length and breadth of Britain in suitable situations. The few exceptions to this general description consist of species nowhere common, but found indiscriminately in high and low situations—a sub-group which has been styled the Scattered Ascending. They all agree, however, in that their occurrences and absences cannot be connected with either colder or warmer thermal means, as in the case of the Montane or Lowland species, but only with the suitability or unfitness of situation presented in each zone. Although not montane, it does not follow they occur in all the zones up to their limit from the lowest. Local conditions limit their range downwards as well as upwards, as is shown by the fact that many of them are elsewhere found much lower as well as much higher than in West Yorkshire. The species in this group in number greatly exceed those of the other two groups put together: there are 680 of them. Three of these Ascending species—*Silene maritima, Armeria maritima*, and *Asplenium lanceolatum*—are found *only* in the Super-Agrarian zone; whilst 108 are found only in the Mid-Agrarian with us, but then our area in the lowest zone of all is so very small. The rest (560) occur either in the first three zones, or in the second and third, or in all four. Of the 132 species found in the Infer-Arctic zone, 90 come under this Ascending category; and of the 450 found in the Super-Agrarian zone, 339 belong to this group; and of the 873 species occurring in the Mid-Agrarian zone, 676 are of this type: 94 of the whole being exclusively Lowland, and 103 Montane in their rôle.

The table on page 43 gives the total number of species observed at various altitudes for the Ascending species, along with the Montanes. By subtracting the number of the latter found at their respective elevations, given in the table on pp. 54-6, the number of Ascending species only, at the different heights, might be got at; but I have never kept notes for them apart, and no particularly useful purpose would be served by giving a separate set of figures now. If the Montane species were excluded, the rate of diminution upwards would show forth as comparatively more rapid; and also, probably, *more irregular*, since accidents of situation are especially powerful in imposing limits upon species of the Ascending type. Here, it only remains to illustrate this irregular altitudinal range by giving a selection from the species of the Ascending group stopping short at each successive elevation. The principle guiding the selection has been to give preference *not* to those rarer species the collector would deem most interesting, but to such as are exceptional, or typical, in their ascensions within our area. Those which are exceptionally high or low in their limits are italicised.

TABLE GIVING THE ALTITUDINAL LIMITATIONS IN WEST YORKSHIRE FOR 420 SELECTED SPECIES OF THE ASCENDING GROUP.

Clearly Limited to Mid-Agrarian Zone.

Stopping upwards at 150 feet.

Ranunculus Lingua. *Nymphæa alba.* Viola canina Bab. *Drosera anglica* (formerly up to 300 yards). Gentiana Pneumonanthe (formerly to 600 feet, but extinct now). Utricularia vulgaris. Sparganium minimum. Carex Pseudo-cyperus (erroneously reported for 1,200 feet). *Carex filiformis.*

Stopping upwards at 300 feet.

Ranunculus trichophyllus. R. sceleratus. Berberis vulgaris. Thlaspi arvense. *Mœnchia erecta.* Dianthus deltoides. Cerastium arvense. Alsine tenuifolia. *Erodium cicutarium.* Astragalus hypoglottis. *Potentilla verna.* Potentilla argentea. Myriophyllum alterniflorum. Bidens cernua. Utricularia minor. *Lysimachia vulgaris.* Narcissus Pseudo-narcissus (as a native). Cladium Mariscus. Carex digitata. Carex riparia. Calamagrostis Epigejos. Lastrea spinulosa.

Stopping upwards at 450 feet.

Nuphar lutea. Nasturtium sylvestre. Lepidium campestre. Scleranthus annuus. Radiola millegrana. *Acer campestre.* Pimpinella magna. Œnanthe crocata. Anthriscus vulgaris. *Cornus sanguinea. Galium tricorne. Galium uliginosum.* Inula Conyza. Picris hieracioides. *Filago minima. Campanula glomerata.* Convolvulus arvensis. Atropa Belladonna. Hyoscyamus niger. Orobanche major. Verbena officinalis. Lycopus europæus. Calamintha Acinos et menthifolia. Ballota fœtida. Echium vulgare. Lamium Galeobdolon. Parietaria officinalis. *Myrica Gale. Salix pentandra.* Salix undulata et rubra. Typha latifolia. Sparganium simplex. Potamogeton eu-natans. Sagittaria sagittifolia. *Alisma ranunculoides.* Butomus umbellatus. Orchis ustulata et pyramidalis. Cephalanthera ensifolia. Iris Pseudacorus. Allium vineale. Colchicum autumnale. Scirpus sylvaticus. Carex pendula. *Catabrosa aquatica.* Glyceria aquatica. *Osmunda regalis. Equisetum hyemale.*

Stopping upwards at 650 feet.

Ranunculus auricomus. Helleborus viridis. Teesdalia nudicaulis. Viola hirta. *Silene maritima. Arenaria serpyllifolia.* Spergularia rubra. Hypericum dubium. Geranium columbinum. Euonymus europæus. *Rhamnus Frangula.* Ulex Gallii. Ononis arvensis. Rosa Sabini. Rosa micrantha. *Rosa arvensis.* Epilobium hirsutum. Helosciadium nodiflorum. Silaus pratensis. Daucus Carota. Conium maculatum. *Galium Mollugo.* Carduus eriophorus. Artemisia vulgaris. Senecio erucifolius.

Inula dysenterica. *Convolvulus Sepium.* Linaria vulgaris. *Veronica Anagallis.* Scutellaria galericulata. Daphne Mezereum. *Cypripedium calceolus.* Tamus communis. Gagea lutea. Schœnus nigricans. Arundo phragmites. Poa compressa. Aspidium angulare.

Stopping upwards at 850 feet.

Nasturtium officinale. Cardamine amara. Polygala eu-vulgaris. Spartium scoparium. Geranium pratense. *Anthyllis vulneraria.* Genista tinctoria. *Trifolium arvense. Ornithopus perpusillus.* Spiræa Filipendula. Potentilla reptans. Rubus discolor, et rhamnifolius, et suberectus, et leucostachys, et carpinifolius, et corylifolius. Pyrus Malus. Cotyledon Umbilicus. Adoxa moschatellina. Viburnum Opulus. Knautia arvensis. Gnaphalium sylvaticum. Eupatorium cannabinum. Petasites vulgaris. *Jasione montana. Andromeda polifolia.* Erythræa Centaurium. Solanum Dulcamara. Nepeta Glechoma. Melampyrum pratense. *Veronica montana.* Lathræa squamaria. Lithospermum officinale. Primula veris. Ulmus suberosa. Salix purpurea et rugosa. Orchis Morio. *Neottia Nidus-avis.* Epipactis latifolia. Malaxis paludosa. *Ophrys muscifera.* Allium ursinum. *Scirpus multicaulis.* Carex muricata. Carex paludosa. Avena pubescens. Sclerochloa rigida. Hordeum sylvaticum. Ophioglossum vulgatum.

Found clearly within Super-Agrarian Zone.

Stopping upwards at 1,000 feet.

Aquilegia vulgaris. *Corydalis claviculata. Hypericum montanum.* Hypericum elodes. Lychnis diurna. *Geranium sanguineum.* Trifolium minus. Rubus *affinis*, et radula, et *thyrsoideus*, et pallidus, et diversifolius, et cæsius. Epilobium parviflorum. Sedum Telephium. Saxifraga granulata. Sanicula europæa. Carduus crispus. *Centaurea Scabiosa.* Hieracium umbellatum. Scrophularia nodosa. [Peucedanum Ostruthium. *Den.*] Origanum vulgare. *Scutellaria minor. Sambucus Valerandi.* Plantago maritima. [Fagus sylvatica, *Den.*] *Paris quadrifolia.* Convallaria majalis. Allium scorodoprasum. Luzula sylvatica. Carex pallescens. *Carex lævigata. Milium effusum.*

Stopping upwards at 1,100 feet.

Hydrocotyle vulgaris. Hypericum hirsutum. Trifolium medium. Veronica Beccabunga. *Calamintha Clinopodium. Anagallis tenella.* Quercus Robur. *Alnus glutinosa. Arum maculatum.* Orchis mascula. Habenaria bifolia. Listera ovata. Carex remota et sylvatica. Festuca elatior. Asplenium lanceolatum (one locality only). Potamogeton rufescens (one locality only).

Stopping upwards at 1,250 feet.

Ranunculus heterophyllus. Stellaria glauca (?). Cerastium glomeratum. *Sagina ciliata.* Ilex Aquifolium. Genista anglica. Prunus spinosa. Sanguisorba officinalis. Rosa spinosissima. Rosa canina (chiefly the forms dumalis, dumetorum, subcristata, and Watsoni at the highest elevations).

Cratægus Oxyacantha. Chrysosplenium alternifolium. Lonicera Periclymenum. Galium Aparine. (Pyrethrum inodorum. *Col.*). *Ligustrum vulgare.* Gentiana Amarella. Lamium purpureum. Littorella lacustris. (Chenopodium album. *Col.*) (Atriplex patula. *Col.*) *Lemna minor. Potamogeton heterophyllus*, et *densus*, et *perfoliatus*, et *lucens*, et prælongus, et *crispus*, et *pusillus. Alisma Plantago.* Blysmus compressus. Scirpus setaceus. *Carex teretiuscula*, et *limosa*, et ampullacea. Phleum pratense. Avena elatior. Cynosurus cristatus. Bromus sterilis. Equisetum limosum.

Stopping upwards at 1,350 feet.

Lychnis Flos-cuculi. Silene maritima. Ulex europæus. Medicago lupulina. *Hippocrepis comosa.* Sedum acre. Helosciadium inundatum. Valeriana dioica. Pimpinella Saxifraga. Sambucus nigra. *Carlina vulgaris. Serratula tinctoria.* Lactuca muralis. Fraxinus excelsior. Veronica scutellata. Rhinanthus Crista-galli. Teucrium Scorodonia. Stachys Betonica. Ajuga reptans. *Plantago media.* Mercurialis perennis. Salix cinerea. Ulmus montana. Habenaria viridis. Gymnadenia conopsea. Carex dioica. Carex curta. Holcus lanatus. Koeleria cristata. Bromus mollis. Scolopendrium vulgare.

Stopping upwards at 1,500 feet.

Caltha palustris. Arabis hirsuta. Peplis Portula. *Hippuris vulgaris. Hedera Helix.* Scabiosa Columbaria. Carduus nutans. Senecio aquaticus. Senecio vulgaris. *Limosella aquatica.* Primula vulgaris. Plantago major. Rumex nemorosus. Polygonum aviculare. Polygonum Bistorta. Urtica dioica. [Pinus sylvestris, *Den.*] Orchis maculata. Allium oleraceum. Carex præcox. Carex fulva. Briza media. Glyceria fluitans. Festuca duriuscula. Ceterach officinarum. Asplenium Trichomanes. Aspidium aculeatum. Botrychium Lunaria.

Stopping upwards at 1,600 feet.

Poterium Sanguisorba. Parnassia palustris. Hieracium boreale. Salix Capræa. *Taxus baccata.* Poa nemoralis. *Brachypodium sylvaticum.* Athyrium Filix-fœmina.

Stopping upwards at 1,700 feet.

Drosera rotundifolia. Fragaria vesca. Taraxacum officinale. Armeria maritima. Betula alba. Corylus Avellana. Salix aurita, et repens. Triglochin palustre. Juncus lamprocarpus. Carex panicea. Agrostis canina. Pteris aquilina.

Clearly reaching Infer-Arctic Zone.

Ascending as high as 1,900 feet.

Anemone nemorosa. Helianthemum vulgare. *Polygala depressa.* Viola palustris. Geranium Robertianum. *Orobus tuberosus. Comarum palustre.* Spiræa Ulmaria. Epilobium montanum. Callitriche verna. Saxifraga tridactylites. Galium verum. Galium palustre. Scabiosa

succisa. Carduus palustris, et arvensis. Hieracium Pilosella. Tussilago Farfara. Veronica Chamædrys, et arvensis. Digitalis purpurea. Pedicularis palustris. Thymus Serpyllum. Potamogeton polygonifolius. Juncus supinus. Carex pilulifera. Molinia cærulea. Aira cæspitosa. Equisetum palustre. Asplenium Ruta-muraria. Cystopteris fragilis.

Ascending as high as 2,000 feet.

Linum catharticum. Oxalis acetosella. Trifolium repens. Lotus corniculatus. Achillea Ptarmica. Eriophorum vaginatum. Juncus effusus. Carex pulicaris. Carex stellulata. Nephrodium dilatatum.

Ascending as high as 2,100 feet.

Viola Riviniana. Alchemilla vulgaris. Apargia autumnalis. Bellis perennis. Hieracium vulgatum. Campanula rotundifolia. Euphrasia officinalis. Eriophorum angustifolium. Carex flava. Aira flexuosa.

Ascending as high as 2,200 feet.

Cardamine hirsuta. Draba verna. Achillea Millefolium. Erica Tetralix. Erica cinerea. Vaccinium Myrtillus. Calluna. Veronica officinalis. Rumex acetosa. Blechnum boreale. Nephrodium Filix-mas.

Ascending as high as 2,300 feet.

Potentilla Tormentilla. Chrysosplenium oppositifolium. Galium saxatile. Luzula multiflora. Carex glauca. Anthoxanthum odoratum. Triodia decumbens.

Reaching 2,400 feet.

Cerastium triviale. Rumex acetosella. Juncus squarrosus. Luzula campestris. Carex vulgaris, et binervis. Agrostis vulgaris. Festuca ovina. Nardus stricta. Lycopodium clavatum.

XIII. *Lowland Species.*—I include in this term those species that within our area are *only* to be found below 300 feet elevation, and that upon the eastern plains only. Most of them, if not quite all, are confined to the lowest levels in other parts of England as well, and nearly all belong to that type styled by H. C. Watson the Germanic-English—to that 46 per cent. of the whole British Flora already alluded to as evincing in horizontal distribution a marked preference for the south of our island, and of which class these truly Lowland species form about 35 per cent.

Within West Yorkshire we have about ninety species thus termable non-ascending. The following list names the principal and more interesting amongst them: nowhere with us ascending above 300 feet in elevation (i.e., at heights with a mean annual temperature of 47° Fahr.), or off the Triassic and Permian tracts and alluvial plains. A large proportion are *hygrophilous* as regards their habitats; and the localities affected by those which are rather xerophilous than wet-loving are probably connected with the hot summers there prevailing: in other words, the species of this class not having preferences for watery places have a comparatively high zero. This is out-shadowed to some extent in the two groups into which these

Lowland species are divided in the table; and those personally acquainted with the local features of the eastern plains will see from the species in each that these groups have been made rather with reference to lithological character of surface and soil than to any material difference of elevation as affecting temperature. The species of the second list as found up to 300 feet are mainly such as love the dry soils associated with our Magnesian Limestone, the gentle swells of which attain to 300 feet on the average; but had it happened that the surface at such a height was mainly watery plateau, and the surface below 150 feet limestone mainly, the two lists would simply have been reversed.

TABLE OF EIGHTY SELECTED EXCLUSIVELY LOWLAND SPECIES OF WEST YORKSHIRE.

Found only below 150 FEET.

Thalictrum flavum. Ranunculus circinatus. R. hirsutus. Barbarea stricta. Nasturtium amphibium. Senebiera Coronopus. Viola lactea. Drosera intermedia. Ulex eu-nanus. Trifolium fragiferum. Vicia lathyroides. Lathyrus palustris. Apium graveolens. Sium latifolium. Œnanthe Lachenalii. Peucedanum palustre. Carduus pratensis. Aster Tripolium. Campanula patula. Veronica triphylla. Mentha Pulegium. Hottonia palustris. Rumex maritimus. Rumex Hydrolapathum. Polygonum minus. Typha angustifolia. Potamogeton mucronatus. Zannichellia palustris. Hydrocharis Morsus-ranæ. Stratiotes aloides. Juncus obtusiflorus. Juncus compressus. Carex axillaris. Carex paradoxa. Carex stricta. Sclerochloa distans. Lastrea cristata. Lastrea Thelypteris. Pilularia globulifera.

Found as high as 300 FEET.

Clematis Vitalba. Anemone Pulsatilla. Myosurus minimus. Turritis glabra. Reseda lutea. Silene noctiflora. Stellaria glauca. Linum perenne. Trifolium scabrum. Rosa systyla. Epilobium roseum. Bryonia dioica. Sison Amomum. Œnanthe fistulosa. Œnanthe Phellandrium. Galium erectum. Asperula cynanchica. Dipsacus pilosus. Viburnum Lantana. Erigeron acris. Carduus acaulis. Lactuca virosa. Specularia hybrida. Chlora perfoliata. Antirrhinum Orontium. Linaria Elatine. Orobanche minor. Calamintha Nepeta. Salvia Verbenaca. Centunculus minimus. Lysimachia Nummularia. Daphne Laureola. Lemna gibba. L. polyrhiza. Ophrys apifera. Spiranthes autumnalis. Scirpus acicularis. Carex distans. Calamagrostis lanceolata. Bromus erectus. Brachypodium pinnatum. Hordeum pratense.

XIV. *Montane or Descending Plants.* The species thus designated comprise those which occur in greatest abundance as individuals, and most numerously as species, on and among our mountains; having their homes on the higher levels preferentially, because there meeting with less competition, whilst encountering climatic influences approaching most nearly

to those of their headquarters in regions outside our area. They lessen in variety of form as we descend from higher levels to lower ones; but with regard to their number, as an ascent is made from the vales to the hill-peaks, the Montanes added to the flora never fully compensate for the ascending species running out because their climatic limits have been touched or their requirements as to soil and situation become obsolete.

The advantages possessed by species of the Montane category over the others in the fight for supremacy over the surface, lie in their greater capability for enduring *cold*, but they are by no means intolerant of warmth if moisture accompany it; and under conditions of soil and situation precluding severe competition on the part of glareal weeds and grasses, they will thrive even more luxuriantly on the plains than upon the peaks. Many of them, relics of some glacial epoch stranded as it were on the retreat of the ice, may be found not merely just existent, but flourishing upon the undrained, water-logged lowland moors and commons.

But although they all agree in being Montane so far as our area is concerned, the group is a compound one, capable of division into two parts if we correlate the species with their 'Types of Distribution' (in the Watsonian sense) horizontally as regards Great Britain and Ireland. They are not all of the true Alpine or Boreal group; and I have, therefore, separated them into two classes—the Montane and the Sub-Montane. In our indigenous flora of over 900 species, 83 are truly Montane and 43 Sub-Montane.

The true Montanes comprise those species, of unmistakably Alpine, Scottish, or Highland type, which not merely with us but elsewhere throughout Britain are characteristic of high latitudes and altitudes—of mountain regions only. Of our 83 species, 39 belong to the *Highland* type of Watson, 25 to the Scottish type, 14 to the 'intermediate' type (those species having their headquarters in N. England and S. Scotland), 4 to the Atlantic type, and only one to the British type of distribution.

The Sub-Montanes, 43 in number, are those species which within our limits are either confined to high or medium elevations, or else do show a preference for the hilly districts, *descending* rather than ascending in presence of relaxations of rigour in climatic conditions; but which in Britain as a whole do not show themselves to be clearly Montane. Montane with us, they are *not* so elsewhere in nearly so great a degree. Of the species brought under this head, none are of the Highland type, only 17 of the Scottish, 15 of the British type, 4 of the Atlantic type (*Helianthemum canum*, *Sedum anglicum*, *Wahlenbergia hederacea*, and *Hymenophyllum Wilsoni*), 6 of the Intermediate, and 1 (*Polygonatum officinale*) of the English type in distribution. The reason these Sub-Montanes are, with us, found principally among the hills, is doubtless the greater humidity of the atmosphere they there find, together with the larger area of surface presenting them with situations suitable for their growth, and most similar (although at higher levels, they not being cold-fearers) to those they normally occupy elsewhere.

Clubbed together these Montane and Sub-Montane species constitute about one-eighth of the Phanerogamic Flora, 12 per cent.—as against one-twelfth in North Yorkshire, which included the lower hill-area of Cleveland; but if Mosses and Hepatics be reckoned in, the proportion is considerably augmented.

These Montane species deserve special prominence in such a flora as ours, and they are therefore accorded a much more minute exposition as to vertical range than has been given to the Ascending species. Certain species not inserted have a rôle not very different from some of those in the Sub-Montane category. Such are *Hypericum dubium*, *Serratula tinctoria*, *Gymnadenia conopsea*, and *Cystopteris fragilis*. In the following table, by means of contrasted columns, each species of each sub-group has its limitations, so far as ascertained, indicated in feet in both an upward and downward direction, and although referring to an area of only 2,760 square miles, it is, I believe, abundantly instructive.

TABLE OF THE EIGHTY-THREE MONTANE AND FORTY-THREE SUB-MONTANE SPECIES OF WEST YORKSHIRE FLORA, CONTRASTING THEIR ZONAL AND ALTITUDINAL RESTRICTIONS.

THE SUB-MONTANE SPECIES IN ITALICS.

LOWEST DESCENT.	Altitude in feet.	UPWARD LIMIT.
	2,400	Cochlearia alpina, Saxifraga stellaris, Saxifraga hypnoides, Empetrum nigrum, Salix herbacea, Carex rigida, Allosorus crispus, Lycopodium alpinum.
	2,300	Rubus Chamæmorus, Sedum Rhodiola, Poa alpina, Epilobium alpinum, Lycopodium Selago.
Epilobium alpinum	2,200	Saxifraga oppositifolia, Saxifraga aizoides.
	2,100	Epilobium angustifolium, Pyrola secunda, *Scirpus cæspitosus*, Lycopodium Selaginoides.
Sedum Rhodiola, Salix herbacea ..	2,000	Trollius europæus, Viola lutea, *Montia fontana*, Epilobium alsinifolium, Vaccinium Vitis-idæa, Asplenium viride.
Carex rigida	1,900	Alchemilla alpina, Gnaphalium dioicum, Polypodium Dryopteris.
Poa alpina	1,850	Ranunculus Lenormandi, Draba incana, Arenaria verna, *Sagina nodosa*, *Pyrus Aucuparia*, Sedum villosum, Galium sylvestre, *Solidago Virgaurea*, *Gentiana campestris*, Vaccinium Oxycoccos, *Menyanthes trifoliata*, *Pinguicula vulgaris*, Listera cordata, *Narthecium ossifragum*.

TABLE OF MONTANE AND SUB-MONTANE SPECIES—*continued.*

LOWEST DESCENT.	Altitude in feet.	UPWARD LIMIT.
Epilobium alsinifolium, Pyrola secunda	1,700	Pyrus rupicola, Primula farinosa, *Juniperus communis*, *Orchis latifolia*, Habenaria albida, Sesleria cærulea, *Avena alpina*, Poa Balfourii, *Polypodium calcareum*, Polypodium Phegopteris, *Hymenophyllum Wilsoni*.
Alchemilla alpina, Saxifraga oppositifolia.	1,600	*Thlaspi occitanum*, *Rosa mollis*, *Geum rivale*, Carduus heterophyllus, Hieracium gothicum, Hieracium pallidum, Hieracium murorum, Myosotis repens.
Saxifraga stellaris, Poa Balfourii ..	1,500	Thalictrum montanum, Draba muralis, Hutchinsia petræa, Galium boreale, Arbutus Uva-ursi, Scirpus pauciflorus, Aspidium Lonchitis, *Nephrodium Oreopteris*.
Arbutus Uva-ursi	1,400	*Actæa spicata*, *Prunus Padus*, Ribes petræum, Hieracium anglicum, Hieracium cæsium, *Populus tremula*, Nephrodium rigidum.
	1,300	Potentilla alpestris, Dryas octopetala, Bartsia alpina, Trientalis europæa, *Wahlenbergia hederacea*, *Carex limosa*.
Dryas octopetala, Rubus Chamæmorus, *Callitriche autumnalis*, Saxifraga hypnoides, Aspidium Lonchitis.	1,200	*Cardamine impatiens*, Geranium sylvaticum, Rubus saxatilis, *Callitriche autumnalis*, Saxifraga umbrosa, Crepis paludosa, Hieracium prenanthoides, Andromeda polifolia, Myosotis sylvatica, *Polygonatum officinale*, *Equisetum sylvaticum*.
Saxifraga aizoides, Bartsia alpina ..	1,100	Meconopsis cambrica, *Sedum anglicum*, Meum Athamanticum, Crepis succisifolia, Hieracium iricum, *Campanula latifolia*, Polygonum viviparum, *Epipactis ovalis*.
Hutchinsia petræa, Lycopodium alpinum, Carex capillaris.	1,000	*Helianthemum canum*, Stellaria nemorum, Saxifraga Geum, Myrrhis odorata, *Symphytum tuberosum*, Rumex aquaticus, *Salix phylicifolia*, *Eriophorum latifolium*, Carex capillaris, Pyrola minor, *Melica nutans*.
Helianthemum canum, Saxifraga umbrosa, Hieracium iricum.	900	*Geranium lucidum*, *Polemonium cæruleum*.
Saxifraga Geum, Trientalis europæa ..	800	Thalictrum flexuosum, Hieracium crocatum, Hieracium rigidum, Melampyrum sylvaticum, *Equisetum variegatum*.
Thalictrum montanum, Potentilla alpestris, *Sedum anglicum*, Hieracium anglicum, *Wahlenbergia hederacea*, Habenaria albida, Allosorus crispus, *Equisetum variegatum*.	700	Circæa alpina, *Vicia sylvatica*, *Festuca sylvatica*, *Salix nigricans*, *Salix laurina*.

TABLE OF MONTANE AND SUB-MONTANE SPECIES—*continued.*

LOWEST DESCENT.	Altitude in feet.	UPWARD LIMIT.
Cardamine impatiens, Viola lutea, Meum Athamanticum, Galium sylvestre, Hieracium pallidum, Hieracium rigidum, Hieracium crocatum, *Polemonium cæruleum*, Polygonum viviparum, *Polygonatum officinale*, Asplenium viride.	600	Pyrola media.
Draba muralis, Thlaspi occitanum, Arenaria verna, Sedum villosum, Crepis succisifolia, *Gentiana campestris*, *Epipactis ovalis*, Melampyrum sylvaticum, *Avena alpina*, Nephrodium rigidum.	500	
Circæa alpina, Hieracium gothicum, *Salix laurina*, *Salix phylicifolia*, *Salix nigricans*, *Orchis latifolia*, Scirpus pauciflorus, *Eriophorum latifolium*, Lycopodium Selaginoides, *Hymenophyllum Wilsoni.*	400	
Meconopsis cambrica, Ribes petræum, Galium boreale, Vaccinium Vitis-idæa, Listera cordata, Sesleria cærulea, *Festuca sylvatica.*	350	
Ranunculus Lenormandi, Trollius europæus, *Vicia sylvatica*, Carduus heterophyllus, Hieracium prenanthoides, Pyrola media, *Myosotis repens*, Primula farinosa, *Polypodium calcareum.*	300	
Cochlearia alpina, *Actæa spicata*, Draba incana, *Geranium lucidum*, Geranium sylvaticum, *Montia fontana*, Rubus saxatilis, Pyrus rupicola, *Solidago Virgaurea*, Gnaphalium dioicum, Hieracium murorum, Hieracium cæsium, Crepis paludosa, *Populus tremula. Juniperus communis.*	200	
Thalictrum flexuosum, Stellaria nemorum, *Prunus Padus*, *Rosa mollis*, *Geum rivale*, Myrrhis odorata, Myosotis sylvatica, *Symphytum tuberosum*, *Melica nutans*, *Equisetum sylvaticum.*	100	
Sagina nodosa, *Pyrus Aucuparia*, Epilobium angustifolium, *Campanula latifolia*, Andromeda polifolia, Vaccinium Oxycoccos, Pyrola minor, *Menyanthes trifoliata*, *Pinguicula vulgaris*, Rumex aquaticus, Empetrum nigrum, *Narthecium ossifragum*, *Scirpus cæspitosus*, *Carex limosa*, Polypodium Phegopteris, Polypodium Dryopteris, *Nephrodium Oreopteris*, Lycopodium Selago.	Below 50	

What a remarkably wide range some species have—what hardy constitutions—what powers of vital adaptation! We see in this, that, no more than with the Ascending species, is it humidity or temperature alone that controls their range. *Cochlearia* and *Empetrum* top of the upward limit column, and descending respectively to 200 and to below 50 feet. Others have nearly as wide a vertical range, e.g., *Epilobium angustifolium*, *Trollius*, *Gnaphalium dioicum*, and *Solidago*.

Some species seem to descend very much lower with us than they do in 'North Yorkshire' (vide Mr. Baker's book), e.g. *Draba incana*, *Rubus saxatilis*, *Hieracium prenanthoides*, and *Polypodium calcareum*.

Three species given as exclusively *low-land* forms in North Yorkshire, ascend very much higher with us than there, viz., *Hippuris vulgaris*, *Limosella aquatica*, and *Potamogeton densus*—all Hygrophilous species, be it noted.

An examination of the zonal facts in the table shews us that 42 species reach into the Upper or Infer-Arctic zone; that 13 only fail in growing above 900 feet—i.e., are restricted to the zone below 900 feet, amid the hills; whilst 71 occur between 900 and 1,700 feet, not overpassing 1,800 feet. Only five species are altogether restricted (confined) to the highest Infer-Arctic zone; and nine have both their highest and lowest limits within the Middle or Super-Agrarian zone, whilst there are but 22 species found *only* confined to the Upper and Middle zones taken together. This seems a very small number—not much over two per cent.—yet if we take into account the great reduction in superficies, and restricted variety of situation in these upper zones, together with their lessened mean temperature, the result is not so very surprising—it is, rather, much what was to be expected. Each fact set down in the table must, however, be held to be capable of correction to some small extent; no natural limit is ever quite a hard and fast line. Any deductions which may be drawn from the *apparent* cessation of a species at a certain height, may be fallacious too. Most of the Alpines rise higher elsewhere out of the Riding, because there finding suitable conditions. We, for example, have no purely calcareous rock much above 2,000 feet, such as is found on Great Shunnor and Mickle Fells.

ANALYSIS OF ZONAL RANGE OF SPECIES.

MONTANE WITH SUB-MONTANE SPECIES.
126=12½ per cent. of whole Flora.

	Descend into	No. of Sp.	Highest Limit in	No. of Sp.
II.	Lower (Mid-Agrarian) Zone	104	Lower Zone	13
III.	Middle (Super-Agr.) Zone	17	Middle Zone	71
IV.	Upper (Infer-Arctic) Zone	5	Upper Zone	43

ANALYSIS OF ZONAL RANGE OF SPECIES—*continued.*

MONTANE SPECIES ALONE. Eighty-three in number.

	Descend into	No. of Sp.	Highest Limit in	No. of Sp.
IV.	Upper Zone	5	Upper Zone	33
III.	Middle Zone	16	Middle Zone	44
II.	Lower Zone	61	Lower Zone	6

SUB-MONTANE SPECIES ALONE. Forty-three in number.

	Lowest Descent	No. of Sp.	Highest Limit in	No. of Sp.
IV.	Upper Zone	0	Upper Zone	9
III.	Middle Zone	1	Middle Zone	27
II.	Lower Zone	42	Lower Zone	7

This tabulated zonal analysis, which concludes this inquiry for the present, may prove useful in some ways; but a careful study of the columnar Montane Range table, especially if an effort be made (*note-book and aneroid in hand in the field!*) to disprove or check any dozen of its individual facts, will impress upon the mind more forcibly than anything else the influences (complex they undoubtedly are) exerted upon plant-distribution by the phenomena embraced by the term Climatology.

XV. *Genesis of the Flora.*—A last inquiry, not strictly a part of either Climatology or Lithology, yet germane to it is,—whence, how, and at what periods came the vegetation now clothing the surface? These problems present many difficulties, and, although they invite speculation by their great interest, it may be that the materials for a correct solution are not yet altogether ours.

We may safely assume, to begin with, that when vegetation was created no region was left wholly without plant life in some form; at any rate, we know that no large tract is without it altogether, and in the *past* the like was, probably, always the case. What has been styled the theory of 'special creation' we may at once dismiss, as that could not have been specially created for it which is not confined to it. Another theory, that of specific centres, advanced first by Prof. E. Forbes, upon the whole fits in best with what we know. This theory was, briefly and broadly expressed, that definite centres were created, whence certain groups of species radiated like the ripples on a pool to circumjacent areas, according as conditions there existing permitted them, by virtue of such constitutional sturdiness and adaptive powers as they severally possessed. Later geological changes, depressions of land, ice-floods, etc., annihilated the species on certain tracts, which, when the conditions again changed, were overrun by other groups of species; but the species of any one period

were not since the Glacial epoch utterly destroyed everywhere, whatever they were before that era.

In West Yorkshire, in our coal-beds, we have geological evidence that various kinds of giant horsetails and huge cycads occurred long ago that are now non-existent anywhere. Since then, at least our present flora has come. The more recent remains of pines, etc., in the sand under the peat, show us that this pre-peat age flora was in type much nearer to our own of the present day.

Dealing with the configuration of the surface as we now know it, for we cannot go back further than that for the origin of our present flora; if we examine carefully into the *range* of our existing species *outside* our limits, we at once find them capable of division into three or four groups, the species involved in each having reached us at three or four successive times—the exact dates it is impossible to fix, but their *order* in time it is possible to do somewhat more than guess at. These may be styled the Northern or Arctic, the Continental or Germanic, the Western or Atlantic, and the Recent or Historic groups.

1. *The Northern Group.* This, within our county limits, includes undoubtedly the oldest existing types, because in part the alpine-arctic element in it, the on-lingering offspring, in unbroken succession, of species that survived the Ice Age flood, of the late Tertiary period, on those mountain peaks that stood up like islands from the glacial sea. This ice-flood reached no greater height in our Riding than 1,450 to 1,500 feet—to the summit of the mural mountain scars of the western dales. Such species as *Silene maritima* on Moughton Fell, *Plantago maritima*, *Cochlearia alpina*, *Aulacomnium turgidum*, *Carex rigida*, etc., survive as representatives of this class. The great bulk of our northern species, however, probably spread to us from adjacent areas not ice-bound, immediately *subsequent* to the eons of this Ice Age; when the country from the low-lying vales of York and Trent to Norfolk, and across to Holland, had some such moss-grown, lichen-clad, sedgy surface as the Lapp and Siberian tundras exhibit now. The Montane species of Thorne Waste, *Empetrum nigrum*, *Carex limosa*, *Cetraria*, and others more universal, like *Galium saxatile* and *Juncus squarrosus*, belong to this second sub-group of northern type—a sub-group now mainly confined to the moorlands, but which for a long time must have covered the whole of Yorkshire. The area of surface representing this period to be seen in an aboriginal condition to-day at the lower levels is very small indeed, parts of Thorne and Riccall Moors and Askham Bogs being almost the only examples.

2. *Continental or Germanic Group.* Other of our species, those of 'Germanic' or 'English' type (so called by Watson because preponderating most markedly in the S.E. and S. of England), afford evidence of the probability of a second diffusion of forms. There was once a land connection between England and the Holstein and Hollandic coasts when (so geology tells us) the shallower southern part of the German Sea was a

marsh reticulated with streams, and by means of this it has been supposed that the English-Germanic element in the flora reached Britain. That element was the next one undoubtedly to reach Yorkshire, probably coming to us through Lincolnshire from the south of England, and it gradually dispossessed of the soil the boreal flora already described, ever creeping up and on, highest (i.e., most successfully) where limestone rock entailed dryer soil, even then doing (favoured by increasing leniency of seasons) almost exactly what species of such type will yet do (with the modern aids of drain-pipe, bill-hook, and lilly-low), namely, waging a successful war for supremacy with species much older in type and harder in cold-resisting capabilities than these Germans, and yet weaker in the long-run from lack of indifference to competition and crowding.

3. *The Western or Atlantic Group.* In these we may see evidences of the time when S.W. Ireland was joined to the Peninsula, and St. George's Channel was not yet: but very *indirect* evidence, of course, so far as our area is concerned ; for although in time—looking at Britain as a whole—the 'Atlantic' is probably the oldest element of all in its flora, yet as Yorkshire species they are probably among the latest arrivals brought to *us*, in part at least, over-sea as spores or seeds by the south-west wind. The prevalence of *Wahlenbergia hederacea* in the sheltered cloughs of the South and West of the Riding I hold to be evidence opposed to the idea that it is really an 'Atlantic' type species at all.

4. Lastly, we may fairly ascribe to human agencies—commerce, agriculture, the Roman invasion, the Danish inroads, and the influence of the Monks—a not inconsiderable and decidedly most recent element in our flora : the class of Glareal Weeds, the Nettles, the Goutweed, the Monk's Rhubarb, the Snake-weed, and several others.

It is thus seen that what we call our indigenous flora is not wholly, or in great part even, aboriginal. Fully two-thirds of it, and perhaps more, is made up of the successors of invaders that have long ousted the original possessors of the soil. The present mingling is of types that exist at no great distance from us, the Boreal on the Scotian and Scanian mountains, the Atlantic type in S.W. England and Western Wales, the English and Continental in South and Mid-England. To the correlative inquiry why some species are found to occur in certain spots and not in others, where the environment *appears* identical, it is no satisfactory answer to suggest that some accident of mountain barrier, or the very chance-like action of an air current must have prevented it. The intrusion theory will sometimes explain some puzzling occurrences of plants in outlying stations at a distance from their present centres : the occurrence of a boreal moss like *Rhacomitrium lanuginosum* on the grit stonework of a sluice at Dyke Marsh, near Thorne, can be accounted for only on the supposition that it came as a spore or protonemal layer with the stone from some quarry in the western hills. A similar instance is that of the Alpine mosses on the Scanian boulders of the North German plains, and another, perhaps, in the

occurrence of *Bryum filiforme* on the ice-brought stones called 'Travellers' about Dunford Bridge.

All our Yorkshire species are Continental also—none are endemic. What conclusion may be drawn from this? Considering how many thousand years we have had *no* land connection, spontaneous differentiation of specific types in the vegetable kingdom does not take place so rapidly as has been asserted. The shade-growing *Carex saxumbra* of the Plumpton grit rocks, is the only Yorkshire species known to me which has not its Continental counterpart, though I believe one or two endemic brambles are known to Mr. Baker. How comes it we can point to so few instances, how be so readily able to correlate our Yorkshire—I might say our British—species to their types, northern, eastern, and western, and find geological discovery hinting, so conformably, the over-land routes by which they reached us, if fertile enduring spontaneous differentiation is ever-acting all around us?

Much remains to be said which I cannot say. If I have succeeded in crystallising from the rich mother-liquor of this great subject (not nearly exhausted yet) some few of the grosser, more obvious forms of fact and theory, of which the beauty and symmetry may be apprehended without profound study, the attempt—suggestive only, as it is meant to be—may perhaps be held to be justified.

LITHOLOGY.

LITHOLOGY.

Although Lithology in the strict sense of the word has no place in a botanical work, the strata of the West Riding, through the soils derived from their abrasion under the action of moisture, exert, really, a profound influence upon the flora : and thus require to be considered at some length in relation to plant distribution.

I. Two Great Rock Types. So long ago as 1849, in an 'Essai de Phytostatique,' M. Jules Thurmann, after a study of the contrasts and agreements afforded by the floras of the granitic Vosges Mountains and the Limestones of the Jura, discerned in the strata of those ranges two great 'types' of rock. To these he gave the somewhat uninviting names of Dys-geogenous (dys = difficult ; geogenous = earth-bearing) and Eu-geogenous (eu = true ; geogenous = earth-bearing), those terms, however uncouth, expressing fairly well those differences in their behaviour, under the disintegrating action of water, which have the most direct bearing upon plant-distribution. M. Thurmann decided, what has never been controverted, that to the *mechanical* properties of a rock when subjected to weather and water—and only very slightly to its *chemical* constitution (as calcareous, aluminous, or siliceous)—was its indirect action upon plant-life due.

Following Thurmann and Baker—who was the first in England to recognise the value of and apply Thurmann's arrangement to the North Yorkshire flora,—I shall adopt the terms ; with this modification, that in West Yorkshire an intermediate kind of rock—the Clay Slate, intermediate both in its mechanical properties and its flora—may be recognised. These varieties call for definition.

1. *Dysgeogenous Rocks.* These are homogeneous in nature, hard or moderately so, very permeable to water because much cracked and fissured, very slightly absorbent, and usually occur in large masses not readily abraded by weather. The soil above such rock (except where a deposit of silicious clay from grits at a higher level has held moisture and formed a bog, as above the limestone 'pavement' of Ingleborough) is dry, scanty, but comparatively fertile. This type of rock is represented in the West Riding by

(1) The Mountain or Scar Limestone.
(2) The Magnesian or Permian Limestone.

2. *The Pseudogeogenous Strata.* These are intermediate in their character between that just named and the next, although including two very different kinds of rock from the geologist's standpoint. They are represented within our area by

(1) The Silurian Clay Slate.
(2) The Yoredale Limestones.

The first stratum is very hard and homogeneous in its mass, as little or less permeable to water than the scar-limestone, but from its dip, fracture, &c., water runs off it to a great extent, and although it has peat above it, its exposed crags and slopes have a scanty, shallow soil, rich and comparatively fertile in certain alpine forms. The mechanical properties (in the botanical sense) of the Yoredale limestone series is due to the fact that whilst its calcareous beds present scar-like faces and masses of limestone almost as pure as that of the thicker 'mountain limestone,' like that homogeneous, little absorbent, and very permeable to fluids, yet these are much separated by intercalations of gritty or bituminous shale, which arrest water percolating down through the porous beds, and hold it. The soils above these strata are, therefore, very variable, but on the whole richer in vegetative forms than the next type, because lacking only the drier-loving among the calcareous-soil species.

3. *Eugeogenous Rocks*. These are very absorbent, and only slightly permeable, and when in beds they hold water above. The matrix in which the hard silicious particles or pebbles are bound up is easily abraded by weather, and an abundant soil is formed by the disintegration. This soil is often light (it has much moisture in it always) over the pure rough grits and Red Sandstones; but over the clay and coal shales is stiff, wet, typically cold, and backward in its vegetable productions. This type of rock is represented in its three main aspects (as to resulting soil) in the West Riding by

(1) The Millstone Gritstone.
(2) The Coal Measures.
(3) The New Red Sandstone.

II. Physical Characters of the Hills of each Type. The three types of rock described determine to a great extent the configuration of the surface—the character of the landscape. The scenery by which each is characterised is as boldly contrasted as the items that make up the flora—to be dealt with presently.

The Scar Limestone region is marked by rounded hills, green with a close rich turf, broken regularly at the edges of the valley by battlemented wall-like cliffs. The Sandstone hills are more irregular and undulating, with irregular edges of rock, scarred and riven by gorges, reticulated with branching gullies sloping from peaty swells of moorland, for the most part clad with heather, bilberry, and ling. The slate hills of the extreme North-West of the Riding are still more irregular in outline, rising by a series of serrations (somewhat masked by soil and barren talus) into distinct peaks. The dales of the scar limestone are narrow, bordered by mural tree-clad scars, over the plates of which cascades fall, or at the foot of which streams gush out, the water having sunk on the plateaux at higher levels through 'swallow' or 'pot' holes. The sandstone dales are wider and shallower, the escarpments of weathered gritstone where present lying higher up the slopes, more unobtrusively because less in proportion to the breadth of the valley than the scar edges of limestone.

III. Lithological Classification of Soils. That the three types of rock sketched determine not only the carving of the surface, but also the soils, and so point the main features of phyto-geography, is not surprising when it is from soil that plants get most of their nourishment. For no plants, so far as I have ever seen, is any particular kind of calcareous earth a necessity; but still, whilst the constitution of the eugeogenous and pseudo-geogenous rocks almost exclusively determines the general character of the floral integers, *moisture* being in their case a common factor, with the flora of soils wholly derived from dysgeogenous rocks it is probable that the reverse is the case in a few instances: there are some few plants that will not bear *wet* at their roots, although even here I consider it is very doubtful if the necessity for a limestone soil is to be laid down to its chemical element calcium. Even *Actæa spicata*, *Gentiana Amarella*, *Ceterach* and *Polypodium calcareum*—three of them forms most characteristically 'xerophilous' (a term to be presently explained)—will, and do, grow off (i.e., away from) limestone, whilst many others, such as *Orchis ustulata* and *Spiranthes autumnalis*, constantly do so under certain well-known conditions.

Thurmann's division of soils in the district he so exhaustively analysed, with its two types of rock, was a triple one: calcareous soils, sand soils, and clay soils, the eugeogenous type of rock yielding the two latter according as they were silicious and less abradable, or aluminous and easily disintegrated. We have no strata in West Yorkshire, and only the Lias shale band of Eastern Yorkshire, precisely answering to the latter class; but the tenacious clays of the Coal measures in the respect of their strength in clay-liking species represent it in a minor degree.

The classification of soils in West Yorkshire I propose, as being those with characteristics mirroring themselves in their floras in a tolerably accurate manner, is as follows:—

(1) Slaty or Pseudo-calcareous—over Clay-Slate.
(2) Calcareous—over the Limestones and Boulder Clay.
(3) Arenaceous—over Gritstone and Trias rocks.
(4) Peaty—over Grit or water-logged sand.
(5) Argillaceous—over Coal Measures, and Yoredale Strata and Boulder Clay.
(6) Alluvial—over Warp and Silt.

The botanical features and restrictions more or less peculiar to these varieties of earth have each to be considered. Thurmann christened the *species* showing marked preferences for one or other of his three kinds of soil by three not inconvenient terms: plants prevalent mainly upon the dry jurassaic-limestone he designated *xerophiles* (dry-loving); those pertaining to the wetter soils of his eugeogenous strata he styled *hygrophiles* (moisture-loving); dividing the latter class into two—*psammophiles* (those typically sand-loving), and *pélophiles* (clay-loving)—those exhibiting in their

distribution preferences for the soils of the softer aluminous shales. For purposes of correlation with the chief soil-conditions of West Yorkshire similar terms—

(1) Xerophilous—on soil No. 2,
(2) Sub-Xerophilous—Slate and Boulder Clay, on soils 2, 1, 3, and 6,
(3) Hygrophilous—on soils 4 and 1,
(4) Psammophilous—on soil 3,
(5) Pelophilous—on soil 5,
(6) Halophilous—on soil 6,

will answer sufficiently well, with this modification—that the Hygrophilous species in West Yorkshire, although typically those of peat over sandstone, are present more or less over the Slaty and Argillaceous soils as well, being markedly less abundant on the dryer calcareous soils only.

IV. Comparisons between Thurmann's List and the Flora of West Yorkshire.—In venturing upon these it must be remembered that the conditions are not equal, and therefore the relationship of many of the species to a certain soil and underlying rock is not always similar in both areas. The contrasts brought to light by a careful examination, are, however, both interesting and instructive.

Of his list of 50 xerophiles—species characteristic in the Jura district of dysgeogenous-rock soils—the flora of all Great Britain possesses only 24 (48 per cent.); whilst of his list of 50 hygrophiles there are as many as 42 (84 per cent.) indigenous or denizen—facts reflecting incontestably the insular (equable and humid) character of the climate.

The corresponding figures (quoted here for the sake of readier reference by Yorkshire students of phyto-geography) given by Baker for North Yorkshire, Trimen and Dyer for Middlesex, and Townsend for Hampshire with Wight, in their respective floras of those counties, are as follow:—

North Yorkshire has of Thurmann's 50 typical *xerophiles*,					8 species,
Middlesex	,,	,,	,,	,,	8 species,
Hampshire with Wight	,,	,,		,,	11 only,

notwithstanding the great area of cretaceous soil in the latter county.

In West Yorkshire, of the fifty given as the commonest species typically restricted to dysgeogenous soils in the Oolitic Jura, we possess only 8, viz.:—

Helleborus fœtidus,	Orchis pyramidalis,
Rosa rubiginosa,	*Mercurialis perennis*,
Calamintha officinalis,	Fagus sylvatica,
Daphne Laureola,	Brachypodium pinnatum.

One is, of course, prepared for differences, but some that we shall come across in our examination are fairly startling. Take that italicized for instance: *Mercurialis perennis* in a list of species having a 'typical' preference for a calcareous district, a *dry*-lover! If rightly included—one may doubt without absolutely denying,—should we not expect that it would in England be a doubly-evident limestone-liker? We can readily understand

a typical moisture-lover over-riding bounds and occurring on calcareous strata, dysgeogenous in type, because of the comparatively moister conditions prevailing there to what is found in Central Europe, but the reverse is truly unexpected and inexplicable except on the supposition that some factor such as mean temperature (no one would expect it to grow on an Alpine peak) had been unaccounted for. In West Yorkshire *Mercurialis* is one of that not small class of positive indifferents, growing in damp hedgebanks on all kinds of soils (least freely perhaps on the *dry* Trias), and seeming to have no likes or dislikes whatever.

Further on in his essay Thurmann goes more fully into the matter by framing two further lists of xerophiles, less common or abundant than those styled 'typical,' but including most of them over again. The first is one of 53 species, 'preferentially dry-loving'; and the second of 98 species, partisans ('*adherentes*') of dry-lovers—these being, I presume, what I should call sub-xerophilous—of which (the two, 151 in number, taken together) 36 occur in North Yorkshire, 29 in Middlesex, and 30 in Hants. Of these, in West Yorkshire we have 35 or 36; but some few of them, e.g., *Crocus vernus* and *Ruscus aculeatus* are not indigenous integers; and, again, 3 or 4 of them are not xerophilous in any strict sense. The list includes many that are such, as *Hypericum montanum*, *Carduus acaulis*, Orchids *ustulata*, *apifera*, *muscifera*, *Ceterach*, and *Polypodium calcareum*, but some of these even do occur in odd stations off dry soil and others, alluded to above, such as *Cerastium arvense*, *Verbascum Thapsus*, and *Melica uniflora*, are quite as much psammophiles, if their stations be any guide, as lovers of dry limestone soils.

We have now to see how far Thurmann's observations upon the flora of the Eugeogenous Triassic and Granitic *Vosges* region hold good when contrasted, similarly to the xerophiles, with that of West Yorkshire.

Thurmann's first list is again one of fifty type *Hygrophiles*. Of these

North Yorkshire has (*teste* Baker) 31 species,
Middlesex (*teste* Trimen) 33 species,
Hants with Wight (*teste* Townsend) 35 species,

whilst West Yorkshire has 35, including

Lepigonum rubrum.	Meum athamanticum.
Stellaria Holostea.	Saxifraga stellaris.
Montia fontana.	Senecio aquaticus.
Hypericum pulchrum.	Senecio sylvaticus.
Hypericum humifusum.	Filago minima.
Sarothamnus scoparius.	Hieracium boreale.
Orobus tuberosus.	Verbascum Blattaria.
Ononis spinosa.	*Digitalis purpurea*.
Lotus major.	Galium saxatile.
Trifolium fragiferum.	*Jasione montana*.
Prunus Padus.	Calluna vulgaris.

Vaccinium Myrtillus.
Rumex Acetosella.
Quercus sessiliflora.
Betula alba.
Alnus glutinosa.
Juncus squarrosus.
Luzula multiflora.
Alopecurus pratensis.
Phragmites communis.
Aira flexuosa.
Aira cæspitosa.
Triodia decumbens.
Nardus stricta.

Those species similarly restricted in West Yorkshire are italicised. It will be seen how few they are; but still the fact that we have 32 out of 50 shows how conspicuously the moisture-preferring species predominate; and the five in the above list which evince a distinct dislike of a calcareous soil are augmented by certain other species within our area which do not appear at all in the drier climate of Central Europe. The explanation why so many species, restricted to the damper tracts in the areas Thurmann studied, are found over-running tracts of dysgeogenous strata with us in West Yorkshire as elsewhere in England, is, plainly, that greater humidity of our insular climate counterbalances the drier soils over the limestones; indeed, that they are not so dry as typical dysgeogenous soils should be is proved by the much smaller number of typical xerophiles that find themselves able to exist upon them, compared with the Continent.

But Thurmann, with regard to this class of moisture-lovers, likewise amplifies his observations by furnishing two additional lists of what may be termed restricted hygrophiles. The first of these consists of 82 *psammophiles*—sand-preferring likers of moisture, of which North Yorkshire has 34, Middlesex 44, Hants 31, and West Yorkshire 35 in its flora, viz.:—

Myosurus minimus.
Teesdalia nudicaulis.
Erysimum cheiranthoides.
Sisymbrium Sophia.
Lepigonum rubrum.
Radiola millegrana.
Montia fontana.
Scleranthus annuus.
Ornithopus perpusillus.
Ulex europæus.
Sarothamnus scoparius.
Ononis arvensis.
Melilotus vulgaris.
Saxifraga granulata.
Galium saxatile.
Scandix Pecten-veneris.
Filago minima.
Senecio sylvaticus.
Thrincia hirta.
Jasione montana.
Verbascum Blattaria.
Myosotis versicolor.
Antirrhinum Orontium.
Rumex Acetosella.
Amaranthus retroflexus.
Salix viminalis.
Populus nigra.
Betula glutinosa.
Juncus squarrosus.
Scirpus setaceus.
Aira præcox.
Aira caryophyllea.
Phragmites communis.
Festuca sciuroides.
Osmunda regalis.

Some of them appear also in Thurmann's typical fifty, and it need hardly be said that with us in West Yorkshire most of them are not restricted to sand, and two or three of the rarer, notably *Myosurus minimus* and *Sisymbrium Sophia*, occur only on soil over calcareous strata rather marly and clayey than sandy; and two or three are only denizens. The disproportionately large number, 44, claimed for Middlesex is due to the inclusion of certain species (e.g., *Xanthium strumarium*; also found in West Yorkshire, but only on 'wool waste'), which cannot rightly be regarded as much more than casuals in that or any other British area.

Thurmann's last sub-division of hygrophilous plants, is one of 39 *pélophiles*—clay-preferring moisture-lovers, of which only 27 are British plants at all. Of these 27 (some again are in the 'general' list) the North Yorkshire flora includes 24, I believe, Middlesex is said to have 24,* Hants with Wight only 18, and West Yorkshire 23, viz:—

Ranunculus Flammula.
Stellaria Holostea.
Alsine tenuifolia.
Hypericum humifusum.
Hypericum pulchrum.
Trifolium fragiferum.
Lotus major.
Lonicera Periclymenum.
Bidens cernua.
Senecio aquaticus.
Tussilago Farfara.
†Inula dysenterica.
†Specularia hybrida.
†Chlora perfoliata.
Lysimachia nemorum.
Centunculus minimus.
Veronica scutellata.
Salix aurita.
Alnus glutinosa.
Alisma Plantago.
Luzula multiflora.
Scirpus acicularis.
Triodia decumbens.

The above is truly the most remarkable list of all, according to insular experience, putting side by side, as it does—† marked clay-preferrers abroad, be it remembered—such plants as *Trifolium fragiferum* (*not* a pélophile at all in England) and *Centunculus minimus* and *Hypericum humifusum* (psammophiles with us) with *Specularia hybrida* and *Chlora*! Nothing could evidence more plainly the diverse—aye, even reverse—relationship to soils and underlying rock exhibited by the same species in Western France compared with Yorkshire. Even the results of the imperfect enumerations given show that there is no similarity or relationship—the formulation of limiting conditions beyond the main one of dryness and wetness is true only for a particular area; and Thurmann's terms dysgeogeneous and eugeogenous, xerophilous and hygrophilous, as expressed botanically, affording precise contrasts for the Jura and the Vosges, are not parallels for England. There can be no parallel where the conditions are different, and therefore it is not

* NOTE.—It may seem strange for the small county of Middlesex to have as many clay-loving species as the much larger areas of N. and W. Yorkshire. The flora of the northern divisions is numerically richer by nearly two hundred species than the south-country ones, but the increase is in the marked infusion of montane elements evident with us. The floras are in other respects much upon a par with each other, Middlesex appearing to be rich in clay-loving integers, and Hants notably poor in sand-preferrers.

possible to exactly correlate 'dysgeogenous' rock and soil with 'xerophilous' species—in the field the facts would at once discredit the attempt, many Jurassaic 'xerophiles' growing also on sand with us, and the Vosgesian 'hygrophiles' largely on dysgeogenous strata. The fact is that we have in West Yorkshire (as also, I suspect, in the Jura) a group of *hygrophiles*—those cannot be considered *xerophiles* which clearly need considerable moisture in the soil for them to flourish conspicuously on it—which prevail mainly or almost exclusively on calcareous soil. Most of them are pelophilous (clay-loving) abroad, and it is in this class that we find species for which, *if any*, the element of calcium is a necessity. This conclusion is allowable where we find the common factor to be the *calcareous* matrix of growth, under the most diverse conditions in other respects. Such West Yorkshire species are *Inula dysenterica*, *Chlora perfoliata*, *Gentiana Amarella*, *Specularia hybrida*, *Neottia Nidus-avis*, and, perhaps, also *Helminthia echioides* and *Triticum caninum*, for they, with us, are plants pertaining obviously to roadsides, stiff fields, damp roadside hedges and ditches, stiff-soiled fields and woods on the Magnesian Limestone and Boulder Clay.

V. Botanical Features of West Yorkshire Soils. Coming now to an exposition of the actual facts as to species-restrictions and preferences discernible in our own area, in connection with rocks and soils, anyone turning back a few pages may see how very imperfectly the species integers under the various heads here given tally with those of Thurmann. The broad distinction between the floras of the two types of rock remain truly, to show that Thurmann's adumbration embodied a truth—a general law, but a law capable of a wide diversity in its expression under the complicating influence of a variety of local conditions, such as altitude, the greater moisture of an insular climate, uniformity of level, &c. Many of the items in M. Thurmann's lists show misinclusions, sadly wide of the mark to the British field-botanist; the species which we are told is a clay-lover abroad being, perhaps, well known to us here as a psammophile solely. The like may be the case with many of the species to be ranged here as characteristic of sand, or peat, or clay, albeit true for the area under consideration, as the accumulated results of actual investigation by many observers are almost bound to be.

1. *Slaty or Pseudo-calcareous (over Clay-slate).* These, overlying hard intractable slaty material, are from the slight disintegration, shallow and damp; but where the hill-slopes are much overlaid with drift, or in the alluvial valley bottoms, the soil is often deep, and more or less argillaceous. Certain Alpine species appear only on these rocks and soils, plus which the botanical character of the vegetation approaches that of argillaceous shales and boulder clay. Alpines descend to a low level, and the ordinary Ascending species of the lowlands are found running out at a lower altitude than on limestone. The Slate-fell district florula is somewhat peculiar: it is a mixed one, marked not so much by hard-and-fast absences as by showing species (of the Ascending type chiefly) in juxtaposition upon

its crags and stream banks, which, elsewhere in the Riding, are more or less sharply defined in their soil preferences. Our West York slate-area is, however, too small to exhibit the fullest expression of features peculiar to the Silurian Slate. The following table, however (adapted from Baker's, in his Northumberland Flora), presents such a type florula—those species actually found on the Slate in West Yorkshire being placed in italics.

Flora of Silurian Slate.

Thalictrum alpinum.
Anemone nemorosa.
Cardamine pratensis.
Viola sylvatica.
Viola palustris.
†*Alchemilla alpina.*
Alchemilla montana.
Pyrus Aucuparia.
Cerastium triviale.
Cerastium alpinum.
†*Circæa alpina.*
Epilobium alsinifolium.
Saxifraga hypnoides.
Saxifraga nivalis.
Saxifraga stellaris.
Saxifraga aizoides.
Saxifraga oppositifolia.
Oxalis Acetosella.
Chrysosplenium oppositifolium.
Galium saxatile.
Sedum Rhodiola.
Sedum anglicum.
Saussurea alpina.
Hieracium chrysanthum.
Jasione montana.
Campanula rotundifolia.
†*Pyrola secunda.*
Vaccinium Myrtillus.
Vaccinium Vitis-idæa.
Euphrasia officinalis.
Thymus Serpyllum.
Rumex Acetosa.
Oxyria reniformis.
Juncus squarrosus.
Juncus triglumis.
Carex stellulata.
Carex rigida.
Anthoxanthum odoratum.
Poa alpina.
Festuca ovina.
Aira flexuosa.
Allosorus crispus.
Hymenophyllum Wilsoni.
Asplenium septentrionale.
Cryptogramme crispa.
Lastrea dilatata.
Cystopteris fragilis.
Lycopodium alpinum.
Lycopodium Selaginoides.
Hedwigia ciliata.
†*Pterogonium gracile.*
Lecidea geographica.

In the above, I think, the mixture of the supposed 'types' is well shown. The West Yorkshire species are in italics; those confined absolutely to our slate-rocks are preceded by the obelisk sign. The Purple Saxifrage is not italicised, because, though it occurs in West Yorkshire in two places, it is *not* upon slate-rock.

2. *Botanical Features of Calcareous Soils.* These, richer in characteristic forms than any other, are marked by the absence of several species which, common even on the mixed soil of the slate area, appear to actually dislike limestone. Such absentees are *Spergularia rubra*, Broom, *Galium saxatile*, *Digitalis purpurea*, *Erica cinerea*, *Vaccinium Myrtillus*, *Empetrum nigrum*, *Betula alba*, *Aira flexuosa*, *Bartramia pomiformis*, etc. This absence is

not absolutely without exception, of course. In certain districts, as near Thorner in walking up Scarcroft Hill to the N.W., or going from Wetherby to Spofforth, as the Permian is left and the Gritstone entered upon, or on leaving the limestone plateau of Malham Moor and walking north on to the sandy-peat, we may actually observe a total change in the flora coincident with the change in rock.

The species having more or less preference for the dysgeogenous limestones (modified by conditions already sufficiently dealt with), are conveniently divisible into three groups: (A) the montane species of xerophiles, (B) the characteristic xerophiles of our area, and (C) other plants, subordinately dry-loving, whose restrictions compel us to place them in a *sub-xerophilous* category.

(A) The Xerophilous Species, forming a typical flora of a Mountain Limestone Hill from base to summit.

Thalictrum montanum.
Actæa spicata.
Hutchinsia petræa.
Arabis hirsuta.
Cardamine impatiens.
Draba muralis.
Draba incana.
Thlaspi occitanum.
Helianthemum canum.
Alsine verna.
Hypericum montanum.
Geranium sanguineum.
Geranium lucidum.
Hippocrepis comosa.
Poterium Sanguisorba.
Potentilla alpestris.
Dryas octopetala.
Rubus saxatilis.
Pyrus rupicola.
Saxifraga tridactylites.
Galium sylvestre.
Scabiosa columbaria.
Hieracium cæsium.
Carlina vulgaris.
Polemonium cæruleum.
Gentiana campestris.
Origanum vulgare.
Thymus Serpyllum.
Calamintha Clinopodium.
Orobanche rubra.
Juniperus communis.
Taxus baccata.
Orchis conopsea.
Ophrys muscifera.
Epipactis ovalis.
Habenaria albida.
Cypripedium Calceolus.
Cephalanthera ensifolia.
Polygonatum officinale.
Carex capillaris.
Carex sylvatica.
Melica nutans.
Avena alpina.
Sesleria cærulea.
Asplenium viride.
Ceterach officinarum.
Cystopteris fragilis.
Nephrodium rigidum.
Polypodium calcareum.
Scolopendrium vulgare.
Bartramia Œderi.
Tortula tortuosa.
Tortula recurvifolia.
Encalypta ciliata.
Ditrichum flexicaule.
Neckera crispa.
Anomodon viticulosus.
Metzgeria pubescens.
Lejeunia calcarea.
Placodium murorum.
Endocarpum miniatum.

Some of these occur on the Permian-lime strata in the east, but the bulk of the fifty are restricted to the calcareous area in the western third of the Riding. A very few (e.g., *Asplenium viride*) occur, but only at a high altitude, on grit.

A very limited number of species appear to be truly xerophilous, in the sense of occurring only on *dry* soil, without bias as to its being silicious or calcareous. Those best exemplifying this in West Yorkshire are:—

Sisymbrium Sophia.
Reseda Luteola.
Cerastium semidecandrum.
Mœnchia erecta.
Trifolium striatum.
Trifolium procumbens.
Anthriscus vulgaris.
Veronica arvensis.
Linaria minor.
Verbascum Thapsus.
Echium vulgare.
Calamintha Acinos.
Plantago coronopus.

Taking the whole country they, perhaps, appear oftener on limestones than sandstones; if so, it is another proof of the felicity of Thurmann's main proposition—that (cæteris paribus) dysgeogenous rocks have driest soils.

(B) The Characteristic Xerophilous Species (exclusive of montanes) of the Limestones in West Yorkshire.

Clematis Vitalba.
Anemone Pulsatilla.
Helleborus viridis.
†Actæa spicata.
†Aquilegia vulgaris.
Reseda lutea.
†Helianthemum vulgare.
Hypericum montanum.
Linum perenne.
Astragalus hypoglottis.
Astragalus glycyphyllus.
Spiræa Filipendula.
Potentilla verna.
Pyrus rupicola.
Caucalis daucoides.
Cornus sanguinea.
Viburnum Lantana.
Asperula cynanchica.
Scabiosa columbaria.
Carduus eriophorus.
Carduus acaulis.
Inula Conyza.
Crepis taraxacifolia.
Cichorium Intybus.
Erigeron acris.
Hieracium murorum.
Ligustrum vulgare.
*Chlora perfoliata.
Gentiana Amarella.
Atropa Belladonna.
Verbena officinalis.
*Cynoglossum officinale.
Calamintha Nepeta.
Calamintha menthifolia.
Salvia Verbenaca.
Lithospermum officinale.
Plantago media.
Taxus baccata.
*Orchis ustulata.
*Orchis pyramidalis.
Gymnadenia conopsea.
Ophrys apifera.
*Spiranthes autumnalis.
Cephalanthera ensifolia.
Convallaria majalis.
Carex digitata.
Melica nutans.
Avena pratensis.

Bromus erectus.
Brachypodium pinnatum.
Hordeum sylvaticum.
Scolopendrium vulgare.
Seligeria pusilla.
Didymodon rubellus.
Encalypta streptocarpa.
Encalypta vulgaris.
Ditrichum flexicaule.
Orthotrichum saxatile.
Hypnum commutatum.
Madotheca platyphylla.

Five (those asterisked) of the above sixty occur, as is well known, on sands by the sea, outside our area; and others (marked with the †) have odd stations on sand within it. The *Spiræa*, the *Gentiana*, and the *Chlora* are *pélophiles*, in the sense of liking a stiff boulder clay soil with us.

(C) West Yorkshire Species of the Sub-xerophilous Class.

*Ranunculus auricomus.
Arabis hirsuta.
Sisymbrium thalianum.
Draba verna.
Viola hirta.
Geranium lucidum.
Hypericum hirsutum.
*Cerastium arvense.
Alsine tenuifolia.
Euonymus europæus.
*Rhamnus catharticus.
Ononis spinosa.
*Anthyllis vulneraria.
Poterium Sanguisorba.
Sanguisorba officinalis.
Rosa spinosissima.
Rosa micrantha.
Saxifraga tridactylites.
Bryonia dioica.
Parnassia palustris.
*Pimpinella magna.
Pastinaca sativa.
*Galium Mollugo.
Galium tricorne.
Dipsacus pilosus.
Carduus crispus.
Carlina vulgaris.
Centaurea Scabiosa.
Senecio erucifolius.
Picris hieracioides.
*Lactuca virosa.
Campanula glomerata.
Specularia hybrida.
Gentiana campestris.
Origanum vulgare.
Calamintha Clinopodium.
Linaria Elatine.
Daphne Laureola.
Fagus sylvatica.
Orchis ustulata.
*Ophrys muscifera.
Neottia Nidus-avis.
Gagea lutea.
Colchicum autumnale.
Carex divulsa.
Carex sylvatica.
Blysmus compressus.
Avena pubescens.
*Koeleria cristata.
Sclerochloa rigida.
Cystopteris fragilis.
Asplenium Ruta-muraria.

Those with the * prefixed have one or more stations on sandy or gritstone soil, such as the drift over the Permian and Trias presents in parts where the surface is composed of mingled gritty and calcareous particles with larger pebbles, allowing water to quickly percolate through; but the whole of them are so very constantly found to skip the retentive-soiled clay and flagstone tracts, that inclusion in the sub-xerophilous category was inevitable. *Parnassia*, *Blysmus*, and *Primula farinosa* are moisture-lovers,

hardly sub-xerophilous species, yet they show a similar lithological restriction to the rest in West Yorkshire.

3. *Hygrophilous Species of Eugeogenous Rocks.* Hitherto the species enumerated in connection with soils have had a more or less close relation with the 'xerophiles' of Thurmann; those it is now necessary to exemplify have a relation to rocks, the main constituent of which is silica, that stands undoubtedly in much closer agreement every way with the Hygrophiles of the same writer. A better name for the whole of them as a class, so far as Yorkshire is concerned, would be *Indifferents*; for, besides being in an immense preponderance over our gritstone area, on account of our more humid atmosphere and soil, with some half score of exceptions, they are very common over the dysgeogenous tracts too. The following table specifies the species which may be regarded from their exclusive prevalence as forming (from base to summit) a

Typical Florula of the Eugeogenous Gritstone Hill.

†Anemone nemorosa.
Cochlearia alpina.
*Corydalis claviculata.
Drosera rotundifolia.
Hypericum pulchrum.
Cerastium triviale.
Polygala depressa.
Orobus tuberosus.
Rubus Chamæmorus.
Galium saxatile.
Lonicera Periclymenum.
Pyrus Aucuparia.
Erica Tetralix.
*Erica cinerea.
Calluna vulgaris.
Myosotis sylvatica.
*Vaccinium Myrtillus.
Vaccinium Vitis-idæa.
Vaccinium Oxycoccos.
*Digitalis purpurea.
Teucrium Scorodonia.
Ulmus montana.
Quercus pedunculata.
*Betula alba.
Alnus glutinosa.
Salix aurita.
Salix capræa.
Empetrum nigrum.
Narthecium ossifragum.
Scilla nutans.
Juncus squarrosus.
Juncus effusus.
Juncus supinus.
Scirpus cæspitosus.
Eriophorum vaginatum.
Eriophorum angustifolium.
Carex pulicaris.
*Carex binervis.
Carex glauca.
Agrostis vulgaris.
Aira flexuosa.
Holcus mollis.
Festuca ovina.
Nardus stricta.
*Blechnum boreale.
*Lastræa dilatata.
Lastræa Filix-mas.
Ceratodon purpureus.
Polytrichum commune.
Bryum nutans.
Hypnum splendens.
Cladonia rangiferina.
Parmelia omphalodes.

The above fifty are about the commonest species on the dale slopes and ridges of unmixed gritstone tracts, some few of them, such as *Rubus Chamæ-*

morus, *Vaccinium Oxycoccos*, *Drosera*, etc., as individuals being the small minority, although frequently recurring, the rest in countless abundance as individuals repeating themselves throughout the vast area of swelling moorland slopes and peaty plateaux. From the above enumeration, however, many of the general hygrophilous species, e.g., Hazel, Ash, *Crepis paludosa*, *Caltha*, *Cardamine sylvatica*, *C. pratensis*, etc., should be excepted, because they are found quite as plentifully in the limestone dales. The species against which the asterisk has been placed are characteristically rare, substantially absent, over the calcareous strata.

Arenaceous Soils and Psammophilous Species. The purely sandy soils do not within our limits vary much except in the amount of water which, as the rainfall increases from east to west, passes quickly through them. A good many of the fifty odd, named in the following table as characteristic of the light barren soil over our triassic and millstone-grit rocks, are plants rather of the western than of the eastern side of England. Where a few of them stray in odd places into localities overlying limestones, it is because of some exceptional condition of the sub-soil there. None of them seem to affect the grits (comparatively few occur on the slate at all) for any chemical property in the soil—as it is possible that a few of the less strictly xerophilous species do in regard to lime,—but for the *soft* water their well-developed finely-divided roots, well fitted for ramifying in the light soil, can absorb.

Myosurus minimus.
*Corydalis claviculata.
Papaver Argemone.
*Teesdalia nudicaulis.
†Turritis glabra. (?)
Coronopus Ruellii.
Sagina ciliata.
Spergula arvensis.
Silene anglica.
*Dianthus deltoides.
Stellaria nemorum.
*Malva moschata.
Scleranthus annuus.
*Spergularia rubra.
Hypericum humifusum.
Radiola millegrana.
*Erodium cicutarium.
†Ononis arvensis.
Sarothamnus scoparius.
*Trifolium arvense.
Ornithopus perpusillus.
*Vicia lathyroides.
Alchemilla arvensis.
*Potentilla argentea.

†Silaus pratensis.
Conium maculatum.
Galium verum.
Artemisia vulgaris.
†Chrysanthemum segetum.
Centaurea Cyanus.
*Filago minima.
Gnaphalium sylvaticum.
Hieracium boreale.
Solidago Virgaurea.
Senecio sylvaticus.
Jasione montana.
Campanula latifolia.
Arbutus Uva-ursi.
*Veronica triphylla.
Linaria minor.
Melampyrum pratense.
Galeopsis ochroleuca.
Teucrium Scorodonia.
Nepeta Cataria.
Orobanche major.
Trientalis europæa.
†*Centunculus minimus.
*Euphorbia platyphylla.

Plantago Coronopus.
Rumex Acetosella.
Littorella lacustris.
Salix repens.
†Tamus communis.
Allium vineale.
Luzula sylvatica.
Juncus bufonius.
Carex lævigata.
Aira præcox.
*Festuca sciuroides.

*Apera Spica-venti.
†Milium effusum.
Sclerochloa distans.
Holcus lanatus.
Hordeum murinum.
Cryptogramme crispa.
Pilularia globulifera.
Lycopodium clavatum.
Equisetum sylvaticum.
Trichostomum homomallum.
Brachythecium albicans.

Those with the asterisk are lowland sand-lovers absent on limestone; those with the obelisk are *pélophiles* also to some extent, and occur on flagstone-rock clays.

4. *Peat and its Vegetation.* Peat is usually found in conjunction with an underlying gritstone rock, and (in the low-lying vale of York) upon a stratum of fine sand saturated with water. It is formed but rarely over clay or limestone. It consists mainly of vegetable humus in a peculiar state—the remains of the small sandstone or hygrophilous species of plants originally preferrent of the rocky stratum before the process (an imperfect carbonisation under pressure) of formation took place, mixed with sand; and it supports a type of florula unique in the vastly abundant individuals of its relatively small number of species. Its flora is distinctly hygrophilous in type, and some of the species are plainly closely allied to the sand-lovers in character, and several of them are found growing indifferently in either matrix. The scarcity of Gramineæ—whose individuals outnumber everything else in our pastures—found on the peat, small in number of plants as well as in number of species, is inexplicable, and more especially since it can and does support an innumerable aggregation of some things. *Yellow*-flowered species are likewise almost wholly absent—a fact first pointed out by Dr. H. F. Parsons. *Utricularia minor* and *Hypericum elodes* are the only ones I call to mind. The list which follows includes the main plants characteristic of peaty soil in the West Riding. In seasons the florula of the Peat is fully three weeks to a month behind that of gravelly or marly tracts at the same level.

Drosera rotundifolia.
Drosera anglica.
Drosera intermedia.
Hypericum elodes.
Rhamnus Frangula.
Comarum palustre.
Sagina nodosa.
Rubus Chamæmorus.
Montia fontana.

Hydrocotyle vulgaris.
Gentiana Pneumonanthe.
Erica Tetralix.
Wahlenbergia hederacea.
Vaccinium Oxycoccos.
Andromeda polifolia.
Veronica scutellata.
Utricularia minor.
Pinguicula vulgaris.

Anagallis tenella.
Myosotis repens.
Empetrum nigrum.
Myrica Gale.
Listera cordata.
Narthecium ossifragum.
Rhynchospora alba.
Scirpus multicaulis.
Eriophorum latifolium.
Eriophorum angustifolium.
Carex limosa.
Carex fulva.
Carex curta.
Carex ampullacea.
Carex vulgaris.
Carex stellulata.

Aira uliginosa.
Molinia cærulea.
Triodia decumbens.
Nardus stricta.
Lastrea Thelypteris.
Nephrodium spinulosum.
Nephrodium Oreopteris.
Lycopodium inundatum.
Lycopodium alpinum.
Sphagnum—several species.
Leucobryum glaucum.
Bryum nutans.
Campylopus pyriformis.
Cladonia rangiferina.
Cladonia uncialis.
Cetraria aculeata.

5. *Boulder Clay and Argillaceous Soils: Botanical Characters.* The soils are of two sorts, and their florulas are only in part interchangeable. The soil over the boulder clay is rather a stiff semi-calcareous mud than a true clay, the subsoil being a kind of breccia, with many rounded pebbles and nodules of limestone, the flora being rather rich, approaching that of chalky districts in general character, with a relative abundance of such plants as the Primrose, Cowslip, *Ballota fœtida*, *Lamium Galeobdolon*, *Acer campestre*, Blackthorn, Hazel, &c., in contradistinction to such species as *Stellaria Holostea*, *Lychnis diurna*, *Malva moschata*, *Genista tinctoria*, *Vicia Cracca*, *Hieracium boreale*, *Solanum Dulcamara*, *Scilla nutans*, Elder, Sallow, Birch, and Coltsfoot, found on the cold, aluminous, or bituminous clays, shales, &c., over the Coal-measures flagstones. These are backward, and agree with the peat flora in one respect—the great number of individuals to a small number of species over a given tract. Numerically, the flora of both tracts is poorer than either that of limestone or sand; its salient feature being that species characteristic of those *dry* soils are here 'conspicuous by their absence.'

The chief Pelophilous (clay-loving) species in the Riding are embraced in this list.

Ranunculus Ficaria.
*Ranunculus auricomus.
Ranunculus Flammula.
Thlaspi arvense.
Turritis glabra.
Stellaria Holostea.
Lychnis diurna.
Lepidium campestre.
Hypericum dubium (?).
Hypericum tetrapterum.

Genista tinctoria.
Lotus major.
Vicia Cracca.
Orobus tuberosus.
Rubus Idæus.
*Spiræa Filipendula.
Epilobium obscurum.
Sanicula europæa.
*Sison Amomum.
*Bupleurum rotundifolium.

Lonicera Periclymenum.
Viburnum Opulus.
*Eupatorium cannabinum.
Hieracium boreale.
Gnaphalium uliginosum.
*Helminthia echioides.
Tussilago Farfara.
Senecio aquaticus.
Bidens tripartita.
Pulicaria dysenterica.
*Chlora perfoliata.
*Gentiana Amarella.
Convolvulus arvensis.
Lysimachia nemorum.
*Primula vulgaris.
Primula veris.
Solanum dulcamara.
Lathræa squamaria.
*Antirrhinum Orontium.
Scutellaria minor.

*Lamium Galeobdolon.
Alnus glutinosa.
*Ballota fœtida.
*Allium oleraceum.
Orchis mascula.
Fritillaria Meleagris.
Paris quadrifolia.
Alisma Plantago.
Scilla nutans.
Juncus bufonius.
Heleocharis acicularis.
Carex pendula.
Carex flava.
Carex pallescens.
Ophioglossum vulgatum.
Equisetum maximum.
Equisetum arvense.
*Barbula cavifolia.
Hypnum filicinum.

Those with the * prefix in the above list are distinct preferrents of the boulder-clay mud soils, but all of them have stations on truer clays. Very few of them in their distribution show any bias for sand, unless Turritis or Sison may be said to do so. Most of the fruticose Rubi are sand-lovers of the lower levels, but *Rubus fruticosus* and *R. carpinifolius* are rather partial to the calcareous marls and boulder-clay soils.

6. *Alluvial Soils. Warp Soils. Halophilous Species.* The complex nature of alluvium, and the presence of a saline constituent in the silt and warp near the outlets of our larger rivers—the result, of course, of the varied character of the strata from the abrasion of which they are derived—eventuates, botanically, in a flora of a mixed kind too, many species of quite contrasting preferences in their normal situations occurring cheek-by-jowl in the oddest way on the banks of the Ouse, Aire, and Don, from Selby down to Goole, from Barnsley down to Thorne, and on the half-reclaimed borders of Thorne and Goole Moors, Marshland, etc. *Campanula latifolia* may be seen in juxtaposition with *Allium oleraceum*, the montane *Epilobium angustifolium* with such hygrophiles as *Sium latifolium*, *Epilobium hirsutum*, and *Rumex maritimus*, not of course actually side by side, but within a short distance in the same area. In addition, about Goole the occurrence of certain Halophilous (salt-loving) plants must be noted; species that in their range throughout Britain show a more or less marked tendency to maritime restriction. The list of species in this sub-maritime, littoral category is not large, of course, and if the tide with its silt and warp did not constantly bring up reinforcements of seeds as well as salt, most of them would doubtless soon disappear.

West Yorkshire Species of the Sub-Maritime Class.

[N.B.—The four italicized are Montane, and the two with the asterisk before them occur *only* on the hills.]

Ranunculus hirsutus.	Plantago Coronopus.
Cochlearia officinalis.	Atriplex littoralis.
**Silene maritima.*	Rumex maritimus.
Spergularia media.	Zannichellia pedicillata.
Trifolium fragiferum.	Potamogeton scoparius.
Apium graveolens.	Triglochin maritimum.
Œnanthe Lachenalii.	Juncus Gerardi.
Carduus tenuiflorus.	Scirpus glaucus.
Aster Tripolium.	Scirpus maritimus.
Glaux maritima.	Carex distans.
Samolus Valerandi.	Carex divisa.
**Armeria maritima.*	Sclerochloa maritima.
Plantago maritima.	Sclerochloa distans.

VI. In concluding this part of the inquiry, if we briefly examine into how far the foregoing soil-restricted groups apportion themselves among the various dispersion types—Northern, Montane, English, Germanic, and Atlantic—we may see at once how the facts find collateral corroboration in what has been advanced as to centres and genesis. Where the noterophilous species (inclusive of some Sub-Montane forms) common to the high moorlands and lowlands, occur at *low* levels, it will surely be on sandy, heathery commons, or wet sandy peat. Such are the general Hygrophiles—*Galium saxatile*, *Epilobium angustifolium*, *Empetrum*, *Montia fontana*, *Erica cinerea*, *Pinguicula*, *Juncus squarrosus*, *Carex panicea*, *Aira flexuosa*, and *Molinia cœrulea*. These are found retaining barren ground they have occupied for centuries and centuries, not because they are cold-lovers, but because they can endure wet, either with or without much relative cold, so long as the soil conditions are such as to prevent the warmth-loving, cold-fearing species from crowding them out.

On the other hand, the Germanic continental type of species are distinct cold-fearers (moisture they many of them like in presence of warmth), and these are always found ascending to their highest levels on calcareous soils —i.e., soils as dry and warm as any we have at all. The annuals among them are many, and *they* are truly heat-lovers, the mean temperature having little to do with their persistence, since they spring up from seed, leaf, bloom, and seed again during the summer season; which is hottest on the eastern plains, whence they originally came. Many of these annuals could not maintain themselves (any more than wheat) in some parts of the west at high-levels, not because of the inability of their seeds to endure cold during germination, as because the summer heat is not long and high enough to allow of the reproductive cycle being perfected.

On the other hand, the Atlantic type species are cold-fearers rather than heat-lovers; they are all perennials, most of them liable to be killed off by severe winters. Most of them are markedly hygrophilous—e.g., *Hypericum elodes*, although some, such as *Meconopsis*, are pseudo-xerophilous.

VII. Geographical Allies. Certain species of plants are seen, after long familiarity with their peculiarities in any area, such as Yorkshire, presenting a considerable variety of conditions, which appear to group themselves naturally together as geographical associates—triplets or quartettes—the integers composing which agree in having an almost identical horizontal distribution, and often a very similar vertical range as well. The species in each group will not always (although it is often so) be found actually side by side; but their faithful adherence to each other in the field, through sub-divided river-basins and over differing strata, skipping places together, monopolising others in company, clearly points to their being stational allies from kindred preferences, finding just what they each require combined in local conditions. In what these preferences exactly consist is not always evident, however, without a close acquaintance with the climatal and lithological minutiæ of the stational facts.

The squads that follow are given as West Yorkshire instances of these climato-stratigraphical alliances. Some of them are borne out by Mr. Baker's observations in North Yorkshire, some are not, but their accuracy for the area in question can be vouched for, despite the fact that some of the bracketed species seem of dissimilar constitution.

Class I.

Actæa spicata, Aquilegia vulgaris, Helleborus viridis, Melica nutans,	and nearly allied, but not ascending so high,	Atropa Belladonna, Gagea lutea, Carex digitata, Hordeum sylvaticum.
Thlaspi occitanum, Arenaria verna, Hutchinsia petræa, Sesleria cærulea,	and near to this group, but descending lower,	Draba incana, Galium sylvestre, Hippocrepis comosa.
Geum rivale, Myosotis sylvatica, Allium Scorodoprasum,	and similar, but not so low or so frequent	Trollius europæus, Carduus heterophyllus, Primula farinosa.

Class II.

Stellaria nemorum, Crepis paludosa, Myrrhis odorata,	Thalictrum flexuosum, Rubus saxatilis, Allium Scorodoprasum.
Sagina nodosa, Parnassia palustris, Blysmus compressus,	Genista tinctoria, Malva moschata, Convolvulus arvensis, Pimpinella magna.

Class II—*continued.*

Cerastium arvense, Centaurea Scabiosa, Echium vulgare,	Anthyllis Vulneraria, Plantago media, Galium verum.
Erigeron acris, Campanula glomerata, Hordeum pratense,	Geranium lucidum, Lactuca muralis, Cystopteris fragilis.

Class III.

Thalictrum flavum, Eupatorium cannabinum Lysimachia vulgaris,	Ranunculus sceleratus, Bidens cernua, Lycopus europæus.
Corydalis claviculata, Lepigonum rubrum, Jasione montana,	Geranium pratense, Sanguisorba officinalis, Polygonum Bistorta.

Naturally, these illustrative groups—not all of those discernible—are made up of forms neither ubiquitous nor very rare, since their occurrence in several places in at least three various stations, and their co-absence in others, is necessary for their acceptance as allies at all. Anyone who tries, from fancy, to *erect* similar groups will find that the difficulty is in proportion to the intimacy of his acquaintance with the area. There are many more species (than seems primâ facie likely to be the case) that will *not* be pressed into any groups at all, by reason of some essential singularity. The following names are suggestive—*Hypericum Androsæmum*, *Rhamnus Frangula*, *Antennaria dioica*, *Trientalis europæa*, *Asplenium viride*. These seem to have no precise allies. The first-named has one or two species, e.g., *Carex pendula*, very nearly corresponding, but they *all* occur in one or two places without the Hypericum, and that is found in one spot without them, so that to truthfully bracket them is impossible. The last-named, again, one would think had two constant associates discoverable; but no! they do not exist: *Arenaria verna* and *Sesleria cærulea* are allies distinctly enough over one large tract, in wet and dry situations, *but* the Asplenium breaks away, revealing the *accident* of the close association on the scar-limestone, by occurring *alone* at a high level on pure grit.

The subject is by no means exhausted, nor indeed is the larger question of phyto-lithology, but requires residence within the division, leisure, and untiring patience for its elucidation. Pursuing the inquiry, we should find ourselves concerned in agencies of dispersion, chemical and mechanical peculiarities of surface strata, soils, and water (for some aerated, for others lime in solution appears a necessity), much more profound than those broadly determining the flora; and there is little doubt the labour would elicit not a few contrasts and correspondences of supreme interest: some of the facts might even throw light upon the antiquity of the types found fused in the British Flora, and the sequence in which they overspread tracts previously denuded (?) of vegetation by slowly acting climatarchic agencies.

BIBLIOGRAPHY.

BIBLIOGRAPHY

OF

WEST YORKSHIRE BOTANY:

1548–1885.

Works or Papers containing references, more or less lengthy, bearing upon West Yorkshire Botany, to which reference has been made in the Flora.

1548.

1. TURNER, WILLIAM. The names of herbes in Greke, Latin, Englishe, Duche, and Frenche, wyth the commune names that Herbaries and Apotecaries use. (Under *Centunculus* (Gnaphalium) and *Taxus* ('Ughe tree') will be found what are probably the earliest allusions to Yorkshire plants.)

1568.

2. TURNER, WM. The secunde parte of William Turner's Herball . . . . imprinted at Collen by Arnold Birchman, MDLXVIII. (Vide fol. 109, respecting Oak-galls at Halifax, the first distinct allusion to a West Riding plant.)

1583.

3. CLUSIUS, CAROLUS. Rariorum Stirpium per Pannonias Hist. Antverhiæ (1583).

1597.

4. GERARDE, JOHN. The Herball or general Historie of Plants. London (1597).

1601.

5. CLUSIUS, C. Rariorum Plantarum Historiæ Antverhiæ (1601). (Vide pp. 98, 99.)

1633.

6. JOHNSTON, T. and GERARD. The Herball, etc., by John Gerard, very much enlarged and amended by Thomas Johnson (1633).

1640.

7. PARKINSON, JOHN. Theatrum Botanicum. . . . Lond. Folio. (Vide pp. 76, 133 (fide Gerarde) recte 133 bis, 286, 392, 510, 534, 727, 755, 756, 889, 1015, 1458.)

1650.

8. HOW, WILLIAM. Phytologia Britannica natales exhibens Indigenarum stirpium sponte emergentiam. Londini . . 1650.

1664.

9. TURNER, ROBERT. Botanologia, The British Physician: or, the Nature and Vertues of English Plants, etc., by R. T. Botanolog. Stud. Lond. 1664. (Vide pp. 148 and 316.)

1666.

10. MERRETT, CHRISTOPHER. Pinax Rerum Naturalium Britannicarum. Lond. (1666).

1670.

11. BUDDLE, ADAM. Hortus Siccus Buddleanus, sive methodus nova stirpium. Mus. Brit. Bibl. Sloan. 2970-2980.
12. RAY, JOHN. Catalogus Plantarum Angliæ. 12mo. Lond.

1677.

13. RAY, JOHN. Catalogus Plantarum Angliæ. Ed. 2. Londini, 1677.

1686—8—1704.

14. RAY, JOHN. Historia plantarum. Vol. i, p. 1—983 (1686). Vol. ii, p. 985—1994 (1688). Vol. iii ad. fin. (1704). London.

1688.

15. RAY, JOHN. Fasciculus Stirpium Britannicarum, etc. Londini. (Dated) Maii 16 (1688).

1695.

16. GIBSON, EDMUND, BISHOP. His Ed. of Camden's Britannia—the lists communicated by John Ray (pp. 768—772).

1696.

17. PLUKENET, LEONARD. Plukenetii Almagestium Botanicum, etc. 4 vols., 4to. Lond. (Quoted as Pluk. Almagest.) Pp. 8, 173, and 251 of Vol. 2 (1696).
18. RAY, JOHN. Synopsis Methodica Stirpium Britannicarum, tum Indiginis, tum in agris cultis, locis suis dispositis. . . . Editio secunda . . . (Lond. 1696). (See p. 242, etc., for the earliest notices of many of the rarer West Riding species.)

1699.

19. BOBART, J. Plantarum Hist. Oxoniensis Universalis. Pars tertia, [Partem hanc tertiam, post auctoris mortem . . . explevit et absolvit.] Bobart, etc. Oxonii, 1699. (Vide pp. 149, 477, 488, 489.)

1711.

20. PETIVER, JACOB. An Account of divers rare plants. Phil. Trans., vol. xxvii, pp. 375—394, (for remarks anent *Draba muralis* in Craven, and the claim of *Oxyria* to be considered a Yorkshire plant, and compare Pluk. Alm., p. 8).

1713.

21. PETIVER, J. Graminum, Muscorum, Fungorum, etc., Britannicorum Concordia. A methodical concordance of British grasses, mosses, etc. (Lond., 1713.)

1718.

22. DERHAM, W. Philosophical Letters. (Correspondence between Mr. Wray (J. Ray), T. Lawson, and Dr. Lister.) . . . published by W. Derham. (Lond., 1718.)

23. LAWSON, THOMAS. Letter to Mr. Ray (dated from Great Strickland, April 9th, 1688), in W. Derham's Philosophical Letters. (Vide pp. 213—223.)

1724.

24. DILLENIUS, J. J. Ray's Synopsis Method. Stirp. Brit., Ed. iii (with additions). (Vide *Melampyrum cristatum*, *Actæa*, etc., etc.)

1741.

25. DILLENIUS, J. J. Historia Muscorum . . . . cum Appendice . . . Oxonii, 1741. (Vide pp. 88, 89, 132, 208, 271, 335, 435, 476, 503, 511, 542, etc.)

1744.

26. WILSON, JOHN. A Synopsis of British Plants, in Mr. Ray's Method. Newcastle-on-Tyne, 1744.

1746.

27. BLACKSTONE, J. Specimen Botanicum quo Plantarum plurium rariorum Angliæ indigenarum . . . Lond. (Hinc vide pp. 3, 8, 10, 14, 15, 19, 25, 26, 27, 28, 34, 35, 36, 44, 45, 52, 54, 67, 68, 70, 74, 75, 77, 79, 80, 81, 86, 87, 88, 89, 96, 97, 105.)

1760.

28. DERHAM, W. Select Remains of John Ray. (Vide p. 146.)
29. MILLER, PHILIP. Gardener's Dictionary.

1762—1778.

30. HUDSON, WILLIAM. Flora Anglica. 8vo. Lond. Ed. i and Ed. ii.

1763.

31. MARTYN, T. Plantæ Cantabrigiensis. . . . To which are added lists of the more rare plants growing in many parts of England and Wales. 8vo. Lond. (Vide Appendix, p. 106.)

1775.

32. BOLTON, JAMES. A Catalogue of Plants growing in the Parish of Halifax (authorship only acknowledged later), forming pp. 729—764 in John Watson's Hist. and Antiq. of the Par. of Halifax. Lond.
33. JENKINSON, JAMES. A Generic and Specific Description of British Plants . . . with notes of their localities. (Some in West Yorkshire.) Kendal, 1775.

1778.

34. ROBSON, STEPHEN. The British Flora. 1 vol., 8vo. York, 1778. (Contains notes of many West Yorkshire localities, but generally without indication of whether original or copied, as many obviously are.)

1782.

35. CURTIS, WILLIAM. A Catalogue of certain Plants growing wild chiefly in the environs of Settle, in York, observed . . . in months vii. and viii., 1782; forming Appendix in vol. i. Flora Londinensis, 1777—1787. (Reprinted in Phytologist, New Series, vol. i.)

1785—1790.

36. BOLTON, J. Filices Britannicæ. . . . Part i, Leeds, 4to, 1785. Part ii, Huddersfield, 1790. The original drawings are in Botanical Department, British Museum.

1787.

37. WITHERING. Syst. Arr. of British Plants. Edition ii. Four vols. 8vo. (Contains many first notices of West Riding Plants.)

1788—1791.

38. BOLTON, JAMES. History of Fungusses. 4to.

1789.

39. GOUGH, RICHARD. Britannia . . . by Wm. Camden. Translated from the edition published by the Author in MDCVII. Enlarged by the latest discoveries by R. G. 3 vols. Lond., 1789. (Vide vol. iii, 98 to 101.)

1790—1814.

40. SOWERBY, J., and SMITH, J. E. English Botany. 36 vols, 8vo. Lond. (*Sedum sexangulare*, *Epipactis rubra*, &c. Contains many earliest notices of West Yorkshire species.)

1792.

41. TEESDALE, R. Plantæ Eboracenses. Part i. Trans. Linn. Soc., vol. ii, pp. 103—125.

1796.

42. WITHERING, J. Syst. Arrangement of British Plants. Ed. iii, 8vo.

1798.

43. TEESDALE, R. A Supplement to the Plantæ Eboracenses. Trans. Linn. Soc., vol. v, pp. 36—95. (Contains notices of many West Riding Plants.)

1800.

44. SALT, JONATHAN. 'Flora Sheffieldiensis' in MS. Dated June 25th, 1800. In the Lib. Lit. and Phil. Soc., Sheffield.

1800—1804.

45. SMITH, J. E. Flora Britannica. In 3 vols. Vols. i and ii, 1800. Vol. iii, 1804.

1801.

46. WITHERING, J. Arrangement of British Plants. Ed. 4. (Vide *Hypochæris maculata*, etc.)

1804.

47. MILLER, EDWARD. Hist. and Antiq. of Doncaster. (Vide sub Nat. Hist., p. 8, where a list is given of 68 plants stated to be found in Doncaster district, including the name '*Elatine alsinastrum*'; and also see pp. 203—205 for list by Dr. Ed. Chorley.)

48. WHITE, Dr. (of York). List of Plants of York and Doncaster in E. Miller's Hist. and Antiq. of Doncaster. 8vo. 1804.

1805.

49. TURNER, DAWSON, and DILLWYN, LEWIS WESTON. Botanist's Guide. 8vo. (Vide pp. 663—744.)

1807.

50. MARTYN, T., B.D., F.R.S. His Edition of Philip Miller's Gardener's Dictionary. (Vide *Bunium bulbocastanum* et *flexuosum*.) 2 vols.

1811.

51. WINDSOR, J., M.D. On the two forms of Thlaspi alpestre. Trans. Linn. Soc., vol. x, p. 196.

1812.

52. EVELYN, JOHN. Sylva . . . with notes by A. Hunter. Fourth edition, with the Editor's last Corrections. York, 1812. (Vide pp. 206—9, 211—12, concerning Sheffield Oaks.)

53. HAILSTONE, SAM. List of Plants of Settle district, in Whitaker's History of Craven. 4to.

1816.

54. WHITAKER, Dr. Loidis and Elmete. Folio.

1817.

55. WOODS, JOSEPH. Monograph of British Roses, in Trans. Linn. Soc., vol. xii, part 1. (References to Settle district Roses, etc.)

1819.

56. SMITH, Sir J. E. On Elatine Alsinastrum. (Vide sub Art. *Elatine* in Rees' Cyclopædia, vol. xii.)

1824.

57. ATKINSON, J. Sketch of the Geograph. Distrib. of Plants in Yorkshire. Mem. Wernerian Society of Edinburgh, vol. [illegible], part i, p. 277, and Philosoph. Mag., vol. ii, p. 211.

1824—1828.

58. SMITH, Sir J. E. The English Flora. 4 vols., 8vo. (Very numerous references to West Yorkshire localities.) The Second Ed. 1828, 1829, and 1830 does not materially differ.

1827—1834.

59. BOHLER, J. Letters on Botany of Sheffield and Roche Abbey district. Published at irregular intervals in the Sheffield Courant, a defunct newspaper, issued first on July 6, 1827, and last on June 13, 1834.

1828.

60. LANGLEY, LARRET. Flora of Rotherham, i.e., a list of plants noticed within 8 miles of Rotherham, by Larret Langley (Brampton Academy, near Rotherham) and Rev. E. Wilson, junr. (of Swinton), in Loudon's Mag. Nat. Hist., ii, 269. Paper dated December 6, 1821.

1829.

61. LOUDON, J. C. Mag. Nat. Hist., vol. ii. See p. 312, under Obituary Notice of Geo. Caley, for remarks anent *Hieracium villosum*.

1830.

62. BREE, W. T. Botanical Frauds. Mag. Nat. Hist, vol. iii, p. 150 (March 1830). Has a bearing upon *Hutchinsia alpina*.

63. SMITH, J. E., and SOWERBY, J. English Botany, 2nd Edition. 12 vols., 8vo, Lond.

64. WITHERING, W. A Botanical Arrangement of British Plants. Ed. vii, 4 vols.

1831.

65. T. E. L. On Rare Plants. Mag. Nat. Hist, vol. iv, p. 71 (1831).

1831—49.

66. SOWERBY, JAMES. The Supplement to English Botany, containing figures of such newly-discovered Plants as are not contained in the 36 vols. finished in 1814. Vol. i, 1831; vol. ii, 1843; vol. iii, 1843; vol. iv, 1849. (Contain the earliest notices of a few W. R. Phanerogams, and a good many of the mosses.) A later fifth vol. contains no reference to West Yorkshire Botany.

1832.

67. APPLEBY, S. Rare Plants in the neighbourhood of Doncaster. Mag. Nat. Hist., vol. v, p. 558 (July 1832).

68. HARGROVE, E. Names of some Plants which grow in the neighbourhood of Knaresborough. Hist. of Castle, Town, and Forest of Knaresborough. Small 8vo. (Vide pp. 308—314.)

1833.

69. HOOKER, Sir W. J. The British Flora.

1835.

70. TURNER, DAWSON. Richardson's Correspondence. Edited by D. T.

71. WATSON, HEWETT COTTRELL. The New Botanist's Guide to the Localities of the Rarer Plants of Britain. Vol. i, Eng. & Wales. (For West Yorkshire localities, vide pp. 174 to 299.)

1837.

72. HOLLAND, J., M.D. Tour of the Don, Sheffield, 2 vols., 12mo. (Contains an account of Jonathan Salt, and incidental allusions to plants found by him and others in Sheffield district.)

73. WATSON, H. C. New Botanist's Guide. Vol. ii, Scotland, and Supplement to England and Wales of vol. i. (Vide pp. 651—658 for notices of West Yorkshire localities.)

1838.

74. CHAMBERS, R. On Cypripedium. Mag. Nat. Hist. vol. ii, N.S., p. 40.

1840.

75. BAINES, HENRY. The Flora of Yorkshire. Demy 8vo, Lond.

76. NEWMAN, ED. History of British Ferns; Lond. 1840. A 2nd Ed.—but not so called on title page—much enlarged, in 1844.

1842.

77. GIBSON, S. Wharfedale Mosses. Phytologist, O. S., vol. i, p. 291.

78. LANKESTER, ED. An Account of Askern, with a Sketch of the Natural History of the Immediate Neighbourhood. 8vo, Lond. (Vide pp. 50—57.)

79. TATHAM, JOHN. On Settle Botany. Phytologist, O. S., vol. i, p. 87.

1843.

80. NEWMAN, EDWARD. County Lists of the British Ferns, and their Allies. Phytologist, O. S., vol. i, p. 449.

1844.

81. BACKHOUSE, JAMES, junr. Notes of a Botanical Ramble in Summer of 1844. Phytologist, vol. i, pp. 1065, 1089, 1126.

82. KING, S. On Halifax District Plants. Phytologist, vol. i, p. 1096.

1845.

83. HASSALL, A. H. History of the British Fresh-water Algæ. 2 vols., 8vo.

84. SPRUCE, RICHARD. List of the Musci and Hepaticæ of Yorkshire (including stations in West Yorkshire). Phytologist, O. S., vol. ii, p. 147.

1847—1851.

85. WATSON, HEWETT COTTRELL. Cybele Britannica. 4 vols, demy 8vo, Lond.

1848.

86. LANKESTER, E. Correspondence of John Ray. 8vo. (Vide p. 197, where Thos. Lawson's Letter to Mr. Ray, of 1688, is reprinted, with the modern synonyms for the old plant-names added by Professor Babington.)

1849.

87. BABINGTON, C. C. Sagina ciliata . . . and its British localities. Bot. Gazette, vol. i, p. 176. (July 1849.)

88. BERRY, ELIHU. List of the rarer Doncaster and Huddersfield Plants. Phytologist, Jan. 1849.

89. BLOXAM, Rev. A. British Rubi in Yorkshire. Phytologist, O. S., vol. iii, p. 325, etc.

90. INCHBALD, P. Some Adwick-le-Street and Doncaster Plants. Paper in Phytologist, vol. iii, 445. (March 1849.)

1849—1850—1851.

91. HENFREY, ARTHUR. The Botanical Gazette, a Journal of the Progress of British Botany. Vol. i, 1849; vol. ii, 1850; vol. iii, 1851. No more published. (Contains various notices of West Yorkshire plants.)

1850.

92. HOWSON, Rev. WM. An Illustrated Guide to the Curiosities of Craven, with . . . a Local Flora, 12mo. Wildman, Settle.

1852.

93. 'The Botany of Ben Rhydding: a Fell-side and Field Ramble in Wharfedale,' by 'Flora' in 'The Ben Rhydding Ariel'—a literary miscellany published August 1852, at Leeds, by Edward Baines and Sons.

1854.

94. BAKER, JOHN GILBERT, F.L.S., and NOWELL, JOHN. A Supplement to Baines' Flora of Yorkshire. Demy 8vo. Lond.

1855.

95. WILSON, WILLIAM. Bryologia Britannica. 8vo. Lond. (Contains various notices of West Riding mosses.)

1855—1858.

96. Windsor, J., M.D. Papers on Settle Botany. Phytologist, New Series, vol. i, pp. 145, 173, 259, 263; vol. ii, p. 12; vol. iii, pp. 423, 464.

1858.

97. Bohler, J. Doncaster Fungi. Phytologist, New Series, vol. iii, p. 198.
98. Irvine, Alexander. The Illustrated Handbook of British Plants. Small 8vo, Lond. (Vide p. 325 and elsewhere.)

1859.

99. Carrington, Benj., M.D., F.L.S. On the Geological Distribution of Plants in some districts of Yorkshire. Rep. Brit. Assoc. for 1858. Sectional Trans., p. 115.
100. Hobkirk, Chas. P. Huddersfield: its History and Natural History. Small 8vo. Huddersfield.

1862.

101. Babington, Charles Cardale, F.R.S. Manual of British Botany. 5th Edition. 8vo, Lond.
102. Carrington, Benj., M.D., and Miall, Louis C. The Flora of the West Riding (exclusive of the Ainsty). Flowering Plants, by L. C. M.; Cryptogams, by B. C. Small 8vo, Bradford.
103. Gissing, T. W. Ferns of Wakefield. 8vo, Wakefield.

1863.

104. Baker, John Gilbert, F.L.S. North Yorkshire: Studies of its Botany, etc. (exclusive of Ainsty). Lond.
105. Moore, T. Note on Trichomanes radicans. Journ. Bot., O. S., vol. i, p. 238.
106. Newbould, Rev. W. W., M.A., F.L.S. Is Hutchinsia alpina, Br., a British Plant? Journ. Bot. (Dec. 1863), vol. i.

1863—1871.

107. Journal of Botany. First Series. Edited by B. Seemann, J. G. Baker, and H. Trimen. (Vide vol. iii, p. 83, etc.)

1863—1872.

108. Syme, J. T. Boswell. English Botany. 3rd Edition, 11 vols., roy. 8vo, Lond. (Many West Yorkshire localities incidentally mentioned.)

1864.

109. Baker, J. G. Review of the British Roses, especially those of the North of England. Naturalist, vol. i, pp. 14, 33, 60, 93, 141, 185.
110. Britten, James, F.L.S. Spontaneous Exotics. Naturalist, vol. i, pp. 183, 205.
111. Dobson, William. Rambles by the Ribble. 1st Series, 2nd ed., 1864. Contains sundry notices of West Riding botany. (Vide pp. 16, 21, 23, 25, 31, 87, 98, 142.)
112. Kenyon, J. E. 'List of Horton Pagnell and Doncaster Plants.' 'Botanists' Chronicle,' August 1864.

113. Miall, L. C. On the Botany of Malham. Naturalist, vol. i, pp. 209 and 246 (Nov. and Dec. 1864).
114. Ib. *Hutchinsia alpina* on Ingleborough? Naturalist, vol. i, p. 51.
115. Roberts, Geo. Plants observed in Summer of 1864 on Magnesian Limestone in Pontefract neighbourhood. Naturalist, vol. i, p. 255 (Dec. 15, 1864).

1865.

116. Britten, James, F.L.S. Spontaneous Exotics. Naturalist, vol. i, pp. 262, 263, 359; vol. ii, p. 46.
117. Gissing, T. W. Materials for a Flora of Wakefield. Naturalist, vol. ii, pp. 6, 32, 56, 80, 94, 128, 147, 165, 184, 212, 232.
118. Hobkirk, C. P. *Lonicera xylosteum* near Huddersfield. Botan. Exchange Club Report for 1865. (Compare with account given in Botan. Exchange Club Report for 1871, p. 15.)
119. Miall, L. C. Ferns and Lichens of Malham. Naturalist, vol. ii, pp. 277 and 377.
120. Roberts, Geo. Plants found on Limestone north of Pontefract, and on Sandstone near Lofthouse. Naturalist, vol. ii, p. 194.
121. Windsor, J., M.D., F.R.C.S. On *Thlaspi alpestre*. Naturalist, vol. ii, pp. 31, 60, 108.

1866.

122. Banks, W. S. Walks in Yorkshire. 12mo, Wakefield. (Contains notices of West Yorkshire plants, pp. 60, 75, 171, etc.)
123. Miall, L. C. Mosses of Malham. Naturalist, vol. ii, p. 263.

1867.

124. Gissing, T. W. Flora of Wakefield. 8vo. Reprinted from Naturalist, 1865-6.
125. Nowell, John. Notes on some Rare Mosses at Todmorden. Naturalist, vol. iii, pp. 1-3.
126. Payne, Dr. (of Wath). 'Flora of the Dearne Valley' in MS., pp. 67. Lent to Author in 1877. Many casuals included and some erroneous records—the names not to be depended upon always, e.g., *Cicuta virosa*, 'Burton Grange,' proved to be *Œ. crocata*.

1868.

127. Hobkirk, C. P. Huddersfield: its History and Natural History. 2nd Edition. (Some of the errors in Flora of Huddersfield in 1st ed. corrected in this.)

1868—1872.

128. Watson, H. C. Compendium of the Cybele Britannica. Three parts and Supplement (1872). Demy 8vo, Lond. (Containing particulars innumerable relating to the vice-comital divisions of Mid-West and South-West Yorkshire.)

1869.

129. Babington, C. C. The British Rubi. 8vo, Lond., 1869.
130. Baker, J. G. Monograph of the British Roses. Journ. Linnean Soc., No. 52, p. 225, etc.

131. Gatty, Rev. Dr. His edition of Hunter's Hallamshire. (Includes a brief account of Sheffield botany. Vide p. 10.)

1870.

132. Bailey, Charles. *Geum intermedium* near Skipton. Botan. Exchange Club Report, 1870.
133. Carr, A. *Mentha rubra* near Sheffield. Botan. Exchange Club Report, 1870.
134. Hooker, J. D., C.B., M.D. Student's Flora of the British Islands. Lond. (Vide p. 32, etc.)

1871.

135. Leighton, Rev. W. A., F.L.S. The Lichen Flora of Great Britain. 8vo. Shrewsbury. Ed. ii in 1872.
136. Trimen, Henry, M.B., F.L.S. *Trichomanes radicans* in Yorkshire, etc. Journ. Bot., O. S., vol. ix, p. 174 and 214.
137. Kynnersley, H. Sneyd. *Mimulus guttatus* in Nidderdale. Botan. Exchange Club Report for 1871.

1872—1873.

138. Lees, F. A., F.L.S. Notes on West Riding Botany. Yorkshire Naturalists' Recorder. (Vide pp. 37, 93, 111, 141, 160, 179, 195, 208.) This periodical also contains the names of numerous West Yorkshire plants observed by the members of the West Riding Consolidated Naturalists' Society at their monthly excursions, quoted in Flora as Y. N. Excurs.

1873.

139. Cash, James. Where there's a Will there's a Way. Lond. 8vo. (Vide included lives of John Just and James Crowther, in connection with *Cypripedium Calceolus*.)
140. Hobkirk, C. P. A Synopsis of the British Mosses. 8vo. Lond. (Contains a few West Riding moss stations not published elsewhere. Vide p. 22, etc.)
141. ——— On the Mosses of the West Riding of the County of York. Journ. Bot., New Series, vol. ii, pp. 327, 358.
142. Lees, F. A., F.L.S. The Peculiarities of Plant Distribution in the Leeds District. Journ. Bot., N. S., vol. ii, p. 67.
143. Windsor, John, M.D., F.R.C.S. Flora Cravoniensis. Manchester, Cave and Sewer (posthumously published).

1873—1874.

144. Watson, Hewett Cottrell. Topographical Botany: in two parts. Part I, 1873; Part II, 1874. (Contains innumerable references to the species of plants found in the vice counties of mid-west and south-west Yorkshire, such in great part resting upon the 'personal authority' of catalogues ticked off for their respective divisions, by J. Tatham, J. G. Baker, F. A. Lees, J. Leyland, etc.)

1874.

145. Babington, C. C., F.R.S., F.L.S. Manual of British Botany. 7th Ed.
146. Lees, F. A., F.L.S. On the Flora of the Leeds and Bradford District. Journ. Bot., N. S., vol. iii, p. 78.
147. ——— On the Flora of the Yorkshire Coalfield. Journ. Bot., N. S., vol. iii, page 144.
148. ——— Report of the Botanical Locality Record Club for 1873. (Vide pp. 11, 12, 15, 17, 18, 19, 27, 28, 31.)
149. Roebuck, W. D. *Butomus umbellatus*, near Leeds. Journ. Bot., N. S., vol. iii, p. 307.
150. Willis, John, Ph. D. On the Flora of the Environs of Bradford. Journ. Bot., N. S., vol. iii, p. 10.

1875.

151. Lees, F. A., F.L.S. Report of Botanical Locality Record Club for 1874. (Vide pp. 40, 41, 48, 53, 60, 63, 65, 69.)
152. ——— On an Alien Species of Rubus found near Huddersfield. Naturalist, New Series, vol. i, p. 56. (This publication also contains many lists of species found at the Yorkshire Naturalists' Union Excursions in 1876 and 1877.)
153. Parsons, H. Franklin, M.D., F.G.S. The Maritime Plants of the West Riding of Yorkshire. Naturalist, N. S., vol. i, p. 115.
154. Pearson, W. H. Yorkshire Hepaticæ. (Extract of West Riding stations from Dr. Carrington's British Hepaticæ, of which only four parts are yet published.)

1876.

155. Foggitt, T. J. *Pyrola media* at Harrogate. Botan. Exchange Club Report for 1875.
156. Lees, F. A., F.L.S. Report of Botanical Locality Record Club for 1875. (Vide pp. 98, 99, 102, 103, 105, 106, 107, 108, 110, 114, 116, 137, etc.)
157. Roebuck, W. D. Account of a Field Day at Aldborough and Boroughbridge (with plant list). Naturalist, N. S., vol. ii, p 55.

1877.

158. Blow, T. B. Report of the Botanical Locality Record Club for 1876. (Vide pp. 160, 163, 165, 168, 187.)
159. Lees, F. A., F.L.S. Recent Moss Discoveries in West Yorkshire. Naturalist, N. S., vol. iii, p. 19.
160. Watts, W. Marshall, B.Sc. A School Flora. 8vo, Wildman, Settle.

1878.

161. Birks, Thos., Jun. Report on Botany, of Goole District, for 1878, prepared by T. B., for the Goole Scientific Society.
162. Davis, J. W., and Lees, F. A. West Yorkshire: An Account of its Geology, Physical Geography, and Botany, Post 8vo. London, 1878. Pp. xl and 414.

163. LEES, F. A., F.L.S. *Hypnum salebrosum* in West Yorkshire. Naturalist, N. S., vol. iii, p. 99.

164. LEES, F. A., F.L.S. Bot. Loc. Record Club Report for 1877. (Vide pp. 207 to 229, 252, 257.

165. LEES, F. A., F.L.S. *Carex capillaris*, L., in Gordale. Journ. Bot., vol. vii, N. S., page 247.

166. MIALL, L. C., F.G.S. The Geology, Natural History, and Antiquities of Craven, in Yorkshire. 4to. J. Dodgson, Leeds. (A reprint of Supp. to 3rd Ed. of Whitaker's Hist. of Craven. 4to, 1877. J. Dodgson, Leeds.)

167. NYMAN, C. F. Conspectus Floræ Europææ. Part I. Orebro (Sueciæ) Typis Officinæ Bohlinianæ. (Vide pp. 78, 83, etc.)

168. TRANSACTIONS OF THE YORKSHIRE NATURALISTS' UNION. Series E. Botanical Report for 1877, by W. Fowler and H. F. Parsons.

169. WESLEY, J. SEBASTIAN, M.B. *Thuidium abietinum* in West Yorkshire. Naturalist, N. S., vol. iii, p. 92.

170. WHITEHEAD, JOHN. *Seligeria tristicha* in Upper Wharfedale.

1879.

171. LEES, F. A., F.L.S. Bot. Loc. Record Club Report for 1878. (Vide pp. 16, 19.)

172. PARSONS, H. F., M.D., F.G.S. Report of the Botanical Section of Y. N. Union for 1878. Trans. Y. N. Union. Series E, pp. 9—50.

1880.

173. LEES, F. A., F.L.S. Bot. Loc. Record Club Report for 1879. (Vide pp. 52, 53, 58, 61-64.)

174. LE TALL, B. B., and WEBSTER, GEO. Askham Bog: its Ferns and Flowers. Reprinted from Nat. Hist. Journal of Friends' School. Vol. iii.

1882.

175. LEES, F. A., F.L.S. Bot. Record Club Report for 1880. (Vide pp. 132, 133, 136, 138, 139.)

176. ROBERTS, GEORGE. Topog. and Nat. Hist. of Lofthouse and its neighbourhood. (Small 8vo, Leeds, 1882.) Vide pp. 374, 389, etc.

177. WEST, WILLIAM. Report of the Botanical Section of the Y. N. Union for 1879. Trans. Y. N. Union. Series E, pp. 64—110.

1883.

178. LEES, F. A. Report on Yorkshire Botany for 1880. Trans. Y. N. Union. Series E, pp. 112—120. (First half of Report. The Concluding portion, pp. 121—124, published May 1884, in Part vii of the Transactions.)

179. LEES, F. A. Bot. Record Club Report for 1881 and 1882. (Vide pp. 188, 193, 196, 230.)

1884.

180. IBBOTSON, H. The Ferns of York, including also Nidderdale. Reprint, with many additions. Edited by B. B. Le Tall, A. R. Waller, and F. T. Walker. Post 8vo, York, 1884 (pp. 24.)

A list of the principal persons who have furnished information, lists, or specimens, bearing on the Flora.

ABBOTT, JAMES, Leeds.
BIRKS, THOS., junr., Goole.
CHEESMAN, W. N., Selby.
FOWLER, Rev. W., Liversedge.
CARR, AMOS (the late), of Sheffield.
EMMET, JOHN, Boston Spa.
INCHBALD, PETER, Harrogate.
IBBOTSON, HY. (the late), York.
JOHNSON, HY., Barnsley.
LEE, P. FOX, Dewsbury.
HICK, THOMAS, Manchester.
NEWBOULD, Rev. W. W. (the late).
LE TALL, B. B., York.
PARSONS, H. F., M.D., London.
ROBERTS, GEORGE, Lofthouse.
SLATER, Rev. H. H., Wellingborough.
SMITH, GERARD E. (the late), Ockbrook.
SOPPITT, H. T., Bolton, Bradford.
STANSFIELD, A., Todmorden.
WESLEY, J. S., Wetherby.
WEBSTER, GEORGE, Holgate, York.
WEST, WILLIAM, Bradford.
WILLIS, JOHN, Bradford.
WHITEHEAD, JOHN, Dukinfield.
WATTS, W. MARSHALL, Giggleswick.
WHITWELL, WM., Wandsworth.

The Principal Herbaria referred to are those of James Abbott (Leeds); Buddle; Dalton, Middleton, etc., in York Museum; W. Kirkby (Leeds); Sloane (Brit. Museum); Salt (Weston Park Museum, Sheffield); W. West (Bradford); and W. Whitwell (Wandsworth, London).

181. LEES, F. A. Bot. Rec. Club Report for 1883. (Vide pp. 9, 12, 13, 15, 17, 21, 27.)

182. SLATER, Rev. H. H., M.A., F.Z.S. The Flora of Ripon and neighbourhood. Transactions Y. N. Union. Part vii. Series E, pp. 125—152 (first half of Flora issued April 1884. Second half, pp. 153—184, issued with part viii in May 1885).

1884—1885.

183. 'THE NATURALIST.' August 1884 to December 1885. Numerous notices of Plants observed at Excursions of the Y. N. Union, etc.

1885.

184. COLLYER and TURNER. Ilkley, Ancient and Modern. Walker and Sons, Otley. Royal 8vo. Contains the Botany of Ilkley Parish, by F. A. Lees.

185. ROBERTS, GEORGE. Topography and Natural History of Lofthouse and its neighbourhood. (Small 8vo, Leeds, 1885.) List of plants at pp. 226—236. (Contains numerous manuscript entries made by Mr. David Dixon, a veterinary surgeon of Wakefield, about the end of last century, in a copy of Berkenhout's Synopsis of British Plants, which he kept as a kind of diary. The book is now in the possession of Mr. Joseph Wilcock, of Wakefield.)

PLAN OF THE FLORA.

PLAN OF THE FLORA.

THE system adopted in the following enumeration of all the Phanerogams and Ferns that are, or have been said to be, constituents of the Flora of the West Riding of Yorkshire, is of course the natural system. The designations of the species, and the arrangement, are substantially in agreement with Watson's 'London Catalogue,' 7th Edition (first issued 1877); the departures made from the names of that list, though intended as improvements (where the classicality is open to grave doubt), or given as more in accordance with our present knowledge, being neither numerous nor important enough to require explanation or defence in this place. The rank accorded in that list to names has not, however, been strictly adhered to, some there lined as species being here reduced to varieties.

The names of the Orders are printed in bold, compact capitals, and the names of species in thick conspicuous antique type for easier reference. Those accepted for the area treated of as Natives, Colonists, or fully-naturalised Denizens are preceded by a consecutive number from *Clematis* to *Equisetum*. Aliens, and plants extinct or erroneously recorded, are not numbered, and are enclosed in square brackets [ ].

Each *Native* species, and some few long-established Colonists and Denizens, are treated after the following system :—

First. The Latin botanical name is given, followed by the English name (if any) commonly applied to it in the Riding. Names not heard therein, such as Fleur-de-luce, Paigle, etc., are omitted. Synonyms have only been given where absolutely necessary.

Second. The ante-Linnean designations used by Turner, Gerarde, Parkinson, Ray, Blackstone, etc., in their quoted works, are next given; but only in those cases where those early writers have left distinct records for the occurrence in the county of the species under notice. References to figures or manual descriptions have been refrained from. They are now seldom made use of, and the species liable from this omission to wrong misinterpretation by the average botanist do not amount to a dozen.

Third. The next paragraph furnishes information respecting the distribution of the species, expressed in two ways. The areal indication as to the vice-comital occurrences, in Watson's *Topographical Botany* (Ed. ii) is first quoted : the figs. 63 standing for South-West Yorkshire (that part of the Riding south of the river Aire and the Leeds and Liverpool Canal); 64 for Mid-West Yorkshire (the area north of the division just stated, and lying to the south of the northern boundary of the Riding from York to Newby Head Inn, and of an artificial line carried westwards from the Inn

along the summit of drainage over Bleamoor to the top of Gragreth—where the counties of Westmoreland, Lancaster, and York meet); and 65 (only given where the Topographical Botany record belongs to a plant found there only) referring to that trapezium-shaped portion of the West Riding which lies to the north of the artificial line mentioned. This small area embraces the basins of the Dee, Clough and Rawthey streams, tributaries of the Lune, and the eminences of Rysell, Baugh fell and Howgill fells. This rather minute description is rendered necessary by this not-to-be-forgotten fact, that Mr. Watson's *North-West Yorkshire* (V.C. 65) thus includes a part of the West Riding that in its phyto-geographical relations is, to all intents, a part of Westmoreland, its geological character (Silurian slate) essentially similar, and its westward flowing streams to no inconsiderable extent the initial ramifications of its river system. The brackets, square and curved, occasionally enclosing one or all of such groups of figures, are also *part of the quotation* from Topographical Botany, meaning, respectively, in Mr. Watson's opinion:—[ ] Plants now extinct, or erroneously recorded; and ( ) Introduced species, not truly wild, within the area expressed by the numerals.

Following this, any marked lithological restrictions (exemplified within the area dealt with) are indicated by the Thurmannic terms (elsewhere explained)—Xerophilous, Psammophilous, Pelophilous, or of Montane type; and the *Range* in an upward or downward direction, where known, and given in feet, shows the limits attained. Such range is *not* a quotation from the zonal information of the *Compendium Cybele Britannica*, but refers solely to the Riding.

Fourth. Into this paragraph a variety of information is condensed. In succession are stated: (1) The category in a scale of 'citizenship' under which the plant comes, i.e., whether a Native indigenous to the soil, a Colonist, a Denizen, or an Alien (or Casual)—the signification of these terms being the same as defined in the *Compendium Cyb. Brit.*; (2) The general or usual habitat, place of abode of the plant, and the sort of situation —dry or damp, exposed or shady—for which it exhibits preferences; (3) The relative commonness or rarity of its recurrence in such stations*; (4)

* For the exact and concise expression of this, the following comparative scale of terms is used.

A. *Very Common*; generally distributed in every part of every district, and plentifully as regards individuals—e.g., the *Wood Violet*.

B. *Common*; i.e., found in every river basin, and frequently recurring; although it may be in no great abundance, and even rare in unsuitable stations—e.g., *Ranunculus Flammula*.

C. *Infrequent*; i.e., hardly generally distributed, rather rare, but still fairly plentiful in several localities—e.g., *Nasturtium palustre*.

D. *Rare*; i.e., found, perhaps, in several scattered stations, but far from generally distributed, and nowhere in any quantity—e.g., *Listera cordata*. Many 'rare' plants have occurred in more localities than some infrequent ones.

E. *Very rare*; found only in one or two or three localities, although in one of these spots it may be plentiful enough—e.g., *Dryas* or *Alchemilla alpina*.

The vital duration of the individual; and (5) The months of the year during which it is usually in a suitable state of maturity for identification and preservation in the herbarium—this, mostly, the time when it is in bloom, but in the case of such genera as *Potamogeton*, *Carex*, etc., the date affixed is that in which it is in a fruiting condition. The statements made under the 1st, 2nd, 3rd, and 5th of the foregoing heads have exclusive reference to the species as an inhabitant of West Yorkshire.

Fifth. This paragraph, under the phrase 'first record,' gives the name of the botanist who first gathered the plant in the West Riding, with date, if known—preference being always given to a printed record, or herbarium specimen; or the author of the work (with date) in which is to be found the earliest notice of the plant as a West Riding one. In the case of Ray, or Bolton, or Salt, the author is usually the discoverer also, but it is not always so: when the earliest notice appears in Merrett, or Parkinson, or Dillenius, some correspondent (such as Richardson) was often the gatherer; in other cases it is uncertain.

Sixth. The stational distribution of the species is next indicated. Where the plant is 'very common,' or in some cases where 'common' with very numerous localities, the word 'general' and a brief account of its restrictions or peculiarities follows the word Distribution. An ancient record or records for a very common plant has, however, often been inserted, if of historical interest. But in the case of the majority of species, the localities are enumerated in this paragraph (or *a selection* of them), giving preference to the more ancient ones, grouped under the capitals M., L., R., Y., N., W., A., C., D., T., indicative by their initial letters of the ten river-basin districts or divisions. The *map* sufficiently defines these drainage districts of the Mersey (tributaries), the Lune (tributaries), the Ribble (with Hodder), the Yore (a partial basin), the Nidd, the Wharfe, the Aire, the Calder (with Colne), the Don (with Dearne), and lastly, the Trent (tributaries). These divisions are strictly natural, so far as that part or those parts of them, within the political boundaries of the Riding, are concerned.

With regard to the stations, where an italicised name follows the detailed locality, such name is the authority. References are added in the case of old records only. Where no name follows a station, the occurrence of the plant in question in such station rests upon the writer's own authority. The mark '!' *after* a record, *before* a name, signifies that the author has verified the station by a visit, and can certify the plant to be still there—if the record be an old one. Where the mark '!' *follows* the name of an authority, it means that a specimen gathered within ten years has been examined; not necessarily, but very often, this specimen rests in the author's own herbarium.

Lastly, follow any observations it is deemed necessary to make for the elucidation of any point of interest connected with the species under consideration: its vulgar name, its history, or what not, undealt with previously.

Species of Foreign origin. Alien waifs and strays have no number affixed, are placed within square brackets [ ], and are dealt with upon no systematic plan.

Ancient Records. Unquestionably John Ray added to the West Yorkshire Flora a greater number of new records (between 1670 and 1688) than any other of the old botanists, and to none does the Riding owe more than to him; but How, Stonehouse, Parkinson, Merrett, Thornbeck, D. Dixon, and others each contributed no mean share towards it. Among the botanists of a century later, Jonathan Salt stands pre-eminent. From the fact that the botanical visitors to the Riding, attracted by the fame of its rarities, placed on record *mainly* their observations concerning such species as were to them novelties, room was left for such residents as Salt, Bolton, Brunton, and Dalton to register faithfully the prevailing integers in the flora. Salt, especially, appears to have most painstakingly noticed the commonest weeds—with the result that to him must be assigned over three hundred of the First Records!

Repetitions of early records by later compilers have only been traced onward by references down to 1805. Copies of quotations have been *noted* whilst working chronologically, but are not always expressed, in order that space might be saved. Many—if not most—of the records in Baines' Flora of Yorkshire relating to West Yorkshire have been traced back and given at first hand. Old records now certainly or presumably (since not re-confirmed within 40 years) extinct, have been enclosed in square brackets [ ]. Erroneous records, or localities thought to probably refer to something else, are invariably enclosed in circular brackets.

THE FLORA.

SIGNS AND ABBREVIATIONS.

A. = Annual. B. = Biennial. P. = Perennial.

* After a locality indicates probable introduction, doubtful indigenity, in the place so asterisked, although it may be wild in the Riding in some, or all, of the other stations given.

* After a species name or an areal number as (e.g.), 63*, indicates suspicion of introduction or non-indigenity in the area concerned.

! After authority = Specimen seen.

! After locality = Seen *in situ* there by author.

÷ = Between: the usual sign in Euclid, etc.

[ ] Enclosing a locality record indicate certain or very probable extinction in that locality.

( ) Enclosing a record indicate that some error of name, or of localisation, is deemed probable. All such records require confirmation.

Dist. = District.

D. Dixon = Particulars of those entries from Berkenhout.

Etc. = and by other authors (of later date); or (where a plant station is concerned) other localities in same district.

Excurs. = Excursion; applied to plants gathered and exhibited at the excursions of the Yorkshire Naturalists' Union.

G. Harr. MS. = Stations in Sheffield district communicated in MS. to Rev. W. W. Newbould by George Harrison.

Hab. = Habitat.

Loc. = Locality.

Hb. or Herb. = Herbarium.

Neighd. = Neighbourhood.

Nom. = Nomen (name).

Pl. = Place, or places.

Rep. = Repeated in.

Spn. = Specimen.

Sta. = Station.

Top. Bot. = Topographical Botany of Watson, 2nd Ed.

Towns. O. B. G. = A copy of the Yorkshire portion of Old Bot. Guide, belonging to Mr. F. Townsend; formerly Mr. Joseph Woods', and containing MS. interlineations by J. Windsor, Jos. Woods, etc.

Y. N. U. = Yorkshire Naturalists' Union.

For the Titles of Books and Authors' names abbreviated in the Flora, reference must be made to the Botanical Bibliography under year of publication.

THE FLORA OF WEST YORKSHIRE.

PHANEROGAMIA.

DICOTYLEDONES.

I. RANUNCULACEÆ.

1. Clematis Vitalba L. '*Hedge Feathers.*'

Area: (63, 64), Top. Bot. Typically Xerophilous. Range: 0—300 feet. Denizen, just possibly native (running out, like *Carduus acaulis*) on Limestone tract of Trent district, but according to *Top. Bot.* comital census, not indigenous north of an irregular curve drawn from the Wash on the east, through Warwick, to the Dee estuary on the west. Bushy and stony ground, hedges and railway cuttings, rare; mainly on Permian tract. P. July-August.

Earliest notice: Blackburn, 1860.

L. Clapham (escaped); *G. Roberts.*

W. By the river, Harewood Bridge (escaped from Inn garden ?). Langwith Wood, Collingham! *J. S. Wesley.* Railway bank, south side of line, half-mile west of Thorparch Station! South-east of Thorparch Station, by railway side; *J. Emmet.* Known to him since 1857 as a Boston Spa plant, in a spot from which it has long since disappeared. Quarry near Scholes; *G. Roberts.*

A. Lane beyond School-house at Meanwood! *J. Abbott* (1867). Garforth; *T. W. Gissing* (1862). Quarry near Allerton Bywater; *G. Roberts* = Old limestone quarry, by cross-roads, half-mile north of Ledstone! *Jos. Blackburn* (1860). Hedge near village of Monk Fryston; *H. F. Parsons.* Ledsham, seed strayed from bush in rectory garden; *W. Whitwell.*

D. Hedge (with *Humulus*) between Conisborough and Mexborough. Levitt Hagg.

T. Firbeck! Kiveton; *Amos Carr.*

[**Thalictrum alpinum** L. Error.

[64] Top. Bot. Giggleswick Scar; Gordale, etc.; *Winch, add. N.B.G.* (1835). Winch surely intended *T. montanum*; and Watson mis-referred the *MS.* additions copied from an interlined Flora Britannica lent him by Mr. Winch (vide p. 5, *N.B.G.*).]

2. Thalictrum minus L. **var. montanum** Wallr.

(*T. calcareum* Jord.)

Area: 64, Top. Bot. Montane. Range: 300—1,500 feet.

Native, mountain limestone crags, rare; the normal, maritime type, *dunense*, is of course absent. P. July.

First record: Ray, 1677.

L. Mill-bank and Friar Wood near Ingleton, *Mr. Thornbeck; Blackst. Sp.* (1746); Id. *Martyn* (1763), Skirwith Wood, near Ingleton; *Curtis, Gough's Cam.* (1789), etc.

N. (Mackershaw Wood, *N.B.G.* Is *T. flexuosum*).

W. Kilnsey; *Curtis* (1782). Arncliffe and Buckden Crags. 'Compton Bank, Wetherby; *P. F. Lee* teste *J. G. Baker*.' I knew the plant here, and always regarded it as the small-leaved race of *majus*—the carpels are more ovoid than in the type of that, but they are much infested with some species of gall-fly.

A. Abundantly on the mountains near Malham and Settle; *Ray, Cat.*, ii, 283 (1677); rep. *Hist.* (1686), *Syn.* (1690). etc. Nothing more common on the rocks about Malham and Wharfe; *R.* in *Gibs. Cam.* (1695). Plentiful in Gordale; *J. Woods, O.B.G.* (1805). Gordale, *near the top; L. C. Miall* (*Nat.* 1865). Malham Cove; *C. Bailey* (1881).

C. A little below Lady Bolles' Bridge, near Heath Hall, 1790; *D. Dixon.*

D. (Askern, *P. Inchbald; Baines.* 'Evidently an error—should be *majus (flexuosum)*—misled by the stunted aspect of the plant.' *P. Inchbald, MS. ad auctor.*).

Obs.—I have great doubts as to the specific distinctness of this mountain form and 'flexuosum.' Both plants grow in Gordale, that on the precipitous slopes and ledges high up, as at Ingleton and on Kilnsey Crag, being small of stature, with ovoid-fusiform carpels, and with leafless sheaths only at the base; whilst lower down, by the stream, undoubted *flexuosum*, with elliptical beaked fruits and stems leafy to base, occurs. But *intermediates* occur with crossed characters, such as a plant (placed by me under the next name) before me, gathered by Dr. Parsons: 7 inches high, squat, and *minus-maritimum* like in facies, but stems leafy to the very base (? from nodal suppression)—leaflets like normal *montanum*, but the striate elliptical-crescentic carpels of *flexuosum.*

3. Thalictrum majus Sm. **a. flexuosum** Reich.!

(*T. collinum* Wallr.)

Area: 64, 65, Top. Bot. Sub-montane. Range: 150—600 feet.

Native; rocky woods, thickets, lanes, and riversides, infrequent. P. July-August.

Earliest specific notices: Rev. W. Wood, 1787; Kenyon, 1800.

L. Wenning banks from Clapdale by Ingleton down to Wennington! near Bentham; *Wm. Kenyon* (about 1800), *Towns. O.B.G.*

R. Arnforth Wood, Long Preston; *T. Nuttall*, in *Windsor.*

Y. Mackershaw Wood (!); *H. H. Slater.* Limekilns, Wicliffe Lane, Ripon; *James Ward* (1839).

W. Arncliffe, 'Eastfield Wood;' *R. Bowman, N.B.G.* (1835). Grassington Woods. Wharfe banks, Ilkley! Bolton Woods! *Geo. Lord.* Dalton Lane, leading from Beilby Grange to West Woods! Thorp-arch; *J. Emmet.* About Collingham.

A. About Skipton; *Rev. W. Wood* (1787). Gordale Scar; *S. Woods, O.B.G.*; *Winch, Add. N.B.G.* Furzy ground near Lotherton! *G. Webster.* Road side, Monk Fryston; *W. N. Cheesman.* Gordale Scar? a plant with stems leafy to the base, 1878, *H. F. Parsons.* Gordale; *W. West.*

D. Askern! *P. Inchbald.*

Obs.—This plant differs much in size, and in facies to some extent also, from variation in the size and cutting of the leaflets. A large form, three or four feet high, with *hollow* hardly striate main stem, is common about Wetherby, but it is not *Kochii* Fr., for the carpels are narrowly elliptical, not in the least approaching ovoid. The small-leaved race of this, above-mentioned, is often called *minus*; and long ago Mr. J. G. Baker said (Miall's Flora, App.) that it agreed better with *T. eminens* Jord. than with *T. flexuosum.* He was writing of the Bolton *Wharfeside* plant, but the Compton bank (vide sub *minus*) and Boston Spa *riverside* plants are similar; being slenderer, with smaller more-lobed leaflets, and a less-branched panicle. Can these be the product of seeds of '*montanum*' washed down from Kilnsey by winter spates, and altered by growth in richer, alluvial soils?

4. Thalictrum flavum L.

Area: 63, 64, Top. Bot. Hygrophilous. Range: 0—300 feet.

Native; watery places in the lowlands, common; but unknown in the more elevated districts. P. June-July.

First record: Bolton, 1775.

(R. Ribble bank, between Settle and Stackhouse; *J. Windsor.*) The large-leaved form of *majus*?

Y. Sharow Mires, Ripon; *H. H. Slater.*

N. Ditch by wood, Bilton Haggs. Wet meadow, right hand, on York Road, half a mile from Knaresborough; *Hargrove*, iii, 1782.

W. Marshy places by river Cock, Seavy Carr, Stutton. Askham Bogs.

A. Milford; *G. Roberts* (1876).

C. In a meadow at Thornhill Brigs, Brighouse; *J. Bolton* (1775).

D. Rotherham district; *L. Langley* (1828). Potteric Carrs, Doncaster! Rother marshes, Treeton, Wales, etc.; *A. Carr!* Goole fields (!) *H. F. Parsons.*

5. Anemone Pulsatilla L. *Pasque Flower.*

Pulsatilla folio crassiore et majore flore; Ray. *Pulsatilla folio cæruleo*; Merr.

Area: 63, 64, Top. Bot. Xerophilous. Range: 100—300 feet.

Native; dry grassy turf of the Permian Limestone-tract ridge, from S. to N. of our area, once common, now almost extinct. P. Mid April.

First record: Merrett, 1666.

Y. Near North Stainley, rare; *R. A. Summerfield* in *Slater's Flora* (1879).

W. Pastures near Aberford; *W. Brunton* (1805). [About one mile north of the road leading from Tadcaster to Bramham; *Winch, Add.* in *N.B.G.* (1835)=Oglethorpe Hills and Plantation; *O. A. Moore.*]

A. In the town Close, Kippax; and near Ledsham Church, *Mr. Dawson; Blackst. Sp.* (1746). About Ledstone Hall; *Ray, Syn.* iii (1724); rep. *Martyn* (1763). Dry hilly ground near Brotherton! *Dr. Wales* (1869).

C. In a close at the bottom of Parkhills, Wakefield, in 1795; and at Pigburn Leys in 1799; *D. Dixon.*

D. Near Pontefract; *Merrett*, 99 (1666); rep. *Gough's Camd.* (1789). Said to abound near Pontefract, 'in a place called the Close, above the lake;' *Dr. Richardson*; *Ray, Syn.* 260 (1724); rep. *Huds.* 237 (1778). Found at Pontefract and on Darrington Leys in the summer of 1791; *D. Dixon.* Darrington Leys, near Ferrybridge; *Baines* (1840). This is near Pontefract, and may be really the same as Ray's station. Smeaton Crag pastures, Went Vale: 'is fast disappearing from this the only locality in which it has been seen in flower for the last four or five years;' *T. W. Gissing* in *Nat.* (1865) and *Flora.* There still in 1870! On the limestone east of Rotherham; *L. Langley* (1828). Heath between Doncaster and Brodsworth; *Pierson* (1798). This locality is Scawsby Leys.

6. Anemone nemorosa L. *Wind Flower. Wood Anemone.*

Area: 63, 64, 65, Top. Bot. Range: 0—1,800 feet.

Native; coppices, hedge banks, and sometimes open pastures, very common. Generally distributed—reported in all the districts.

First record for Riding: Bolton, 1775.

Obs.—The bloom is often a fine purple—and very occasionally of a pale blue (*G. Webster*). In West Yorkshire this is often a meadow species; I remember *fields* at the foot of Woodhouse Ridge, in the Meanwood valley, within a mile and a half of busy streets, white with it. That was about 1855 to 1860.

[**Anemone apennina** L. Alien.

In a wood near Otley, 1843; *Garnett* (*Phyt.* i, 650). Probably an Escape from a garden. I have no other record.]

[**Anemone ranunculoides** L. Alien.

Sub-spontaneous in one or two places in parks, as at Shireoaks (*A. Carr*), and in plantations (perhaps introduced with soil at the root of young trees) at Whitwell, near Worksop (*T. P. Lewis*).]

[**Adonis autumnalis** L. Alien.

Area: '[63],' Top. Bot.

A sporadic Casual in corn and seed-fields, occasional, but for a season or two only.

W. Garden, Wetherby—a weed.

A. Old quarried ground, Brotherton (1867). Oat-field, Myles Hill, Meanwood (1866-8).

D. Field near Wales, two or three specimens; *A. Carr.*

T. Sandy cornfield near Thorne (1870).]

7. Myosurus minimus L. *Mousetail.*

Area: 64, Top. Bot. Range: low levels only.

Colonist, sandy or clayey fields, very rare. A. June.

N. York (and Doncaster) neighbourhood; *White* (1804): *Bot. Guide* (1805). =In fields at Staveley; *H. Baines.*

W. Holgate near York; *Baines.* Acomb; *G. Webster.* Sandy field west end of Thorner (1870)!

8. Ranunculus heterophyllus Fr.

Area: 63, 64, Top. Bot. (sub *R. diversifolius*). Range: 50—1,250 feet.

Native; streams, ponds, and dikes, common, especially in its immature form of *Drouetii.* P. May-August.

[R. Near the new boat-house at Giggleswick Tarn; *J. Windsor.* Tarn since drained.]

W. Askham Bog dikes (!) *G. Webster.* [Forms with variously dissected floating leaves, some agreeing with *R. fissifolius* Schk., others with *R. rhipiphyllus* Bast.]

A. Malham Tarn (!) *W. Todd.* Form with almost orbicular fan-cut leaves approaching *rhipiphyllus.*

Note.—In the aquatic Ranunculi which follow I have felt compelled to depart from the names and arrangement of the London Catalogue; and I confess that I have found myself unable in many cases to assign to their right places some dozen or two earlier records for segregate forms—names having (as the specimen bases I have been able to examine have shown) been used in such various senses and applied to such widely-different plants. My own namings of ten years back have undergone revision with the rest. The more forms I have examined and watched *in situ*, the nearer have I arrived at a profound conviction of the truth of the grouping given by Sir J. D. Hooker in the 3rd Edition of the *Student's Flora.* The main alteration I would make in Hooker's arrangement would be in grouping *R. Drouetii* (which has a *hispid* receptacle invariably) under *R. heterophyllus*, along with *R. triphyllos*, keeping the well-marked *R. trichophyllus* (with its compressed carpels, very small starlike flower, and rigid habit) apart, as probably distinct. I constantly find plants of our northern streams and ponds, passing one year unchallenged by 'authorities' as *Drouetii*, developing later into indisputable restricted *heterophyllus*, with fan-like floating leaves and larger blooms! None who have watched the true *trichophyllus* of the brackish stells by the sea in Durham, North Lincoln, and South Essex, through all their stages, could mistake *R. Drouetii* for it. *Inland*, the latter is infinitely the commoner plant, *trichophyllus* having an affinity with *circinatus* not sufficiently recognised, and, like it, very rarely indeed (if ever?) producing floating leaves. The plant so often passing as '*trichophyllus*,' with largish petals, and peduncles as long as the leaves, these having *somewhat rigid*, divaricated capillary segments, is usually some form of *flori-*

D. Dam near Bradfield.

Obs.—The type and its form *Drouetii* are distinctly more *montane* than the *peltatus* forms.

STATUS **submersus** Bab.

Frequent; usually May to June.

W. Askham Bogs; *G. Webster*!

A. Osier beds, Brotherton. Cattle-plash in field by road between Scholes and Seacroft. In the river Aire, near bridge on Carlton Road, Skipton, July 1884; *W. Whitwell*! With fan-like, rigid, submerged leaves, product (under altered conditions) of the Malham Tarn plant twenty miles higher up the river?

D. Goole drains.

Obs.—This group, perhaps, includes some plants that would in some seasons develop into *R. floribundus*, but without floating leaves it is difficult to separate the, as yet, undifferentiated forms.

VAR. **Godronii** Gren.

W. Ditches at Askham Bog; *G. Webster*. [Erroneously referred by me to *R. trichophyllus* instead of *R. Drouetii*, in Bot. Rec. Club. Rep. 1877.]

D. Potteric Carrs; *P. F. Lee* (1883)! Rossington drains. Ditch by Don, between Sprotborough and Doncaster; *A. Carr*! teste *C.C.B.*

FORMA **Drouetii** Schultz.

Area: 63 (Carr), 64, Top. Bot.

Native; pools (often on peat), drains, etc., common, and with a vertical range coinciding with its type. P. June-August.

---

bundus before floating leaves have been produced, or where some local condition has prevented their development altogether, such as would fall under the *sphærospermus* of Hiern's Key, or *submersus* of Bab. Man. The submersed capillary segments are not always flaccid and collapsing, nor are those of *Drouetii*, both in some hard waters becoming incrusted and semi-rigid.

The lineal arrangement I substitute for that of London Catalogue will therefore be as follows:—

1. **R. HETEROPHYLLUS** Angl. Auct.
   a. **heterophyllus** Fries.
      forma *triphyllos* Wallr.
      f. *rhipiphyllus* Bast.
      f. *fissifolius* Schrank.
         status *submersus* Bab.
   b. **Godronii** Gren.
      f. *Drouetii* Schultz.
2. **R. PELTATUS** Fr.
   a. **peltatus** Fr. (*truncatus* Dum.).
      f. *elongatus* Hiern.
      f. *penicillatus* Dum.
         st. *pseudo-fluitans* Hiern.
   b. **floribundus** Bab.
      st. *sphærospermus* Boiss.
      st. *cæspitosus* Thuill. (*terrestris.*)
3. **R. MARINUS** '(Fr.)' Hooker.
   a. **Baudotii** Godr.
      f. *marinus* Fr. (propriè.)
   b. **confusus** Godr.
      f. *salsuginosus* Hiern.
4. **R. FLUITANS** Lamk.
   f. *Bachii* Wirtg.
5. **R. TRICHOPHYLLUS** Chaix.
   f. ? *radians* Revel.
   f. *confervoides* Fr.
6. **R. CIRCINATUS** Sibth.

L. Howgill, peaty pools and rills on the slopes.

Y. Queen Mary's Dubb and other pools by the Yore at Ripon (!) *H. H. Slater*, etc.

N. In a plash on the 'Stray,' Knaresborough.

W. Askham Bog dikes. Ditch, west side of race-course ings, Wetherby.

A. River Aire, Bell Busk; *P. F. Lee*—etc.

D. Askern ditches!

T. Ditch near Kiveton Canal; *A. Carr*! Pools, Lindrick Common; *A. Carr*!

9. Ranunculus peltatus Fr. (including **truncatus** Dum.).

Area: 63, 64, 65, Top. Bot. Range: 0—1,000 feet.

Native; slow and quick streams, ponds and ditches, common. P. May-August.

First record: Slater, 1881.

Y. Pilmoor Carr near Sharow, etc.; *H. H. Slater*!

N. Plash on the Stray, Knaresborough.

W. Mill-race, Leathley Valley.

C. Ditch, Horbury; *P. F. Lee*—a form approaching *elongatus*. Pond, Howley, near Batley; *P. F. Lee*—approaching the next form.

D. Near Sheffield; *A. Carr*!

FORMA **penicillatus** Dum. (*pseudo-fluitans* Newb.).

Y. In the Yore and Skell at Ripon.

N. In the Nidd, above and below Knaresborough.

W. In the Wharfe at Pool (!) *J. Abbott*. Linton. By Thorparch Bridge, in a water cut! (Long called *fluitans.*)

A. Swift reaches of river Aire above Skipton.

D. In the Rother, near Whiston; *A. Carr*! Stream near mill, Lindrick Valley; *A. Carr*!

VAR. **floribundus** Bab. The most frequent form.

M. Rare. Pool near Delph.

Y. Ponds near Ripon, not uncommon (!) *H. H. Slater*.

W. Ponds near Askham Bog; *G. Webster*! About Ilkley, Pool, Wetherby, etc.

A. Black-hill Dam, Adel.

C. Stanley, 1867; *T. W. Gissing*.

D. Campsall. Barnsley. Pond near Shire-green, Sheffield; *A. Carr*.

T. Laughton. Waleswood; *A. Carr*.

STATUS **cæspitosus** Th. (terrestris), i.e., a succulent solely-capillary leaved form, growing on mud, has occurred; the leaves closely resemble the submersed ones of *Baudotii* or *confusus* in colour and furcation.

A. By dried up pool, Roundhay limehills.

10. Ranunculus fluitans Lam.!

Native; deep, swiftly-running water, very rare. P. July.

First (and only) reliable record: Baker, 1854.

(R. In the mill-race, Settle, halfway between the bridge and the railway station; *J. Windsor*. The old station on the Mid Lancaster line of rails. Not to be seen in 1870).

Y. & N. In the river Swale, above York; *J. G. Baker*.

A. (In the river Aire, Saltaire; *W. West*. Perhaps rather *penicillatus*).

D. (In the river Rother; *G. Harr. Hb.*, MS. note in *Newb. Baines*; but Carr's specimens were all *penicillatus*.) Goole; in river Ouse, detached plants washed down; *T. Birks*.

Obs.—I have never seen any plant from within the limits of the West Riding which, when examined closely, had a *glabrous* receptacle. The Wharfe plant of Bolton, Pool, etc., referred to *Bachii* by some, in years back, when the forms of all the Batrachium section were less understood than now, is all *penicillatus*.

11. Ranunculus trichophyllus Chaix.

Native; still shallow dikes and drains in the low country, very rare. P. July-August.

First record: the Author, 1872.

W. Seavy Carr dikes, Stutton (1880-2).

A. Osier-bed dikes in the Brotherton water-meadows (1869). Water-cut near Selby (1875)—a small rigid plant.

D. Boggy ditches between Askern and Campsall: a rather doubtful plant, rigid and small flowered, but light green, persistently without floating leaves, but may prove to be a depauperate state allied to *Drouetii* or *cæspitosus*. The receptacle is markedly hairy. Drains, Goole fields; *H. F. Parsons*. [I have not seen the plant from this locality.]

Obs.—This species is nearest to *R. circinatus* in habit, local preferences, and in the fact that it very rarely (if ever) forms any floating leaves—the plants classed with it, or rather reported as it, when they do produce floating leaves are at once seen to be either *R. Godronii* or typical *R. heterophyllus*, i.e., they are *Drouetii* forms whilst without floating leaves.

12. Ranunculus circinatus Sibth.

Area: 63, 64, Top. Bot. Range: 0—300 feet.

Native; streams and peaty dikes, rare, and at low levels only. P. July.

First record: Newbould, 1852. Cyb. Brit., iii, 521.

R. (Waste ground by the Ribble, below the Long Stream; *J. Windsor*, *Flora*. Probably an error.)

Y. In the canal, Ripon, *Miss Morton*; *H. H. Slater*, *Flora*.

W. In a brook crossing beneath the road from Harewood to Weardley, a mile or so west from Harewood Bridge! Askham Bog dikes (!) *G. Webster*.

A. Drains by river Aire, Brotherton! *Dr. Wales*.

C. Coxley Dam (1885); *P. F. Lee*!

D. Campsall! Dyke, Black Carr, Doncaster! *J. Abbott*. Potteric Carr! *O. A. Moore*, *W. W. Newbould*, etc. Whitgift; *H. F. Parsons*.

T. Dykes on south border of Thorne Moor.

13. Ranunculus Lenormandi Schultz.

Area: 63, 64, 65, Top. Bot. Sub-Montane. Range: 350—1,800 feet.

Native; pools and rill-sides, in elevated districts. P. May-August. Common in the moorland tracts, rare in the more cultivated.

First notice: Newbould, 1852.

L. Howgill, Baugh Fell.

R. Blea Moor.

Y. Moors west of Kirby Malzeard.

W. Rombalds Moor, Ilkley. Barden Tower Moor.

A. Watery hollow on Malham Moor peat plateau, with *Hippuris*, etc., at 1,250 feet. Adel Bog and Dam, low end (perhaps gone). Baildon Moor and Esholt Common; *W. West*

C. Norland Moor! *H. F. Parsons* (1877) [Small form simulating *tripartitus*]. Wibsey; *W. West*. Salterhebble; *J. Backhouse*, in *Windsor's Flora*.

D. About springs and rills, Woolley Edge; *W. West*. Wharncliffe; Bell Hagg; *G. Harr. MS.* Dunford Moors. Ecclesall; *W. W. Newbould* (*Cyb. Brit.*, iii).

14. Ranunculus hederaceus L.

Area: 63, 64, 65, Top. Bot. Range: 0—1,800 feet.

Native; on wet mud, shallow ditches and pools, common. P. May-August.

First notice: Salt, 1800.

Generally distributed—reported for all the districts.

The VAR. or STATE **homœophilus,** floating on deep water, is occasional, and has been misreported as *Lenormandi*. The best-marked examples I ever saw were in deep water in a pool at Bardsey, and at Castle-shaw, Wragstone Clough, Mersey tributary district.

15. Ranunculus sceleratus L. '*Blistercup.*'

R. palustris rotundifolius Ger. = *R. sceleratus*, fide C. C. Babington.

Area: 63, 64, Top. Bot. Range: Low, 0—300 feet.

Native; ditch and pond sides, on stiff clay, infrequent; on peat or sand, rare. P. June-August.

First record: Lawson, 1688.

[N. Long flat, Knaresborough; *E. Hargrove* (1832). Not visible in 1868.]

W. Hobmoor brick ponds, etc., near York. [By the Wharfe between Pool and Arthington. Gone.] Drain, in field near Stockeld Park gates. Knavesmire, York; *W. Whitwell.*
A. Burton Salmon and Hillam; *G. Roberts.* By the Aire below Methley; *J. Abbott.* Ferrybridge district; *Y. N. Excursion* (1873). Frizinghall, near Shipley; *J. W. Carter* teste *H. T. Soppitt.*
C. Sandal; *P. F. Lee.*
D. By Robin Hood's Well, nigh Wentbridge (1688); *T. Lawson*; *Derh. Phil. Lit.*, p. 221 (1718). Askern; *Geo. Harr. Hb.* Rossington. Marsh near Sprotborough! *W. Todd.* Near Rotherham; *W. W. Newbould* (1844). Aldwarke; *G. Harr. MS.* Goole fields; *H. F. Parsons.*

16. Ranunculus Flammula L.
Area: 63, 64, 65, Top. Bot. Range: 0—2,000 feet.
Native; ditches, pool-sides and wet grassy places, common. P. June-August.
First record: Salt, 1800.
M. In this district; *J. Whitehead.*
L. Sedbergh. Bentham. Garsdale.
R. Settle; *J. Windsor.* Slaidburn. Wigglesworth.
Y. Yore district, frequent.
N. Boggy field, right hand, in Forest Lane, Knaresborough; *E. Hargrove* (1832).
W. A. Wharfedale and Airedale, common.
C. Birkenshaw; *J. Emmet.* Norland; Southowram, *L. C. Miall.* Todmorden; *A. Stansfield.* Colne basin, frequent; *C. P. Hobkirk.*
D. Rotherham district; *L. Langley* (1828). Common about Sheffield; *J. Salt* (1800). Waleswood; *A. Carr.* Treeton and Catcliffe; *G. H., MS.*
T. Maltby Woods.

Forma **pseudo-reptans,** a state rather than a variety, in wet, gravelly or boggy-sandy ground on the borders of dams. It reverts to the more succulent and upright type in richer adjacent soil. I have seen extreme *pseudo-reptans* on the gravelly margin of Malham Tarn; and less-marked states from Widdale Fell Tarns, Gorple Water, shore of Black Hill Dam, Adel, and boggy ground near old dam at Bradfield (*A. Carr*).

17. Ranunculus Lingua L. *Great Spearwort.*
Area: 63, 64, Top. Bot.
Native; marshes and wet thickets, rare. P. July-August.
First record: White, 1804.
[R. Giggleswick Tarn; *J. Windsor.* Gone—if ever there?]
Y. Near Ripon; *W. Brunton* (1805). Sharow Mires (!) *H. H. Slater.* Marsh near Copgrove; *J. Dalton.*

W. Askham Bogs (!) *Baines.* [Bramham, 1863; *J. Emmet.* Now gone?]
[A. Adel Bog, near Leeds; *L. C. Miall*, etc. Formerly—now gone.]
D. Doncaster neighbourhood; *White* (1804). Potteric Carrs! *S. Appleby* (1831).

18. Ranunculus auricomus L.! *Goldilocks.*
Area: 63, 64, 65, Top. Bot. Pelophilous. Range: 50—900 feet.
Native; woods and shady banks, rather preferring calcareous or stiff clayey soil, common. P. May.
First record: Bolton, 1775.
R. Kendall's Fell, Kelcowe, etc., near Settle; *J. Windsor.* Slaidburn.
Y. Clotherholme, Mackershaw (!) Studley; *H. H. Slater.* Copgrove, *H. F. Parsons.*
N. Longwalk Wood, Knaresborough (!) *Hargrove* (1832). Hunsingore.
W. Bramhope staircase, in plenty, on grit (!) *W. W. Newbould.* Wood at Roundhay limehills (!) *W. Todd.* Wetherby, etc.
A. Skipton woods; *W. Whitwell.*
C. Rastrick, Hipperholme, etc.; *J. Bolton.* Near Lofthouse; *G. Roberts.* Kirklees; *Y. N. Excurs.* (1873).
D. Road to Darnall from Catcliffe; *G. Harr., Herb.* Conisborough dist. (!) *Y. N. Excurs.* (1873). Sheffield; *J. Salt.*
T. Maltby Woods. Sandbeck Park.

The State **depauperata** with abortive petals occurs with the type, chiefly in barren calcareous soil.

19. Ranunculus acris L. *Buttercup.*
Area: 63, 64, 65, Top. Bot. Range: 0—1,800 feet.
Native; fields and banks, very common, with a wide vertical range. P. June-August.
First specific notice: Salt, 1800.
Generally distributed; in all the districts.
Our West Yorkshire plant is almost wholly the form **vulgatus** Jord. with creeping rootstock; but I have seen **Boræanus** Jord., with very narrow leaf segments and erect rootstock occasionally.

20. Ranunculus repens L. '*Toad-teather.*'
Area: 63, 64, 65, Top. Bot. Range: 0—1,500 feet.
Native; wet fields and waste ground, ditch banks, roadsides, and as a weed in neglected gardens. Common. P. May-July.
First record: Salt, 1800.
Distribution general; all the districts.

21. Ranunculus bulbosus L. *Buttercup.*
Area: 63, 64, 65, Top. Bot. Range: 0—1,250 feet.
Native; meadows and old rich pastures, very common at lower levels, rarer on the hills, and hardly ascending above the plateau of Malham Tarn and Conistone Moor.

First record: Bolton, 1775.
General in distribution—reported from all the districts.

Status **flore pleno:** in some fields about Well Head, near Halifax; *J. Bolton.*

22. Ranunculus hirsutus Curtis.
Area: 63, (64), Top. Bot. Sub-maritime in its preferences.
Native; waste ground near tidal influence, very local. A. July-August.
First record as a Native: Parsons, 1876.
[R. Cornfield, Settle, 1855; *J. Tatham*, 'one plant.' A Casual only, there.]
D. Goole fields; by Hook Road, near Goole! *H. F. Parsons* (1876).
T. Near Thorne; *H. F. Parsons.* Casual here?

23. Ranunculus parviflorus L.
Area: 63, Top. Bot.
Colonist, or perhaps rather a Casual; arable land, very rare.
First record: Dalton, 1835.
(Y. Hackfall, *Rev. J. Dalton*; *Hook. Brit. Flo.*, vol. iii (1835). I fear doubtful, since not quoted later; and station an unlikely one.)
D. Grass-grown limestone débris by a potato patch near cottages at the foot of the Sprotborough quarries, near Conisborough; 1870!

24. Ranunculus arvensis L. '*Burr-Crowfoot.*' '*Hedgehogs.*' '*Hell-weed.*'
Area: 63, 64, Top. Bot. Range: 0—500 feet.
Colonist; corn and turnip fields, on light soils common, on stiff clays rare.
First record: Salt, 1800.
('Everywhere,' *Miall* and *Carr. Flora.* By no means the case, but probably more abundant formerly: its presence in quantity is said to be an indication of bad farming.)
R. Field, Wigglesworth. No entry for Settle in *Flora Cravoniensis.*
Y. Markington, so plentiful in 1881 as to give fields a yellow colour; *H. H. Slater.* Ripon.
N. Knaresborough; *E. Hargrove* (1832). Hunsingore. Spofforth.
W. Acomb. Wetherby district.
A. Meanwood. Eccup. Selby.
C. Lightcliffe and Hipperholme; *H. Baines.*
D. Sheffield, Tinsley Park, etc.; *J. Salt.* A common weed on warp soil about Goole; *H. F. Parsons.*
T. Marshland; *H. F. Parsons.*
No record for Lune district, and not marked for that of Mersey by Mr. Whitehead, whose list forms the basis of my information in regard to that area.

25. Ficaria verna Huds. *Pilewort.*
Area: 63, 64, 65, Top. Bot. Range: 0—1,000 feet.
Native; shady banks and open grassland, very common in the type (*divergens* F. Schultz.)
Area general—all the districts.
First record: Bolton, 1775.

Var. **incumbens** F. Schultz (*R. calthæfolia* Reich.), a luxuriant variety of watery places (if not a hybrid with *Caltha*?), only once seen, in a ditch with *Caltha*, by the tannery dams of Weetwood bottoms, near Leeds.

26. Caltha palustris L. '*Water Blobs.*' '*Marsh Marigold.*'
Area: 63, 64, 65, Top. Bot. Range: 0—1,500 feet.
Native: stream sides and marshes, common, with a wide range. P. March-May.
Distributed generally—all the districts.
First records: Bolton, 1775; Salt, 1800.
It is often reduced in size high up on the moors.

Var. '**minor** Mill. Dict.' is recorded for 'Todmorden' in *Baines.* Mr. Newbould has in his copy a MS. note: 'Further investigation must show if a plant found near Endcliffe Wood is a large state of *C. minor.*' I know nothing further as to either of these.
The narrow non-contiguous petalled *Guerangerii* Bor. I have never seen within West Yorkshire boundaries.

27. Trollius europæus L. *Globe-flower.* '*Boxes.*'
Ranunculus globosus; Ger.; Ray.
Area: 63, 64, Top. Bot. Montane. Range: 2,000 to 300 feet.
Native; in rocky moist woods and stream banks, rare. P. June.
First records: Gerarde, 1597, for Yorkshire generally; Ray, 1670, for West Riding.
In most places of Yorkshire; *Gerarde* (1597); Id.; *How, Phyt.* 102; 'divers places;' *Merr.* 102; near water-courses, etc., plentifully; *Gibs. Cam.*
L. Friar Wood and Scar Close, below Ingleton, towards Chapel-le-Dale, *Mr. Thornbeck*; *Blackst. Spec.* 80, 1746; Id., *Martyn* (1763); Skirrith Wood, Ingleton; *Gough's Cam.* (1789). Sedbergh. Dentdale. Feizor Wood; *L. C. Miall.*
R. Between Settle and Hinkel-haugh, etc.; *Ray, Cat.*, 259 (1670). Settle and Ribble banks; *J. Windsor.*
Y. Ripon; *W. Brunton.* Near Copgrove; *J. Dalton.* Clotherholme and Burton Leonard; *H. H. Slater.*
W. Wharfedale, many plants from Buckden downwards by Grassington; *W. Wood* (1798). Barden and Bolton Woods (!) as far down as Ilkley.

A. Malham Cove; *D. Turner* (1805). Rombalds Moor, a mile above Eldwick; *H. Baines.*

C. Mossleden and Ripponden; *J. Bolton* (1775). [Turner Clough, Barkisland. Gone.]

Obs.—The country-children's name of 'Boxes' or 'Butter Boxes' has clearly been given from some fanciful resemblance of the overlapping curled-in petals shutting in the butter-coloured stamens, as they do, to the lid of a box. The name is local, of course, and, so far as I know, confined to Ribblesdale, Garsdale, and Dentdale in the West Riding.

[**Eranthis hyemalis** Salisb. Alien.

Subspontaneous.

Y. Skell banks, Ripon, occasional. Near Copgrove; *Y. N. U. Exc.*, 1885.

N. Hedgeside, Long Flat, Knaresborough; *W. Kirkby* !

W. Swinnow Park plantation, naturalised; *J. Jackson* ! Plantations bordering Archbishop's Palace grounds, Bishopthorpe, York; *W. Whitwell.*

A. Plantations in Cowick Park, near Snaith; *W. Coates* through *W. Kirkby* !]

28. Helleborus viridis L. *Setterwort.*

Area: 63, 64, Top. Bot. Range: 50—500 feet.

Native; woods, thickets, and hedge-banks, mostly on calcareous soil, rare. P. April-May.

First record: Turner, 1664.

Helleborus. There are accounted eight kindes of this Hellebore, some whereof grow . . . about Lancashire and Yorkshire; *Turn., Bot.*, 148 (1664).

L. In a meadow at Wharfe in Craven; *H. Baines* (1840).

Y. Fountains Abbey; *W. Brunton* (1805).

N. Knaresborough, in an orchard at the Abbey; *E. Hargrove.* Banks opposite the Mill; *J. Dalton.* Birkham Wood.

W. On a bank side in Kidhall Lane, near Leeds, *Mr. Dawson*; *Blackst. Spec.* (1746); *Martyn* (1763); *Gough's Cam.* (1789); etc. Still there! Langwith Scrub, Collingham! *J. Abbott.* The Wilderness Wood, Wetherby. Tadcaster, pasture behind the Church; *H. Baines*; still there; *J. Emmet.* Aberford; *Rev. W. Wood* (1798). Becca Banks, Aberford!

D. Woods between Warmsworth and Conisborough; *J. Bohler*; *Baines.* Campsall; *G. Roberts* (1872). Conisborough Castle; *Jos. Woods* (1838); Id.; *J. Salt* (1800), 'am doubtful whether it be a native there;' *Salt, Flo. S.* [Can Salt have known only the station in the moat of the keep, and been unaware of it in Levitt Hagg, not a mile away?] Conisborough; *L. Langley* (1828). Lime and clay soils west of Doncaster; *S. Appleby* (1831). Old moat near Hook Church; *H. F. Parsons.* Probably alien in this last spot.

T. Roche Abbey, bank at low end of Laughton Pool, etc.

29. Helleborus fœtidus L. *Bear's-foot.*

Area: '64, Tatham sp.,' Top. Bot.

Denizen: banks and thickets, rare, much less certainly indigenous than *H. viridis*, but naturalised. P. March-May.

L. Rocky pasture above Feizor; *H. Baines* (1840); *F. Howson*, 1850; rep. *L. C. Miall*, etc., 'probably indigenous.' Oxenber near Austwick; *R. Clapham.*

R. Near the old tan-pits at Langcliffe; *J. Windsor.*

Y. Copt Hewick and Norton Conyers; *H. H. Slater.*

N. Bushy ground on cliffs, above the gardens, a little to the east of Knaresborough town! *W. Kirkby.*

[A. Under Malham Cove, 1839; *L. C. Miall* in *Nat.* i, 211 (1865). Ledstone Park, *Mr. Schofield* (1869); *G. Roberts.*]

D. Lanes at Campsall; *R. Teesdale* (1798), and *O.B.G.* Limestone tract west of Doncaster; *S. Appleby.* Smeaton, Went Vale; *T. Birks* (1878). Woods between Warmsworth and Conisborough, with a habit very different from the cultivated plant; *H. Baines* (1840). On and beneath rocks in Levitt Hagg Wood; *G. Harr.* (1863). Conisborough district still, in 1873, *Y. N. Excurs.*

30. Aquilegia vulgaris L. *Columbine.* '*Fools'-cap.*'

A. sylvestris, C.B.; Ray.

Area: 63, 64, Top. Bot. Xerophilous. Range 100—1,000 feet.

Native: woods and thickets by streams, mainly on limestone (then usually blue, but on grit pale pink); rare, although locally plentiful. P. June-July.

Earliest notices: How, 1650; Salt, 1800.

Yorkshire; *Ray, Cat.*, ii, 26; ii, 25 (1670).

L. Ingleborough lower scars; *R. Gough.* Wherf scars.

R. Stainforth Scars and Kelcowe; *J. Windsor.* Mountain woods near Settle; *H. Baines.*

Y. Mackershaw Wood (!) *H. H. Slater.* Fountain's Abbey; *W. Brunton.*

N. Scotton banks, Knaresborough.

W. By the Wharfe, near Grassington. Near Ilkley. Headley and Jackdaw Crag Wood, near Tadcaster. Thorparch. Aberford. Norwood bottoms, Thorner. (Damp land near Lindley Wood.)

A. Various parts of Craven; *L. C. Miall.* Adel bridge; *H. Baines* (1840). Thicket by stream a little on Adel side of the Seven-Arch Aqueduct, Meanwood (1867—70), on grit, with pale pink flowers.

C. In Catholis Clough, below Hartley Royd, probably a garden outcast; *A. Stansfield* (1877). [Shibden; field opposite Copley Hill, Norland. Gone.] (Stanley, near Wakefield, one plant—a stray; *Geo. Roberts.*)

D. Many about Brodsworth and Hamp[o]le Woods (near Doncaster); *How., Phyt.*; Id., *Merr. Pin.*, 9. Rotherham dist.; *L. Langley.* Conisborough! *G. Roberts* (1873); *H. F. Parsons.* Levitt Hagg Wood; *G. Harr., MS.* In Old Park, Sheffield; *J. Salt* (1800).

T. Maltby. Between Anston and Kiveton Park.

Obs.—Folly-flower or Fools'-cap from the shape of each spurred petal—suggestive of the Jester's head-dress.

[**'Aquilegia alpina.** Ingleborough,' *Martyn* (1763); rep. *Gough's Cam.* (1789). A dwarfed, but large-flowered form of *A. vulgaris*, such as grows in the fissures of the elevated limestone pavements, was intended by this, not the Linnean, continental *alpina.*]

[**Delphinium Ajacis** Reich. Alien.

Casually with crops, but fugitive, not colonising.

Y. Near Ripon; *W. Brunton, O.B.G.* (1805). Waste ground by Yore, outcast (1870).

N. Cornfield, North Deighton.

A. Kippax; *Atkinson* in *Miall.* Lime quarry, Brotherton.

D. Balby; *G. E. Smith.* Sandy cornfield, near Thorne.

Obs.—I have referred the above old records to this species as being the *Consolida* of English authors, but in any case the plant may have been true Linnean *Consolida*—although that is much rarer than *Ajacis.*]

[**Delphinium Consolida** L. Alien.

Casual in corn and seed fields.

W. Field at Thorparch (1870).

C. Casual, near Hirst's Corn Mill, Huddersfield; *Yorks. Nat. Rec.* 128.]

31. Aconitum Napellus L. *Monkshood.*

Area: '(64)' Top. Bot.

Denizen: stream sides and woods; rare, long-enduring, but not wild. P. July.

First record: Hailstone, 1815?

R. Island in Ribble below Settle; *J. Tatham.*

Y. Spa Gill, 1881, *Mrs. Berkeley*; *Slater's Flora.* Near the Plaster Pits between Ripon and Mickley; *W. Grainge.*

W. Wharfe banks, Bolton Bridge, and below Ilkley; *J. Willis.* Farnley; *H. F. Parsons.* North side of Wharfe above Thorparch; *S. Hailstone.* Still there (near Flint Mill), 1880.

A. Hambleton Haugh! *T. Birks.*

[**Aconitum Lycoctonum** L. Alien.

Naturalised in the wooded swamp to the back of the old ruined Hall at Bramham Park!]

32. Actæa spicata L. *Bane-berry.*

Christophoriana; Ger., Ray. *Aconitum racemosum*; Ray.

Area: 63, 64, Top. Bot. Xerophilous. Montane. Range 1,400—200 feet.

Native; rocky, bushy places, normally at some elevation, but comes down to a very low level in Aire district. Herb. L. Mid May-June.

First record: Ray (1670).

L. Chapel-le-Dale pastures, Ingleton, *Mr. Thornbeck*; *Blackst. Spec.* (1746); *Martyn* (1763). Clefts of rocks at the foot of Ingleborough, between Chapel-in-the-Dale and Meer Gill; *J. Windsor.* Near Clapham, in a wood; *Martyn* (1763). Ingleborough; *Id.* Fissures in limestone pavement at the foot of Ingleborough! *Stokes* (1805).

(N. Nidderdale; *H. H. Slater, Flo. Ripon.* Why inserted in that Flora, or where the station is, I do not know. Perhaps it is 'Birkham Wood, Knaresborough, *Maurice Lee*, oral report,' with '!' instead of '?' appended in *Yorks. Nat. Rec.*, 142. I never saw the plant there myself.)

W. Mr. Witham showed me this growing in Haselwood-woods, near Sir Walter Vavasour's park-pale in Yorkshire, but not plentifully; *Ray, Cat.*, 71 (1670); *Hist.* (1688). Still there, 1867; *Jos. Blackburn*! Rocky scrub of scars above Cowside Beck (!); between Darnbrook and Arncliffe; *Jos. Woods, MS.* in *Towns. O.B.G.* (1838). Rocky wood between Arncliffe and Kilnsey. Boston Spa Woods. Thorparch Woods; *W. Wood* (1805). Aberford. Beech wood at Barwick-in-Elmet. Hesleden Gill (!) *J. Windsor.*

A. Woods near Allerton Bywater, bordering Kippax Park, near road from Kippax to Fairburn! *Jos. Blackburn* (1868, plentiful). Crevices of limestone tables between Malham and the Tarn (!) *J. Windsor.* Among the Shrubs by Malham Cove, *T. Lawson*; *Ray, Fascic.* 22 (1688); *Syn.* i, 95 (1690). By brook below Malham Cove (! very sparingly); *J. Windsor* (1867).

C. Liley Wood, Upper Whitley, near Mirfield! *E. Leyland* (1812); *Baines* (1840). [On coal-measures; *here*, perhaps, introduced? 'nearest the farm-house above the edge of the wood'; *Hobk. B.E. C. Rep.*, 1871.]

[**Pæonia officinalis** Retz.

Pœonia mas vulgaris; Ger.

'Found growing in Mr. Field's Well Close, in Darfield: which, though farre from any house, I beleeve it came first out of a garden with some dung. *Mr. Stonehouse*'! *How. Phyt.*, 95 (1650); rep. *Merr.*, 96. Ray omitted this from 2nd Cat., and in preface gives reasons, and includes it and many others in the Indic. Plantarum Dubiarum of his Synopsis.]

II. BERBERACEÆ.

33. Berberis vulgaris L. *Barberry.*

Area: 63†, 64, Top. Bot. Range: Low.

Native; hedges and thickets, rare; often planted but still clearly indigenous in certain places on the Permian tract. Shrub. May-June.

First record: Salt, 1800.

R. Kilcorse Wood, near Settle; *Baines* (1840). Rathmell; *J. Windsor.*
Y. Hedges, Clotherholme, rocks at Fountain's Abbey, etc., near Ripon(!) *H. H. Slater.* Copgrove. (Hackfall*.)
N. Long-flat, Knaresborough; *E. Hargrove* (1832).
W. Dalton Lane, between Beilby Grange and West Woods, Bramham! *J. Emmet.* Rigton. Hedges, Hob Moor and Acomb, near York; *H. Baines.* (Bolton Abbey†.)
A. Ledstone! *G. Roberts.* Fairburn; *Baines.* Near Pontefract; *Baines.* Between Kippax and Ledsham. Ledsham; *W. Whitwell.*
C. Oulton and Rhodes Green, not wild there; *G. Roberts.*
D. Bushy scar between Hampole Stubbs and Doncaster, and hedges, Hampole; *G. E. Smith.* Askern; *G. Harr.* Between Knottingley and Womersley, plentiful; *G. Roberts.* About Conisborough; *J. Salt* (1800); *G. Roberts.* Levitt Hagg.
T. Maltby. Letwell. King's Wood.

Obs.—This does not ripen its fruit well with us, except in the Trent district.

[**Epimedium alpinum** L. Extinct Introduction.

Found by that accurate Botanist, Dr. Richardson, in Bingley Woods, and not sparingly; *Blackst. Spec.* (1746). In woods, Bingley; *Martyn* (1763); *O.B.G.* (1805). Introduced to deceive Dr. Richardson; *Smith* rep. *Baines.* One plant in an old wall at Bramham Park, doubtless introduced; *John Emmet.*]

III. NYMPHÆACEÆ.

34. Nymphæa alba L. *White Water-lily.*

Area: 63, 64, Top. Bot. Range: Low (with us).

Native; ponds and slow waters, very rare in a truly native state. P. Late June-July.

First record: Hargrove, 1782.

[R. Giggleswick tarn. Gone.]
Y. Ripon; *G. Roberts.* Musterfield pond (!), near Sleningford; *W. Grainge.*
[N. In a pond near Collin Bridge, Follifoot; *E. Hargrove*, Ed. iii, 1782.]
W. Pond at Thorparch, along with *N. lutea*, where they have grown for over 20 years; *J. Emmet.*
[C. Top of Westgate Common, Wakefield, 1791; *D. Dixon.*]
D. Barnsley; *Miall, Flo.* (1862). Ackworth; *Miall.* Potteric Carrs; *E. Miller* (1804); Id., *S. Appleby* (1831).
T. Near Rossington. Thorne Waste; *H. F. Parsons.*

35. Nuphar luteum Sm. '*Brandy Bottle.*'

Area: 63, 64, Top. Bot. Range: 0—450 feet.

Native; ponds and slow rivers or dikes, infrequent. P. July.

First record: Salt, 1798.

R. In river Ribble (still parts) between Settle and Long Preston; *J. Windsor.*
Y. Ripon, near the Rifle range; *H. H. Slater.*
N. Easedyke, near Wighill! *J. Emmet.*
W. Drains by Wharfe, near Tadcaster. Pond in pasture between Thorparch Woods and the river Wharfe; *J. Abbott.* In the river Cock near Grimston.
D. In river Went, Smeaton; *G. Roberts.* Potteric Carrs; *E. Miller* (1804) etc. Rotherham district; *L. Langley* (1828); *A. Carr* (1870). Near Conisborough Ferry, Cadeby Common, July 1798; *J. Salt.* River Rother; *G. Harr.*

IV. PAPAVERACEÆ.

[**Papaver somniferum** L. Alien.

Casually, in fields, or on rubbish outcast from gardens, and without permanence.

A. Hawksworth; *W. West* (1879). Canal side, Woodlesford; *G. Roberts.*
C. Sandal; *G. Roberts.*
D. Gravel pits in Campsall Park; *E. Lankester* (1842).]

36. Papaver Rhœas L. *Poppy.*

Papaver Rhœas fl. purp. saturo; Merr., Ger.

Area: 63, 64, Top. Bot. Range: 0—600 feet.

Colonist; waste and arable land, rail-banks, etc., common. A. June-August.

First record: How, 1650.

Distribution general—all the districts, but in Mersey, Lune, and elevated districts, sporadic and fugitive.

'In mine own Rie at Darfield, *Mr. Stonehouse*'; *How, Phyt.*, 88; *Merr. Pin.*, 90.

'*Pap. Rhœas flore albo.* Amongst wheat betwixt Pontefraict and Ferry-bridge, *Mr. Stonehouse*'; *How, Phyt.*, 87 (1650). A var., or really *somniferum*? *Cf. supra.*

37. Papaver dubium L. (*P. Lamottei* Bor.)

Area: 63, 64, 65, Top. Bot. Range: 0—600 feet.

Colonist; fields and waste ground, such as railway banks and sidings; common up to 500 feet, rare above that. A. May-August.

First record: Baines, 1840.

Distribution—all the districts, but local in some.

Y. Blois Hall, near Hewick; *H. H. Slater* (1881).
D. Several places in Ecclesall (with white juice); *W. W. Newbould.* Sheffield (1859), *Newb.*; *Bab. Flo. Camb.*, 302.

Obs.—The yellow-juiced variety, *P. Lecoqii*, I have not seen in Yorkshire.

38. Papaver Argemone L.

Area: 63, 64, Top. Bot. Range: 0—300 feet.

Colonist; light-soiled cornfields, common on the Trias, frequent on limestone, very rare on grit or coal-measures. A. June-August.

First record: Salt, 1800.

Y. N. W. Common in the Ainsty between York and Boroughbridge. Copgrove. Wetherby.
A. Cornfield near Saltaire; *L. C. Miall.* Selby.
C. Rothwell (1869), and near Lofthouse; *G. Roberts.*
D. Doncaster; *Geo. Harr., Hb.* Near Booth-Ferry; *H. Baines.* Quarry at Crooks Moor; *A. Carr* (1876). Cornfields near the Manor, Sheffield; *J. Salt* (1800). Manor farm fields, Sheffield; *A. Carr.* Near Sheffield; *W. W. Newbould.* Cornfield, Rawcliffe; *H. F. Parsons.* Thorne. Barnby Dun; *G. Roberts.*
T. Hatfield; *G. Roberts.*

[**Papaver hybridum** L. Casual.

Area: 63 (Lees), Top. Bot.

One plant only seen, in dry calcareous cornfield between Pontefract and Knottingley, July 1871, with *Bupleurum rotundifolium.*]

39. Meconopsis cambrica Vig.

Area: 64, Tatham 'wild,' (63, 65), Top. Bot. Montane. Range: 900—100 feet.

Denizen; doubtfully indigenous; rocks about waterfalls and lower down by streams, very rare. P. Late June-July.

First record: Tatham, 1832.

L. Near Sedbergh; *J. Carr, MS.*, in *Towns. O.B.G.* About Wethercote Cave; *Jos. Woods* (1838); *B. Carrington.* Near village of Feizor; *J. Tatham* (*spn.*, 1832), *N.B.G.*
R. About Stainforth and Stackhouse; *J. Windsor.* Giggleswick; *W. Howson* (1850). By the river side below Settle; *H. Baines* (washed down from higher up).
Y. Near Mickley, Hackfall; *H. H. Slater.*
W. Heseltine Gill, Arncliffe. Bolton Wood (hedges, etc.), near the Abbey (!) *J. Abbott.* Thorparch, river side; *L. C. Miall* (washed down). [Wood between Clifford and Bramham, at the left hand, bottom of hill†; *J. Emmet.* Gone by 1880.]
[C. Wheatley, near Halifax: introduced and was lost; *Miall, Flo.*]

40. Chelidonium majus L. '*Sallenders.*'

Area: 63, 64, Top. Bot. Range: 0—600 feet.

Denizen; hedge-banks and thickets, but seldom very far from habitations; common. P. End of May—July.

First record: Bolton, 1775.

L. Sedbergh. Dentdale.
R. Near Settle! *J. Windsor.*
Y. Fountains Abbey walls; *H. H. Slater.* Hackfall; *Mrs. Berkeley.*
N. Scotton and Bilton Banks, by Nidd, in bushy, rocky pl., here looking very wild and associated with *Myrrhis* and *Geranium sylvaticum.*
W. At the foot of steep rocks in the wood to the west of Boston Spa, etc.
C. Halifax; *J. Bolton.* Walton Common, 1804; *D. Dixon.* Near Huddersfield; *C. P. Hobkirk.*
A. D. Frequent in the more rural parts. Ledsham; *W. Whitwell.*

V. FUMARIACEÆ.

41. Corydalis lutea DC.

Denizen; now well established, by old walls, on old bridges, etc., but rare.

First record: W. Brunton, 1800.

R. Right-hand side of highroad from Langcliffe to Settle, nearly opposite the Well; and Holmhead, near Langcliffe High Mill; *Mr. Gough*; *With.*, Ed. 7 (1830).
Y. All about Fountains Hall, Studley (!) *W. Brunton* (*Fl. Brit.*, ii, 749). Fountains Abbey; *Brunton* and *Dalton* in *O.B.G.*; *H. H. Slater.*
N. Old wall in Abbey ground, Knaresborough; *G. Roberts* (1884).

42. Corydalis claviculata DC.

Area: 63, 64, 65, Top. Bot. Arenophilous. Range: 150—800 feet.

Native; turf banks and rocky, bushy places and woods, on grit rock and sand, common; on limestone very rare. A. June-August.

First record: Bolton, 1775.

L. Bentham.
R. Peart's Crags, near Settle, and south end of Craven Bank, Giggleswick; *J. Windsor.* Bowland Knotts.
Y. Hackfall.
N. Hookstone Crags, Harrogate; *E. Hargrove* (1832). Plumpton. Brimham Rocks. Grimbald Crag, Knaresborough. Scotton Banks. Birk Crag; *J. Emmet.*
W. Arncliffe Wood; *H. Baines.* Otley Chevin. Addingham.
A. Bingley Wood (!) Nab Wood, Apperley Bridge. Stonehall, Rawden (*Rev. W. Wood* in *With.*, Ed. 7, 1830), etc., in Airedale. Thickets on Adel Moor, near bridge! Bramley Fall Wood. Kirkstall Forge; *H. Baines.* Hambleton Haugh. Brayton Barf, Selby. Morton Bog; *H. F. Parsons.*

C. Hebden Valley! Elland Wood, *C. Eastwood*; *G. Roberts*. Skircoat, Salterhebble, Warley, etc., near Halifax; *J. Bolton*. Primrose Hill, Huddersfield; *Hud. N. H.* Luddenden Foot; *Eastwood*.
D. About Wharncliffe Lodge; *J. Salt* (1800). Wyoming Brook; *G. Harr. Hb.* Bradfield Moors; *G. H. MS.* Rotherham district sandstones; *L. Langley* (1828).
T. Thorne Waste; *H. F. Parsons*.

43. Fumaria Boræi Jord.
Colonist; arable land, waste ground, etc.; very rare. A. July.
First record: Todd, 1869.
[R. 'Not seen near Settle;' *J. Windsor*. *Rumour* of aggregate occurring, alluded to.]
Y. Bishop Monckton, near Ripon!
N. Harrogate; *W. Todd* (1869)!
W. Ilkley. Harrogate and Thorparch; *R. Teesdale* (Linn. Trans., 1798, sub nom. aggreg.).
A. Meanwood! Yeadon; *B. Carrington* (1860) sub nom. *capreolata*.
D. Near Conisborough; *T. Birks*!
T. Tickhill ('*pallidiflora*'); *T. Birks* (1878).

44. Fumaria confusa Jord. (*F. agraria*, Mitt.)
Area: 63, Top. Bot. A. July-August. Range: apparently Low.
Colonist; hedges and field-borders, distribution ill ascertained. Dried herbarium examples are very difficult to name with certainty, and my own gatherings at Ripon, Ilkley, Meanwood, etc., placed under *Boræi* because so determined at the time of collection (when I did not, possibly, rightly understand the distinctions) may well in part belong to this segregate. Briggs declares it (*confusa*) to be the prevalent form about Plymouth; More makes the same remark for the Isle of Wight; Trimen says *muralis* is the most frequent near London; whilst Hooker applies the term 'Common' to *Boræi* only; and Baker gives *Boræi* as the most usual form of N. Yorkshire. All three splits are too rare in West Yorkshire for me to feel I have seen material enough for any trustworthy expression of opinion.
First record: Salt, Herb. spn., 1802.
D. Rotherham dist.; *L. Langley* [sub nom. aggreg.]. Id.; *Amos Carr*! Kexborough, near Barnsley; *H. Johnson* (1885)! Sheffield; *Baker* in *Miall* (sub aggregate name—*capreolata*). Near Sheffield (1802); *Salt, Herb.*, is *F. agraria*; *W. W. Newbould, MS.* in his copy of *Baines' Flora*. Hedge of a garden at Catcliffe in 1862, but destroyed in 1864; *G. Harr., MS.*

45. Fumaria muralis Sonder.
Area: '63. Newb., Bab. 64, Baker,' Top. Bot. A. May-Sept.
Colonist; fields and waste ground; very rare. Range: Low.
First record: Carrington (1860).

Y. Rubbish heap near Bishopton, Ripon—one plant, 1880; *H. H. Slater*.
W. Ilkley; *B. Carrington* teste *C. C. Babington* (1860).
D. Field near Kimberworth; *A. Carr* teste *C. C. B.* Sharrow Bank, Sheffield; *W. W. Newbould*.

[**Fumaria parviflora** Lam.
Casually in fodder-vetch field, near Oil Mills, Meanwood Valley (1876).]

[**Fumaria densiflora** DC. (*F. micrantha* Lag.)
Area: 63,* Top. Bot.
Arable land, very rare; a Casual rather than a permanent integer in the flora.
First (only) record: the Author.
Field between Knottingley and Brotherton (1870).
Obs.—A ramose procumbent form of *F. officinalis* has once or twice been misreported as this [*vide Y. N. U. Trans., Series E*, p. 144 (1882)].

46. Fumaria officinalis L. '*Fumitery*.'
Area: 63, 64, 65, Top. Bot. Range: 0—900 feet.
Colonist; waste and cultivated ground, gardens, etc., very common. A. May-September.
Earliest notice; Bolton, 1775.
Generally distributed; recorded for all the districts, but only ascending as high as field cultivation reaches.

Var. **densiflora** Gren.
Ripon; *H. H. Slater*! and fields at Priest Hill, Wetherby.
Obs.—Known as 'Fumitery' to many unlearned country folk: can this term really be a corruption of *fumus-terræ*, earth cloud or smoke, from the grey glaucous hue of the curling foliage, resembling masses of cloud or smoke wreaths?

VI. CRUCIFERÆ.

47. Raphanus Raphanistrum L.
Area: 63, 64, Top. Bot. Range: 0—600 feet.
Colonist; in arable fields, common. A. June-August.
First record: Salt, 1800.
Distribution general—found in every district.
Obs.—Varies in having pale yellow or whitish lilac-streaked petals—the latter the least frequent.

[**Raphanus sativus** W.
Has occurred on manure heaps, etc., and by footpath side at Methley (*G. Roberts*).]

[**Rapistrum orientale** DC.
Casual. Sandal, *G. Roberts*.]

48. Sinapis arvensis L. *Charlock*.
Area: 63, 64, 65, Top. Bot. Range: 0—600 feet.
Colonist; corn, potato, and turnip fields, very common. A. June-September.
First record; Salt, 1800.
Distribution general—found in every district.

Var. **S. orientalis** Murr.
Sandal; *G. Roberts*: a Casual, with other exotics (1885).

49. Sinapis alba L. *Mustard*.
Area: 63†, 64, Top. Bot. Range: Low.
Colonist, but fugitive and casual-like; cornfields and rubbish heaps, quite rare. A. July-August.
First record: Backhouse, in Baker's Flo. Supp., 1854.
W. Ilkley; *B. Carrington*. Near York; *J. Backhouse*.
A. Brotherton! Yeadon; *B. Carrington*. Methley (in clover); *G. Roberts*. Field between Moortown and Eccup, with *Centaurea Cyanus*! Roundhay; *W. West*.
C. Lofthouse; *G. Roberts*.

50. Sinapis nigra L. *Black Mustard*.
Area: 63, 64, Top. Bot. Pelophilous. Range: Low.
Native; cornfields occasionally, but chiefly on river and stream banks; infrequent. P. June-August.
First record: Baker, 1854.
W. Wharfe side by railway bridge, Linton!
A. Canal side, Shipley. By the Aire at Snaith; *H. F. Parsons*. Brotherton! Skipton; *W. Whitwell*! etc.
C. Cornfields at Hebden Bridge; *Miall, Flo.* Spen banks, near mills, Heckmondwike.
D. By the Don, Doncaster. Parkgate; *A. Carr*.

51. Brassica polymorpha Syme. **(B. campestris L.).**
Perfoliata siliquosa; Ger. 536. *Perfoliata sive B. campestris*; Park. 580.
First record: How, 1650.
'Codded Thorow-waxe: at Darfield, Mr. Stonehouse'; *How, Phyt.*, 90; Merr. Pin. (1666) 92.
Range: 0—450 feet?
I consider the three following are best regarded as forms of one species.

F. 1. **Brassica Rutabaga** DC. *Tuberous-rooted Swede*.
Colonist; lowland fields and waste ground (occasionally), only as a relic of cultivation. A. September-October.
I have seen it myself in several districts, chiefly lowland ones, but there are few published records. It appears mainly late in the year, after the turnips have been munched off by sheep.

F. 2. **Brassica Napus** L. *Rape* or *Coleseed*.
Not wild, and rare, in fields, because rarely grown as a crop. A. July.
First record: uncertain, because doubtful whether or not the plant under this name mentioned by Salt (1800) really referred to this—more likely it was the Turnip.
Y. This District; *H. H. Slater, Flora of Ripon*.
W. Field—a crop—between Tadcaster and Wighill.
D. Near Sheffield, 'in cornfields'; *J. Salt, MS.* (1800).

F. 3. **Brassica sylvestris** Wats. *Navew*. (Rapa L.—? *Wild Turnip*.)
Common and generally distributed on the banks of the larger streams up to 150 yards, and in field borders and lanes, but rarer there. Sometimes Annual only, I think; but I cannot distinguish clearly between the truly indigenous *sylvestris* and the strays which result from seed of the cultivated Turnip. The root-leaves are rather evanescent, but are not glaucous, and are hispid, just as in Mr. H. C. Watson's Thames-bank seedlings distributed some years ago. B. May-September.
First record: uncertain.
R. Settle ings; *J. Windsor*. Ribble bank, near Wigglesworth.
N. Fields, etc., about Cowthorpe.
W. Ilkley.
A. By the Aire, above and below Leeds. Snaith; *H. F. Parsons*. Ferrybridge. Bingley and Cottingley; *W. West*.
C. Near Huddersfield. By the Calder, near Castleford.
D. By the Don, near Conisborough.
Obs.—In North Yorkshire the common plant of the stream sides is *Rapa*,—*Napus* is quite uncertain,—*Rutabaga* cultivated only; *J. G. Baker*.

[**Brassica Cheiranthus** Vill. Alien.
'[64],' Top. Bot.
Introduced with ballast or grass seed, but for some years pretty well established on a made embankment west of the Workhouse at Wetherby! (W.) Reported in Yorks. Nat. Recorder (128, 1873) from waste ground near Hirst's Corn Mill, Huddersfield (C.).]

[**Eruca sativa** Lamk. Casual.
Quarry and waste ground at Sandal; *Geo. Roberts* (1885-86)!]

52. Diplotaxis tenuifolia DC.
Area: 63, Top. Bot.
Denizen; old walls, very rare. P. June-September.
First record: Baines, 1840.
A. On old walls, Pontefract Castle and Church (!) *Baines* (1840). Still there in 1885; *G. Roberts*.

[Diplotaxis muralis DC.
Area: 63†, Top. Bot.
Casual, hardly colonising; on waste ground, very rare. A. August.
D. About Goole; *H. F. Parsons*.]

53. Sisymbrium officinale Scop.
Area: 63, 64, Top. Bot. Range: 0—600 feet.
Native; by road sides, waste ground, or field borders in dry soil, common, but most so on calcareous, least so on flagstone-rock soils. A. July-August.
First record: Salt, Flo. Sheff., 1800.
Distribution general—except Mersey district. Very many localities, from Ingleton and Settle downwards.

54. Sisymbrium Sophia L. *Flixweed*.
Sophia chirurgorum; Ger.; Park.; Ray.
Area: 63, 64 Tatham Cat., Top. Bot. Sub-xerophilous. Range: 0—450 feet.
Colonist (but in some of its reported stations it can have been a casual only, I think); waste and dry ground, calcareous mainly, rare and never in any quantity. A. July-August.
First record: Salt, 1800.
In agro Eboracensi; *Ray, Hist.*, 812 (1686).
R. By the old high-road from Settle to Long Preston and near Swinden (lower down the dale); *J. Windsor*; rep. in *Howson, Flo.* No recent observance.
Y. Great Ouseburn; *H. Ibbotson, MS.*
A. Knottingley; *G. Roberts*! On wool-waste by Boothroyd's Mill, Meanwood Valley! 2 plants.
C. Sandal, field near Three Houses; *G. Roberts*!
D. Near Aldwarke bridge; *Salt, Flo.* (1800). Sprotborough (1869).

[Sisymbrium Irio L. Casual.
Sporadic, but fugitive, and very rare. Old walls at York; *Jos. Blackburn, spn.* (1850). In 1869, a few plants at Pontefract, growing at the foot of walls near the Castle, with *Diplotaxis tenuifolia* and *Cheiranthus Cheiri*.]

[Sisymbrium junceum M.B. Casual.
Waste ground at Sandal; *G. Roberts*. Named at Kew; not seen by me.]

[Sisymbrium orientale L. (*Columnæ* Jacq.) Casual.
Waste ground at Sandal; *G. Roberts* (1885)!]

55. Sisymbrium Alliaria Scop. *Sauce-alone*.
Area: 63, 64, 65, Top. Bot.
Native; in hedges, very common. B. May-July.
First record: Bolton, Cat., 1775.
Distribution general; recorded for every district.

56. Erysimum orientale Br.
Casual, or Colonist; stiff-soiled arable fields, potato mostly, very rare.
W. Field near Healaugh, one or two plants (1880).
D. Waste ground, Goole; *H. F. Parsons* (1875).
C. Sandal, field near Three Houses; *G. Roberts*!

57. Erysimum cheiranthoides L.
Area: (63), Top. Bot.
Casual, colonising; waste sandy and cultivated ground; very rare. A. July-September.
First record: G. Harrison, Herb., 1862.
W. About York, on rail-banks. By river below mill, Wetherby.
A. Near Milford; *G. Roberts*.
D. Cultivated ground, Goole; *H. F. Parsons* (1876). Conisborough; *Geo. Harr. Herb.* Cornfield at Greenside; *Amos Carr*. Barnby Dun; *H. Johnson*!

58. Hesperis matronalis L. *Dames'-Rocket*.
H. Pannonica inodora; Park.; Ray.
Denizen; river banks in bushy, sandy places; rare; persisting at Bolton, gone from many of its other stations. In many places, on waste heaps, etc., as a garden escape. P. June.
First record: Ray, 1670.
R. 'Was found and communicated to me by my often-remembered ingenious Friend Mr. Martin Lister, in the beck that parts Yorkshire and Lancashire, as you goe from Westby in Craven to Pendle Hill'; *Ray, Cat.*, i, 164; *not* in Ed.ii. Between Stackhouse and Stainforth; *W. Howson* (1850).
[Y. Near Ripon.] Does not seem to be there now; *H. H. Slater* (1883).
N. Spofforth; *J. Emmet*.
W. By Wharfe, east side, between Barden and Otley (!) *J. G. Baker*. In several places lower down river to Boston Spa.
A. Gargrave; *T. Simpson*. Road between Rylstone and Calton; *H. Baines*. Brookside, near Kiddall Hall; *S. Hailstone* (1808). Between Bell Busk and Airton, 1882; *W. Whitwell*.
[C. Hollins Wood, Luddenden; *S. King*.]
[D. Hellerby Wood, Doncaster; *H. Baines* (1840).] [By Earnshaw's Warping Drain, Goole; *H. F. Parsons*.]

59. Cheiranthus Cheiri L.! *Wallflower*. '*Hand-flower*.'
Denizen; thoroughly naturalised—the small yellow-petalled form (*C. fruticulosus*) on rocks and old walls. P. May-June.
First record: Hargrove, 1782.
R. Castlebar rock, Settle (!) *R. Clapham*; *Miall's Flora*; rep. *Windsor's Flora*. Introduced by J. Tatham about 1837, from seed off the old City-walls, York, sent by his son; *W. Whitwell* teste *Mrs. Silv. Thompson*.

N. Church walls and Abbey rocks, Knaresborough; *Hargrove* (Ed. 3, 1782).
W. Bolton Abbey walls! Rocky bank, north side of the bridge over the Wharfe at Thorparch (!) *J. Emmet*.
A. Growing abundantly on an old wall at Tanshelf, and also on the ruins of Pontefract Castle, 1790; *D. Dixon*. Pontefract Castle, etc. On bridge at Methley, 1870; *G. Roberts*. Skipton Castle walls.
T. Roche Abbey.
Obs.—Formerly (and probably still) the cut blooms were brought to the Leeds Market for sale in great bunches, and known and spoken of then as 'Hand-flowers—a penny a bunch.'

60. Cardamine amara L.! *Bitter-cress*.
Area: 63, 64, Top. Bot. Hygrophilous. Range: 0—850 feet.
Native; wet woods, ditch-sides, and by streams; infrequent, but locally plentiful. P. End of May-July.
First record: Bolton, 1775.
L. Tatham's Wood, Ingleton. Bentham.
R. Ribble banks, from Horton downwards to Bolton-in-Bolland. Runley Bridge Mill, Settle.
Y. Little Nunwick, marshes. Sharow Mires; *H. H. Slater*. Ripon; *W. Brunton* (*Sir J. E. Smith, E.B.*).
N. Brookside near the viaduct, Pannal. Spofforth, etc.
W. Askham Bogs. Newton Kyme; *H. Baines* (1840). Pool to Wetherby, by Wharfe. Boston Spa, 1859 and since; *J. Emmet*.
A. Weetwood bottoms, Meanwood (!) *J. Abbott*. Aire banks, above Leeds (!) *Rev. W. Wood* (1805). Canal side, Shipley; *W. West*.
C. Warley Clough, etc., Halifax; *J. Bolton* (1775). Barnsley (!) *G. Roberts, H. Johnson*, etc.
D. Conisborough dist.; *Y. N. Excursion* (1873). Rotherham district; *L. Langley*. By the Don between Doncaster and Conisborough. Beeley Wood (*Salt Herb.*), Wadsley Forge (*J. Salt*), Endcliffe Wood (*W. W. Newbould*), and other places near Sheffield.

61. Cardamine pratensis L. *Milkmaids*. '*Ladies' Smocks*.'
Area: 63, 64, 65, Top. Bot. Range: 0—1,900 feet.
Native; damp pastures and marshy places, common, and to be found close to the large towns. P. Early May-June.
Generally distributed—reported from all the districts, and reaching nearly to 2,000 feet on Ingleborough and east side of Whernside.

62. Cardamine hirsuta L. (Bab.).
Area: 63, 64, 65, Top. Bot. Range: 100—2,200 feet.
Native; rocks and old walls, mostly dry, or not so wet as the spots affected by the following species; rather rare. A. April-July.
First notice: Salt, 1800.
M. Greenfield valley; *J. Whitehead*.

N. Old walls at the Abbey, Knaresborough; *E. Hargrove* (1832); not in the Eds. of 1782 or 1809.
D. Near Sheffield; *Herb. Salt*; 'some of Mr. Salt's specimens seem starved *C. sylvatica*, others on a separate sheet look like *C. hirsuta*,' *W. W. Newbould, MS.*

63. Cardamine sylvatica Link. (*C. flexuosa* With.)
Area: 63, 64, 65, Top. Bot.
Native; wet woods, ditches, shady gills, etc., common. B. or P. May-August.
The first record for the present segregate, as distinct from *eu-hirsuta* Syme., is difficult to ascertain: probably 'C. sylvatica, Endcliffe Wood, *Geo. Harrison, MS.*' (1863) is the first, unless Mr. Newbould himself visited Endcliffe Wood and published the plant of that station prior to receiving Mr. Harrison's *MS.* Salt did not know it apart from No. 62.
Reported for all the districts.
Endcliffe Wood, Sheffield (1863); *W. W. Newbould*, '*sylvatica* only.'

64. Cardamine impatiens L.
C. impatiens vulgo Sium minus impatiens; Ger., Ray.
Area: 64, 65, Top. Bot. Sub-montane. Range: 600—1,200 feet.
Native; rocky woods, on limestone; rare, but locally abundant. A. June-July.
First record: Ray, 1677.
L. Douker Gill, Sedbergh; *G. Pinder*, 1853. Howgill.
R. Among the stones under the scars near Wharf, a village three miles distant from Settle; Ray Cat., ii, 51. [In Ed. i, p. 54 (1670), Ray wrote: 'I never yet met with this in England.'] *Hist.* 815 (1686), *Syn.* i, 104; etc. About Settle; *R. Teesdale* (1798). Linn Gill, abundant! Kelcowe below the greater cave, Cavehole Wood, about Giggleswick Scar Limekilns, Attermire Cove, etc.; *Windsor*.
A. Malham Cove; *W. Kirkby*! (Weetwood, *A. Shipley*, in *Miall*.)

[Cardamine bellidifolia L. Incognit. Error.
C. pumila Bellidis folio Alpina; G., 260. *Plantula Cardamines alterius æmula*; Clus. Park., 828. . . . In variis locis juxta Rippon ad monasterium ejus; *Merr. Pin.*, 20 (1666). In connection with this, Ray (Cat. Ang. ed. i (1670), p. 38, sub *Barbarea muralis*, 'Wall-cress with Daisie leaves,' has the following: 'Hanc olim habui pro Cardamine pumila Bellidis foleo Clus.,' etc. 'If Ray, Gerarde, and Parkinson mistook this plant, surely Merrett is not to be blamed for following the best botanists of his day'; *W. W. Newbould*. I am sorry it does not grow there; *Turritis hirsuta* appears to have been mistaken for it; *W. Brunton* in *O. B. G.* (1805).

65. Arabis Thaliana L.

Area: 63, 64, 65, Top. Bot. Range: 50—1,500 feet.

Native; walls, dry banks, and fields; infrequent over our area as a whole, but locally common on the limestones. A. April-May.

First record: Salt, 1800.

L. Dentdale. Ingleton. Sedbergh.

Y. Ripon (!) *H. H. Slater.*

N. About Knaresborough. Cowthorpe.

W. Ilkley. Harewood. Wetherby. Collingham.

A. Skipton. Awes Scar Rocks, Malham, etc. Hut Green, Kellington; *H. F. Parsons.* Knottingley; *G. Roberts.*

D. At Upperthorpe, Sheffield; *J. Salt* (1800). Crooks Moor, Sheffield; *A. Carr.* Conisborough.

[**Arabis stricta** Huds.

What Merrett had from Rippon cannot be determined; *Sir J. E. Smith, E. Bot.* t. 614. Was it not *A. hirsuta? W. W. Newbould.*]

66. Arabis hirsuta Brown.

Barbarea muralis; J. Bauh.; Ray. *Erysimo similis hirsuta planta*; P.

Area: 63, 64, 65, Top. Bot. Sub-xerophilous. Range: 100—1,500 feet.

Native; on rocks, dry banks, and old walls, common on, rare off, the limestones. B. End of May-August.

First record: Merrett, 1666.

L. Between Ingleborrow hill and Settle, among ye rocks; *Merr.* (1666) MS., 14. Many places!

R. Many places in and about Settle; *Jos. Woods, MS.* Ribble side, Kelcowe, etc.; *J. Windsor.* Many other places.

Y. Many places near Ripon; *W. Brunton* (1805).

N. Grimbald Crag pasture (!) *E. Hargrove* (1832).

W. Thorparch; *H. Baines* (1840). Old wall at Acomb, 1880; *W. Whitwell.*

A. In Craven everywhere; *Rev. W. Wood* (1805). Malham. Gordale. [Kirkstall Abbey, up to 1867; now gone.] Skipton; and 'The Vale,' Ledsham; *W. Whitwell.*

C. Stanley, near the canal; *G. Roberts.*

D. Askern. Smeaton Crags! Goole—casually; *H. F. Parsons.* Conisborough Castle (!) 1838; *Jos. Woods' MS.* in *Towns. O.B.G.*

67. Arabis perfoliata Lam. (= **Turritis glabra** L.).

Area: 63, 64, Top. Bot. Arenophilous. Range: low.

Native; road-side banks and hedges on sand or gravel, rare. A. June-July.

First record: Salt, 1787.

Y. Rail-bank hedge between Hutton Conyers and Melmerby. Cornfields near Ripon; *W. Brunton* (1800).

N. By road (right hand) between Green Hammerton and Boroughbridge. Between Copgrove and the river Nidd, in hedge bank.

D. Near Darfield about 1787; *Salt*; *Baines' Flo.* Ed.; *spn. G. Harrison Hb.* about 1862. Found still on bank by road to Darfield from Barnsley in 1884; *H. Johnson*! Rotherham district; *L. Langley* (1828); *N.B.G.* (1832).

T. Between Auckley and Finningley; *G. E. Smith.* Tickhill; *T. Birks, junr.* (1878).

68. Barbarea vulgaris Brown. '*Yellow-cress.*'

Area: 63, 64, Top. Bot. Range: 0—750 feet.

Native; stream sides, ditch banks, etc., common in the lower lands, rare among the hills above 400 feet. B. or P. May-July.

First record: Salt, 1800.

L. Near Sedbergh; *Pinder.*

R. Ribble side near Settle; *J. Windsor.* Summit of Castleberg rock; *Miall's Flora.* Wigglesworth. Sawley.

In the lower parts of the other districts very common.

Obs.—'A form with adpressed pods occurs occasionally'; *Baker, Supp.* I only know *B. stricta.*

69. Barbarea arcuata Reich.

Native; stream banks, very rare. P. June.

First (only) record: the Author.

A. Stream bank by Eddison's Farm, Adel bridge (1870). This plant has an elongated raceme in fruit, with slender arched spreading siliques and long seeds.

70. Barbarea stricta Andrz.

Area: Non. rec., Top. Bot. Range: 0—300 feet.

Colonist; spreading and increasing along the water-ways. P. June-August.

First record: Borrer, discov., 1842; Baker, Supp., 1854.

W. Ouse banks about York and down to Selby! Sherburn; *Y.N.U. Excurs.* (1883).

A. By Aire and canal below Leeds, towards Castleford. Methley; *O. A. Moore.*

C. Between Sheffield and Halifax; *Bab. Man.* By canal at Stanley, near Lofthouse (!); *G. Roberts.* Dewsbury and Thornhill; *P. F. Lee*!

D. Conisborough; *O. A. Moore.* Doncaster (!) and Barnby Dun; *Id.* By canal, Barnsley! Bolton ings, Barnsley; *E. Johnson.* Thorne.

Obs.—An increasing species of comparatively recent introduction, like *Potentilla norvegica*—probably first brought with timber for the coal-pits.

[**Barbarea præcox** Br. Alien.

On rubbish by waysides and occasionally in sandy fields amid grain or clover, but not permanent.

R. Several places near Settle; *J. Windsor.*

D. Field at Oughtibridge; *Geo. Harr. MS.* (?)]

71. Nasturtium officinale Br. *Water-cress.*

Area: 63, 64, Top. Bot. Range: 0—900 feet.

Native; watery places about springs and in clear slow streams, on gravel or sand, common. P. June-August.

First record: Bolton, 1775.

Distribution general—recorded from every district, although in some (e.g., Mersey) *not* common, and not ascending very high.

Obs.—The narrow-leafleted small form (*N. microphyllum* R.) is the product of too dry conditions—it bears the same relation to the type that *Valeriana Mikanii* does to *V. sambucifolia.* It is rare. *N. siifolium* I have not seen from West Yorkshire.

72. Nasturtium sylvestre Br.

Area: 63, 64, Top. Bot. Hygrophilous. Range: 50—450 feet.

Native; river-banks and ditches, infrequent, rather local. P. June-September.

First record: Salt, 1798.

R. Waste ground by Ribble, below Birkbeck's Weir, Settle; *J. Windsor.*

Y. Ure banks, near Ripon; *W. Brunton* (*O.B.G.*); *Baines.* Boroughbridge.

N. Ouse banks between Myton and York, at intervals.

A. Below Leeds; *Rev. W. Wood*; *Flo. Brit.*, ii. (1800) = Riverside, below Ferrybridge, 1806 (handwriting unknown); *Herb. Hailstone.* Aire bank, by osier-beds, below Brotherton (1869).

C. Canal side, Dewsbury; *P. F. Lee*!

D. Jordan Dam, four miles below Sheffield; *Herb. Salt* (1798). Rotherham, Catcliffe, Conisborough, Wath-upon-Dearne, Rotherham; *G. Harr. MS.* By the Doncaster Canal, Barnsley; *H. Johnson* (1885). Armin; *H. F. Parsons.* By river, near Thorne Quay.

73. Nasturtium palustre DC. (**terrestre** Br.).

Area: 63, 64, Top. Bot. Range: 0—300 feet.

Native: wet, waste, muddy, or clayey ground by ponds, etc.; infrequent. A. June-September.

First record: Salt, 1800.

Y. About Boroughbridge.

W. Ainsty; *G. Webster.* Healaugh; *J. Jackson.*

A. By Boothroyd's mill-reservoir, Meanwood Valley, Leeds! Near Bingley; *L. C. Miall.* Hawksworth; *W. West.*

C. Canal side, Dewsbury; *P. F. Lee.*

D. Near Sheffield; *Salt, Herb.*

74. Nasturtium amphibium Br.

Area: 63, 64, Top. Bot. Hygrophilous. Range: 0—300 feet.

Native; river-banks and pool-sides, with much the same distribution as *N. sylvestre*, but more *local*, and yet, where it does occur, more abundant, preferring muddy and peaty (rather than sandy) situations. P. July-August.

First record: Salt, 1800.

N. In the river Crimple near Newsome Bridge, Spofforth.

W. Askham ditches. Moss-Carrs ponds, between Wetherby and Bilton.

A. By Aire, from Methley to Castleford. Pond, Carlton Bridge, near Snaith; *H. F. Parsons.*

D. Near Doncaster; *Herb. Salt.* Rotherham, Ecclesfield; *G. Harr. MS.* Dikes on south of Thorne Moor. Goole Moor-dikes.

T. Treeton; *A. Carr.*

75. Armoracia rusticana Baumg. *Horse-radish.*

Raphanus rusticanus; Ger., Park. *Sylvestris sive Armoracia*; Ray.

Denizen, well established but never perfecting seed; ditches, field-corners, river-banks, infrequent. Often an obvious escape. P. June.

R. Ribble banks, Settle.

W. Wharfe banks, Otley to Tadcaster.

A. 'At D. Martinus Lister sponte natam copiose invenit.' Upon the banks of Skipton-beck and elsewhere in Bolland; *Ray, Cat.*, 261 (1670); *Hist.*, 818.

A. C. Aire and Calder banks about Castleford, etc.

D. Doncaster neighbourhood; *Dr. White* (1804). By Don, Sprotborough! *H. Johnson, A. Carr*, etc.

Obs.—Before How no attempt seems to have been made to distinguish wild from introduced plants.

76. Cochlearia officinalis L., **b. alpina** Wats. *Scurvy Cress.*

Cochlearia; Ray; Park. *Thlaspi aliud lunatis foliis sive Magorum Arabum*; Lob. P. 848; Merr. *C. grœnlandica* Sm.

Area: 63, 64, 65, Top. Bot. Range: 50—2,400 feet.

Native; mountain rocks and rillsides, and also (washed down and more luxuriant) on the banks of the larger streams; common in the river-basins of Lune, Ribble, Ure, Wharfe, and Aire, rarer in Calder and Don, unknown in the Trent tributary district. B. and P. Mid-May to August. I believe our West Yorkshire plant, notwithstanding the size it attains at low levels in rich soil, to be all of the alpine variety or form.

First record: Gerarde (1597).

L. Upon a great hill . . . called Ingleborrough Hill; *Gerarde, Hb.* (1597). Hath been also found growing upon Ingleborough Hill . . assured me by a worthy gentlewoman, Mrs. Thomazin Tunstall; *Park. Theat.*, 286 (1640). Near rills . . . upon Penyghent and Ingleborough Hills; *Ray, Cat.*, 75 (1670); *Hist.*, 822. On the top of Penigent Hill, in Yorkshire, under the stones; *Merr. Pin.*, 118 (1666). Upon Penigent; *Huds.*, ii, 283 (1762). Ingleborough; *Martyn* (1763). Helk's Wood; *Baines.*

R. By the river Ribble, near Settle; *Gough's Cam.* (1789), for the first time as *var. grœnlandica* expressly! Ribble banks, Settle Crags, etc.; *J. Windsor.*

N. Above Pateley Bridge; *W. Brunton.*

W. Woods near Bolton Abbey, and Wharfe Banks to Ilkley; *Rev. W. Wood* (1805). Thorparch! Hob Moor, York, *H. Baines*. There still, 1881; *W. W. Newbould*. Newton Kyme; *J. Emmet*.

A. Malham; *W. West*. By Aire, above Skipton.

D. By Don side, Conisborough; *Amos Carr*. And by Ouse, above Goole.

Obs.—Ray remarks, in *Gibs. Cam.* (1695), of this plant that 'upon Penigent and Ingleborough Hills . . . by reason of the coldness of the air it is so little that it hath been taken for a distinct species.' Merrett's plant could hardly have been aught else, as only it and *Draba verna* of the Cruciferæ occur on the summit of Penyghent Hill.

77. Draba verna L. *Whitlow Grass.*

Paronychia vulgaris; Ger.; Ray.

Area: 63, 64, 65, Top. Bot. Range: 0—2400 feet.

Native; rocks, roofs, wall-tops, and dry sandy fields, common. A. July-August.

Reported from all the districts, but very rare in that of Mersey, and decidedly scarce in those parts of Calder and Aire districts which overlie the coal formation and are thickly populated. Not in Bolton's Halifax Catalogue (1775); but given as 'common' by Salt in Flo. Sheff. (1800), which will most likely be the earliest record for the Riding.

VAR. **brachycarpa** Jord.

Area: 63, 64, 65, Top. Bot.

In the elevated gritstone districts commoner than the long-podded type. I have notes and records of it in four districts, Lune, Ribble, Wharfe, and Aire.

78. Draba muralis L.

Thlaspi veronicæ folio; Pk. *Bursa pastoris major loculo oblongo*; R.

Area: 64, 65, Top. Bot. Range: 600—1,600 feet.

Native; dry limestone rocks and walls, very rare, abundant locally. A. Late April to June.

First record: Ray, 1670.

(L. Near the Moorcock Inn, Hawes Junction, 1879, 'in a rocky gill on right of road going up towards Baugh Fell from a point at Garsdale Head a little to the west of the railway station;' *H. F. Parsons*.)

R. On limestone rocks about Dale Head, near Penyghent; *J. Windsor*. = Rocks near Peter Castle, Silverdale Head; *J. Tatham, MS.*

W. Arnber Scar, near Arncliffe in Littondale; *Curtis* (1782); *Gough's Cam.* (1789). Id. misprinted 'Amber Scar, Littledale'; *O.B.G.* Kilnsey Crag, eastern end; *Rev. W. Wood*, rep. *J. Windsor*.

A. In the way from Malham to the Cove, among the stones by the rivulet; *Martyn* (1763). About Malham, plentifully still! Awes Scar, Malham Tarn (!) *J. Windsor*.

Mr. Ray found this on the mountain sides in several parts of Craven; *Petiv., Phil. Trans.* (1711). Id., *Ray, Cat.*, i, 50; ii, 47 (1670); *Hist.*, 790 (1686).

79. Draba incana L.

Paronychia Gnaphalii facie; Merr. *Leucojum s. Lunaria*; Pluk.; Ray in Cam. *Lunaria contorta major*; Ray.

Area: 64, Top. Bot. Montane. Sub-xerophilous. Range: 1,850—200 feet.

First record: Merrett, 1666.

Native; on limestone rocks and banks among the hills; common on the scar limestones, rare off them.

L. Garsdale. Dentdale Crags. Ingleton. Lower crags of Whernside. At Clapdale, in Yorkshire . . . on the rocks; *Merr.* 90. Ingleborough; *Ray, Cat.*, 297 (1670); *Gibs. Cam.*; *Martyn*; etc.

R. Rocks about Settle; *Gough's Cam.* (1789); *J. Windsor* (!). Penyghent. Hinklehaugh (= Ryeloafhill); *Ray, Cat.* Moughton Fell. Rocks above the Ebbing and Flowing Well; *D. Turner* (1805).

W. Buckden. Arncliffe. Addingham; *L. C. Miall* [?] In an old limestone quarry by the Leeds Road, rather more than a mile from Tadcaster; *H. Ibbotson*.

A. Malham. Gordale (!) *Rev. W. Wood*; *O.B.G.*

Obs.—Small specimens are often well-nigh without cauline leaves, and there is also a variety or stunted form which has an untwisted pod, shorter and more tumid than the type, which I believe to have given rise to the reported occurrence of *D. rupestris*. I have also seen the straight-podded form labelled *muralis* in herbaria.

[**Draba rupestris** Br. Incognit.

'[64],' Top. Bot.

Ingleborough; *Herb. Hooker* (1810). 'The Ingleborough plant gathered by Sir Wm. Hooker, and referred to by Syme, E.B., i, is a small state of *D. incana*;' *Hook., Stud. Flo.*, p. 32.]

80. Alyssum calycinum L.

Area: not treated of in Top. Bot. Range: Low.

Colonist; constantly recurrent in clover and seed fields, but rarely in any quantity. A. July.

First record: Smith, 1845.

Y. Field near the Yore, Ripon; *Mrs. Berkeley* (1880).

W. Sicklinghall; *J. S. Wesley*. In a rough field half a mile to north-east of Sicklinghall.

D. Cantley, clover fields; *G. E. Smith* (1845).

Obs.—This class of Colonist appears to need a fallow of one or two years for its development.

[**Alyssum incanum** L.

Casual, hardly colonising—i.e., not permanent; clover and 'seed' fields, very rare. A. July.

W. Field on north side of road near Flint Mill, going from Thorparch to Walton, with *Anthemis tinctoria*, in plenty in 1870.]

[**Alyssum maritimum** Lam.

Alien—an outcast of gardens, on waste ground or rubbish heaps, rare. Seen under such circumstances at Meanwood (A. District), once in a sandy field at Acomb (W.), and on the waste ground by Thorne Quay (D.).]

81. Camelina sativa Crantz.

Casually, but fugitive, in flax fields, oat fields, and on waste ground, infrequent. A. July-August.

W. By the river just below the Flour Mill at Wetherby and Tadcaster.

A. Scarcroft, near the Hall. Pontefract; *H. Baines*. Methley (in clover); *G. Roberts*. [Meanwood, in rubbish by the Oil Mill; now gone.]

C. Stanley; *G. Roberts*. Old quarry, Sandal; *G. Roberts*.

D. Arable fields on Low Ellers farm, below Doncaster; *G. E. Smith*. Rotherham district; *L. Langley*. Near Rotherham; *H. Baines*. Goole; *H. F. Parsons*.

[**Neslia paniculata** Desv. Casual.

Clover field at Methley; *G. Roberts* (*Nat., Dec.* 1885). Field, Sandal; *P. F. Lee* (1886).]

82. Thlaspi arvense L.

Area: 63, 64, Top. Bot. Range: Low.

Colonist; arable fields on gravels or clays, infrequent. A. June-August.

First record: Baines, 1840.

R. (Old wall between Settle and the bridge; *H. Baines*.) Settle; *J. Bates* in *Miall's Flora*.

Y. Bishop Monckton; Givendale, Sharow (!); and road-side, near Littlethorpe; *H. H. Slater*.

W. Collingham, near the sand-quarries.

A. Garforth; *G. Roberts*. Near Drax; *H. F. Parsons*.

C. Fields between Salterhebble and Copley; *Miall's Flora*. Sandal; *G. Roberts* (1886).

D. Doncaster; *H. Johnson*. Between Reedness and Swinefleet; *H. F. Parsons*!

[**Thlaspi perfoliatum** L. Error.

'[64],' Top. Bot.

The York plant of Merrett, Nicolson in Camden, etc., was *T. alpestre*, the confusion arising from the Oxfordshire *Thlaspi* being regarded as identical with the York plant; both are given under the same heading of *Thlaspi perfoliatum minus* by Ray in Syn., Ed. iii, 305, 306. Ray has 'on most limestone pastures in Westmorland and Cumberland,' which Hudson (Flo. Ang., 282) mis-condenses into 'on the moist limestone pastures in Yorkshire, Westmorland, and Cumberland.']

83. Thlaspi alpestre L. var. **occitanum** Jord.

Thlaspi fol. globulariæ; Ray.

Area: 64, Top. Bot. Montane. Range: 1,600—500 feet.

First records: Merrett, 1666; Ray, 1670.

Native; elevated rocks and shale-banks, rare, locally abundant, closely associated with the spoil-heaps of the lead mines. P. May-July.

L. On the south side of Ingleborough; *Ray, Syn.*, Ed. iii, 305. On Ingleborough; *Woodward* in *O.B.G.* (1806).

R. On the pastures above the Ebbing and Flowing Well, a mile from Settle; *Ray, Cat.*, 296 (1670); etc. Found there, i.e., Stackhouse Scar, by J. Carr, in 1801; *J. Windsor, MS.*

W. Limestone pastures above Grassington; *J. S. Wesley* (1877).

A. At a place called Coppermine, on the left hand . . . from Settle to Malham, *Mr. Thornbeck*; *Blackst. Spec.*, 97 (1746); *Martyn* (1763). In many places between Settle and Malham; *Ray, Cat.*, ii, 284. Within half a mile of the Tarn; *W. Curtis* (1782). Upon all the heaps of earth that have been dug out of the old lead mines between Malham and Settle; *J. Richardson in litt.* (1737) *to Rauthmell*. Earthy banks and foot of walls in Malham village, and down by stream nearly to Airton.

[**Thlaspi minus, Clusii**; G. 268. *T. perfoliatum minus*; Ray.

In the pastures above the Ebbing and Flowing Well, two miles from Grislewick, in stony ground amongst the grass; *Merr., Pin.*, 118 (1666). This belongs here—the plant is *T. occitanum*, although the name given by Merrett, and quoted by Ray, is now usually given as an old synonym for *T. perfoliatum* L.]

84. Iberis amara L.

Area: '(64),' Top. Bot.

Casual; in 'seed' fields, and on railway banks, quite rare, and clearly *not* native to our area, as in the south of England. A. July.

R. [Waste ground below Birkbeck's Weir, Settle, doubtfully indigenous; *J. Windsor*.] Roadside between Giggleswick and High Ridge; *Id.*

W. Rough field, Sicklinghall; *J. S. Wesley*. Thorparch; *J. Emmet*.

C. Rail-bank, Bottom Boat, Stanley; *G. Roberts*!

(D. River side near Sheffield, etc.; *Miall's Flora*. I cannot find the authority.)

Obs.—Probably the plants reported as *I. amara* are not always that species.

85. Teesdalia nudicaulis Br.

Area: '63 Denny ms., 64,' Top. Bot. Range: 100—1,000 feet.?

Native; sandy fields, waste quarried ground, and on mountain slopes, rare. A. End of May-July.

First record: Salt, 1800.

L. How Gill, near Sedbergh; *J. Carr, MS.* (about 1815).
Y. South of Hutton Moor; *W. Brunton* (1805).
N. Grimbald Crag, Knaresborough, a very few plants (1870).
W. Sandy fields, etc., Hobmoor, near York; *W. Kirkby.* Ainsty; *G. Webster.*
[C. Near Mirfield, on the road from Halifax to Wakefield; *H. Baines.* Sought here in vain, 1870.]
D. Gravel pit near Whitley Bridge; *W. Fowler* and *H. F. Parsons*! East of Smeaton! Crook's Moor, Sheffield, plentiful; *Salt* and *Hb. Salt* (*Flo. Brit.*, ii, 693).

86. Hutchinsia petræa Br.
Lepidium petræum O. F.
Area: 64, 65, Top. Bot. Montane. Range: 1,600—1,000 feet.
Native; on limestone crags among the hills, rare, with a restricted range, although abundant enough locally. A. April.
First record: Hooker, 1824.
L. Middle Fells of Whernside, in both Chapel-le-Dale and Kingsdale. Ingleton Fells, above Southerscales; *R. Clapham.*
W. Lime scars of Langstrothdale, about Cam and Outershaw.
A. On the higher part of the farthest east cleft in Awes Scar, Malham Tarn; *W. J. Hooker*; *J. Windsor.* Towards the east end of a ridge by which one may pass over Malham Cove.

[Nasturtium montanum tenellis et dissectis foliis siliquosum.
Mountain Cresse, with a finely-cut leaf. Found by Th. Willisell on a broken mountain, about the midway between Settle and Ingleborough. Of this I have seen only some young tender plants; *Ray, Cat.*, Ed. i, 219. Not in Ed. ii. Ray manifestly did not consider this worth further notice. Was it not either the preceding species or *Cardamine impatiens*?]

[Hutchinsia alpina Br. Ambiguity.
Said to have been found on Ingleborrow; *Bab., Man.*, Ed. iii (1881). Ingleborough, Yorkshire, but very uncertain; *Wats.* in *Comp. Cyb. Brit.*, 481 (1870). Reported from Ingleborough; never verified; *Hooker, Student's Flora.* Can its seeds have been wilfully sown by 'the party of botanists, making a tour in the neighbourhood of Ingleborough,' of whom the Rev. W. T. Bree wrote in *Mag. Nat. Hist.*, iii, p. 150 (1830)? 'A list of the species attempted to be introduced was kept,' but only one of these was mentioned to Mr. Bree—*Scilla campanulata.* Rev. W. W. Newbould has written exhaustively on the matter in Seemann's Journ. of Bot., i, p. 359 (1863); and compare also Vol. iii, 92, and xviii, 375.]

87. Capsella Bursa-pastoris L. *Shepherd's Purse.*
Area: 63, 64, 65, Top. Bot. Range: 0—1,250 feet.
Native; waste ground at foot of walls, roadsides, etc., very common.
First notice: Bolton, 1775.
Distribution general—in every district, but so intimately associated with surfaces altered by human agency, that except about the mountain roads, and barns and dwellings, it is rare amongst the hills.

[Lepidium latifolium L. Extinct Casual.
Area: (64), Top. Bot. A species of maritime preferences.
'Yorkshire, *Dr. Richardson*'; *Sm., Eng. Flora* iii, 165. Rocks near the Church at Knaresborough; *Archd. Pierson* (1798). Plompton; *R. Teesdale* (1798). Rocks at Brompton, near Knaresborough; *Baines.* Never confirmed; Brompton a misprint for Plumpton. Knaresborough, upon the rock going down to the river Nidd down the (flax) factory steps; *E. Hargrove* (1782). Not there 1869-70-7. 'On a rock on the left going out of Kirkgate, Knaresborough, *Baines'* variation (1840).]

[Lepidium ruderale L. Casual.
Has been noticed once or twice on ballast at Goole (*H. F. Parsons*), and about Thorne Quay (!) and at Carr Pit, Huddersfield (*T. H. Bartlam*).]

[Lepidium sativum L. Alien.
A casual, on rubbish and manure heaps outcast from gardens; and once seen in a flax field near Wighill.]

88. Lepidium campestre Br.!
Area: 63, 64, Top. Bot. Range: 50—600 feet.
Native; fields, road- and rail-banks, infrequent and seldom in any abundance, mostly on sand or clay. B. July-August.
First record: Leyland in Baines, 1840.
R. About Settle; *J. Windsor.*
Y. Ripon. Not uncommon there; *H. H. Slater.*
W. Quarry near Tadcaster. Between Pool and Otley!
C. Upper Foot in Midgley, Halifax; *J. Leyland.* Luddenden Foot; *Eastwood.* Dalton; *C. P. Hobkirk.*
D. Askern! Doncaster neighbourhood; *White* (18[illegible]4). Machon bank, Sheffield, and Ecclesfield; *W. W. Newbould.* Banks of Rother at Woodhouse; *G. Harr., Hb.* Waste ground in Thorne village. Edge of Moors, Goole; *H. F. Parsons*!

89. Lepidium Smithii Hook.!
Area: 63, 64, Top. Bot. Range: 200—900 feet.
Native; rocky banks, waste spots, with rather a preference for dry calcareous soil, rare, but the commoner species among the hills. P. July-August.

First record: Carrington, 1862.
L. About Ingleton.
R. Near Giggleswick; *B. Carrington.*
W. Hedge bank, Copmanthorpe.
A. Skipton, by road to Grassington!
D. Between Pontefract and Smeaton Crags; *W. Todd*!

[Lepidium Draba L. Casual.
Fugitive, not naturalised, in fields and on railway-banks (ballast-brought), occasional.
Y. Field near Ripon, one plant (1870).
A. Clover fields, Brotherton.
C. Waste ground at Sandal; *G. Roberts.*
D. By the rail-bank near the Sprotborough quarries.]

90. Senebiera Coronopus Poir. (*Coronopus Ruellii* Gaert.)
Area: 63*, Top. Bot.
Native; waste sandy ground by way sides, very rare; one of the widely-distributed species unknown over by far the greater part of our area. It almost shows sub-maritime preferences with us. A. June-August.
First record: Salt, 1800.
Y. Cornfield, sparingly, Hewick Bridge; *H. H. Slater* (1880). Weedy patch, Mackershaw Wood; *H. H. Slater.*
W. Foot of walls, Holgate, near York; *H. Baines.* Acomb; *H. Ibbotson.*
A. Drax; *H. F. Parsons.* By Ouse, near Selby.
C. (Near Lofthouse? *G. Roberts*—the author has doubts.) Hirst's Mills, Huddersfield*; *T. H. Bartlam.*
D. Brought to Conisborough meeting of W. R. Consol. Nat. Soc. (1873) from 'Don Valley.' Goole and Marshland dist.; *H. F. Parsons.* About Brightside; *Salt, Flo.* (1800).

[Senebiera didyma Pers. Casual.
Area: 63, 64, Top. Bot.
Very rare; only found in one place. Goole (on waste ground where ballast is thrown); *H. F. Parsons*!]

VII. RESEDACEÆ.

91. Reseda lutea L.
Area: 63, 64, Top. Bot. Xerophilous. Range: 0—350 feet.
Native; limestone quarries, rough waste ground, cultivated field-borders, etc. Common upon but almost restricted to the Permian tract. B. June-July.

First record: Langley, 1828.
(R. Craven limestone; *L. C. Miall.* Seen by others? Not given as a Settle plant by *J. Windsor.*)
Y. (Rail-banks near Masham; *H. H. Slater, Ripon Flora.* Perhaps out of West Riding boundaries.) Ripon!
W. Wetherby. Tadcaster. Thorparch.
A. Brotherton and Knottingley quarries and fields. Pontefract; *T. W. Gissing.* Ferry Fryston; *Y. N. Excurs.* (1873). Skipton (Gargrave Road), 1884; *W. Whitwell.*
D. Rotherham (limestone) dist.; *L. Langley.* Smeaton (!) *W. West.* Conisborough. Sprotborough quarries. Askern; *Geo. Harr., Hb.*
T. About Firbeck, Letwell and Kiveton.

92. Reseda Luteola L. *Woad. Weld.* '*Yellow-weed.*'
Area: 63, 64, Top. Bot. Range: 0—600 feet.
Native; rough waste ground, rail-banks, limestone quarries, etc., common, but sparsely distributed on calcareous soils, rare on the coal measures. B. June-August.
First record: Bolton, 1775.
R. Kelcowe, etc., Settle; *L. C. Miall.*
L. Rivulet side below Clapham bridge; *J. Windsor.* Rail-siding, Ingleton.
Y. Quarry moor, Wicliffe lime-hills, etc., near Ripon (!); *H. H. Slater.* By the river Skell; *Miss M. Plues.*
N. Knaresborough; *E. Hargrove* (1832). Plumpton *Id.* (1809).
W. Thorparch (!); *J. Emmet.* Ainsty; *G. Webster.*
A. Brotherton. Ferrybridge. Garforth. South Milford.
C. Dirtcar Green and Sharleston Green, 1791; *D. Dixon.* Colne bank, Huddersfield; *S. King.* Halifax district; *Bolton, Cat.* Stanley; *G. Roberts.*
D. Askern. Aldwark and Levitt Hagg, Conisborough (!); *G. Harr. MS.* 'Common in uncultivated places'; *Salt, Flo.* 1800). But Salt's Sheffield Flora area was a wide one. Only noticed on the Doncaster side of Sheffield; *W. W. Newbould.*
T. Anston; *Geo. Harrison, MS.* Kiveton; *A. Carr.*

[Reseda alba L. Alien.
Hardly naturalised—always the white-petalled but un-ornamental plant occasionally seen in cottage gardens—and in all its stations doubtless little more than an escape or outcast from gardens.
R. Near Settle; *A. Irvine, H.B.P.*, 729.
N. On rocks on the cliff at Knaresborough (!); *W. Kirkby.* Seemed pretty well established here in 1869-72.
A. About Milford and Sherburn.
C. Oatfield at Stanley; *G. Roberts.* This was *R. odorata* L.
D. Near Balby, Doncaster.]

VIII. CISTACEÆ.

93. Helianthemum vulgare Gaert. *Rock-rose.* '*Sun-flower.*'

Area: 63, 64, 65, Top. Bot. Xerophilous. Range: 50—1,900 feet.

Native; dry banks and rocks, common on the limestones, very rare on the coal measures and red sandstone. P. June-August.

First record: Salt, 1800.

L. Helks Wood, Ingleton; *H. Baines.* All about the limestone scars of Kingsdale and Chapel-le-dale.

R. Kelcowe, Lord's Wood, etc. Settle; *J. Windsor.* Many places.

N. Grimbald Crag pastures! Knaresborough; *E. Hargrove* (1809).

W. Arncliffe. Grassington scars. On sandy alluvium by river side above Bolton Bridge! Linton, and between Collingham and Wetherby. Etchell Crags, Scarcroft. Thorparch.

A. Roundhay lime-hills (nearest locality to Leeds). Malham Cove, etc. Ledsham, etc.

D. Rotherham district; *L. Langley* (1828). Conisborough (!); *J. Salt* (1800). Roman ridge, Doncaster; *W. W. Newbould.*

T. Maltby. Roche Abbey (!); *Geo. Harrison, MS.* Kiveton. Anston; *A. Carr.*

94. Helianthemum canum Dunal.

Area: '64, Lees,' Top. Bot. Xerophilous. 'Range: 400—450 yards, nearly;' *L. C. Miall.*

Native; mural mountain limestone scars, very rare. P. July.

Earliest notice: Miall, 1862.

A. Malham, near the Cove; *Miall* and *Carrington's Flo.*, p. 6 (1862). 'Malham Cove!' *Miall* in *Nat.*, i, 212 (1865). Given also—name only—in a list of species illustrative of the 'Flora of Malham,' at p. 14 of an appendix to a reprint of Whitaker's 'History of Craven,' entitled 'The Geology, Nat. Hist., and Pre-Hist. Antiquities of Craven in Yorkshire,' by L. C. Miall (1878). I believe the record has never been confirmed by any botanist beyond myself. The plant grows on the step-like ledges of rock near the top of the face of the Cove, with a S.W. aspect; but reaching it is somewhat risky for all but the cool-headed.

(D. Conisborough limestone district; *Y. N. Excurs. Report* (1873) in *Y. N. Recorder*, p. 205. An error of designation.)

IX. VIOLACEÆ.

95. Viola palustris L. *Bog Violet.*

Viola martia palustris, sive V. montana; Gesn.; How. *V. rubra striata Eboracensis*; *Park.* 755.

Area: 63, 64, 65, Top. Bot. Hygrophilous. Range: 0—1,900 feet.

Native; swamps and wet hilly pastures, preferring hilly districts, yet generally distributed, and with a wide range. P. May-June.

First record: Parkinson, 1640.

M. Greenfield valley; *J. Whitehead.*

L. Common about Bentham, Ingleton, Kingsdale, Dentdale, Sedbergh, Howgill, Ingleborough, Helwith Moss, Austwick; *R. Clapham.*

R. In Giggleswick tarn; *W. Curtis* (1782); *Gough's Cam.* (1789), etc. Cocket Moss, Ryeloaf Hill, etc., about Settle; *J. Windsor.* Hodder Valley, Ribblehead, etc.! Horton; *W. Howson.*

Y. Hackfall, Kirby Malzeard, etc.

N. Wood under Brimham rocks; *J. S. Wesley.* Aketon bog field. Plumpton. Birk Crag, Harrogate; *J. Emmet.*

W. Askham Bog. Bramham, 'Black fen' and West Woods. Ilkley; *W. West.* Barden. Uncommon in the cultivated districts over the Permian strata.

A. Malham Moor. Esholt, Yeadon, Bingley; *W. West.* Alwoodley; *T. Scatcherd.* Adel Bog. Sharpah, Skipton; *W. Whitwell.*

C. Oakenshaw Moss and Roadeshall Woods; *Rich.*; *Ray, Syn.*, iii (1724). Lofthouse; *G. Roberts.* Norland Moor, etc., Halifax; *J. Bolton* (1775). Birkby, Huddersfield; *C. P. Hobkirk.*

D. Master Stonehouse, a reverend minister of Darfield, . . assured me he had found a kinde of Wilde Violet neere unto his habitation, whose leaves were rounder and thinner than of others, and the flowers reddish with sadder veines therein; *Park., Theat.*, 756. Found by Mr. S. in Yorkeshire; *How, Phyt.*, 129; *Ray, Hist.*, 105 (1688). Ackworth; *T. W. Gissing.* Rawcliffe Rabbit Hills; *H. F. Parsons.*

T. Thorne Moor, south side.

96. Viola odorata L. *Sweet Violet.*

Area: 63, 64, Top. Bot. Range: 0—600 feet.

Native; hedge-banks in woods and by road-sides, common; but I believe that in *many* of its stations it is *not* indigenous, although clearly so in the warmer parts of the area. P. Late March to May.

First record: Bolton, 1775.

L. Between Ingleton and Bentham.

R. Near Catterall Hall, Langcliffe Hall, etc.; *J. Windsor.* Near Settle bridge; *W. Howson.*

Y. Common about Ripon; *H. H. Slater.*

N. Scotton and Bilton, Knaresborough, Cowthorpe.

W. Grassington. Bolton Abbey. Collingham. Wilderness banks, Wetherby (!); *J. Jackson.* Thorparch (!); *J. Emmet.* Aberford. Stutton. Bolton Percy; *W. Whitwell.*

A. Belbank, Bingley; *L. C. Miall.* [Saltaire; *Id.*] Allerton, Ledstone, Lingwell Gate; *G. Roberts.* Broughton, near Skipton; *W. Whitwell.*

C. Wakefield district, frequent; *T. W. Gissing.* Chevet; *Id.*

D. Ackworth. Smeaton! Conisborough. Town Cross Lane, between Sheffield and Grimethorpe; *J. Salt.*

T. Common about Roche Abbey (!), Maltby, Firbeck, and near Worksop; *A. Carr.*

Obs.—The white form not uncommon, the lilac one rare. I know no West Yorkshire plant in regard to which it is more difficult to decide between indigenous and non-wild stations.

97. Viola hirta L.

Area: 63, 64, Top. Bot. Sub-xerophilous. Range: 50—900 feet.

Native; woods and rough bushy banks, common on the limestones, rare off them. P. April-June.

First record: Knowlton, 1805.

L. Lower Scars of Kingsdale and Clapdale. Feizor Wood; *R. Clapham.*

R. Kilcove, Major Springs, etc., near Settle; *J. Windsor.*

Y. Copgrove; *J. Dalton*; *O.B.G.* Mackershaw Wood, etc. (!) Ripon, *Mr. Brunton*; *Id.*

N. Knaresborough, many places.

W. Norwood bottoms, Thorner. Thorparch; *Knowlton*; *O.B.G.* (1805) Wetherby. Collingham! Aberford. Common all over the Permian tract in this district.

A. Malham Cove; *J. Windsor, MS.* Gordale (!); *W. West.* Kippax.

(C. Near Halifax; *L. C. Miall*).

D. Smeaton Crags, near Norton. Wentbridge; *T. W. Gissing.* Conisborough. Doncaster. Warmsworth.

T. Roche Abbey. Maltby Woods.

Var. **sciaphila** Koch (a cordate-leaved glabrous *hirta*). Very rare; wood by road between Aberford and Barwick, June 1871.

Hyb. **permixta** Jord. With both parents only, on banks under beeches between Barwick and Aberford! Some individuals assignable to either species.

Obs.—This species differs much in hairiness, in relative length of petiole to leaf; and varieties with narrow, pale petals are not very rare in dry, barren, open places, but *calcarea* Bab. I have not seen.

98. Viola sylvatica Fries. (a. *riviniana* Reich.)

Area: 63, 64, 65, Top. Bot. Range: 0—1,900 feet.

Native; woods, rocks, and hedgebanks, very common. P. March-July.

First record: Bolton, 1775.

Distribution general; recorded from every district.

Very common about Sheffield; *Salt, Flora* (1800); sub nom. '*canina*,' of course.

Obs.—Small compact forms of this with large yellow-spurred flowers, such as occur in dry open spots, have been frequently mis-reported as *eu-canina*, through the confusion brought about by the name *flavicornis.*

Var. **Reichenbachiana** Bor.

Area: 63, 64, Top. Bot. Range: 100—600 feet.

Native; damp, light, or marly soil in woods on both the limestones, rare, locally abundant. P. May-June.

First record: G. E. Smith, 1840?

Y. Wood, Mackershaw, near Nunwick, etc., always in *shady*, not in open places. The reverse of this is stated by H. H. Slater in his Ripon Flora.

W. Sweep Hill Woods, Wetherby! Thorparch. Aberford.

A. Beech Woods at Ledsham, and Parlington Park!

T. Maltby Wood, Kingswood. Cantley (!) *G. E. Smith, MS.*

99. Viola canina Bab.

Area: 63 Leyland spn., 64, Top. Bot. Range: 50—250 feet.

Native; moist, sandy, and peaty ground; very rare. P. April-June.

First record: Lees, 1871.

Y. Wet rushy field near Blois Hall, Copt Hewick (!) *H. H. Slater.*

W. Among furze, Askham bog, a single tuft (1871). Bit of boggy ground outside wall of White Car Wood, S.W. corner of Bramham Park, a few tufts (1880-1).

A. Morton bog near Hambleton (!) *W. Todd.* (Heaton Woods and wood near Luddenden; *C. Eastwood*; sub nom. *flavicornis*, but I feel certain some small form of *Riviniana* was meant).

(C. Near Halifax, frequent; *Baines*; sub '*flavicornis*,' but neither this nor *lactea* really occur there. *Forster's flavicornis* meant?)

D. 'Thorne Waste'; *Jos. Blackburn*; confirmed by *H. F. Parsons.* Camblesforth and Drax; *H. F. Parsons* (1876). ('Cantley'; G. E. Smith's handwriting, *Herb. Brit. Mus.*; plant *Reichenbachiana*?)

100. Viola lactea Sm.

Area: '63 Parsons,' Top. Bot. Range: Low.

Native; moist, sandy, furzy, or turfy ground, very rare. P. May-July.

First record: Backhouse in Baines, 1840.

D. Rawcliffe 'Rabbit Hills'; *H. F. Parsons*, teste *H. C. Watson* (1877). [Very like the Cornish *intermedia*—receding from *eu-lactea* in the direction of *flavicornis* Sm.]

T. Thorne Moor, *J. Backhouse*; *Baines.*

[Viola stagnina Kit.

[63] Top. Bot.

'I am informed by my friend, James Backhouse, that the Thorne Moor violet (Flora, p. 15) may probably belong to this species. It merits further investigation'; *Baker, Supp. to Baines' Flora.* I think not, although I have never myself seen it there *in situ*: a scrappy specimen Joseph Blackburn, of Leeds, gave me, as gathered there, was certainly either *eu-canina* Bab. or the intermediate form of *lactea* Watson discriminated—in my opinion the former. I consider it very unlikely indeed that the true *stagnina* occurs.]

101. Viola tricolor L. *Field Pansy.*
Jacea, sive herba Trinitatis; Merr.
Area: 63, 64, Top. Bot. Range: 0—600 feet.
Colonist; cultivated fields, as a weed in gardens, etc. Common in tilled ground in the lowlands, yet rather uncertain in its stations from one year to another, and rare among the hills, where its perennial form, *lutea*, takes its place. B. or A. May-July.
First record: Merrett, 1666.

Y. Ripon, river-sides and waste ground, 'abundant'; *H. H. Slater, Flo. Ripon* [?]. The var. not referred to, for that is separately mentioned, but I have not found *eu-tricolor* in such stations at all commonly.
A. This mightily beautifieth the border of an hedge in our own garden at Methley, where it groweth spontaneously, *Mr. Witham*; *Merr.*, 65. Fields at Newlay, Cookridge, etc. Burn and Brayton; *H. F. Parsons.*
In all the other districts, but not permanent in its localities.

VAR. **arvensis** Murr.
Area; 63, 64, Top. Bot. Range: 0—900 feet.
Colonist; tilled fields, very common, reported from all the districts, and reaching as high as field cultivation.
First record: uncertain.

102. Viola lutea Huds.! *Mountain Pansy.*
V. montana lutea grandiflora; Ray.
Area: 63, 64, 65, Top. Bot. Montane. Range: 450—1,800 feet.
Native; hilly pastures, confined to the N. W. half of Riding, but there frequent. P. June-July.
First records: Lawson, 1718; Ray, 1724.
In montosis sept. v. q. agri . . Eboracensis; *Ray, Cat.*, 318 (1670).

L. Ingleton Fells. Between Clapham and Bowland Knotts.
R. Settle, in the way to Malham; *Ray, Syn.* iii (1724); *Martyn.* About Attermire cliff, near Settle; *W. Curtis* (1782); rep. *Gough's Cam.* (1789). Mountain pastures, Settle (!); *J. Windsor.* Penyghent; *W. Whitwell.*
Y. Sawley Moor and Kirby Hill Moor; *W. Brunton, O.B.G.*
W. Mountain pastures of Langstrothdale. Kilnsey district; *W. Curtis.* Grassington. Upper part of Washburn Valley; *Y. N. U. Excurs.* (1883).
[C. Hallifax; *Ray, Syn.* iii, 365; *Martyn* (1763). Warley, Illingworth, Halifax; *J. Bolton.*]
A. By Malham Cove; *T. Lawson*; *Derh. litt.* 223 (1718). High pastures about Malham (!) *Rev. W. Wood*; *O.B.G.* (1803).

X. DROSERACEÆ.

103. Drosera rotundifolia L. *Round-leaved Sundew.* '*Moorgloom.*'
Ros solis folio rotundo; Ray.
Area: 63, 64, 65, Top. Bot. Range: 0—1,700 feet.
Native; peat bogs and Sphagnum swamps, common in the western half, rare in the eastern half of the Riding. P. July-September.
First record: Dixon, 1799.
'In Yorkshire it is called Mŏrgloom;' *Petiv. ms.* in *Ray, Cat.* ii, 254 (1677), in his copy now in Brit. Mus. (Vide *Primula farinosa*).

M. Greenfield Valley bogs; *J. Whitehead.*
L. Baugh Fell, Howgill. Ingleborough plateau. Bleamoor. Helwith moss; *W. Howson.*
Y. Sawley Moor.
N. Ramsgill; *J. S. Wesley.* Hookstone Crags, Harrogate Moor (!); *E. Hargrove* (1809).
W. Rombalds Moor, Ilkley. Fewston. Rare in the cultivated districts. Not yet found in Askham Bogs; *B. B. Le Tall.*
A. Malham Tarn Bog. Baildon Moor; *L. C. Miall.* Adel Bog, near Leeds.
C. Growing in the shrogs near the Lodge, on the Wakefield side of Patterdale, near Kirkthorpe, 1799; *D. Dixon.* Todmorden Moors! *A. Stansfield.* Blackstone Edge; *J. Eastwood.* Norland Moor; *L. C. Miall.* Holmfirth Moorlands; *C. P. Hobkirk.*
D. Doncaster; *White* (1804). Potteric Carrs; *Appleby* (1831). Bog near Cliff Rocher; *J. Salt* (1800). Redmires Bogs, Sheffield; *W. W. Newbould.* Goole Moors. Thorne Waste.

104. Drosera anglica Huds.!
Rorella longifolia perennis; Ray.
Area: 63, Top. Bot. Range: 0—1,600 feet.
Native; spongy bogs, rare, perhaps extinct at its one or two elevated stations. P. July-August.
First record: Ray, 1688.

C. Top of Warley Moor, and in plenty along with the round-leaved sort on Blackstone Edge, about a mile from the road on the right hand, going from Halifax to Rochdale; *Bolton, Cat.* (1775). Robin Hood bed, Blackstone Edge, formerly, failed to find it two years ago; *A. Stansfield* (1877).
D. & T. Between Doncaster and Bawtry, by *Tho. Willisel*; *Ray, Hist.*, 1100 (1688); *Syn.*, 3, 356 (1724). Potteric Carrs; *S. Appleby* (1831). Peat bogs near Doncaster, *Rev. H. Davies*; *O.B.G.* Thorne Waste (*Jos. Woods, MS.*) and Goole Moor!

HYB. VAR. **obovata.**
D. Thorne Moor, *Mr. J. Hardy*; *W. W. Newbould, MS.*

105. Drosera intermedia Hayne.!
Area: '63 [64],' Top. Bot. Range: under 50 feet.
Native; peaty moors at low levels only, very rare, but locally abundant. P. July-August.
First record: Salt, 1800.

N. (Harrogate; *W. Brunton* in *O.B.G.*; rep. in *Miall, Flo.* Never confirmed.) [Staveley Carrs; *Archd. Pierson, O.B.G.* Now gone through drainage of Carrs.]
(W. Askham bog. An error through admixture of specimens.)
(C. Blackstone Edge, *C. Eastwood*; *L. C. Miall.* An error.)
D. Between Doncaster and Bawtry; *Ray*; *O.B.G.* Turner refers Ray's record quoted above under *D. anglica* to this species. Both occur, but there was some confusion also. Near Doncaster; *White* (1804). Mossy bog near Bawtry; *J. Salt* (1800). Thorne Waste (!) fine and abundant; *J. Backhouse*; *Herb. Kew.*
T. Goole Moor (!); *H. F. Parsons.*

XI. POLYGALACEÆ.

106. Polygala vulgaris L. *Milkwort.*
Area: 63, 64, Top. Bot. Range: 50—500 feet.
Native; dry pastures and banks, mainly on calcareous soil, infrequent. P. May-August.
First notice: for the true segregate, uncertain.

R. Kelcowe, etc., Settle; *J. Windsor.*
N. Knaresborough (!); *E. Hargrove* (1832); not in 1809 Ed.
W. Wetherby! Thorparch, Jackdaw Crag, Stutton, etc.
A. Fields between Scarcroft village and Brandon. Ferrybridge district; *Y. N. Excurs.* (1873).
D. Smeaton Crags. Conisborough.
T. Roche Abbey. Kiveton; *A. Carr*!

107. Polygala oxyptera Reich.
Area: 63, 64, Top. Bot. Range: 0—300 feet.
Native; dry fields, with a preference for sand, very rare. P. June-August.
First record: Lees, 1876.

N. Sandy pasture field by Nidd, below Knaresborough, 1877; *T. J. Foggitt*!
W. Marked for Ainsty by *G. Webster.*
T. Bank between Sandbeck Park and Firbeck (1876).

108. Polygala depressa Wender. *Milkwort.*
Area: 63, 64, 65, Top. Bot. Range: 0—1,900 feet.
Native; hilly pastures, heaths, etc.; very common among the hills, rarer in the cultivated districts. P. May-September.
First notice: Bolton, 1775, but sub nom. *vulgaris* of course.

M. Greenfield valley; *J. Whitehead.*

L. Common on moory ground.
R. Common.
Y. Kirby Malzeard!
N. Pateley Bridge, Harrogate, etc.—the common form.
W. Ilkley, Rombalds Moor slopes. Washburn valley. Askham bog; *W. Whitwell.*
A. Malham Moor. Skipton.
C. Skircoat and Warley Moor (!) *J. Bolton* (1775).
D. Common on heaths and pastures; *J. Salt* (1800).
No record for Trent district, but it doubtless occurs.

XII. CARYOPHYLLACEÆ.

[**Dianthus Armeria** L. Extinct Casual.
Area: '63, Newb. ms.,' Top. Bot.
I cannot accept this species as a permanent integer in our West Yorkshire flora.
D. Brightside, Sheffield, *Mr. Bohler*; *W. W. Newbould, MS.* in *Baines Flora.*]

109. Dianthus deltoides L.
Area: '63, 64 Biden sp.,' Top. Bot. Range: 100—200 feet.
Native; sandy, rocky, or grassy ground, very rare. P. July-August.
First record: G. E. Smith, 1840.

Y. Fountains Abbey, *Mr. Winch*; *With. Arr.*, Ed. 7 (1830). This doubtless referred to the Fringed Pink we now refer to *D. plumarius.*
[A. Meanwood (1854), *Biden.* 'VAR. *glaucus.*' An escape.]
D. In turf about Cantley vicarage; *G. E. Smith.* [Dry gravelly bank east of Sprotborough station, 1871. Perhaps here brought with rail-ballast from native stations to the east; not there in 1881.]

110. Dianthus plumarius L. *Clove Pink.*
Denizen: old walls, well established in one place. P. July-August.
Y. Fountains Abbey walls, abundant.
[W. On a rock in Thorparch Wood; *O. A. Moore.* Destroyed; *J. Emmet.*]
T. Tickhill Castle; *Herb. Tofield.*

[**Dianthus Caryophyllus** L. Incognit.
The Thorparch, Fountains Abbey, and Studley Woods plants of the *O.B.G.* and *Baines' Flora* were all *D. plumarius.*]

[**Dianthus cæsius** Sm. *Cheddar Pink.* Alien.
Window-ledges of Nine Altars Chapel, Fountains Abbey, in 1870. No doubt planted.]

[**Dianthus barbatus** L. *Sweet Williams.* Alien.
Only where deliberately planted, and now extinct I believe.
Y. On a marle bank in Studley Woods; *W. Brunton*; *O.B.G.* (1805).
W. In Thorparch Woods; *S. Hailstone.*]

111. Saponaria officinalis L. *Soapwort.*
Area: 63*, 64*, Top. Bot. Range: 0—450 feet.
Denizen, well established; river banks, rare, but locally abundant. P. August-September.
First record: Bolton, 1775.

L. Austwick; *Miall's Flora.*
R. Giggleswick, near the further end of Roper's Walk; *J. Windsor, MS.* in *Towns. O.B.G.* School yard, and small field by Mitchell Lane; *J. Windsor, Flora.*
N. Nidd side above Goldsborough Mill (!) *E. Hargrove.* In plenty on sandy north bank of Nidd by foot bridge over river near Cowthorpe Oak. Between Scriven and Knaresborough. Ouse banks below Myton (washed down from higher up).
W. By Wharfe, Collingham!
A. By Aire, near Castleford. Near Snaith, *Rev. W. Wood*; *O.B.G.* (1805). River side between Shipley and Esholt, Kirkstall; *Miall.*
C. [Hepton Bridge, Ripponden, etc.; *J. Bolton, Cat.* (1775). Now gone?] Bank above river Calder, near Mirfield; *C. P. Hobkirk* (1886).
D. Doncaster neighbourhood; *White* (1804). Rotherham; *L. Langley.* By Don, Conisborough. Banks of the river below Sheffield; *J. Salt* (1800). Aldwarke; *Geo. Harr.*

112. Saponaria Vaccaria L.
Casual; rail-banks, sandy 'seed' fields, etc.; occasional, but seems to be increasing.

W. Railway bank, Wetherby (!) *J. S. Wesley* (1879).
A. Methley, in clover, frequent; *G. Roberts.*
C. Sandal (waste ground and in a field); *G. Roberts*! Lofthouse, one plant; *G. Roberts.*

113. Silene inflata Sm. (*S. Cucubalus* Wib.).
Papaver spumeum sive Behen; Park.; How. *Cucubalus Behen* L.
Area: 63, 64, Top. Bot. Range: 0—700 feet.
Native; dry banks and waste ground, abundant on the calcareous Permian strata, common on grits and sands, rare in the higher lands. P. June-August.
First records: How, 1650; Bolton, Cat., 1775.
Distribution general; recorded from every district.
In Womersley field, Mr. Stonehouse; *How, Phyt.*, 87 (A *var.* 'foliis auratis elegans').

VAR. **puberula** Jord.
Common about Ripon, Wetherby, and other places, where it is the prevailing form; on *sandstone* rare. Dawson Turner first discriminated this variety, but gave no name to it; *O.B.G.*, 432.

114. Silene maritima With.
Area: '64 Tatham,' Top. Bot. Range: 800—1,000 feet.
Native; mural limestone scars; very rare. P. June-July.
First record: H. Baines, 1840.

L. (Whernside; *Windsor*, 'report.') On an almost inaccessible limestone cliff above the village of Wharf, 5 miles N.W. of Settle; *Baines* (1840) = Moughton Fell Scar, south side (!) *Howson, Buxton* et aliis. (Kilnsey Crag, reported; *Windsor's Flora.*)

Obs.—Moughton, the sole station in the Riding of this maritime species, is, according to geologists, an ancient sea-cliff, against which the waves at one time broke.

Silene conica L. Casual.
Occurred on waste ground near Hirst's Corn Mill, Huddersfield (*J. Tindall, Yorks. Nat. Rec.*, 128, 1873), out-thrown with screenings of grain. Just possibly the plant was really *S. conoidea.*]

[**Silene conoidea** L. Casual.
An Alien introduced with seed.
C. Field, Three Houses, Sandal; *G. Roberts*!]

115. Silene anglica L.
Area: 63, 64, Top. Bot. Range: Low.
Colonist; cornfields and dry sandy ground, rare; but, perhaps, not seldom overlooked. A. July-August.

N. Cornfield near Knaresborough!
A. Clover-field near Ferry-Fryston.
C. [Casual—'dried up bed of a weir, Luddenden, one plant'; *S. King*; a flour mill introduction probably.] With *S. conica* near Hirst's Corn Mill, Huddersfield; *J. Tindall* (1873). Disused quarry at Sandal; *G. Roberts*!—the type *gallica.*
D. Cantley, Doncaster; *G. E. Smith*!

116. Silene nutans L.
Area: 64, Top. Bot. Alt.: 300 feet.
Extinct Native; Casual in its latest discovered station; very rare. P. July-August.

[N. In the rocks on the Abbey Plain; *E. Hargrove, Hist.* (Ed. 3, 1782). Rocks about Knaresborough; *Archdeacon Pierson*; *O.B.G.* (1805). Shelvings of the rock near the Rock House; *E. Hargrove* (1832). Extinct.]
W. Bramham Park, 'dry grassy knoll near furze-bushes, not far from the foot-road across the park,' very sparingly; *A. Crebbin* and *W. West* (1879)! Perhaps a casual here, though how introduced into unbroken turf seems a mystery.

M

117. Silene noctiflora L.
Area: 63, 64, Top. Bot. Range: 50—300 feet.
Colonist; corn and other tilled fields, common on the Permian tract, much rarer on the sandstones. A. June-August.

Y. Near Ripon, in plenty; *W. Brunton.* Copgrove; *J. Dalton*; *O.B.G.* Aldborough!
N. Near Knaresborough and Goldsborough.
W. In cornfields near Wetherby (!) *Gough's Cam.* (1789); *Sir T. Frankland, O.B.G.* Thorparch; *W. Wood* in *O.B.G.* Frequent still about Wetherby (!) and Bramham; *J. Emmet.*
A. Milford.
C. Lofthouse, casually, grain-brought; *G. Roberts.* Netherton, Dewsbury*; *P. F. Lee*!
D. Doncaster; *L. C. Miall.* [Stainton; *G. E. Sm.* in *Miall.* This is in North Riding.]

[**Silene Armeria** L. Alien.
A garden escape.
R. River side near Settle; *J. Tatham.*]

[**Gypsophila porrigens** Boiss. Casual.
Old quarry at Sandal; *G. Roberts*! (1885). A curious Caryophyll, with ovate, hairy, opposite leaves like a *Lychnis*, and a dichotomous panicle of small flowers, with long, wavy hair-like pedicels.]

118. Lychnis vespertina Sibth.
Area: 63, 64, Top. Bot. Range: 0—450 feet.
Native; fields and dry banks, on sandy or calcareous soil, but rare beyond the eastern half of the Riding. B. Late-June to September.
First record: Baines, Flora, 1840?

R. Settle, occasionally; *J. Windsor.*
Y. Ripon, common; *H. H. Slater.*
W. Wetherby (!) and Thorparch, common; *J. Emmet*! Ainsty; *G. Webster.*
A. Bradford valley; *L. C. Miall.* Airedale generally, frequent! Common about Garforth, Castleford, etc.
C. Mirfield. Wakefield; *T. W. Gissing.* Lofthouse; *G. Roberts.*
D. Askern. Thorne (on sand). Goole.
T. Letwell. Waleswood; *A. Carr.*

Obs.—The form with pale-pink flowers (Annual?) is to be met with occasionally among clover or on stubbles.

119. Lychnis diurna Sibth. '*Wake-Robin.*'
Area: 63, 64, 65, Top. Bot. Range: 0—1,250 feet.
Native; woods, banks, hedges, very common. P. May-July.
Distribution general—reported from every district.

120. Lychnis Flos-cuculi L. '*Ragged Robin.*' '*Thunder-flower.*'
Area: 63, 64, 65, Top. Bot. Range: 0—1,300 feet.
Native; marshy meadows and pastures, stream-sides, very common, and with a rather wide range. P. Mid-June to August.
First record: How, 1650.
Distribution general—found in every district.
['ARMORARIA FLORE PLENO. Double-flowered wild Willians, in Netherwood Hall grounds, in Darfield Parish, *Mr. Stonehouse*'; *How, Phyt.*, ii (1650); Id., *Merrett, Pin.*, p. ii (1666). 'A chance form of *Lychnis Flos-cuculi.* See *Sm., Eng. Flo.*, ii, p. 327'; *C. C. Babington.*
['ARMERI PRATENSIS Ger. FLORE ALBO . . by Penigant; *T. Lawson* to *Ray* (1688) in *Derham, litt.*, 213 (1718).' *Dianthus Armeria*, fide *C. C. Babington*; *W. W. Newbould, MS.* in *Flo. Baines*; but I believe the plant actually observed by Penyghent to have been *Silene maritima.*]

121. Lychnis Githago Lam. *Corn Cockle.*
Area: 63, 64, Top. Bot. Range: 0—600 feet.
Colonist; arable fields, chiefly among grain, common in the lower lands, rare indeed west of Skipton. A. June-August.
First record: Salt, 1800.
Distribution general—specific records from every district except Lune and Mersey.

122. Mœnchia erecta Sm.
Sagina erecta Huds.
Area: 63, 64, Top. Bot. Range: 100—300 feet.
Native: banks and waste ground, very rare, confined to gravelly soil, in the Riding. Mostly gone from its older stations.
Earliest notice: Bolton, 1775.

N. Grimbald's Crag, Knaresborough, sparingly.
[C. The Old Way on Oak's Green near Rastrick, on Sodhouse Green near Illingworth, lane-sides in Skircoat, etc.; Bolton (1775).]
D. [Crook's Moor, Sheffield, 1798; *Herb. Salt.*] Near Rotherham; *L. Langley* (1828). About the Wickersley flag quarries.

123. Cerastium semidecandrum L.
Area: 63, 64, Top. Bot. Arenophilous. Range: 0—900 feet.
Native; on dry sandy ground, wall tops, etc.; rare. A. April-May.
First record: Salt, 1800.

R. Swarth Moor.
Y. Ellershaw, Ripon; *H. Baines.* (Waysides, *abundant*; *H. H. Slater, Flo.*! Some misapprehension?).
N. Knaresborough, about Grimbald's Crag.
W. Sandy lanes about Poppleton and Acomb; *H. Ibbotson.*

A. Hawksworth, on sand; *L. C. Miall.* Malham (!) with *Thlaspi alpestre*; *J. Windsor.*
C. Heath, near Wakefield; *T. W. Gissing.*
D. Doncaster; *H. Baines.* Crook's Moor, Sheffield; *Salt, Herb.*

124. Cerastium glomeratum Thuill.
Area: 63, 64, 65, Top. Bot. Range: 0—1250 feet.
Native; stony banks, wet pastures and road-sides, arable land, infrequent. A. or B. May-July.
First record: Salt, 1800.
L. Garsdale Head. Ingleton.
R. Settle, common; *J. Windsor.* About Wigglesworth.
Y. About Ripon; *H. H. Slater.*
N. Cowthorpe.
W. Bardsey. Thorparch!
A. Meanwood.
C. Frequent; *T. W. Gissing.*
D. Crooks Moor, Sheffield; *J. Salt.*

Obs.—Notwithstanding the summary way this is dismissed in previous floras, although I daresay it is really generally distributed, it is not one-tenth as abundant a species as *C. triviale.* Being an unornamental species, it attracts neither interest nor attention. I have no record for Mersey or Trent districts.

125. Cerastium triviale Link.
Area: 63, 64, 65, Top. Bot. Range: 0—1,800 feet.
Native; fields, roadsides, spoil-heaps of lead-mines, rocks, etc., very common.
First record: Salt, 1800.
Distribution general; in every district.

Obs.—The Annual, pentandrous form, with dark-green leaves and smaller petals, is common on wet heaps of débris about the lead-mines on Malham and Conistone Moors; it is usually a sea-shore plant.

[**Cerastium latifolium** Sm. Incognit. Misnomer?
Thorparch, *T. Scatcherd*; *Miall's Flora* (1862). I have no doubt Mr. Scatcherd reported this, using the term *latifolium* in a loose sense; intending *glomeratum* to be understood.]

126. Cerastium arvense L.
Area: '63, 64 Kingsley spec.,' Top. Bot. Range: 0—400 feet.
Native; dry, sandy, or calcareous fields and banks, infrequent. P. May-June.
First record: Salt, 1800.
Y. Yore banks above Ripon (!) 'Var. *suffruticosum* L. By Yore; *H. H. Slater.*

N. Braham Hall Farm rocks, Spofforth! Near Cowthorpe.
W. Thorparch (!); *J. Emmet.* Holgate, York; *W. Whitwell.*
A. Hambleton. Kellington; *G. Roberts.* Burton Salmon (!) *Id.* Garforth!
C. Heath, near Wakefield; *T. W. Gissing.*
D. Cornfields about Doncaster; *J. Salt.* Among bushes by road from Cantley Church to Bessacarr; *G. E. Smith.* Armthorpe! Goole fields and Rawcliffe Rabbit Hills; *H. F. Parsons.*
T. Roche Abbey; *Geo. Harrison.*

127. Stellaria aquatica Scop.
Area: 63, 64, Top. Bot. Range: 0—300 feet.
Native; wet hedge-banks and sandy places by streams, rivers, etc., rare, but locally plentiful. P. July-August.
First record: Salt, 1800.
N. In lane going from Close-houses to Cowthorpe!
W. Clifford; *T. Scatcherd.*
A. By the Aire near Methley! and Calder banks at Castleford (*G. Roberts*) and Wakefield (*T. W. Gissing*). Snaith; *H. F. Parsons.*
D. Rotherham district; *L. Langley* (1828). By the Dearne near the Sewage Fields (!) *H. Johnson.* Don banks below Rotherham; *J. Salt*=near Conisborough Ferry; *Herb. Salt.* By Don at Aldwarke Carr; *A. Carr* (1875).

128. Stellaria nemorum L.
Alsine montana, folio smilacis instar, fl. laciniato; Ray.
Area: 63, 64, Top. Bot. Range: 50—900 feet.
Native; bushy places, by streams and rivers, wet hedge-banks, infrequent. P. Late May to July.
First record: Bolton, 1775.
L. Trow Gill, top of Clapdale Woods; *J. Windsor.* Sides of Wethercote Cave; *Dawson Turner, O.B.G.*
R. Hollin Hall Woods, Settle; *J. Windsor.* Rathmell; *W. Howson.*
Y. Hackfall (!) and Studley; *W. Brunton.* Mackershaw; *H. H. Slater.* Clotherholme; *Miss Morton.* Yore bank above Ripon Bridge, amid bushes.
N. Ditch-bank by stile in the 'Oak' field, Cowthorpe! Pateley Bridge.
W. Bolton. Woodhall! Washburn Valley. Thorparch; *Borrer*; *Jos. Woods, MS.* in *O.B.G.*
A. In Bingley parish, *Richardson*; *Ray, Syn.* iii, 347 (1724). [About Leeds, plentifully; *W. Wood, O.B.G.* Not now within six miles.] Hawksworth; *W. West.* Shipley to Apperley; *W. West.* Milford Junction; *W. Whitwell.*
C. Pennant Clough, Great House Clough, Eastwood, and near Stanally in Harley Wood, Todmorden; *A. Stansfield.* Kirklees Woods;

Y. N. Excurs. (1873). Near Halifax, very abundant; *J. Bolton.* Heath, Wakefield; *T. W. Gissing.* Rastrick and Fixby; *C. P. Hobkirk.*
D. Upperthorpe; *J. Salt* (1800); and Whiteley Wood, Sheffield; *J. Hardy.* Oughtibridge, shady banks; *A. Carr* (1875). Wharncliffe.

129. Stellaria media With.
Area: 63, 64, 65, Top. Bot. Range: 0—1,800 feet.
Native; road-sides, woods, and waste sandy places, very common.
First record: Salt, 1800.
Generally distributed—in every district.

Var. **neglecta** Weihe.
Woods and shady hedge-banks, common.

Var. **Boræana** Jord.
Dry sandy places, common.

Var. **umbrosa** Opitz.
Area: '64,' Top. Bot.
Woods, very rare, or by-passed often as *neglecta* from not examining the seeds.
W. Not far from Bramhope, near Otley; *W. W. Newbould.* Woodhall banks by Wharfe, 6 miles lower down the river!

130. Stellaria Holostea L.
Area: 63, 64, 65, Top. Bot. Pelophilous.
Native; hedges and woods, common.
Distribution general—found in every district.

131. Stellaria glauca With. (*S. palustris* Retz.; Ehrh.)
Area: 63, 64, Top. Bot. Range: 0—300 feet.
Native; marshes, lowland boggy ditch-sides, etc., rare. P. Late May-July.
First record: Salt, 1800.
[R. Giggleswick Tarn edge; *J. Windsor, MS.* in *Towns. O.B.G.* Gone, if ever there.]
Y. Copgrove, common; *J. Dalton, O.B.G.* (1805). Sharow Mires near Ripon (!) *W. Brunton*; confirmed by *H. H. Slater* as late as 1883.
W. Knavesmire, York; *H. Baines.* Askham Bogs.
A. (Bog near Malham Tarn; *B. Carrington* in *Miall.* Doubtful; too high a level.) Marshy place by Swillington Bridge; *Rev. W. Wood.* Selby; *H. Baines.* [On east side of Ouse, I believe.]
D. Aldwarke by the Don (1798); *Salt, Herb.* There still in 1875; *A. Carr.* Sand holes, Swinefleet Common; *T. Birks*!
T. Thorne Waste; *H. F. Parsons.*

132. Stellaria graminea L.
Area: 63, 64, Top. Bot. Range: 0—900 feet?
Native; grassy banks and ditch-sides, common. P. May-July.
First record: Salt, 1800.
Distribution general, recorded for all the districts; but by no means common above 450 feet or so, in the higher parts of the dales.

133. Stellaria uliginosa Murr.
Area: 63, 64, 65, Top. Bot. Range: 0—1,800 feet.
Native; watery, grassy places and boggy thickets common. A. June.
First record: Salt, 1800.
Distribution general—recorded for every district.

134. Arenaria trinervia L.
Area: 63, 64, Top. Bot. Range: 0—600 feet.
Native; shady moist hedge-banks, woods, etc., common, but nowhere obtrusively abundant. A. May-August.
First record: Salt, 1800.
M. Greenfield Valley; *J. Whitehead.*
R. Near Catterall Hall, and Kendall's Gill, Settle; *J. Windsor.*
Y. Ripon, common; *H. H. Slater.*
W. Wharfedale, frequent from Grassington to Stutton. In the Ainsty, at Bilbrough (!) and other places; *W. Whitwell.*
A. Skipton! Bingley and Esholt; *W. West.* Adel. Ledsham; *W. Whitwell.*
C. Wakefield district, common; *T. W. Gissing.* Huddersfield, common; *C. P. Hobkirk.*
D. Rotherham district; *L. Langley.* Doncaster; *H. Baines.* Sheffield; *J. Salt.*
T. Lane-banks, Firbeck and Sandbeck Park! Roche Abbey; *Geo. Harrison.*

Obs.—Normally this has inconspicuous petals,—the reverse of *Ranunculus auricomus*, but they are occasionally as long as the sepals.

135. Arenaria serpyllifolia L.
Area: 63, 64, Top. Bot. Range: 0—1,800 feet.
Native; dry fields, walls, rocks, amid corn, etc., very common in most districts, but on the Coal measures infrequent. A. June-September.
First record: Salt, Flo. MS., 1800.
Distribution nearly general—recorded for every district except that of Mersey.
Our plant is *sphærocarpa.* I am not sure I have seen true *leptoclados* in the Riding.

136. Alsine verna Bartl.

Auricula muris pulchro flore, fol. tenuissimo, T.B.; Ray.

Area: 63? 64, 65, Top. Bot. Montane. Range: 350—1800 feet.

Native; limestone crags and about the spoil-heaps of the lead-mines, and as an Adventive on river-banks, rare, locally abundant. P. June-August.

First record: Ray, 1677.

L. Ingleborough, foot of the hill towards Skirreth; *Thornbeck* in *Blackst. Sp.* (1746); *Martyn* (1763). Field between Stackhouse Borrins and Feizor; *J. Windsor*.

R. On the mountains about Settle; *Ray, Cat.*, ii, 35 (1677); *Hist.* 1033 (1688); *Martyn*; etc. High Hill; *J. Windsor*. Hinklehaugh; *R. Teesdale*. Still plentiful thereabout.

Y. In Yore (adventive) on shifting bank below Hackfall, a dozen plants or more in 1871.

W. Outershaw Crags. Cold Conistone. About the lead-mines on Grassington Moor. Barden Tower. Strid rocks, Bolton. Addingham, by river (adventive).

N. Greenhow Hill near Pateley Bridge (!); *W. Brunton*.

A. By the lead-mines between Malham and Settle (!); *J. Windsor*. Malham and Malham Moor.

[**Alsine laricifolia** Cr. Error. Misnomer.

'About the lead-mines at the bottom of Ingleborough, towards Horton'; *Bolton, MSS.*, 1768; *Hailst. A. verna* was intended, the *laricifolia* of Lightfoot's *Flora Scotica* being that stunted or debased form we now call *Gerardi*.

I have had the Continental *A. laricifolia* Crantz sent to me as *A. verna*, localised from the Yore side at Hutton Conyers; *J. G. Baker* in *North Yorkshire* (1863). Both banks of the Yore near Hutton Conyers are in the West Riding of Yorkshire; but this species of the Alps, Apennines, and Pyrenees could hardly have been washed down the Yore from the Upper Wensleydale Crags. Probably it came from a garden, like the Tanfield *Erinus*.]

137. Alsine tenuifolia Crantz.

Area: '63, 64 Baker,' Top. Bot. Range: 50—300 feet.

Native; limestone walls and rocks, sandy ground, rare. A. June-August.

First record: Salt, 1800.

Y. Copgrove gravel-pits; *J. Dalton*. Fountains Abbey walls (!); *W. Brunton*.

N. Knaresborough!

W. On the railway ballast, between the lines, a mile west of Wetherby, in abundance (!); *J. S. Wesley*. Limestone quarry west of Tadcaster; *H. Wetherill* in *Supp. Flo. Yorks.*

(C. Cullingworth, since 1848 only occasionally found; *Miall's Flora.*)

T. About Maltby; *Salt, MS. Flora*. Maltby Wood-Lea Common, on the rocks; *G. E. Smith* (1845). There in 1878!

138. Sagina eu-apetala E.B.

Area: 63, 64, Top. Bot. Range: 0—300 feet.

Native; wall-tops and dry banks, rare—quite unexplainably so in the West Riding. A. May-September.

First record: uncertain.

Y. Ripon; *H. H. Slater*.

A. Masonry-coping of canal, Gilstead to Bingley! cum *W. West*.

D. Wall-top, Went-vale (1870).

T. Roche Abbey. Maltby.

139. Sagina ciliata Fries.

Area: 63, 64, Top. Bot. Range: 0—1,250 feet.

Native; dry sandy banks and fields, rare. A. June-August.

First record: Gibson, 1850.

Y. Ripon district; *H. H. Slater*—not seen by me.

W. Aberford (1870).

A. Malham Tarn shore! Garforth and Selby; *Gibson* teste *C. C. Babington*.

D. Gravel pits near Armthorpe, Doncaster!

140. Sagina procumbens L.

Area: 63, 64, 65, Top. Bot. Range: 0—2,000 feet.

Native; damp sandy ground, wet mossy rocks, walls, garden paths, etc., very common. A. May-September.

First record: Salt, 1800.

Generally distributed—reported from all the districts.

141. Sagina subulata Wimm.

Area: [61? 62], Top. Bot. Range: at 1,250 feet alt. in District A.

Native: wet sandy-boggy ground, very rare. P. July.

First record: Lees, 1873.

A. Shingly sandy beach, south side of Malham Tarn, very sparingly with *S. ciliata*, *Scirpus setaceus*, etc., in 1870. Confirmed in 1876. The record, from memory, in Yorks. Nat. Recorder, August 1873, by the writer, for Arncliffe Clouder, proved erroneous in 1876.

Obs.—It is very singular that with so many apparently suitable stations for it, this plant should be so very rare in Yorkshire, but its distribution as shown in Top. Bot. is rather peculiar—Lincoln, Yorkshire (except doubtfully), Lancashire, Durham, and Westmoreland being missed.

142. Sagina nodosa Meyer. *White Sandwort.*

Area: 63, 64, 65, Top. Bot. Range: 100—1,500 feet.

Native; damp ground in sand or peat, infrequent, though often abundant enough locally. P. July-August.

First record: Curtis, 1782.

M. Head of Greenfield Valley. Denshaw.

L. Ingleton. Garsdale.

R. Bogs and hills about Settle; *W. Curtis* (1782); *Gough*; etc. Penyghent; *W. Whitwell*.

W. Bramham Moor; *H. Baines*. Jackdaw Crag quarry, Stutton! Bolton and Ilkley (!); *B. Carrington*.

A. Malham Tarn beach! Hawksworth; *L. C. Miall*. Gravelly shore, Adel Dam, near Leeds; and Blackmoor, rare even in 1867, perhaps now gone. Grassington road, Skipton, 1884; *T. W. Edmondson*!

T. Marshes about Roche Abbey; *J. Salt*.

143. Spergula arvensis L., **a. sativa** E.B. *Spurrey.*

Area: 63, 64, Top. Bot. Range: 0—900 feet.

Colonist; sandy, tilled fields and waste ground, very common, and as high as cultivation reaches. A. June-August.

First record: Salt, 1800.

Generally distributed—all the districts.

Var. **vulgaris** Boenn.

W. Field (on sand) near Ingbarrow, Wetherby!

144. Spergularia rubra Fenzl. *Red Sandwort.*

Area: 63, 64, Top. Bot. Arenophilous. Range: 0—500 feet.

Native; sandy waste ground, rocky lanes, fields, etc., rare. A. June-August.

First record: Salt, 1800.

N. Plumpton and rocks about Braham Hall Farm, Spofforth.

W. Near York (and Doncaster); *White* in *Miller* (1804).

A. Lanes about Wyke and Brandon and Shadwell. Lane between Adel and Eccup; *W. West*. About Whitham's flag-quarries, by the railway line at Horsforth Wood-side (!) *H. Barber*. Sandy fields about Hambleton.

C. Sandy lanes, Farnley Tyas, near Storthes Hall, Huddersfield; *P. Inchbald*. Stocksmoor, Dewsbury; *P. F. Lee*! Longwood; *C. P. Hobkirk*. Lee Moor, near Wakefield, 1871; *G. Roberts*.

D. Denby Dale; *Y. N. Excurs.* (1872). About Brightside, Sheffield; *J. Salt*. Armthorpe, etc. Doncaster!

145. Spergularia salina Presl.

Area: '63,' Top. Bot. sub *S. neglecta* Syme. Range: 0—10 feet.

Casual or Adventive, on tidal banks of Ouse from Goole downwards to Adlingfleet.

D. Goole; *H. F. Parsons*!

Obs.—I suspect that this plant, which is hardly glandular, and has seeds without any wing, is not true Syme's *neglecta*, but a succulent saline-mud state of *Spergularia rubra*: the brackish water favouring absorption and checking evaporation from the vegetable cells, whence distension results.

XIII. ILLECEBRACEÆ.

146. Scleranthus annuus L. *Knawell.*

Area: 63, 64, Top. Bot. Arenophilous. Range: 0—300 feet.

Native; sandy banks and fields, infrequent, but abundant locally.

First record: Salt, 1800.

Y. Ripon, occasional; *H. H. Slater*.

N. Spofforth! Plumpton.

W. Wetherby! Healaugh. Acomb fields.

A. About the sand quarries, Meanwood Wood, near Leeds (!) *J. Abbott* (1869). Common about Hambleton and Selby.

D. 'Common'; *Salt, Sheff. Flora*. Doncaster. Cantley.

T. Letwell.

XIV. PORTULACACEÆ.

147. Montia fontana L. *Water Chickweed. Blinks.*

Area: 63, 64, 65, Top. Bot. Range: 100—1,800 feet.

Native; watery places on peat or sand, about springs, etc., common. A. May-July.

First record: Salt, 1800.

M. Greenfield Valley; *J. Whitehead*. Denshaw.

L. Garsdale. Baugh Fell.

R. Several places about Settle; *J. Windsor*. Bleamoor and Ribblehead.

N. Ramsgill; *J. S. Wesley*.

W. Ilkley Moors. Askham 'flat swamp'! Beamsley and Barden Moors.

A. Shipley and Eldwick; *W. West*. Adel bog! Bradford and Heaton; *L. C. Miall*. Skipton; *W. Whitwell*.

C. Fixby; *C. P. Hobkirk*. Norland Moor! Wakefield, 'common'; *T. W. Gissing*.

D. Potteric Carrs; *E. Miller* (1804). Rotherham district; *L. Langley* (1828). Frequent about Sheffield, often in dry places; *J. Salt*.

Obs.—The form or state *rivularis* is not unfrequent in rills on the moors. Wherever it is found water has been, but it often dries up by the time the plant is in seed.

[**Claytonia perfoliata** Don. Alien.

Casually in heathy-sandy ground, rare, and fugitive.

W. Sandy field at Acomb, near York.

A. About Rider's nurseries (among Rhododendrons, etc.) at Moortown, near Leeds.]

[**Claytonia alsinoides** Sims. Alien.

M. Naturalised in a wood at Marsden; *W. West*.

A. Copse near Morley; *P. F. Lee*. Garden stray?]

[XV. ELATINACEÆ.

Elatine Alsinastrum Endl. Error.

First recorded by Dr. White in 1804, as a plant of the York and Doncaster neighbourhoods, in Miller's Hist. and Antiq. of Doncaster, number 38 in list there given, and *not* as one of those 'most rare'; but a reference to the article *Elatine* by Sir J. E. Smith in Rees' Cyclopædia, vol. xii (1819), and to Smith's English Flora, vol. ii, pp. 243, 244 (1824), show that Dillenius was the one originally in error, and that, 'as is proved by his herbarium at Oxford,' *Centunculus minimus* was the plant intended; which does still grow at Maltby, near Doncaster, and (outside our boundaries) near York.]

XVI. HYPERICACEÆ.

148. Hypericum Androsæmum L.

Area: '63, 64† Tatham,' Top. Bot. Range: 300—600 feet.

Native; woods and thickets, rare, of 'Atlantic' type. P. July.

First record: Bolton, 1775.

L. Wood on the way from Thornton Force to Bentham.

A. Pontefract, 1791; *D. Dixon.* About Kirkstall Abbey, rare; *Rev. W. Wood* (1805). Broadstone Wood, Saltaire; *L. C. Miall.* In Shipley (!) or Eldwick Glen; *H. Denny.*

C. Elland, Sowerby, Woodhouse, etc., near Halifax; *J. Bolton* (1775); rep. *Robson* (1805). North Dean and High Royd Wood, Warley; *Leyland.* Near Hebden Bridge; *S. Gibson.*

D. In Cooke Wood, near Sheffield; *Salt, Flora* and *O.B.G.* (misprinted Crooks Wood in *Baines*).

[Hypericum calycinum L. '*Rose of Sharon.*'

Alien. In open woods and thickets, always an intentionally or accidentally introduced plant, but thoroughly naturalised and abundant in (T.) Shooter's Hill plantations, to the south-east of Rossington! cum *W. Fowler* (1878). Reported also for (N.) Plumpton and (W.) Grimston.]

[Hypericum hircinum L. Alien.

R. Near Settle; *Irvine, H.B.P.* (1858).

A. An escape; in lane by cottages between Shadwell and Wyke (1869-71).]

149. Hypericum perforatum L. *St. John's Wort.*

Area: 63, 64, 65, Top. Bot. Range: 0—900 feet.

Native; woods, bushy hedge- and railway-banks, common. P. July-August.

First record: Salt, 1800.

Distribution general; occurs in every district, but most prevalent upon sandstone. Rare in the Mersey area.

Var. **angustifolium** Bab. (*microphyllum* Jord.).

This variety is rather rare, or at least I have only seen it in the heathy-sandy districts about (N.) Plumpton and Hookstone Crags, Harrogate; Long Marston; and (A.) Hambleton. I have received it also from (D.) near Sheffield (*Amos Carr*).

150. Hypericum dubium Leers.

Area: 63, 64, Top. Bot. Range: 50—450 feet.

Native; bushy places by watercourses (mainly) with a peculiar distribution, rare, although plentiful enough in certain areas. P. Mid July-August.

First record: Salt, 1800.

L. Near the Wenning stream, west of Ingleton; *Isaac Hindson.*

R. Bushy places, Bolton-by-Bolland, gorge of Ribble!

Y. In a stone quarry at Hutton, near Ripon; *W. Brunton* in *O.B.G.* But I fear the station is outside the Riding. By the Ouse below Boroughbridge, *adventive.*

(A. Bradford district; *J. Willis* in *Journ. Bot.*, 1874, p. 15.)

D. By the Don above Sheffield, various places; *J. Salt* (1800). Still by the Don in many places (!); *A. Carr* (1876). Near Brightside; *Deakin* in *Herb. Salt.*

151. Hypericum tetrapterum Fries.

Area: 63, 64, 65, Top. Bot. Range: 0—1,250 feet.

Native; damp banks of ditches, hedges, streams, common. P. July-August.

First record: Gerarde, 1597.

Distribution general; recorded from every district.

'Square St. John's grasse, groweth plentifully in the north part of England, especially in . . . Craven'; *Gerarde* (1597).

152. Hypericum humifusum L.

Area: 63, 64, Top. Bot. Arenophilous. Range: 0—600 feet.

Native; dry sandy fields and banks, common, yet by no means regularly distributed, especially infrequent on limestone. P. July-August.

First record: Salt, 1800.

R. Settle, common; *J. Windsor.*

Y. Ripon, not uncommon; *H. H. Slater.*

N. Cattall. Sandy fields, Braham Hall, Spofforth.

W. York neighbourhood; *White* (1804). Acomb, etc.; *G. Webster.*

A. Sandy banks, Alwoodley, near Leeds. Brayton Barf, near Selby.

C. Halifax; Bretton Park; *G. Roberts.* Heath, Wakefield; *T. W. Gissing.*

D. Doncaster; *White* (1804). Crook's Moor, Sheffield; *Herb. Salt.* Don bank, Wadsley; *Geo. Harr. Hb.* Hessel, near Ackworth; *Gissing.*

153. Hypericum pulchrum L.

Area: 63, 64, 65, Top. Bot. Pelophilous. Range: 0—1,500 feet.

Native; heathy or clayey moorland banks, common. P. July.

Earliest notice: Salt, 1800.

Distribution general; occurs in every district.

Obs.—A very abundant and beautiful flower, but one with which I have never been able to connect any local tradition; nor has it in Yorkshire any common name beyond that applied indifferently to it, *H. perforatum* and *H. hirsutum.*

154. Hypericum hirsutum L.

Area: 63, 64, 65, Top. Bot. Sub-xerophilous. Range: 50—1,250 feet.

Native; bushy banks and woods, on dry, mostly calcareous, soil; common. P. June-August.

First record: Salt, 1800.

L. On limestone about Ingleton and Clapdale.

R. Settle, common; *J. Windsor.*

Y. Fountains Abbey and Mackershaw.

N. Birkham Wood, etc. Knaresborough!

W. Arncliffe, Grassington, Bolton, Collingham, Thorparch, etc., common! Askham Lane, near York; *H. Baines.*

A. In Craven; *Baines.* About Draughton, Skipton! Shipley; *W. West. Adventive* here? Huddlestone quarry, near Micklefield (!); *Baines.* Brotherton. Ledsham (!) *W. Whitwell.*

D. Thicket near Brightside; *Herb. Salt.* Wentbridge. Warmsworth. Conisborough, on the Permian limestone, abundant!

T. Maltby, Roche Abbey, Anston, etc.; *A. Carr!*

155. Hypericum montanum L.

Area: 63, 64, Top. Bot. [Unknown apparently in 65.] Range: 50—950 feet.

Native; dry calcareous-soil banks, rare. P. July-August.

First records: Bolton, 1775; Salt, 1800.

L. Sykes' (Helk's) Wood, near Ingleton; *Gough's Cam.* (1789); rep. *var. auct.*

Y. Mackershaw and Fountains Woods; *T. Simpson* (1850). Lane near Stainley; *Id.* Thieves' Gill and Kirby Malzeard; *H. H. Slater, Flora of Ripon.*

N. Knaresborough! *E. Hargrove* (1809).

W. Collingham bank. Bilton Haggs. Thorparch; *H. Baines.*

A. Malham Cove; *J. Nowell.* Gordale; *L. C. Miall*; *W. West.* Garforth; *T. W. Gissing.* Huddlestone quarry; *H. Baines.* There still in 1877; *W. Whitwell.* Fairburn (!) *G. Roberts.*

C. Windle Royd, Warley, Elland Edge, etc.; *J. Bolton.* Gone now; could Bolton have erred?] Roadside, Whitley*, 1886, one or two plants; *C. P. Hobkirk.*

D. & T. Doncaster neighbourhood; *White* (1804). Rotherham district; *L. Langley* (1828). Brodsworth; *O. A. Moore.* Roche Abbey to Maltby, on right hand; *G. E. Smith.* Quarries near Letwell. In an old lane two miles south of Sheffield (1873); none visible 1876, probably eaten off by cattle; *A. Carr.* Cadeby Common (1801); *Herb. Salt.*

156. Hypericum elodes L.

Area: 63, 64, Top. Bot. [Not known for 65.] Range: 0—1,200 feet.

Native; spongy bogs, rare, decreasing. P. August-September.

First record: Salt, 1800.

M. Gulliver's Swamp, Greenfield Valley (!); *J. Whitehead.*

N. Hookstone Crags (near Pannal); *H. Baines.* Gone?

A. Hawksworth and Cottingley Moors; *Miall's Flora.* Adel bog (*H. Denny*, 1835), south (Reformatory) side of stream. There still in 1869 (!) with *Scutellaria minor*, only a small patch; perhaps now gone. = Blackmoor; *H. Baines.* [Headingley Moor; *H. Baines*—now gone.]

D. In a bog near Cliff Rocher, above Sheffield; *Salt, Flora* (1800). Refound, 'head of Rivelin Valley,' by *A. Carr* in 1876-7. Rawcliffe; *H. F. Parsons* (1877).

XVII. MALVACEÆ.

157. Malva moschata L. *Musk Mallow.*

Area: 63, 64, Top. Bot. Noterophilous. Range: 0—900 feet.

Native; hedges, woods, and stream-banks, preferring a sandston soil, common. P. June-August.

First record: Bolton, 1775.

L. By road between Ingleton and Kirby Lonsdale (!); *J. Windsor.*

R. Above Settle bridge; *J. Windsor.*

Y. Copt Hewick, Ripon, and Boroughbridge; *H. H. Slater.*

N. Near Grimbald bridge, Knaresborough; *E. Hargrove.* Between Bilton and Cowthorpe; *H. Baines.*

W. Harewood. Ilkley. Bardsey. Ingbarrow, Wetherby. Boston Spa and Collingham (and a white variety which proved constant in cultivation); *J. Emmet.* Near York; *H. Baines.* = Ainsty; *G. Roberts.* Bolton Bridge; *L. C. Miall.* Washburn Valley.

A. Skipton. Bingley and Roundhay; *W. West.* Adel. Alwoodley. Bolton near Bradford, and Apperley; *L. C. Miall.*

C. Between Loscoe Grange and Normanton Common, 1794; *D. Dixon.* Skircoat, etc., Halifax; *J. Bolton.* Warley; *R. Leyland.* Lofthouse; *G. Roberts.* [Milnsbridge; *C. P. Hobkirk*—now gone]. Copley; *L. C. Miall.*

D. About Sheffield; *Salt, Flora.* Rail-bank, Monk Bretton, Barnsley. Wentbridge; *H. Baines.* Near Car House, Doncaster; *L. Langley.*

158. Malva sylvestris L. *Mallow.* '*Mauls.*'

Area: 63, 64, Top. Bot. Range: 0—600 feet.

Native; waste places, lane-sides near villages and field-borders, common on the limestones, rarer than the last on sand. P. June-September.

First record: Bolton, 1775.

R. Giggleswick and Upper Settle; *J. Windsor.*
Y. Ripon, common; *H. H. Slater.*
N. Scriven and Knaresborough.
W. Wetherby. Walton. Stutton. Thorparch.
A. Brotherton, etc. Sherburn; *W. Whitwell.*
C. Sowerby Bridge, Warley side, and Ovenden; *J. Bolton* (1775). Wakefield, common; *T. W. Gissing.*
D. Brightside, Sheffield; *J. Salt* (1800). Don banks above Sheffield; *G. Harrison.* Kirk Smeaton; *W. West.*
T. About Braithwell and Firbeck.

159. Malva rotundifolia L.

Area: 63, 64, Top. Bot. Range: 0—450 feet.

Denizen; hedge and wayside banks, always near habitations, infrequent. P. July-September.

First record: Salt, 1800.

R. 'Said to have been found at Thornton near Settle'; *J. Windsor.*
Y. Ripon, common; *H. H. Slater.* Bishop Monckton!
W. Rigton-bank foot, near Bardsey.
A. About Pontefract! Between Micklefield and Sherburn.
C. Heath; Hiendley; *T. W. Gissing.*
D. Brightside, plentiful; *J. Salt* (1800).

[**Malva parviflora** L. Alien.

On wool-washings refuse, with other exotics, on waste ground near Boothroyd's Mill, Meanwood! In a clover field at Methley; and waste ground, Sandal (*G. Roberts*); Near Hirst's Corn Mill, Huddersfield in 1873 (*J. Tindall*).]

[**Malva borealis** Wallm. Alien.

Found in plenty on waste ground near Old Goole (Flour) Mill by *T. Birks, junr.*! Grain-introduced? 'Growing where Mallows have been plentiful for many years.']

[**Althæa officinalis** L.

'[64]' Top. Bot. Incognit.]

XVIII. TILIACEÆ.

[**Tilia grandifolia** Ehrh. Alien.

Occasionally in parks, plantations, etc., but always planted, and mostly of recent date; no trees of large size—the largest I recollect is in a park near Firbeck (T.); some fair-sized ones about Wighill (N.) and Goldsborough (N.).]

160. Tilia intermedia DC. *Linden.* '*Lime-tree.*'

Denizen; plantations, road-sides, etc., very common. Tree. July.

This has been long planted—and as such is to be found in every district, though rare in M. and L.,—but does not reproduce itself by seed, spontaneously, with us, and so has no real claim to inclusion in the Flora, any more than Larch or Spruce.

161. Tilia parvifolia Ehrh.

Area: '(64)' Top. Bot. Range: 200—300 feet.

Denizen; rocky wood-slopes on limestone, very rare, but *possibly* a native. Tree. July.

First record: Jos. Woods, 1838.

W. Woodhall (planted).
D. Above Conisborough!
T. Roach Abbey; *Jos. Woods* (6th August, 1838), *MS. in Towns. O.B.G.*

XIX. LINACEÆ.

162. Radiola Millegrana Sm.

Linum Radiola; Mill., etc.

Area: 63, 64, Top. Bot. Range: 0—300 feet.

Native; sandy or sandy-peaty ground, liable to inundation in the winter; scarce. A. July.

First record: White, 1804.

Y. Sawley Moor, near Ripon. Not uncommon near Ripon, often overlooked; *H. H. Slater, Flora.*
A. Shore of Blackhill Dam, Adel, near Leeds, in plenty (!); *J. Abbott.* Wet sandy lane-side near Wike, by Shadwell.
D. Near Doncaster; *White* (1804). Rawcliffe; *H. F. Parsons*!

163. Linum catharticum L. *Purging Flax.*

Area: 63, 64, 65, Top. Bot. Range: 0—2,400 feet.

Native; dry banks, pastures, etc., very common. A. July.

First specifically West Riding record: Bolton, 1775.

Generally distributed—in all the districts.

164. Linum perenne L.

Area: 64, Top. Bot.

Native, but more like a Colonist in some respects; fields and bare stony ground on limestone soil, very rare and uncertain. P. but also A. End of June.

Y. Mackershaw Wood, near Ripon; *W. Brunton*; *O.B.G.* (1805). Broken ground near Wicliffe Lane in open bare spots where the ground had been cleared; present in 1869, not to be seen in 1870.
(W. Other stations between Leeds and York; *Miall's Flora*, p. 8. Some error surely?)
A. Near Ledstone Lodge; *H. Baines, Flora* (1840).

[**Linum usitatissimum** L. *Flax.* Alien.

An impermanent Escape from cultivation, or a Casual on railway-banks, fields, etc. Not unfrequent, but seldom twice in the same place.

Y. Near Copgrove . . . had been cultivated; *J. Dalton*; *O.B.G.*
W. Rail-bank, Thorparch. Fields about Walton.
A. Rail-banks, Headingley station!
C. Woolley Edge, etc.; *T. W. Gissing.*
D. About Goole, Thorne, Sheffield.]

[**Linum angustifolium** Huds.

A. Horsforth; *B. Carrington, M. and C. Flora.* A wool- or grain-brought waif?]

XX. GERANIACEÆ.

165. Geranium sanguineum L. '*Bloody Cranesbill.*'

Area: 63, 64, Top. Bot. [Not known in 65.] Range: 200—900 feet.

Native; rocky bushy places, on limestone only within our area, rare; a decreasing species from transference into gardens, being so handsome. P. July.

First record: Merrett, 1666.

L. On Ingleborough hill; *Merr.* 46. In Friar Wood, also at Thwait near Clapham, *Mr. Thornbeck*; *Blackst., Spec.* (1746). Friar Wood, Ingleton; *Martyn* (1763).
R. About Settle; *Gough's Cam.* (1789). Kelcowe, Lord's Wood, etc.; *J. Windsor.*
N. On the cliff of Knaresborough, bushy ground above the gardens (!); *Pierson* (1805); on the rocks . . . and in Abbey fields in the richest profusion; *E. Hargrove* (1809).
W. Rocky woods about Arncliffe and Kilnsey. Grassington Wood; *Rev. W. Wood* (1805).
A. About Malham; *Gough's Cam.* Malham Cove; in considerable quantity (!); *P. Inchbald* (1876). Gordale (!); *Jos. Woods* (1805).

166. Geranium phæum L.

Germanic Type. Range: 50—350 feet.

Denizen: thickets, hedge-banks, and by streams, rare and not native. P. June.

First record: Thornbeck, 1746.

L. Meadow at Feizor; *R. Clapham.* = In a field three miles west of Settle; *J. Tatham*; *S. Hailstone's list.* About Ingleton, *Mr. Thornbeck*; *Blackst. Sp.* (1746). Clapham; *Martyn* (1763). Near Settle; *Dawson Turner* (1805) = Rose-tree Bank Wood; *J. Windsor.* Many confused repetitions of these two localities, from *Richard Teesdale* downwards.
N. Scriven Park, Knaresborough; *Archd. Pierson* (1805).
A. Baildon Bridge; below the corn mill, Shipley, and Castlefield, Bingley; *L. C. Miall.* Rawcliffe; *H. F. Parsons*!
D. Limestone tract and warp lands west of Doncaster; *S. Appleby* (1831) = Sprotborough Broats; *O. A. Moore*, in *Baines.* Tofield introduced this here, I think.

167. Geranium sylvaticum L. *Wood Geranium.*

G. batrachioides montanum; Ray.

Area: 63, 64, 65, Top. Bot. Montane. Range: 200—1,200 feet.

Native: thickets, river-banks, and pastures, among the dales common, elsewhere rare. P. Mid May-July.

First record; Ray, 1670.

L. In pratis montosis . . Eboracensis copiosé; *Ray, Cat.*, p. 131 (1670); *Hist.*, p. 1062. About Ingleton; *R. Teesdale.*
R. About Settle; *Gough's Cam.* (1789). Mill Island, Major Springs, meadows near Stainforth, etc.; *J. Windsor.* Linn Gill.
Y. Hackfall! Clotherholme; *H. H. Slater.*
N. Near Pateley Bridge; *W. Brunton* in *O.B.G.* Bilton and Scotton banks, by the Nidd, above Knaresborough!
W. Buckden. Littondale. Bolton Abbey; *Rev. W. Wood.* = Woods and hedges near the Wharfe at Bolton; *Knowlton* in *O.B.G.*

[**Geranium nodosum** L. Extinct Alien.

Leeds; *H. Denny MS.* and 'Washerlane' or 'Waterham' near Halifax; *S. Gibson* to *H. C. Watson*; *N.B.G.* (1835). No further information. Perhaps taken from a garden, certainly but an outcast if not. Grades of indigenity were of little account at that time: Bolton in his Halifax Catalogue, sub *Geranium pyrenaicum*, tells how that plant at Skircoat originated from 'seeds brought from Bingley, and scattered here and there.']

168. Geranium pratense L. *Meadow Cranesbill.*

G. batrachioides; Ger.; Ray.

Area: 63, 64, Top. Bot. Range: 0—1,200 feet.

Native; river-banks and pastures, irregularly distributed, locally abundant on gritstone and sand. P. June-August.

First record: Ray, 1724. In pratis montosis . . . Eboracensis; *Ray, Syn.* 3 (1724).

L. Ingleton; *Jos. Woods* (1805).
N. Knaresborough! *E. Hargrove* (1832; not in 1809 Ed.).
Y. Ripon, abundantly (!); *W. Brunton* (1805).
W. Otley (!); *Jos. Woods*. Arthington. Wetherby. All along Wharfe valley from Buckden to Newton Kyme.
A. Malham. Skipton. Between Bingley and Keighley; *Martyn* (1763). River-side at Newlay and down to Kirkstall Forge, within four miles of Leeds. Near Leeds; *W. Wood* in *O.B.G.* Ferrybridge, by Aire; *Y. N. Excurs.* Oulton; *G. Roberts.*
D. Doncaster neighbourhood; *White* (1804). Goole; *H. F. Parsons*!

Obs.—With pure *white* flowers at Pool, Skipton (*W. Whitwell*), and elsewhere, occasionally.

169. Geranium pyrenaicum L. (*G. perenne* Huds.)
Area: 63, 64, Top. Bot. Range: 50—300 feet.
Denizen; road-side banks and river-banks, rare. P. Late June-July
First record: Bolton, 1775.
N. Lane, Flaxby, near Knaresborough, near cottages (1869).
W. Thorparch Wood, immediately behind the Hall, *J. Houlton*; *J. Jackson*! Aberford, by the Toll-bar; *H. Ibbotson.* Holgate and Acomb; *H. Baines.* Perhaps gone—not marked by *G. Webster* for Ainsty.
A. Between Bingley and Keighley *by the river*; *Huds., Flo. Ang.*, 303 (1778). *In meadows and pastures* between B. and K.; *Gough's Cam.* (1789).—Although given under *G. pyrenaicum*, the hab. cited makes one suspect that *Edward Forster* meant *G. pratense.* In such spots near B. the last-named is abundant, whilst the former is confined to a few places on the river and canal bank.
[C. Lower Willow-Hall, Skircoat; *J. Bolton.* Introduced by him. = Pyenest lane; *H. Baines.*]
D. Road-side between Campsall and Barnsdale; *E. Lankester* (1842).

170. Geranium molle L. '*Mother-o'-millions.*' '*Dove's-foot.*'
Area: 63, 64, 65, Top. Bot. Range: 0—1,250 feet.
Native; fields, road-sides, etc., very common.
Earliest notice: uncertain. Salt, 1800?
Generally distributed—in every district.

171. Geranium pusillum L.
Area: 63, 64, Top. Bot. Range: 0—450 feet.
Native; dry banks and fields, on limestone chiefly, infrequent. A. June-August.
First record: Dixon, 1791.
This does not appear to occur on the western scar-limestone.
Y. Near Copgrove, *Rev. J. Dalton*; *O.B.G.* Near Boroughbridge; *H. Baines.*

N. Abbey rocks, Scriven lane, etc., near Knaresborough! Hunsingore.
W. Collingham. Thorparch; *J. Emmet*! Ainsty, near York; *G. Webster.* Copmanthorpe.
A. About Garforth, Sherburn, Ferrybridge. Burr; *H. F. Parsons.* Selby. Ledstone, 1886; *G. Roberts.*
C. Close near Sandal Castle Hill, and near the bottom of Milnthorp Green, 1791; *D. Dixon.* Heath, Woolley, Oakenshaw; *Gissing*!
D. Norton and Wentbridge. Near Doncaster; *Salt's MS. Flo.*, sub nom. *rotundifolium* (1800).

[**Geranium rotundifolium** L. Incognit. Misnomer.
Area: [63, 64, insufficiently certified], Top. Bot.
This species is named in the list in Hargrove's Guide (1832 ed.—not in 1782 ed.) along with four others—still there—as to be 'all found upon the rocks about the Abbey' at Knaresborough. The four, *and* also *G. pusillum*, all occur there yet; and I believe *G. pusillum* to have been intended by the '*rotundifolium*—Round-leaved Crane's-bill.'
Dalton's Copgrove record in *O.B.G.* certainly meant *G. pusillum.*]

172. Geranium dissectum L.
Area: 63, 64, Top. Bot. Range: 0—900 feet.
Native; dry fields, waysides and waste ground, common. B. June-September.
First record: Herb. Salt, 1799.
Distribution general—recorded from every district, localities too numerous to require mention.

173. Geranium columbinum L.
Area: 63, 64, Top. Bot. Range: 0—650 feet.
Native: dry fields and banks, preferring sand, infrequent, although often locally abundant. A. or B. June-July.
First record: Salt, 1799.
Yorkshire, mountain meadows; *Gough's Cam.* (1789).
Y. Not uncommon about Ripon; *H. H. Slater.*
N. Lane from Spofforth to Ribston.
W. Ilkley. Otley. Thorparch; *H. Baines.* Flint Mill, banks! Ainsty; *G. Webster.*
A. Near Bingley Seven Arches; *L. C. Miall.* Skipton! Foot of Malham Cove; *J. Windsor.*
D. Rotherham district; *L. Langley* (1828). Dearne valley; *H. Johnson.*
T. Cornfield, Roche Abbey, 1799; *Herb. Salt.*

174. Geranium lucidum L.
Area: 63, 64, 65, Top. Bot. Sub-montane. Range: 200—900 feet. A. May-August.
Native: rocks, walls, and stony banks.
First record: Curtis, 1782.

L. Frequent in Dentdale. Ingleton.
R. On stone fences about Settle; *W. Curtis* (1782); *J. Windsor*; *Gough's Cam.*; etc. Slaidburn.
Y. Kirby Malzeard. Hackfall. Hutton Moor; *H. H. Slater.*
N. Knaresborough (!) *H. Baines.*
W. Barden. Bolton. Burley; *L. C. Miall.* Ilkley. Boston, by the river; *J. Emmet—adventive.*
A. In Craven, common; *H. Baines.* Malham! Skipton. Between the Aire and Gilstead, and near Bingley (! and *W. West*) *L. C. Miall.*
C. Warmfield field, 1790; *D. Dixon.* Disused quarry at Sandal*; *G. Roberts.*
D. Wentbridge and Purston; *T. W. Gissing.* Rotherham district; *L. Langley* (1828).

175. Geranium Robertianum L. *Herb Robert.* '*Stinking Bob.*'
Area: 63, 64, 65, Top. Bot. Range: 0—1,800 feet.
Native: hedge-banks, old walls, stony places, and woods, very common. A. or B. May-September.
First records: Bolton, 1775; Old Park, near Sheffield, 1799; Salt.
Distributed generally—every district.

Obs.—In the hill districts, where they grow together, this species hybridises with *G. lucidum*, producing handsome, but ephemeral (because sterile) individuals.

176. Erodium cicutarium L'Hérit. *Stork's-bill.*
Area: 63, 64, Top. Bot. Range: 0—300 feet.
Native or Colonist: sandy fields and waste ground, often introduced with crops, and then frequently fugitive.
First record: Bolton, 1775.
Y. 'Not uncommon about Ripon'; *H. H. Slater.*
W. Low-Mill (flour), Boston Spa, 1879, one plant; *J. Emmet.* A large plant thought to be *E. moschatum.* Ainsty, sandy fields; *G. Webster.* Field by Sweep Lane, Wetherby!
A. Manningham; *L. C. Miall* (Casual in North Park Road). Hambleton. Kellington; *G. Roberts.*
C. Near Farnley; *Hudd. Nat. Hist.* Halifax, 'seed-field'; *W. West.* Adwalton and Skircoat Moor, etc., near Halifax; *J. Bolton* (1775). Probably long gone from these places.
D. Rotherham district; *L. Langley* (1828). Brightside, plentiful; *J. Salt* (1800). Barnby-on-Don, 1886; *G. Roberts.*
T. Rossington; *Geo. Harr. Herb.* A very large plant which has often probably been mistaken for *E. moschatum*; *W. W. Newbould, MS*

[**Erodium moschatum** L'Hérit. Incognit. Lost Alien?
G. moschatum; Ger.; Park.
R. A. In Craven common; *Ray, Cat.*, Ed. i, p. 132 (1670). But Ray does not seem to have seen it growing himself, for in his *Synopsis* (Ed.

i, 155; ii, 217; and iii, 358) the versions are:—'I have from Dr. Lister that *in Craven* it is very common.' In *Gibs. Cam.* the version is 'Dr. Lister is my Author.' In *Ray, Hist.*, vol. ii (1688), p. 1057, we read—'in . . . regione *Craven* dicta quæ pars est occidentalis eboracensis comitatus *sponte* et copiosé provenit.' Now, if *both* really wild, and abundant from 1670 to 1724, it would not now have totally disappeared from a district that has altered little, and in which the other salient rarities (noticed by Ray himself), with scarcely an exception, still survive.

The record is repeated and variously noted down by mere copyists as late as Smith and Baines; and it may fairly be questioned whether some luxuriant form of *E. cicutarium* was not the plant really found. Against this, however, is the fact that the Herb. Buddle specimen is undoubtedly *moschatum.* It may, of course, be said that the indigenous integers in the florula of any tract are gradually changing, and that a re-placement by alien, but acclimatising, species is and has long been going on. This is indisputably true, but the tendency in the change is for the species of longer duration—the Perennials, the Shrubs, etc.—to disappear, from alterations effected in physical and climatic conditions by human agencies, such as draining, and their places to be taken by (if anything) shorter-lived or drier-soil loving races, and not vice versâ. An annual arenophilous *Erodium* can upon no hypothesis be considered as a *likely* indigenous item in the primitive flora of an elevated lime and slate district. Smith in Eng. Flo. (iii, 230), appears rather to question the indigenity of the plant, for he remarks on its having long been cultivated in gardens for its scent.

(W. Thorparch; *Rev. W. Wood* in *O.B.G.* Wood was doubtless mistaken in this as in other things.)
D. Near Sharrow Moor (Sheffield); *Salt, Flo. Sheff.* This plant in the Herb. at Weston Park is only *E. cicutarium.*
C. Leeds road, Huddersfield; *Hudd. Nat. Hist.*, Ed. 2. A wool casual].

[**Erodium ciconium** Willd. Alien.
In a field at Three Houses, Sandal (C.); *G. Roberts*! A garden Escape?]

[**Erodium cygnorum** Nees. Alien.
A 'wool' introduction, occasionally found where combings or waste has been out-thrown from mills.
A. Guiseley; *Rev. G. E. Pinder.* Yeadon; *B. Carrington.*
C. In the Leeds road, Huddersfield; *Hudd. Nat. Hist.* (sub nom. *E. moschatum*).]

[**Erodium malacoides** DC. Alien.
A Casual on wool waste (A.) near Boothroyd's mill, Meanwood Valley, Leeds; and (C.) Leeds road, Huddersfield (*Hudd. Nat. Hist.*).]

177. Oxalis Acetosella L. *Wood Sorrel.*
Trifolium acetosum vulgare; Ray.
Area: 63, 64, 65, Top. Bot. Range: 0—2,000 feet.
Native; shady hedge-banks, on leaf-mould in woods, very common P. April-May.
Earliest record: Ray, 1724.
Generally distributed—reported from every district.
Obs.—The purple-flowered form was noticed by Dr. Richardson 'in a hollow lane betwixt Northowram and Halifax'; *Ray, Syn.* (1724). It appears to occur chiefly on coal-measures strata.

178. Oxalis corniculata L.
A Colonist; or Casual rather, since fugitive, in sandy fields; very rare. A. June.
First record: Baker, 1863.
W. In a sandy field near Acomb, pretty plentiful in 1870; not to be seen in 1879. A weed in gardens, occasionally, as at Holgate, near York; *J. G. Baker.*
C. Occasionally in gardens about Wakefield; *Gissing's Flora.*

179. Impatiens Noli-me-tangere L. '*Touch-me-not.*'
Persicaria siliquosa; Ger. *Balsamine lutea*; Ray.
Area: 63*, 64*, Top. Bot. Range: 200—450 feet.
Denizen; shady moist ground by streams and woods, rare, possibly indigenous in the shady cloughs at the S.W. corner of Ribble basin. A. June-July.
First notice; Merrett, 1666.
R. Wooded gorge of Ribble, Bolton-by-Bowland. Bashall Woods, Browsholme.
Y. Banks of Skell, Studley; *W. Brunton.* Foot of old walls, Fountains Abbey; *R. Teesdale* (*Linn. Trans.*, 1798).
W. Harewood Park (plantation near the gates on the Leeds road)!
A. Near Bingley, Yorkshire; *Merr.* 92 (1666). Juxta Bingley, *Dr. Richardson*; *Ray*, iii, 317.
D. Wood at Rockley, near Barnsley; *H. Johnson.*

180. Impatiens parviflora DC.
Casual, but a rather frequent, and easily naturalising one. A. July.
C. Near Lumb-bank Mill, above Mytholmroyd (quite naturalised); *A. Stansfield* (1877). Halifax; *W. West* (1881). By stream, Bretton, near Dewsbury; *P. F. Lee*!

[**Impatiens fulva** Nutt.
This also has occurred once or twice, a product of seeds outcast with garden rubbish; being handsome it is often grown in gardens.
C. Near Elland (1865); *W. Guthrie* (*Nat., N. S.*, p. 82).]

[RUTACEÆ.

'**Ruta sylvestris minima,** etc. The wilde are found . . . on the hils of Lancashire and Yorke'; *Gerarde, Hb.*, 1072; *Ger. Em.*, 1256. 'I think there is some mistake here. Gerarde's plant is a true *Ruta*, I think'; *C. C. Babington.* Not mentioned by Ray in his Indiculus Plantarum Dubiarum, appendix to Syn., Ed. iii. Parkinson in his Theat. Bot., p. 133, says: 'How should we beleeve Gerard who saith some of them grow on the hills in . . . Yorkshire, you may easily gesse by their abiding our winters in gardens, for being of so hot and burning qualities, and growing in hot countries, not enduring any cold; how should they be naturall to our climate, especially the more northerly parts?']

XXI. ILICACEÆ.

181. Ilex Aquifolium L. *Holly-tree.* '*Hollins.*'
Area: 63, 64, 65, Top. Bot. Range: 50—1,000 feet.
Native; woods and hedges, common, but *rare* in the limestone dales above 600 feet. Shrub or Small Tree. May-June.
Earliest specific notice of it as a W.R. plant: Salt, 1800.
Distribution general; occurs in every district.

XXII. CELASTRACEÆ.

182. Euonymus europæus L. *Spindle-tree.* '*Witch-wood.*'
Area: 63, 64, Top. Bot. Range: 0—600 feet.
Native: woods, hedges, and thickets, infrequent. Small tree. May-June.
First record: Hargrove, Hist., 1782; but although not stated in that work, I believe the credit of most, if not all, of these records from 1780 to 1800 is due to the Rev. J. Dalton.
L. Near Feizor; *R. Clapham.*
R. Road-side over the High Ridge near Giggleswick; *J. Windsor.* Crow-nest Wood, Settle; *H. Baines.*
Y. Hackfall; and Studley Park; *W. Brunton*; *O.B.G.*
N. Knaresborough (Cliff-wood!), near Priory Gate; *E. Hargrove* (Ed. iii, 1782). Goldsborough.
W. Collingham. Wetherby. Thorparch! *J. Emmet.* Ilkley; *W. West.*
A. Road-side, Kirby Malham; *J. Tatham.* Pontefract! Allerton, Shipley, and Baildon; *W. West.* By Aire below Hunslet Sewage Works! Below Castleford; *G. Roberts.* Sherburn; *W. Whitwell.*
C. Lofthouse; *G. Roberts.*

D. Rotherham district; *L. Langley* (1828). Rossington; *G. Harrison, Hb.* Conisborough. Near Sharrow Moor, Sheffield; *Salt, Sheff. Flo.* Could it have been wild here? *W. W. Newbould.* Near Doncaster.
Obs.—Mr. Roberts remarks, however, that he thinks the Yorkshire 'Witch-wood' is the Wiggan or Mountain Ash.

[**Staphylea pinnata** L. Alien.
Staphilodendron; Plinii. *Nux vesicaria*; Trag. Dod. Ger.
In some woods on the further side of Yorkshire; *How, Phyt.*, 119 (1650); *Merr., Pin.*, 83, 'in the farthest parts.'
In sepibus prope Pontemfractum, non tamen ita copiosé ubi spontaneam assuere ausim; *Ray, Cat.*, 293 (1670); *Hist.*, 1681. Pontefract; *Martyn* (1763). Hedges near Pontefract, 1795; *J. Berkenhout.*
Ray, it seems from the above, *doubted* the nativity of the Bladder-nut tree at Pomfret (loc. cit. and Preface), a fact no botanical antiquary can fail to be glad of, as one more collateral vindication—if any is needed, indeed—of his acuteness and judgment. Hailstone, in 1805, declares it 'truly indigenous,' an instance of his bad judgment, especially after Ray's remark.]

XXIII. RHAMNACEÆ.

183. Rhamnus catharticus L. *Purging Buckthorn.*
Area: 63, 64, Top. Bot. Range: 0—750 feet.
Native; hedges and wood-borders, infrequent, hardly rare. Small Tree. May-June.
First record: Hargrove, 1782.
L. Ingleton; *P. F. Lee*!
R. Kelcowe and Lord's Wood, Settle; *J. Windsor.*
Y. Ripon hedges, Mackershaw Wood; *W. Brunton.* Copgrove, common; *J. Dalton.*
N. Knaresborough, wood by Abbey gates (!) *E. Hargrove*, 1782 Ed., etc.; rep. *Pierson* in *O.B.G.* (1805).
W. Collingham (!), by road towards Harewood; *J. Abbott.* Cross lane, near Sweep plantation, one mile east of Collingham.
A. Gordale; *W. West.* Monk Fryston; *Rev. W. Wood*; *O.B.G.* Kippax! Garforth; *T. W. Gissing.* Hedge at Milford, 1886; *G. Roberts.*
C. Langley Wood, Castleford, and Lofthouse†; *G. Roberts.*
D. Rotherham district; *L. Langley* (1828). Conisborough; *Salt, Flo.* (1800); *Y. N. Excurs.* (1873). Went-vale. Doncaster; *H. Baines.*

184. Rhamnus Frangula L. *Alder Buckthorn.*
Area: 63, 64, Top. Bot. Range: 0—300 feet.
Native; damp peaty thickets, rare, and in the lowlands only. Small Tree. May-June.
First record: Gough, 1789.

N. Hedge, Moss-carrs, between Bickerton and Wetherby; and plantation, Bilton Haggs (!); *W. Todd.*
W. In Thorparch Woods; *Gough's Cam.* (1789). Askham Bogs!
A. ('Malham, etc.;' *Miall* and *Carr. Flora*, p. 92.) Adel bog, by stream near Adel bridge, and lower down near the Aqueduct!
C. (Langley Wood, Lofthouse; *T. W. Gissing*—but with *doubt* expressed.) It is in Thorp Wood, a little higher up; *G. Roberts.*
D. Rotherham district; *L. Langley* (1828). Ecceshall Wood; *Salt, S. Flora* (1800). Levitt Haggs; *Geo. Harrison, Hb.* On the bushy oases of Thorne Waste!
T. Roche Abbey; *Dr. R. Deakin, Hb. Sheff. Lit. and Phil. Soc.*

XXIV. SAPINDACEÆ.

185. Acer Pseudo-platanus L. *Sycamore.* '*Plane-tree.*'
Area: (63, 64, 65), Top. Bot. Range: 0—1,250 feet.
Denizen: self-sown often, woods, hedges, etc., very common. T. May.
Distribution general; to be found in every district.
In hedges (in Yorkshire) not uncommon; *Gough's Cam.* (1789). 'Common;' *Salt* (1800).

186. Acer campestre L. '*Cat-oak.*' *Maple.*
Area: 63, 64, Top. Bot. Range: 0—600 feet.
Native: woods, hedges, stream-sides, common. T. May-June.
First record: Salt, 1800.
Records of its occurrence in M. and L. are desiderata.
Y. Ripon, common; *H. H. Slater.*
N. Frequent from Knaresborough eastward.
W. Ilkley (!) *W. West.* Common, and very fine from Boston Spa eastwards by the river. Knavesmire, York; *W. Whitwell.*
A. Windhill, Shipley; *W. West.* About Meanwood quarries, and Adel! Kippax. Selby.
C. Frequent, but scarcely wild, about Wakefield; *T. W. Gissing.*
D. Pontefract. Ackworth. Conisborough.
T. Roche Abbey; *G. Harrison, Hb.* Common about Sheffield; *J. Salt* (1800).

XXV. LEGUMINIFERÆ.

187. Ulex europæus L. *Furze. Whin. Gorse.*
Area: 63, 64, 65, Top. Bot. Range: 0—1,350 feet.
Native: stony-slopes, commons, common. Shrub. January-April.
First record: Bolton, 1775.
Generally distributed, reported from every district, but not common above 1,000 feet, and less frequent on the scar-limestone than on the gritstone.

188. Ulex Gallii Planch.! *Autumnal Gorse.*

Area: 63, 64, Top. Bot. Range: 0—900 feet.

Native; sandy or peaty moors, infrequent, but locally abundant. Shrub. August-October.

First record: uncertain; G. E. Smith probably, but he does not seem to have put the segregate-name on record in print prior to the Author's record on p. 81 of the *Journal of Botany* for 1874.

M. Greenfield Valley; *J. Whitehead.* Denshaw!

R. Tosside Chapel.

N. Harlow Heath. Hookstone. Plumpton. Long Marston Moor.

W. Ilkley Moor slopes and Otley Chevin. Rombald's Moor; *T. W. Edmondson*!

A. Farnhill Moor, Skipton, at 900 feet or nearly so; *W. Whitwell*! Shipley, Baildon Moor; *W. West.* Bradford Moor. Adel Blackmoor! Yeadon Moor; *B. Carrington.*

C. Todmorden district; *A. Stansfield.* Greetland (!), Norland Moor; *L. C. Miall.* Whitley; *H. F. Parsons.* Stocks Moor, Horbury; *P. F. Lee.*

D. Doncaster race-course! About Cantley; *G. E. Smith, MS.*

189. Ulex nanus Forst.!

Area: '[63, 64 doubtful]' Top. Bot. Range: 20—200 feet.

Native; barren common-land, very rare with us. Shr. September.

First record: Baines, 1840.

D. On the race-course at Doncaster; *H. Baines*; teste *Hy. Ibbotson*, etc.

T. Sandy lanes south by east of Rossington! (With *Ulex Gallii.*)

190. Genista anglica L. '*Cat Whin.*'

Area: 63, 64, 65, Top. Bot. Range: 0—1,300 feet.

Native: heaths and stony moors, infrequent (now)—a decreasing species as moors are brought under cultivation. P. June-July.

First record: Bolton, 1775.

R. Long Gill, Settle; road-side, Rathwell Moor; *J. Windsor.*

Y. Brimham Rocks; *Miss Morton.*

N. Tockwith; *G. Roberts.* Harrogate.

W. White Car Bog, at S.W. corner of Bramham Park! Knavesmire, York, 1877, still tolerably plentiful; *W. Whitwell*, etc.

A. Adel Blackmoor. Morton Bog, near Hambleton! *T. Birks.* Rawcliffe Rabbit-hills; *H. F. Parsons.*

C. Greetland, Warley, Sowerby Moors; *J. Bolton*, 1775. Crosland Moor; *C. P. Hobkirk.*

D. Cadeby Common; *J. Salt.*

191. Genista tinctoria L. *Greenweed.*

Area: 63, 64, Top. Bot. Pelophilous. Range: 0—600 feet.

Native; in stiff-soiled fields and banks, infrequent, plentiful locally. P. July.

First record: Bolton, 1775.

L. About the villages of Wharf and Clapham, abundantly; *H. Baines.*

R. About Giggleswick; *J. Windsor.*

Y. Fields about Fountains Hall, Studley, Bishop Monkton, and Wormald Green! Hutton Conyers, rail-banks; *H. H. Slater.* Galphay Cow-mires; *Miss Morton.*

W. Gibson's Scrub, south end Askham Bog! *H. Ibbotson.* Lane, near Flint Mill, Wetherby! Ilkley Wells (1828); *D. Smith.* Plentiful in rough pastures between Wood Hall and the 'Banks House' Farm (!) near Sicklinghall; *J. S. Wesley.*

A. Shipley; *Miall.* Hawksworth; *W. West.* Monk Fryston; *G. Roberts.* Seacroft, near Leeds; *J. Emmet.* Ledsham, 1877; *W. Whitwell.*

C. Sharlstone, Hell Lane, Royston; *T. W. Gissing.* Wibsey, Low Moor (!); *W. Fowler.* Between Crigglestone and Haigh! Fixby Edge, Storthes Hall Moor; *Hudd. Nat. Hist.* Rastrick and Elland; *J. Bolton* (1775). Hay-beck, near Woodkirk; *G. Roberts.* Ryhill, Heindly; *G. Roberts*!

D. About Sheffield; *J. Salt* (1800). Don banks above Sheffield; *Geo. Harrison, Hb.*

192. Sarothamnus scoparius Koch. *Broom.*

Area: 63, 64, 65, Top. Bot. Arenophilous. Range: 0—1,250 feet.

Native; sandy or gravelly soil, common, but very local upon calcareous soil. Shrub. May-June.

First record: How, 1650.

Distribution general—recorded from every drainage district, and with a wider range in West than in North Yorkshire, ranging from Blea-moor and the rail-bank at Dent Head down to Bawtry and Finningley at a very low level.

[**Genista flore albo.** Broome with a white flower at Wombwell Head in Darfield Parish, *Mr. Stonehouse*; *How, Phyt.*, 45 (1650). The White Spanish Broom was not introduced until 1690 (v. Loudon), but I never saw a white variety of the British species.]

193. Ononis spinosa L. (*O. campestris* Koch. *O. antiquorum* L.)

Area: '63 Moore ms., 64 Tatham,' Top. Bot. Range: 50—300 feet.

Native; dry fields and road-sides, on Permian strata only, quite rare, although locally plentiful. Small Shrub. July.

First record: uncertain; perhaps Gissing, 1865.

(L. Austwick; *W. Howson* in *Flora of Craven*, 1850. But W. H. only meant a Restharrow *with spines.*)

(Y. Near Thieves' Gill, Ripon, in small quantity—the following species abundant there; *H. H. Slater* in *Ripon Flora.* From this I suspect only the spiny form of *O. arvensis.*)

W. Fields about Wetherby and Bramham in a few places!

A. Lanes about Brotherton, Burton Salmon, Pontefract.

C. Wakefield district, 'frequent'; *Gissing's Flora* in *Nat.*, 1865. But some spot on the limestone about Pontefract was probably meant, if, indeed, the spiny form of *O. arvensis* was not confused with it.

D. Warmsworth.

T. About Firbeck. Kiveton; *A. Carr*!

194. Ononis arvensis Auct. (*O. repens* Koch.) *Arrest-harrow.*

Area: 63, 64, Top. Bot. Range: 0—600 feet.

Native; on dry banks, preferring sand, stream-sides, etc., common, although to a certain extent local. Shrub. June-August.

First record: Bolton, 1775.

Distribution nearly general—no record for Mersey district. Greta banks—Ribble side—banks of Yore, Nidd, Wharfe, Aire, and Calder from Methley and Castleford downwards—Ouse and Don banks, common. Road-side bank near Adel Dam. Stream-bank between Fenay and Woodsome; *C. P. Hobkirk.*

Forma **spinosa** I have noticed by the Ouse above Selby, and on Scarcroft Hill (sandy road-side), near Thorner. The Austwick plant of Flora Cravoniensis is it, likewise.

195. Anthyllis Vulneraria L. '*Twins.*'

Area: 63, 64, Top. Bot. Range: 0—1,300 feet.

Native; dry banks, rocky places, etc., on a calcareous, and sometimes on a gravelly, soil, infrequent. P. June-August.

First record: Salt, 1800.

L. Kingsdale lower scars.

R. Kelcowe, Lord's Wood, etc., near Settle; *J. Windsor.*

Y. Ripon, common; *H. H. Slater.*

N. Bleach grounds, Abbey plain, Knaresborough; *E. Hargrove* (1832). Rail-banks north of Starbeck station.

W. Meadows below Barden; *J. G. Baker.* Wetherby. Thorparch and Tadcaster (!); *H. Baines.*

A. Gordale; *W. West.* [Bingley, Bolton, near Bradford; *L. C. Miall.*] Brotherton, etc. Methley rail-bank, on coal-shales; *G. Roberts.*

D. Pontefract. Smeaton; *T. W. Gissing.* Askern; *Geo. Harrison.* Conisborough. Rotherham district; *L. Langley* (1828). About Sheffield; *J. Salt* (1800).

T. Firbeck. Kiveton Park; *G. Harrison.*

196. Medicago sativa L. *Lucerne.*

Alien, not naturalising readily; railway-banks, fields, common in the lowland districts, rare off rail-banks above 300 feet. P. July.

First record: Salt, 1800—'in several pastures near Sheffield.'

Distribution general—has been found in every district, if its Casual stations by railway-banks and sidings be included.

[**Medicago falcata** L. Alien.

Introduced with wool or grass seeds? Has appeared the last few years.

A. Bradford, in one spot; *H. T. Soppitt* and *W. West* (1885).

C. Waste ground at Sandal, near Wakefield; *G. Roberts*!]

[**Medicago orbicularis** Willd. Alien.

The curious-fruited Medick, sometimes grown in gardens, was found on waste ground at Sandal (C.) by *Mr. G. Roberts* in 1886! It may have come with grass-seed or foreign hay, but is more likely to have been the result of a stray garden seed.]

197. Medicago lupulina L. *Medick.* '*Nonsuch.*

Area: 63, 64, Top. Bot. Range: 0—900 feet.

Native: fields and grassy banks, very common, but for the most part keeping in or near cultivated ground.

Distribution general—in every district.

First record: Salt, 1800.

[**Medicago lappacea** Desr. (*M. pentacycla* DC.)

Casual, hardly colonising, but frequently recurring in certain districts. On waste ground where mill waste is out-thrown. A. July.

On 'wool-waste' heaps, Boothroyd's Mill, Meanwood Valley!

Var. **denticulata** Willd.

M. Greenfield Valley, on mill waste; *J. Whitehead.*

A. By Canal, Woodlesford; and waste ground, Fairburn, 1886; *G. Roberts*! By Boothroyd's Mill, Meanwood!

C. Whitley Willows; *Hudd. Nat. Hist.* Sandal; *P. F. Lee*! Dewsbury; *P. F. Lee*!

D. Mark Lane, Sheffield district, 1878; *Amos Carr*!

Var. **apiculata** Willd.

W. Near Cawood; *G. Roberts* (1884)!

C. Near Lofthouse; *Id.* (1886).]

[**Medicago maculata** Sibth. (*M. arabica* All.)

Area: '(63),' Top. Bot.

Casual (in our area), appearing under similar circumstances to the preceding, but not seeming to naturalise.

A. Yeadon; *B. Carrington* (1858). By Boothroyd's Mill, Meanwood Valley (1876)!

C. Whitley Willows; *Hudd. Nat. Hist.*]

[**Medicago ciliaris** Willd. (*M. intertexta* W.) *Hedgehog Medick.* Casual; on 'wool-waste' heaps, Boothroyd's Mill, Meanwood Valley!]

[**Medicago minima** Lamk. Casual.

Has occurred under the same circumstances as the two preceding; and, further, a plant possibly this but reported as '*Melilotus minima*' is said in Yorks. Nat. Recorder (128) to have occurred near Huddersfield.]

198. Melilotus officinalis Willd. (*M. altissima* Thuill.) *Melilot.*
Area: 63†, 64†, Top. Bot.
Colonist; fields, waste ground, railway-banks, etc., common; has become much more prevalent during the last ten years.
First records—these alone are interesting now: Bolton, 1775. 'Near Conisborough' (1798), *Herb. Salt*; (*not M. arvensis*; *W. W. Newbould*). 'Common about Conisborough'; *Salt, MS. Flora* (1800). Doncaster neighbourhood; *E. Miller* (1804).
Distribution general now—along the lines of rail and water traffic—recorded from every district. I saw it flowering in 1885 on the rail-bank near Dent station at over 1,000 feet elevation (2 or 3 plants).

199. Melilotus alba Lam. (*M. leucantha* Koch.)
Area: (63, 64), Top. Bot.
Colonising Alien: in fields, on waste ground, railway banks, and by canals, tenderer and less common than the preceding, but still not rare, although often fugitive in its stations. A. or B. July.
First record: Baines' Flora, sub *M. officinalis*, 1840.
N. The cliff at Knaresborough, 1869!
W. Near Bishopthorpe; *H. Baines*, p. 30.
A. Silsden; *L. C. Miall.*
C. Lofthouse (1883, 10 feet high); *G. Roberts.* Dewsbury; *P. F. Lee.*
D. Doncaster. Goole.

200. Melilotus arvensis Wallr. Alien.
Casual or Colonist, much rarer than the preceding. A. July.
W. Rail-bank, Wetherby!
A. Sidings, Milford Junction (1879).
C. Balne (properly Balm) Lane, Wakefield (1856); *T. W. Gissing.*

201. Melilotus parviflora Lam. Alien.
Colonist; spreading and establishing itself better than the two preceding (or is it oftener reinforced?); on waste ground, near flour-mills chiefly.
W. Tadcaster. Arthington, rail-bank (1869)! Cawood; *G. Roberts* (1884).
A. Woodlesford, Methley, and Castleford; *G. Roberts*!
C. Lofthouse and Sandal; *G. Roberts*! Near Hirst's Corn Mill, Huddersfield; *J. Tindall* (1873). Horbury, and Turnip field, Dewsbury; *P. F. Lee*! Lofthouse, every year since 1882; *Id.*
D. Amongst oats, Goole Fields (*T. Birks*); Conisborough. Barnsley! Doncaster.
Obs.—Mr. Roberts remarks that to his knowledge it ripens its seed, and is *not* dependent on flour mills. Brought with clover seed.

[Trigonella ornithopodioides DC. Incognit. Extinction?
Area: [64 authority insufficient], Top. Bot.
Trifolium Ornithopodii siliquis. Bird's Foot Trefoil. Found by Tho. Willisel among the corn half a mile from Tadcaster, as one goeth up the hill towards Sherburn in the road to Doncaster; *Ray, Cat.*, Ed. ii, p. 291 (1677); *Syn.* i (1690), and *Syn.* iii, 331. Probably a Casual, at any rate never seen since, if not confused with *Ornithopus*, which is common now in the district indicated. Record repeated by *Hudson, Gough, Withering, O.B.G., Smith*, etc., and later, unquestioned, by *Baines.*]

[Trigonella Fenu-græcum L. '*Fenugreek.*' Alien.
Old quarry at Sandal (C.), sown with eastern grass or clover seed, 1886; *G. Roberts*! Not grown in the district for fodder, being an uncertain crop.]

[Trigonella gladiata Bieb. Alien.
With the preceding in 1886; *G. Roberts.*]

[Trigonella monspeliaca L. Alien.
On shingle by river Wharfe below the flour mill, Tadcaster, in 1880—one plant! Out-thrown with screenings of South European grain.]

202. Trifolium pratense L. *Red Clover.*
Area: 63, 64, 65, Top. Bot. Range: 0—1,500 feet.
Native; pastures, banks, etc., very common, but most frequent as a forage plant. P. May-August.
First record: Salt, 1800.
Generally distributed—every district.

203. Trifolium medium L.
(*T. alpestre* Huds., Gough, etc.)
Area: 63, 64, Top. Bot. Range: 0—1,250 feet.
Native; hedge and rail-banks, bushy places and heaths, infrequent. P. July.
First record: Bolton, 1775.
Distribution general—in every district, but sparingly, rendering it worth while to cite localities.
M. Greenfield Valley; *J. Whitehead.*
L. In Skirwith Wood (near Ingleton); *W. Curtis*; *Gough's Camd.* (1789).
R. Bowland and Tosside. Rail-banks, Hellifield.
Y. By the Yore near Sharow; *H. H. Slater.*
N. Spofforth. Harrogate.
W. Collingham. Ilkley. Barden.
A. Skipton. Bingley and Baildon; *W. West.* Meanwood, etc. Malham Moor!

C. Illingworth, Rastrick, etc.; *J. Bolton.* Todmorden. Ossett and Allerton; *T. W. Gissing.*
D. River-side, Sheffield; *J. Salt.* Wharncliffe, etc.
T. Waleswood; *A. Carr.* Finningley.

[Trifolium maritimum L.
One Casual plant (!) found at Old Goole, by road-side where ballast was often laid, by Rev. T. Dickin in 1878.]

[Trifolium ochroleucon Huds. Casual.
Area: '(64),' Top. Bot.
In clover and seed-fields, rare.
Y. Clover-field, by Ripon, a single specimen; *W. Brunton* in *O.B.G.* (1805).
C. Sandal; *G. Roberts*!
D. Field near Barnsley, one or two specimens only; *H. Johnson* (1880).]

[Trifolium incarnatum L. Casual.
Occasionally cultivated as a fodder crop; and stray examples have occurred at Thorparch (!); Ripon; Roundhay; Milford and Kippax (*G. Roberts*!); Barnsley, etc.]

[Trifolium resupinatum L. Casual.
Has occurred, sparingly, several times.
W. Cawood; *G. Roberts*! Tadcaster!
A. Methley (in clover); *G. Roberts.* Boothroyd's Mill, Meanwood!
C. Waste ground at Sandal, with other exotics; *G. Roberts.* Balne Lane, Wakefield; *T. W. Gissing.*]

204. Trifolium arvense L. *Hare's-foot Clover.*
Area: 63, 64, Top. Bot. Arenophilous. Range: 0—900 feet.
Native; sandy banks and fields, infrequent, but locally common. A. July-August.
First record: Salt, 1800.
Y. Blind pit on Hutton Moor; *H. Baines.*
N. Spofforth (!) *T. Scatcherd.* Braham Hall fields. Near Poppleton, 1877; *W. Whitwell.*
W. Near Rufforth. Ainsty; *Hy. Ibbotson.* Acomb Green, 1866; *W. Whitwell.*
A. Otley Chevin; *B. Carrington.* Knottingley; *T. W. Gissing.* About Hambleton.
C. Sandal; *G. Roberts.* Brighouse; *L. C. Miall.*
D. Near Rotherham; *Herb. Salt* (1799). Heck; *H. F. Parsons.*

205. Trifolium striatum L.
Area: 63, 64, Top. Bot. Range: 0—450 feet?
Native; sandy banks and fields, infrequent. A. June-July.
First record: Salt, 1800.
Y. Near Boroughbridge!

N. Fields and lane-sides at Spofforth and Braham, three and a half miles from Wetherby!
A. Brayton Barf.
D. Near Doncaster; *Salt, MS. Flora.* Armthorpe! Cantley; *G. E. Smith.* (Between Bowbridge and Canklow Mill, five miles from Sheffield; *H. Baines.* Error? A spn. of *T. medium* with the foregoing localisation is in *Herb. Salt.*)

206. Trifolium scabrum L.
Area: 64, Top. Bot. Range: 0—300 feet.
Native; sandy banks, very rare, but perhaps by-passed for *T. striatum* occasionally.
First record: Brunton, 1805.
Y. Ellershaw Hill, near Ripon; *W. Brunton* in *O.B.G.* (1805).
W. Sandy lane between Acomb and the Ouse; *H. Baines.*

[Trifolium subterraneum L. Incognit.
Hilly barren pastures, in Warley and . . . Skircoat, Halifax; *J. Bolton, Cat.* (1775). What Bolton meant is uncertain, I think.]

207. Trifolium hybridum L. (incl. **T. elegans** Savi). '*Alsike.*'
Denizen; now well established, and spreading, at low levels. P. June.
First record: Tatham, 1862.
Probably general in its distribution, but no actual record for Lune and Mersey districts.
R. About Settle; *J. Tatham* (1862).
Y. About Ripon and Boroughbridge.
N. Spofforth. Cowthorpe. Nun-Monkton.
W. Common about Wetherby, Kirk Deighton, Bramham, etc.
A. Garforth. About Selby. South Milford.
C. Sandal.
D. Doncaster.
T. Waleswood; *A. Carr.*

208. Trifolium repens L. *White Clover.*
Area: 63, 64, 65, Top. Bot. Range: 0—1,800 feet.
Native; old pastures, heaths, and road-sides, very common. P. June-September.
First records: How, 1650; Salt, 1800.
Generally distributed—every district.
The proliferated state '*T. flore viridi foliaceo elegans*,' 'In mine own orchard at Darfield,' appears to have been found by Mr. Stonehouse (*How, Phyt.*, 124) in 1650; commented on by *Merrett* in his *Pinax*, p. 120.

[Trifolium spumosum L. Alien.
On waste ground in field, and in a disused quarry near Sandal, 1886 *G. Roberts*! Brought with foreign clover seed?]

209. Trifolium fragiferum L.

Area: 63, 64, Top. Bot. Sub-maritime. Range: 0—200 feet.

Native; grassy, sandy places, very rare, only in the central plain, once an arm of the sea.

First record: Baines, 1840.

Y. In an old quarry, near Wicliffe Lane, Studley; *H. H. Slater*.

W. (Oglethorpe; *T. Scatcherd* in *Miall's Flora*.) Knavesmire, York; *H. Baines*.

A. Banks of river Aire, Newland; *H. F. Parsons*.

D. Alluvial ground at Adwick-le-Street in 1844; *P. Inchbald*. By the Dutch river, Goole; *H. F. Parsons*!

Obs.—This Trefoil usually shows distinctly marine preferences, but Reichenbach remarks that it inhabits also 'deep clayey ground.' In its Yorkshire distribution it has a fellow in *Sclerochloa distans*.

210. Trifolium procumbens L.

Area: 63, 64, Top. Bot. Range: 0—900 feet.

Native; dry banks, fields of sown grass, etc., common, but mostly on a dry gravelly or sandy soil. A. June-August.

First record (distinctly for W. Riding): Miall's Flora, 1862.

Distribution general; all districts except Mersey.

[**Trifolium agrarium** L.

Occasionally in clover fields, introduced with other seed. I have it from (Y.) Sharow; (N.) Harrogate; (A.) Brotherton; (C.) Mirfield; and (D.) near Thorncliffe, Sheffield (*A. Carr*).]

211. Trifolium minus Relhan.

Area: 63, 64, 65, Top. Bot. Range: 0—900 feet.

Native; fields, road-sides, and on rocks, very common, but not ascending very high. A. June-August.

First record: Salt, 1800, sub nom. '*filiforme*.'

Distributed generally—reported from every district.

212. Trifolium filiforme L.

Area: 63, 64, Top. Bot. Range: Low.

Native; dry sandy places, very rare. A. June-July.

First record: H. Ibbotson, 1858.

(R. Foot of Giggleswick Scar; *J. Windsor*. Some error; there is small *T. minus*.)

(Y. Ripon, waste ground, common; *H. H. Slater*, *Flora*. An error!)

W. Ouse banks, Clifton Ings, etc., near York; *H. Ibbotson*!

D. Smeaton Crags, Went-vale, 1869. Rawcliffe Rabbit-hills; *H. F. Parsons* (1878).

Obs.—No Yorkshire plant has been more confused and misreported than this. Salt's MS. Flora entry, 'common on heaths, etc.,' referred to the preceding species. ('Frequent'; *Miall's Flora*!)

213. Lotus corniculatus L. *Common Bird's Foot.* '*Yellow Clover*.'

Area: 63, 64, 65, Top. Bot. Range: 0—1,900 feet.

Native; on dry sandy, heathy, or calcareous banks, very common. P. June-September.

First record: Salt, 1800.

Distribution general—reported from every district, and in none scarce.

Obs.—The Dale farmers say sheep thrive on this; but I scarcely think they care for it. What is meant is that stock thrive best on pasture-land where it is plentiful, which means that the soil is a tolerably dry one. It will not grow on the ill-drained rushy slopes that are so productive of foot-rot and mal-nutrition.

214. Lotus tenuis Kit.

Area: '64 Foggitt', Top. Bot. Range: 0—300 feet.

Colonist; gravelly grassy banks, rail-banks, sown meadows, etc., infrequent. Not met with in old untouched meadow or pasture land. P. June.

I am not quite clear whether this species is of comparatively recent introduction in West Yorkshire, or a true native.

First record: Foggitt, 1875.

R. Rail-bank, Hellifield! Casual?

N. Field at Harlow, Harrogate; *T. J. Foggitt*! Called an 'upland pasture,' but clearly 'refreshed' with grass seeds—and *Crepis nicæensis* grew with it.

W. Rail-banks and seed-fields about Wetherby, Thorparch, Collingham.

D. Rail-bank, Sprotborough (!) *W. Todd*. Ryhill; *H. Johnson*!

215. Lotus uliginosus Schk. (*L. major* Sm.)

Area: 63, 64, Top. Bot. Pelophilous. Range: 0—1,250 feet.

Native; moist or peaty bushy places, common. P. July-August.

First record: H. Baines, 1840, for York generally; for West Riding uncertain.

M. Greenfield Valley; *J. Whitehead*.

R. Settle, common; *J. Windsor* (*Phyt.*, 1855-8).

Y. Ripon; *H. H. Slater*. Kirby Malzeard!

N. Scotton banks, etc., Knaresborough. Aketon, near Spofforth! Harrogate; *G. Harr*.

W. Wetherby. Ilkley. Woodhall! Askham Bogs; *B. B. Le Tall*!

A. Malham Moor. Skipton. Shipley; *W. West*.

C. Todmorden. Wakefield; *T. W. Gissing*. Coxley Valley; *P. F. Lee*.

D. Bushy dike-sides about Thorne.

Obs.—The hairy form (*pilosus* Beeke) is commoner than the glabrous one throughout Yorkshire.

216. Astragalus Hypoglottis L.

Glaux Dioscoridis; Ger.; Ray. *Astragalus incanus parvus*; Pluk.

Area: 63, 64, Top. Bot. Xerophilous. Range: 0—300 feet.

Native; dry grassy banks and road-sides, rare. P. June-July.

First record: Ray, 1670.

In comitatu Eboracensi; *Ray*, *Hist.*, 939 (1686).

W. About Hazelwood (near Aberford) in Yorkshire; *Ray*, *Cat.*, i (1670); *Martyn*; *G. Roberts*. Thorparch; *Rev. W. Wood* (*O.B.G.*). Stubbing Moor, Cross roads, above Rigton! Road-side between Collingham and Thorparch (!); *J. Emmet*. Jackdaw Crag, Stutton (!); *O. A. Moore*.

A. By road between Sherburn and Huddlestone quarry; *H. Baines*. Kippax lime-hills! Glass-Houghton; *G. Roberts*. Ledsham Rectory Field; *W. Whitwell*.

D. Beck Hill, near Hampole; *P. Inchbald*. Smeaton Crags! = Went hill and vale; *T. W. Gissing*. Doncaster; *Martyn*; rep. *Gough's Cam.* Non procul Doncastria; *D. Richardson*; *Ray*, *Syn.* (1724). Rotherham district; *L. Langley*. Pigburn fields, Doncaster; *O. A. Moore*. Brodsworth; *Deakin* in *Herb. Salt*.

217. Astragalus glycyphyllus L. '*Wild Liquorice*.'

Area: 63, 64, Top. Bot. Range: 0—300 feet.

Native; hedges and thickets, on calcareous soil, infrequent, and local though in a good many localities. P. June-July.

First record: Bolton, 1775.

Y. Hackfall; *W. Brunton*. Common near Sleningford; *J. Dalton*; rep. *H. Baines*. Hedge near the Rifle Range, Ripon; *H. H. Slater*. ? = Hedge near Queen Mary's Dubb!

N. Near Flaxby; *R. Teesdale* (*O.B.G.*) = Bridle-road between Flaxby and Allerton Park (!); *E. Hargrove* (Ed. iii, 1782).

W. Hedge on slope above Whitwell House Farm, Linton (!); *J. Jackson*. Thorparch (!); *H. Baines*.

A. Monk Fryston; *W. Wood* in *O.B.G.* (1805). Burton Salmon; *G. Roberts*. Lane between Lumby and Ledsham; *W. Whitwell* (1881).

[C. Above Mixenden and Warley; *J. Bolton*. Can this have been wild there; or is the flora of the country undergoing a gradual change, of which this and other observances of Bolton's are a proof?]

D. Roman ridge, Hampole; *P. Inchbald*.

T. Roche Abbey; *J. Bohler*. Anston, Lindrick, Kiveton Park, frequent; *Geo. Harrison*.

[**Robinia Pseudacacia** L. *The False Acacia Tree.* Alien.

Always a planted tree—a native of North America, but occasionally found in situations to all appearance wild; as, for example, in an old quarried depression nigh a pool (marked Hell Hole on the Ordnance Map) near the road between Spofforth and Crosper in district N.]

[**Glycyrrhiza vulgaris** Ger. *Common Liquorice.*

It is planted . . . about Pontfraict in Yorkshire, which is commended for the best; *Ray*, *Cat.*, i, 135 (1670). Ray seems to have been the earliest botanist who distinguished rigidly between wild and introduced species.]

218. Ornithopus perpusillus L. *Bird's Foot.*

Area: 63, 64, Top. Bot. Range: 0—1,050 feet.

Native; dry sandy fields, banks and waysides, infrequent and local. A. May-June.

First record: Bolton, 1775.

Y. Road from Ripon to Galphay; *H. H. Slater*.

N. Quarry near Plumpton Bar (!); *E. Hargrove* (1809). Crosper fields, Spofforth.

W. Ainsty sand fields; *G. Webster*. Tadcaster; *H. Baines*. Otley Chevin; *B. Carrington*.

A. Baildon Green; *W. West*. Hambleton. Adel, near the Roman Camp!

C. Skircoat, etc., Halifax; *J. Bolton*. Honley, rare; *Hudd. Nat. Hist.* Heath and Oakenshaw; *T. W. Gissing*.

D. Rotherham district; *L. Langley* (1828). Crook's Moor; *J. Salt* (1800). Pollington; *H. F. Parsons*.

[**Ornithopus ebracteatus** Brot. Casual.

(63, S.W. York; Casual), Top. Bot.

C. Canal bank by Oaks Farm, Barnsley; *Elihu Berry*, *Phyt.*, vol. iii, 336. I know nothing further of this, but such occurrences were thought more singular and worthy of preservation 30 years ago than we now know such ephemeral things to be.]

219. Hippocrepis comosa L.

Ferrum equinum Germanicum siliquis in summitate; Ray.

Area: 63 (sine personal authority), 64 Tatham, Top. Bot. Montane only in Yorkshire. Range: 500—1,500 feet.

Native; on the ledges of limestone rocks in the western dales, infrequent, locally plentiful. P. July.

First record: Ray, 1724.

L. Wherf (Wharf near Austwick); *Ray*, *Syn.*, and *Martyn*.

R. Settle; *Ray*, *Martyn*, etc. Attermire Scars above Settle; *H. Ibbotson*. Kelcowe, Giggleswick Scars, Stackhouse; *J. Windsor*.

A. Near Malham, on calcareous rocks, copiously (!) *Ray*, *Syn.*, iii (1724), *Martyn*, etc. Gordale!

W. Kilnsey; *W. Curtis*.

220. Onobrychis sativa Lam.

Area: '(63, 64),' Top. Bot. Range: Low.

Denizen; long established in a few places on the Permian limestone, but casual only in most of its recorded localities. P. June-July.

First record: Salt, 1800.

W. Wetherby and Thorparch; *H. Baines.*

A. Burton Salmon; *G. Roberts.* = Between Tadcaster and Ferrybridge; *H. Baines* (1840). Kippax. (Saltaire, Shipley, Burley; *L. C. Miall.* Casual here.)

(C. Huddersfield and Brighouse; *L. C. Miall.*)

D. Between Rotherham and Maltby; *J. Salt* (1800). Adwick-le-Street, foot of rail-banks! Road-side cutting, Wentbridge; *H. F. Parsons*!

T. Field-banks, Letwell, perhaps a relic of cultivation.

Obs.—Mr. Baker, in Supp. to Baines' Flora (1854), gives this as an unquestioned Native for Yorkshire, but on very slender grounds, I think. His matured opinion (1886), however, is—'I know of no thoroughly satisfactory station.' It may be indigenous on the chalk wolds; it certainly is not, however, in West Yorkshire. Formerly it was much cultivated, and where introduced into suitable stations is very tenacious.

221. Vicia hirsuta Koch. *Wild Tare.*

Area: 63, 64, Top. Bot. Range: 0—600 feet.

Native; cornfields, amid clover, sandy hedge-banks, common. A. July-August.

First record: Bolton, 1775.

Distribution general—to be found in every district.

222. Vicia tetrasperma Mœnch.

Area: 63, 64, Top. Bot. Range: 0—300 feet.

Native; bushy hedge-banks and fields, on gravelly soil, rare. A. July.

First record: Salt, 1800.

Y. Ripon, not uncommon; *H. H. Slater.*

W. About Spofforth and Wetherby!

C. Near Luddenden; *S. King.* Heath, near Wakefield; *T. W. Gissing.*

D. Border of Goole Moors; *H. F. Parsons*! Doncaster; *Y. N. U. Trans.* (1879). Near Sheffield; *J. Salt.* Woodhouse; *G. Harr., Hb.*

[Vicia (Ervum) Lens DC. Alien.

This has occurred, as a Casual, in seed-fields, where wool waste has been thrown out, and near flour mills. Tadcaster, and Boothroyd's Mill, Meanwood.]

223. Vicia Cracca L. *Hedge Vetch.*

Area: 63, 64, 65, Top. Bot. Range: 0—1,250 feet.

Native; hedges and thickets, on all soils almost, very common. P. July.

First record: Bolton, 1775.

Distribution general—reported from every district.

224. Vicia sylvatica L. *Wood Vetch.*

Area: 63, 64, Top. Bot. Range: 0—600 feet.

Native; rocky woods and thickets, rare. P. July-August.

First record: Gough, 1789.

L. In Skirwith Wood, near Ingleton; *Gough's Cam.* (1789). = 'Tatham's Wood, Ingleton'; *H. Baines.* Lower Dentdale.

R. Birkwith, near Horton; *W. Howson.*

Y. Old wood, Hutton Moor; *J. Dalton* (1805)—perhaps a little outside the Riding.

A. Plentifully on the edge of the rivulet just above Hanlith Bridge, near Kirby Malham (!); *J. Windsor.* = Right-hand side of road by the Aire side, half-way between Hanlith Bridge and Seale Gill.

C. Hareley Wood, Stanelly Clough, Todmorden; *A. Stansfield* (1877). Woods, Chevet.

D. Rocky wood, Sprotborough, 1870.

225. Vicia sepium L. *Bush Vetch.*

V. fol. subrotundo brevi obtusi mucronato; Ray.

Area: 63, 64, 65, Top. Bot. Range: 0—1,200 feet.

Native: railway and hedge-banks and bushy places, very common, but not often in any quantity in one spot. P. May-July.

First record: Ray, 1696.

In agro Eboracensi, *D. T. Robinson*; *Ray, Syn.*, ii (1696), p. 188.

Generally distributed—reported from every river-basin.

226. Vicia sativa L. *Fodder Vetch.* '*Tares.*'

Casual, but constantly recurring because constantly reinforced by re-sowing in varying places. I do not think it is anywhere naturalised within our area. With us always a waif of field cultivation, and where it does stray to a hedge or bank, being unfitted to hold its own in competition, it never lasts long; common within the lower zone of cultivated land, very rare above 500 feet.

First record: uncertain; Salt's in Fl. Sheff., 'Dry meadows and pastures,' probably referred to what we now call *V. segetalis.*

Distribution nearly general.

227. Vicia angustifolia Roth. **a. segetalis** Thuill. *Wild Vetch.*

Area: 63, 64, Top. Bot. Arenophilous. Range: 0—750 feet.

Native: sandy lanes, banks, and fields, common. A. May-July.

First record: Salt, 1800, sub nom. *lathyroides.*

Distribution nearly general—every district except Mersey.

The Form **Bobartii** Forst., with very narrow upper leaflets, is not uncommon in dry sandy situations, but it passes by insensible degrees into the type.

228. Vicia lathyroides L.

Area: '64,' Top. Bot. Range: Low.

Native; in grassy sandy soil, rare, some seasons intermitting, as it likes a warm moist season, and in arid ones does not germinate. A. April-May.

First record: Baines, 1840.

Y. Mickley Burrows, Ripon (!) *H. Baines.*

N. Sandy moory lanes at Poppleton (*H. Ibbotson*!) and Moor Monckton!

W. Sand pits at Acomb (and sandy lanes about the Kennels!) and lane leading to the Ouse; *H. Baines.* (Thorparch; *T. Scatcherd.* An error—*V. Bobartii* was probably meant.)

(A. Field borders, Thornton; *L. C. Miall.* One of the very numerous errors of Miall and Carrington's Flora.)

(D. 'On dry parts of Crook's Moor, Sheffield'; *J. Salt, Flora, MS.*, but perhaps the plant was starved *V. angustifolia*, as is the '*V. lathyroides*' of Herb. Tofield, 4to Coll., according to Mr. Newbould.)

[Vicia bithynica L. Incognit.

Near Doncaster, *Mr. Tofield*; *Curtis*; *Gough's Cam.*; *et aliis.* The Old Bot. Guide of 1805, and Baines in 1840, repeat the record amplified, as if universal almost in the area indicated, to 'gravelly cornfields, pastures, and ditch-banks,' but there has been no confirmation, and I do not know of a professed Doncaster specimen in any Herbarium except Tofield's own, in the hands of Dr. Younge, of Sheffield, in 1800; cfr. *Sm., Flo. Brit.*, ii, 774.]

[Lathyrus Aphaca L. Casual.

Area: (64), Top. Bot.

This—a native in the south—has occurred as a fugitive waif as follows:—

Y. Cornfield at Monckton, near Ripon; *W. Brunton* in *O.B.G.* (1805).

W. Cornfield at Cawood, a few plants only; *G. Roberts*!

A. Apperley; *R. Dewhirst* and *H. T. Soppitt.*]

[Lathyrus hirsutus L. Alien.

Found casually at Marley, near Bingley, 1885; *J. T. Walsh* teste *H. T. Soppitt* and *W. West.*]

229. Lathyrus pratensis L. *Vetchling.* '*Wild Pea.*'

Area: 63, 64, 65, Top. Bot. Range: 0—1,250 feet.

Native; moist ditch-banks, fields, open coppices, very common. P. July-August.

First record: Salt, MS. Flora of Sheff., 1800.

Distribution general—in every district.

[Lathyrus sylvestris L. Incognit.

Area: (63, 64), Top. Bot.

Reported by the W.R. Consol. Nat. Society, but I believe in error, and also given by *J. E. Kenyon* (*Bot. Chron.*, August 1864) for Hooton Pagnell. I am uncertain under what conditions this has occurred; but, as an Escape from gardens, the Everlasting Pea (*L. latifolius*), closely allied to it, has been found occasionally and perhaps has been mistaken for the less ornamental smaller-flowered wilding.]

230. Lathyrus palustris L.

Area: '63' and '64' (sine authority), Top. Bot. Range: Low.

Native; bushy boggy places, very rare. P. August.

First record: G. E. Smith, 1845.

Yorkshire, wet meadows; *Gough's Cam.* 'Frequent in the Co.'; *Huds.*

(A. Near Leeds; *Rev. W. Wood*; *Bot. Guide* (1805). Heslington probably meant—that was 'near' enough for Wood.)

D. Near York (and Doncaster); *Dr. White* (1804). White's list was for *both* areas, and he, too, probably only knew the old Heslington station near York, but in E. Riding. Marsh below Brampton (S.E. of Doncaster); *G. E. Smith* (1845). Border of Thorne Moor, bushy dike-side, S.W. angle, 1872!

231. Orobus tuberosus L. '*Wood Pea.*' '*Peasling.*'

Area: 63, 64, 65, Top. Bot. Range: 0—1,500 feet.

Native; on grassy and bushy banks, woods, etc., very common. P. April-June.

First record: Salt, 1800.

Distributed generally—in all the districts.

The writer can remember this growing in 1860-64 on Woodhouse Ridge and in Batty Wood, with *Asperula odorata* and *Lamium Galeobdolon*—all now gone, within a mile and a half of Leeds.

The Var. **tenuifolius** is not uncommon, usually growing with the type. Records of it from every district except Trent and Lune.

[Cicer arietinum W. The *Chick Pea.* Alien.

Occurs casually on wastes amid refuse from flour-mills, etc. Has so occurred at Mirfield, Huddersfield (*J. Tindall*), Tadcaster, Aysgarth, Cawood, etc., to the author and his correspondents since 1865; but probably existing from a very early date. Introduced into England about 1548. I know not to what else to refer the entry in Merrett, 94 (1666): '*Pisum in extremitate clavatum.* At Bingley, Yorkshire, called Chick Peas.' Probably its cultivation was attempted soon after its introduction. These early writers were not accustomed to distinguish wild from cultivated plants very rigidly, and one finds Merrett, How, and others mentioning the cultivation of Liquorice (*Glycyrrhiza glabra*), a plant of the Pea tribe, at Pontefract on exactly the same terms.]

[**Scorpiurus subvillosus** DC. Alien.

Has occurred twice or thrice, the scorpoid, noduled, ribbed, and prickly pods entangling in wool like burrs, and so coming to be out-cast with the dressings; fugitive, of course.

C. Wool-refuse, near Huddersfield; *J. Tindall* (1878)! Batley; garden of Dr. Stuart's house dressed with wool-waste as manure (1885)!]

XXVI. ROSACEÆ.

232. Prunus spinosa L. *Sloe. Blackthorn.*

Area: 63, 64, 65, Top. Bot. Range: 0—900 feet.

Native; hedges and thickets, common, although only fruiting well in the warmer districts, as about Bishop Monckton. Shrub. April.

First record: Bolton, 1775.

Distribution general—occurs in every district.

233. Prunus insititia L. *'Bullaster.'*

Area: '64 Tatham,' Top. Bot. Range: 0—300 feet.

Denizen; hedges and thickets, very rare. Small tree. April-May. Fr. August.

First record: Salt, 1800.

R. Stainforth, near Settle; *W. Howson* (1850).

W. Lanes about York. Thorparch; *H. Baines*. Bramham.

A. Pontefract; *H. Baines*.

D. Sprotborough (!) *Geo. Harrison, Herb.* 'Common in hedges'; *J. Salt, Flora. P. spinosa* not meant, for it is lined also.

T. About Tickhill and Sandbeck, fruiting well.

[**Prunus domestica** L. *Common Plum.* Alien.

Although this gradates into *P. insititia*, going back when neglected, it has no claim to be included in our Flora as anything but an always planted alien.]

234. Prunus Avium L. *Wild Cherry.*

Area: 63, 64, Top. Bot. Range: 0—600 feet.

Denizen; coppices, rocky woods, hedgerows, common in the lowland districts, rare above 300 feet in the gills of the dales. Tree. May.

First record: Salt, Flora, 1800, sub nom. *Cerasus*.

Distribution general—occurs in all the districts, but *usually* as an evident introduction. No record for Mersey district.

R. Bolton-in-Bowland. Winskill Wood, Settle; *W. Howson, Flora.*

Y. Studley, Kirby Malzeard, Hackfall, etc.

N. St. Francis Chapel Woods, Stockeld. Plumpton. Scotton banks.

W. Many places from Bolton Woods downwards by Harewood, Woodhall to Tadcaster and Stutton.

A. From about Gargrave downwards to Castleford. Ledsham.

C. Sharlestone, Barnsley, Huddersfield, etc.

D. Rossington. Sandall! Wharncliffe. Wardsend, Sheffield; *J. Salt.*

T. Sandbeck and Roche Abbey Woods.

235. Prunus Cerasus L. *'Morella.'*

Area: 64, Top. Bot. Range: 0—500 feet.

Denizen; copses and hedges, rare. Small Tree. May-June.

First record: Dixon, 1790, if his plant was not really *P. Avium*; Salt's was.

W. Two or three trees in a fir-wood on the other side of the lane to the blacksmith's shop at Scarcroft! Thorparch, one tree!

A. Bolton Wood, Bradford; *L. C. Miall.* Rawden Woods (!); *B. Carrington.*

C. In Mr. Smyth's Wood at Heath, often in fruit, 1790; *D. Dixon.* Grimescar Wood, Huddersfield; *C. P. Hobkirk.* Walkroyd and Wheatley; *L. C. Miall.* Battye-ford, near Mirfield; *H. F. Parsons*!

236. Prunus Padus L. *Bird-cherry. Heggberry*, or *Heckberry. Cerasus avium nigra racemosa*; Ger.; Ray.

Area: 63, 64, 65, Top. Bot. Sub-montane. Range: 1,400—50 feet.

Native; thickets, stream-sides, and gills, common among the hills, but rare in the low country. Tree. May-June.

First record: Ray, 1670.

In occidentali montosa comitatus Eboracensis *Ray, Cat.*, i, 64 (1670); *Hist.* and *Syn.* In the mountainous part of the West Riding of this county; *Ray* in *Gibs. Cam.* (1695).

L. Sedbergh; *Pinder.* Rawtheydale and Dentdale, many places! Thornton Force. Greta banks.

R. Kelcowe, Clayton's Wood, and many places (!) near Settle; *J. Windsor.* Bolton-by-Bolland. Slaidburn.

Y. Oldfield Wood, Hackfall; *W. Brunton.* Dunkirk Wood, Sleningford; *J. Dalton.*

W. Woods in Wharfedale, plentifully; *Rev. W. Wood*; *O.B.G.* Buckden. Arncliffe. Bolton Woods! Nesfield. Ilkley. Farnley; *H. F. Parsons.* Washburn Valley above Leathley (!); *J. Abbott.*

A. Airedale; *W. Wood* in *O.B.G.* Shipley, etc., and Bolton Outlane, near Bradford; *H. Baines* and *L. C. Miall.* Rawden Woods; *B. Carrington.* Less frequent in Airedale than in Wharfedale or Ribblesdale.

C. Calder Vale; *Rev. W. Wood* (1805). Hebden Valley! (1879). About Todmorden, many places; *A. Stansfield.* Kirklees Wood; *Y. N. Excurs.* (1873). Honley and Bradley; *L. C. Miall.* Haw Wood near Wakefield; *T. W. Gissing.* North Dean, Norland, Copley, etc.; *J. Bolton* (1775). Meltham; *Hudd. Nat. Hist.*

D. Boggy wood near Rossington! cum *W. Fowler.* Rotherham district; *L. Langley.* Askern (!); *H. F. Parsons.*

Obs.—Whether the north-country name for this beautiful and abundant constituent of the dale woods was or was not really derived, by excentric adoption, from Heggdale in Eden-dale (where it is plentiful), as Parkinson implies (Theat. Bot., 1505), it is certain that everyone in the western dales of Yorkshire calls it Heggberry tree, and not one in a hundred would know what was meant if asked whether the Bird-cherry grew about or not. But against Parkinson's view is a statement that the Wild Cherry (*P. Avium*) is in Scotland known as Hagberry. A rough bushy uncultivated place, and also a furrowed peat moss, are in Yorkshire indifferently styled *haggs*.

[**Spiræa salicifolia** L. Alien.

A doubtless-planted shrub, for which one or two stations are on record.

N. Hedge between Green Hammerton and Knaresborough; *R. Teesdale* (1798) = Near Thornville Royal, in the lane between that place and Green Hammerton, a large plant, but may have been brought with quicks from a garden when the hedge was originally planted; *Rev. J. Dalton* in *O.B.G.* (1805).

W. Briery Wood, Addingham; *L. C. Miall.*

C. Near Huddersfield; *L. C. Miall.*]

237. Spiræa Ulmaria L. *Meadow-sweet.*

Area: 63, 64, 65, Top. Bot. Range: 0—1,500 feet.

Native; swamps, ditch-banks, and moist meadows, very common. P. June-July.

First record for Riding: Bolton, 1775.

Distributed generally—in every district.

238. Spiræa Filipendula L. *Dropwort.*

Area: 63, 64, Top. Bot. Range: 50—900 feet.

Native; on dry grassy turf on calcareous rocks, and in clay or sand on river-banks, rare.

First record: Bolton, 1775.

Y. Banks of Yore above Ripon—adventive? 'Now getting very scarce'; *Slater's Flora.* Copt Hewick; *J. S. Hurst*! (1880).

W. Kilnsey Crag; *W. Curtis* (1782). Pastures by Wharfe at Thorparch; *H. Baines.* Now rare, if not quite gone.

[C. Warley, Halifax; *J. Bolton, Cat.* (1775): now gone? A stray when there?]

D. Doncaster neighd.; *Dr. White* (1804). On Smeaton Crags, Went-vale!

T. Lindrick; *Amos Carr.*

239. Agrimonia Eupatoria L. *Agrimony.*

Area: 63, 64, Top. Bot. Range; 0—1,200 feet.

Native; dry banks, thickets, grassy way-sides, common, especially on clayey sandstone soils. P. June-July.

First record: Salt, 1800.

Distribution general, but for Mersey district.

Obs.—I consider it very singular that no one should yet have turned up *Agrimonia odorata* in our area. It should occur about Grassington or Kettlewell. It may be known by the swoln fruiting calyx containing two seeds.

240. Sanguisorba officinalis L. *Great Burnet* *'Redheads.'*

Area: 63, 64, Top. Bot. Range: 0—1,200 feet.

Native; moist meadows, pastures, and stream-banks, infrequent, mainly (and most abundant) on limestone; also on sand, but hardly on shale. P. Late June-August.

First record: Salt, 1800.

Yorkshire, common in pastures; *Gough's Cam.*

L. Dentdale. Ingleton.

R. Common about Settle; *J. Windsor*; *Jos. Woods, junr.* (*O.B.G.*).

Y. Ripon; *W. Brunton.* Common; *H. H. Slater.* Copgrove; *J. Dalton* in *O.B.G.*

W. Upper Wharfedale. Wetherby. Stutton. Church Fenton, etc.

A. About Leeds, abundant; *Rev. W. Wood* in *O.B.G.* (1805). All down Aire valley from Gargrave. South Milford. Pontefract; *Herb. Hooker.* Selby; *H. F. Parsons.*

C. Frequent below Wakefield; *T. W. Gissing.* Darfield and Burton Grange; *H. Johnson*! Lofthouse; *G. Roberts.* Horbury; *P. F. Lee.*

D. Potteric Carr; *E. Miller* (1804). Between Sheffield and Rotherham; *J. Salt.* Barnby Dun! Ouse bank between Goole and Hook; *H. F. Parsons.*

T. Kiveton Park; *G. Harrison.*

Obs.—No records for Nidd or Mersey. Surely there?

241. Poterium Sanguisorba L. *Salad Burnet.* *'Sot-herb.'*

Area: 63, 64, Top. Bot. Sub-xerophilous. Range: 0—1,500 feet.

Native; dry banks, rock ledges and fields, mainly on the Permian and scar-limestone strata, there common; rare elsewhere. P. June-July.

First record: Langley, 1828.

L. Lower Dentdale scars. Ingleton.

R. Kelcowe, etc. Settle; *J. Windsor.*

Y. Ripon! 'Common'; *H. H. Slater.*

N. On the Permian limestone in Nidd area, common

W. Wharfe side, Barden! (*J. G. Baker*), Grassington! and adventive down at Ilkley. On the Permian strata.

A. On the Permian strata.
C. Wakefield, 'frequent' (to the east of the town, meant); *T. W Gissing*.
D. Roman ridge, Hampole (!); *P. Inchbald*. Rotherham district; *L. Langley* (1828). Conisborough. Near the river Don at Aldwarke; *G. Harrison*.
T. Anston and Kiveton Park; *Geo. Harrison*. Roche Abbey!

Obs.—The chewed leaves have a cucumber-like flavour, and in some places, as about Ripon, have been in repute (infused in water) as a cure for, or alleviant of, drunkard's thirst.

Mr. Roberts has favoured me with the following note, which I insert (as the facts are new to me) without endorsing the supposition made. I fancied that *Poterium muricatum* was the fodder plant formerly sown. Whether the Castleford plant is that or not I cannot say. 'About the middle of last century 'Burnet' was much cultivated in meadows for forage and hay. But it soon declined; some of our plants may be the relics of cultivation. Below Castleford there is a field nearly all Burnet, but very little in the other fields.'

[**Poterium muricatum** Spach. Casual.

This has been found once or twice by myself and others, always among clover or lucerne.

A. Field at Brotherton (clover). Amid lucerne in field by tannery at Meanwood.
T. Waleswood; *Amos Carr*!]

242. Alchemilla arvensis Scop. *Parsley Piert.*

Area: 63, 64, Top. Bot. Arenophilous. Range: 0—1,200 feet.

Native; dry fields and bare sandy ground, common on the Trias and drier sandstones, but quite rare over considerable areas on the limestones and shale. A. June-September.

First record: Salt, 1800.

Distribution general; all districts reported except Mersey.

An *unlikely* station is (R.) High-hill, near Settle, common; *J. Windsor*. Lower crags of Attermire, Settle; *W. Whitwell*.

243. Alchemilla vulgaris L. *Bear's-Foot. Lady's Mantle.*

Alchemilla minor, PARK. (538) *quinquefolia*; How, Phyt.

Area: 63, 64, 65, Top. Bot. Range: 0—2,100 feet.

Native; fields and banks, very common. P. June-August.

First record: How (1650).

Distribution general—reported from all the districts, but 'very scarce in the Ainsty' (*H. Ibbotson*). Noticed by *W. Whitwell* in the Ainsty, near Acomb.

In Wombwell cow-pastures, and many other places of Yorkeshire, *Mr. Stonehouse*; *How, Phyt.* p. 3 (1650). '*Not* at Wombwell' [but a station in a lane north of Stamford given for it]; *Merrett, Pinax*, p. 3

(1666). In . . pascuis . . Eboracensibus; *Ray, Hist.*, 1660. In mountainous places . . as Yorkshire; *Ray, litt.* in *Derh. litt.*, 18 (1667). Halifax; *J. Bolton* (1775).

Obs.—Cattle like it, and it is said to make the milk they give richer and less watery—it is mildly astringent and diuretic.

VAR. **montana** Willd. is rather a state than a fixed variety. It is the more common form among the hills and at high levels, but is not rare in the lower lands.

244. Alchemilla alpina L.

A. alpina pentaphyllos; Ray.

Area: 65, Top. Bot. Montane. Range: 1,900—1,600 feet.

Native; mountain crags of Silurian slate, very rare. P. August.

First record: Ray (1690).

In montibus Eboracensis; *Ray, Syn.*, i (1690); rep. *O.B.G.* (1805), etc.

L. Howgill Fells! = Cautley Crags; *Rev. G. Pinder*. Crags in Dentdale; *G. Pinder* (1845).

245. Potentilla Fragariastrum Ehrh. *False Strawberry.*

Area: 63, 64, 65, Top. Bot. Range: 0—1,250 feet.

Native; dry, shady, or open banks, very common. P. March-May.

First record: Salt, 1800.

Distribution general—in every district.

246. Potentilla verna L.

Pentaphyllum incanum repens; Ray, Cat. *Potentilla opaca*; Huds.

Area: 63, 64? Top. Bot. Xerophilous. Range: 100—300 feet.

Native; on dry Permian limestone soil, rare and very local. P. May.

First record: Ray, 1670.

A. In pastures about Kippax; *T. Willisel*; *Ray, Cat.*, 235 (1670); *Syn.* i; *Martyn* (1763); etc. And still there 1870! About Ledstone Park, abundantly; *Rev. W. Wood* in *O.B.G.* (In Craven; *G. Roberts* (1869): spn. not seen.)
D. Darrington Leys, near Ferrybridge; *H. Baines* = (?) Near Knottingley; *P. Inchbald*. Smeaton Crags. Womersley (!) *P. Inchbald*; *W. Todd*.

247. Potentilla alpestris Haller fil.

P. verna; Huds. (Confounded with *P. verna* prior to 1700.)

P. aurea; Flora Brit.; O. Bot. Guide.

Area: 64, Top. Bot. Montane. Range: 600—1,500 feet.

Native; bushy and stony scars on the limestone, rare. P. June-July.

First record: Martyn, 1763.

R. In barren pastures, Giggleswick; *Martyn* (1763); *Hudson* (1778); sub nom. *Pot. verna*. Top of Lord's Wood, Settle, in the corner adjoining Kelkoe; *J. Windsor*. Stainforth Scars. Bottom of the wood at the end of Moughton Fell. Dale Head; *R. Buxton* = Silverdale Head; *J. Tatham* [i.e., head of glen 4 miles N.N.E. of Great Stainforth].

W. Kilnsey Crag; *W. Howson*.
A. Rocks in the glen on the right-hand side between Grisedales and Malham Tarn; *J. Windsor*. Rocks above Gordale Scar; *W. Brunton* (*O.B.G.*); *J. Windsor, spn.* (1865); *W. West, spn.* (1880)! Rocks of the Cove, Malham (!), and Malham Tarn rocks; *W. West*!

248. Potentilla Tormentilla Scop. *Tormentil.*

Area: 63, 64, 65, Top. Bot. Range: 0—1,800 feet.

Native; heathy and grassy banks, very common, with a very wide range. P. June-July.

First W. R. record: Salt, 1800.

Distribution general—occurs in every district.

249. Potentilla procumbens Sibth.! (*Tormentilla reptans* L.)

Area: 63, 64, Top. Bot. Range: 50—450 feet.

Native; dry heathy-sandy banks, waysides, etc., infrequent, or, rather, local. P. July-August.

First record: Bolton, 1775.

R. Roadside between Settle and Wigglesworth (!) plentifully; *J. Windsor*.
N. Lanes, Marston Moor!
W. Lanes between Bilton and Thorparch (!) *H. Baines*.
A. Banks, Hawksworth to Baildon (!) *W. West*. Wayside, Horsforth (!) *J. Abbott* (1875). Bradford neighbourhood; *S. Hailstone* (1815).
C. About Halifax; *J. Bolton, Cat.* (1775).
D. Near Oughtibridge; *Geo. Harrison*.

250. Potentilla reptans L.

Area: 63, 64, Top. Bot. Range: 0—600 feet.

Native; waysides and sandy banks and pastures, common, but rare above 600 feet. P. June-September.

First record: Salt, 1800.

M. Marked by *J. Whitehead* for this district.
R. By-road between Rathmell and Wigglesworth; *J. Windsor*.
Y. N. and W. Common below 300 feet.
A. Leeds, sparingly; *Rev. W. Wood*. Common.
C. Brighouse; *Dr. Smith* (*O.B.G.*). Common.
D. Common; *J. Salt* (1800).
T. Waleswood, Kiveton, etc.; *A. Carr*.

251. Potentilla anserina L. *Silver Weed. Wild Tansy.*

Argentina; Ger. *Pentaphylloides, Argentina dicta*; Ray.

Area: '63, 64,' Top. Bot. Pelophilous. Range: 0—900 feet.

Native; grassy waysides, goose-greens, wet clayey pastures, common and generally distributed. P. July.

First record: Ray, 1686.

Old records are: Circa Settle; *Ray, Hist.*, 617 (1686); and about Halifax; *J. Bolton* (1775).

252. Potentilla argentea L.

Area: 63, 64, Top. Bot. Range: 0—300 feet.

Native; on rocks of gritstone and sandy banks, rare, but locally in plenty.

First record: Salt's Herb., 1802.

N. On the rocks at Plumpton; *E. Hargrove* (1809 et Eds. seq. ad 1832. Hist. Knaresborough); first brought from there to *S. Hailstone* [about 1815]; *Baines' Flora* (1840). Nearly gone. On rocks in fields at Braham Hall and Crosper, above the Crimple, plentiful! (1879-81).
[C. Heavy soils, Halifax; *J. Bolton, Cat.* What had he in view?]
D. Cinder Hills, near Wardsend, 1802; *Herb. Salt*. On the Triassic sand about Cantley; *G. E. Smith*. Armthorpe and Bessacarr!

[**Potentilla fruticosa** L. Incognit.

'(63)' Top. Bot.

The English name Perennial (or Shrubby) Cinquefoil appears in Miller's list of Potteric Carr plants (1804). This cannot be the origin of the later record, since the tract referred to is not the same.

Limestone tract west of Doncaster; *S. Appleby*, 1831 (*Mag. Nat. Hist.* and *New Bot. Guide*)—doubtless introduced as an ornamental shrub. Tofield lived at Balby and introduced plants; and in 1870 I noted *Potentilla fruticosa* in the shrubberies of Melton Hall.]

253. Potentilla norvegica L.

Denizen; well established, spreading and now ineradicable, on the banks and in the masonry of canals and rivers. A. and B. June-July.

A. Masonry of old dam above Smithy Mills, Meanwood (introduced). In the crevices of stone-work of Leeds and Liverpool Canal at Armley! *J. Abbott*—known for 20 years. By canal at Woodlesford (!) and from Altofts to Methley; *G. Roberts*.
C. By the Aire and Calder Co.'s Canal connecting Wakefield with the river Aire, about Bottom Boat, etc., Stanley, in the utmost profusion; known here since 1856. Calder banks, near Castleford; *G. Roberts*. Dewsbury; *P. F. Lee* (1886). Huddersfield; *C. P. Hobkirk* (1886).
D. Canal-bank masonry near Snaith; *T. Birks*.

[**Potentilla hirta** L. Alien.

Found as a Casual or garden escape on road-sides.

R. Near Settle, July 1878; *W. N. Cheesman*!
T. Hatfield Moor, by road-side near Lindholme, near centre of moor; *H. F. Parsons*!]

[**Aremonia agrimonioides** DC. Incognit. Misnomer.

'Pretty abundant by the canal running from Stanley to the Calder'; *T. W. Gissing* in *Nat.* and *Flora of Wakefield*. But the plant was *Potentilla norvegica* L.]

254. Comarum palustre L.

Pentaphyllum, vel potius Heptaphyllum, flore rubro; Blackst.

Area: 63, 64, 65, Top. Bot. Range: 0—1,500 feet.

Native; spongy bogs, swamps, and peaty ditches, infrequent in the cultivated districts; common in the hill-swamps. P. July.

First record: Blackstone, 1746.

L. '. . . and other places near Ingleton, Yorkshire'; *Blackst., Spec.* (1746). Bowland Knotts.

R. In Giggleswick Tarn; *Gough's Cam.* (1789). Wigglesworth. Hellwith Moss; *W. Howson.*

Y. Sharow Mires, Pilmoor Carr, etc., near Ripon; *H. H. Slater.*

N. Ramsgill; *J. S. Wesley.* In marshy field behind Aketon Bleach Works!

W. Barden Moor. Ilkley. Washburn Valley. Askham Bogs!

A. Adel Bog! Adel Lower Dam, under the alders! Eldwick; *W. West.* Rawcliffe Rabbit Hills; *H. F. Parsons.*

C. Warley Clough; *J. Bolton* (1775). Middle of a small marsh at Stanley; *G. Roberts*!

D. Jordan Dam, Sheffield Moors; *J. Salt* (1800). Goole Moor; *H. F. Parsons.* Doncaster Carrs; *E. Miller* (1804); *S. Appleby* (1831).

255. Fragaria vesca L. *Strawberry.*

Area: 63, 64, 65, Top. Bot. Range: 0—1,500 feet.

Native; dry warm banks and rocks, common. P. May-June.

First notice as a West Riding plant; Bolton, 1775.

Distribution general—recorded for every district; but, as was to be expected in the cultivated areas, about the manufacturing towns, etc., it is growing local and even scarce.

256. Fragaria elatior Ehrh.

Denizen or Alien; shady banks, very rare. P. Early June.

W. Ditch-side, Hobmoor, York; *H. Baines.* Probably a 'casual naturalisation of the cultivated Strawberry'; *J. G. Baker, Supp.* 62.

A. In the Well-House Plantation, Meanwood. This was the true plant, but doubtless introduced with the young trees, or with manure? Wood border, near the tramway by road from Skipton (behind Castle), to the Hawbank rock!

257. Rubus Idæus L. *Rasp. Raspberry.*

Area: 63, 64, 65, Top. Bot. Pelophilous. Range: 0—1,500 feet.

Native; damp woods, swampy thickets, etc., common, and often in great plenty locally. P. Mid May-June.

Distributed generally—over every district.

I give the three oldest records: recent ones number over a hundred!

W. Grass-wood, near Kilnsey; *W. Curtis* (1782); *Gough's Cam.* (1789).

C. Steps Wood, Warley, Halifax, etc.; *J. Bolton* (1775).

D. Sharrow Moors, Sheffield; *J. Salt* (1800).

RUBI FRUTICOSI.

[**Rubus fruticosus** L. *Bramble. Blackberries. 'Blegs.' 'Bumble Kites.'*

Range: 0—750 feet.

Hedge-banks, lanes, thickets, heaths, woods. June-August.

The 'Babingtonian' forms will be treated seperately.

In West Yorkshire this lowland type, as elsewhere, is *par excellence* characteristic of the Infer- and Mid-Agrarian Zones; none of its forms ascend anywhere in Yorkshire above 900 feet, and in West Yorkshire I know of no station above 750 to 800 feet, even for '*affinis*' forms of *rhamnifolius, Koehleri*, or *Radula* (the segregates that ascend highest); whilst above 450 feet fruticose Brambles are noticeably rare.

I need only add that all Mr. Carr's specimens of Rubi were gone through by the Hon. J. L. Warren; and also many of my own gathering, as well as all Mr. Carr's, by Professor Babington. Only such names as *affinis, discolor, leucostachys, umbrosus, Radula, pallidus*, and *corylifolius*—the species well known to me—have been set down upon my own responsibility. At the last possible moment (20th September, 1886) the names, etc., have been modified and revised by the light of Prof. Babington's 'Notes on British Rubi: with special reference to the list in Lond. Cat., Ed. 8.']

258. Rubus suberectus Anders.

Area: 63, Top. Bot. Range: 0—600 feet.

Native; wet thickets and bushy places on peat, rare.

First record: G. Anderson, 1824.

M. Denshaw Clough, near stream by Junction Inn!

Y. Dallow Gill, near Ripley (!) *G. Anderson* in *Smith's Eng. Flora*, ii, 406.

N. Scotton Banks, Knaresborough (*R. 'fissus?'* fide *Bloxam*).

A. Malham; *Baines' Flora* (1840). Bog between the two Adel Bridges over the Meanwood Beck, Adel! Bog by the Aqueduct Arches, Scotland Mills, a mile lower down the same stream! Weetwood Bottoms, still lower down stream!

C. Hebden Valley bottom; *Y. N. U. Excurs.* (187[illegible]).

D. Cantley Vicarage plantation; *G. E. Smith.* Wharncliffe; *A. Carr.*

Obs.—The above records are some of them, probably, really for *R. fissus*, and one, perhaps, *R. plicatus*; but as they stand on record for type *suberectus*, where not seen by myself I dare not place them under either segregate.

259. Rubus fissus Lindl.

Area: 63, Top. Bot.

Boggy, bushy places, commoner than the preceding, but fewer records for it.

M. Greenfield Valley; *J. Whitehead.*

R. Boggy ground about Wigglesworth!

W. Askham Bog; *G. Webster.* Scarcroft and Norwood bottoms!

(A. Wood near Bingley; *W. West*, spn. (No. '28,' 1884), perhaps *R. hemistemon*).

D. Rossington Carr! cum *W. Fowler.*

T. Bushy border of Thorne Waste. Ecclesfield, Sheffield; *A. Carr*!

260. Rubus plicatus W. & N.

Area: no record, Top. Bot.

Damp bushy places, with a tendency to a montane restriction in West Yorkshire, rare.

L. Near the Wennington stream in a boggy thicket.

R. Bushy places in wooded gorge of Bolton-in-Bowland.

Y. Hackfall!

N. Nidd bank, south sides below Knaresborough!

W. Arncliffe; *H. Baines* (1840). Wood near stream, Kilnsey! Langwith Scrub, Collingham! Askham Bogs; *J. G. Baker* (1862)!

A. Malham (!) *L. C. Miall.*

[**Rubus hamulosus** P. J. Müll. et Lefévre, '*Pollichia*,' v, 16-17. Forma sub judice. 'Mr. S. Gibson sent it from Hebden Bridge, Yorkshire, as the "*R. rhamnifolius* ε. *macilentus* Lees?" It was once called a fastigiate form of *R. rhamnifolius*. . . . Probably [this Bramble] will have to be distinguished specifically'; *C. C. Babington*, Journ. Bot., vol. xxiv, N.S., p. 218 (1886). Unknown to me and therefore not numbered. A form of *R. plicatus*; *J. G. Baker.*]

261. Rubus Lindleianus Lees.

Area: 63, Top. Bot.

Native; banks, hedgerows, and thickets, especially in a sandy soil or one approaching to peat, common, and apparently a substantially endemic form of *R. rhamnifolius.*

M. Mersey district; *J. Whitehead.*

N. Marston Moor!

W. Dringhouses! Askham confines. Brandon, near Shadwell. Collingham. Etchell Crags, Scarcroft!

A. About Hambleton. Allerton; *W. West.* Cookridge to Horsforth.

C. Near Halifax! *H. Baines.* Between Ossett and Wakefield; *L. C. Miall.* Dewsbury; *P. F. Lee.* About Liversedge; *W. Fowler*!

D. Bell Hagg, Sheffield; *Bab., Brit. Rubi.* Thorne Moor borders!

262. Rubus rhamnifolius W. & N. (incl. *cordifolius* Bab. non Baker).

Area: no record, Top. Bot. Range: 0—900 feet.

Native; banks, etc., in sandy districts, the type rare.

W. Chandler's Whins, Askham Bog; *G. Webster.* Hedges near Walton, Thorparch!

A. Meanwood, lane-side from Orphanage to Scotland Mills. Allerton; *W. West*! Morton Bog; *T. Birks* (sub nom. *incurvatus*).

C. Near Halifax; *R. Leyland.* (I expect *R. affinis* was the plant.)

D. Goole Moors; *H. F. Parsons.* Manchester Road-side, near Hollow Meadows above Sheffield; *A. Carr.*

263. Rubus affinis Auct. Angl. et Génév. non W. & N. (*R. montanus* Wirtg. teste Baker, sed non C. C. Bab.)

Area: 63, Top. Bot. Range: 0—600 feet.

Hedges and open, stony thickets, the prevailing form in the dales, replaced by *R. Lindleianus* in the lowland sandy-heathery districts.

M. Saddleworth district.

L. Howgill and about Sedbergh.

R. Slaidburn. Settle vicinage.

Y. Kirby Malzeard. Fountains Abbey.

N. Pateley Bridge. Harrogate. Kirby Overblow.

W. Addingham. Ilkley. Otley. Leathley. Askham; *G. Webster.*

A. Adel. Weardley. About Skipton Rock, etc. Cookridge. Hawksworth; *W. West.* Bingley; *Id.* Baildon; *Id.*

C. Marsden! Hebden Bridge!

D. Hooton and Rotherham; *A. Bloxam*; *Bab., Brit. Rubi.* Ravenfield, abundant; *A. Bloxam* in *Phyt.*, iii, 325. Goole Moor! Thorne Waste border! Rivelin; *A. Carr*!

T. Finningley (!) *W. Fowler.*

Var. **laciniatus** Willd.

A denizen, which has escaped from gardens and become naturalised. It is by its stem, prickles, pink petals, panicle, and habit clearly a cut-leaved variety of the plant (cf. *Baker, Journ. Bot.*, vol. 24, p. 7) we in England have been calling *R. affinis.*

W. Lane near Ben Rhydding, Ilkley. (Roots brought from there into the field-garden at Meanwood Lodge have retained their characters during 15 years.)

C. Wessenden Valley, 'known for 7 or 8 years'; *Rev. W. Fowler* and *R. Jessop* (1875). Near Cook's Study, and at Dunford Bridge; *C. P. Hobkirk*; *Nat.*, 1875, p. 57.

264. Rubus rusticanus Merc., C. C. B. (**R. discolor** Auct. Angl.)

Area: 63, 64, Top. Bot. Sub-xerophilous. Lowland. Range: 0—750 feet.

Native; lanes and hedges, abundant in low country on dry soils, rarer on cold wet soil, quite unknown in the higher hill districts. P. August.

On record for all the districts.

A curious plant with broad over-lapping cordate leaflets, Mr. Bloxam marked 'A form of *discolor*.'

265. Rubus pubescens W. & N. (C. C. Bab.).
Area: no record in Top. Bot. Range: 0—750 feet.
Hedges, a rare form in Yorkshire, and of distinct facies; it is what has until just lately passed as *R. thyrsoideus* Wimm.
W. About Boroughbridge.
D. Hollow-Meadows at head of Rivelin Valley; *A. Carr*, sub nom. *thyrsoideus*.

266. Rubus leucostachys Sm. (Incl. *vestitus* Bab.).
Area: 63, Top. Bot.
Native; open or shady hedge-banks and thickets, common, seldom in great quantity, but very frequently recurring.
N. Harrogate! Sandy lane at Spofforth.
W. Bilton. Wighill in Ainsty. Barwick-in-Elmet.
A. Skipton Rock. Bingley! Rider's Wood, Cross roads, Meanwood. Scotland Mills, Meanwood! Hawksworth; *W. West*. Cookridge.
C. Near Halifax; *H. Baines*. Shibden and Huddersfield; *L. C. Miall*. Dewsbury; *P. F. Lee*.
D. Bell Hagg, Sheffield; *Bab.*, *Brit. Rubi* (1869). A frequent form about Sheffield; *A. Carr*.
T. About Kings Wood, Roche Abbey!

267. Rubus Salteri Bab. (1846). (*R. sylvaticus* W. & N. teste Focke.)
A rare West Yorkshire form, approaching robust *R. carpinifolius* in facies.
R. Thicket near Tosside Chapel, towards Wigglesworth!
W. Chandler's Whins, near Askham Bogs; *G. Webster*. = '*R. calvatus* Blox.', Quaker's Wood, Dringhouses; *Id.*, *Y. N. U. Trans.*

268. Rubus carpinifolius Bab.
Area: 63, Top. Bot. Range: Low in West Yorkshire.
Native; open hedge-banks on a dry soil, rare. I have long considered this plant (well known to me in Worcestershire and Staffordshire) to be a form of *R. rhamnifolius*—the passage to the woodland *R. umbrosus*. Prof. Babington in 1886 admits that some of his specimens 'may belong to other named forms, perhaps *R. vulgaris* W. and N.,' but of that as a British plant the Professor says he knows nothing.
Y. Wicliffe Lane, Ripon!
A. Ledstone. Micklefield, on the limestone. Hambleton; *T. Birks, jun.*
C. Hebden Bridge, *S. Gibson*; *C. C. Babington*, *Brit. Rubi*, 140 (1869). (Shibden; *L. C. Miall*. Probably *R. umbrosus*).
Obs.—I am unable to vouch for all the records given above being the true *R. carpinifolius*, with *narrow* panicles. I fear some of them belong rather to *R. umbrosus* of Arrh., and others to *R. vulgaris* W. & N., the terminal leaflet being very cuspidate with narrowed base, and the felting underneath the leaflets being very dense. (The foregoing sentence—showing my doubts—was written long before seeing Prof. Babington's paper in Journal of Botany, July 1886, but I let it stand.—20, ix, 86.)

269. Rubus villicaulis W. & N. (*R. vulgaris* Lindl.).
Damp, rather shady thickets, rare. This has a peculiar unmistakable *facies*, with its broad, dull-green flat leaves and large compound panicle. With *R. Salteri* and *R. carpinifolius* it forms a sort of passage to *R. rhamnifolius*.
Y. Studley Royal, Ripon, bank near Fountains Hall!
W. Askham Bog border; *G. Webster*.
C. Haybeck, Dewsbury; *P. F. Lee*! A plant between this and *R. macrophyllus*.
D. Roadside, near Staveley Coal Pit; *Amos Carr*!

270. Rubus macrophyllus W. & N.
Range: 0—300 feet.
Woods and shady banks, the type rare, the var. *umbrosus* common.
N. Poppleton; *G. Webster*. Stockeld Park Woods!
W. Shady places between Tadcaster and Wighill Park! Askham Bogs; *G. Webster* (and var. *glabratus*). Woodhall Woods, Linton.
C. Dewsbury; *P. F. Lee*.
D. Wharncliffe Woods; *A. Carr*.

Var. **amplificatus** Lees.
Area: 63, Top. Bot.
D. Hooton Cliff; *A. Bloxam*; *Bab.*, *Brit. Rubi*.

Var. **umbrosus** Auct. Angl. (**R. Maassii** Focke teste C. C. Bab., 1886.)
Area: 63, Top. Bot. Range: 0—450 feet.
One of the commonest woodland brambles in the Riding.
D. Bell Hagg, Sheffield; *Bab.*, *Brit. Rubi*. Symmingthwaite, W. York (*J. G. Baker*!); *Id.* Hooton Cliff, Ravenfield; *A. Bloxam* (*Phyt.*, iii, 325), sub nom. *carpinifolius*; but Mr. Baker remarks 'what Bloxam meant by *carpinifolius* was always *umbrosus*.'
Distribution general; I have seen it myself in every district except Mersey.
Obs.—Hitherto this has been universally named *R. umbrosus* Arrh., and considered—as is here done—a variety of *R. macrophyllus*; but now Prof. Babington in referring it to *R. Maassii* seems to consider it specifically distinct, and further, the glabrous-leaved var. of *macrophyllus*, on record above, is suggested as possibly being a smooth-leaved form of *Maassii* also. There is clearly a close relationship betwixt all these Rubi of the *macrophyllus* group, and I believe them all forms of one ver-species.

271. Rubus mucronatus Blox. (**mucronulatus** Bor. ?).
Hedges and thickets, rather a local form, having a well-marked facies.
Y. Lane-hedge between Bishop Monckton and Ripon!
A. Lane near Park gates at Meanwood! teste *C. C. Babington*.
D. Hooton Cliff, near Doncaster; *A. Bloxam* (1859).

272. Rubus festivus Müll. & Wirtg. (nom. teste J. L. Warren).
A rare form discriminated from amongst Carr's Sheffield district gatherings by the Hon. J. L. Warren. Mr. Baker, in referring British forms to Nyman's names, places this under *R. Lejeunii* Bab., in same group as *R. infestus* W. & N. Prof. Babington in Bot. Rec. Club Rep. (1877), p. 237, considered it not really distinguishable from *R. mucronulatus*; but now in his latest 'Notes on British Rubi' he inclines apparently to Focke and Baker's opinion, placing it in the series after *Lejeunii* among the *Glandulosi*; although Wirtgen classes it (as above) with the *Spectabiles* group of *Villicaules*.
A. Allerton, Bradford; *W. West*.
D. Wharncliffe Wood and Endcliffe Wood, Sheffield; *A. Carr*!

273. Rubus Sprengelii Weihe (cum *Borreri* Bell-Salt).
Area: 63, 64, Top. Bot.
Woods and coppices, occasionally in the open, rare—a pretty and well marked form.
M. Greenfield Valley (!) *J. Whitehead*.
N. Harrogate; *Bab.*, *Brit. Rubi* (1869). Scotton banks, Knaresborough.
W. Bolton Woods!
A. Cookridge Wood, Horsforth. Scotland Mills Wood, Meanwood.
C. Hebden Bridge, *S. Gibson*; *Bab.*, *Brit. Rubi*. Coxley Wood; *P. F. Lee*!
D. Sheffield; *W. W. Newbould*; *Bab.*, *Brit. Rubi*. Denaby Wood, Mexborough; *A. Bloxam*, *Phyt.* iii, 325. Endcliffe Wood; *A. Carr*.

274. Rubus rosaceus Weihe (incl. **R. hystrix** Weihe).
Area: 63, Top. Bot.
A rare, well-marked form, in open woods.
R. Brushwood near road, Tosside Chapel!
Y. Aisenby; *Bab.*, *Brit. Rubi*. About Boroughbridge!
W. Hedges near Askham Richard; *J. G. Baker*, *North Yorkshire* (1862).
A. Hambleton; *T. Birks, jun.* (1878).
D. By the Rother, near Beighton; *A. Carr*. Mark Lane, Ringinglow; *A. Carr*. Loxley, near Sheffield; *Bab.*, *Brit. Rubi* (1869).
T. Roche Abbey Wood; *A. Bloxam*, *Phyt.* iii. Near Lindrick; *A. Carr*.

[**Rubus pygmæus** Weihe. Incognit.
Quaker's Wood, Dringhouses; *G. Webster*. I believe the plant here, named *R. pygmæus* at first by some authority, was afterwards decided to be *R. calvatus*.]

275. Rubus Bloxamii Lees.
Area: '64 Webster,' Top. Bot.
A beautiful, well-differentiated form, but a rare one with us, showin a partiality for sandy soils.
W. Acomb Wood; *G. Webster*.
A. Thicket by old quarry, South Milford!
D. Upper Bradfield, Sheffield; *A. Carr*.

276. Rubus scaber Weihe & Nees.
D. Lane, Lower Bradfield, and Cliff Rocher; *A. Carr*, spns. teste *C. C. Babington*.

277. Rubus echinatus Lindl. (*R.* '*rudis* Weihe' of Bab., Brit. Rubi.)
A rare form with us, and with a distinct physiognomy; partial to sandy soils.
N. Near the Crimple, Pannal Valley!
W. Hedges between Ilkley and Otley.
T. Between Tickhill and Sandbeck Park!

278. Rubus Radula Weihe.
Area: 63, Top. Bot.
Open hedge-banks and stony slopes, common; one of the most frequent of the open hedgerow forms.
W. Sherburn. Barwick-in-Elmete. Wetherby. About Ilkley.
A. Church Fenton. [About the Delph quarries, Woodhouse Ridge (!); *J. Abbott*.] Baildon; *W. West*.
C. Stocks-moor, Dewsbury; *P. F. Lee*.
D. Loxley, near Sheffield; *Bab.*, *Brit. Rubi*. Cliff Rocher and Scraith Wood; *A. Carr*!
T. Hooton Rectory; *A. Bloxam*. Kiveton; *A. Carr*.

Var. **denticulatus** Bab., Brit. Rubi, 195.
D. Valley of the Loxley, near Sheffield; *W. W. Newbould*—the true var.—*Bab.*, *Brit. Rubi*, p. 225.
Obs.—Prof. Babington now says this is a doubtful plant seeming to connect *R. Radula* with *R. rosaceus*, 'not yet identified with any described species, and should it prove distinct' he would 'like to call it *R. Newbouldii*.'

279. Rubus Koehleri Weihe.
Woods and thickets, common in one form, rare in the rest.
A. Allerton; *W. West*. Baildon; *Id.*
C. Coxley Valley; *P. F. Lee*.
D. Mark Lane, near Ringinglow, Sheffield; *A. Carr*.

Var. **pallidus** Bab. non Weihe.
Area: 63, Top. Bot.
Native; the commonest of our woodland and shade brambles; in fact it is *Koehleri* less robust and well developed from growing with less light. Generally distributed—I have it from every district.

Var. **infestus** W. and N.
C. Hebden Bridge; *Bab.*, *Brit. Rubi*. Dewsbury; *P. F. Lee*.
D. Rivelin Valley; *A. Carr*.

280. Rubus fusco-ater Weihe.
Area: 63, 64, Top. Bot.
N. Goldsborough; *Bab.*, *Brit. Rubi*. Poppleton; *G. Webster*.

A. Brayton Barf! 'A var. near *Briggsii*'; *C. C. Babington*. Meanwood Wood-side!
C. Coxley Valley; *P. F. Lee*.
D. Near Sheffield, *Newbould* (off type); *Bab.*, *Brit. Rubi*. Loxley, near Sheffield; *Id.* Lane at Upper Bradfield, and Endcliffe Road, Broomhall; *A. Carr*.

281. Rubus diversifolius Lindl. et Génévier (*R. dumetorum* Warren.).
Hedgerows and open thickets on the sandstone chiefly, rare—a glandulose *R. corylifolius*, *R. tuberculatus* forming a passage between them.
Y. About Ripon.
W. Healaugh, in the Ainsty. Barwick-in-Elmete.
C. Dewsbury district; *P. F. Lee*.
D. Hutton, near Waith (? Wath); *A. Bloxam*, off type; *Bab.*, *Brit. Rubi*. Rivelin Valley, near flour mill; *A. Carr*.

282. Rubus Guntheri Weihe (= *R. flexuosus* M. & L., No. 463, Lond. Cat., 8th ed.).
(C. Near Hebden Bridge; *E. Lees*, *Phyt.*, iv, 930.)

283. Rubus Lejeunii Weihe.
A. Lotherton Moor; *G. Webster*.
D. Wood near Rotherham; *A. Carr*. Cliff Rocher; *A. Carr*. This was queried *R. pygmæus* by C. C. Babington.

284. Rubus humifusus Weihe; Bab., Br. Rubi (*R. saxicolus* Müll.).
D. Wharncliffe Wood; *A. Carr*!—So named by C. C. B. when going through Carr's Sheffield Rubi, but I see little or no difference between it and the *R. festivus*, No. 272, from same locality. I believe it to be simply a more glandulous form of that. *R. saxicolus* is placed very near it in the lineal series of Prof. Babington's latest Notes on British Rubi; and although only two West Yorkshire stations can be cited for it, in many parts of England it would seem to be not uncommon, if the 'abundance of British specimens so named by Focke' in Herb. Génévier be anything to judge by.

FORMA **'atro-rubens** Wirtg.' (Sub *R. foliosus* W.).
D. Cliff Rocher, Loxley-dale, and Wharncliffe Wood; *A. Carr*. This name, which I had placed here, and which being *sub judice*, it is hardly worth while to remove and treat as a species, Prof. Babington now drops entirely as being synonymous with *R. foliosus* Weihe.

285. Rubus Bellardi Weihe (*R. hirtus* W. & N., in part).
Area: 63, 64, Top. Bot.
Very rare in West Yorkshire. I believe we have only one 'species'—the old *glandulosus*—though Prof. Babington I see now places *dentatus* as a var. of *Bellardi*, and *rotundifolius* as a var. of *hirtus*.
A. Hambleton Hough (Var. '*hirtus*').
C. Whitley; *H. F. Parsons*.

VARS. **dentatus** Blox. and **rotundifolius** Blox.
D. Loxley, near Sheffield; *Bab.*, *Brit. Rubi*, 253. Wharncliffe Wood; *A. Carr*! Roadside to Canklow Mill, near Rotherham; *A. Carr*.

286. Rubus corylifolius Sm. *Hazel-leaved Bramble*.
Area: no record in Top. Bot. Range: 0—450 feet.
Native; the commonest hedgerow and open-lane bramble next to *R. discolor* (*rusticanus*) in the lowlands under 200 feet; and very variable; the typical form *sublustris* Lees is commonest, but the var. *purpureus*, which Prof. Babington now identifies with *fasciculatus* Müll., is not uncommon.
Distribution general—in all the districts, but quite rare in Mersey and Lune, i.e., the two western ones.
R. Settle; *Bab.*, *Brit. Rubi* (*sublustris* Lees) = Cavehole Wood; *J. Windsor*.
W. Thorparch; *H. Baines*. Ainsty; *J. G. Baker*.
A. Baildon; *W. West*. Roundhay! Skipton. Ledsham. Selby.
C. Horbury; *P. F. Lee*. East of Barnsley!
D. About Sheffield; *A. Carr* (two forms).

287. Rubus scabrosus Müll. (**R. tuberculatus** Bab.).
A rather local hedgerow form; off *R. corylifolius*, in the direction of *diversifolius*—its glandulous representative.
W. Ainsty hedgerows, frequent; *J. G. Baker*. Askham (!); *G. Webster*.
A. Hedges about Selby.
D. Hollow meadows, Sheffield; *A. Carr*.

288. Rubus Balfourianus Blox.
N. Hedgerow near Knaresborough, towards Scriven!
W. Acomb; *G. Webster*.

289. Rubus cæsius L. *Dewberry*.
Area: 63, 64, Top. Bot. Range: 0—750 feet.
Native; hedge-banks and rocky ground, mostly on limestone, infrequent, but locally abundant.
First record: uncertain.
L. Lower Dentdale. Ingleton. Cavehole Wood, Feizor (!); *J. Windsor*.
R. Giggleswick Scar, Kelcowe; *J. Windsor*. Gisburn.
Y. Sleningford. Ripon. Copgrove.
N. Flaxby. Cowthorpe. Ribston, etc. On the limestone common.
W. Grassington. Bolton Woods. Common on the Permian tract from Wetherby, Norwood Bottoms, Thorner, and Bramham to Aberford and Barwick. Ainsty, frequent; *J. G. Baker*. Near Marston; *W. Whitwell*.
A. Airton. Gargrave; *W. West*. Common on the limestone about Micklefield. Sherburn.
C. Wakefield; *T. W. Gissing*. About Featherstone.
D. Hooton Cliff; *A. Bloxam*. Conisborough. Warmsworth. Went-vale.
T. Roche Abbey; *A. Bloxam*. Kiveton: *A. Carr*. Common about Firbeck, Lindrick, etc.!

290. Rubus saxatilis L.
Rubus alpinus humilis, J. B.; Ray.
Area: 63 Leyland spn., 64, 65, Top. Bot. Montane. Range: 1,200—200 feet.
Native; rocky gills and stony woods, infrequent. P. June-July.
First records: Parkinson, 1640; Ray, 1670.
In Yorkshire; *Park.* (1640). In the north-west part of Yorkshire; *Ray*, *Cat.*, 269; *Syn.* i, etc.
L. Mill-bank, Ingleton; *Mr. Thornbeck*; *Blackst. Spec.* (1746); *Martyn* (1763). Sykes' and Helks' Wood; *Dawson Turner* in *O.B.G.* Feizor Wood; *H. F. Parsons*.
R. About Settle; *Gough*; *Curtis* (1782); etc. Lord's and Cavehole Woods; *J. Windsor*. Linn Ghyll and Douk Gill. Stackhouse Borrins, Kelcowe, etc.; *H. Baines*.
Y. Mackershaw Wood; *J. Dalton*; 'there on ordinary turf' (!) *H. H. Slater*. Hackfall (!) *W. Brunton*, *O.B.G.*
W. Grass Wood and Kilnsey Crag; *W. Curtis* (1782); *Wood* in *O.B.G.* Bolton Woods! Barden.
A. Gordale; *J. Nowell*.
C. Turner Clough, Barkisland, and Ogden Clough; *R. Leyland*.

[**Rubus arcticus** L. Imposition.
Sent to Mr. Watson as having been gathered on a West Yorkshire moor; but probably originated in a Lancashire Botanic Garden. *R. arcticus* can hardly by any possibility have been one of the *two* Knotberries mentioned by Parkinson.]

291. Rubus Chamæmorus L. *Cloudberry*. *Knotberry*. *Knoutberry*. *Vaccinia nubis*; Park. *Chamæmorus*; Ger.; Ray.
Area: 63, 64, 65, Top. Bot. Montane. Range: 1,200—2,400 feet.
Native; heathery peat moors from 1,600 feet upwards, common there, very scarce at a lower level, but descending in one or two spots to 1,250 feet or thereabout. P. late June-July. Fruit from last week of July to mid August—soon eaten by Grouse, Plovers, etc.
First record: Gerarde, 1597.
L. First made knowne unto us by Thomas Hasket, a painefull Chirgion, and simplist of Lancashire; but Dr. Lobel found it on Ingleborough hill . . . as Mr. Bradshaugh . . . did likewise; *Park. Theat.* (1640). Upon Ingleborough hils, plentifully . . . . where the clouds are lower than the tops of the same all winter long, whereupon the people of the country have called them Cloudberries, found there by . . . Master Hesketh; *Gerarde*, *Herb.* 1091 and 1368 (misprinted 1386).
R. Ingleborough; *Martyn*. Whernside. Baugh Fell. Hinckelhaugh, near Settle; *Ray*, *Cat.*, i, 68; ii, 66, 67 (with remarks to show that two previously considered species are only one); *Ray*, *Hist.*,

653, 654, and compare *Clusius*, *Rar. Pl. Hist.*, 118 (1601) = Ryeloaf Hill; *J. Windsor*. Penyghent—N.W. part, draining into Ribble; *W. Whitwell* (1884).
N. Great Whernside, ridge at head of the How Stean Beck!
W. Moors above Outershaw and Cam Fells. Buckden Gavel. North end of Penyghent and hills near Cowside; *J. Windsor*. Fountains Fell.
C. Mossleden Moor Ridge, for two miles; *J. Bolton* (1775). [Bouldsworth Hill; *R. Teesdale* (1775), but the summit ridge about 1,600 feet, where it chiefly grows, is within Lancashire boundary.] Rishworth Moor! Blackstone Edge.
A. Bog by Malham Tarn; *H. F. Parsons*.

Obs.—Some of the peasants call the berries Knotberries, and, asked why, one smocked ancient at once replied, 'Because they grow only on the Knots'—the eminences rising above the general level of the fell plateaux (some 1,500 feet) which divide the river basins, being called Knots; a British word signifying a craggy protuberance. At p. 802 of Gibson's Camden (1695), however, we have a very different account. Which is the true one who shall decide, when doctors disagree? Archdeacon Nicholson there (loc. cit.), says, 'Instead of Gerard's mistaken name of Clowdberry, the northern peasants call it Cnoutberry, and have a tradition that the Danish king Knut, being (God knows when) distressed for some time in these vasts, was relieved *by feeding upon* these dainties.' A hardly defensible piece of etymology, with the additional suspicious coincidence attaching to it that one of our feathered winter visitors—the *Tringa canutus* L.—is said to have got its name from the circumstance that the same royal personage was accustomed to feed on it also. Both berry and bird are delicious *bonnes bouches*—properly cooked!

292. Geum urbanum L. *Herb Bennet*.
Area: 63, 64, 65, Top. Bot. Range: 0—1,000 feet.
Native; hedges and woods, very common. P. June-August.
First record: Bolton, 1775.
Generally distributed—in all the districts.

293. Geum intermedium Ehrh.
Area: 64, Top. Bot. Range: 0—900 feet.
Native; damp thickets or hedge-banks, with the preceding and the next, rare, or rather occasional. Hybrid P. June-July.
First record: Brunton, 1805.
R. Ingleton.
Y. Mackershaw Wood; *W. Brunton* in *O.B.G.*
N. Wet wood by Nidd, under Birkham Wood, Knaresborough.
W. Bolton Bridge; *J. Windsor* (1867). By Wharfe, Thorparch, etc.; *J. Emmet*.
A. Skipton, and between Skipton and Bolton; *C. Bailey* (1871). Lane above Tarn House, Skipton; *W. Whitwell*, 1884.

294. Geum rivale L. *Avens.*

Caryophyllata montana purpurea; Ger. Em.; Ray.

Area: 63, 64, 65, Top. Bot. Sub-montane. Range: 1,600—100 feet.

Native; road-sides, woods, and stream-banks, everywhere amongst the hills, common by the larger rivers down into the low country, rare on the coal-measures and lowland areas away from the rivers. P. June-July.

First notice: Ray, 1670.

L. Circa Settle, Ingleton, etc., frequentissimé; *Ray, Cat.*, i, 59 (1670), ii, 56; *Hist.*, 607. Dentdale. Sedbergh. Howgill. Garsdale.

R. Settle; *Ray* in *Gibs. Cam.* (1695); *Martyn* (1763); *Gough* (1789); etc. About Settle, common (!); *J. Windsor.* Ribblehead. Slaidburn. Bolton-in-Bolland.

Y. Copgrove; *J. Dalton.* Hackfall. Near Ripon, common; *H. H. Slater.*

N. Long Walk, Knaresborough; *E. Hargrove* (1782). Scriven Park; *Id.* (1809). Scotton banks.

W. About Leeds; *Rev. W. Wood.* [As usual with Wood's records, 'about' means within 12 miles or so.] Buckden. Bolton Woods. Ilkley. Collingham. Boston Spa.

A. Skipton, in great abundance; *W. Wood* in *O.B.G.* (1805); *W. Whitwell*, 1882-84. Malham (!); *W. West.* Near Adel, 1872; *G. Roberts.*

C. Mossleden, Ripponden, etc.; *J. Bolton* (1775).

D. Doncaster district; *Dr. White* (1804). Wentbridge, 1872; *G. Roberts.*

295. Dryas octopetala L.

Area: 64, Top. Bot. Montane. Range: 1,350—1,200 feet.

Native; on dry, exposed turf over limestone rock, one station only. P. Late June-July.

First record: Curtis, 1782.

(R. Near Settle; *Dr. Fell*; *Bot. Guide* (1805)—the Arncliffe station doubtless meant. N.B.—Fell's record has the pen run through it by Joseph Woods in Townsend's copy of O.B.G. formerly in his possession.)

W. On Arncliff Cloud, a mountain near Arncliff in Littendale; *W. Curtis* (1782); *Gough's Cam.* (1789). In the N.W. of Yorkshire, in Craven; *Rev. W. Wood*; *Sm., English Flora* (1824). Still on Arncliffe (the 'Hawkswick' of the Ordnance Map) Clouder; in plenty up to 1870.

Obs.—The sepalar variations in a series of Arncliffe specimens assisted J. E. Gray (cf. Ann. Nat. Hist., xv, p. 139, 1845) to prove that the var. *depressa* Mackay, had no constant character.

[**Rosa cinnamomea** L. Extinct Alien.

In the wood in Aketon pasture near Pontefract; *Mr. Salisbury*; *O.B.G.* Spn. in *Herb. Smith.* Not now to be found there; *Mr. Sabine*; *Eng. Flora*, ii, 373. Doubtless planted like *Staphylea.*]

[**Rosa alpina** L. Alien.

R. Near Settle—surmise only; *J. Windsor, Flora Cravoniensis*, p. 27.

W. Shrubberies at Boston Spa, east end, and one plant in altered ornamental grounds by Wharfe, near Jackdaw Crag, without a doubt planted.]

296. Rosa spinosissima L. '*Cat-rose.*'

Area: 63, 64, Top. Bot. Range: 0—1,250 feet?

Native; hedges, roadsides, and thickets, locally common on the limestones. P. June.

First record: Baines, 1840.

L. Crow Nest Wood, N.W. of Settle; *H. Baines*=Cavehole Wood; *J. Windsor.*

R. Kelcowe, etc.; *J. Windsor.*

Y. Sleningford, Ripon; *H. H. Slater.*

N. Road-side between Harrogate and Pannal; *E. Hargrove* (1809).

W. Wood near Etchell Crags, Scarcroft! Flint-mill, Wetherby! Towton-field and Newton Kyme; *G. Roberts.* Bramham! Collingham; *J. Emmet.*

A. Gordale; *W. West.* Ledsham (!) *G. Roberts.* Kippax! Malham (moor crags) and Airton; *H. Baines.* Roadside between Garforth and Aberford; *H. Baines.*

T. Roadside near Firbeck.

No record for D. Surely not absent on the Permian tract therethrough? I have some cloudy recollection of having seen this somewhere near Pontefract, but can find no voucher. Yet it seems strange that no record exists for it for Went-vale or Conisborough. I have not myself seen this at a higher elevation than 900 or 1,000 feet in West Yorkshire.

Obs.—This is *the* rose on Towton field, according to Mr. Emmet.

297. Rosa involuta Sm. (*R. Sabini* Woods.)

Area: 64, Top. Bot. Range: 0—600 feet.

Native; woods and thickets, rare and local. Shrub. July.

First record: Hailstone, 1808?

R. Specimens sent to me from *Settle* by the Misses Thompson, as *Rosa Sabini*, I believe correctly named; *W. Whitwell.*

L. Helks Wood, Ingleton (!) *J. Tatham.* Has been called '*Doniana.*'

Y. Near Boroughbridge; *Geo. Webster.*

W. [Oglethorpe Ings (S.E. of Boston Spa); *S. Hailstone*; extinct; *J. Emmet.*] Hill above Collingham, on the road to Harewood; *S. Hailstone*—sub nom. *spinosissima var.*; *Baines' Flora.*

A. On Roundhay 'lime hills,' near Leeds! (Near Kildwick and Skipton; *Miall's Flora.* Small form of *R. mollis*?)

D. Field near Greenside, Sheffield; *Amos Carr*!

298. Rosa mollis Sm. *Red Rose. Apple Rose.*

R. pomifera major; Park. Parad.; Ray. (*R. villosa*, L. Hb.).

Area: 63, 64, 65, Top. Bot. Range: 0—1,500 feet.

Native; hedges and thickets, common amongst the hills, rare in the low country. Shrub. June.

First record: Ray, 1669.

Eboracensi territorio copiosé; *Ray, Cat.*, 266 (1670); *Hist.*, p. 1472 (1688). 'The Rose with the large prickly fruit . . . I have observed in many places of Yorkshire; *Ray* in *Derh. litt.*, p. 51 (1669).

L. Very common in Dentdale, Garsdale, about Ingleton, etc.

R. Many places near Settle (*J. Windsor*), Slaidburn, Wigglesworth, etc.!

Y. Mackershaw, etc., near Ripon (!) *W. Brunton.* Hackfall (!) *H. H. Slater.*

N. About Scotton, Pateley Bridge, etc. Harrogate.

W. Grass Wood, near Kilnsey; *W. Curtis* (1782); *Gough's Cam.* (1789). Common down to Ilkley. Washburn Valley.

A. Malham, Airton, Skipton! Adel!

C. Warley, etc., Halifax; *J. Bolton* (1775). Todmorden. Hebden Valley!

D. Doncaster neighbourhood; *Dr. White* (1804). Sheffield Manor; *J. Salt.* Frequent about Sheffield; *Amos Carr.*

299. Rosa tomentosa Sm.

Area: 63, 64, 65, Top. Bot. Range: 0—1,250 feet.

Native; hedges, and in natural woods, common; but rare in some parts of the low country, and for Mersey district no record. Shrub. June.

First record: Brunton, 1805.

L. Common enough, but less so than *R. mollis.*

R. Common near Settle, in Bolland, etc.

Y. Hedges near Ripon; *W. Brunton*; *Herb. Turner.* Studley; *J. Dalton* in *O.B.G.*

N. Pateley Bridge, Scotton Banks, Birkham Wood, Knaresborough, etc.

W. Common from Barden downwards to Thorparch.

A. Malham! Gargrave, Bingley, and Eldwick; *W. West.* Camblesforth, near Drax; *H. F. Parsons.*

D. Common about Sheffield—many localities (!); *Amos Carr.*

T. Frequent (!) *A. Carr.*

300. Rosa rubiginosa L. *Sweetbriar.*

Rosa sylvestris odora Eglenteria; Ger.

Area: 63†, 64†, 65, Top. Bot. Range: 0—450 feet.

Doubtfully Native, but clearly alien in most stations; thickets on limestone and hedges, rare. Shrub. July.

First record: How, 1650.

R. Kelcowe, Settle; *J. Windsor.*

N. Hedge near the river Crimple, Harrogate*; *P. Inchbald.*

W. Rigton Hill top.

C. Dixon's Scar, Sowerby Bridge; *J. Bolton* (1775). On Sandal Castle Hill, 1799; *D. Dixon.*

D. In divers hedges by Darfield, but with more smooth shining leaves than the garden kind. This is like that of Park., pag. 1017; but that the leaves are sweet as well as the flower, *Mr. Stonehouse*; *How, Phyt.*, 105; *Merr., Pinax*, p. 105. Lane near Storrs, Sheffield; *A. Carr*! Conisborough; *O. A. Moore* (1840).

301. Rosa micrantha Sm.

Area: 63, 64, Top. Bot. Range: 0—500 feet.

Native; hedges and thickets, rare, confined to the limestone. Shrub. July.

First record: Woods, 1838.

R. Lodge Gill, Settle; *W. Howson.*

N. Scotton banks, Knaresborough (!).

W. Hedges near Bramham (*J. Backhouse*), Collingham, and Boston Spa; *H. Baines.* Norwood bottoms, Thorner!

D. Roche Abbey; *Jos. Woods* (1838).

302. Rosa cryptopoda Baker. (*R. inodora* Fr.)

Area: '63 S. King,' Top. Bot. Range: uncertain.

Native; hedges, very rare. Shrub. July.

C. Luddenden neighbourhood; *S. King* (1862). I do not know if anyone has gathered this since. Said to be allied to *R. sepium*—although allocated to *R. Borreri* in previous West Riding floras. It must be very rare, for out of many hundred West Yorkshire gatherings I have not seen one with 'short, naked peduncles, round fruit, and leaflets distinctly glandulose beneath.'

303. Rosa canina L. *Dog-rose.* '*Hep-tree.*' '*Choups*'—the hips.

Area: 63, 64, 65, Top. Bot. Range: 0—1,350 feet.

Native; hedges, thickets, woods, and scars, very common. Shrub. Mid June-July.

First record: Salt, 1800.

Generally distributed; in every district.

My experience (with Mr. Baker's assistance in determining the names) together with that afforded by gatherings made by Amos Carr, G. Webster, W. Todd, and a few others, of the distribution of the forms, is as follows:—

Var. **lutetiana** Leman.

Common in hedges, seen from every district except Mersey, but not so abundant as var. *dumalis.* Intermediates, neither well marked *dumalis* nor purely simply serrate-leaved long-fruited *lutetiana*, are confusingly prevalent.

Var. **dumalis** Bechst.

Very common—more abundant than any other dog-rose in the lowlands; still prevalent among the hills, but there merging into *subcristata*, just as *lutetiana* merges into *Reuteri* under like conditions.

Noted in every district.

Var. **biserrata** Mer. (incl. *vinacea* Baker).

Not rare; but near the preceding—of which it is a more highly-specialised form.

W. Woodhall and Langwith, Collingham!

A. Wall-side by road, Adel Dam; *W. Todd.*

C. Hebden Valley!

D. Hedges near Crosspool, Scraith Lane, Bell Hagg Crags, etc., all near Sheffield; *A. Carr.*

Var. **urbica** Leman.

Very common in hedges, especially in the lowland limestone districts.

Noted in districts L., R., Y., N., W., A., C., D., T.

Var. **frondosa** Steven. (**dumetorum**, Herb. Jos. Woods).

Rare.

N. Near Rufforth; *G. Webster!*

W. Holgate, near York; *G. Webster.*

A. Brickfield lane, Adel!

T. Roche Abbey! teste *J. G. Baker.* Road-side at Wales; *A. Carr!*

Var. **arvatica** Baker.

Rather rare.

W. Harewood! Between Acomb and Dringhouses; *G. Webster.*

A. Malham!

C. Ogden Clough; *W. Todd!*

D. Rother side below Woodhouse Mill, Sheffield; *A. Carr.*

Var. **dumetorum** Thuill.

Rather common.

W. Between Kilnsey and Grassington! By stream, Norwood bottoms, Thorner!

A. Brickfield lane at Adel!

Var. **tomentella** Leman. (cum *obtusifolia* Desv.).

Local; seems to prefer a limestone, or at least a dry soil.

N. Hedge on road to Scriven! Near Rufforth; *G. Webster.*

W. Holgate, near York; *G. Webster.*

A. Roundhay. Scrub at Weetwood, near Headingley! Ledsham!

D. By the R. C. Cemetery, Rivelin Valley, and Oughtibridge; *A. Carr!*

T. Near Lindrick Common; *A. Carr.*

Var. **Andevagensis** Bast.

This—which is simply var. *lutetiana* with a few setæ on the fruit stalk—is now and then to be met with where the variety of which it is a state is found.

Var. **aspernata** Déségl. (**verticillacantha** Baker).

Rather common.

W. Collingham. Bramham. Hedge at Holgate; *G. Webster.*

A. Bell Busk (1870).

D. Redmires; by Whiteley Wood Dam, Cliff Rocker, etc., near Sheffield; *A. Carr.*

Var. **Koscinciana** Besser. (**collina** Jacq.).

Local, but not rare; best developed with us in the warm rocky limestone districts.

Y. Studley Woods; *J. G. Baker.*

N. Between Goldsborough and Green Hammerton.

W. Norwood Bottoms, Thorner. Holgate; *G. Webster* (sub nom. *collina*).

D. Ecclesfield, road to Sheffield, near the toll-gate; *A. Carr!* Hedges near Guilthwaite; *Id.*

Var. **decipiens** Dumort.

Rare.

Y. Hedge near Ripon towards Sharow, and hedge near Kirby Malzeard!

Var. **Reuteri** Godet.

Common amongst the hills.

Noted in L., R., Y., N., W., A., and D. districts.

Var. **subcristata** Baker.

Common in the hilly districts.

Noted in L., R., Y., W., and A. districts. Probably in Don, too; but Carr's specimens contained no plants clearly exhibiting the persistent ascending sepals, combined with glabrous, doubly-serrate leaflets.

Var. **Hailstoni** Baker.

Local.

R. Ribble bank near Hellifield!

D. Near Sheffield; *Amos Carr!*

Var. **implexa** Gren.

Rare and local.

L. Near Old Wennington Bridge: spot where gathered perhaps a little outside the boundary of the West Riding.

D. Went-vale, near Pontefract; *Jas. Abbott!*

Var. **Watsoni** Baker.

Rare.

W. Fields between Acomb and Dringhouses; *G. Webster.*

Var. **celerata** Baker.

Rare.

A. Adel Moor, near Church, four miles from Leeds! (teste *J. G. Baker*).

Var. **Borreri** Woods. (*R. inodora* of Flora Crav. and Miall's Flora.)

Rare, and much confused, I am afraid, from misunderstanding certain forms of *R. tomentosa* with hooked prickles. No great reliance is to be placed on any of the records except Backhouse's and Carr's.

L. Road-side near Sedbergh; *J. Backhouse.*

R. Lodge Dingle, and Cave-hole Wood near Settle; *J. Windsor.*

A. Near Shipley; *Miall & Carr., Flora.*

D. Lane between Meadow Hall and Grange Lane Station; *A. Carr.*

Var. **Bakeri** Déségl.

W. Hedges, Holgate, near York; *G. Webster.*

And perhaps elsewhere, but in my experience Dog-roses with leaflets distinctly glandulose beneath are the very rarest of the Yorkshire Bakerian forms.

304. Rosa stylosa Desv. (**systyla** Bast.).

Area: 63, Top. Bot. Range: 100—200 feet.

Native; open thickets, very rare.

(C. Near Skircoat; *Bates* in *Miall's Flora.* Some error. The plant *arvensis*?)

D. Levitt Hagg, near Conisborough (1871)! cum *W. Todd.*

305. Rosa arvensis Huds.

Area: 63, 64, Top. Bot. Range: Low. Not above 450 feet.

Native; warm hedge-banks, common at low levels, *rare* above 300 feet. P. June.

First record: Salt, 1800.

Distribution general—higher localities only here given.

M. Saddleworth district; *J. Whitehead.*

L. Lower Bentham.

R. Rathmell, near Settle; *W. Howson.* About Sawley!

Y. & N. Common, but not known where it ceases going up valleys.

W. Addingham, and eastwards, growing commoner as river descends.

A. Hawksworth; *W. West.* Bradford; *H. Baines.* Common in Lower Airedale.

C. Todmorden; *A. Stansfield.* Halifax; *H. Baines.* Dewsbury; *P. F. Lee.*

D. Near Sheffield Manor; *J. Salt* (1800). Common east of Sheffield and Barnsley.

T. Throughout this district.

Obs.—I have never seen any satisfactory specimen of the long-urnshape-fruited var. *bibracteata* from a West Yorkshire station, although W. Todd found an intermediate form approaching it, near Tadcaster.

306. Cratægus Oxyacantha L. *Hawthorn.* '*May.*'

Area: 63, 64, 65, Top. Bot. Range: 0—1,400 feet.

Native, but mostly planted where now seen; hedges, thickets, scars, etc., very common. Shrub. May-June.

First record: Salt, 1800. 'The monogyna most common with us'; *Flora of Sheffield.*

Generally distributed; in every district.

Obs.—The fruits seem to be universally known as Haws.

> How oft, my slender slice of pocket store consumed,
> I've fed on scarlet hips and stony haws.—*Cowper's Task.*

Var. **oxyacanthoides** Thuill.

Rare, in hedges and wood-borders, and I fancy mostly planted. It flowers a fortnight later than *C. monogyna* and *C. kyrtostyla.* Hybrids occur likewise.

W. Hedge near Kirk Deighton. Hedge at Stutton.

[A. Headingley 'Old Gardens,' Leeds—now villa plots.]

D. Levitt Hagg and Sprotborough. Conisborough!

T. Maltby Wood, by road-side! Roche Abbey Vale and Sandbeck Park!

Var. **monogyna** Jacq.

This is the vastly-prevailing form. Salt in 1800, however, was aware that a *two* or *three* styled plant occurred: see his note above (under type form).

Var. **kyrtostyla** Fing.

Common.

I have noted it in districts Y., N., W., A., D., and T. Most of the *old* gnarled thorns in parks and pastures belong to this, if the hairy peduncles and reflexed styles constitute a sufficient distinction.

Var. **laciniata** Ster.

Occasional in hedges, but not flowering nearly so freely as the other forms.

I have seen it in districts R., Y., N., W., A., D., and T. I have remarked it to be a sparing bloomer not only in Yorkshire, but also in Worcestershire, Lincolnshire, and Berkshire.

307. Pyrus Aria Hook. *White Beam.*

Area: 63, 64, Top. Bot. Range: 0—450 feet.

Denizen; woods and plantations in the lower country, rare. Tree. Late May.

Woods in Yorkshire; *Gough's Cam.* (1789).

A. Plantation near the Park gates, Meanwood! Cookridge.

C. Kirklees Woods†; *Y. N. Excurs.* (1873). Stock moor, near Dewsbury*; *P. F. Lee!* Near Boothtown, and Gledholt, Huddersfield; *L. C. Miall.*

Obs.—I hold this, *eu-Aria* of E. B., to be always a planted tree in Yorkshire.

308. Pyrus rupicola Syme. (*P. scandica* Baker, Supp.). *Wild Beam.*
Area: 64, 65, Top. Bot. Range: 1,400—100 feet.
Native; limestone scars and rocky mountain woods, rare. Tree. May-June.
First record: Hargrove, 1782, but of course under the name of *P. Aria*, since the present segregate was not then differentiated.

L. Dentdale Scars. Ingleton. Crow Nest Wood, N.W. of Settle; *H. Baines*. Clapdale and Giggleswick Scars; *H. Ibbotson*.
R. Stainforth Wood. Linn Gill.
N. On the edge of the Abbey rocks, Knaresborough; *E. Hargrove* (1782 et seq. eds.); *R. Teesdale* (1798) (!). Rocky woods, Thorparch; *Archdeacon Pierson* (1805); and Boston Spa!
W. East-field Wood, Arncliffe, rock fissures (!) *J. Windsor*. Kilnsey Crag.
A. Malham Cove (!) Rocks above Gordale, and on 'High Mark.'
D. Went-vale.

[**Pyrus torminalis** Ehrh. Alien.
I hesitate to include this as a Native, as its occurrences have been too few, not to say apocryphal; and Yorkshire is clearly beyond its range as an undoubtedly indigenous plant in England.

D. In old Park near Sheffield; *Salt's Flora* (1800).
T. Maltby; *T. Birks, jun.*, in *Goole Sci. Soc. Recorder's Report* for 1878.

If the names were correctly applied in these two instances, I hold it probable that the plants were introduced by some forest-tree nurseryman.]

[**Pyrus fennica** Sm.
Area: '64*,' Top. Bot.
Alien, hardly Denizen; in plantations occasionally, but always a nurseryman's introduction.
Noted in N., W., A., and C. districts.]

309. Pyrus Aucuparia Gaert. *Mountain Ash. 'Wiggen.' Roan-Tree.*
Area: 63, 64, 65, Top. Bot. Range: 0—1,800 feet.
Native; woods, thickets, and scars among the hills, common. T. May.
First record: Salt, 1800.
Distribution general—in every district.

Obs.—This species and *Salix cinerea* ascend in sheltered gills among the north-west hills higher than any other woody shrub. This is the 'Witch-wood' of some parts of Yorkshire according to Mr. Roberts, but I never heard it so spoken of myself; Wiggen may, however, be a corruption of 'Witchen.'

310. Pyrus communis L. *Wild Pear.*
Denizen; in hedges or wood-borders, very rare and with no claim to be called indigenous. Tree. Early May.

A. One tree near Adel, towards Smithy Mills, Meanwood! = 'About a mile from Adel Church towards Leeds'; *A. Shipley* (in *Miall's W. R. Flora*, 1862).

311. Pyrus Malus L. *Crab-tree.*
Area: 63, 64, 65, Top. Bot. Range: 0—900 feet.
Native; hedges and thickets, in the low country common, among the hills rarer.
First record: Salt, 1800.
Distribution general—in every area.

XXVII. LYTHRACEÆ.

312. Lythrum Salicaria L. *Loosestrife.*
Area: 63, 64, Top. Bot. Range: 0—1,250 feet.
Native; wet hedge- and ditch-banks, and pool-sides, frequent. P. August.
First records: Hargrove, 1782; Dixon, 1790.

[R. Giggleswick Tarn; *J. Windsor*—now gone: tarn drained.]
Y. Sharow Mires, etc., Ripon; *H. H. Slater*.
N. Half a mile from Knaresborough in wet meadows on right side of York road; *E. Hargrove* (1782). Stockwell Lane, Knaresborough; *Id.* (1809 and 1832 eds.).
W. Bickerton road, Wetherby! Askham Bogs. Sherburn and Church Fenton! Ouse-side, Middlethorpe Ings; *W. Whitwell*.
A. Malham Tarn; *J. Tatham*. At an unusual altitude, and not seen by me, but the recorder was too good and careful a botanist for the record to be doubted. Fields near river and by canal at Steeton; *S. Hailstone* (1805). Milford! [Meanwood Valley under Myles Hill.] Pontefract.
C. Plentiful at Silcoats, 1790; *D. Dixon*.
D. Potteric Carrs; *E. Miller* (1804). Rotherham; *L. Langley*. Doncaster Car; *J. Salt*, etc. Ackworth; *T. W. Gissing*. Askern. Barnsley.
T. Dike-sides near Bawtry and Rossington!

[**Lythrum hyssopifolia** L. Alien.
'(64 casual or erroneously reported),' Top. Bot.

N. Near Harrowgate; *Rev. Archdeacon Pierson* in *O.B.G.* (1805).
C. Casual at Whitley Willows, 1858; *Hudd. Nat. Hist.* There in 1868.]

313. Peplis Portula L. *Purslane.*
Area: 63, 64, Top. Bot. Range: 0—900 feet.
Native; shallow water or on wet mud, infrequent. A. July-August.
First record: Salt, 1800.

R. Rathmell, Greenhippings, etc., near Settle; *J. Windsor*.
A. On mud, Adel lower dam!
C. Wibsey; *W. West*. Wakefield, frequent; *T. W. Gissing*.
D. Goole fields; *H. F. Parsons*. Ditch near Womersley (!); *W. Todd*. Crook's Moor; *J. Salt*; thereabout still; *A. Carr* (1876).

XXVIII. ONAGRACEÆ.

314. Epilobium angustifolium L. *'Rose-bay.'*
Lysimachia speciosa quibusdam Onagra dicta siliquosa; J.B.; Ray.
Area: 63, 64, 65, Top. Bot. Montane. Range: 0—2,100 feet.
Native; woods, mountain cliffs, and low moors, common, but not indigenous in many of its stations. P. July.
First record: Gerarde, 1597.

L. Foalfoot, Ingleborough; *T. Pennant* in *Hailstone*. Howgill Fells. Garsdale. Near Meer Gill at foot of Ingleborough; *J. Windsor*.
R. Linn Ghyll, Ribblehead! Rail-bank in several places between Horton and Ribblehead stations. North part of Scar near Attermire Cove; *J. Windsor* = Langcliffe Scars near Victoria Cave!
Y. Sawley; *W. Brunton* in *O.B.G.* (1805). Quarry moor near Ripon, *Mrs. Berkeley*; *H. H. Slater*. Morcar Wood, Copgrove; *J. Dalton*.
W. In Grass Wood, near Kilnsay; *Gough's Cam.* (1789).
A. Scars and swallow-holes between Malham and Kettlewell and Malham and Settle! Deep glen above Gordale, among rocks, betwixt Gordale and Grisedales; *S. Hailstone*; *J. Windsor*.
C. On the side of Dirtcar Green, 1791; *D. Dixon*. North Dean; *W. West*. Skircoat and Elland; *J. Bolton* (1775). [Ardsley Wood; *G. Roberts*.] Haw Park—a natural wood—near Walton; *G. Roberts*.
D. In Yorkshire, neere unto a close called the cow-pasture, in a place called Hooke; *Gerarde*, *Hb.* (1597). [Hook, near Goole, on the uncultivated moor-borders, near by which places it is still, but decreasingly of course, a conspicuous feature in the vegetation.] Near Sheffield; *J. Ray*, *Cat.* (1670); *Hist.*, 860; *Gough's Cam.* (1789); etc.

VAR. **brachycarpum.** *French Willow.*
Not rare, the usual plant in *suspicious* stations, hedges and plantations near villages, etc.; much grown in gardens.

315. Epilobium hirsutum L. *'Codlins-and-Cream.' Willow-weed.*
Area: 63, 64, Top. Bot. Range: 0—600 feet.
Native; watery places, very common in the lowlands, rare in the dales above 450 feet, but ascends to Settle. P. July.
First record: Bolton, 1775.
Distribution general—occurs in every district.

Obs.—This species hybridises with *E. parviflorum* (Barnsley; *H. Johnson*!)

316. Epilobium parviflorum Schreb.
Area: 63, 64, 65, Top. Bot. Range: 0—1,200 feet.
Native; moist boggy places, ditches and stream-sides, common. B. July.
First record: Bolton, Cat., 1775, sub nom. *hirsutum*.
Distribution general—occurs in every district.
'Marsh on Crook's Moor, Sheffield, 1794'; *Herb. Salt*.

Obs.—There are two forms of this, one small and very hairy, in peat bogs mostly; the other nearly glabrous, tall and wand-like (*E. rivulare* Wahl.) in very wet ditches. It will hybridise with *E. hirsutum* and *E. obscurum*.

317. Epilobium montanum L.
Area; 63, 64, 65, Top. Bot. Range: 0—1,900 feet.
Native; damp ground in woods, ditches, or gardens, very common, ascending very high by shaded rills on the mountains. P. June-September.
First record for Riding: Salt, 1800.
In every district.

318. Epilobium roseum Schreb.
Area: 63, 64, Top. Bot. Range: Low.
Native; shady ditch-sides, and as a weed in nurseries and gardens, rare. P. August.
First record: Baines, 1840.

W. York; *H. Baines*. Garden at Bilbrough! Gardens at back of Thorner main street, and by stream there! Damp ground near Ilkley; *G. S. Gibson* (1852).
A. Nursery Gardens, Meanwood Valley (!) and Woodhouse Carr.

319. Epilobium tetragonum L. (Bab.).
Area: No record in Top. Bot. Range: Low.
Native; ditch-sides, but the true long-podded plant really very rare. P. August.
First record: Harrison Herb., 1862.

(A. Bingley; *W. West*.)
D. Near Woodhouse Mill, Sheffield; *Geo. Harrison, Herb.* ['This does not seem *E. obscurum*'; *W. W. Newbould*.]
T. Wet ditch near Bawtry (1876)!

320. Epilobium obscurum Schreb.
Area: 64, Top. Bot. Range: 0—900 feet.
Native; wet places, ditch-banks, garden ground, etc., common. P. July-August.
First record: 'Near Park Furnace (1798)'; *Herb. Salt.*
Distribution general; every district.
The '*E. tetragonum*; not uncommon' of Craven Flora meant *obscurum.*

321. Epilobium palustre L.
Area: 63, 64, 65, Top. Bot. Range: 0—1,800 feet.
Native; swamps and peat bogs, common, except in the more cultivated lowlands. P. July-August.
First record: Baines, Flora, 1840.
Distribution general—occurs in every district.
M. Greenfield Valley; *J. Whitehead.* Wragstone Clough.
L. & R. Common on the peat moors.
Y. & N. Frequent in the upper parts of these river-basins. Near Rufforth; *H. Baines.*
W. Askham Bogs.
A. On all the moors, Malham, Adel Bog, etc.
C. Todmorden. Ogden Clough. Heath; *T. W. Gissing.*
D. Doncaster Carrs. By Rother, near Treeton; *A. Carr.*

A VAR. **virgatum** Fr., stands on record from 'road-side near Sedbergh'; *J. Backhouse.*

322. Epilobium alsinifolium Vill.
Area: 65, Top. Bot. Range: 2,000—1,700 feet.
Native; in tufts (semi-evergreen) by the spring heads on the mountain plateaux, very rare. P. August.
First record: the Author, unless Curtis's plant of 1782 recorded as *E. alpinum* was really this.
L. Very sparingly on Howgill Fells west to north side! By one spring on the N.W. slope of Ingleborough.
Obs.—The few West Yorkshire specimens I have seen in situ are much smaller in all their parts than the Micklefell and Teesdale, form which is var. *E. anceps* Fr.

[Epilobium alpinum L. Incognit.
'[64 error],' Top. Bot.
Near Settle; *W. Curtis* (1782). Sides of rivulets in the mountainous parts of the West Riding; *R. Teesdale* (1798). I have a memorandum of having found this on the highest ridge, nearly as far as is accessible, on the right hand descending to Foalfoot, Ingleborough, but whether this or a small form of *E. montanum* I cannot now determine; *J. Windsor, Flora.* A small flowerless (young) plant which I gathered by a rill head high up the west slope of Ingleborough, and queried in list in 'West Yorkshire,' p. 263, is probably *E. alsinifolium.*
Obs.—Curtis' second station, 'about Buckbarrow Well,' inserted in *Yorkshire* list of O.B.G., is in Westmoreland, north of Kendal. Mr. Baker informs me that the species there, too, is *alsinifolium*, so in all probability his Settle plant was the same.]

323. Œnothera biennis L. *Evening Primrose.*
Colonising Alien; sandy fields and waste ground, rare and hardly permanent; usually introduced with seed crops.
First record: Appleby, 1831.
W. A few dozen plants in a seed-field (with *Anthemis tinctoria*), near Flint-mill, Wetherby, 1870.
A. Rail-bank, Shipley; *W. West.* Ouse banks, above Selby.
C. Calder side, Bottom Boat; *G. Roberts.*
D. On the sand-lands east of Doncaster: *S. Appleby* (1831). Sandy fields at Cantley (*G. E. Smith, MS.*).
T. Finningley!

324. Circæa lutetiana L. *Enchanter's Nightshade.* '*Philtrewort.*'
Area: 63, 64, 65, Top. Bot.
Native; damp shady places, especially where the ground has been newly disturbed or trees felled, common and mostly abundant.
First records: Bolton, 1775; Dixon, 1790.
Distribution general—in all the districts.
M. Greenfield Valley; *J. Whitehead.*
L. Howgill. Wennington and Bentham.
R. Settle, common; *J. Windsor.* Slaidburn! Sawley; *R. Brown.*
N. Long Walk, Knaresborough; *E. Hargrove* (1782). Grimbald Crag, Scriven Lane, etc., Knaresborough; *Id.* (1832).
W. Bolton. Bilton Haggs (small form), abundant. Wetherby! Bramhope staircase! *J. Abbott.* York neighbourhood; *H. Baines.*
A. Malham Cove! Bingley, Shipley, etc.
C. Near the foot-bridge leading to Heath, 1790; *D. Dixon.* Todmorden, Hebden Bridge, etc.! Horbury, Ossett, etc.; *T. W. Gissing.*
D. Rotherham; *L. Langley* (1828). Sheffield; *J. Salt* (1800). Endcliffe Wood; *G. Harrison.* Plentifully about Little Matlock; note by *J. H[olland]* in *Herb. Salt.*

VAR. **intermedia** Ehrh.
Area: 63 Leyland spn., 64, Top. Bot. Range: 500—600 feet.
Native; shady rocky bushy places, very rare. P. July-August.
First notice: Bolton, 1775.
A. Malham Cove; *J. Windsor* (about 1812). 'I now doubt if the plant I found is more than a slight variety of *C. lutetiana*'; *J. Windsor* (1873). [About Leeds; *Mr. Woodward*; *With.* 7th ed. (1830).]

C. Warley Clough; behind Scars-head, Norland; Cliff Scar Wood, Rastrick, etc.; *J. Bolton, Cat.* (1775). The foregoing sub nom. *alpina*, but Leyland's specimen shows this species to have been intended. Perhaps extinct now—no recent records. The 'Ripponden and Luddenden' records of *Baines' Flora* relate to the same places, and the same plant, as the foregoing.

325. Circæa alpina L.
Area: '65; also 63 64 insufficiently vouched,' Top. Bot. Montane. Range: 900—600 feet.
Native; rocky bushy places.
First record: Dixon, 1793.
M. Reported for head of Greenfield Valley, but never seen by me; *J. Whitehead* (1877).
L. Howgill; rocky shady bank by stream!
(A. & C. [About Leeds, *W. Wood*; *Withering* (1787). Probably erroneous; not there now at least; and referred in later eds. of Withering to *intermedia.*] Sowerby Bridge, Luddenden and Ripponden; *H. Baines.* Malham Cove; *Miall's Flora.* But the plant is *intermedia.* Near Low Laithes, Wakefield, 1793; *D. Dixon.*)
D. Rivelin; *Amos Carr*! 'Foot of the Cemetery at Crookes' (1876)!

XXIX. HALORAGACEÆ.

326. Myriophyllum verticillatum L.
Area: 63, 64, Top. Bot. Range: Low.
Native; slow streams and dikes, rare. P. July.
First record: Salt, 1800.
Y. Ponds near Ripon; *W. Brunton.* Sharow Mires; *H. H. Slater.* (In the river Ure; *Miss Morton.*)
A. Milford; *G. Roberts.*
D. Potteric Carrs, Doncaster (!) *S. Appleby* (1831). 'Car near Doncaster, July 1800'; *Herb. Salt.*

327. Myriophyllum spicatum L.
Area: 63, 64, Top. Bot. Range: 0—300 feet.
Native; still waters, rare. P. June.
First record for the segregate: Harrison, 1862.
R. Birkbeck's Weir and Beggar-wife's Bridge, Settle; *J. Windsor.*
W. Mill-cut, Washburn Valley!
A. Yeadon ponds; *W. West.* River Aire at Carlton Bridge, Skipton; *T. W. Edmondson* (1886)!
C. Kirklees district; *Y. N. Excurs.* (1873).

D. Potteric Carrs; *S. Appleby* (1831) = Doncaster! *H. F. Parsons.* Near Attercliffe Coal Mines; *J. Salt.* Near Leighton; *Geo. Harrison, Herb.* ('Not *alterniflorum*'; *W. W. Newbould.*)
Obs.—Not being critically distinguished from *alterniflorum* in Baines' Flora, the earlier records therein, and that of Appleby, cannot be held to have referred to restricted *spicatum.*

328. Myriophyllum alterniflorum DC. *Milfoil.*
Area: 63, 64, Top. Bot. Range: 0—900 feet.
Native; streams and canals, infrequent, but the commonest of the genus, and ascending highest. P. May-July.
First record: Miall's Flora, 1862.
R. In the Ribble at Wigglesworth!
N. Knaresborough; *L. C. Miall.* Nidd at Cowthorpe.
W. Askham Bog ditches! River Wharfe, and mill-cut, Thorparch!
A. Dikes at Sherburn! Selby 'Cut' canal!
C. Storthes Hall Wood, near Huddersfield; *C. P. Hobkirk.* Canal near Halifax!
D. Barnsley Canal.

329. Hippuris vulgaris L. *Mare's-tail.*
Area: 63, 64, Top. Bot. Aquatic. Range: 50—1,500 feet.
Native; slow streams, ponds, mill-dams, etc., infrequent. P. July.
First record: Bolton, 1775.
R. Branch of Ribble, $1\frac{1}{2}$ miles S.W. of Settle; *H. Baines.* Near Birkbeck's Weir; *J. Windsor.*
Y. Aldborough; *Archd. Pierson.*
N. Staveley Carrs and Corn-mill Dam at Boroughbridge; *Archd. Pierson* in *E. Hargrove* (Eds. 1782 and 1809). Pond in Belmont Wood, Knaresborough; *Id.* (1832).
W. Bardsey pool! Mill dam, Stutton; *H. Ibbotson.*
A. Drying-up clay-bottomed plash on 'High mark,' Arncliffe Moor! Milford; *G. Roberts.* River and canal at Kirkstall; *J. Abbott.* Mill dam between Micklefield and Sherburn! In the mill-stream is a barren floating form, wholly submersed and quite flaccid; which led to the notice of the variation in Bab. Man., Ed. 8; *W. Whitwell.* Aberford; *H. Baines.*
C. Old House mill-dam, Norland, and Mixender; *J. Bolton, Cat.* (1775).
D. Rotherham district; *L. Langley* (1828). Potteric Carrs; *S. Appleby* (1831). Eastoft; *H. F. Parsons.* Askern; *G. Harrison.* Askern Pool; *C. P. Hobkirk* (1886).
T. In a muddy pond at Roche Abbey; *Salt* (1800). Pools at Laughton, etc.

330. Callitriche vernalis Kütz. *Water Starwort.*

Range: 0—1,250 feet.

Native; ponds and slow streams, rare, or confused with deep water states of the two succeeding, which probably are not distinct.

The distribution of this—it has a *shallow* sinus to the turgid, sharply-winged fruit—is ill ascertained. The old records for *verna* cannot safely be assigned to this; so that it is probable the following three stations are the earliest records for the restricted species.

A. Malham Tarn. Dikes at Sherburn.
T. Pool at Roche Abbey.

331. Callitriche platycarpa Kütz. *Mud Starwort.*

Area: 63, 64 (by implication), Top. Bot. Range: 0—1,800 feet.

Native; shallow waters, wet mud, peaty or sandy rill-sides, the commonest species and with the widest range.

First record: Baker in Supp. Flora Yorks. (1854).

In all the districts; from the ditches of Marshland by Goole, Thorne, Askham Bogs, Adel Moor, etc., to Blea-moor and Baugh Fell tarns—the highest stations in which it has been observed.

Obs.—I include *C. stagnalis* Scop.—the mud state—with this.

332. Callitriche hamulata Kütz. (incl. *pedunculata* DC.).

Area: 63, 64, Top. Bot. Range: 0—450 feet.

Native; sandy and muddy-bottomed slow streams, infrequent. P. July.

First record: Salt, 1800; sub nom. *autumnalis* (Hook.).

R. In the Ribble, near Wigglesworth.
Y. Near Ripon!
W. Dike, Wetherby Ings, border near railway.
A. Ditch, near Fairburn!
C. Ditch, Batley; *P. F. Lee.*
D. Doncaster Carr Dikes! In the Rother; *A. Carr*! Sheffield; *Salt, Flora.*
T. Stones Dike, near Kiveton; *A. Carr.*

333. Callitriche autumnalis L.

Area: 64, Top. Bot. Range: 1,250 feet.

Native; lakes, very rare. P. August.

First record: the Author.

(Y. Near Ripon, on magnesian limestone; *Miall's Flora.* Error, the *autumnalis* Hook. was meant = *C. hamulata*; but what has magnesian lime to do with its occurrence?)

A. Malham Tarn! Not seen *in situ*; but a matted mass of what was undoubtedly this, I gathered near the outlet of the tarn, up-cast on its shingle near the south sluice end, in 1870!

XXX. CUCURBITACEÆ.

334. Bryonia dioica L. *Bryony Vine. 'Mandrake.'*

Area: 63, 64, Top. Bot. Range: 0—300 feet.

Native; in hedges on the magnesian limestone band running through districts T., D., A., W., N., Y., there common; much rarer on the sand of the Trias, quite absent over the scar-limestone, and almost unknown on the grit, but occurs in a few spots about Acomb and Nether Poppleton. Mr. Hobkirk adds 'also in district C.'; and there is a station given in *Miall* and *Carrington's Flora*—'Wood Head'—which I am unable to locate; but most certainly over nineteen-twentieths of the Calder and Colne basins, as of those of the Upper Don, Aire, and Wharfe, upon coal shales and grits, it is a characteristic Absentee.

First record: Baines' Flora, 1840.

XXXI. GROSSULARIACEÆ.

335. Ribes Grossularia L. *Gooseberry.*

Denizen; bird-sown in hedges and thickets, common as single bushes, rarer in wild situations on scars. Shrub. May-June.

First record: *Baines*, 1840.

L. Rock-clefts between Chapel-le-Dale and Meer Gill, by the foot of Ingleborough; *J. Windsor.*
R. Ditch-side, Settle Ings, and on High-hill above middle of Castleberg; *J. Windsor.*
Y. Ripon; *H. H. Slater.*
N. Hedges, frequent.
W. Scars about Grassington and Barden! Occasional all over the district in the lowland hedgerows.
A. Frequent, from Malham Scars downwards.
C. Occasional; *T. W. Gissing.*
D. Conisborough! Askern.
T. Occasional; *A. Carr.* Roche Abbey, apparently wild; *H. Baines.*

Obs.—Some of the above records, notably those for Ingleton and Grassington Scars, are for the smooth-fruited form (*R. Uva-crispa* L.) which is less doubtfully the product of bird-sown seeds; it is said to be the *wild* European type, hairless-fruited like the currants.

336. Ribes alpinum L.

R. alpinus dulcis; Ray.

Area: '63? Dill R.S., 64 Tatham,' Top. Bot. Range: 0—1,250 feet.

Native on precipitous limestone-rock slopes, but in most of its woodland and hedgerow stations planted; rare. Shrub. May-June.

First record: Dodsworth in Ray, 1688.

In agro Eboracensi invenit, *D. Dodsworth*; *Ray, Hist.* (1688); *Syn.*, etc.

L. Sedbergh and Dent; *G. Pinder.* Thornton. Clapham; *H. F. Parsons.*
R. Stainforth Wood; *J. Tatham.* Near Settle; *H. Baines.*

Y. Studley and Hackfall; *W. Brunton* in *O.B.G.* Fountains Abbey.* By Markington Beck, one bush; *J. S. Tute.*
N. Dropping Well Wood, Knaresborough; *E. Hargrove* (1809 et sequ.). Bilton Banks.
W. Bolton Woods. Outershaw, Langstrothdale!
A. Between Gordale and Malham in rocky clefts (!) *J. Windsor.* In hedges near Bradford, plentifully; *D. Richardson* in *Ray Syn.*, iii (1724); rep. *Huds. Flora Angl.* (1778). Shipley and Wilsden; *L. C. Miall.* About Bradford, *Mr. Wood*; *With.*, ii, 243 (1787). Apperley Bridge; *Rev. W. Wood*; *O.B.G.*
D. Woods near Camp's Mount, Askern; *E. Lankester* (1840); not wild.
T. Woodland banks about Roche Abbey, but not wild there.

No records for C. and M. districts.

337. Ribes rubrum L. *Red Currant.*

R. vulg. fructu rubro; Ger.; Ray.

Range: 0—900 feet.

Denizen; thickets and river-banks, infrequent, bird-sown or escaped. Shrub. June-August.

First record: Ray, 1690.

In sylvis in septentrionale com. Eboracensis parte; *D. Johnson*; *Ray, Syn.*, i, 221 (1690).

L. Dentdale, by stream near Sedbergh.
R. Wood near Langcliffe Mill and by Ribble banks (*J. Windsor*) as far as Bolton-in-Bolland! ('Truly wild, *Tatham*'; *H. Baines.* But *petræum* intended.)
Y. By Yore side above Ripon bridge, etc.
W. Many places by Wharfe from Bolton to Thorparch or lower.
A. Occasional.
C. Ardsley, an escape; *G. Roberts.*
D. Occasional. Conisborough. Went-vale.

338. Ribes petræum Sm. *Rock Currant.*

R. vulgaris fructu dulci; Clus. H., 120; Ray.

Area: 64, 65, Top. Bot. Montane. Range: 1,400—350 feet.

Native; on rock ledges and stony thickets on the scar-limestone, rare. Shrub. May.

First record: Ray, 1724.

R. Near Settle, 'Bracken-brow'; *Ray, Syn.*, iii, 456 (1724); *Martyn* (1763). Linn Gill. Douk Gill Scar.
W. Littondale Scars. Buckden. Near Burnsall, truly wild; *L. C. Miall.* Wharfe-side at Bolton; *J. G. Baker.*
A. Rock crevices between Gordale and Malham Tarn on left; *J. Windsor.* Above Gordale (!); *J. Nowell.* Malham Cove Scars; *C. Bailey*! Rocky scar above Malham Tarn (!); *J. Windsor*,

339. Ribes nigrum L. *Black Currant.*

Area: not treated of in Top. Bot. Range: 50—900 feet.

Denizen; damp woods, thickets, and hedges, rare, mostly an escape. Shrub. May-June.

First record: Irvine, 1858.

L. Clefts of the limestone pavement at the foot of Ingleborough; *Alex. Irvine.* North side of Ingleborough, apparently indigenous; *L. C. Miall.*
R. Birkbeck's Weir and Settle Ings; *J. Windsor.*
W. Wharfe-side, odd plants, from Bolton (*J. G. Baker*) to Newton Kyme! Askham Bog, in the 'Far Wood'; *B. B. Le Tall.*
A. Aire-side, Kirkstall; *A. Shipley.*
C. Lingwell Gate, Ardsley; *G. Roberts.*
D. Rossington Carr Woods!

XXXII. CRASSULACEÆ.

340. Sedum Rhodiola DC. *Rose-root.*

Rhodia radix; Park.; Ger.; Ray.

Area: 64, Top. Bot. Range: 2,300—2,000 feet.

Native; high exposed grit or slate rocks; rare. P. July.

First record: Gerarde, 1597.

L. Howgill slate fells, north side between Cautley and Black Force. Ingleborough Fels, neere unto the brooke sides; *Gerarde, Hb.* 426 (1597); rep. *Merr.*, 104. Ingleborough; *Ray, Cat.*, 264 (1670), etc.; *Martyn* (1763); 'rock on the summit of'; *Woodward* in *With.*; *Hailstone.* Ingleborough and Hardknot in Yorkshire; *Ray, Syn.*, i, 99. Upon the mountains of . . . Ingelborough, oftentimes on the very ruggiest places and most dangerous of them, scarce accessible, and so steepe, that they may soone tumble downe that doe not very warily looke to their footing; *Park., Theat.*, 727 (1640).
R. West side of Penyghent (!)

341. Sedum Telephium L. *Orpine.*

Area: 63, 64, Top. Bot. Range: 50—1,500 feet.

Native in both its forms; hedgebanks, on sandy or gravelly soil in the south-east of the Riding (*S. purpurascens*—the broad-leaved form) and on limestone scars, in the north-west portion of the Riding (*S. Fabaria*—the narrow-leaved form); both forms rare. P. August.

First record: Dixon, 1791.

L. Sedbergh. In and about village of Wherf; *H. Baines.*
R. Linn Gill. Douk Gill, near Horton! Wood and cliff under Winskill two miles north of Settle; *H. Baines* = Richardson's Scar, Settle; *J. Windsor.*

W. Acomb, hedge-banks; *H. Baines*. Road-side by Station, Poppleton, introduced from garden in 1877; *W. Whitwell*.

A. In a small glen on right of road at turn to go over fields to Gordale from Malham; *S. Hailstone* (1815). Malham Cove; *J. Windsor*. [About Bradford; *H. Baines*; and Meanwood Wood stone quarries! formerly]. Kippax; *Loidis and Elmete*, 1816. Hedges, etc., on the Trias about Hambleton and Gateforth! cum *W. Todd*.

C. On Walton Common, 1791; *D. Dixon*. Near Fenay Bridge, Huddersfield—a garden escape; *Hudd. Nat. Hist.*

D. Hedges near Grimethorpe; *J. Salt*.

342. Sedum villosum L.

Sedum purpureum pratense; Ray.

Area: 64, Top. Bot. Montane. Range: 1,900—500 feet.

Native; wet shales and grit rocks, or on peat, rare. Late June-July.

First record: Ray, 1677.

L. Ingleborough, on the moist rocks going towards Horton-in-Ribblesdale, where peat is got; *Ray, Cat.*, ii, 270 (1677); *Hist.*; *Syn.*; *Martyn* (1763); *Huds.* (1778). Rawthey Dale shales; and on the Dent slope of Widdale! Swarth Moor (!) *J. Windsor*. Bentham Moor; *J. Windsor*. Close by Wethercote Cave; *Woodward* in *O.B.G.*

R. Near Settle; *Martyn* (1763). In et juxta montem Hincklechew prope Settle; *Dr. Richardson*; *Ray*. Long Preston Moor; *J. Windsor*. Foot of Rye-loaf Hill—the same as the Hinklehaugh of old writers. Roadside between Settle and Penyghent; *Winch.*

W. Moor near Oughtershaw, Langstrothdale; *Rev. W. D. Fox*. Cowside Beck, wet grits; *J. Tatham*; N.B.G. Fountains Fell; *Jos. Woods, MS.* in *Towns. O.B.G.*

A. Malham Tarn bog (N.B.G.) and between Kirkby-Malham and Settle. Hartside Fells; *R. Teesdale* in *O.B.G.*

[**Sedum album** L. Alien.

On old walls of cottage gardens, etc., noticed in Went-vale and near Tickhill, but with no claim to inclusion.]

[**Sedum dasyphyllum** L. Alien.

On old wall, but only where planted. I have seen it at Bramham, near Thorner, at Bilbrough, etc.]

343. Sedum anglicum Huds.

S. minimum non acre flore albo; Ray. *S. min. fl. mixto ex albo et rubro*; Merr.

Area: 65, Top. Bot. Range: 600—1,500 feet.

Native; wet rocks and rill-sides on silurian slate of the N.W., very rare.

First record: Merrett, 1666.

L. Howgill Fells, north aspect over towards Black Force! On the west of Ingleborough hill in the mud of the hollow-topped stones; *Merr.*, *Pin.*, iii (1666). On rocks, Ingleborough and Clapham; *Gough's Cam.* (1789). Query—lost? Not confirmed for 100 years.

[N. Near Harrogate; *R. Teesdale, Linn. Trans.* (1798).

C. Walls near Storthes Hall, Huddersfield—introduced; *P. Inchbald*; *Hudd. Nat. Hist.*

Obs.—Curiously enough not quoted for any West Riding station in *Baker's Flo. Supp.*, although old authors were worked up. I have seen it only in the station first given.

344. Sedum acre L. *Stonecrop. 'Golden Moss.'*

Area: 63, 64, 65, Top. Bot. Range: 0—1,800 feet.

Native; old walls, sandy banks, limestone rocks, common; scarce only in the manufacturing districts. P. June-July.

First record: Baines, 1840.

Distribution nearly general—every district except M.

Ascends to the limestone scar-ridges of Whernside and Cam Fell.

[**Sedum sexangulare** L. Alien.

Banks and old walls, very rare, and always an escape? P. July.

(L. Ingleton, July 23rd, 1864; *W. Todd, MS.* list of specimens in his herbarium. But naming perhaps not correct.)

W. Old wall, near cottages between Thorparch and Collingham; planted doubtless.

A. Just below Malham Cove, on left hand; *J. Windsor* (1801). Doubt expressed (*Flora*, 1873) if plant of that station be true *S. sexangulare*. I have thrice failed to find aught but *S. acre* by the Cove, but on a grassy-sandy bank of the Aire beck, close by Airton, in 1870, I gathered a trifid-cymed plant, with base of leaves spurred, which I consider to be the true thing. It does not follow it is wild, since occurring only *below* the town of Malham, from rockeries in which other plants (e.g., *Lamium maculatum* and *Linaria Cymbalaria*) have escaped and become naturalised, and if those, then possibly this *Sedum* also.]

345. Sedum reflexum L.

Denizen; old walls of gardens, and on rocks in ornamental grounds where planted; not native. P. August.

First record: Dixon, 1790.

R. Castleberg Rock, Settle, etc.; *J. Windsor*.

Y. About Fountains Abbey.

W. Moor End, Boston Spa!

C. Sandal Church walls, 1790; *D. Dixon*.

D. Walls near Carbrook; *J. Salt, MS. Flora*. Wentbridge. Kirk-Smeaton. Goole; *H. F. Parsons*.

T. Roche Abbey.

[**Sedum rupestre** Huds. Alien.

Seen on rocks, escaped from adjoining gardens, at Knaresborough, and also seen at Slaidburn.]

[**Sempervivum tectorum** L. *Houseleek*. Alien.

On thatched roofs, old walls of cottage gardens, etc., but only where planted, and showing no disposition to spread itself, as *Sedum reflexum* occasionally will, beyond the roof or wall on which it is acclimatised.]

346. Cotyledon Umbilicus L. *Navelwort.*

Area: '63† Leyland spn.,' Top. Bot. Range: 450—900 feet.

Denizen; perhaps native in south-western cloughs of Don district, but nowhere else; on rocks and masonry near villages and reservoirs, rare. P. June.

First record: Salt, 1803.

C. Cragg Valley, Mytholmroyd! and above Catholis Dam, Harley Wood, near Todmorden, probably planted; *A. Stansfield* (1877). Luddenden Valley; *R. Earnshaw*. [Formerly at Greetland and Warley.]

D. Nether Bradfield; *J. Salt* (1803). Still on walls near stream at Bradfield; *Ed. Birks* (1876).

T. Abbey brook cloughs, south border Don district in Yorkshire, but physico-geographically part of Derbyshire Derwent district; *R. A. Gatty*; *A. Carr*.

[**Cotyledon lutea** Huds. Incognit.

'Walls and rocks in the West Riding, *Mr. Tofield*'; *O.B.G.* 'Mr. Fairburn informed me that the stock of this plant, now in Chelsea Garden, originated from a Yorkshire root introduced by Mr. Hudson.' *Dawson Turner*: note in *O.B.G.*]

XXXIII. SAXIFRAGACEÆ.

347. Saxifraga oppositifolia L. *Purple Saxifrage.*

Sedum alpinum ericoides, cæruleum; C. B.; Ray.

Area: '64 Rylands spn.,' Top. Bot. Montane. Range: 2,200—1,600 feet.

Native; exposed alpine rocks, very rare, locally plentiful, but only on Yoredale-rock strata with us. P. May.

First record: Ray, 1677.

L. Ingleborough, ad latus septentrionale copiosissimé; *Ray, Cat.*, Ed. ii, 269 (1677), et *Syn.* i; *Martyn* (1763); 'non ad summitatem nec septentrionem, sed ad latus occidentem versus; *Ray, Syn.*, iii 353 (1724). Still plentiful there on the vertical western rocks!

R. On Pennigent; *W. Curtis* (1782); *Gough's Cam.* (1789). Cliffs of Penyghent; *W. Whitwell*; one flower still in bloom on August 4th 1884.

(A. Limestone rocks near Malham; *Miall's Flora*. Error.)

348. Saxifraga stellaris L.

Area: '64, 65 Pinder ms.,' Top. Bot. Montane Range: 2,400—1,500 feet.

Native; wet rocks and boggy rill-sides on the higher hills, scarce. P. July.

First record: Teesdale, 1798.

L. Moor above Cautley Spout; *J. Nowell*. Ingleborough (!), various places about Ingleton; *R. Teesdale, Linn. Trans.*; rep. *O.B.G.* (1805). On wet rocks on north and west side by rills, but rare. I have not myself met with it there; *Windsor's Flora*.

R. Hinklehaugh, and various places circa Settle; *R. Teesdale* in *Linn. Trans.*

W. Penyghent, both slopes! North-east of Pennigent; *H. Baines*.

A. Malham Cove; *R. Teesdale*; *O.B.G.*

[**Saxifraga Geum** L.

Denizen; on shady rocks, rare. P. June-July.

First record: Hailstone, about 1844.

L. Near Weathercote Cave, Chapel-le-dale, naturalised (!) *J. Backhouse* in *Baker* Flo. Supp. (1854).

Y. Grounds at Studley and Fountains Abbey (!) not wild; *H. H. Slater*.

W. Planted on a rock in Thorparch Woods; *S. Hailstone*. In several places thereabouts now (1880).]

349. Saxifraga umbrosa L. *London Pride.*

Area: 'Alien?' Wats. Comp. Cyb. Brit. Range 300—1,000 feet.

Native—the wild crenate-leaved type; wet sheltered rocky places, rare, but locally plentiful. The garden form naturalised. P. Late June.

First record for wilding—Bingley, 1800; for Denizen—Teesdale, 1798.

R. Linn Gill! *Rev. J. Howson* in 1840; *Tatham* (*Phyt.*, i, 267); loc. rep. *Howson's Guide* (1850). 'About 20 yards above the waterfall in Linn Ghyll'; *J. Windsor* in *Flora Cravon* (1855-8).

W. Betwixt Arncliffe and Horton in Craven; *Rev. W. Bingley* (*Flo. Brit.* and *Sm.*, *E. F.*, 1824), i.e., same as 'Heselden Gill, in profusion . . . the typical form of the species . . . possessing strong claims to be considered indigenous'; *C. C. Babington* (1884) in *Ann. Nat. Hist.*, xiii, p. 465. Truly native in the Heseltine locality, a deep ravine at the foot of Pennygent and Fountains Fell—fide *John Tatham* (*Phyt.*, i, 267, 1842); *C. C. Babington* (*Phyt.*, i, 845); *W. Borrer* (*Phyt.*, ii, 430). Heseltine Gill is the Penyghent Gill of the map. Hazleton Gill, Littondale; *F. Brunton*, *O.B.G.* (1805). First found by Mr. Brunton (in 1792) who pointed out the place to Mr. Bingley, on whose authority it stands in Flora Britannica; *D. T.*, *O.B.G.* In a shady place under a rock by the side of the rivulet in Hezleden Gill, about a quarter of a mile from the house, 1806; *J. Windsor.* Banks, Thorparch and Boston Spa Woods; *Sir T. Frankland* and *Dr. White*, *Sm.*, *E. Flora.* Denizen only here. Wetherby; *Hook.*, *Brit. Flora*, Ed. i, 192 (1830).

A. Reddins Gill, near Keighley, Mr. Knowlton; *R. Teesdale* (1798); rep. *O.B.G.*

C. Woods at Storthes Hall, covering large tracts, but not wild; *P. Inchbald.*

350. Saxifraga aizoides L.

Sedum minus alpinum luteum; Ray, S. ii. *S. alpinum floribus luteis*; C. B. Pin. *Sedum sive illecebra fol. oblongis floribus luteis maculosis*; Merr. [Dill. refers this in Ray Syn. iii to *Sedum villosum*, but the plant was *S. aizoides.*]

Area: 64, 65, Top. Bot. Montane. Range: 2,200—1,100 feet.

Native; wet rocks, and in peat by rills on the mountains, rare. P. July-August.

First record: Merrett, 1666.

L. On the north side of Ingleborough hill, near a bog by the side of an underground river; *Merr.*, iii (1666); *Ray*, *Cat.*, 279 (1670); *Syn.*, i., etc. Ex humidis montis Ingleborough jugis collectum *D. T. Lawson*; *Bobart*, 477 (1699). Ingleborough Hill, north side; *Thornbeck* in *Blackstone* (1746); Id. 'about the rivulets'; *Martyn* (1763). Still there! Very sparingly on Yoredale limestone escarpment near *S. oppositifolium*; *H. F. Parsons.* Chapel-le-Dale; *W. Howson.*

R. On banks of the rill above Hunt Pot (Thund Pot), west side of Penyghent; *W. Whitwell* (1884).

351. Saxifraga tridactylites L. '*Whitlow Grass.*'

Area: 63, 64, 65, Top. Bot. Range: 50—1,800 feet.

Native; damp old walls and rocks, common in the calcareous districts, rare in the purely gritstone ones. P. April-June.

First record: Baines, 1840.

L. Very common in Garsdale, Dentdale.

Y. Sharow; *H. H. Slater.* Mickley.

R. Settle, common; *W. Howson*, etc. Slaidburn.

N. Upper Nidderdale down to Cowthorpe.

W. Bolton. Ilkley. Arncliffedale. Outershaw!

A. Malham (!) *W. West.* Skipton. Snaith; *H. F. Parsons.*

D. Conisborough Castle; *H. Baines.* Womersley and Smeaton (!) *T. W. Gissing.*

T. Roche Abbey Valley, near the mill! Anston Vale; *A. Carr.*

No record for Mersey trib. district, or for Calder Valley.

352. Saxifraga granulata L.

Sax. rotundifolia alba; Ray. *S. alba*; Ger.

Area: 63, 64, Top. Bot. Range: 100—1,000 feet, perhaps to 1,500 feet.

Native; rocky ground on limestone, damp marly pastures and banks, infrequent; and peculiar in its restrictions in W. Riding, not appearing saliently on the red sandstone as in some other parts of England, and quite absent over large areas on the coal measures and millstone grit. P. May.

First record: Merrett, 1666.

L. Lower limestone scars of Dentdale. Kingsdale.

R. Lord's Wood, High Hill, etc., Settle; *J. Windsor.*

Y. Near Sleningford Mill, Ripon, and rail-banks near Hutton Conyers!

N. By Crimple, near Harrogate; *P. Inchbald.*

A. Malham pastures and scars. Field, Eccleshill, Bradford*; *W. West.*

D. Frickley Park†; *H. Johnson.*

A 'double-flowered' variety (*Sax. alba fl. pleno*) is mentioned in *Merrett's Pinax*, p. 109 (1666), as having occurred 'on the west side of Ingleborough Hill,' referring to which record *Blackstone* (*Spec.* 88) says, 'It is found in the same place as mentioned by Dr. Merrett; also in the pastures towards Bleamoor, in a place called Sleights, *Mr. Thornbeck.* Hortorum elegans est ornamentum.' Referred to also in *Ray Syn.*

Obs.—This is frequently grown in gardens, on rockeries, etc., under the name of the *Pill-Saxifrage.* Its root nodules have great vitality, and (being often out-thrown with rubbish by gardeners) make it one of the plants which it is not always safe to assume to be truly wild, where found at low levels, in parks, lane-sides, and fields near houses especially.

[**Saxifraga decipiens** Ehrh. Incognit.

North side of moss at end of Malham Tarn; *Windsor*, *Flora Crav.*, 34. No further information, but it was probably a form of *S. hypnoides*, perhaps *S. sponhemica* (which, although not yet detected, should occur). Certainly, but it runs into *hypnoides* very much! *J. G. Baker.*]

353. Saxifraga hypnoides L. *Mossy Saxifrage.* '*Eve's Cushion.*'

Sedum alpinum trifido folio; C.B.; Ray.

Area: 64, 65, Top. Bot. Montane. Range: 2,400—900 feet.

Native; wet rocks and banks in the hill-country, frequent. P. Mid May-July.

First record: Ray, 1670.

L. Dent and Rawthey Dale Crags. Whernside Crags. In Dale Pastures near Ingleton; *Mr. Thornbeck*; *Blackst.*, *Sp.* 89 (1746)=Chapel-le-Dale pastures, Ingleton; *Martyn* (1763). On Ingleborough; *Ray*, *Cat.* (1670); *Hist.*, 1043 (1688).

R. In Craven abundantly; *Rev. W. Wood* in *O.B.G.* About Settle, many places; *Curtis*, *Cat.* (1782); *J. Windsor.* Linn Gill. Horton. Lord's Wood. Moughton Scars.

A. By Malham Cove; *Mr. Lawson*; *Ray*, *Syn.*, i, 151 (1690), and rep in Eds. ii and iii; *Derh.*, *litt.*; *R. Teesdale* (1798); etc. There still, 1882. Malham Moor. Leeds; *H. Denny*; *N. B. Guide* (1832) sub nom. *S. hirta*; but, of course, the localisation was not meant to be precise. It does not grow wild within 30 miles of Leeds.

N. Near Pateley Bridge; *W. Brunton*, *O.B.G.* (1806).

W. Outershaw, Langstrothdale. Arncliffe Clouder; *H. Ibbotson.*

[C. Warley Clough; *J. Bolton.* Long gone.]

354. Chrysosplenium oppositifolium L.

Saxifraga aurea Dodonæi; Blackst.

Area: 63, 64, 65, Top. Bot. Range: 50—1,800 feet.

Native; springs, wet banks, rocks, and rill-sides, common. P. April.

First record: 'Shadowie moist rockes, in Yorkshire'; *Johns. Ger.*, 842 (1633). Later records: Blackstone, 1746; Martyn, 1763.

Distribution general—recorded for every district.

355. Chrysosplenium alternifolium L.

Area: 63, 64, Top. Bot. Range: 50—1,800 feet.

Native; shady rocks by rills, rill-sides in woods, etc.; infrequent. P. May.

First record: Blackstone, 1746.

L. On the north side of Ingleborough hill, cum priore [*C. oppositifolium*] cujus varietas esse videtur; *Blackst. Spec.*, 90 (1746). Ingleton. Garsdale!

R. Settle Ings, Peart's Crags, etc., near Settle; *J. Windsor.*

Y. Studley, equally as common as *C. oppositifolium*; *W. Brunton.*

N. Scotton banks, Knaresborough!

W. Bolton Woods! Wetherby; *J. S. Wesley.* Washburn Valley!

A. Bingley; *Martyn* (1763); *Hudson*; *Gough's Cam.* (1789); etc. Near Apperley Bridge; *Rev. W. Wood* in *O.B.G.* Skipton Woods; *W. Whitwell* (1883).

C. Step's Wood, Warley, etc., Halifax; *J. Bolton.* Other places near Halifax! Woodsome, near Huddersfield; *C. P. Hobkirk.*

D. Rill-side, Nether Hallam, Sheffield, 1802; *Herb. Salt.* 'Near Pilley Green, Barnsley'; *H. Payne.*

T. Roche Abbey; *H. Baines.*

356. Parnassia palustris L.

Area: 63, 64, 65, Top. Bot. Range: 50—1,550 feet.

Native; wet boggy places, and also on the limestone in quarries about springs, etc., infrequent, yet not rare, and often locally plentiful. P. July-August.

Earliest notice: Gerarde, 1597.

Very plentifully in . . . Craven, in Thornton fields, and at Doncaster; *Gerarde*, *Hb.*, 692.

L. & R. Not uncommon in wet spots on the scars in Garsdale, at Ingleton and Settle; *J. Windsor.*

N. Near St. Robert's Well, Knaresborough; *E. Hargrove* (Ed. 1782 et seq.). Perhaps gone here. Oakdale Farm, Harrogate, and near Starbeck (!); *W. Grainge.*

W. Kilnsey; *J. Tatham.* Bolton Woods, Valley of Desolation; *L. C. Miall.* Between the Far Wood and the Tadcaster road at Askham Bog; *B. B. Le Tall.* Plentiful about Kirk Deighton, Thorparch, Barwick-in-Elmete, etc. Bramham; *H. Baines.*

A. Glass Houghton; *G. Roberts.* Gordale; *W. West.* Malham Moor; *J. Nowell.* Edge of Tarn (!) *W. Whitwell.* [Adel Bog (!); *A. Shipley.* Perhaps now gone.]

D. Askern! Potteric Carrs; *E. Miller* (1804); *S. Appleby* (1831). Rotherham district; *Larret Langley* (1828). Doncaster; *Gerarde* (vide supra).

T. Near Roche Abbey; *J. Salt.* Quarry, Hooton Roberts; *W. W. Newbould.*

Obs.—In Miller's list of English names of Doncaster plants (1804) this is entered as 'Grass upon nasus': a curious misprint (?) for its book name of Grass of Parnassus, which I never heard applied to it by the rustic. About Wetherby, where it is very abundant in certain spots, the children gather it by handfuls and call it '*Lady Smocks*'—a local misnomer, of course.

XXXIV. UMBELLIFERÆ.

357. Hydrocotyle vulgaris L. *White-rot.* *Marsh Pennywort.*

Area: 63, 64, Top. Bot. Range: 0—1,100 feet.

Native; swamps and lowland bogs, very common, but not ascending high. P. June.

First record: Salt, 1800—'Crooks Moor, Sheffield.'

Distribution general—recorded from every district.

Greenfield Valley swamps (*J. Whitehead*), Ilkley Moor slopes (*W. West*), and Lawkland Moss (*J. Windsor*) in the west, and Ryhill (*G. Roberts*) in the eastern half of our area are the highest stations I know of.

358. Sanicula europæa L. *Sanicle.*

Area: 63, 64, Top. Bot. Range: 0—1,000 feet.

Native; rocky, or light-soiled woods, infrequent (hardly to be called common). P. Late May-July.

First record: Bolton, 1775.

L. and R. Frequent in the mountain woods of lower Dentdale, Ingletondale, Clapdale, Settle (*J. Windsor*), etc. Linn Gill.

Y. Woods, abundant; *H. H. Slater.*

N. Scotton ! etc. Long Flat Lane, Knaresborough, and several other places; *Archd. Pierson* in *Hargrove* (1782).

W. Bolton ! Grassington. Ilkley; *W. West.* Harewood. Wetherby. Bramham Woods.

A. Skipton ! Esholt; *W. West.* Rare near Leeds, i.e., within six miles. Ledsham and Ledstone Park; *W. Whitwell.*

C. Dixon's Scar, Sowerby, Warley Clough; *J. Bolton* (1775). Wakefield; *T. W. Gissing.* Langley Wood; *G. Roberts.* Woodsome and Storthes Hall; *P. Inchbald.*

D. Doncaster and York neighbourhoods; *Dr. White* (1804); *E. Miller.* Rotherham district; *L. Langley* (1828). Old Park, near Sheffield; *J. Salt.* Conisborough. Went-vale.

T. Maltby Woods.

[**Cicuta virosa** L. *Cowbane* or *Water Hemlock.* Incognit. Extinct, but mostly errors?

Area: [63, 64 on old or insufficient authority], Top. Bot.

[N. 'Pond near the Park Lawn, Knaresborough'; *E. Hargrove* (1832, not in Eds. of 1782 or 1809). No pond left 1868.]

(W. Hasling Ford, near the Wharfe; *Miall's Flora* (1862); only *Œ. crocata* there in 1870, with some *Œ. Phellandrium* in a marsh.)

[T. By the river at Scrooby Bridge, near Bawtry; *Sir T. G. Cullum*: *O.B.G.* Perhaps out of Riding.]

(D. Burton Grange, Barnsley; *Payne's Flora* (1867); this turned out to be *Œ. crocata*, but called 'Cowbane' locally! Potteric Carrs; *Miller's list* (1804), under English name of 'Water Hemlock.' *C. virosa* has never been found there by any reputable botanist, as far as records go, and as several other large fine-leaved aquatic Umbellifers do grow there, I presume it likely that one of them was the plant Edw. Chorley (who assisted Dr. Miller in drawing up the list) had in mind.)]

[**Eryngium campestre** L. Incognit. Error of Localisation.

'*Eryngium vulgare Camerarii.* Common Eryngo. Below Melling, Yorkshire, plentifully; *Mr. Thornbeck*'; *Blackstone* (1746); copied into *Martyn* (1763); was put by Hudson to *E. campestre*; but whatever the plant was is immaterial so far as Yorkshire is concerned, since Melling is in North Lancashire, by the river Lune. In *Gough's Camden* (1789), however, the record is varied to 'sea meadows at Millington,' the only place of that name I know being near Pocklington, in the East Riding, and far from the sea.]

359. Apium graveolens L. *Celery.*

Area: '63, 64,' and '[63, 64 inland locs.]' Top. Bot. Range: 0—50 feet, as a wilding.

Native; salt marshes and dike-sides where the influence of the tide is felt, rare; confined to the S.E. corner of the Riding. B. July.

First record for wild plant: Parsons, 1876; but Hargrove (in 1782 Ed.) gives it as growing (doubtless an Escape) 'about the ponds in Staveley Town.'

A. Armin; *H. F. Parsons.*

D. About Ousefleet, Reedness, Goole fields; *H. F. Parsons*!

(An Escape in its other rural stations, and fugitive under such conditions; chiefly by canals. Noticed at Settle Bridge, Sherburn, Ledstone, Saltaire, Bingley, Hambleton Dike near Selby—in this station 'not unlikely to be native' [*H. F. Parsons*], Staveley, etc.)

360. Helosciadium nodiflorum Koch. *Water Parsnep.*

Area: 63, 64, Top. Bot. Range: 0—600 feet.

Native; ditches, shallow slow streams, etc., common. P. July.

First record: Salt, 1800.

Distribution nearly general; every district except M., but quite rare among the hills.

A large broad-leaved succulent plant, growing in dikes at Sherburn, and between Keighley and Haworth, has been more than once misreported as *Sium latifolium.* A sub-repent form grows on the tidal bank of the river Ouse between Goole and Hook.

The VAR. **repens** of Baines' Flora was probably applied to a mere procumbent form of *nodiflorum*; at any rate the ditch plant found 'by the road from Knaresborough to Ripley' near Scriven Park Lodge, which I have examined, is only that, and not true *repens* of Koch, such as occurs in Skipwith 'Line-ponds' in the East Riding.

361. Helosciadium inundatum Koch.

Area: 63, 64, Top. Bot. Range: 0—1,350 feet.

Native; pools and slow waters, rare. P. June.

First records: Gough, 1789; Salt, 1800.

W. Near York; *Gough's Cam.* = Askham Bog dikes, plentiful!

A. Yeadon Dam; *W. West.* Pool on Malham Moor, with *Hippuris.*

C. Sharleston; *T. W. Gissing.* Canal, Dewsbury (in inky-black water); *P. F. Lee*!

D. Doncaster neighbourhood; *Dr. White* (1804). Marsh on Doncaster Raceground, 1797; *Herb. Salt.* Sand-holes, Swinefleet Common; *T. Birks* (1878). Near Attercliffe Coal-mines; *J. Salt.* Rawcliffe; *H. F. Parsons.* By the river Don, Brightside; *A. Carr*!

[**Petroselinum sativum** Hoffm. *Parsley.* Alien.

An escape from cultivation, which has occurred occasionally in fields and on rubbish. Kellington; *H. F. Parsons.* Darnall; *A. Carr.* It seems to be well established, in quantity, only in one place: on the steep cliff-face of a disused limestone quarry, at four cross roads, half a mile north of Ledstone village. Has been there many years, perhaps the result of its having been cultivated in a field above the quarry. *Clematis* occurs near it.]

[**Petroselinum segetum** Koch. Casual.

Area: '63 (without personal authority, but unquestioned),' Top. Bot. Grass field, Hillam, near Pontefract; *G. Roberts, Nat.*, 1865, p. 256.

This occurred doubtless as a field Casual, introduced with seeds in the same way as *Papaver hybridum, Melampyrum cristatum*, etc., have occurred; but a consideration of its distribution in England at once shows that it is very unlikely to be an indigenous integer in the florula on the same footing as in the west and south. I have never seen a specimen gathered in Yorkshire. Mr. Gissing, as a namer of species, was, I am sorry to say, not altogether reliable. Mr. Roberts, however, wrote in 1877 respecting the above: 'The late Mr. Gissing, whom I assisted in the compilation of the 'Flora of Wakefield,' named nearly all the plants I collected in 1865-6. If you turn to the Flora . . . you will see that he has not entered it at all. I have referred to my note-book of that year, and I find this entry—'P. segetum, *one* specimen.'']

362. Sison Amomum L. *Honewort* or *Stonewort.*

Area: 63, Top. Bot. Range: Low.

Native, but just running out; hedge-banks on calcareous gravels and clays, rare, but locally abundant. A. August.

D. Lanes all about Adwick-le-street; *E. Lankester* (1840). Kirk Smeaton; *Y. N. U. Trans.* (*Series* E., p. 20). East Cowick; *H. F. Parsons*!

363. Ægopodium Podagraria L. *Goutweed.* '*Ashweed.*'

Area: 63, 64, 65, Top. Bot. Range: 0—600 feet.

Native, but one of a class like *Verbena*, etc., mainly affecting the vicinage of dwellings; common in cultivated districts, rare above 300 feet, among the hills. P. July.

First records: about Hepton- [Hebden] bridge, and Trimmingham in Skircoat; *Bolton*, 1775. In a close near the footbridge at Heath, near the river, 1790; *D. Dixon.*

Distribution nearly general; every district except M., but is rare in L.

364. Carum Carui L. *Caraway.*

Colonising Denizen; field-borders, way-sides and warped land, infrequent. B. July.

First record: Brunton, 1805.

Y. By the road-side on Sawley Moor; *W. Brunton* = Near the Black's Head above Grantley; *J. Dalton* in *O.B.G.* (1805).

W. Railway-bank, under Grimston demesne, with *Cichorium Intybus*!

D. On warp lands about Goole; *T. Birks*! Rail-bank, Grange Lane *A. Carr.* By King's Causeway, Ousefleet; *H. F. Parsons*!

365. Bunium flexuosum With. *Earth-nut.* '*Pig-nut.*' '*Yer-nut.*'

Area: 63, 64, 65, Top. Bot. Range: 0—1,200 feet.

Native; meadows, pastures, banks, very common. P. May-July.

First record: Salt, 1800.

Distribution general—in every district.

[**Bunium Bulbocastanum** K. Incognit.

'Worsbrough, near Barnsley; *Mr. E. Cockshut*; *With., Bot. Arr.*, Ed. 2, p. 276. Probably not true *Bulbocastanum*, but the state of *B. flexuosum* then so named. Omitted from Withering, Ed. 3. See Martyn's Ed., Millar's Gardener's Dict.'; *W. W. Newbould.*]

366. Pimpinella Saxifraga L. '*Burnet Kex.*'

Area: 63, 64, 65, Top. Bot. Range: 0—1,500 feet.

Native; pastures, lanes, and rocks, common.

First record: Gough, 1789.

M. Mersey tributaries district; *J. Whitehead.*

L. Very common in the dales of the Greta, Wenning, Dee, and Rawthey!

R. In limestone-rock fissures about Settle; *Gough's Cam.* (1789); etc.

Y. & N. Quite general.

W. Very common from Buckden downwards.

A. Common in western half; rare just about Leeds. Hawksworth; *W. West.*

C. Wakefield district; *T. W. Gissing.* Storthes Hall; *C. P. Hobkirk*; Hungerhills, Dewsbury; *P. F. Lee.*

D. & T. Common generally, but rare in the alluvial district around Goole.

367. Pimpinella major Huds. (*P. magna* L.).

Area: 63, 64, Top. Bot. Range: 0—450 feet.

Native; wood-borders, hedge and lane-sides, rather unequally distributed than rare, but locally abundant, by no means restricted to a 'chalky or limestone soil' as Sir J. E. Smith asserts.

First record: Gough in Cam., 1789.

R. Ribble banks below Settle bridge; *J. Windsor.* Betwixt Settle and Rathmell, and near Newton-in-Bolland, plentifully; *Jos. Woods, MS.* in *Towns.O.B.G.* Sawley.

Y. Near Ripon; *Gough's Cam.* (1789); rep. *O.B.G.* Lanes N.W. of Ripon, abundant. Every hedge near Copgrove; *J. Dalton* in *O.B.G.*

N. About Knaresborough; *R. Teesdale, Linn. Trans.* (1798); rep. *W. Wood*; *O.B.G.* (1805). Pateley. Nunmonckton Park; *J. E. Leefe*; N.B.G.

W. Ainsty lanes. About York, common, on sand and gravel. Sparingly about Pool and Leathley. Spofforth Hill, Wetherby, locally abundant. Collingham.

A. Skipton; *W. Whitwell.* Baildon; *W. West.* Hambleton; *G. Roberts.* By river Ouse below Selby; *H. F. Parsons.*

C. Low Moor near Bradford; *W. West*!

D. Goole; *H. F. Parsons.*

T. About Maltby; *G. E. Smith.* Firbeck.

VAR. **dissecta** Wallr.
Rare.
D. Road-side between Catcliffe and Treeton; *Amos Carr*!

368. Sium latifolium L.
Area: 63, 64, Top. Bot. Range: 0—100 feet.
Native; in and by still waters, rare. P. July-August.
First record: Rev. W. Wood, 1805.
A. [Burton-Salmon; *G. Roberts*. Ascertained to be an error—was rightly entered as for the next species in *Gissing's Flora*.] Sedge Lane, near Drax; *H. F. Parsons*! (Hollin-beck, Keighley; *L. C. Miall*. Leeds; *H. Denny*; N.B.G. (1832), 654. Both errors.)
D. Jerry's Lane, Goole; *T. Bunker* and *T. Birks*. Near Whitgift on banks of Yorkshire Ouse; *Rev. W. Wood, O.B.G.* (1805). Still plentiful in dyke by King's Causeway, Whitgift; *H. F. Parsons*.
[T. Everton Common, 1803; *Herb. Salt*. Near Bawtry: I think just out of Riding.]

369. Sium angustifolium L. '*Water-Kex*.'
Area: 63, 64, Top. Bot. Range: 0—450 feet.
Native; in dikes and ditches, common. P. July-September.
First record: Salt, 1803.
L. Near Sedbergh; *H. Baines*.
R. Settle Ings; *J. Windsor*. Rathmell.
Y. Near Hewick Bridge, Ripon; *H. H. Slater*. And other places!
N. Cowthorpe! Boroughbridge.
W. Pool to Otley; *L. C. Miall*. Harewood. Wetherby. Thorparch. Towton. Askham Bogs.
A. Rawden Crag Wood; *L. C. Miall*. Weetwood bottoms. Fairburn. Burton-Salmon; *G. Roberts*.
D. Pontefract. Adwick-le-Street. Doncaster. Wadsley and Ecclesfield; *A. Carr*.
T. Roadside between Bawtry and Everton (July 1803); *Herb. Salt*.
Obs.—Many other localities. In Yorkshire many of the hollow-stemmed Umbel flowers are called Kexes or Kecksies. Wm. Marshall, an agricultural writer about 1787, spells it *Kecks*.

370. Bupleurum rotundifolium L. '*Hare's Ear*.'
Area: 63, 64? Top. Bot. Range: 0—300 feet.
Colonist, decreasing, and often perhaps only a Casual; cornfields, very rare, and not permanent. A. July.
First record: Dalton, 1805.
Y. Ripon; *W. Brunton*. Every ploughing field near Copgrove; *J. Dalton, O.B.G.* (1805).
A. Waste ground, Frizinghall; *H. T. Carter* and *H. T. Soppitt* (1885).
C. Sandal (*Mr. Cooper*); *G. Roberts*.
D. Rotherham district; *L. Langley* (1828). Goole ballast; *H. F. Parsons*.
T. Cornfields near Maltby; *G. E. Smith*.

371. Œnanthe fistulosa L.
Area: 63, 64, Top. Bot. Range: 0—300 feet.
Native; marshes and ditches, infrequent. P. July.
First records: Bolton, 1775; Salt, 1800.
Y. Sharow Mires, Ripon; *H. H. Slater*. Pillmoor Car, near Blois Hall. About Boroughbridge.
W. Marshy pasture by a rill, Wetherby Ings! *J. Jackson*. Stutton *G. Roberts*. Askham Bogs! Hob Moor, York; *W. Whitwell*.
A. Ferrybridge district; *Y. N. Excurs.* (1873).
C. [Warley Clough, by Ripponden brook, etc., near Halifax; *J. Bolton* (1775). Now gone.] Beside the mill at the end of Horbury Bridge, 1792; *D. Dixon*. Heath, Sharlestone, etc.; *T. W. Gissing*.
D. Dikes of the 'Marshland' district, Goole; *H. F. Parsons*. Near Conisborough, frequent; *J. Salt* (1800).
T. Near Kiveton Station; *A. Carr*.

372. Œnanthe Lachenalii Gmel.!
Œ. peucedanifolia; Old Floras. *Œ. pimpinelloides*; O.B.G.
Area: '63 O. Moore spn.,' Top. Bot. Range: 0—150 feet.
Native; marshes and dike-sides, rare. P. July-August.
First record: Salt, 1800.
Y. Ponds and ditches near Ripon; *W. Brunton*. Wet places, Copgrove, very common; *J. Dalton* in *O.B.G.* (1805). One plant only by Yore below Hewick Bridge; *H. H. Slater* (1883).
W. Dikes by railway line near Towton, plentiful!
D. Potteric Carrs, Doncaster (!); *J. Salt*; *O.B.G.* 'Marshland' dikes, Goole! *H. F. Parsons*; *T. Birks*; etc.

373. Œnanthe crocata L.! '*Cowbane*.' *Hemlock Dropwort*.
Œnanthe cicutæ facie Lobelii; Park.; Ray.
Area: 63, 64, Top. Bot. Range: 0—450 feet.
Native; in swamps and ditches, common. P. July.
First record: Ray, 1670.
In Eboracensi comitatu; *Ray, Cat.*, 223 (1670); *non* in septentrionalibus . . . Westmorlandia scilicet et Eboracensi provincia; *Ray, Hist.*, i, 441 (1688).
R. New Hall, near Settle; *H. Baines*. Wigglesworth. Rathmell; *W. Howson*.
Y. Ripon, Ure side (!) *H. H. Slater*.
W. Wharfe banks from above Ilkley. Common at Newton Kyme.
A. Dike-side near Aire, Castleford! Bingley, Baildon, and Hawksworth; *W. West*. Adel Lower Dam! Silsden; *H. Baines*. Near the canal, Woodlesford; *G. Roberts*. Near Carlton; *H. F. Parsons*.

C. Dirtcar, 1792; *D. Dixon*. Near Storthes Hall, Farnley Tyas; *P. Inchbald* (1860). Kirklees district; *Y. N. Excurs.* (1873). Hebden valley! Stanley canal near Heath Bridge; *T. W. Gissing*. Coxley Valley; *P. F. Lee*.
D. Dearne bank at Darfield; *J. Salt*. Burton Grange, Barnsley! Conisborough.

374. Œnanthe Phellandrium Lam. '*Horsebane*.'
Area: 63, 64, Top. Bot. Range: 0—300 feet.
Native; ponds and ditches, infrequent, locally common.
First record: Salt, 1800.
Y. Pilmoor Carr, near Sharow (!) *H. H. Slater*.
N. Pool near Bilton Haggs (!) *W. Todd*.
W. Pond near Haslingford, Pool! Askham Bog ditches!
A. Newton near Castleford; *G. Roberts*. South Milford.
C. Warmfield, New-Miller Dam; *T. W. Gissing*.
D. Marsh below Aldwark Corn Mill (1798); *Herb. Salt*. Goole dikes; *H. F. Parsons*! Pond between Doncaster and Conisborough!

[**Œnanthe fluviatilis** Coleman. Incognit.
This was mis-reported to me from waste ground, Hirst's Corn Mill, Huddersfield, in the Calder district, in mistake—the species being immature *Carum Carui*: it is said to have occurred with other Casuals, such as *Caucalis daucoides*, *Silene conoidea*, *Melampyrum cristatum*, etc. (Vide *Yorks. Nat. Recorder*, 1873, p. 128.)]

375. Æthusa Cynapium L. *Fools' Parsley*.
Area: 63, 64, 65, Top. Bot. Range: 0—900 feet.
Native; fields and garden ground, very common, and occurring as high as field cultivation reaches.
First record: Salt, 1800.
Generally distributed—in every area.

[**Fœniculum vulgare** Gaert. *Fennel*. Alien.
This, as an escape from gardens, has occurred at Hawksworth (*W. West*), waste ground at Conisborough (!), railway bank near Stutton (!), and near Sharow (*H. H. Slater*); but has no claim to inclusion as a permanent integer in the West York Flora. There is one old record: York and Doncaster neighbourhood; *Dr. White* (1804): about which, however, I know nothing further. Occasionally as an escape about Askern; *C. P. Hobkirk*.]

[**Ligusticum scoticum** L. *Scottish Lovage*. Casual.
This has occurred once (on colliery débris?) in a '*wet field* near Grange Lane Station,' near Sheffield; found there by the late *Amos Carr*. Specimen seen—a very unusual Casual.]

376. Silaus pratensis Bess.
Area: 63, 64, Top. Bot. Range: 0—900 feet.
Native; pastures and grassy road-sides, on gravel or sandy soil chiefly. P. July-August.
First record: Salt, 1800.
Distribution nearly general.
L. Austwick, Lawkland, and Clapham; *J. Windsor*.
Y. Bridge Hewick, Hutton Conyers rail-banks; *H. H. Slater*.
W. Harewood! Thorparch. Towton! Askham district!
A. Gargrave. Skipton rock! Steeton; *W. Whitwell*.
C. Hungerhills, Dewsbury; *P. F. Lee*.
D. Conisborough. Went-vale!
T. Roche Abbey; *Herb. Salt* (1798).

377. Meum Athamanticum Jacq. '*Bawd-Money*' (Ray, Syn., 3, 207).
Meum, *Meu*, or *Spignel*; Ger., 1052. *Meum vulgatius*; P., 888.
Area: '63, 64? error, 65 Pinder ms.,' Top. Bot. Range: 600—1,200 feet.
Native; stony mountain pastures, very rare.
First records: Parkinson, 1640; Turner, 1664; Merrett, 1666.
Wilde in the North countries of this Land as in . . Yorkeshire; *Park.*, 889 (1640). It grows in Yorkshire; *R. Turn.*, 316 (1664). Mountainous parts of the West Riding, sparingly; *Ray* (1670); rep. *Teesdale* (1798).
L. Sedburg, Yorkshire; *Merr., Pin.*, 77 (1666): confirmed *Pinder* (1850). About Low-gill. Pastures near Black-Burton; *I. Hindson*.
R. Pastures, Newby Head, very sparingly.
C. On the banks of a moist meadow, Scamonden, *Richardson*; *Ray, Syn.*, iii, 208; *Martyn* (1763); etc. In some rough pastures about half a mile above Katty's Well in Warley, on the right-hand side of the brook; *J. Bolton* (1775).
Obs.—I have thrice failed to re-gather this in the old Calder valley stations. Parkinson's record, though not localised as in the W. Riding like Merrett's, was founded upon the same observance by the Lancashire 'simplist,' Thos. Hesketh, cotemporary of Gerarde.

378. Angelica sylvestris L. '*Holy-ghost*.' '*Ghost Kex*.'
Area: 63, 64, 65, Top. Bot. Range: 0—1,350 feet.
Native; moist woods and wet thickets, ditch-sides, common.
First record: Bolton, 1775.
Warley Clough, etc.; *J. Bolton*. Potteric Carrs; *E. Miller* (1804).
Generally distributed—reported from every drainage district.

379. Peucedanum palustre Mœnch. (*Selinum p.* L.) *Water Parsnip*.
Area: '63 O. A. Moore spn.,' Top. Bot. Range: Low.
Native; marshes and banks of dikes in the low country, rare: a decreasing species. P. July-August.

(A. Hollin Beck, Bingley (*Stephen Hartley*) in *Miall's Flora*; error, some *Œnanthe* was taken for it).

D. Potteric Carrs; *E. Miller* (1804); marshes near Doncaster; *Tofield* (1762) in *Huds.*; *Gough's Cam.* (1789). ['But a friend writes me from Doncaster that the best botanist in that neighbourhood is of opinion that it does not grow there, and that Mr. Tofield's specimen came from a distance; *Mr. Wood*'; *With.*, Ed. ii, 282. Mr. Wood must have been misinformed. I have myself seen the plant growing in Potteric Carr, where it looked quite as wild as in Burwell Fen. Has not Mr. Wood made several other mistakes about West Riding Botany? *W. W. Newbould, MS.* Yes, Mr. Wood recorded *Circæa alpina*, *Mentha sylvestris*, etc., wrongly, unless his localisations 'near Leeds,' etc., be interpreted very loosely or widely indeed; *F. A. Lees.*]

T. Thorne Moor, south side, 1872! = 'Lane near Whittaker's plantation'; *T. Birks* (1883)! Re-observed by *H. Johnson* in 1882. 'Was abundant in 1844'; *P. Inchbald.* Low moors between Goole and Thorne; *Rev. W. Wood*; *O.B.G.* (1805). In great plenty in low wet moors near Whitgift; *Withering*, Ed. vii, 372 (1830). By New Torne Dyke near Wroot.

Obs.—'Water-Parsnep, or Milk Parsley as some hath it, groweth in marshes near unto Doncaster'; *Tofield.* Parsnip, I suppose, because the orbicular-oblong fruits, when ripe, are very like those of the common Parsnip. Records show that about the beginning of the century not only was it common in South Lincolnshire and Yorkshire swamps, but that it occurred in many places on the west coast of Lancashire—Southport (where I saw it still in 1870), Pilling, Flookburgh, etc. 'Modern confirmation is wanted,' as Mr. Baker writes in his recent 'Lake District Flora'; but the lack of this, from veritable extinction, cannot be held to cast any doubt upon its having occurred—the records are too numerous and circumstantial. One, *not* quoted in the 'Flora of the Lake District,' occurs as a MS. note (in Dr. Windsor's hand) in *Townsend's copy* of *O.B.G.*; it runs: 'in a ditch on the right-hand side of the road, in a level field between Witherslack and Kendle, *J. Winsor.*' Why inserted in the *Yorkshire* list of Woods's copy and not in the Westmoreland list, I cannot conjecture.

380. Peucedanum Ostruthium Koch. '*Felon-wort.*'

Denizen; waste ground in rural lanes, field-corners, near barns, etc., in the hillier tracts; rare. P. July-August.

First record: the Author, 1878.

W. By barns, etc., Outershaw, Langstrothdale. Near an old barn, Farnley near Otley; *H. F. Parsons.*

Obs.—A decoction of this is still in use by hinds, farmer-vets., etc., as a drench for cattle, to promote secretion of milk in cases of arrested lactation, and in some other disorders.

381. Pastinaca sativa L. *Wild Parsnip.*

Area: 63†, 64, Top. Bot. Range: 0—600 feet.

Native; dry sandy or calcareous banks and fields, infrequent. B. July-August.

First record: Salt, 1800.

No record for districts M., L., R., and Y.

N. Rail-banks, Cattall to Knaresborough!

W. Dry banks and fields and rail-banks, frequent on the Permian tract. Wetherby. Thorparch.

A. Brotherton. Kippax.

D. Common on the limestone from Conisborough to Wentbridge!

T. Near Dia-Carr Wood, Roche Abbey, 1803; *Herb. Salt.* Common in this district.

382. Heracleum Sphondylium L. *Hog-Weed*

Area: 63, 64, 65, Top. Bot. Range: 0—1,250 feet.

Native; fields, banks, and thickets, very common. B. July-August.

First record: Ray, 1724.

Distribution general—in every district.

The VAR. **angustifolium** L. is not infrequent, especially on dry gravelly soil. Ray records it '*Sphond. majus aliud laciniatis foliis*, at the Town's-end at Settle' in *Syn.* iii, 205 (1724). It also occurs about Boroughbridge, Wighill, Wetherby, Selby, Cantley (*G. E. Smith*), and Sheffield (*A. Carr*).

383. Daucus Carota L. *Wild Carrot.*

Area: 63, 64, Top. Bot. Range: 0—750 feet.

Native; gravelly and calcareous fields and stony banks, river-sides, common. B. July-August.

First record: Bolton, 1775.

Distribution general; has occurred in every district—from Ingleton, Settle, Skipton, Bingley, down to sandy fields on the borders of Thorne waste.

'Adjoining *Duel Lane*, Warley, Halifax'; *J. Bolton* (1775). Warmfield Field, plentiful, 1816; *D. Dixon.*

384. Caucalis daucoides L.

Area: 63, 64, Top. Bot. Range: 0—200 feet.

Colonist; cornfields, very rare, and probably seldom permanent; more a Casual than anything else. A. July.

First record: Withering, 1796.

Y. Near Ripon; *W. Brunton* in *O.B.G.*

N. Knaresborough; *H. Baines.*

W. Near Thorparch, *Mr. Aikin*; *Withering*, Ed. 3, ii, 287 (1796); rep. *R. Teesdale* in *O.B.G.* Tadcaster, one plant by flour mill, 1870.*

C. Casual, out-thrown with grain screenings, near Hirst's Corn Mill, Huddersfield; *Yorks. Nat. Rec.*, p. 128.

D. Rotherham district; *L. Langley* (1828).

T. In a calcareous cornfield near Roche Abbey; *Herb. Salt.* (July 17—rest of date obliterated; but about 1798—1800, doubtless.)

385. Torilis infesta Spreng.

Area: '63 and 64 (without personal authority),' Top. Bot. Range: not ascertained.

Colonist or Casual; dry calcareous banks and cornfields, very rare. A. July-August.

Y. Cornfields near Ripon; *H. Baines.*

(C. Frequent; *T. W. Gissing* in *Flora of Wakefield.* Not so; some misnomer probably.)

T. Fields near Waleswood Colliery; *A. Carr.*

Obs.—I have never seen this—or a specimen of it gathered in the Riding.

386. Torilis Anthriscus Gaert. '*Red Kex.*'

Area: 63, 64, 65, Top. Bot. Range: 0—900 feet.

Native; hedge-banks and fields, in a dry soil. A. July-August.

First record: Salt, 1800.

Distributed generally, in every district.

387. Torilis nodosa Gaert.

Tordylium nodosum; Huds.

Area: 63, 64, Top. Bot. Range: 0—300 feet.

Native; light-soiled arable fields, sandy wayside banks, etc., rare. A. June.

First record: Bolton, 1775.

W. Dringhouses; *H. Baines* = Banks by bridge over railway there! Fields, Priest Hill, Wetherby!

A. About Pontefract and Knottingley! Field at Methley; *G. Roberts.*

C. Lightcliffe and Southowram; *J. Bolton* (1775). Oakenshaw; *T. W. Gissing.* Ardsley and Sandal (!) *G. Roberts.*

D. Smeaton; *H. Johnson!* Roman ridge, Hampole; *P. Inchbald.* Brightside; *J. Salt* (1800). Tinsley, the Rivelin, Darnall, etc.; *A. Carr.*

388. Chærophyllum Anthriscus Lam. (*Anthriscus vulgaris* Pers.).

Area: 63, 64, Top. Bot. Range: 0—500 feet.

Native; dry banks and roadsides, on a gravelly or calcareous soil, infrequent. A. June.

First record: Bolton, 1775.

R. Near Giggleswick School; *J. Windsor.*

Y. Ripon; *H. H. Slater.*

N. Green Hammerton and Knaresborough! Wighill.

W. Spofforth to Wetherby. Thorparch.

A. Snaith; *H. F. Parsons.* Sherburn.

C. Hedges about Warley and Skircoat; *J. Bolton* (1775).

D. Near Wadsley Forge; *J. Salt.* Doncaster; *J. Willis.* Balby! Askern.

389. Chærophyllum sativum Lam. (*A. Cerefolium* Hoffm.)

Denizen; dry hedge-banks (formerly cultivated), rare. A. July.

N. Cliff (above gardens) at Knaresborough!

D. Bank by hedge near Christ's Church, Doncaster; *J. Willis!*

390. Chærophyllum sylvestre L. '*Cherrill.*' '*Kex.*' '*Cicely.*' '*Orchard Bun*'—a 'bun' being, vulgarly, any hollow-stemmed Umbellifer.

Area: 63, 64, 65, Top. Bot. Range: 0—900 feet.

Native; hedge-banks and meadows, very common. P. May-June.

First record: Salt, 1800.

Distribution general; in every district.

391. Chærophyllum temulum L. *Chervil* or '*Cherrill.*'

Area: 63, 64, Top. Bot. Range: 0—1,200 feet.

Native; damp hedge-banks and thickets, common. B. June-August.

First record: Salt, 1800.

Distribution general; found in every district, but not so abundant as the preceding.

392. Myrrhis odorata Scop. '*Cicely.*' '*Anise.*' '*Cow-weed.*'

Cicutaria vulgaris, J.B.; Ray. *Scandix odorata*, Huds.; Teesd.

Area: 63, 64, 65, Top. Bot. Montane. Range: 1,100—100 feet.

Native; streamsides, woods, field-banks, etc., common. P. May-June.

First record: Curtis, 1782.

M. Mersey tributaries district; *J. Whitehead.*

L. Garsdale, Dentdale, Rawtheydale, Ingleton, etc., very common.

R. About Settle; *W. Curtis* (1782). Many places about Settle (!) *J. Windsor.* Ribble banks; *H. Baines* (1840).

Y. Hackfall; *W. Brunton*; *O.B.G.* (1805). Yore banks (!) *J. Dalton*; *H. H. Slater*; etc.

N. Dropping-well Wood, Knaresborough; *Teesdale, Linn. Trans.* (1798).

W. At the entrance to Kilnsey; *Gough's Cam.* (1789). Common all down the dale, by Bolton and Ilkley to Newton Kyme.

A. Many parts of Craven; *W. Wood*; *O.B.G.* Airedale to below Castleford! Pretty common near Leeds, and certainly wild; *Mr. Wood*; *With.* (1787). [Mr. Wood's term *near* Leeds, used here and for *Circæa alpina*, etc., clearly included a considerable area—say 20 miles. The nearest spot where the *Circæa* grows is quite that distance. But the *Myrrhis* has occurred on the Aire banks at Kirkstall Forge, only four miles away from Leeds Bridge. Bingley; and Kippax; *H. Baines* (1840).

C. Hebden Valley! Calder banks, North Dean (!) *J. W. Davis.* Meadows between Morton and Rushworth; *Dr. Richardson* (*O.B.G.*) and *Herb. Buddle.* Calder banks to Castleford! Megg's Scar, Ripponden; *H. Baines* (1840).

D. Denby Dale; *Y. N. Excurs.* (1872). Don banks above Sheffield; *J. Salt, Herb.*, June 1799. Doncaster neighbourhood; *Dr. White* (1804). Aldwarke; *A. Carr.* Whiteley Wood; *G. Harrison.*

Obs.—Dillenius must have known of this in Yorkshire, since his contemporary, Dr. Richardson, did, and a specimen exists in Herb. Buddle, but he does not mention any special localities in the 3rd Ed. of Ray's Synopsis (1724): why is not clear, unless it was at that time (1720-30) so abundant, or (in his view) so generally distributed a plant as to merit only the same treatment as Wild Chervil or Shepherd's Needle. Even Hudson in 1778 only mentions Westmoreland, Cumberland, and Lancashire.

393. Scandix Pecten-Veneris L. *Shepherd's Needle.* '*Crake-needle.*'
Area: 63, 64, Top. Bot. Range: 0—900 feet.
Native or Colonist; in fields, common. A. June-September.
First record: Salt, 1800.
Distribution general; has occurred in every river-basin, but above 600 feet in elevation it becomes quite rare.

[**Coriandrum sativum** L. *Coriander.*
Occurs as a Casual, product of stray seeds, generally on railway-banks, rare in W. Yorkshire because so rarely cultivated.

W. Wetherby, one plant, 1880.

C. Field at Sandal (in 1886) with *Erodium Cygnorum* (misnamed *E. ciconium* on page 183); *G. Roberts.*]

394. Conium maculatum L. *Hemlock.* '*Humlocks.*'
Area: 63, 64, Top. Bot. Range: 0—900 feet.
Native: dry hedge-banks, thickets, roadsides, etc., infrequent, preferring a sandy or gravelly soil. B. July.
First record: Salt, 1800.

L. Between Bentham and Wennington; *Isaac Hindson.*

R. Near Settle; *J. Windsor.* Near Gisburn.

Y. Rocks at Fountains Abbey (!) *H. H. Slater.* Boroughbridge.

N. Hunsingore. Cattall.

W. Pool and Harewood! Wothersome. Bardsey.

A. Bingley; *W. West.* Fairburn; *G. Roberts.* Near Knottingley; Castle Hill, Pontefract; *W. Whitwell* (1884).

C. Wakefield, frequent; *T. W. Gissing.* Bottom Boat; *G. Roberts.* Calder Valley, Dewsbury; *P. F. Lee.*

D. By canal, Swinton; *A. Carr.* Brightside; *J. Salt.* Goole; *H. F. Parsons.*

T. About Tickhill and Sandbeck Park. Finningley.

395. Smyrnium Olusatrum L. *Alexanders.*
Area: '63*, 64*, denizen,' Top. Bot.
Denizen; woods, foot of old walls, banks of hedges, etc., rare.
First record: Rev. W. Wood, 1787.

W. Under the walls of York; *Mr. Wood*; *With.*, 310 (1787). Hedge bank, near Wighill Church! Thorparch; *H. Baines.*

A. Pontefract Castle; *J. Willis* (1864).

D. Doncaster neighbourhood; *E. Miller, Hist. Donc.* (1804).

XXXV. ARALIACEÆ.

396. Hedera Helix L. *Ivy.* '*Hyvin.*'
Area: 63, 64, 65, Top. Bot. Range: 0—1,250 feet.
Native; tree trunks, banks, and rocks, very common.
First record: Salt, 1800.
Distribution general; in every district.

XXXVI. CORNACEÆ.

397. Cornus sanguinea L. *Dogwood.*
Area: 63, 64, Top. Bot. Range: 0—750 feet.
Native; hedges and thickets, on dry, mostly calcareous, soil. Shrub. June.
First record: Salt, 1800.

Y. Copt Hewick; *H. H. Slater.* Bishop Monckton.

N. Staveley; *H. H. Slater.* Common from Knaresborough to Hammerton!

W. Arthington eastwards to Tadcaster, common. Also still further eastward, in the Ainsty; *W. Whitwell.*

A. Airton to Gargrave (!) *W. West.* On the Permian tract, plentiful.

C. Wakefield, frequent; *T. W. Gissing.*

D. Rotherham district; *L. Langley* (1828). Hedges near Jordan Dam; *J. Salt* (1800).

T. Very abundant.

[**Cornus suecica** L. Incognit. Some error.
'C. herbacea, Dwarf Honeysuckle, on Pickering Moor, *and near Doncaster*'; *Gough* in *Camden's Brit.* (1789). I know no more of this.]

[**Cornus alba** Willd. Alien.
Plentiful in plantations about Wighill Park and Bilton; also (always planted) in wet spinneys about Rossington; plantation bordering Malham Tarn (*H. F. Parsons*) and young plants up-springing in the neighbouring bog!]

XXXVII. LORANTHACEÆ.

398. Viscum album L. *Misseltoe.*
Area: '(63) 64,' Top. Bot. Range: Low.
Native originally; Alien in most of its *remaining* stations, perhaps.
First record: Ray, 1677?
In woods and old orchards in the vale of York, not very unfrequent; *H. Baines* (1840).

Y. Orchard trees, Vicarage Garden, Kirkby Malzeard! *Rev. F. D. Horner.* Introduced.

N. On ash trees in a field called Gunrith, near Goldsborough, also on crab trees near Rudding Hall; *E. Hargrove* (Ed. of 1809). Orchard, Cowthorpe.

W. [Woods about Nun Appleton (*J. G. Baker*, 1863); 'on Crab-trees and Poplars,' destroyed by depredations (in 1880). Newton Kyme and Thorparch Ings; *J. Emmet*; gone in 1880. 'Once at Stutton—now gone (1881)'; *Old Inhabitant.* About Wetherby, here and there, but not wild; *J. S. Wesley.* Thorn trees at Woodhall (Mr. Scott's)*!

A. Growing on several trees at Clubcliffe [Clump-cliff], near Methley, 1793; *D. Dixon.* Drax Abbey, on tall Black Poplars; *H. F. Parsons.* Apple trees at Steeton Hall; *W. Whitwell* (1881).

C. Stanley* (1870); *G. Roberts.* Walton Hall! Garden tree, St. John's, Wakefield; *T. W. Gissing.*

D. [Si quis Viscum in Quercû crescentem videre desiderat, Sheffeldiam adeat, à qua non procul inveniri eum à D. Fr. Jessop certior factus sum; *Ray, Cat.*, ii, 307 (1677); 'but has it not been said afterwards that *Hedera* was the plant meant? *W. W. Newbould, MS.*]. Rockley, nr. Barnsley; *Payne.* Planted. On a hawthorn, near Beighton; *Hb. Salt.* Near Adwick-le-Street; *W. W. Newbould.*

Obs.—Dr. H. F. Parsons kindly writes anent a word used above, 'Why 'Alien'? On lofty trees such as those at Drax Abbey it is not likely to have been planted intentionally or by human agency. Probably it has been sown by birds; but is not that one of the natural agencies by which plants are dispersed?' Yes, it may be supposed to be 'natural' in the case of a parasitic epiphytal shrub which probably seldom spreads from tree to tree in any other way; but it still remains a fact that if the existing stations of *Viscum* in W. Yorkshire be fully sought into, in most it will be found that the plant was introduced or re-introduced by human hands.

XXXVIII. CAPRIFOLIACEÆ.

399. Adoxa Moschatellina L. '*Muskweed.*'
Area: 63, 64, 65, Top. Bot. Range: 0—900 feet.
Native; warm shady hedge-banks, woods and thickets, common P. April-May.

First record: Salt, 1800.

M. Mersey tributaries district; *J. Whitehead.*

L. Bentham! Dentdale.

R. Settle, many places; *J. Windsor.* Slaidburn.

Y. About Ripon, Studley, Sleningford (*W. Grainge*), etc.

N. Cowthorpe, Hunsingore, etc., frequent.

W. Otley, Collingham, Thorparch, Tadcaster, frequent. Grassington. Bolton Bridge; *W. Whitwell.*

A. Esholt Woods! Kellington; *H. F. Parsons.* Rylstone; Skipton; *W. Whitwell.*

C. Elmley Woodhouse; *W. Rushforth.*

D. Hugset Wood, Barnsley; *T. W. Gissing.* Rotherham district; *L. Langley* (1828). Grimethorpe, Fulwood, Sheffield; *J. Salt* (1800). Wharncliffe Wood; *A. Carr*! Old Park Wood; *Id.* Endcliffe Wood; *W. W. Newbould.*

T. Roche Abbey. Wales; *A. Carr.*

400. Sambucus nigra L. *Elder.* *Bore-tree.*
Area: 63, 64, 65, Top. Bot. Range: 0—1,250 feet.
Native; hedges, thickets, common. Tree. Mid-June.
Generally distributed—in every district.
The cut-leaved Var. **laciniata** is often found as a planted shrub. Kelcowe, east of Cove (*J. Windsor*); Thorparch; near Sheffield (*A. Carr*); etc.

401. Sambucus Ebulus L. *Danewort.* *Dwarf Elder.*
Area: '63† Leyland, 64,' Top. Bot. Range: 0—600 feet.
Denizen; hedges and road-side wastes, very rare, perhaps mostly an Alien—it used to be planted and used for dyeing blue (teste *W. Brunton, O.B.G.*). Herb. P. August.
First record: Dixon, 1791.

L. Austwick; *W. Howson* (1850).

R. Wall-side near Close House, Settle; *J. Windsor* (about 1806).

Y. [Near Ripon, lane close to Leeming turnpike gate (*J. Dalton, O.B.G.*) is out of the Riding. It should not have been included in the list at p. 290 of 'West Yorkshire.'] Copgrove, 1870.

[W. Boston Spa churchyard; now gone. Plants of it were removed into the hedge of Dr. T. Scatcherd's garden at the time of the alterations to the church which led to its destruction in what was possibly not a native station to begin with.]

A. Growing plentifully at Castleford, June 1791; *D. Dixon.*

C. Rail-bank by a bridge about half way between Liversedge and Cleckheaton! *W. Fowler* (1876). Between Horbury and Wakefield.

D. Silkstone; *H. Johnson*—not wild. River-bank one mile below Sheffield; *J. Salt.*

402. Viburnum Opulus L. *Gueldres Rose.* '*Snowball-tree.*'
Area: 63, 64, Top. Bot. Range: 0—750 feet.
Native; hedges and thickets, common, ascends to above Horton in Ribblesdale (*W. Howson*). Shrub. May-June.
First record: Salt, 1800.
Distribution general; found in every district.

403. Viburnum Lantana L.
Area: '63, (64),' Top. Bot. Range: 0—300 feet.
Denizen, often merely planted; hedges, rare. Shrub. May.
First record: Lankester, 1840.
W. Scrub at Bilton Haggs, north of Wighill Park.
D. The Brakes, Askern; *E. Lankester* (1840). Wharncliffe; *Geo. Harrison, Hb.*
T. Hedges and thickets about Firbeck and Letwell.

404. Lonicera Caprifolium L.
Denizen: woods and hedges, rare. Shrub. May-June.
First record: Langley, 1828.
Y. Studley (!) *L. C. Miall.*
W. Thorparch Wood (!) *J. Emmet*; but planted.
D. Rotherham district; *L. Langley* (1828).
T. Sandbeck Park.

405. Lonicera Periclymenum L. *Woodbine. Honeysuckle.*
Area: 63, 64, 65, Top. Bot. Range: 0—900 feet.
Native; hedges, wet woods, thickets, common. June-September.
First records: Bolton, 1775; Salt, 1800.
Distribution general; found in every district.
Obs.—A variety 'with flowers smaller and light green' in a hedge near Knaresborough; *T. E. L.* in *Mag. Nat. Hist.*, iv, 71 (1831), was perhaps the next species, if not some monstrosity.

406. Lonicera Xylosteum L. *Fly Honeysuckle.*
Denizen, alien originally; in woods, parks, etc., rare. Shrub. July.
First record: Hailstone, 1815?
Y. Studley Woods. Near Boroughbridge; *Y. N. U. Excurs.* (1876).
W. Thorparch Woods; *Mr. Hailstone*; *W. A. Bromfield*; *G. E. Sm., MS.*
C. Hedge at Birkby, one mile from Huddersfield; 'been there 30 years'; *C. P. Hobkirk* (1871); now destroyed (1886). Jacky Wood, Ardsley, 1882; *G. Roberts.*

[**Lonicera alpigena** L. *Twin-berry.* Alien.
A. In shrubbery coppice near Well-house, Far Meanwood, planted !]

[**Lonicera etrusca** Sav. Alien.
W. Hedge, Holgate, York, planted of course; *Baker, North Yorkshire*, 242.]

[**Symphoricarpus racemosus** Michx. '*Snowberry.*'
An Alien. Shrub. Plentiful, and quite naturalised (self-propagating) on islands in, and thickets by, the Wharfe below Thorparch Bridge.]

XXXIX. RUBIACEÆ.

407. Galium boreale L.
Area: '63 without pers. authority, 64,' Top. Bot. Montane and of the Scottish type. Range: 1,500—350 feet.
Native; bushy and rocky places, by rivers and rills, rare, and restricted in Yorkshire (curiously enough) to river-basins running eastwards. P. July-August.
First records: Blackstone, 1746; Curtis, 1782.
W. Alpine wood between Kilnsey and Kettlewell on the left side of the road; *W. Curtis.* On the hills between Kilnsey and Grass-wood; *Gough's Cam.* (1789). Langstrothdale. Cow-side Beck. Top of Kilnsey Crag; *J. Windsor.* Rocks about the 'Strid,' Bolton Abbey; *W. Wood* (!) *With.* (1792); rep. *Hailstone* (1805). 'Arncliffe (1841), *J. B. Wood*,' *Herb. Kew.*
A. At the east end of Malham Tarne; *Mr. Thornbeck*; *Blackst., Spec.* (1746); *Martyn* (1763); rep. with misprint 'Come' for Tarn; *O.B.G.* and error rep. in *Baines' Flora.* Still by the Tarn in 1870—8. By the wall-side next the foot-road betwixt Capon Hall and Malham Water-Houses; *J. Windsor, MS.* (1801). Rocks above Gordale; *Rev. W. Wood.*

408. Galium cruciatum With. *Honeywort. Crosswort.*
G. religiosum of some old Lists.
Area: 63, 64, 65, Top. Bot. Range: 0—900 feet.
Native; hedge-banks and thickets, common. P. April-May.
First record: Salt, 1800.
Distribution nearly general; no record for Mersey district, but hardly absent therefrom?

409. Galium verum L. *Yellow Bedstraw.* '*Maiden's Hair.*'
G. luteum; Ger.; Park.; Ray. '*Cheese-Rening*'; Ray.
Area: 63, 64, 65, Top. Bot. Range: 0—1,800 feet.
Native; dry sandy or calcareous banks, common. P. July-August.
First records: How, 1650; Salt, 1800.
Distribution nearly general; every district except Mersey.
Obs.—Called in Yorkshire, Mayden's haire; *How, Phyt.*, 45; rep. *Ray, Cat.*, 126. And to this day, in some parts of Wharfedale, still so styled.

410. Galium erectum Huds.
Area: '[64],' Top. Bot. Range: 50—500 feet.
Native? dry banks, very rare. P. July.
First record: Baines, 1840.
W. Hobmoor, near York; *Baines' Flora.*
D. Churchyard at Fulwood; *A. Carr, spn.*! The plant is rightly named, but I fear some introduction with *grass seed.*

411. Galium Mollugo L. *White Bedstraw.*
Area: 63, 64, Top. Bot. Range: 0—750 feet.
Native; dry hedge-banks, quarries, broken thickety ground, in frequent, but on the Permian lime tract abundant. P. July-August.
First record: Salt, 1800.
L. Ingleton, rare. Lower Dentdale.
R. Several places near Settle; *J. Windsor.* Rimmington. Horton-in-Ribblesdale, by station; *W. Whitwell* (1884); possibly adventive there.
Y., N., W., A., D., T. Abundant on the Permian belt crossing these divisions.
W. Roadside near Nesfield and Addingham. Catterton, etc.; *W. Whitwell.*
A. Skipton.
C. Ouslewell Green, Lofthouse, on Flagstone; *G. Roberts.*

412. Galium saxatile L. '*Rockstraw.*'
G. montanum; Huds.
Area: 63, 64, 65, Top. Bot. Arenophilous. Range: 0—2,200 feet.
Native; rocks, banks, heaths, and stony pastures, very common, but, although not absent, replaced to a great extent on the scar limestone by *G. sylvestre*, and on eu-geogenous gritstone quite a salient feature of the wilder slopes.
First records: 'Yorkshire, on rocks, common,' Gough's Camd., 1789; Crook's Moor, Salt, 1800.
Generally distributed—in every district.

413. Galium sylvestre Poll.
G. pusillum; O.B.G.
Area: 64, Top. Bot. Sub-xerophilous. Montane. Range: 1,850—350 feet.
Native; on calcareous mountain-banks and rocks, rare, but locally very abundant. P. July.
First record: Jos. Woods, 1805.
L. Ingleton; *O. A. Moore.* Kingsdale. Wethercote Cave. Yoredale Scars of Dentdale, Garsdale, Chapel-le-Dale, etc.
R. About Settle, everywhere; *Jos. Woods, MS.* in *Towns. O.B.G.* On sand by Ribble (adventive), near Wigglesworth! Penyghent: *W. Whitwell.*
W. Buckden, Arncliffe, Kilnsey, Conistone Scars, etc., and (adventive) on rocks by Wharfe in Bolton Woods!
A. Pastures above Gordale; *J. Windsor.* Malham Moor, Awes Scar, etc., common. Tarn House Lane, Skipton; *W. Whitwell* (1884)!

[**Galium spurium** L. Casual or misnomer?
Clifton and Elland Edge; *G. L. Lister*; *Y.N.U. Trans.*, 1878. I was unable to obtain sight of any specimen. It is usually a waif of flax fields.]

414. Galium palustre L.
Area: 63, 64, 65, Top. Bot. Range: 0—1,800 feet.
Native; wet grassy or peaty places, ditches, marshes, etc., common.
First records: Salt, 1800; Miller, 1804, for 'Potteric Carrs.'
Distribution general—in every district.

Var. **elongatum** Presl. Rare.
W. Askham Bogs.
A. Pool-sides, South Milford!
C. Stanley, by the canal!
D. Doncaster Cars; *Geo. Harrison, Hb.* (as *G. palustre*). Thorne Moor, dike-sides!

Var. **Witheringii.** Common in bogs; very often mistaken for *G. uliginosum*, which, however, dries *green.* Noted in every Drainage District except Trent.

415. Galium uliginosum L.
Area: 63, 64, Top. Bot. Range: 50—1,250 feet.
Native; bushy or grassy marshy places, rarely on peat, rare. P. July-August.
First record: Baines, 1840.
(Y. Ripon, two stations; *H. H. Slater.* Both doubtful.)
W. Askham Bog. Ingbarrow marsh-field, near Stockeld!
A. Malham Tarn Bog! Bushy north-east side of Adel Lower Dam, near Leeds!
(C. Wakefield, frequent; *T. W. Gissing* (?).)
D. In moist ground, *frequent*; *J. Salt.* Did he know the true plant? Denby Dale; *Y.N. Excurs.* (1872).

416. Galium Aparine L. *Goose-grass. Cleavers* '*Hairough.*'
Area: 63, 64, 65, Top. Bot. Range: 0—1,300 feet.
Native; hedge-banks, thickets, and arable land, very common. A. June-August.
Distribution general; in every district, but rare above limit of cultivation.
Obs.—First Yorkshire record I have is one by *Dr. White*, of York (dated July 13th, 1788), in Miller's Hist. and Antiq. of Doncaster (1804). *Gal. Aparine* 'grows in almost all our hedges . . . and it has been quite the fashion here to take the expressed juice as a certain cure for the Scurvy. The Dean (Dr. Fountayne) was cured by it,' etc.

417. Galium tricorne *With.*
Area: 63, 64, Top. Bot. Range; 0—300 feet.
Native; cornfields over limestone or drift, restricted to eastern third of Riding, but frequent there. A. July-August.
First record: Wood, 1792.

N. Flaxby near Knaresborough, in oat-field!
W. About Thorparch and Wetherby (!) *W. Wood*; *With.*, ed. ii (1792). Priest Hill, Kirk Deighton.
A. Cornfields about Monk Fryston; *Wood*; *With.*, Ed. ii. *Woodward* given as the authority in *O.B.G.*
C. Field at Three Houses, Sandal, amongst other aliens, 1886; *G. Roberts*!
D. Near Conisborough Castle; *Rev. W. Wood*; *O.B.G.* Field near the Don, Rawmarsh; *A. Carr.* Doncaster; *P. Inchbald.* Robroyd, Barnsley; *H. Johnson*!
T. Firbeck.

418. Asperula odorata L. *Woodruffe.*
Area: 63, 64, 65, Top. Bot. Range: 0—1,000 feet.
Native; woods and shady hedge-banks, frequent, on lime shales or grit, in the Western half of the Riding, rare in the Eastern half, and almost unknown on the Trias sand. P. May-June.
First record: Bolton, 1775.
M. Greenfield Valley; *J. Whitehead.*
L. Ingleton. Dentdale. Clapdale; *W. Howson.* Linn Gill.
R. Settle, several places; *J. Windsor.* Sawley.
Y. Common about Ripon; *H. H. Slater.*
N. Knaresborough, Long Walk, and near 'Dropping Well'; *E. Hargrove* (1782 ed. et seq.). Scotton Banks.
W. Bolton. Grassington. Harewood. Wetherby. Thorparch.
A. Skipton Woods. [Batty Wood, Woodhouse Ridge, Leeds, common in 1860—gone now.]
C. Woodhouse Wood, Skircoat; Allen's Wood, Norland; *J. Bolton, Cat.* (1775). Wakefield, frequent; *T. W. Gissing.* Soothill, Dewsbury; *G. Roberts.* Near Huddersfield; *H. Ibbotson*!
D. Old Park Wood; *Herb. Salt.* Eccleshall, Cliff Rocher, etc.; *A. Carr.*
T. Common in woods about Roche Abbey and Maltby.

419. Asperula cynanchica L. *'Quinzy-wort.'*
Area: 63, 64, Top. Bot. Xerophilous. Range: 0—300 feet.
Native; dry calcareous, grassy or stony ground, rare. P. July-August.
First record: Hargrove, 1782.
Y. Near Ripon; *H. Baines.* Old quarried ground off Wicliffe Lane.
N. Knaresborough, foot of rocks on Abbey plain; *E. Hargrove* (1782 ed. et seq.).
W. Linton Common, abundant. Cross roads, Stubbing Moor, near Rigton. Jackdaw Crag Quarry, Stutton! Thorparch, several places; *J. Emmet*! Jackdaw Crag, Boston Spa; *W. Whitwell.*
A. (Malham; *E. Hailstone* (1805); rep. *J. Windsor, Flora Crav.*). Old quarry between Micklefield and Huddlestone. Old quarry near Aberford, 1885; *G. Roberts.* Ledsham (1878) and rocks on roadside, Newthorpe; *W. Whitwell* (1880).

D. Went-vale (!) *Mrs. Watson.* On the Roman Ridge (Ermine Street), Doncaster; *P. Inchbald* (1844). Brodsworth; *Dr. Deakin.*
T. About Roche Abbey, and rocky field-corner near Firbeck!

[Asperula arvensis L. Casual.
C. Between Brighouse and Sowerby Bridge; *S. Gibson* (1850).]

420. Sherardia arvensis L. *'Madderwort.'*
Area: 63, 64, 65, Top. Bot. Range: 0—900 feet, as high as field cultivation.
Colonist; fields and waste ground, very common in all suitable situations. A. June-October.
First record: Salt's MS. Flora Sheffieldiensis, 1800.
Reported from all the districts, but of course scarcer in the hill regions, where enclosed or arable lands are the exception, and grazing pastures the rule. On dry calcareous soils the flowers are not unfrequently white.

XL. VALERIANACEÆ.

[Centranthus ruber DC. *Red Valerian.* Alien.
Naturalised on wall-tops and on rough ground in old limestone quarries about the villages of Brotherton, Firbeck, etc.
Growing near the ruins of Red Hall, near Newton, 1790; *D. Dixon.* Verified at Red Hall in 1874; *J. Wilcock.*]

421. Valeriana dioica L.
Area: 63, 64, 65, Top. Bot. Range: 0—1,500 feet.
Native: marshy, grassy, or peaty ground, common. P. Late May-June.
First record: Salt, 1800.
M. Swamps in the district; *J. Whitehead.*
L. Ingleton. Ribblehead.
R. Settle Ings, Kendall's Gill, etc.; *J. Windsor.* Rathmell. Site of Giggleswick Tarn; *W. Whitwell.*
W. Frequent from Buckden downwards. Clifford, 1849; *S. Hailstone.*
A. Below Malham Cove; *W. West.* Shipley.
D. Askern; *Y. N. Excurs.* (1873). Rotherham district; *L. Langley* (1828). Fulwood; *J. Salt.* Edlington Wood!
T. Maltby Woods!

[1. **Nardus Celtica.** 2. **Nardus Celtica repens.** The first and seconde, if my memorie faile me not, do growe in a field, in the north part of England, called Crag close, and in the foote of the mountaine called Ingleborough Fels; *Gerarde, Hb.*, 921; *Ger. Em.*, 1080. I do not know to what to refer the foregoing. '*N. celtica* = *Valeriana celtica* L.—a Swiss Alpine. Probably a mistake

of Gerarde' [*C. C. Babington*]. 'Part of the preface to Ray's Catalogues (Eds. i and ii) gives reasons for Ray's excluding several plants mentioned by previous authors. . . . One or two relate to Yorkshire, e.g., if I mistake not, two species of *Nardus*, *Pæonia*, etc.'; *W. W. Newbould.* Dillenius gives both in his *Indiculus Plantarum Dubiarum*, at end of his ed. of *Ray's Syn.* (1724), remarking only 'Gerard saith these two grow about Ingleborough hill in Yorkshire': *How* (*Phyt. Brit.*, 79-80) likewise copies Gerard's entries. *V. dioica*, however, grows commonly in the stations mentioned by Gerarde.]

422. Valeriana officinalis L. *Valerian. 'All-Heal.' 'Catslove.'*
Area: 63, 64, Top. Bot. Range: 0—1,000 feet.
Native; damp hedge-banks, ditches, etc., common. P. July-September.
First record: Salt, 1800.
Distribution general—found in every district.

VAR. **Mikanii.** This dry-soil state, for I can regard it as nothing more, is not rare in dry or stiff soil in limestone woods. It occurs in (Y) Mackershaw Woods, (T) Maltby Woods, (W) Thorparch Woods, (A) about Micklefield and (D) in Edlington Wood near Conisborough; but sometimes, where a *dry* stretch of soil slopes gradually down to a ditch or stream, numerous intermediates may be observed in every stage of transition, from luxuriant succulent elder-like leaved plants by the water, to others a foot high at most, with linear deeply-cut coarsely serrate leaflets, in the driest situation.

423. Valerianella olitoria Mœnch. *Lamb's Lettuce.*
Area: 63, 64, Top. Bot. Range: 0—750 feet.
Native; in cornfields, dry hedge-banks, and rocks, common. A. May-July.
First record: Salt, 1800.
L. Low warm scars about Ingleton! Weathercote Cave.
R. Stainforth Scars, and Kelcove, Settle; *J. Windsor.*
Y. Ripon, not uncommon; *H. H. Slater.* Fountains Abbey rocks!
N. Rocks above gardens, on the cliff at Knaresborough!
W. Ilkley. Wetherby. Tadcaster. Near Holgate, York; *W. Whitwell* (1877).
A. Knottingley.
C. Wakefield; *T. W. Gissing.*
D. Rotherham district; *L. Langley* (1828).
T. Kiveton and Anston Stones; *A. Carr.*

[Valerianella carinata Lois. Casual.
Field near Hundsworth; *Amos Carr.* Probably fugitive.]

424. Valerianella Auricula DC.
Area: '63 G. E. Sm.,' Top. Bot. Range: Low.
Colonist, or Casual? Fields, very rare. A. July.
D. Cornfields at Cantley; *G. E. Smith.* Warp-soil cornfield, Goole fields!

[Valerianella eriocarpa Desv. Casual.
'A specimen gathered at Banner Cross, Sheffield, is perhaps this plant'; *W. W. Newbould, MS.* in his copy of *Baines' Flora.*]

425. Valerianella dentata Koch. *Corn Valerian.*
Area: 63, 64, Top. Bot. Range: 0—750 feet.
Colonist; cornfields, arable land, etc., infrequent. A. July-August.
First record: Dalton, 1805.
R. Field at Rimmington.
Y. & N. Fields near Staveley, Bishop Monckton, etc.; *J. Dalton*; *O.B.G.*
W. Thorparch and Wetherby; *H. Baines.* Bickerton. Catterton; *W. Whitwell.*
A. Adel. Meanwood.
D. Rotherham district; *L. Langley* (1828). Tinsley; *Geo. Harrison, Hb.*
T. Firbeck. Kiveton; *A. Carr.*

XLI. DIPSACEÆ.

426. Dipsacus sylvestris L. *'Tea-sell'* (Miller's List).
Area: 63, 64, Top. Bot. Range: 0—300 feet.
Native; truly so, but running out and therefore much less frequent in Yorkshire than south of Derbyshire. 'I have not found it north of Derbyshire and Nottinghamshire'; *Mr. Wood* in *With.*, Ed. ii, 138. Waste bushy ground, hedge and field-banks, chiefly on the Triassic and Permian tract of the S.E. third of the Riding. B. September.
First record: Dixon, 1792.
W. About Towton, Church Fenton, etc.
A. Lane between Castleford and Fryston, 1792; *D. Dixon.* Snaith. Knottingley station; *J. Willis* (1867); still plentiful in 1885; *G. Roberts.*
D. About Goole; *H. F. Parsons.* Conisborough; *A. Carr.* Doncaster; *E. Miller* (1804).
T. Firbeck and Kiveton, sporadically. Anston; *A. Carr.* Roche Abbey; *J. Salt.*

[Dipsacus fullonum Mill. Alien. *'Tassell.'*
Occasionally (a relic of cultivation) on hedge-banks, etc., but growing rarer as the Teazel (superseded by machines) is less and less grown. It has not become naturalised and has been found almost exclusively in the lower catchment areas of the Aire, Wharfe, and Don.
A. Bank of river Aire, near Snaith (casual); *H. F. Parsons.*
D. Rotherham district; *L. Langley* (1828).]

427. Dipsacus pilosus L. *Shepherds' Staff.*
Area: 63, 64, Top. Bot. Range: 0—300 feet.
Native; woods and dry bushy banks, mainly on calcareous soil, rare. B. August.
First record: Pierson in Hargrove's History, 1782.
Y. Fountains Abbey; *Rev. J. Dalton* (1805); found '*lately* under the rocks there,' by Miss Morton; *H. H. Slater, Flora* (1884).
[N. Under the [Knaresborough] Castle rock; *Hargrove, Hist.*, 1782 Ed.; omitted from 1809 and later eds., possibly because it had become extinct; *Archd. Pierson*; rep. *O.B.G.* (1805) with name of Archd. Pierson as the authority].
(W. Burley, between Otley and Skipton; *Rev. W. Wood*; *O.B.G.* Never seen by others—is this another of Wood's mistakes?)
D. 'Warp' lands near Doncaster; *S. Appleby* (1831). Near the river Don about one mile below Conisborough, July 1800; *Herb. Salt.* Road-side, Little Smeaton; *P. F. Lee.*
T. Roche Abbey; *J. Salt.*

428. Scabiosa succisa L. '*Devil's-bit.*' '*Blue-caps.*'
Area: 63, 64, 65, Top. Bot. Range: 50—1800 feet.
Native; damp pastures, common, most frequent amid the hills. P. August-September.
First record: Bolton, 1785.
M. Greenfield Valley; *J. Whitehead.* Denshaw.
L. & R. Very common on the upland pastures.
Y. & N. Abundant in upper parts of these dales.
W. Common from Outershaw downwards.
A. Common throughout the division.
C. 'About Halifax'; *Bolton, Fungi*, i, xv (in 1780).
D. Frequent. Very common; *J. Salt.* Redmires, etc.; *A. Carr.*
T. Roche Abbey grounds; *Geo. Harrison, Herb.* Kiveton; *A. Carr.*
Obs.—'We have found it in Yorkshire with double flowers'; *R. Teesdale* in *O.B.G.* In what sense *double*? I have often looked for, but never found any tube-flower such as *Dipsacus* or *Lonicera* with a trace of duplication. Mr. Baker, however, remarks that 'Phyllody of stamens occurs in various species of Scabiosa.'

429. Scabiosa Columbaria L.
Area: 63, 64, Top. Bot. Sub-xerophilous. Range: 0—1,650 feet.
Native; dry limestone-banks in the low country, scars and banks in the hill districts; common on the limestones, rare off them. P. June-July.
First record: 'Yorkshire'; Gough's Camd., 1789.
L. & R. Common on the lower scars.
Y. & N. On the belt of Permian strata.
W. From Buckden and Arncliffe to Bolton. Collingham to Newton Kyme. Bramham; *W. Whitwell.*

A. Gordale; *W. West.* Malham! Skipton rock! Kippax. Ledstone. Brotherton. Pontefract; *J. Willis.* Hillam; *G. Roberts.* Fairburn, 3 feet high; *W. Whitwell.*
D. Rotherham district; *L. Langley* (1828). Conisborough. Sand lands east of Doncaster; *S. Appleby* (1831).
T. Common on the limestone-soiled portion of this tract. Roche Abbey; *W. W. Newbould.* Firbeck, etc.

430. Scabiosa arvensis L. '*Great Blue-caps.*'
Area: 63, 64, Top. Bot. Range: 0—600 feet.
Native; gravelly fields and banks, rather preferring sand. P. July.
First record: Salt, 1800.
Distribution general; has occurred in every district, but does not rise nearly so high as the others of the genus.
A pure white variety by canal, Skipton; *W. Whitwell.*

[**Cephalaria syriaca** Schrad. Casual.
C. Waste ground at Sandal; *Geo. Roberts* (1885)!]

XLII. COMPOSITÆ.

431. Onopordon Acanthium L. *Cotton Thistle.* '*Pigleaves.*'
Area: 63, 64*, Top. Bot. Range: intermediate?
Denizen or Casual? road-side banks and dry fields, very rare. B. August.
First record: Salt, 1800.
W. Thorparch Woods, most probably introduced by the late Mr. Hadfield; *J. Emmet.*
A. Peckfield; *Geo. Roberts.* Booth Ferry, Goole, casual, only once seen; *H. F. Parsons*!
C. Waste ground, Sandal, 1886; *G. Roberts*!
D. Near the road-side between Barlby and Doncaster, July 1800; *Herb. Salt.* Rail-bank near Woodhouse Junction; *A. Carr*! Three or four plants in a sandy field near Thorne Quay (1870).

432. Silybum Marianum Gaert. *Milk Thistle. Virgin's Thistle.*
Denizen; in fields and waste ground, never lasting very long: usually the outcast from some garden. B. July.
First record: Salt, 1800.
Y. Near Kirby Malzeard, many plants in a field and lane near a farm-house. Ellershaw Hill, near Ripon; *H. Baines.* Near Boroughbridge; *H. Baines.*
W. Acomb, near York; *H. Baines.*
D. Rotherham district; *L. Langley* (1828). Brightside; *J. Salt.*

433. Carduus tenuiflorus Curt. *Sand Thistle.*
Area: '63, 64 Tatham Cat.,' Top. Bot. Range: Low.
Native; in sandy fields or waste ground, along the Triassic vale of the Riding, with other sub-maritime species; beyond this tract rare and alien. A. and B. July-August.
First record: Gissing, 1862.
(Y. Ripon, *not uncommon*; *H. H. Slater.* Some error.)
W. Aberford; *Gray*; *L. C. Miall.*
A. Field at Methley; *G. Roberts.* Sandy seed-field, Brayton-Barf!
C. Old quarry at Sandal; *G. Roberts*! Wakefield, frequent; *Gissing's Flora*=Heath; *T. W. Gissing* in *Miall's Flora.*
D. Goole (on ballast); *H. F. Parsons.* Rossington!

434. Carduus nutans L. *Musk Thistle.*
Area: 63, 64, Top. Bot. Range: 0—1,250 feet.
Native; dry fields and banks, infrequent. A. or B. June-August.
First record: Salt, 1800.
R. Attermire Scars pastures.
Y. & N. Ripon, not uncommon (!) *H. H. Slater.* Knaresborough. Poppleton.
W. Wetherby and Bramham! Boston Spa; *J. Emmet.*
A. Gordale Scar pastures! (1870) and about Malham. Hambleton. Garforth! Kippax; *T. W. Gissing.* Methley; *G. Roberts.*
C. Heath, near Wakefield; *T. W. Gissing.*
D. Kirk Smeaton; *W. West.* Rotherham district; *L. Langley.* Askern. Hatfield; *G. Roberts.* Sandall Beat; *G. Harrison, Hb.* Brightside; *J. Salt.*
T. Firbeck. Tickhill. Lindrick; *A. Carr.*

435. Carduus crispus L. (incl. *acanthoides* L.). *Welted Thistle.*
Area: 63, 64, Top. Bot. Range: 0—600 feet.
Native; dry banks and fields, common on the limestone tract of the east; rare elsewhere.
First record: Baines, 1840.
L. Ingleton, rail-bank!
R. Sawley; *R. Brown.*
W. Ilkley; *W. West.* Abundantly from Collingham east to Tadcaster.
A. Gargrave; *W. West.* Garforth. Thorner to Barwick. Oulton; *G. Roberts.* Brotherton to Pontefract, common.
D. Went-vale. Pontefract. Hampole.
T. Common in this district; as elsewhere, varying much; among which two forms, one with small crowded heads (*C. polyanthemos* K.) and another with broader leaves and larger heads (*C. acanthoides* L.), can be distinguished.

Obs.—Our early Yorkshire botanists seem to have thought this too common to be worth special mention: neither Salt nor Bolton notice it in their lists for Sheffield and Halifax, but this, perhaps, was because just about those towns it is absent.

HYBRID. **nutanti-crispus.**
This was found by Mr. A. Carr on Lindrick Common with *C. nutans* and *C. crispus*, and I gathered a similar intermediate in old quarries at Brotherton.

436. Carduus lanceolatus L. *Bur Thistle.*
Area: 63, 64, 65, Top. Bot. Range: 0—1,500 feet.
Hedge-banks, field-corners, and waste ground, very common. B. July.
First record: Salt, 1800.
Distribution general—common in every district.

437. Carduus eriophorus L. *Woolly Plume Thistle.*
Area: 63, 64, Top. Bot. Xerophilous. Range: 0—300 feet.
Native; thickets, open woods, and waste dry ground upon the Permian strata only; rare. B. July-August.
First record: Salt, 1800.
Y. Near Ripon; *W. Brunton*=Quarry Moor, Mackershaw (!) *H. H. Slater.* Near Copgrove; *J. Dalton, O.B.G.* (1805).
W. About Wetherby; *Rev. W. Wood*=Lanes about Etchell Crags and Stubbing Moor, Rigton!
A. Milford, 1886; *G. Roberts.*
(C. Copley Woods; *L. C. Miall.* Surely some error?)
D. Brockerdale, Wentbridge. Levitt Hagg, Conisborough.
T. Near Maltby (!) *G. E. Smith.* By the high-road side near Beckhouse, about one mile from Roche Abbey, July 1799; *Hb. Salt.* 'About Brookhouse'; *Salt, MS. Flora.*

438. Carduus palustris L.
Area: 63, 64, 65, Top. Bot. Range: 0—1,800 feet.
Native; marshy, rushy fields, and moorland pastures, common. B. July.
First record: Salt, 1800.
Distributed through all the districts, but not very frequent in the more highly cultivated ones now.

439. Carduus pratensis Huds.
Area: 63, 64, Top. Bot. Range: 0—100 feet.
Native; damp boggy pastures and thickets, rare. P. June-July.
First record: Rev. W. Wood, 1805.
W. Marshy fields bordering Askham Bog on the east (!) *H. Baines.*
(A. Leeds; *H. Denny, MS.* in *N.B.G.*)

(C. Near Castleford; *Miall's Flora.* 'Huddersfield, common'; *Hudd. Nat. Hist.*, Ed. i, in error for *arvensis.*)

D. Between Goole and Thorne, with *Selinum palustre* and *Myrica Gale*; *Rev. W. Wood* in *O.B.G.* Potteric Carrs, Doncaster, about Great Eaa plantations!

T. Dysker Wood, near Firbeck; *Herb. Salt*!

440. Carduus heterophyllus L. *Melancholy Thistle.*
Cirsium Britannicum; Clusius. *C. B. Clusii repens*; J. B.; Ray. *Cirsium Anglicum*; Lobel.
Area: 63 without personal authority, 64, Top. Bot. Montane. Range: 1,500—300 feet.
Native; mountainous woods and pastures, rocky stream-sides, etc., common in the N.W. half of the Riding; rare in the S.E. half. P. June-July.
First record: Clusius, 1583.

M. By the Tame Stream below Wragstone Clough, Denshaw!

L. Provenit in pratis ad radices montis Engleborow . . . in comitatu Eboracensi; *Clus. Stirp. Pann. Hist.* (p. 655); *Clusius, R.P. Hist.* (1601). Ad montium radices Eboraci; *J. Bob.*, 149 (1699); *Ray, Cat.*, i, 74 (1670). In the medowes at the foot of Ingleborow hill, by *Thos. Penny*; *Johns. Ger.*, 1184—who quotes Clusius and gives his figure. Skirreth Wood; *Gough's Camd.* Clapham; *H. F. Parsons.* Clapdale; *W. Whitwell.*

R. Stackhouse Borrins, Wood near Clayton's Bridge, etc., near Settle; *J. Windsor.* Linn Gill. Slaidburn. Fields near Giggleswick; *Gough's Cam.*=Meadows near Feizor; *J. Windsor.*

N. Near Pateley Bridge; *W. Brunton, O.B.G.*

W. Langstrothdale and Littondale, common. Grassington. Barden. [Near Bishopthorpe in the lane between that place and York; *J. Dalton* in *O.B.G.* ? *pratensis.*]

A. Near Malham Town, in the way from Airton to Leeds; *Mr. Dawson*; *Blackst. Spec.* (1746); *Martyn* (1763). Bushy boggy place by the stream a little above the Seven Arches aqueduct at Adel, near Leeds. Dying out, seldom flowering, 1867-76. Perhaps not there now—used to be the place nearest the town of Leeds (4½ miles) in which it occurred.

C. Between Halifax and Blackstone Edge; *W. Wood, O.B.G.*

441. Carduus acaulis L.! *Dwarf Thistle.*
Area: 63, Top. Bot. Xerophilous. Range: Low.
Native; pastures and waste broken ground on the limestone tract, very rare, but locally abundant, finding its northern limit on the east side of Britain in the extreme south of the Riding. P. August.
First record: Salt, 1800.

(C. Heath, Garforth, etc.; *Gissing's Flora* (1867). Never confirmed, and I fear some error.)

[D. Cadeby Common (north side of Don, opposite) Conisborough; *Herb. Salt* (July 1800).]

T. Lindrick Common (Kiveton side!), near Anston, plentiful; *Amos Carr* (1877).

442. Carduus arvensis Curt. *Common* or *Corn Thistle.*
Serratula arvensis Linn.
Area: 63, 64, 65, Top. Bot. Range: 0—1,800 feet.
Native; fields and waste ground, ubiquitous in the two lower zones (to 900 feet), scarce above. P. July-September.
First record: Salt, 1800.
Generally distributed—all the districts.

Obs.—For this, as for many other common plants, I have not found any specifically *West Riding* record earlier than Bolton's Cat. or Salt's MS. Flora. With reference to the Linnean synonym quoted above, a curious note (worth repetition) occurs in a rare old book entitled 'The Rural Economy of Yorkshire' (8vo, 2 vols., London, 1788), by Wm. Marshall—an 'Agricultural Writer.' 'Let no voluminous writer pretend to perfect accuracy. Linnæus, whose system is a wonderful excition of the human mind with respect to accuracy of arrangement, appears to have made an evident mistake in the classification of this common plant. How he could be induced to tear it from its natural family *Carduus* and force it into that of *Serratula* may now be difficult to be ascertained. I retain the name, but protest against the propriety of it. The *Linnean names* are now gone forth throughout all nations; and whoever changes them is speaking a language unknown to *universal botany.*'

VAR. **setosus.** Alien, occasional on heaps of wool-refuse in the clothing districts.

443. Carlina vulgaris L. '*Everlastings.*' *Carline Thistle.*
Area: 63, 64, Top. Bot. Sub-xerophilous. Range: 50—1,250 feet.
Native; old turf, mainly over limestone rock, frequent, but hardly common, and not very abundant in any one spot. P. July-September.
First record: Bolton, 1775—'Warley and Elland Park.'
Distribution general—occurs in every district from Feizor and Chapel-le-dale, High-hill, Settle (*J. Windsor*), Malham Moor and Gordale (*W. West*) down to the Yoredale shales of Diggle valley (*J. Whitehead*).

444. Arctium minus Schkuhr. *Burdock.*
Area: 63, 64, Top. Bot. Range: 0—1,250 feet.
Native; woods, waysides, hedges, and waste ground, very common. P. July-August.
First records: Bolton, 1775; Salt, 1800.

Distribution general—is found in every district. It varies much in the woolliness of the heads, but these are always small, sessile or nearly so, and clustered in a racemose kind of way, very unlike the following species.

445. Arctium intermedium Lange. (*A. pubens* Bab. olim.)
Bardana major altera; Ger. *B. major lanuginosis capitulis*; P.(Ray).
Area: 63, 64, Top. Bot. Range: 0—450 feet.
Native; road-sides and hedge-banks, rare and local, on Permian limestone mostly. B. August.
First record: Ray, 1670.
Yorkshire, frequent; *Baker, Supp.* to *Baines' Flora* (1854).

Y. Fountains Abbey.

W. Road and hedge-sides, Spofforth hill, near Wetherby, 1881! Near Towton! Near York; *H. Baines.*

A. About Skipton (!) *Ray, Cat.*, i, 39 (1670); *Hist.*, i, 332 (1686)!

D. Ecclesall, near Sheffield; *W. W. Newbould* teste *C. C. Babington* (*Ann. N. H.*, vol. 17, 1856).

T. About Kingswood, Firbeck! Near Kiveton station; *A. Carr.*

Obs.—Ray's plant may have been only woolly-headed *A. minus*, but I have myself seen *A. intermedium* there; and Mr. Whitwell a form intermediate between it and *minus*, differing only in the less woolly anthodes and less white-felted underside of leaf, in the Gargrave Road near Skipton.

446. Arctium majus Schkuhr.
Area: 63, 64, Top. Bot. Range: 0—200 feet.
Native; field-borders and waste ground, rare. B. August.
First record: the Author, 1878.

R. Near Sawley Abbey; *R. Brown.* A little doubtful.

W. By the moat, Healaugh Priory Farm, 1880!

A. Field-border near Selby!

D. Waste ground by railway south of Doncaster, 1877!

447. Serratula tinctoria L. *Saw-wort.*
Area: 63, 64, Top. Bot. Range: 50—1,500 feet.
Native; rocky and heathy places, boggy thickets, etc., infrequent. P. July.
First records: Hargrove, 1782; Dixon, 1790.

L. Between Ingleton and Kirby; *J. Windsor.* Greta Valley; *P. F. Lee*! Wood, Thornton Force; *J. Willis.* Bentham! Clapham; *W. Howson.*

R. Giggleswick Scars; *J. Windsor.*

Y. Heathy ground, Kirby Malzeard!

N. Knaresborough, on the left hand by the cartway going down the Abbey field to the Mill; *E. Hargrove* (1782 and 1809 eds.; not in the later editions). Cliff top, Knaresborough (!) *W. Kirkby* (1869). Road between Allerton and Whixley; *J. Jackson*!

W. Arncliffe Clouder! Askham Bog, bordering field (!) *H. Ibbotson.*

A. Plantation at Fryston, 1790; *D. Dixon.* Oat-field, Eccleshill; *L. C. Miall.* Micklethwaite; *Id.* Malham Tarn bog; *P. F. Lee* (*Var.* approaching *monticola*, but scarcely worth separating varietally).

D. Hedge near Norton Lees; *J. Salt.* Sheffield, hedge near Grayson's farm, Bellhouses; *A. Carr*! Silkstone Common; *H. Johnson.*

448. Centaurea nigra L. *Hard-heads. Horseknops.*
Area: 63, 64, 65, Top. Bot. Range: 0—1,500 feet.
Native; meadows, pastures, wayside banks, very common. P. June-August.
First record: Salt, 1800.
Generally distributed—occurs in every district.
The radiate form is very rare: I have noticed it only in the valley from Maltby to Roche Abbey, and in quarries near Conisborough. *C. decipiens* we have not in West Yorkshire at all.

[**Centaurea Jacea** L. Incognit.

W. Wetherby; *Miall's Flora.* The radiate form of *C. nigra* mistaken for it? Still it is a likely species to occur as a Casual in fields laid down with foreign grass-seed.

A. Limestone tract near Leeds; *H. Denny, Cat., N.B.G.* No doubt Mr. Denny, who was not a critical botanist (H. C. Watson, suo opinione), meant radiate *nigra.*]

449. Centaurea Scabiosa L. *Great Knapweed.*
Area: 63, 64, Top. Bot. Range: 0—900 feet.
Native; waysides, thickets, and open woods, on dry, mostly calcareous, soil. P. July-August.
First record; Bolton, 1775.

L. Ingleby's Brows, Austwick; *J. Windsor.* Ingleton.

R. Several places near Settle; *J. Windsor.*

Y. & N. On the Permian belt, common.

W. Arncliffe; *W. Howson.* Grassington—down Wharfe to Newton Kyme. Wetherby, Stutton, Bramham, etc., abundant.

A. Malham and Gordale. Skipton! By Otley Road beyond Far Headingley. Garforth. Newthorpe, near Sherburn; *W. Whitwell.*

C. Warley; *J. Bolton* (1775). Wakefield, frequent; *T. W. Gissing.*

D. Conisborough; *J. Salt.* Rotherham district? *L. Langley* (1828).

T. Frequent about Tickhill, Roche Abbey, Lindrick, etc.

450. Centaurea Cyanus L. *Blue-Bottle.* '*Bottle-of-Sorts.*'
Cyanus; Park. *Cy. flore cæruleo medio purpureo*; Merr.
Area: 63, 64, Top. Bot. Range: 0—500 feet.
Colonist: sandy fields, infrequent. A. July-August.
First record: Merrett, 1666.

R. Settle, occasionally; *J. Windsor.*
Y. Occasional about Ripon; *H. H. Slater.*
N. Spofforth!
W. Rare—about Walton and Acomb.
A. Hawksworth; *W. West.* Moortown, near Leeds!
C. Warmfield; *T. W. Gissing.* Dewsbury; *P. F. Lee.* Lofthouse and Stanley, uncertain and infrequent. At Sandal amongst aliens; *G. Roberts.*
D. In Stonebersley field, Yorkshire; *Mr. Witham*; *Merr.*, 32. Common near Sheffield; *J. Salt.* Manor Fields; *G. Harrison, Hb.*; *A. Carr.*
T. Sandy field below Rossington towards Bawtry.

[**Centaurea Calcitrapa** L. *Purple Star-thistle.* Alien.
Area: '(63),' Top. Bot. Occurring occasionally on refuse out-thrown from seed warehouses, woollen mills, etc.; but often what is named this is really *C. melitensis.*
C. Near Hirst's Corn Mill, Huddersfield; *J. Tindall.* Cornfield at Brighouse; *W. West.*]

[**Centaurea solstitialis** L. *Golden Star-thistle.* Alien.
An introduction with grass-seed mainly, but sometimes in clover, fugitive, not naturalising.
A. Dry pastures about Fryston; *A. Aiken, O.B.G.* (1805).
C. Sandal, waste ground in two places; *G. Roberts*!]

[**Centaurea melitensis** L. (=*collina*).
Casual, introduced with grain or wool from South Europe, often called *solstitialis* a few years back, but now occurring with increasing frequency, especially near flour-mills. Flower small and yellow.
W. By flour-mill, Wetherby; and at Tadcaster.
A. Clover field at Methley; *G. Roberts* (1885).
C. Dewsbury; *P. F. Lee*! Sandal; *G. Roberts.*
D. Waste ground near Old Goole Mill; *T. Birks*!]

[**Microlonchus Duriæi** Sp. (*Centaurea muricata* Willd.).
Alien—a South French and Spanish Centaurea: brought with seed? Waste ground, old quarry, Sandal; *P. F. Lee*!]

[**Centaurea montana** L. Alien.
A garden escape; was found in the disused quarry at Sandal by Mr. Roberts; and is occasionally seen on rubbish heaps in lanes, fugitively.]

[**Atractylis (Acarna) cancellata** Willd. Alien.
Clover-field at Methley with many other Casuals (*Geo. Roberts*). A South European Thistle-like Composite, doubtless intoduced with seeds of fodder plants. For a full list see *Naturalist* forDecember 1885, p. 391.]

451. Chrysanthemum segetum L. *Corn Marigold.*
Area: 63, 64, Top. Bot. Range: 0—450 feet.
Colonist; sandy or clayey cornfields, infrequent, but often in great abundance when it occurs at all. A. June-September.
First record: Leyland, 1831.
R. Rathmell and Halton; *J. Windsor.* Tosside Chapel!
Y. Near Ripon; *Mrs. Berkeley.*
N. Kirkby Overblow and Spofforth!
W. Thorparch, Tadcaster, and Wetherby; *H. Baines.* Ouse bank, Middlethorpe; *Richard Whitwell* (1863). Boston Spa; *W. Whitwell* (1866); rare; *J. Emmet.*
A. Saltaire; *J. Willis.* Eccup! Warmfield and Pontefract; *T. W. Gissing.*
C. Brighouse, Lightcliffe, etc., abundant; *R. Leyland.* Dewsbury; rare and uncertain.
D. 'Common'; *J. Salt.* 'I have not seen more than one or two specimens in the neighbourhood of Sheffield'; *W. W. Newbould.* Field by footpath to Ecclesfield; *A. Carr.*
T. Fields on moor border to south-east of Thorne!

452. Chrysanthemum Leucanthemum L. *'Ox-eye.' Moon Daisy. Dog Daisy. 'Margarets.'*
Area: 63, 64, 65, Top. Bot. Range: 0—1,400 feet.
Native; fields, banks of railways and roads and wastes, common. P. Mid June.
First record: Salt, 1800.
Distribution general; found in every district as high as cultivation reaches. I once saw it by a barn at Middle Houses on the plateau of Malham Moor.
Obs.—In some parts of Yorkshire the common Daisy is styled the 'Dog Daisy.'

[**Stuartina Muelleri** Sonder.
A West Australian Alien, doubtless introduced with wool. Found by Messrs. J. Shaw and Allen Godward on waste ground at Linthwaite, Colne Valley, in September 1886; and named at Kew.]

453. Matricaria Parthenium L. *Feverfew.*
Area: 63, 64, Top. Bot. Range: 0—600 feet.
Denizen; or Alien, but pretty well established in hedges near dwellings, in many wide-apart localities. P. July.
First record: Salt, 1800.
L. & R. Lawkland and Catteral Hall; *J. Windsor.* By farms, Bentham!
Y. Sharow, Ripon; *H. H. Slater.*
W. Walton, Wighill, Rigton, Thorparch, etc. Bolton Abbey ruins; *H. Baines.*

A. Allerton, Baildon, Idle, etc.; *L. C. Miall.*
C. Huddersfield and Halifax; *L. C. Miall.* Wakefield, but always a garden escape; *T. W. Gissing.*
D. Conisborough; *G. Harrison.* About Sheffield; *J. Salt.* Barnsley.
T. About Firbeck and Roche Abbey.

454. Matricaria inodora L. *'Mayweed.'*
Area: 63, 64, 65, Top. Bot. Range: 0—900 feet.
Native: fields and waste places, very common. A. July-October.
First record: uncertain, not named by Salt in his manuscript *Flora Sheffieldiensis.*
Distributed generally—reported from every district.

455. Matricaria Chamomilla L. *Corn Chamomile.*
Area: 63, 64, Top. Bot. Range: 0—300 feet.
Colonist; corn and other arable fields; common. A. August-September.
First record: Salt, 1800.
N. Fields about Goldsborough and Flaxby!
W. Thorparch; Thorner; Stutton—common in the calcareous fields. York.
A. Garforth and Micklefield. Woodlesford.
C. Wakefield, frequent; *T. W. Gissing.* Sandal quarry; *G. Roberts*! Barnsley.
D. Plentiful about Conisborough; *J. Salt.* Askern. Goole fields; *H. F. Parsons.* Thorne Quay!
Obs.—Sometimes lacking ray-florets, FORMA **discoidea.** Found so at Wetherby! Methley (*G. Roberts*), Thorne Quay! and Manor farm coal-pits (*A. Carr*!).

456. Tanacetum vulgare L. *Tansy.*
Tanacetum; Ger.; Ray.
Area: 63, 64, Top. Bot. Range: 0—450 feet.
Native: sandy river-banks and waste bushy places, locally common. P. August.
First records: In Eboracensis comitatûs occidentali parte, etc.; *Ray, Cat.*, 293 (1670). Bolton, 1775.
R. Settle; *J. Windsor.* And lower down the river too!
Y. & N. Common on banks of main streams from Ripon and Knaresborough eastward.
W. Wharfe banks from Bolton downwards to junction with Ouse. Ouse bank below York; *W. Whitwell.*
A. Aire banks, Shipley, Apperley, Horsforth, Methley, etc. Allerton.
C. Skircoat, etc.; *J. Bolton.* Calder banks from Sowerby.
D. Don banks from above Sheffield (*J. Salt*) to Goole (*H. F. Parsons*).

457. Anthemis Cotula L. *Maither* or *Mather. Stinking Chamomile.*
Area: 63, 64, Top. Bot. Range: 0—450 feet.
Colonist; arable fields, common. A. July-August.
First record: Salt, 1800.
Y. Ripon (!) *H. H. Slater.*
N. Knaresborough. Goldsborough. Hunsingore. Wilstrop.
W. Acomb! Thorner and Wetherby, common. Ilkley! Thorparch!
A. Skipton; *W. Whitwell*! Castlefield, Bingley; *L. C. Miall.* Calverley.
C. Wakefield; *T. W. Gissing.* North Dean; *W. West.* Huddersfield; *L. C. Miall.*
D. Shire Green, Heeley, etc.; *A. Carr.* Womersley; *W. West.* Goole fields; *H. F. Parsons.* Barnby-Dun; *G. Roberts*!
T. Lindrick and Whiston; *A. Carr.*

458. Anthemis arvensis L. *'Dog-finkel.'*
Area: 63, 64, Top. Bot. Range: 0—300 feet.
Colonist; arable land, infrequent, almost rare. A. June-July.
First record: Miall's Flora, 1862.
Y. Hutton Moor, Bridge Hewick, and Sharow; *H. H. Slater.*
N. Field near Goldsborough!
W. Thorparch; *J. Emmet.* Wetherby and Bardsey!
C. Halifax and Huddersfield; *L. C. Miall.* Horbury Bridge; *P. F. Lee*! Near Lofthouse and Sandal; *G. Roberts.*
T. Near Wales; *A. Carr.*

[**Anthemis nobilis** L. *Chamomile.* A Casual. Escaped from cultivation or brought with grass seed or ballast. Gone in most, if not all, of its stations.
W. Thorparch; *T. Scatcherd.*
C. Whitley Willows, one or two plants (1858); *Hudd. N. Hist.* On Clifton Common and Blackstone Edge; *J. Bolton* (1775).
D. Brightside; *J. Salt.* Near Sheffield; *G. Harrison.*]

[**Anthemis tinctoria** L. Casual.
An increasing Casual; reinforced, not naturalising.
W. Field (with *Alyssum incanum*) near Flint Mill, between Wetherby and Thorparch!
A. In clover at Knottingley! Cornfield, Bingley; *W. West.*]

459. Achillea Millefolium L. *Yarrow. 'Milfoil.'*
Area: 63, 64, 65, Top. Bot. Range: 0—1,800 feet.
Native; banks, heaths, pastures, very common. P. July.
First record: Salt, 1800.
Generally distributed—occurs in every district.
Obs.—The ray-florets vary in hue from pure white to the deepest purple-red.

460. Achillea Ptarmica L. *Goose-tongue.* '*Sneeze-wort.*'
Area: 63, 64, 65, Top. Bot. Range: 0—1,800 feet.
Native; river-banks, stony pastures, and heaths, common, although seldom in great quantity in any one spot. P. July-September.
First record: Salt, 1800.
Generally distributed—occurs in every district, but near the larger towns, in highly cultivated areas, it is not very frequent. I can recollect it growing on Woodhouse Moor, Leeds, about 1860-62.

[**Achillea tomentosa** L. Alien.
In grassy turf, Bramham Park, in front of the ruined Hall (September 1879), doubtless of garden origin—out-thrown with rubbish.]

[**Achillea tanacetifolia** All. Lost Alien.
Found on a rough hilly bank near Ringing Low, five miles from Sheffield, in July 1844, by Mr. John Hardy. Considered 'indubitably indigenous and not an escape from cultivation' by its finder (cf. *C. C. Babington* in *Ann. Nat. Hist.*, xix, 137, 1847), accompanied as it was stated to be by *Vaccinium Vitis-Idæa*, and growing amongst 'a profusion of *Lastrea Oreopteris*'; but we judge differently now. 'An escape'; *Bab.*, *Man.*, v, viii. Extinct in 1849; *W. Borrer* (*Cyb. Brit.*, iii, 464). 'Now rare or almost eradicated . . . has been carried into several gardens at Sheffield'; *J. Hardy* in litt. (1854); *Baker, Supp. Flo. Yorks.*]

[**Achillea serrata** Sm. (*decolorans* Schr.). Incognit.
? Leeds; *Denny*, *Cat.*; *N.B.G.* Manifestly not worth further notice.]

461. Artemisia Absinthium L. '*Wormwood.*'
Area: 63†, 64†, Top. Bot. Range: 0—300 feet.
Denizen; road-sides and waste ground near habitations, rare. P. August.
First record: Salt, 1800.
Y. Littlethorpe; *H. H. Slater.* Kirby Hill, Boroughbridge; *H. Baines.*
N. Hunsingore, Marston, Overton; *H. Baines.*
W. Thorparch; *T. Scatcherd.*
A. Yeadon; *B. Carrington.*
C. Sandal; *T. W. Gissing.* Haigh; *H. Payne.*
D. Armthorpe; *H. Baines.* Conisborough; *J. Salt.* By the Don, Sheffield; *A. Carr.*

462. Artemisia vulgaris L. '*Mugwort.*'
Area: 63, 64, Top. Bot. Range: 0—900 feet.
Native; hedges and thickets, common generally, although rare in some parts of the Calder and Aire valleys. P. July-August.
First record: Salt, 1800.
Distribution nearly general—every district except Mersey.

[**Calotis cuneifolia** DC. **Calotis hispidula** F.M.
Two 'wool refuse' Aliens, natives of West Australia, found near Linthwaite, Colne Valley, in September 1886, by Messrs. Godward and Shaw. Composites, in facies between Gnaphalium and Artemisia.]

463. Filago germanica L. *Cudweed.* '*Hen and Chickens.*'
Area: 63, 64, Top. Bot. Range: 0—900 feet.
Native or Colonist; in sandy or gravelly fields, very common. A. July.
First records: Salt, 1800; 'Brightside (1799)'; *Herb. Salt.*
Distribution general—reported from every river-basin.
Obs.—There is a form branching from the root, and with broader leaves than the type, which has been called *F. spathulata* Presl. Such is a specimen in *Herb. Kirkby* from Gates' Wood, Knaresborough.

464. Filago apiculata G. E. Sm.
Area: 63, Top. Bot. Range: Low.
Colonist; sandy fields, rare and local. A. July-August.
First record: G. E. Smith, 1842.
A. Sandy fields about Hambleton Hough!
D. In clover and wheat fields, Cantley; *G. E. Smith.* Lane, Bessacarr, below Doncaster. Field at Rossington!

465. Filago minima Fries.
Area: 63, 64, Top. Bot. Range: 0—450 feet.
Native; dry places in fields, on sand-rocks, etc., infrequent. A. June-August.
First record: Salt, 1800.
N. Hookstone quarries, near Pannal! Rocks in fields near Plumpton and Braham Hall, Spofforth!
W. Acomb. About Otley Chevin and Almias Cliff.
A. Very common about Hambleton.
C. Heath, Wakefield; *T. W. Gissing.*
D. Crook's Moor, Sheffield; *J. Salt.* Armthorpe, Doncaster. Hooker Hill, near Wentworth; *H. Payne.*

466. Gnaphalium uliginosum L.
Area: 63, 64, 65, Top. Bot. Range: 0—900 feet.
Native; fields and bare ground by road-sides where water has stood, common. A. July-September.
First record: Salt, 1790. 'Cornfield near Sheffield'; *Herb. Salt.*
Generally distributed—occurs in every district.

467. Gnaphalium sylvaticum L.
Area: 63, 64, Top. Bot. Range: 0—1,200 feet.
Native; dry pastures, heathy banks, infrequent. P. July-August.
First record: Bolton, 1775.

M. Greenfield district; *J. Whitehead.*
L. About Mason Gill, and Bowland slope towards Clapham station.
R. Meerbeck Woods, Giggleswick, etc., Settle; *J. Windsor.*
Y. Pasture near Blow (Blois) Hall; *W. Brunton* (1805). Above Clotherholme; *H. H. Slater.*
N. Plentiful in sandy fields about Crosper and Braham Hall Farm, Spofforth.
W. Buckden! Wharfedale; *J. G. Baker.* Field adjoining Copmanthorpe Wood; *H. Ibbotson.* Washburn Valley; *P. F. Lee.* Ilkley; *H. F. Parsons*!
A. Ledstone Park. Fields about Brayton Barf and Selby.
C. Skircoat Moor-end, near Halifax; *J. Bolton* (1775). Storthes Hall, Huddersfield; *P. Inchbald.*
D. Sheffield, frequent; *J. Salt.*

[**Gnaphalium margaritaceum** L. Alien.
Waste sandy ground by streams, rare, and not in any sense wild.
A. Fields near Skipton, in plenty, but probably not wild; *H. Baines* (1840). Spn. collected by Tatham, 1850, and spn. collected by Webster, 1880, in *Herb. Whitwell*! Scotland Mills; *A. Shipley* (1850). Here it came out of Speight's garden! Near Shipley not uncommon; *L. C. Miall.*]

468. Gnaphalium dioicum L. '*Catsfoot.*'
G. montanum album; Ger. Em.; Ray.
Area: 63, 64, 65, Top. Bot. Montane. Range: 1,800—100 feet.
Native; stony pastures and rocky places, mainly on limestone, rare. P. June-July.
First records: How, 1650; Ray, 1670; Bolton, 1775.
M. Dry hills about Saddleworth; *J. Whitehead.*
L. & R. In Crummack pastures, Craven; *Blackst. Sp.* (1746). On the sides of Ingleborough Hill; *Ray, Cat.*, i, 136 (1670), ii, 135; *Ray, Hist.*; *Syn.* i; *Martyn* (1763). Giggleswick Scars; *J. Windsor*; and on site of Tarn, 1884; *W. Whitwell.*
Y. Copgrove; *J. Dalton.* Kirby Malzeard!
W. Near Thorparch; *W. Wood* in *O.B.G.* = Cross-roads, Stubbing Moor, Rigton! Turf of Jackdaw Crag Quarry, Stutton (!) *J. Emmet.* Aberford; *Loidis and Elmete* (1816).
A. In the meadows between Leeds and Ledstone Hall, Yorkshire, plentifully, *Mr. Dawson*; *Blackst. Spec.* Ledsham Hall; *Martyn* (1763); *G. Roberts*, 1878-83!
C. Plentiful at Burkwood, near Altofts, 1790; *D. Dixon.* Mount, Stiperden Cross; *A. Stansfield* (1877). Norland and Warley; *J. Bolton.* Ogden Clough; *W. Todd.*
D. 'Fl. rubro . . . in great plenty mixed with the white, on a great mountainous place on the left hand of the way as one goes from

Pontefract to Womersly, and almost directly opposite to Stapleton Town on the right hand; *Mr. Witham*; *Merr.*, *Pin.*, 47. Found in Scosby Leas neer Donkester, *Mr. Stonehouse*; *How*, *Phyt.*, 48; *Merr.*, 47. Hoober Hill, near Wentworth; *H. Payne.* Cliff Rocher and Bradfield; *J. Salt* (1800). Conisborough, bank near river Don (1873); *G. Roberts.*
T. Wood Lea, Maltby, in the turf; *G. E. Smith.*

469. Senecio vulgaris L. '*Groundsel.*' '*Simson.*'
Area: 63, 64, 65, Top. Bot. Range: 0—1,350 feet.
Native; fields and garden ground, very common, but not found above habitations with their accompanying garths or gardens.
First record: Salt, 1800. As so often happens, the early writers doubtless *saw* this species in the West Riding, but it being so common generally, have left no record of such observance.
Generally distributed.

470. Senecio sylvaticus L.
Area: 63, 64, Top. Bot. Arenophilous. Range: 0—900 feet.
Native; sandy hedge-banks and waste ground, railway embankments, gravel pits, etc.; common, but rare on limestone. A. July.
First record: Bolton, Cat., 1775—'Warley, etc.'
L. About Burton and Bentham.
R. Settle; *J. Windsor.* Tosside Chapel. Cocket Moss; *W. Howson.*
Y. Near Swinton; *H. H. Slater.* Hutton Moor.
N. Common about Spofforth, Marston Moor, Rufforth, etc.
W. Wike and Weardley. Acomb. Bilton.
A. Clayton Heights, Bradford; *L. C. Miall.* Hambleton. Selby.
C. Dungeon Wood, Huddersfield; *Hudd. Nat. Hist.* Ossett, Ardsley, etc.; *T. W. Gissing.*
D. Sheffield; *J. Salt.* Dovecliffe; *G. Harrison.* Thorne. East side of the Dearne Valley. Wooldale; *J. Willis.* Rawcliffe; *H. F. Parsons.*
T. Tickhill, Bawtry, and Rossington.

471. Senecio viscosus L.
Area: 63, [64], Top. Bot. Range: 0—300 feet?
Casual; rail-banks and waste ground, rare; often mis-named or robust *S. sylvaticus* reported for it. A. July-September.
First record: Salt, 1800.
(L. 'Frequent'; *Howson's Craven Flora.* Some mistake.)
(A. Limestone tract near Leeds; *H. Denny, N.B.G.*)
C. Wrose Hill, near Bradford; *L. C. Miall.* Milnsbridge; *C. P. Hobkirk* (1858). Rail-bank, North Dean Station! cum *J. W. Davis.* Railway-bank near Crigglestone; *H. Payne.*
D. ('Common'; *Salt, Flora.* Did Salt know the true thing?) Near Barnsley; *H. Johnson*!

Obs.—Truly, the above form a fine 'chapter of errors'; Mr. Hobkirk's Milnsbridge plant, my own rail-bank Casual, and Mr. H. Johnson's—also a Casual—are probably the only records certainly accurate as to name.

[Senecio squalidus L. Alien or Misnomer?
Near Thorparch; *T. Scatcherd* in *Miall's Flora.*]

472. Senecio erucifolius L.
Area: 63, 64, Top. Bot. Sub-xerophilous. Range: 0—450 feet.
Native; hedges and thickets on a dry soil, infrequent. P. July.
First record: Bolton, 1775.

L. Near Ingleton.
R. Runley Bridge and Kelcowe, Settle; *J. Windsor.*
Y. Wicliffe Lane, Fountains, etc., Ripon! Givendale; *H. H. Slater.*
N. Knaresborough (*Mrs. Berkeley*), Flaxby, Scriven, etc.
W. York; *H. Baines.* On the limestone about Wetherby, Thorparch, Aberford, Thorner, etc., very common.
A. Skipton (!) *W. Whitwell.* Airton; *Id.* Hawksworth; *W. West.* Frystone. Castleford; *G. Roberts.* Ledsham; *W. Whitwell.*
C. Near Halifax; *J. Bolton* (1775). Park Hills, 1790; *D. Dixon.*
D. Sheffield; *J. Salt.* Ecclesfield to Greenside; *A. Carr.*
T. Finningley! Firbeck!

473. Senecio Jacobæa L. *Ragwort.* '*Dogstanders.*' '*Seggrum.*'
Area: 63, 64, 65, Top. Bot. Range: 0—1,300 feet.
Native; road-sides, hedge-banks, and commons, very common. P. July-August.
First record: Salt, 1800.
Distribution general—a too-abundant dreadful weed, neither useful nor ornamental!

Obs.—A hybrid between this and the preceding was found by Mr. W. Whitwell near Airton; and, later, shown to me on a limestone bank above Skipton. It has the characters of both parents in a modified form, but is a handsomer plant than either. A form lacking ray-florets (*S. flosculosus* Jord.) occasionally occurs on waste ground; I have a suspicion, from its affecting the neighbourhood of docks and railway sidings, that it may be an alien ballast-casual.

474. Senecio aquaticus Huds. '*Water Seggrum.*'
Area: 63, 64, 65, Top. Bot. Range: 0—1,250 feet.
Native; wet meadows, stream-sides, wet commons, very common. P. July.
Earliest West Riding notice: Salt, 1800.
Generally distributed—reported for every district.

Obs.—A form or variety with much divided lower leaves (*S. pinnatifidus* G. et G.) is occasionally to be met with in very wet places, where the plant has had to grow through standing water.

[Senecio palustris DC. Incognit. 'Conyza foliis laciniatis; sive Conyza Helenites. A stone's cast from the east end of Shirley Poole neere Rushie-moore belonging to Mr. Darcy Washington. In Yorkeshire, *Mr. Heaton*'; *How, Phyt.*, p. 30; rep. *Ray* in *Gibs. Cam.* (1695); changed to 'Shirty Pool near Rushby' in *Gough's Camden.* I cannot find anything more as to this. The locality named so explicitly is about two miles to the south of Askern, and the district is even yet a very marshy one; but apparently no one has found the plant there since Heaton's time.]

475. Senecio saracenicus L. *Worm-wort.*
Area: untraced out in Top. Bot. Range: 0—450 feet.
Denizen, long known, and probably introduced for its real or supposed virtues in rural cattle pharmacy; wet waste places and field corners, rare. P. August.
First record: Ray, 1724.

L. River-side, Sedbergh; *Rev. G. Pinder.* In the meadows near Ingleton, *Mr. Thornbeck*; in *Blackst., Spec.* 105; rep. *Martyn.* In the village of Ingleton; *H. Baines.* Below the bridge at Ingleton. Clapham; *Martyn* (1763); rep. *Huds.* (1778).
A. Weetwood slope, Meanwood, near Leeds, in a field on left, a little past Smithy Mills towards Adel! By stream, Weetwood Bottoms! By the river Aire at Snaith; *H. F. Parsons.*
C. Bierley, Bradford; *H. Baines.* Hallifax, towards Richay, in the road on the left hand; *Ray, Syn.*, iii (1724); rep. *Martyn* (1763). 'The Halifax plant is certainly *S. sarracenicus*, Linn . . *R. A. Salisbury MSS.* in his copy of *With.*, Ed. ii, p. 914. Now (June 1871) belonging to me'; *W. W. Newbould.* Soothill Wood, Batley; *Neville* (1873), *Y. Nat. Rec.*, 220. Mirfield and Deighton; *T. H. Bartlam.*

Obs.—This, like Ragwort, is said to be an aromatic bitter, and was formerly used as an anthelmintic for cows that would not fatten, and horses suspected of worms.

476. Doronicum Pardalianches L. *Leopard's Bane.*
Area: undicated in Top. Bot. Range: 0—300 feet.
Denizen; well established in woods, osier-holts, thickets, etc., infrequent. P. July.
First record: Manby, 1805.

Y. Sandy places on islets, and among osiers, on the Yore near Little Nunwick, well established, and *looking* perfectly wild!
N. Near the World's End, Harrowgate; *Manby* in *O.B.G.* (1805). Hedge near Hospital! Plantation on the Sicklinghall Road, Wetherby (*J. S. Wesley*), and thicket by Stockeld Park gates!
W. Harewood, plantation on right of road, near lower gates, in going from Moortown! Denton, near Otley; *H. F. Parsons.*

A. Ledsham, for many years, in plenty; *W. Whitwell.*
C. Storthes Hall Woods! *P. Inchbald.* Near Soothill Wood, Batley; *Smith*; *T. W. Gissing.*
D. Quarry at Aldwarke; *G. Harrison.* Near Sprotborough (1845) yearly extending; *P. Inchbald* (1876). Hexthorpe; *T. Birks* (1878).

477. Bidens cernua L.
Area: 63, 64, Top. Bot. Range: 0—100 feet.
Native; by canals and ponds, rare. A. July-August.

Y. Ripon Common; *Miss Morton.* Gone or nearly so; *H. H. Slater.*
A. Frizinghall; *W. West.* Marsh by the 'Cut,' Selby!
C. Canal, North Dean (!) *W. West.* Stanley; *G. Roberts.* Heath and Sharlestone; *T. W. Gissing.*
D. Campsall! Worsborough reservoir; *Payne's MS. Flora.* Ditches by the river Rother near the dam; *A. Carr.* Swinefleet, by the warping drain; *H. F. Parsons*!

478. Bidens tripartita L.
Area: 63, 64, Top. Bot. Range: 0—300 feet.
Native; by slow streams, canals, ponds, infrequent. A. August-Sept.
First record: Bolton, 1775.

R. Rathmell; *J. Windsor.*
A. By the Aire from Hunslet downwards! Leeds and Liverpool Canal at Apperley (*W. West*) and Skipton. Frizinghall; *J. Willis* (1871).
C. Stanley canal; *T. W. Gissing.* Sandal; *G. Roberts.* North Dean; *W. West.* 'Not found in Halifax Parish'; *H. Baines.*—Because Leyland could not find it? It does grow there. Ball Plash, Brighouse; *J. Bolton, Cat.* (1775). Hemsworth Dam; *J. Willis.* Pond near Huddersfield, formerly, place now built over; *C. P. Hobkirk.*
D. Rotherham; *L. Langley* (1828). Goole fields; *H. F. Parsons.* Canal, Barnsley; *H. Payne.* Between Wadsley Forge and Wardsend; *J. Salt* (1800) = By the Wardsend dam; *A. Carr* (1876).

479. Inula Helenium L. *Elecampane.*
Area: comital allotment unattempted in Top. Bot. Range: 0—450 ft.
Denizen; 'native in N. Yorks.,' *J. G. Baker*; thickets, hedges, and rough waste ground; now very rare (dying out?), but formerly commoner, because much planted for use as an expectorant. P. August.
First records: Brunton and Dalton, 1805.

Y. Pastures above Fountains Abbey; *W. Brunton* (1805). Near Copgrove and Newby; *J. Dalton, O.B.G.*
N. Broken ground of field by the siding at Wilstrop, near Marston (!) *J. S. Wesley.* Goldsborough; *L. C. Miall.*
W. Hedge between Thorparch village and the Church (!) *J. Emmet.*
A. Orchard hedge, Scurff Hall, Drax; *H. F. Parsons* (1877).
D. On strong clay soil and low 'warp' lands below Doncaster; *S. Appleby* (1831).

480. Inula Conyza DC. '*Spikenard.*'
Area: 63, 64, Top. Bot. Xerophilous. Range: 0—300 feet.
Native; thickets and dry hedge-banks, mainly on Permian limestone, infrequent. B. July-August.
First record: Dixon, 1790.

Y. Mackershaw Wood, Ripon! [Not noted by *H. F. Slater.*]
W. Thorparch; *H. Baines.* Jackdaw Crag, Boston Spa; *W. Whitwell.* On the rocks in wood at Flint Mill, Wetherby
A. Near Castleford and Pontefract! Brotherton to Burton Salmon! Fairburn; *G. Roberts.* [Gargrave; *L. C. Miall.*]
D. Went-vale, Womersley; *J. Willis.* Brockerdale! Woodhouse Common, 1790; *D. Dixon.* Conisborough.
T. Roche Abbey; *J. Salt*! Firbeck and Letwell! Anston; *P. F. Lee*!

481. Inula dysenterica L. *Fleabane.*
Area: 63, 64, Top. Bot. Pelophilous. Range: 0—600 feet.
Native; damp fields, road-sides, etc., on a stiff soil, common. P. August.
First records: Bolton, 1775; Salt, 1800.

M. Mersey tributaries district; *J. Whitehead.*
L. About Black Burton.
R. Near Ebbing and Flowing Well, and Meer beck, Settle; *J. Windsor.* Gisburn. Long Preston; *R. Clapham.*
W. Common about Wetherby, Barwick-in-Elmete, Thorparch, Clifford.
A. Bolton by Bradford, and field by Trench Wood, Shipley; *L. C. Miall.* Ferry-Fryston district; *Y. N. Excurs.* (1873). Seacroft and Kidhall; *G. Roberts.* Methley; *Id.* Carlton; *H. Payne.*
C. Wakefield, frequent; *T. W. Gissing.* Ossett and Batley; *Id.* Greetland and Skircoat; *Bates.* Halifax; *J. Bolton.* Rotherham district; *L. Langley* (1828).
D. Barnby Dun. Endcliffe Wood; *G. Harrison.* Bolton-on-Dearne; *H. Payne.* Between Sheffield and Brightside; *J. Salt.*
T. Firbeck. Kiveton.

[Inula Pulicaria L. Incognit.
Vide *Baines, Supp.*, p. 89. Reported also in *Yorks. Nat. Recorder* for Conisborough (1873), but in error for *I. dysenterica*, as was acknowledged.]

[Asteriscus aquaticus Less. (*Buphthalmum aquaticum* L.) Alien.
Sent with other aliens from the disused quarry at Sandal, 'where I saw garden rubbish being shot,' by *Mr. P. F. Lee.* A South European plant, allied to the *Inuleæ* and *Belliæ*: how introduced it is difficult to say.]

482. Bellis perennis L. *Daisy.* '*Bairnwort.*'
Area: 63, 64, 65, Top. Bot. Range: 0—1,800 feet.
Native; fields, waysides, etc.; very common up to 1,250 feet, above that rare. Mr. Baker gives 2,400 feet as the height to which it ascends on Micklefell; but in West Yorkshire I doubt if it occurs anywhere quite up to 1,800 feet.
First record for West Yorkshire: Salt, 1800.
Occurs in every division.

483. Erigeron acris L.
Area: 63, 64, Top. Bot. Xerophilous. Range: 0—300 feet.
Native; dry banks and fields, mostly on limestone, infrequent. B. July.
First records: Merrett, 1666; Salt, 1800.
Y. Quarry Moor, Ripon; *H. H. Slater.* Quarries, Burton Leonard!
W. Bramham (!) *H. Baines.* Linton Common, Wetherby! Thorparch; *J. Emmet.*
A. Ledstone and Kippax; *Rev. W. Wood.* Garforth! Knottingley; *T. W. Gissing.* Kellington; *H. F. Parsons.*
D. Rotherham district; *L. Langley* (1828). Wentbridge! Adwick-le-Street; *W. W. Newbould.* Hampole; *P. Inchbald.* Conisborough. Wentworth; *H. Payne.*
T. Roche Abbey; *J. Salt* (1800).
Obs.—Conyza acris flore albo . . . and in Yorkshire, *Mr. Witham*; *Merr., Pin.*, 29 (1666). I can hardly think *E. canadensis* can have been meant by this.

484. Aster Tripolium L.
Area: no record in Top. Bot. Halophilous. Range: 0—10 feet.
Native; on saline tidal mud on banks of rivers and drains. P. August-September.
First record: Parsons, 1875.
A. By the Ouse to Aire mouth.
D. Goole, and by the Ouse and warping drains past Reedness to Trent Falls, the rayless form (*discoideus*) occasionally intermingled; *H. F. Parsons.*

[**Aster paniculatus** Lam. (*A. salicifolius* Scholl. *A. salignus* W.).
Aster lævis L. (*A. lævigatus* Willd. *A. Novi-Belgii* W.?).
Aliens.
The first of these—a rather small-flowered, narrow-leaved plant—occurred to *Amos Carr* on the banks of the Don at Wardsend, and by the Loxley above Sheffield, in a naturalised state; whilst the second has been gathered semi-wild on waste ground in a quarry at Sandal by *Geo. Roberts* and *P. F. Lee.* It also grows (or grew) by the Wharfe below Tadcaster, and by Ouse above Selby. Both are alike only im-

permanent garden Escapes, both frequently (*A. lævis* much the most) cultivated under the name of Michaelmas Daisy, less for their intrinsic beauty than for their showy hardihood at a season when almost all other herbaceous perennials have died down.]

[**Hemizonia pungens** T. & G. Alien.
A Californian weed, on waste ground at Goole; *T. Birks.* 'There many years—flowering in 1880.' Named by Prof. Asa Gray.]

[**Hemizonia fasciculata** T. & G. Casual.
Near Hirst's Corn Mill, Woodhouse, Huddersfield; *J. Tindall.* Blooming in October 1880. A Californian weed brought with grain.]

[**Madia glomerata** Hook. Casual.
Near Hirst's Corn Mill, Huddersfield; *J. Tindall* (1880). With the preceding.]

485. Solidago Virga-aurea L. *Golden-rod.*
Virga-aurea angustifolia serrata, sive Solidago Saracenica; Ray.
Area: 63, 64, 65, Top. Bot. Sub-Montane. Range: 1,850—200 feet.
Native; thickets, rocks, and hedge-banks, infrequent. P. August.
First record: Dillenius in Ray, 1724.
M. Greenfield Valley; *J. Whitehead.*
L. & R. Common throughout these districts.
Y. Kirby Malzeard. Hackfall! Wicliffe Lane, Mackershaw; *H. H. Slater.*
N. Scotton banks, Knaresborough!
W. Bolton. Ilkley. Woodhall. Thorparch (*J. Emmet*), on rocks in woods!
A. Skipton, Bingley; *Ainley.* Adel Moor. Hambleton Haugh!
C. In the road . . . from Halifax towards Richlay [Keighley] *Mr. Newton*; *Ray, Syn.*, iii, 177 (1724). Abundant near Halifax (!) *H. Baines.* Wakefield: *T. W. Gissing.*
D. Don banks, Sheffield (*J. Salt*). Rivelin Valley; *A. Carr.* Bradfield! Rotherham district; *L. Langley* (1828). Conisborough! Barnsley.

VAR. **cambrica** Huds. Mountains in the West Riding; *R. Teesdale.* Common on and about Ingleborough; *D. Turner* in *O.B.G.* (1805). Foal Foot, Ingleborough (!) *J. Windsor.* The usual form on high rocks.

486. Tussilago Farfara L. *Cleats. Coltsfoot. Foalsfoot.*
Area: general; 63, 64, 65, Top. Bot. Range: 0—1,800 feet.
Native; banks and fields, pelophilous naturally, and most abundant on stiff wet soils, but more indifferent to the physical constitution of what it grows in than any other plant. Very common up to 1,200 feet, but following hand-made roads as high as 1,800 feet. P. March-April.
First specific record: Salt, 1798.
In all the districts.

487. Petasites officinalis Mœnch. *Butterbur.* '*Wild Rhubarb.*'
Area: 63, 64, 65, Top. Bot. Range: 0—750 feet.
Native; damp thickets, river-banks, etc., the *male* plant very common and recorded from all the districts. (*P. vulgaris* Desf.).
First record: Bolton, 1775, and first for *hybrida.*

PLANTA FEM. **P. hybrida** Hook. is the female plant, but some specimens show abortive stamens also, as has been pointed out by Mr. Nicholson (*Journ. Bot.*, xxii, 251). This female plant, everywhere much rarer than the male, grows with the male plant. Brunton, in 1800, appears to have known it was the fertile form, for in *O.B.G.* he remarked in a Note: 'This plant . . . always bears seed; but *T. Petasites*, which is still more plentiful, never.'
L. Stream-side, Helk's Wood, Ingleton; *B. Carrington* (1862).
Y. Ure banks near Ripon (!) *W. Brunton.* Near Sleningford; *J. Dalton.*
N. Boroughbridge; *Mr. Borrer*; *Jos. Woods, MS.* in *Towns. O.B.G.*
W. Near the river Wharfe, between Ilkley and Bolton (!) *R. Teesdale.*
A. Skipton Woods and Aire side, Cononley; *W. Whitwell.* By Aire, below Castleford.
C. Abundantly by the Calder at North Dean, and lower down near Halifax (!) *J. Bolton* (1775); damp wood at Fenay and clough at Thunder Bridge, near Huddersfield; *Hudd. Nat. Hist.*

488. Petasites fragrans Presl. *Sweet Butterbur.*
Denizen; in plantations and thickets, rare, but establishes itself readily. P. March.
First record: the Author, about 1870.
W. In the wood at Thorparch, on rocky banks near the river, half way between Flint Mill and the Bridge, in plenty!

[**Petasites albus** Gaert. Alien.
Damp shady places, introduced by man.
A. Near Lower Bents, Wilsden, partly destroyed (in 1862); *L. C. Miall.*
C. In a quarry in the wood north of Storthes Hall, 1843, over a considerable area; but evidently a stray; *P. Inchbald, MS.* (1876).]

489. Eupatorium cannabinum L. *Agrimony.*
Area: 63, 64, Top. Bot. Range: 0—600 feet.
Native; damp thickets, stream-banks, etc., infrequent. P. August.
First record: Hargrove, 1782.
R. Kelcowe, Kendall's Gill, etc., Settle; *J. Windsor.*
Y. Near Ripon, not uncommon; *H. H. Slater.* Hackfall!
N. On the rock at the Dropping Well, Knaresborough (!) *E. Hargrove* (1782 et seq. eds.).
W. Norwood Bottoms, Thorner! *Jos. Blackburn.* Marsh at Leathley, Wharfedale! Barwick. Collingham! Ditch by road-side near jungle, Askham Bogs!

A. Milford Junction and Fryston; *G. Roberts.* Burton Salmon. Shipley Glen; *H. T. Soppitt.*
C. Wakefield, frequent; *T. W. Gissing.* Not very; *G. Roberts.*
D. Conisborough, frequent; *J. Salt.* Womersley, deep ditches; *G. Roberts* (1871). Askern; *H. F. Parsons.*
T. Old quarries near Kiveton; *A. Carr.*
Obs.—I never heard *this* species called '*Hemp*' Agrimony by the peasantry—only by more or less cultivated collectors of plants, herbalists, etc.

490. Cichorium Intybus L. *Chicory.* '*Blue Succory.*'
Area: 63†, 64, Top. Bot. Range: 0—300 feet.
Denizen (also as a Colonist); in chalky fields and rocky limestone railway-banks, rare. P. July-August.
First record: Hargrove, 1782.
Y. Givendale, by the Yore (one plant); *H. H. Slater* (1880).
N. In a lane leading from Grimbald Bridge to Ribston; *E. Hargrove* (1782). Knaresborough, on the Wetherby road beyond Grimbald Bridge bar! (*Id.*; eds. of 1809 and 1832). Five or six plants in 1868.
W. In plenty on the rocky limestone banks of railway one and a half miles west of Wetherby station (!) *J. Jackson.* Rail-banks near Stutton. Sandy fields near York; *H. Baines.* Church Fenton, Thorparch, and Collingham; *L. C. Miall.*
A. Saltaire; *L. C. Miall.* Gilstead Lock; *Id.* The 'Green Lane,' Ledsham, several plants; *W. Whitwell.*
C. Lofthouse, a stray plant; *G. Roberts.*
D. Rail-bank, Goole; *H. F. Parsons.* Conisborough, frequent; *J. Salt.* Rother banks; *G. Harrison.* Dearne Valley, casually; *H. Payne.*

[**Cichorium Endivia** Don.
Casual. Clover-field at Methley (*G. Roberts*); Dewsbury (*P. F. Lee*); Goole; Sheffield; etc. Debased garden outcasts.]

491. Lapsana communis L. *Nipplewort.*
Area: 63, 64, 65, Top. Bot. Range: 0—900 feet.
Native; hedge-banks, very common. A. July-September.
First record: Salt, 1800.
Generally distributed—to be found in every district.

[**Hypochæris glabra** L. Incognit.
Area: '63, 64' (sine personal authority), Top. Bot.
W. Thorparch; *S. Hailstone.* No one has found it since.
D. On the common close to the Inn at Bank-top, near Barnesley; *Rev. W. Wood*; *With.*, ed. ii (1787); rep. *O.B.G.* (1805). I have both sought this (in the above locality) and further information, in vain.]

492. Hypochæris radicata L. *Cat's-Ear.*
Area: 63, 64, 65, Top. Bot. Range: 0—1,250 feet.
Native: meadows, pastures, and rough banks, very common. P. July.
First record: Salt, 1800.
General in its distribution—occurs in every district.

[**Hypochæris maculata** L. Error.
Area: '[64], Maculate vars. of Hieracium have been repeatedly mistaken for this plant,' Top. Bot.
(R. Near Attermire Cave; *J. Windsor, MS.* Near Ottermine Cove, Settle; *Mr. Caley* in *O.B.G.*; from *R. Teesdale*, 1798.)
(A. Rocks in Gordale; *J. Windsor, MS.* About Gordale and Malham Cove; *Mr. Wood*; *With.*, ed. ii, 859 (1787); omitted from ed. iii; rep. '*Dr. Smith*'; *O.B.G.*)
Obs.—Windsor himself mistook *H. Gibsoni* for this up to about 1838, but learnt his mistake prior to the publication of his *Flora Cravoniensis* issued after his death in 1873.]

493. Leontodon hirtus L.
Area: 63, 64, Top. Bot. Range: 0—300 feet.
Native; on sandy and sandy-heathy ground, rare. P. August.
First record for certain for West Riding: the Author's? No previous flora has cited any stations, dismissing the species as one vaguely occurring in 'fields and moors.'
(R. High Hill near Settle; *J. Windsor.* Probably an error.)
N. Sandy field, Wilstrop! Braham Hall rocks and Hookstone Crags!
W. Near York; *H. Baines.* Acomb fields.
A. Brayton Barf (!) *W. Todd.*
(D. Rother bank near Woodhouse; *Geo. Harrison.* But is the plant small *Apargia autumnalis*? *W. W. Newbould.*) Waste ground, Thorne Quay.

494. Leontodon hispidus L.
Area: 63, 64, Top. Bot. Range: 0—1,800 feet.
Native; fields and banks, common, on the limestones abundant. P. July.
First record; Salt, 1800.
Generally distributed—occurs in every district.

495. Leontodon autumnalis L. *Hawkbit.*
Area: 63, 64, 65, Top. Bot. Range: 0—1,250 feet.
Native; in fields, roadsides, and river-banks, etc., very common. P. August.
First record: Salt, 1800.
Distribution general—occurs plentifully in every area, but at or about 600 feet is largely replaced by the var. with shaggy phyllaries. A many-phyllaried, rayless monstrosity is occasionally found.

VAR. **pratensis** Koch. Range: 300—1,800 feet. This is the more common form on the hill-pastures of the dales, and occurs as low down as the dry hills about Ledstone—perhaps lower, but if so it will probably be as an adventive on river-banks.

[**Arnoseris pusilla** Gaertn. Incognit.
'*Hyoseris mascula* Male Swine's Cichory: in some barren fields in Yorkeshire, *Mr. Stonehouse*'; *How, Phyt.*, 61 (1650). No notice of this is to be found in Baines' or Baker's Yorkshire Floras; yet it may really have occurred as a Casual, in same way as *Crepis setosa* and *Melampyrum cristatum* were found.]

496. Picris hieracioides L.
Area: 63, 64, Top. Bot. Xerophilous. Range: 0—300 feet.
Native; dry or damp clayey calcareous banks, rare. B. July-August.
First record: O. A. Moore (1849).
W. Spofforth Hill, etc., Wetherby. Scarcroft Hill, Thorner! Thorparch; *O. A. Moore.* On Mr. Handley-Taylor's farm, Askham Bryan; *H. Ibbotson.* Jackdaw Crag Quarry, Stutton!
A. Field between the North Lane and Park at Headingley; *A. Shipley* in *Baines*—was probably a Casual here.
D. Doncaster! Smeaton Crags! Hampole Ridge.
T. Quarries about Roche Abbey, Firbeck, etc.

497. Helminthia echioides Gaert. *Ox-tongue.*
Area: 63, 64, Top. Bot. Pelophilous. Range: 0—450 feet.
Native; in stiff clayey fields and on ditch-banks, rare. B. July-September.
First record: Baines' Flora, 1840.
W. Lane behind Wood Hill, east of Wetherby. Thorparch; *H. Baines.*
A. By road between Selby and Hut Green; *H. Ibbotson* (1838). Field at Meanwood!
C. Lofthouse; *G. Roberts.* Haybeck near Ardsley (among clover and *Cuscuta*); *P. F. Lee*!
D. Clay Lane, Ousefleet; *T. Birks.*

498. Tragopogon pratensis L. '*Go-to-bed-at-Noon.*' *Goat's-beard.*
Area: 63, 64, Top. Bot. Pelophilous. Range: 0—750 feet.
Native; meadows, hedge-banks, and railway-banks, common on stiff clayey or calcareous soils. B. June.
First record: Hargrove, 1782.
Distribution general—occurs in every district.
R. Near Settle, frequent; *J. Tatham.*
Y. Ripon.
N. Long Flat, Knaresborough (*major*!); *E. Hargrove* (1782 et eds. seq.).
W. Spofforth. Arthington!

A. Windhill, Apperley, and Eccleshill; *W. West.* Rail-bank, Headingley station!
C. Wakefield, frequent; *T. W. Gissing.*
D. By canal near Attercliffe; *A. Carr*!

VAR. **minor** Fr. This form is commoner than the type, but the records have not been kept apart by precedent recorders, and I am unable to rightly assign them. I have seen *minor* myself in all the districts except Mersey and Lune.

499. Taraxacum officinale Wigg. *Dandelion.* '*Pissabed.*'
Area: 63, 64, 65, Top. Bot. Range: 0—1,900 feet.
Native; meadows, road-side banks, sandy places, etc., very common. P. April-August.
First record: Bolton, 1775.
Distribution general; plentiful in every district.

VAR. **lævigatum** DC.
Sandy places; no doubt not rare on *dry* banks and rocks, but I have noticed it only on the Triassic sandstones of the eastern quarter of the Riding. Dearne valley; *H. Payne.*

VAR. **palustre** DC.
Area: '63, 64,' Top. Bot.
Not uncommon. Ripon Common; *W. Brunton.* Copgrove marshes; *J. Dalton* (1805). Morton Bog, Hambleton.

VAR. **erythrospermum** Andrz.
Area: '64,' Top. Bot.
Rare in sandy places. Rocks above Gordale Scar on the way to Malham Tarn; *W. West*! By the Greta stream, Ingleton; *P. F. Lee*!

500. Lactuca virosa L. *Wild Lettuce.* '*Goat Lettuce.*'
Area: 63, 64, Top. Bot. Range: 0—300 feet.
Native; walls of ruins, limestone cliffs, and sandy fields, quite noticeably rare in West Yorkshire. B. August-September.
First record: Hargrove, 1782.
[R. Mis-referred to this species instead of the *L. muralis* in *Miall's Flora*, for Settle district.]
Y. Fountains Abbey—one plant on rocks!
N. [On most of the rocks at Knaresborough; *E. Hargrove* (1782); on the Abbey rocks; *Id.*, eds. of 1809 and 1832; none visible 1869.] By the road-side going from Rufforth to Acomb (!) *Geo. Webster.* Skip-bridge (near Kirk Hammerton); *J. G. Baker.*
W. In a stone quarry at Thorparch; *Mr. Wood*; *With.*, ed. ii, 835 (1787). Precipitous banks of Wharfe at Boston Spa; *W. Todd*! Steep bank of road near the Bobbin Mill by Thorparch Bridge!
C. Old quarry at Sandal (on arenaceous sandstone strata), 1885; *Geo. Roberts*!

[**Lactuca saligna** Huds. Incognit.
'*The Least Wild Lettuce.* Duel Lane, Cliff Hill, Warley, Halifax'; *J. Bolton* in *Cat.* 1775, must have referred to *L. muralis*, I think. At any rate there is no such plant there now.]

501. Lactuca muralis Fresen. *Wall Lettuce.*
Area: 63, 64, 65, Top. Bot. Range: 0—1,250 feet.
Native; shady hedge-banks and rocks, mainly calcareous, common.
First record: Salt, 1800.
M. Diggle valley, on wet Yoredale limestone shales, sparingly.
L. Common on all the damp shady limestone scars in the gills of Garsdale, Dentdale, Kingsdale.
R. Settle, several places (!) *J. Windsor.* Slaidburn. Sawley; *R. Brown.* Attermire Scar fissures; *W. Whitwell.*
Y. Fountains Abbey! Hackfall!
N. Knaresborough!
W. Wetherby, Collingham, Thorparch, and Stutton, chiefly on shady crags by the river Wharfe. Bolton Woods! Addingham; *W. West.*
A. Skipton, chiefly on damp walls; *W. Whitwell.* Bingley and Esholt; *W. West.*
C. Wakefield, common; *T. W. Gissing.* Not so—the Permian belt to east was referred to. Hebden Bridge.
D. Denby Dale; *Y. N. Excurs.* (1872). Cliff Rocher; *J. Salt.* Conisborough!
T. Roche Abbey Valley!

502. Sonchus oleraceus L. '*Swine-thistle*' (teste *G. Roberts*).
Area: 63, 64, Top. Bot. Range: 0—900 feet.
Native; fields and waste banks by roads, common. A. July.
First record: Salt, 1800.
General in its distribution, but not quite so abundant as *S. asper.* Occurs in every district.

503. Sonchus asper Hoffm.
Area; 63, 64, Top. Bot. Range: 0—1,250 feet.
Native; cultivated and waste ground, on rocks, etc. A. July-August.
First record: Geo. Harrison, 'Rossington, 1862,' as Baines and Miall give no specific West Riding stations.
Distribution general; in every district, and ascending higher than *S. oleraceus.* Seen on Malham Moor.

504. Sonchus arvensis L. *Sow-thistle.*
Area: 63, 64, Top. Bot. Range: 0—600 feet.
Colonist; cultivated fields, common. P. August.
First record: Salt, 1800.
Distribution general; in every district.
The robust glabrous form, without the glands on the peduncles, I have gathered only once—in a Swede turnip field east of Wetherby.

Obs.—This is the species, well known of farmers, which is called Sow-thistle in West Yorkshire—'Sow,' because it sows itself so annoyingly alike in corn and turnip fields. They, the farmers, don't know *S. oleraceus* as a troublesome weed at all.

505. Crepis taraxacifolia Thuill.

Area: 64†, Top. Bot. Range: Low.

Colonist, mostly fugitive; rail-banks, seed fields, etc., rare. A. Aug.

First record: the Author, 1869.

W. On rubble at foot of rail-bank, Thorner! Wetherby.

A. Knottingley.

C. Sandal (!) *G. Roberts.*

T. Sandy field, Rossington.

[Crepis (Borkhausia) rubra Link. Casual.

In the disused quarry at Sandal, with so many other aliens, 1886; *G. Roberts*!]

[Crepis setosa Haller fil. Casual.

C. Field at Ryhill near Walton, 1885; *H. Johnson*!]

506. Crepis virens L. *Meadow Hawk's-beard.*

Area: 63, 64, 65, Top. Bot.

Native; dry banks, meadows, etc., common, but almost absent over certain areas about Leeds, etc. A. June-August.

First record: Salt, 1800; 'Crook's Moor, 1799'; *Herb. Salt.*

Distribution general; but only as a Casual on railway-banks and such like places in M. and L.

[Crepis nicæensis Balb.

Casual, but seeming established: introduced with grass-seed.

N. In a field at Harlow Hill, Harrogate (!) *T. J. Foggitt* (1876).]

[Crepis biennis L.

Area: [64], Top. Bot.

Casual, in our area of rarer occurrence than *C. taraxacifolia.*

Y. Clover field near Sharow, Ripon!

W. Thorparch; *S. Hailstone.*

(C. Ivy House Clough near Luddenden; and Ovenden Moor side; *Miall's Flora.* Very unlikely stations for the plant—if rightly named—to occur in as a Casual, unless in some clover-fields, if there were any. Probably *C. paludosa* was intended.)]

507. Crepis hieracioides W. & K. (*C. succisæfolia* Tausch.).

Area: 64, Top. Bot. Montane. Range: 1,200—600 feet.

Native; damp thickets and banks of streams on the Scar-limestone only, rare. P. July.

First record: Windsor, 1811.

L. By the Greta stream, near Ingleton!

R. Major Spring's Wood, Settle; *J. Windsor* (1811). Stainforth Wood (!) *J. Tatham.* [The first record mis-referred to Ingleton, and Tatham wrongly quoted as authority in *Miall's Flora.*]

W. Hesletine Gill (head of Littondale); *W. Borrer.* Arncliffe Dale, bushy places by stream below village of Arncliffe!

508. Crepis paludosa Mœnch. '*Yellow or Marsh Succory.*'

Hier. montanum Cichorei folio nostras; Ray. *Hieracium paludosum* Huds. and Sm.

Area: 63, 64, 65, Top. Bot. Montane. Range: 150—1,500 feet.

Native; damp shady woods and stream-sides, common.

Earliest record: Richardson in Ray, 1724. The record in Gibson's Camden, 1695, refers to *Burnley*, outside the county.

In Craven abundant; *Richardson*; *Ray, Syn.*, iii, 166; rep. *Gough* in *Cam.*; rep. *Withering*, ed. ii (1787); etc.

M. Wet 'cloughs' of Mersey district; *J. Whitehead.*

L. Very common in Dentdale, Garsdale, etc. Ingleton.

R. Many places near Settle (!) *J. Windsor.* Douk-gill Scar (!) *W. Whitwell.*

Y. Studley; *Dawson Turner* (1805). Hackfall! Skell and Laver banks, Ripon; *W. Brunton.*

W. Buckden Woods. Arncliffe. Heselden Gill. Bolton.

A. Malham Tarn border! Shipley Glen; *W. West*!

C. Hebden Bridge (!) *A. Stansfield.*

D. Silkstone Fall; *H. Payne.* Wharncliffe! Cliff Rocher! *A. Carr.*

509. Hieracium Pilosella L.

Area: 63, 64, Top. Bot. Range: 0—1,800 feet.

Native; dry fields and banks, preferring calcareous soil. P. July.

First record: Bolton, 1775.

M. Marked as occurring in this district by *J. Whitehead.*

L. & R. Frequent on the dry banks under the scars.

Y. Ripon, common; *H. H. Slater.*

N. Knaresborough; *E. Hargrove* (1832).

W. Very frequent on the limestone tracts. Bolton! Collingham! Bramham. Etchell Crags, Thorner, Tadcaster, etc.

A. Skipton. Kippax lime-hills. Roundhay lime-hills! Brotherton.

C. Heath, Wakefield; *T. W. Gissing.* Field near Mixenden; *J. Bolton* (1775). Huddersfield; *C. P. Hobkirk.*

D. & T. Common on the Permian tract. On sand at Rossington!

[Hieracium aurantiacum L. Alien.

R. Naturalised near Settle; *J. Backhouse.* Still at Settle, 'lane-side, about a mile to the south of the town,' 1881; *Miss Thompson* in *Herb. Whitwell.* I have never seen the spot, and *Dr. Windsor* in his *Flora Cravoniensis* omits any mention of it.

Y. Swinton (1881); *Mrs. Berkeley*; an Escape; *H. H. Slater.*

C. Reported from Illingworth, near Halifax; *Y. N. Recorder*, p. 35 (1873), but Mr. R. Earnshaw informed me, later, in correspondence, that the plant exhibited at the meeting came off a rockery.]

510. Hieracium anglicum Fries.

Area: 64, Top. Bot. Montane. Range: 1,800—700 feet.

Native; on mountain rocks, rare. P. July-August.

First record: Windsor, 1813.

L. High up the west slope of Ingleborough (!) *J. Windsor.* 'Foot of Whernside, *Horsfield* and *Percival* (1866), a var.; *J. Windsor.*

R. (Mill Island; *J. Windsor*?) Rocks near Attermire Cave (1813); *J. Windsor.* On the high terraces of Penyghent, and about Stainforth; *W. Whitwell* (1884)!

W. Heseltine Gill, Littondale (!) *J. Backhouse.*

A. Several places upon the scars both about Settle and Malham; *J. G. Baker.* Gordale (1865); *J. Windsor.* Awes Scar, Malham Tarn; *P. F. Lee*!

Obs.—The upward limit of this is given as 400 feet too low in the table on page 55, but until Mr. Whitwell sent me specimens, I had not seen it from the high west scars on Penyghent.

[Hieracium Lawsoni Sm.

A. Gordill; *Dr. Richardson*; *Sm., Eng. Flora* (Ed. 1824 and 1829). The species which occurs there is now referred to *H. anglicum.*

C. Farnley Fields, Huddersfield! *J. Bartlam, Yorks. Nat. Recorder*, p. 120 (1873). A misnomer, of course.]

511. Hieracium iricum Fries.

Area: no record in Top. Bot. Montane. Range: 1,100—700 feet.

Native; on limestone rocks by water, very rare. P. August.

First record: the Author?

(L. At the foot of Whernside; *Horsfield* and *Percival* (1866). A plant said to be doubtful—probably a form of *H. anglicum.*)

A. On low plates of limestone by the stream flowing from Gordale Scar. The few plants seen were quite as well developed as the Winch Bridge *H. iricum* I made acquaintance with in 1873-6; but I look on it as not at all improbable that *H. iricum* is itself only a luxuriant specialised form of *H. anglicum*; a view borne out, perhaps, by the fact that it occurs only at the lower elevations and in less exposed situations, but always within the *anglicum* range. Its 'lowest descent' is given 200 feet too high in the table on page 55.

[Hieracium oreades Fr. Incognit.

Vide *Dill.* in *Ray, Syn. Stirp.* (1724), iii, 169, No. 8. The Gordale *H. oreades* must evidently be referred to *H. pallidum*; *J. G. Baker* in *Miall's Flora* (1862). Mr. Baker adds (in proof, November 10th, 1886) that this statement is condensed from a note at p. 44 of Backhouse's *Monograph*, which reads 'Both of these plants [specimens of *pallidum* and *anglicum* on same sheet, labelled by Dillenius, in Herb. Sherard] may *now* be found in the district referred to—*in loco declivi Gordil prope Malham Cravoniæ vicum.*' I should rather approximate it to one of the many puzzling forms of *H. cæsium*, perhaps that called *Gibsoni*, which is a feature of the Craven limestones, whilst *H. pallidum* is more partial to trap or slate rock.]

512. Hieracium pallidum Fr.

Area: '64 Tatham spn.,' Top. Bot. Range: uncertain.

Native; on limestone rocks, very rare indeed. P. July.

I insert this in deference to the opinion of *J. Windsor* as given in his *Flora Cravoniensis*, p. 45, and *Mr. Backhouse's* and *Mr. Baker's* statement quoted above; but I myself, after several searches, and a general examination of all the forms from the lime-rocks of Craven, do not believe that true *H. pallidum* occurs in West Yorkshire at all: I consider the Craven lime-rock plants to be all forms of *H. cæsium*, a belief somewhat borne out by Mr. Baker's additional note (November 10th, 1886)—'I never myself gathered *pallidum* at Gordale, only *Gibsoni*, *anglicum*, and *cæsium*'; *J.G.B.*

A. Gordale (one plant), 1865; *J. Windsor.* Gordale (1866); *W. Horsfield.* Malham Cove; *H. F. Parsons.*

513. Hieracium Gibsoni Backh. (*H. hypochæroides* Gibs. *Phytol.*, i, 907).

Area: '64 Tatham spn.,' Top. Bot. Montane. Range: 600—1,600 feet.

Native; among limestone talus and on dry exposed rocky ledges, rare. P. July-August.

First record: Gibson, 1844.

L. Helk's Wood, Ingleton!

R. 'Near Settle, Yorkshire; *Mr. Caley, With.*;' *Sm. Fl. Brit.*, ii, 833 (1800). Amongst débris at the foot of Giggleswick Scars; near Attermire Cave; *J. Windsor, MS.* in *Towns. O.B.G.* (all sub nom. *Hypochæris maculata*).

A. Awes Scar above Malham; *J. Windsor.* Gordale Scars; *J. Backhouse* and *J. Tatham*; *J. Windsor* in 1865.

Obs.—This has yellow styles, and it runs into *H. maculatum.* It occurs in plenty upon Yoredale limestone in Summer Lodge Gill, Swaledale, and in one or two places in Wensleydale. It is closely allied to *H. cæsium.* Its range in the table on pp. 55-6 is given as that of *H. pallidum.*

514. Hieracium murorum L.; Fr., Backh. *Golden Lungwort.*
H. macrocaulon hirsutum folio rotundiore; Ray.

Area: 63, 64, 65, Top. Bot. Montane. Range: 1,600—200 feet.

Native; on scars of limestone, and rocky banks, rare, but locally in plenty. P. June-August.

First record: for true plant, E. Hargrove, 1782.

L. & R. Common on the lower scars of Dentdale, Chapel-le-Dale, and Kingsdale! Cavehole Wood, Giggleswick Scar, etc.; *J. Windsor.* Lime crags above Settle; *J. Tatham*; etc. N.W. crags of Penyghent; *W. Whitwell*!

Y. Hackfall! Walls of Fountains Abbey; *J. G. Baker.*

N. On the ruins of the Castle, Knaresborough; *E. Hargrove* (1782). Rocks, Abbey gates; *Id.*, eds. of 1809 and 1832.

W. Littondale! Grassington. Buckden Scars. Bolton Woods, in variety; *J. G. Baker.* On Almias Cliff; *E. Hargrove* (1809). Rock by river at Boston Spa!

A. In Gordil, near Malham; *Richardson*; *Dill.*, *Ray*, *Syn.*, iii, 169 (1724); *Martyn* (1763)—the first record in point of time, but according to Backhouse the type sheet of this in the herb. of Dillenius is made up of specimens of *anglicum*, *pallidum*, and *vulgatum.* Malham Cove and Scars! Skipton!

D. (Old Park, Sheffield; *J. Salt* (1800). Beauchief Abbey; *G. Harrison.* I doubt both these being what we now know as *H. murorum.* I visited Beauchief Abbey with Mr. Carr in 1877, and *H. vulgatum* was only to be seen there then.)

515. Hieracium cæsium Fries.

Area: '64 Backhouse sp.,' Top. Bot. Montane. Range: 200—1,450 feet.

Native; on limestone scars, not in shade, rare. P. Late June-July.

First record: Tatham, 1854.

L. Dentdale Scars. Kingsdale.

R. Castlebar rocks, etc., near Settle; *J. Tatham.* Limestone terraces of Penyghent! Douk Gill!

Y. Fountains Abbey; *J. G. Baker.*

W. On the lower lime scars about Buckden! Wharfedale; *J. G. Baker* in *Miall's Flora.*

A. About Gordale Scars and Malham Cove in plenty!

516. Hieracium vulgatum Fries. (*H. sylvaticum* Sm.) *Hawkweed.*

Area: 63, 64, 65, Top. Bot. Range: 100—1,800 feet.

Native; on rocks or old walls, common. P. June-August.

First record: Merrett, 1666.

Distribution general—recorded for every district.

M. Mersey district, rocks and shale banks (!) *J. Whitehead.* And in localities in the other areas too numerous to make them worth mentioning.

FORM **maculatum** Backh. Mon. '*Lungwort.*'
Pulmonaria maculosa nigra; Merr. *H. murorum* γ; Huds. *H. maculatum* Sm., E.B. t. 2121.

L. 'Under ye rocks betwixt Ingleborough hill and Settle. It hath a large leaf spotted with black spots, 'tis of a redish colour underneath'; *Merr.* 99 (1666). This form is still quite a feature of the scar-limestone rocks of West and North Yorkshire. It is not *Gibsoni*, which has *yellow* styles with *few large* heads.

R. Lime crags above Settle, 1884; *W. Whitwell*!

W. A variety with the leaves spotted with red, on the rocks in Grass Wood [near Grassington], and at Gordale; *Gough's Cam.* (1789).

A. Goredale, near Malham; *Withering*, ed. ii (1787). Mallā, *Dr. Richardson*; *Herb. Sherard*, No. 4896, sub nom. *Hieraciū alpinū.* Gordale Scar; *D. Turner* in *O.B.G.* But *R. Teesdale* found it unvarying under cultivation (*O.B.G.*; note, p. 707), from which I think *H. Gibsoni* was wholly or in part what he intended. High Close Scar, Malham Tarn; *P. F. Lee*!

VAR. **rubescens** Backh.

R. & A. Giggleswick and Gordale; *J. Backhouse.* Attermire Crags, 1884; *W. Whitwell*!

C. Storr's Hill, Ossett; *P. F. Lee*! A large form which I can only refer to this.

Obs.—This species varies greatly in physiognomy according to local conditions. In woods it assumes the form known as *nemorosum*—a luxuriant shade-grown, rather leafy plant: on exposed, high rocks it assumes a squat, robust appearance, simulant of *H. murorum*, and is then the *montanum* of some authors. In cultivation both merge into large, ordinary *vulgatum.* As bearing upon the frequent maculation of the foliage in one form of this species, Mr. Whitwell avers that in his experience wherever it occurs the leaves of the Dandelion are black-spotted in the same manner. I had not noticed the correspondence.

517. Hieracium gothicum Fries.

Area: '64 Tatham spn.,' Top. Bot. Montane. Range: 1,600—400 ft.

Native; scars and rocks by streams, very rare, or perhaps seeming so because ill-understood and taken for rigid, stunted forms of the preceding.

L. Wenning bank near Bentham! (approaching what I regard as *H. strictum* Fr.). Sparingly in the small crevasses of the limestone pavement on Ingleborough.

(C. Hebden Bridge; *Y. N. U. Trans., Series* E, p. 75. An error—it was a form of *H. sylvaticum*!)

Obs.—Mr. Baker in Supp. Flora Yorks. remarks that Fries' true plant is allied to *H. vulgatum* rather than *H. tridentatum.* My Yorkshire specimens were referred doubtingly to a rigid narrow-leaved form of *H. vulgatum* by Mr. Watson! I do not understand the species at all well myself.

518. Hieracium tridentatum Fries.

Area: 64, 65, Top. Bot. Range: 300—1,200 feet.

Native; bushy places by streams among the hills, mostly on limestone; rare. P. July-August.

First record: Irvine, 1858.

L. Dentdale and Garsdale thickets. Wenning banks from Wethercote to Bentham.

R. Linn Gill. Ribble-side thickets down to Rathmell! Stainforth; *A. Irvine.* Douk Gill Scar, Horton; *W. Whitwell*!

N. Scotton banks, Knaresborough!

W. About Buckden and down Wharfe to Bolton Woods!

[No observance of this in the upper *Aire* valley.]

Obs.—This is not a rare species among the western dales, but has been confused with *H. vulgatum* in that leafy form which passes as *H. nemorosum.*

519. Hieracium prenanthoides Vill.

Area: '64 Tatham spn.,' Top. Bot. Montane. Range: 1,200—300 ft.

Native; damp stony thickets by streams, rare and local. P. July-August.

First record: Tatham, 1850.

R. Stainforth Force; *J. Tatham*!

Y. Rocky and sandy spots on the Yore bank at Hackfall! Well established in clumps (!), but normally Alien there, i.e., the result of roots washed down by the river in spate. [It is abundant by the Yore two miles above Hawes, and odd clumps occur down to Askrigg, but on the intervening stretch of thirty miles it has not got a footing.]

W. Hesletine Gill; *J. Tatham.*

A. Malham Cove; *H. F. Parsons.*

520. Hieracium umbellatum L.

Area: 64, Top. Bot. Range: 0—450 feet.

Native; sandy banks of roads and streams, heaths, commons, etc., rare. P. July-August.

First record: uncertain.

R. By bridge between Settle and Giggleswick; *J. Tatham.* Ribble bank near Hellifield! Tosside.

Y. Blind (sand) pit at Hutton, near Ripon; *H. Baines.*

N. Hedge-banks about Marston.

W. In the sandy 'jungle' (south-west end) of Askham Bogs! cum *J. Abbott.* Sandy banks of Ouse from York down to Selby! or lower?

A. Aire bank near Carlton; *H. F. Parsons*! Brayton Barf.

C. Whitley and Dungeon Wood; *Hudd. Nat. Hist.*

D. Hatfield Moor; *H. F. Parsons.* Rail-bank near Rother below Catcliffe, and Redmires; *A. Carr.*

T. Finningley and about Bawtry!

521. Hieracium crocatum Fries.

Area: '64 Lees cat.,' Top. Bot. Range: 1,500—300 feet.

Native; on rocky banks by streams, quite rare. P. Late July.

First record: Tatham, 1854.

R. Attermire Cove (near Victoria Cave!); *J. Tatham*

A. Malham! Glen above Gordale; *Horsfield* and *Percival* (1865).

522. Hieracium boreale Fries.

Area: 63, 64, 65, Top. Bot. Range: 0—1,400 feet.

Native; on railway banks (chiefly), stream-sides, hedge-banks, on shale, sand and grit; and, less common, on limestone scars. P. August-September.

First record: Salt, 1800.

M. Mersey district, rail-banks, plentiful; rocks, scarcer!

L. & R. Frequent in thickets, and on river banks throughout both districts.

Y. Hackfall. Yore banks, Newby!

N. By Nidd at Pateley Bridge, Scotton Banks, etc. Cowthorpe.

W. Buckden! Bolton Woods, and by Wharfe down to mouth!

A. Very common from Keighley to Normanton. Meanwood Wood. Eccup. Snaith.

C. Wakefield, frequent (!) *T. W. Gissing.* Horbury; *P. F. Lee.* Cleckheaton, etc.!

D. Old Park Wood; *J. Salt.* On rail-banks, etc., about Sheffield (*A. Carr*) and Barnsley. Conisborough. Swinton! Woolley Edge! Rare in low-lying district about Goole; *H. F. Parsons.*

T. Finningley and Tickhill!

[**Hieracium villosum** L. Incognit. Error.

Clefts of rocks at Mur [Meer] Gill, at the foot of Ingleborough; *Mr. Caley*; *O.B.G.* On the second ridge of rocks on the north-east side of Ingleborough; *J. Windsor, MS.* An unsuccessful search made (with T. W. Simmonds) for the Meer Gill plant; *J. Windsor.*

The plant gathered on Ingleborough by Mr. Geo. Caley was pronounced '*villosum*' by 'lynx-eyed' 'Jemmy' Dickson, but, of course, it turned out not to be that species. Cfr. Withering, Arr., Ed. iii; et cfr. Mag. Nat. Hist., ii, 312 (1829). Mr. Watson, in Top. Bot. (ii, 232), remarks that it is *H. anglicum* which has occasionally been reported under the name of *H. villosum.* 'Probably both Mr. Caley's plant and my own are to be referred to *H. anglicum*': *J. Windsor, Flora.*]

[Xanthium strumarium L. Alien.

A Casual, occurring on shoddy-waste, with which the burrs have been outcast, but more rarely than the next-named *Xanthium*.

A. Boothroyd's Mill, Meanwood Valley !

C. Waste ground, railway goods station, Dewsbury ; *P. F. Lee, spn.* !]

[Xanthium spinosum L. Alien.

Casual on rubbish and wool waste near shoddy mills, etc., getting rather common, although not often flowering and never perfecting seeds.

A. Hunslet ! Meanwood Valley, Boothroyd's Mill !

C. Batley ; *P. F. Lee* (1879). Morley (1866) ; *G. Roberts*. Huddersfield.]

[Ambrosia artemisiæfolia L. Alien.

A Casual, in a clover field at Methley ; *G. Roberts* (1885).]

XLIII. CAMPANULACEÆ.

[Lobelia Dortmanna L. Incognit.

A pseudo-record from memory of a valueless kind is my only authority for giving this a place in a list of West Riding species. Dr. Windsor, in *Flora Cravoniensis* (p. 55), says he thought he saw the leaves (only) at the bottom of Giggleswick Tarn (now drained). Probably the leaves were those of *Littorella lacustris*.]

523. Jasione montana L. '*Blue-cap*.'

Area : 63, 64, 65, Top. Bot. Range : 0—750 feet.

Native ; on sandy banks, heaths, and furzy wastes, infrequent. B. July.

First record : Blackstone, 1746.

L. By the roadside going from Ingleton, Yorkshire, to Kirby Lonsdale, Westmoreland, just at the division of the counties of York and Lancaster ; *Mr. Thornbeck* ; *Blackst. Spec.* (1746). Between Ingleton and Kirby Lonsdale ; *Martyn* (1763) ; *Gough's Camden* (1789). On grit rocks about Masongill, still there 1870 !

R. By road between Rathmell and Wigglesworth ; *J. Windsor*. Tosside !

Y. Quarry Moor and Hutton Moor, Ripon ; *H. H. Slater*. Brimham Rocks !

N. Pateley Bridge ; near Harrogate ; Plumpton, and about Braham Hall. Crosper, and St. Francis' Chapel, Spofforth ! Marston Moor.

W. Weardley !

A. Eccup Lane and Adel Moor Top ! About Gateforth and Brayton Barf !

C. Near Hardcastle Crags, Hebden Valley ; *J. Willis* ! Near Halifax ; *Bates* in *Miall's Flora*. Todmorden ; *A. Stansfield*. Norland Moor ; *H. F. Parsons*. Storrs' Hill, Ossett ; *P. F. Lee* ! Chevet ; *H. Payne*.

D. Near Sheffield Park ; *J. Salt*. Ecclesall ; *W. W. Newbould*. Between Heeley and Manor Coal-pit ; *A. Carr*.

524. Campanula glomerata L. *Little Throatwort*.

C. pratensis flore conglomerato ; C. B. Pin. ! Ray.

Area : 63, 64, Top. Bot. Xerophilous. Range : 0—300 feet.

Native ; dry banks and pastures, common on the Permian tract, very rare off it, and in such situations as on alluvial soil by river-banks, where it is luxuriant as to size and adventive. P. July-August.

First record : Dillenius in Ray, 1724.

Y. Fields by Yore, above Ripon Bridge ! Other places near Ripon ; *H. H. Slater*.

N. Abbey Fields, Knaresborough (!) *E. Hargrove* (Eds. of 1782 to 1832).

W. 'Flore albo . . . juxta Bramham' ; *Dr. Richardson* ; *Ray, Syn.*, iii, 277 (1724). Rail-cutting near Stutton ! Thorparch, Wetherby (!) Bramham (!) and Tadcaster ; *H. Baines*. Middlethorp Ings, York, 1864 ; *R. Whitwell* ; *Herb. Whitwell* !

A. Brotherton. Fairburn (with lilac corollas) ; *G. Roberts*. Kippax ; *T. W. Gissing*. Garforth. Ouse banks (adventive), on sand, from Selby downwards to mouth of Aire (on warp) (!) *H. F. Parsons*. Micklefield ; *W. Whitwell*.

D. Rotherham district ; *L. Langley* (1828). Campsall ; *E. Lankester*. Askern. Adwick-le-Street ; *W. W. Newbould*. Went-vale (!) *T. W. Gissing*. Near Cadeby and Sprotborough (1801) ; *J. Salt*. Conisborough ; *J. Willis*.

T. Firbeck to Letwell !

[Campanula Trachelium L. Alien.

Casually in one spot. Other records open to doubt. It is *unlikely* to be a native on the Permian tract of the West Riding, looking to its range in England.

(N. Given under title of '*Camp. Trachelium*, Great Throatwort,' as growing in the neighbourhood of Knaresborough, in *Hargrove's Hist.* (1832 ; not in eds. of 1782 or 1809) ; but no locality assigned to it beyond 'Hedges.' Doubtless *C. latifolia—not* given in the list—was intended.)

(A. 'Saxton' ; *G. Roberts, MS.*, but perhaps an error).

(C. Kirkthorp [near Wakefield], 1796 ; *D. Dixon, MS.* in *Berkenhout*. Probably the next species intended, it being the *Trachelium* of old writers).

D. By the tram-road from Manor Coal-pits, Sheffield ; *A. Carr*, spn. ! An adventive here, clearly.]

525. Campanula latifolia L. *Bellflower. Great Throatwort*.

Trachelium majus Belgarum, sive giganteum ; Ger. *C. maxima* ; Ray, Syn.

Area : 63, 64, 65, Top. Bot. Sub-Montane. Range : 0—1,100 feet.

Native ; hedges and thickets, common ; although rather uncommon in some of the more populous parts of Airedale and Calderdale. P. July-August.

First records : Johnson's Gerarde, 1633 ; How, 1650.

Everywhere among the mountains ; *Ray* in *Gibs. Cam.* (1695). In divers places of Yorkeshire ; *How, Phyt.*, 122 (1650) ; rep. *Merr.*, 119 (1666). In the mountainous parts of Yorkshire ; *Ray, Cat.*, 309 (1670) ; *Hist.* (1688).

L. Ingleton. Dentdale.

R. Crummock Lane, near Catteral Hall, etc., near Settle ; *J. Windsor*. Ribble bank ; *Jos. Woods, junr.*

Y. Copgrove ; *J. Dalton*. Ripon ; *W. Brunton*. Common about Ripon still.

N. Pateley Bridge. Scotton Banks. Knaresborough !

W. Bolton to Thorparch, frequent ; *H. Baines*. Middlethorp Ings, York ; *R. Whitwell* (1865) in *Herb. W. Whitwell*.

A. In the yeere 1626 I found it in great plenty growing wilde upon the banks of the river Ouse . . . . as I went from Yorke to visite Selby ; *Johns. Ger.*, 450 (1633). At Malham, on the rocks by the Whern ; *Ray, Syn.*, iii, 276 (1724). About Leeds, abundant ; *Rev. W. Wood, O.B.G.* [not now much within six miles of the town]. Common from Skipton downwards to Selby. Between Snaith and Rawcliffe ; *H. F. Parsons*.

C. Halifax Woods ; *H. Baines*. Fixby ; *Hudd. Nat. Hist.* Bretton ; *A. F. Gissing*.

D. Near Doncaster ; *Dr. White* (1804) ; *E. Miller*. Went-vale ! *T. W. Gissing*. Rotherham district ; *L. Langley* (1828). Limestone tract west of Doncaster ; *S. Appleby* (1831). Near Treeton and Rother bank ; *J. Salt* ; rep. *A. Carr*.

T. Tickhill ! Roche Abbey.

526. Campanula rapunculoides L.

Area : '(63), (64),' Top. Bot. Range : 0—300 feet.

Denizen ; roadside hedges and plantations, but I think always an Escape. P. August.

First record : Backhouse, 1830.

Y. Sleningford ; *T. Simpson*.

N. Plumpton Park, near Ripley ; *L. C. Miall*.

W. Bramham Woods ; *H. Baines*.

A. Ditch by road between Skipton and Broughton ; *W. Whitwell* (1884) !

D. By roadside, half-mile from Darrington towards Wentbridge ; *J. Backhouse* in *Hooker's Brit. Flora*, Ed. i (1830).

Obs.—In the Supplement to *Baines' Flora* (1854) this is said to be 'doubtless a true native' in the Darrington and Bramham stations : in 'North Yorkshire' the dictum is 'Alien.' Mr. Watson in *Cybele* does not accept it as indigenous to Scotland even.

527. Campanula rotundifolia L. *Harebell. Hairbell*.

Area : 63, 64, 65, Top. Bot. Range : 0—1,800 feet.

Native ; heathy and sandy waysides, fields, and on rocks, common. P. July.

First record : Salt, 1800.

Distribution general—occurs in every district.

Obs.—Sometimes the flowers are milk-white. One old record for such is 'Field between Acomb and Hobmoor' ; *S. Hailstone*.

[Campanula persicifolia L. Alien.

A naturalised plant by the river Wharfe in ornamental grounds at Thorparch (!) *S. Hailstone*.]

528. Campanula Rapunculus L. *Rampion*.

Area : (63), Top. Bot. Range : At 100 feet.

Denizen ; waste ground, very rare. B. August.

D. Amongst brushwood, upon hills of limestone débris near the quarries at Sprotborough, west of Doncaster (1871).

Obs.—The plant was in plenty over an area of 4 or 5 square yards, semi-tuber rooted, with a close, not patently-branched inflorescence, but the calyx segments were in many cases toothed at the base. The records given for the next species perhaps related in part to the same plant ; that passing for *C. patula* forty years back, which I consider to be certainly (and, as positively, non-native) *C. Rapunculus*.

529. Campanula patula L.

Area : 63, 64, Top. Bot. Range : uncertain.

Denizen ? on dry rocky banks, very rare. B. July-August.

First record : Langley, 1828.

(Y. Ripon, occasionally ; *Miall's Flora*. I do not understand the 'occasionally,' as the wild plant is not uncertain or fugitive in its habitats, whilst the Rampion is, because mostly the result of roots out-thrown from gardens into lanes, etc., with rubbish : why was not a single station thereabouts indicated ?)

[N. Name, but no localities, in *E. Hargrove's List* (1832). Near Knaresborough, several places ; *H. Baines*. Baines seems to have copied Hargrove, but I know of no specimens in herbaria, and no confirmatory record of recent date seems to exist.]

D. Rotherham district ; *L. Langley* (1828). Limestone tract west of Doncaster ; *S. Appleby* (1831). Between Doncaster and Selby in 1838, exact place not remembered ; *H. Ibbotson, MS.* On limestone near Doncaster ; *H. Baines*. Between Swinton and Doncaster, on the pathside over a bridge, about 1845 ; *Gerard E. Smith*.

[Phyteuma orbiculare L. Incognit.

Mr. Watson had a specimen sent to him (vide *Cyb. Brit.*, ii, 143) localised from 'Sheffield.' Probably some admixture of specimens or the *Phyteuma* sent as representing a memory of something else, possibly *Jasione*, seen near Sheffield.]

530. Specularia hybrida A.DC.
Area: 63, 64 Baker spn., Top. Bot. Range: 0—300 feet.
Colonist: dry-soiled arable fields, mainly over limestone, rare. A. August.
First record: Gough in Cam., 1789.
Yorkshire, in corn-fields; *Gough's Cam.* (1789).

Y. Near Copgrove; *H. Baines* (1840).
W. Thorparch (!) and Bramham; *H. Baines*. Field by Jackdaw Crag, Boston Spa; *W. Whitwell* (1866)!
A. Near the limekilns at Kippax; *Rev. W. Wood* (1805); rep. *Whitaker* (1816).
D. Doncaster neighbourhood; *Dr. White* (1804). Rotherham district; *L. Langley* (1828).

Obs.—It has been pointed out that the 'Thorparch' of *Baines' Flora* often means Boston Spa, on the south side of the river Wharfe, the old ecclesiastical parish embracing both places. In the case cited above, however, the author has seen *Specularia* on the north side of the river.

531. Wahlenbergia hederacea Reich. *Ivy-leaved Bellflower.*
Campanula cymbalariæ foliis; Ger.; Ray.
Area: 63, 64? Top. Bot. Range: 300—900 feet.
Native; damp shady banks in the warm cloughs of the hills, rare. P. July-August.
First record: Ray, 1670.

A. Moist meadows near Gargrave; *H. Baines*. Seen since?
C. Rill-edge, Barkisland Moor, Ripponden side; *J. Bolton* (1775) = Cob Clough, Ripponden. There still, 1869. Near Halifax; *J. Bolton*; *O.B.G.* Holme Clough above Holmbridge. Near Farnley Tyas (*R. Jessop*) and Thurstonland; *C. P. Hobkirk*. Wessenden Valley above Marsden (!) *W. Fowler*. [Near Halifax; *Gough's Cam.* (1789).]
D. I have also found it in the North, as about Sheffield; *Ray, Cat.*, i, 52; Ed. ii, 49; *Hist.*, 741 (1688). '*In watery places* about S.'; *Ray* in *Gibs. Cam.* (1695), etc. Still rather plentiful in three or four shaded cloughs on the moors west of Penistone, about Ewden Brook, Abbey Brook, Bradshaw Clough, and Dunford Bridge.

XLIV. ERICACEÆ.

532. Vaccinium Oxycoccos L. (*Oxycoccus palustris* Pers.) *Cranberry.*
Area: 63, 64, 65, Top. Bot. Montane. Range: 1,900—10 feet.
Native; spongy swamps and peat bogs, common on the moors; rare now from drainage at the lower levels. P. June-August.
First record: Blackstone, 1746.

M. Greenfield Valley swamps, etc.; *J. Whitehead*. Gulliver's Swamp! Wragstone Clough.
L. On Baugh Fell, Bleamoor, and north slope of Bowland Knotts!
R. About Settle; *W. Curtis* (1782); rep. *Gough*; etc. Settle Mosses, not uncommon; *J. Windsor*. Cocket Moss; *W. Howson*. Rathmell Moor.
Y. Blacksmith's Car, Copgrove; *J. Dalton*. Bogs near Ripon, so plentifully that the berries are brought to market in large quantities; *W. Brunton* in *O.B.G.* (1805).
N. Angram. Nidderdale Moors. [Cranberry Car, near Hunsingore, up to A.D. 1850, now gone.]
W. On all the moors from Outershaw to Rombald's Moor!
A. Malham Tarn Bog. On Blackmoor, Leeds; *Mr. Dawson*; *Blackst. Spec.* (1746); *Martyn* (1763); rep. '*Rev. W. Wood* in *O.B.G.*' [Addle Bog, down to 1867—perhaps gone now.]
C. Halifax Moors; *J. Bolton*. Meg's Dyke, beyond Scamonden, Todmorden (!) *A. Stansfield*.
D. [Rotherham district; *L. Langley* (1828).] Potteric Carrs; *S. Appleby* (1831). Sheffield High Moors; *J. Salt* (1800). Bell Hagg Moors; *A. Carr*. By the 'Wells' on Thorne and Goole Moors (!) abundantly; *H. F. Parsons*, *T. Birks*, etc.

533. Vaccinium Vitis-idæa L. *Whortleberry.*
Vaccinia rubra Buxeis foliis; Park. *Vitis-idæa semper virens fr. rubro*; Ray.
Area: 63, 64, Top. Bot. Montane. Range: 2,200—350 feet.
First record: Parkinson, 1640.
Native; gritstone rocks, stony moors, and peat bogs on the mountains, common on all the higher tracts of moor. P. June-July.

M. On rocks at head of all the Mersey tributary streams; Seal-Bark, Buckstones, etc.!
L. Baugh Fell. Ingleborough; *Woodward* in *O.B.G.*
R. Ryeloaf Hill, etc., near Settle; *J. Windsor*. In Yorkshire; *Park.* 1458 (1640); *inter Ericas*, Yorkshire; *Ray, Cat.* 309 (1670); *Hist.*
Y. Brimham Rocks and Grantley Lakes, Sawley Moor; *W. Brunton*. Dallow Gill; *J. Dalton*, *O.B.G.*
N. Near Harrowgate; *R. Teesdale*, *Linn. Trans.* (1798).
W. Otley; *Rev. W. Wood*. Otley Chevin. Barden Moor; *W. Whitwell*.
A. Malham Tarn Bog. Sharpah, Skipton; *W. Whitwell*. Rumbolds Moor. Addle Bog, by stream between Millfall Cottage and Nanny Bridge, up to 1870. Moseley Wood, Horsforth (!), in plenty; *J. Abbott*.
C. About Halifax; *W. Wood*.
D. Sheffield High Moors; *J. Salt*. Rivelin, Bradfield, and Redmires Moors; *A. Carr*.

534. Vaccinium Myrtillus L. *Bilberry. Blaeberry. Whimberry.* '*Blegs.*'
Area: 63, 64, 65, Top. Bot. Range: 0—2,200 feet.
Native; rocky woods and stony moors, on gritstone mainly, common; but decreasing in the lower levels from enclosure of commons, etc. P. May-June.
First record: Ray, 1688.
'In ericētis, cum priore, . . Eboracensibus'; *Ray, Hist.*, 1487-89 (1688). Abundant in Haw Park [near Wakefield] in 1792; *D. Dixon*.
Distribution nearly general; every district except Trent, from which it is possibly absent.

535. Arctostaphylos Uva-ursi Spreng.
Vaccinia rubra foliis myrtinis crispis; Merrett.
Area: 63 Leyland sp., Top. Bot. Montane. Range: 1,800—1,500 ft.
Native; on mountain gritstone crags, very rare.
First records: Merrett, 1666; Salt, 1799.

M. About Seal Bark Rocks, Greenfield Valley head; *J. Whitehead*.
C. [Four miles from Heptenstall, near Widdop, on a great stone by the river Gorple; *Merr. Pin.*, 123 (1666); rep. *Rev. W. Wood* as authority in *O.B.G.* (1805). Heptenstall Eaves, formerly, now eradicated; *Baines' Flora*.] High Moor, near the 7th milestone to the left of road leading from Sheffield to Grindleford Bridge, 1799; *Herb. Salt*. This means Ankirk Moor.
D. Bare rocks above Abbey Brook to the north of Black Tor, in small quantity; *A. Carr* and *F. A. Lees* (1877).

536. Andromeda polifolia L. '*Bog-bell.*'
Rosmarinum sylvestre minus; Park. *Sedum palustre nostras Arbuti*; Ray.
Area: 63, 64, Top. Bot. Montane. Range: 1,500—50 feet.
Native; peat moors and mosses, now very rare, although abundant locally. P. June-July.
First records: Parkinson, 1640; Ray, 1724.
In divers places of Lancashire and Yorkshire bordering these; *Park.*, 76 (1640).

L. Wennington Moss [perhaps just out of Riding]; *Dawson* in *Blackst.* (1746); *Martyn* (1763). Austwick Moss; *R. Clapham*. Near Ingleborough; *Gough's Cam.* (1789).
R. Cocket Moss, three miles S.W. of Settle; *W. M. Watts* (1879)! In tolerable plenty in the osier ground between Fordale and Helwith Bridge; *J. Windsor*.
[A. Blackmoor, near Leeds; *W. Wood* in *O.B.G.* Addle Bog; *T. W. Gissing*.]
C. Blackstone Edge, in the way from Halifax to Rochdale in high and swampy place; *Ray, Syn.*, iii, 472 (1724). Near Keb-cote, etc., on Stansfield Moor, Todmorden; *A. Stansfield*. Hills above Keighley; *Knowlton* (1805). Moss above Meltham to south-west. Ogden Clough, about Halifax; *Mr. Crowe*, *O.B.G.*

[D. Potteric Car; *E. Miller's* list (1804); *S. Appleby* (1831).] Now gone from there, but remaining still on not distant moors.
T. Hatfield Moor, near Wroot (*A. Carr*), and abundantly yet on Thorne and Goole Moor—'Thorne Waste.'

Obs.—A decreasing species, certainly gone from its Adel and Halifax stations, but only within the last thirty years.

537. Erica Tetralix L. *Bell-heather.*
Area: 63, 64, 65, Top. Bot. Range: 100—2,200 feet.
Native; spongy bogs and peat moors; common on the hills, rare at low levels. P. Mid July-August.
First record for West Riding: Salt, A.D. 1800; but the 'Heath' incidentally mentioned as growing upon Ingleborough Hill by *Gerarde* (*Herball*, 1091) may have in part referred to this species, which is even commoner among the Lings than the Purple Heather.
Distribution nearly general; every district except Trent, and in that also if the isolated Abbey Brook area draining into the Derbyshire Derwent be regarded as belonging to the same river-basin.

Obs.—A curious form, with the tubular corolla cleft to the base into five petals, is occasionally found. I have gathered it on Blea-moor, and received it from Mr. T. J. Foggitt, who collected it on Harlow Heath, Harrogate, in 1877.

538. Erica cinerea L. *Purple Heath.* '*Crow-ling*' (Marshall).
Area: 63, 64, 65, Top. Bot. Range: 50—1,700 feet.
Native; dry sandy moors and stony hill-slopes; common, more abundant locally than *E. Tetralix*, but not so generally distributed. P. July-October.
First record: Gerarde, 1597.

M. Mersey district moors (!) *J. Whitehead*.
L. 'Upon Ingleborough hils'; *Ger. Em.*, 1273.
R. Near Giggleswick Tarn; *J. Windsor*. Moors about Tosside and Slaidburn.
Y. Hutton Moor, near Ripon; *H. H. Slater*. Kirkby Malzeard.
N. Pateley Bridge, Harrogate, and Brimham.
W. Bolton and Ilkley Moors. Fewston.
A. Otley Chevin. Adel moor.
C. Halifax, Huddersfield, and Rishworth Moors. Heath and Outwood; *T. W. Gissing*. Chevet.
D. Woolley Edge! New Miller Dam; *T. W. Gissing*. Penistone. Rotherham district; *L. Langley* (1828). High moors, Sheffield; *J. Salt* (1800). Endcliffe Wood; *W. W. Newbould*. Rivelin and Bradfield Moors; *A. Carr*.

539. Calluna vulgaris 'Salisb.' (*C. Erica* DC.) *Ling. Heather.*
Area: 63, 64, 65, Top. Bot. Range: 0—2,300 feet.
Native; on peaty moors and heaths, very common. P. Mid July-August.
Earliest notice; Gerarde, 1597.
'Upon Ingleborough hils'; *Gerarde, Herball*, 1091; *Ger. Em.*, 1273.
Distribution general; occurs in every district.

Obs.—The nearest spot to Leeds in which this now grows is, I believe, on the south rail-bank in the cutting between Manston and Crossgates stations on the N. E. Railway.

Var. **incana.**
Not common, and chiefly on low sandy moors, as about Spofforth, Selby, etc.

[**Pyrola rotundifolia** L. Incognit.
Pyrola rotundifolia major, officinale; B. Pin., 191; Blackst., 79.
Pyrola nostras vulgaris; Park., 510.
Area: '[63],' Top. Bot.
I doubt if true *P. rotundifolia* occurs at all in West Yorkshire. P. Late July.
Northwards, in Yorkeshire; *Park., Theat.*, 510 (1640).

(Y. Near the Waterfall at Lord Grantley's Lakes on Sawley Moor; *W. unton* in *O.B.G.* (1805). This was *P. media*, not differentiated by Brunton.)

(W. In Black Fen Wood, Bramham, and Weetwood, not far from Leeds, plentifully, *Mr. Dawson*; *Blackst., Sp.* (1746). Black Fen Woods [solus]; *Martyn* (1763).) '(Tennant's Wood, near Kilnsey; *R. Clapham*)'; *Windsor's Flora.* Erroneously quoted by *Windsor* under *P. rotundifolia.* Curtis first recorded it as *minor*, but very likely Surgeon Clapham (who was not a critical botanist) merely used the specific name loosely as *a* round-leaved *Pyrola*, not at all intending to assert that the plant was what we now understand by the name.

C. Near Hallifax, by the way leading to Kighley; *Ray, Cat.*, 256 (1670); *Martyn* (1763). (At North-bridge, half a mile from Halifax, plentifully; *Merr.*, 99 (1666).) Woods near Halifax; *Rev. W. Wood*; *O.B.G.*

Obs.—Most, if not all, of the old stations given above for *P. rotundifolia vera* belong to *P. media*, but the two species were not at that early date clearly distinguished from each other. I have never seen a specimen of *eu-rotundifolia* from within the Riding.]

540. Pyrola media Swartz.! *Winter-green.*
Area: '63 Leyland spn., 64,' Top. Bot. Range: 100—300 feet.
Native; shady stony moors and wooded slopes, where the soil is wet and stiff; rare, often on steep slopes where landslips have taken place. P. July.

First record: Ray, 1690.

Y. Sawley Moor, Grantley demesne! Woods at Studley, *Mr. Winch*; *Jos. Woods, MS.* in *Towns. O.B.G.*

N. Harlow Heath, Harrogate (!) *T. J. Foggitt* (1876).

C. 'On the moors south of Heptenstall in the way to Burnley in great plenty for near a mile's riding'; *Ray, Syn.* i, 176 (1690); *Martyn*, etc.; still thereabouts (Horsebridge Clough!), but now sparingly. Woods above Mytholm Church; *A. Stansfield* (1877). Storthes Wood, in several places flowered abundantly in 1857, uncertain in its blooming; *P. Inchbald.* 'Ray's plant, found plentifully about Halifax, is *P. media*, of which I have received fine specimens from Mr. R. Leyland, of that town, who first detected the error'; *Sir J. E. Smith, English Flora*, iv, 266. Yes, the Heptonstall species is really *media*, but the North Dean Wood one is not. *Someone* jumped to the conclusion that after what Smith wrote it must all be *P. media.*

541. Pyrola minor L.! *Lesser Winter-green.*
Area: 63 Leyland spn., 64, Top. Bot. Range: 0—900 feet.
Native; stony- or peaty-soiled woods, mostly under larch or fir trees, infrequent. P. July.
First record: Gerarde, 1597.

L. In Craven . . . . in a close called Cragge Close; *Gerarde, Hb.* (1597). It grows in . . . . Craven, as Gerarde affirmes; *How, Phyt.* 100. In a wood between Clapham and Ingleton; *Martyn* (1763); *Gough's Cam.*; *W. Curtis*; *R. Teesdale*; etc. Wood at Thornton Force; *T. W. Simmonds*; *Jos. Woods* in *Towns. O.B.G.*

Y. Hackfall Wood! [sub nom. *rotundifolia*, but the sp. is *minor*] *Gough's Cam.* (1789). Hackfall; *R. Teesdale, O.B.G.* [rightly given as *minor*]. Mackershaw Wood (!) *H. H. Slater.*

W. In Hazelwood Woods, near Sir W. Vavasour's park! *Ray, Cat.*, 256 (1670); etc. Bramham Woods! Tennant's Wood, near Kilnsay; *W. Curtis* (1782) in *O.B.G.* Fir plantation near road between Arncliffe and Kilnsey (1870)! Quaker's Wood, Dringhouses; *G. Webster*!

C. Hudhole Wood, near Scout Harley Wood, Todmorden; and above Lumb-bank, Mytholm; *A. Stansfield.* Wood at North Dean (Copley end), near Halifax! (published wrongly as *P. media*); *J. Willis, spn.* (1876)! Several places near Halifax; *Ray*; *O.B.G.*

D. Thorncliffe, near Wortley; *H. Payne.* Ditch banks, Sandall Beat Wood, east of Doncaster, 1879! Edlington Wood, *T. Tofield, junr.* (1771); *Herb. Tofield*, sub nom. *rotundifolia*; *W. W. Newbould.* Rotherham district; *L. Langley.*

T. Maltby Woods! *G. E. Smith.*

542. Pyrola secunda L.
Area: '65 Pinder spn.,' Top. Bot. Range: 2,000—2,200 feet.
Native; mountain turfy banks and rock ledges, very rare. P. August.
First certain record: Pinder, 1850.

L. Cautley Crags, Howgill Fells; *Rev. G. Pinder.* 'At a high elevation, and the sheep crop it.' Black Force, White Pike, Howgill, very sparingly, but the true plant; blooms very scarce!

(W. Near Sir W. Vavasour's Park, Hazelwood; *Martyn* (1763), who seems to have erroneously referred Ray's record to *P. secunda.* Ray remarks in his *Cat.*, p. 256, Ed. ii, p. 246, 'Mr. Witham shew'd me a sort of *Pyrola*, with a lesser and more pointed leaf than the common, having the flowers also lesser, and standing thicker together, which I suspect to be a different species. It nearly resembles *Pyrola folio mucronato serrato*, and grows in the like places, e.g., shady woods.' But Ray nowhere says the Hazelwood plant is *positively* that—whilst what grows there now is *P. minor*—although in *Hist. Plant.*, 1234 (1688), he says 'Hanc esse suspicet quam in sylva Haselwood.' *R. Teesdale* in *O.B.G.* (1805) says 'I believe Ray's plant will prove to be *P. minor*, as I have received this from Haslewood, and am told no other is now to be found there.')

543. Monotropa Hypopitys L.
Area: '63, 64 'Cramound',' Top. Bot. Range: 100—150 feet?
Native; under beeches, in calcareous soil, very rare. P. July.
First record: Salt, 1805.

Y. Studley Woods; *H. Ecroyd Smith*; about 1860, specimens seen by me whilst fresh; *J. G. Baker.*

W. Bramham Park Woods; *H. Baines.* In the wood at Jackdaw Crag Quarry, near Stutton; *O. A. Moore* (1839).

D. Campsall Park; *Ed. Lankester* (1842).

T. Wood near Everton; *Mr. Salt*; *O.B.G.* (1805).

Obs.—I have not been able to discover *who* first gathered this plant at *Bramham*: was it J. Backhouse, senior, between 1830 and 1838?

XLV. JASMINACEÆ.

544. Fraxinus excelsior L. *Ash.* '*Esh.*' '*Keys*'—the seed vessels.
Area: 63, 64, 65, Top. Bot. Range: 0—1,250 feet.
Native; woods, hedges, and scar-thickets among the hills, very common. Tree. May.
First record: Salt, 1800.
Distribution general—in every district.

The uni-foliolate State, **heterophylla** Vahl., is quite rare. I only know of two trees, one in Kings-wood near Roche Abbey, the other in ornamental grounds near the Abbey at Fountains Hall, Studley; but *O. A. Moore*, in *Baines, Supp.*, says, 'much cultivated about Doncaster.' I have an idea that it is always, with us, a nurseryman's planted tree.

545. Ligustrum vulgare L. *Privet.*
Area: 63, 64, Top. Bot. Range: 0—900 feet.
Native; although only alien in *most* of its stations; woods, hedges, and thickets, infrequent. Shrub. June.
First record: Gough, 1789.

L. & R. Giggleswick Scars; *W. Howson.* Other scar-thickets about Feizor and Cavehole Wood!

Y. Mackershaw Woods (? planted here).

W. Grass Wood, Kilnsey; *Gough's Cam.* (1789). Woods at Thorparch; planted.

A. Gordale Scar thickets (!) *J. Windsor*; *G. Roberts*; *L. C. Miall*; etc.

D. Went-vale, Melton Woods, and Conisborough.

T. Maltby, Firbeck, Kiveton, Tickhill—abundant in this district.

Obs.—It may be found in very numerous stations in plantations, etc. (besides those above named), but in none of these is it indigenous.

XLVI. APOCYNACEÆ.

[**Vinca major** L. Alien.
Not, in my opinion, naturalised sufficiently to merit inclusion in a Yorkshire flora.

W. Wharfe banks near Wetherby; *W. Brunton* in *O.B.G.* (1805).

C. Storthes Hall Woods; *P. Inchbald* (*Phyt.*, iii, 332).

D. Rotherham district; *L. Langley* (1828). Conisborough; *J. Salt, MS.* (1800). Road-side at Loversall, probably a garden outcast; *H. Baines.*]

546. Vinca minor L. *Periwinkle.*
Area: 63, 64†, Top. Bot. Range: 0—450 feet.
Denizen; woods and banks, well established in many spots. P. May-June.
First record: Bolton, 1775.

Y. Ripon; *J. Dalton* in *Smith's English Flora*, i, 339.

N. Belmont Wood, Knaresborough (!) *E. Hargrove.*

W. Bolton Woods and Thorparch Woods.

A. Weetwood, near Leeds! Gascoigne Wood, Milford; *G. Roberts* (1885) Carlton, near Snaith; *H. F. Parsons.*

C. [Wood belonging to Willow-hall, Skircoat; *J. Bolton* (1775).] Plentiful in a plantation at Thornes, 1790; *D. Dixon.* Hebden Bridge; *R. Leyland.* Storthes Hall Woods; *P. Inchbald.*

D. Rotherham district; *L. Langley* (1828). Summer Hill Estate, Rossington! Cliff Rocher, and rail-banks near Nunnery Coal-pits, Sheffield; *A. Carr.*

T. Roche Abbey Woods (!), abundantly; *G. E. Smith.*

XLVII. GENTIANACEÆ.

547. Erythræa Centaurium Pers. '*Centaury.*'

Area: 63, 64, Top. Bot. Range: 0—600 feet.

Native; dry stony, heathy, or calcareous-soiled pastures and banks, common. A. July-August.

First record: Salt, 1800.

M. Mersey tributaries district; *J. Whitehead.*

L. Greta bank, Ingleton; *P. F. Lee!* Garsdale and Dentdale.

R. Kelcowe, Stackhouse, etc., near Settle; *J. Windsor.*

Y. Thieves' Gill, Mackershaw, Quarry Moor, etc., Ripon; *H. H. Slater.*

N. Hookstone Crags, and Spofforth.

W. Otley Chevin! Common on the limestone about Collingham, Bramham, Thorparch, and Stutton. Askham Bogs.

A. Field by road near Adel Dam. Lanes, Wike and Brandon. Ledstone Park. Frequent on the Permian tract about Brotherton and Pontefract.

C. Hebden Valley; *Y. N. U. Excurs.* (1879). Woolley Edge (!) Wakefield, frequent; *T. W. Gissing.*

D. Rotherham district; *L. Langley* (1828). Pilley and Royston; *H. Payne.* Rother banks; *G. Harrison.* On the limestone tract, common. Goole; *H. F. Parsons.*

T. Sandbeck Park! Ravenfield, Kiveton; *A. Carr.*

548. Erythræa pulchella Fries.

Area: no record, Top. Bot. Range: 200 feet.

Native; dry sandy pastures, very rare. Given as 'common' under *E. Centaurium* in *Miall's Flora*, regarding it as a variety; but it is, I consider, distinct. Slender individuals of *E. Centaurium* have been taken for it, from not observing that the apparently-pedicelled lateral flowers are really sessile between two small leaves, the lengthening from the axil below being a lateral stem.

First record: the Author?

W. Sandy pasture to north-west of Etchell Crags near Scarcroft.

T. Sandy lane-side between Finningley and Bawtry.

549. Chlora perfoliata L. *Yellow-wort.*

Area: 63, 64, Top. Bot. Xerophilous. Range: 0—300 feet.

Native; on stiff calcareous or clayey soil, infrequent. A. June-July.

First record: Teesdale, 1798.

Y. Mackershaw Wood, Studley [=Wicliffe] limekilns, etc, near Ripon; *W. Brunton* (1805). Burton-Leonard limekilns and about Copgrove; *J. Dalton*, in *O.B.G.* Quarry moor, and field near Thieves' Gill (1881); *H. H. Slater.*

W. About Thorparch; *Rev. W. Wood* in *O.B.G.* Wood by railway half-mile west of Thorparch station (!) *J. Emmet.* Collingham bank! Wetherby! Huddlestone Quarry; *F. Whitwell.*

A. Plentiful at Frystone (1801); *D. Dixon.* Kippax; *Rev. W. Wood.* All about Ledsham, Ledstone Park, and Newthorpe. Between Sherburn and Aberford; *H. Baines.* Castleford; *A. W. Gissing.*

C. Horbury,* near lime-burning pits, brought with limestone; *P. F. Lee!*

D. Between Ferrybridge and Doncaster; *R. Teesdale* (1798). Doncaster neighbourhood; *Dr. White* (1804). Rotherham district; *L. Langley.* Adwick and Hampole; *P. Inchbald.* Askern; *Walker.* Levitt Hagg, Conisborough! Near Sheffield; *D. Turner* in *O.B.G.*

T. Maltby; *Jos. Woods, MS.* in *Towns. O.B.G.* Roche Abbey; *J. Salt.*

550. Gentiana Pneumonanthe L. '*Marsh Gentian.*'

G. palustris angustifolia; B. P.; Ray. *Pneumonanthe*; Ger.

Area: 63, 64 Kingsley spn., Top. Bot. [The Kingsley spn. was gathered in Heslington Fields, and therefore mis-referred by Watson to Mid W. instead of S.E. Yorks.] Range: 0—450 feet.

Native; heathy-sandy moors, rare, and decreasing. P. August.

First record: Ray, 1724.

In Eboracensi in ericetis humidioribus passim; *Ray, Cat.* (1670); In agro Eboracensi; *Hist.* 269; *Syn.* i.; etc.

[L. A quarter of a mile beyond Clapham, in a field going the middle way to Engleton; *Mr. Newton*; *Ray, Syn.* iii. 274 (1724); *Derh. Litt.*, 222 (1718). It is lost between Clapham and Ingleton, where Mr. Newton observed it, as Mr. Thornbeck informs me; *Blackst., Spec.* (1746).]

W. Kilnsey; *J. Tatham.*

A. Morton bog, near Hambleton! *H. F. Parsons.* Rawcliffe Rabbit Hills; *H. F. Parsons.*

D. *Prope* Doncaster; *D. Tofield*; *Huds.* (1778); *Dr. White* (1804); etc. =Potteric Carr; *H. Baines.* Bog near Bawtry; *Mr. Keirshort*; *G. E. Smith.*

T. Hatfield Common, *Dr. Maton*; *O.B.G.*

551. Gentiana Amarella L. *Felwort. Gentian.*

Area: 63, 64, Top. Bot. Range: 0—1,800 feet.

Native; dry banks, pastures, and mountain-turf; common, but mainly over limestones. A. August-September.

First record: Pierson (?) in Hargrove, 1782.

On lime hills . . . Yorks.; *Gough's Cam.*

M. Hills about Delph and Greenfield (on Yoredale shales); *J. Whitehead.* [I think this will prove to be the var. *uliginosa.*]

L. Dentdale. Ingleton. Feizor and Austwick; *R. Clapham.*

R. High pastures about Settle, Giggleswick lime-kilns, Kelcowe (1806), etc.; *J. Windsor.* Attermire Slopes; *W. Whitwell* (1884).

N. Hilly pastures, Knaresborough (!) *E. Hargrove* (Ed. 1782, et seq.).

W. Common on the Permian tract about Wetherby, Thorparch, Bramham, Barwick, etc.

A. Stockdale and Malham pastures; *J. Windsor.* Roundhay lime-hills. Kippax. Ledstone. Newthorpe; *W. Whitwell* (1880).

D. Rotherham; *L. Langley* (1828). Conisborough; *J. Salt* (1798). Went-vale; *T. W. Gissing.*

T. Kiveton; *A. Carr.* Maltby. Firbeck. Anston.

VAR. **uliginosa** Willd.

This flowers in May and early June. It has a 4-cleft corolla, occurs about Halifax! Ovenden (*R. Earnshaw!*), Delph (? *J. Whitehead*), in moist elevated pastures, and more plentifully about Pontefract, and evinces none of that partiality for calcareous soil which characterises the type.

[**Gentiana germanica** Willd. Extinction or Misnomer?

'York,' Top. Bot.

L. A specimen found at the foot of Ingleborough Hill, Yorkshire, 1816, *H. Shepherd*, in *Herb. W. Whitwell*, which spn. came to him from Mr. J. Tatham's coll. through Mr. Silvanus Thompson, is queried *germanica.* It seems only a large unbranched *Amarella.* No other specimen in the remaining *Herb. Tatham.* Mr. H. Shepherd was the Curator of the Liverpool Botanic Gardens, 1843.

Y. Ripon; *W. Brunton*; cfr. *Hook., Brit. Flora*, Ed. v (1842).]

[**Gentianella minima Bavarica** Cam. P., 407. By Pontefract plentifully, in *Ray, Syn.*, iii, *Indic. Pl. Dub.*; I believe was the vernal variety of *G. Amarella*, mentioned above, but proof positive is, of course, impossible.]

552. Gentiana campestris L. *Field Gentian.*

Area: 63 sine pers. auth., 64, Top. Bot. Range: 1,800—500 feet.

Native; mountain pastures, rare. A. September.

First record: Hudson, 1778.

R. About Giggleswick Tarn, High Ridge, etc., near Settle; *J. Windsor.*

Y. Dallow Gill (*Herb. Kew*) and field behind Dalton Lodge at Copgrove; *Rev. J. Dalton* in *O.B.G.*

A. On 'High Mark,' between Malham and Arncliffe! *L. C. Miall.* [Boggy field adjoining Horsforth beck, 4 miles from Leeds;

Hudson.] About Gordale, *Rev. W. Wood*; *O.B.G.* (Glass Houghton; *G. Roberts.* Perhaps the var. of the preceding.) Through ignorance of the calycine variability of *G. Amarella*, and a too rigid adherence to book characters in naming gatherings, I know *G. Amarella* has been several times reported as *eu-campestris.*

D. Road-side near Redmires, Sheffield; *A. Carr!*

T. Old quarries near Kiveton; *A. Carr.*

553. Limnanthemum peltatum Gmel. **(Villarsia nymphæoides** Vent.).

Denizen or Native. Introduced? Still waters. P. July-August.

First record: Whitwell, July 23rd, 1881.

W. In plenty in brick-ponds at Hob Moor, near York (!) *W. Whitwell.* The ponds were being drained or, at any rate, the level of the water greatly reduced, and, as Mr. Whitwell thinks, that circumstance led to the abundant blooming in 188 . The plants, from their extensive roots in the mud, etc., were clearly of some years growth when I examined the ponds in August of the same year. The plant is one that will live in deep water long, but prefers shallow water. Mr. G. Webster writes (November 1886) that Mr. Backhouse has had the *Villarsia* growing in a tank in the Holgate Nursery Ground (near half a mile away) for ten years, but he does not admit the possibility of any transfer having taken place, nor could birds have carried the seeds, as they ripen under water. There is no drainage into the ponds from the nursery. I should, unhesitatingly, have written 'Alien, undoubtedly planted' above, in place of Denizen, with its implication of obscurity as to how first introduced, but for a statement in *Smith's English Flora*, i, 275 (1824), where, referring to the occurrence of the *Villarsia* in Castle Howard Lake, Sir J. E. Smith says, 'brought thither *from near York* by Mr. Teesdale; *Sir T. G. Cullum, Bart.*' The statement is a singularly suggestive one—botanists and others did not *introduce* species native in one part to others where they were not so, much, if at all, prior to 1800—except, perhaps, herbs having a medical use.

Introduced into a pond by the Wharfe, and into a slow dike on west side of Wetherby Racecourse.

554. Menyanthes trifoliata L. *Bogbean. Buckbean.* '*Threefold.*'

Area: 63, 64, Top. Bot. Range: 0—1,700 feet.

Native; in marshes and spongy bogs, infrequent. P. June-July.

First record: Salt, 1800.

L. Marsh between Clapham and Bowland Knotts, near Keasdon brook!

R. Several places about Settle; *J. Windsor.* Horton in Ribblesdale; *W. Howson.*

N. [Pond in Belmont pasture, Knaresborough ; *E. Hargrove.* Gone]. Marshy field at Ingbarrow, near Stockeld (!) *J. S. Wesley.* Aketon bleach-works field !
Y. Whitemere, Hutton Moor ; *H. H. Slater.* Swinton ; *Mrs. Berkeley.*
W. Bog above Buckden, on moor near tarn ! By reservoir above Conistone. Marsh at Leathley. Ilkley, near 'Old Well.' Stutton (!) *G. Roberts.* Askham Bogs (!) *H. Ibbotson.*
A. Malham Tarn edge (!) *W. West.* Skipton end of Rumbold's Moor ; *W. Whitwell.* [Adel bog—now gone.]
D. Stream near Stamper's Hill, Barnsley ; *H. Payne.* Potteric Carrs ; *E. Miller* (1804). About Conisborough ; *J. Salt* (1800). Redmires, Sheffield ; *A. Carr.* Thorne Waste ; *H. F. Parsons.*

XLVIII. POLEMONIACEÆ.

555. Polemonium cæruleum L. *Jacob's Ladder.*
Valeriana græca ; Ger. 1076. *P. vulgare cær. et alb.* ; Ray, Syn.
Area : 64, Top. Bot. Range : 450—1,200 feet.
Native on the scar limestone, but Alien in many stations ; bushy, damp thickets, rare. P. June.
First record : Merrett, 1666.
L. Near Settle and Ingleton ; *R. Teesdale* in *O.B.G.* Near Ingleborough ; *G. Roberts.*
R. Giggleswick, as a waif, and fields south of Settle ; *Misses Thompson.* On the hill between Stockdale and Malham ; *Herb. Tatham.*
Y. White form, near Fountains Abbey ; *Mrs. Berkeley* in *H. H. Slater's Flora.* Not native there.
N. Bushy cliff, Brown Hill Quarry, Plumpton Cross-roads ; not wild.
W. Arnber Scar [Littondale], and in a low wood in Conistone 'Dib' [dip or gill] ; *Gough's Cam.* (1789). Arncliffe Wood, near Mount Grace (!) *H. Baines.* Thicket near high-road, Cote Hill, opposite Harewood Park south gates ; escape ?
A. On the rocks betwixt Maw-water Tarn and Mawam-co, where the highest rock standeth round like a castle ; *Merr.* (1666). Malham Cove, in a wood (!) and at Cordil or the Wern ; *Ray, Syn.* iii, 288 ; *Martyn* (1763). In Carleton Beck [there is a Calton near Airton, and a Carleton Beck near Skipton, but which is intended I do not know. It does not *now* grow by the Aire tributary], in the falling of it into the river Aire ; *Ray, Cat.* i, 310 (1670)—doubtless plants washed down by the stream in spate ; and in the 2nd. Ed. (1677) the older known stations are added for the first time by Ray*—'but more plentifully, both with a blue flower and

* The reason they were not inserted in the 1670 Ed. is explained in a letter from Mr. Ray to Dr. Lister, dated August 22nd, 1670, in reply to one of Mr. Lister's of date June 4th, in which he [Ray] sent, post haste, particulars of the Cove Wood locality.

a white, about Malham Cove . . . . and also at Cordill or the Whern' (p. 299). Gordale, plentifully ; *W. Brunton* (1805) in *O.B.G.* 'Just above the natural bridge there' ; *J. Windsor, MS.* in *Towns. O.B.G.* Aire-side above Kirby Malham, 1882 ; *W. Whitwell.* Still occurs at Malham Cove and Arncliffe !
[C. Lane near Skircoat Green, Halifax ; *J. Bolton* (1775).]

[Gilia capitata Don. Casual.
A waif from gardens or screenings of grain.
A. Methley ; *G. Roberts.*
C. By the canal, Mirfield ; *P. F. Lee* !]

[G. (Navarrettia) intertexta Steud. Casual.
D. Waste ground near Old Goole (flour) Mill ; *T. Birks, junr.* !]

[Leptosiphon luteus Benth. Alien.
A Californian annual. On rubbish heap in a lane behind Park Hill, Wetherby (1880) !]

XLIX. CONVOLVULACEÆ.

556. Convolvulus arvensis L. *Field Bindweed.*
Area : 63, 64, Top. Bot. Range : 0—450 feet.
Native ; fields, rail-banks, way-sides, frequent ; restricted to calcareous or warm gravelly soil as a rule, but common by cinder roads and on spoil-heaps in many parts of the colliery districts, because such places are porous, and furnish artificial conditions at variance with their lithological surroundings.
First record : Bolton, 1775.
Distribution peculiar ; no record for districts of Lune, Ribble, or Mersey—i.e., none of the western river-basins.
Y. Ripon—a troublesome weed ; *H. H. Slater.* Copgrove.
N. Knaresborough, etc.
W. Wetherby ; *J. Jackson.* Boston Spa. The Ainsty ! *G. Webster.*
A. Roundhay limehills. Pontefract. Burley and Apperley ; *W. West.*
C. Frequent among corn about Lightcliffe, Brighouse, etc. ; *J. Bolton.* Cleckheaton. Kirkheaton ; *C. P. Hobkirk.*
D. Conisborough. Grimethorpe and Darnall, near Sheffield ; *A. Carr.*
T. Waleswood ; *A. Carr.* Kiveton. Roche Abbey ; *J. H. Aveling.*

557. Convolvulus sepium L. *Great Bindweed. Hedge Convolvulus.*
Area : 63, 64, Top. Bot. Range : 0—500 feet.
Native ; hedges and thickets, perhaps occasionally an introduced plant, as about Settle ; its range in West Yorkshire is wider than that given by Baker for North Yorkshire. A common plant in the lowlands. P. July-August.

First record : Bolton, 1775.
M. Hedges in Mersey district ; *J. Whitehead.* [Wild here ?]
R. Lane near Hanlith Bridge, Settle, etc. ; *J. Windsor.*
Y. Ripon district, common ; *H. H. Slater.*
N. Hedgerow near Grimbald Bridge, Knaresborough ; *E. Hargrove* (1832). Hunsingore !
W. Common in Wharfedale from Bolton to Tadcaster. Ilkley. Arthington. Stutton.
A. Near Headingley Station. Moortown to Roundhay.
C. Wakefield, common ; *T. W. Gissing.* Plentifully in the hedges near the river Calder, Halifax ; *J. Bolton.* Huddersfield ; *C. P. Hobkirk.*
D. Potteric Carr ; *E. Miller* (1804). Conisborough. Barnsley ; *H. Payne.* Rotherham district ; *L. Langley* (1828). Sheffield and Brightside, 1798 ; *Herb. Salt.* By the river Rother ; *A. Carr.*
T. Frequent from Tickhill to Roche Abbey. Kiveton ; *A. Carr.*

558. Cuscuta europæa Murr.
Area : '63, [64],' Top. Bot. Range : Low.
Colonist or Casual ; on herbaceous plants, mostly vetches in cultivated fields, very rare. A. July-September.
First record : Cramond, 1854.
N. Field near Ouseburn ; *J. Cramond* in *Supplement to Baines' Flora.*
C. (Lofthouse (on *Galium*) ; *G. Roberts* ; *T. W. Gissing.* Perhaps this was *C. Epithymum*). On tares in a fodder-vetch field near Lofthouse ; *G. Roberts,* spn. !
D. Barmby Farm, near Barnsley, on vetches ; *H. Payne.* On nettles near the Don, by Mexborough !
T. Roche Abbey ; *J. H. Aveling* in *Hist. Roche Abbey* (1870).

559. Cuscuta Epithymum Murr.
Area : No record in Top. Bot. Range : Low.
Native ; on heath, furze, bedstraw, etc., very rare. A. June-August.
First record : Payne, about 1870 ?
Y. Furzy bank above the Yore, near Little Nunwick, Ripon !
D. Cawthorne (on ling) ; *H. Payne.*

560. Cuscuta Trifolii Bab. *Clover Dodder.*
Area : 63, 64, Top. Bot. Range : Low.
Colonist ; on clover, and occasionally amid corn crops, rare. A. July-August.
First record : Baines in Baker's Supp., 1842.
Y. Near Ripon Bridge.
W. On clover near York ; *G. Webster.* Amid corn in field near Walton, north of Thorparch ; *J. Abbott* ! Tadcaster, in 1842 ; *H. Baines* in *Baker's Supp.* (1854).

C. Clover-field, Haybeck, West Ardsley ; *P. F. Lee* ! Lofthouse, 1879 ; *G. Roberts.*
D. Near Barnsley ; *H. Johnson* ! Field between Tinsley Park and Catcliffe ; *A. Carr.* Clover-field, Goole ; *H. F. Parsons* !

[Cuscuta Epilinum Weihe.
Casual—very rare.
On flax near Darley Hall, Dearne Valley ; *H. Payne.*]

L. SOLANACEÆ.

561. Solanum Dulcamara L. *'Bitter-Sweet.' 'Nightshade.'*
Area : 63, 64, Top. Bot. Range : 0—500 feet.
Native ; damp hedge-banks and swampy thickets, common, but with a low range. P. June-August.
First record : Bolton, 1775.
M. Greenfield and Saddleworth districts ; *J. Whitehead.*
L. Lawkland Green ; *J. Windsor.* Bentham !
R. Giggleswick, etc. ; *J. Windsor.* Wigglesworth. Slaidburn !
Y. Ripon, common ; *H. H. Slater.*
N. Long Flat, Knaresborough ; *E. Hargrove.*
W. Common from Bolton downwards. Rare at Grassington !
A. Frequent throughout this district from Airton to the Ouse.
C. Halifax, 'not common here' ; *J. Bolton* (1775). Wakefield, common ; *T. W. Gissing.*
D. Sheffield ; *J. Salt* (1800). By the Loxley and Rivelin ; *A. Carr.*
T. Thorne. Roche Abbey ! Kiveton ; *A. Carr.*

562. Solanum nigrum L.
Casual ; fugitive, and quite rare. A. July-August.
First record : Salt, 1800.
W. Near Tadcaster ; *H. Baines.* A garden weed, Wetherby ! Acomb ; *H. Baines.*
A. Leeds ; *H. Denny, MS.* (*N.B.G.*) Rawcliffe ; *H. F. Parsons.*
D. Have only seen it in gardens about Sheffield ; *J. Salt.* High Hoyland, casually ; *H. Payne.* Sandy ground, Thorne Quay, 3 plants !
T. Roadside between Tickhill and Bawtry ; *Herb. Salt.* Roche Abbey.

[Lycium barbarum L. *'Tea-tree.'*
Alien. Planted, but persisting remarkably in exposed situations.
W. On rocks near Thorparch Bridge ! Hedge on left going up Spofforth Hill beyond the Union Workhouse, Wetherby.
A. Little Fenton ; *H. F. Parsons.*
T. About Firbeck.]

[Physalis Alkekengi L. Casual. Incognit.
Solanum vesicarium Halicacabum Trag. Hath been observed by Mr. Jervase Dickson, a Physitian, in an hedge by Bently neer Doncaster; *How, Phyt.* 115 (1650).]

563. Atropa Belladonna L. *Dwale. Deadly Nightshade.*
Area: 63, 64, Top. Bot. Range: 0—300 feet.
Native; waste broken ground, thickets, and woods; infrequent and intermittingly. Herb. P. June-July.
First records: Hargrove's Hist., 1782; Gough's Camden, 1789.
Y. On and under rocks, Fountains Abbey (!) *H. Baines.*
N. At Knaresborough, plentifully; *Gough's Camd.* (1789). In the road from the corn-mill to the Abbey; *E. Hargrove* (1782). Now scarce; *Id.*, eds. of 1809 and 1832. There in 1881; *H. H. Slater.* Again in 1884; *G. Roberts.*
W. In Haselwood; *Gough's Camd.* In York and Doncaster neighbourhoods; *Dr. White* (1804). Boston Spa, under the far wood beyond Jackdaw Crag (!) *H. Baines.* Bramham Moor. Thorparch Woods, about the old quarries! Road between Aberford and Becca Lodge. Sherburn churchyard, north side; *W. Whitwell* (1880).
A. Growing luxuriantly at Weldon, between Castleford and Frystone, 1792; *D. Dixon.* Wood by Roundhay lime hills (!) *W. Todd.* Sherburn Churchyard. Waste hollow by road, half-mile beyond the 'Pointer Inn,' Peckfield, going towards Sherburn!
D. Marr churchyard and Askern; *H. Baines.* Conisborough; *J. Salt* (1798).
T. Firbeck churchyard, abundant formerly, now gone; *J. H. Aveling* in *Hist. Roche Abbey* (1870).

564. Hyoscyamus niger L. *Henbane.*
Area: 63, 64, Top. Bot. Range: 0—500 feet.
Native, I think, but often impermanent and Casual-like in its irregular appearances. On waste ground, stony wayside banks and thickets, rare, often in suspicious stations, but mostly on limestone soils within the Riding. A. or B. July-August.
First record: Bolton, 1775.
L. Ingleton; *W. Howson* (1850).
R. In three or four places about Settle; *J. Windsor.* Very intermittent.
Y. Road near Clotherholme (one plant) 1870! Near Boroughbridge; *H. Baines.* And again in 1876 (*Rev. R. W. Marriott*).
N. Abbey Plain, Knaresborough, near the mill; *E. Hargrove* (1832). One plant, 1869.
W. Collingham, steep bankside under the lee of a garden, 1880-3! Had occurred previously in same district to *J. Emmet, J. Jackson,* etc. Thorparch; *H. Baines.*

C. [Gibbet Hill, above Halifax; *J. Bolton* (1775).] On Sandal Common, at Agbrigg, and at Altofts, adjoining to Burkwood, 1791; *D. Dixon.* Lane near Sandal Castle, two plants (1856); *T. W. Gissing.*
D. Rotherham district; *L. Langley* (1828). Brightside; *J. Salt.* Hooton Pagnell; *J. E. Kenyon* (1864). Went-vale Kirk Smeaton); *T. W.* and *A. F. Gissing.*
T. Formerly at Stone, near Roche Abbey, now gone; *J. H. Aveling* (1870).

[Datura Stramonium L. *Thorn Apple.*
Alien; occurring casually on waste ground or in patches of cultivated ground. A. August.
Y. Rubbish heap near Ripon (1870)! In 1881, in some abundance, in a potato patch, Skelgate, Ripon; *H. H. Slater.*
C. One plant in Dr. Bacon's Croft, 1791; another near the dog-house at Lupset; *D. Dixon.* Near Chald House, Wakefield, a score plants; *T. W. Gissing* (1865).]

LI. SCROPHULARIACEÆ.

565. Verbascum Thapsus L. *Mullein. Flannel-leaf.*
Area: 63, 64, Top. Bot. Range: 0—600 feet.
Native; dry fields and bank-sides, common, but erratic, and rarely more than a plant or two in one spot. B. July-September.
First record: Salt, 1800.
R. Near Crow Nest, Settle; *W. Howson* (1850).
Y. Wicliffe Lane and Fountains.
N. Knaresborough! Plumpton and Spofforth!
W. Aberford; *G. Roberts.* Bolton Abbey, and road-side near Beamsley; *J. G. Baker.* Wood at Thorparch; *J. Willis.* Bramham.
A. Brotherton and Burton Salmon! Between Pontefract and Castleford; *T. W. Gissing.* Brayton Barf; *H. F. Parsons.* Micklefield.
C. Lofthouse and Morley (casuals); *G. Roberts.*
D. Doncaster neighbourhood; *Dr. White* (1804); *E. Miller.* Rotherham district; *L. Langley* (1828). Near Conisborough Castle (1798); *J. Salt.*
T. Maltby! Old quarries near Kiveton; *A. Carr.*

[Verbascum pulverulentum Vill. (*V. floccosum* W. & K.) Incognit.
Recorded once, as below; but whether correctly named I do not know, as the occurrence has never been confirmed.
Rotherham district; *L. Langley* in *Loudon's Mag. Nat. Hist.*, ii, 269 (1828).]

[Verbascum Lychnitis Huds. Lost Alien?
[Hoary, white-flowered Mullein; in dry pastures the year after they have been fallow. In several fields about Lower Willow Hall in Skircoat; about the slate-pits on Elland-Edge, and in some fields about Illingworth; *J. Bolton, Cat.*, 1775.] Was this really an Alien in 1775; or was it one of those Natives, now going or gone, which evidence the truth of what M. Christ asserts, that, apart from eradication by drainage and the plough, indigenous vegetation, especially as regards forest ground, throughout the globe, is perpetually changing? Held by Mr. Watson to be very much of an Alien throughout Britain, its native stations—stony hills—in the Belgian coal-field are very similar to those about Halifax indicated above, and to those places in which it yet occurs on the south border of the Staffordshire Coal Measures. To be erratic in their appearances and disappearances is, however, somewhat a characteristic of all the Mulleins in indigenous areas, with always a tendency towards springing up most plentifully in soil newly turned, or thrown up from a depth.
A flowerless Verbascum found in the old quarry at Sandal by Mr. P. F. Lee in 1886—probably a garden waif—appears, judging from its leaves, to come nearest this species. Singularly enough D. Dixon has a note, in his copy of Berkenhout so often referred to, that he saw this White Moth Mullein as well as the next species 'in a quarry somewhere near Wakefield.']

[Verbascum Blattaria L. Alien.
Occasional, as Escapes usually, but always fugitive.
N. Near Green Hammerton; *H. Baines.*
W. Bank of road below a garden, outside Thorner village towards Scar croft, 1869.
C. In 1764 several plants in a tenter field . . . Ovenden, near Halifax, but not seen in any other place; *J. Bolton* (1775). In Agbrigg Quarry, 1790; *D. Dixon.*
T. Walls of Tickhill Castle; *Dr. Maton*; *O.B.G.* (1805).]

[Verbascum virgatum With. Alien.
Has occurred once at least; 'a casual straggler from garden cultivation' (*Supp. Flo. Baines,* 180); and sometimes confused with *V. Blattaria.*
W. Island in river below Thorparch Bridge, 1866, one large plant—not seen since, I believe; *W. Whitwell.*]

[Verbascum nigrum L. Casual.
Area: '(63),' Top. Bot.
I cannot admit this into any enumeration for West Yorkshire; although it is indigenous in Mid and South England, with us it is of less frequent occurrence than the preceding even, of course because less grown in gardens, being smaller flowered and not so handsome.

Y. Near Robin Hood's Hill, Studley; *Miss Morton* in *Slater's Flora.*
A. Near Frizinghall Railway Station, on the embankment (two years); *W. West*!]

[Erinus alpinus L. Lost Alien.
In abundance in the old bed of the river Yore at Tanfield; *Bab., Man.*, Ed. vi, p. 251, and Ed. viii, p. 261. Not there now.]

566. Scrophularia Balbisii Horn. *Water Betony* '*Bishop Leaves* (Gerarde, 1597).
Area: 63, 64, Top. Bot. Range: 0—600 feet.
Native; wet thickets, river and stream-sides, infrequent. P. July.
First records: Hargrove, 1782; Salt, 1800, sub nom. agg. '*aquatica.*'
L. By the Wenning, Bentham!
R. Near the Ribble, Wigglesworth!
Y. By Yore, Ripon, *not* common; *H. H. Slater.* By stream, How Hill; *Id.* Newby!
N. Stockwell and most other lanes, near Knaresborough; *E. Hargrove* (Eds. 1782, et seq.). Cowthorpe.
W. Wharfe side, Bolton (*not* Ehrharti)! Ilkley, Pool, Collingham, Boston Spa, etc. Askham Bog ditches!
A. Adel and Meanwood! Gargrave and Eshton; *J. Windsor.*
C. Wakefield district, common; *T. W. Gissing.* Barnsley!
D. Potteric Carrs; *E. Miller* (1804). Don banks above Sheffield; *A. Carr*! Rother banks below Catcliffe; *Id.* Goole; *H. F. Parsons.*
T. Roche Abbey valley.
Obs.—This is not common in the valleys among the hills, but seems there to be, to some extent, if not wholly, replaced by *S. Ehrharti.*

567. Scrophularia Ehrharti Stev.
Area: 64, Top. Bot. Range: 300—600 feet.
Native; bushy places by water, rare. P. July-August.
First record: J. Backhouse, about 1849.
L. Thornton-in-Craven, 1850, *J. Tatham*; *Herb. Whitwell.* By the Wenning, west of Clapham!
R. Roadside near Gisburn; *J. Windsor* (?) Salterforth, near Settle; *J. Backhouse.*
W. Wharfe bank, Ilkley (!) *J. G. Baker.* Langwith and Woodhall. Collingham!
A. Gargrave; *W. Howson, Flora.* Skipton and Gargrave; *J. Backhouse.*

568. Scrophularia nodosa L.
Area: 63, 64, 65, Top. Bot. Range: 0—900 feet.
Native; damp hedge- and ditch-banks, and by streams; common. P. July-September.
First records: Bolton, 1775; Hargrove, 1782; Salt, 1800.
Distribution general; occurs in every district.

569. Scrophularia vernalis L.
Range: 0—300 feet.
Denizen; hedge-banks and lane-sides, rare. B. May-June.
First record: Appleby, 1831.
Y. Wood near Fountains Abbey; *H. Baines*. (Near James Carter's House, Masham—not wild; *H. H. Slater*. A little outside the Riding?)
D. Balby (!) 'Just under walls of Tofield's garden, from which it escaped?' *J. Hardy*. There still in 1873; *Y. N. Excurs.* Limestone tract west of Doncaster; *S. Appleby* (1831). About Cantley Hall; *G. E. Smith*.

570. Digitalis purpurea L. *Foxglove.*
Area: 63, 64, 65, Top. Bot. Arenophilous. Range: 0—1,250 feet.
Native; field- and hedge-banks, stony woods, old quarries, etc., on sand and peat over grit- and flag-stones chiefly, but common, abundant locally, very rare on limestone. B. or P. Late June-July.
First records: Gerarde, 1597; Bolton, 1775; Salt, 1800.
'Those with white flowers do grow naturally in . . . . Craven, in a field called Cragge Close'; *Gerarde, Hb.* 697.
Distribution general; occurs in every river-basin, but restricted in the way indicated above.

[**Antirrhinum majus** L. *Snapdragon.* Alien.
Found upon old walls, ruins, etc., but without any claim to inclusion as a Native, although it might be classed as a Denizen, since it maintains its ground.
Y. Fountains Abbey (!) *Miss M. Plues*.
A. Skipton Castle walls (!) Pontefract!
T. Tickhill Castle!]

571. Antirrhinum Orontium L.
Area: no record in Top. Bot. Range: Low.
Casual or Colonist; corn-fields on gravel or limestone, and as a weed in cultivated soil; very rare. A. July-September.
[W. Sherburn district; *P. F. Lee* (1883) in *Naturalist*. Error—it was *Linaria minor*.]
D. A weed in gardens at Stainborough, near Barnsley; *H. Johnson*! Field near Manor Coal-pits; *A. Carr*!
T. Cornfields about Firbeck, Letwell, etc.!

572. Linaria Cymbalaria Mill. *'Aaron's Beard.' 'Mother-o'-Millions.'*
L. hederacea fol. glabro, seu Cymb. vulgaris; Ray.
Denizen; well established on old walls, masonry of canals, etc., common.
First record: Dillenius, 1724.
In agri Eboracensis; *Dr. Richardson*; *Ray, Syn.*, iii, *282 (1724).

R. Walls near Settle; *H. Baines*.
Y. Ripon, common; *H. H. Slater*. Fountains Abbey!
N. Common about Knaresborough on rocks and walls.
W. Bolton Abbey ruins (!) *W. West*. Old walls abutting on the river Wharfe at Wetherby, firmly established!
A. Walls of Skipton Castle; *W. Howson* (1832). Still there, 1884; *W. Whitwell*. Wall at Cottingley; *W. West*. Waterside masonry of Leeds and Liverpool Canal in many places about the old bridges, etc., from Skipton to Methley! Pontefract; *W. Brunton* (*O.B.G.*). Castle walls, etc., common there still!
C. On a wall at Grange Hall, 1790; *D. Dixon*. Wakefield; *T. W. Gissing*. Stanley; Walton; and Salterhebble; *G. Roberts*. Woodsome; *C. P. Hobkirk*.
D. 'Dr. Richardson remarked it in the quarries about Darfield in Yorkshire'; *Martyn's* Ed. of *Miller's Dict.* Wentbridge.
T. Roche Abbey.

573. Linaria Elatine Mill.
Area: 63, 64†, Top. Bot. Range: 0—300 feet.
Native; cornfields, rare; perhaps a Casual only in some of its stations, introduced with seed corn. A. July-September.
First record: Dixon, 1791.
W. Thorparch: *O. A. Moore* (1837). Boston Spa, 1866, in a quarry; *W. Whitwell*. Catterton; and in cornfields on Jos. Mawson's farm at Acaster; *Hy. Ibbotson*.
A. Knottingley (!) *T. W. Gissing*. Near Pontefract; *J. Willis*. Monk Fryston and Drax; *H. F. Parsons*. Birkin; *G. Roberts*.
C. Sharlston Field, 1791; *D. Dixon*.
D. Campsall; *E. Lankester*. Rotherham district; *L. Langley*. Hooton Pagnell; *J. E. Kenyon* (1864). Adwick; *P. Inchbald*. Redbrook, near Barnsley; *H. Payne*. Fields near Ecclesfield Road from Wadsley Bridge; *A. Carr*! Near Page Hall, Sheffield, 1799; *Herb. Salt*.
T. Hedge-side, Maltby; *O. A. Moore*. Fields about Kiveton.

[**Linaria spuria** Mill.
[63] Top. Bot.
A Casual; met with once or twice only, in cornfields, probably grain-introduced.
A. Knottingley; *T. W. Gissing*.
C. Plentiful in Sharlestone Field, 1791; *D. Dixon*. Did not Dixon rather intend the preceding species?]

[**Linaria purpurea** L. Alien.
On old walls bordering gardens, rare and tender, although sowing itself when once introduced.

A. At Upper Meanwood, in a stone quarry by Walsh's house, for many years, but perhaps now gone.
C. Bradford Road, Huddersfield, among débris, in a waste place left after delving for flagstone, near where the gardens of (the now pulled down) Longwood House stood; *W. Guthrie* in *Nat.* (1864) i. 175.
T. Tickhill Castle!]

574. Linaria vulgaris Mill. *'Butter and Eggs.'*
Area: 63, 64, Top. Bot. Range: 0—750 feet.
Native; hedge-banks, field-borders, etc., by preference on gravelly or sandy soil, but not absent on limestone, common. P. July.
First record: Bolton, 1775.
Distribution nearly general; really common in every district except that of Lune, but no record at all for it in that of the Mersey tributaries.
The highest station I know of is that by the river Loxley at Bradfield, where Mr. A. Carr showed it to me.

Status **Peloria.**
Waste ground near Ripon; *W. Brunton* (1802).

Linaria italica Trev. Alien.
This occurred persistently for several years on waste ground (now villa sites) at Headingley near Leeds, formerly a portion of the public zoological gardens. Now gone, but gathered there by J. Abbott as late as 1869.]

575. Linaria minor Desf.
Area: 63, 64, Top. Bot. Range: 0—500 feet.
Native; on calcareous or sandy fields, infrequent, but often more of a Casual than aught else on railway lines, etc. A. June-September.
First records: Dalton and Wood, 1805.
Y. Copgrove cornfields; *J. Dalton*. Ellershaw Hill and Bondgate Green by Ripon; *W. Brunton*. Rail-banks between the sleepers, etc., near Hutton Conyers; *H. H. Slater*.
N. Between Bilton and Cowthorpe; *H. Baines*.
W. Thorparch; *O. A. Moore*. Boston Spa. Railway-track, Wetherby to Spofforth! Etchell Crags, Scarcroft! Road-side, Bramham!
A. About Leeds, not common; *W. Wood* in *O.B.G.* Frizinghall; *W. West* (Casual here). Monk Fryston; *H. F. Parsons*. Kippax lime-kilns; *H. Baines*. In a pea-field, Kippax; *G. Roberts*. Knottingley; *T. W. Gissing*. Micklefield station; *W. Whitwell*.
C. Lofthouse; *G. Roberts*. Warmfield; *T. W. Gissing*.
D. Smeaton Crags; *E. Lankester* (1840). Rotherham district; *L. Langley*. Levitt Hagg; *G. Harrison*. Redbrook, Barnsley; *H. Payne*. New road by a newly-made dam at Loxley; *A. Carr*. Whiston, Sheffield; *Id.*
T. Maltby, in August 1838; *Jos. Woods, MS.* in *Towns. O.B.G.* Rail sides, Kiveton; *A. Carr*.

[**Linaria supina** Desf. Alien.
An Alien on a recently made railway bank at Arthington (1873); introduced with ballast?]

576. Mimulus luteus L. *'Monkey-flower.'*
Denizen, well established; boggy rill-sides, stream-banks, and watery places, infrequent. P. Late June-August.
First record; Baker, 1850.
R. Bog at Horton-in-Ribblesdale; *J. Backhouse*; rep. *W. Howson*.
Y. In the Skell at Studley (1850), and about Fountains Abbey (1852); *J. G. Baker* in *Supp.* to *Baines' Flora*.
N. By the beck, Ramsgill: *H. S. Kynnersley* (1871). Re-seen by *J. S. Wesley*! (1877).
W. Plentifully by stream running down from Hawkswick Clouder into Arncliffe.

Forma **guttatus** DC.
The plant seen by stream on a wild moor in Upper Nidderdale by *H. Sneyd Kynnersley* (*Bot. Exch. Club. Rep.* 1871) was reported as this—at best a slight variety, perhaps of garden origin: specimens sent to me later (1877) by *J. S. Wesley* had not spotted corolla-lips. I have therefore ventured to place the record under *M. luteus*.

577. Limosella aquatica L. *'Mudwort.'*
Area: 63, 64, Top. Bot. Hygrophilous. Range: 20—1,500 feet.
Native; muddy margins of pools and shallow streams, rare. A. July.
First record: F. A. Lees, 1867.
A. Drying-up pool (not a boggy bottom) with *Hippuris*, on 'High Mark,' Malham Moor (1870). On the drying-up mud at the south end of Adel Lower Dam, near Leeds, in great plenty 1867, scarce in 1868 and 1869, abundant again in 1870!
D. Rawcliffe 'Rabbit Hills,' by pools! *H. F. Parsons* (1877). Area being drained—perhaps now gone. 'Several cart-loads of potato tops have been put down, covering up the place where it grew'; *T. Birks* (1878).
Obs.—The 'Bolton beck' station of *Baines' Flora*, pp. viii and 79, is not in the West Riding; but referred to Bolton-on-Swale. The York neighbourhood station of *White* (1804) related to a locality east of the city.

578. Veronica hederifolia L.
Area: 63, 64, Top. Bot. Range: 0—1,250 feet.
Native; sandy or loamy cultivated ground, rubbish heaps, etc., very common. A. April-August.
First record: Salt, 1800.
Distribution general; recorded from every district.

'Near Malham Tarn'; *J. Windsor* (1,250 feet) is the highest station I know of; and, singularly enough, not this only, but *many* of the Scrophulariaceæ extend considerably higher in West Yorkshire than the range assigned to them in North Yorkshire by Mr. Baker. Is the reason to be found in the fact that North Yorkshire has no limestone plateaux, habitable as is Malham Moor, at so high a level? 'Yes, and the same things are much lower at the Lakes than in North Yorkshire'; *J. G. Baker.*

579. Veronica polita Fries. *Garden Speedwell.*
Area: 63, 64, Top. Bot. Range: 0—900 feet.
Native; in cultivated ground, very common. A. May-September.
First record: Salt, 1800.
Generally distributed, occurring in every area; and ascending into potato patches, etc., high up the Hodder Valley. It will often flower far into winter, and under snow.

580. Veronica agrestis L.
Area: 63, 64, 65, Top. Bot. Range: 0—1,350 feet.
Native: in cultivated ground, very common everywhere, ascending to Middle Houses on Malham Moor. A. April-September.
First record: Salt, 1800.
General in its distribution; in every district.

581. Veronica Buxbaumii Ten.
Area: 63, 64, Top. Bot. Range: Low.
Colonist: in sandy cultivated fields, among seeds, clovers, grain, etc., of comparatively recent introduction (Yorkshire observations of it date from 1841; it is not in Baines' Flora); now common, and still spreading. A. April-October.
First W.R. record: G. E. Smith, 1846.
Y. Near Ripon!
W. Near York; *O. A. Moore* (1841). Steeton Hall farm, Tadcaster; *W. Whitwell*! Ainsty Fields, Thorparch.
A. Kippax (!) *J. Blackburn.* Apperley Bridge; *J. Willis.* Pontefract; *J. Willis.*
C. Newton, Wakefield, and Darton; *T. W. Gissing.* Storthes Hall; *P. Inchbald.* Sandal: *G. Roberts.*
D. Near Ecclesfield and Woodhouse Junction; *A. Carr.* Goole fields; *H. F. Parsons.* Near Thorncliffe: *A. Carr.* Doncaster!
T. Maltby; *G. E. Smith* (1846). Rossington!

582. Veronica triphyllos L.
Area: 63, 64, Top. Bot. Arenophilous. Range: Low.
Native; sandy fields, rare. A. May-June.
First record: Tofield in Hudson, 1778.
N. Sandy bank near Nether Poppleton; *H. Wetherill.* Abundant there and very large, 1879; *W. Whitwell.*

W. Hobmoor, Acomb, etc., near York (!) *S. Thompson*; *G. Webster*; etc.
D. About Cantley and Armthorpe (!) *G. E. Smith.* Near Thorne; *C. Gray.*
T. In cornfields near Rossington; *Tofield, Huds.* (1778); *Gough's Cam.* (1789).

583. Veronica arvensis L.
Area: 63, 64, 65, Top. Bot. Range: 0—1,800 feet.
Native; on dry banks, sandy fields, wall-tops, railway ballast, etc., very common. A. May-June.
First record: Salt, 1800.
Generally distributed; occurs in every district

584. Veronica serpyllifolia L.
Area: 63, 64, 65, Top. Bot. Range: 0—2,000 feet.
Native; sandy banks and pastures, cultivated ground, and rocks, common. P. June.
First record; Salt, 1800.
Distribution general; in every district, and to be found high up on the rocky slopes of Whernside, Cam, and Ingleborough Fells.

[**Veronica peregrina** L. (*V. romana* L.) Alien.
A North American, linear-leaved plant, found casually in cultivated ground. Has occurred once in West Yorkshire—with other waifs in the old quarry at Sandal (*G. Roberts*)!]

[**Veronica repens** DC.
A Corsican plant, naturalised (or naturalising) at York (vide *Babington's Manual*, Eds. vii. and viii.) in market-garden ground.]

585. Veronica officinalis L. *Speedwell.*
Area: 63, 64, 65, Top. Bot. Range: 0—2,200 feet.
Native; hedge-banks, heathy ground, woods, common. P. May-June.
First record: Bolton, 1775.
Distributed generally, throughout every district.

586. Veronica Chamædrys L. *'Bird's Eye.'*
Area: 63, 64, 65, Top. Bot. Range: 0—1,800 feet.
Native; grassy hedge-banks, thickets, and pastures; common. P. May-June.
First record: Bolton, Cat., 1775—various places near Halifax.
Distribution general; occurs in every district.

587. Veronica montana L.
Area: 63, 64, Top. Bot. Range: 0—900 feet.
Native; on loamy banks in woods and thickets, rarely in the open, infrequent. P. June
First record: Gough, 1789.
M. Greenfield district; *J. Whitehead.*
L. Lawkland woods; *W. Howson.*

R. High Ridge Wood, and Hollin Hall Wood, Settle; *J. Windsor.* Catterick Force; *W. Whitwell.* Slaidburn!
Y. Hackfall (!) Studley, Mackershaw Woods; *W. Brunton, O.B.G.* Morcar Wood, Copgrove; *J. Dalton, O.B.G.*
W. In Stockeld Woods; *Gough's Cam.* (1789). Wood by pool east of Bardsey! Bolton Woods! Thorparch; *H. Baines.*
A. Skipton Woods; *W. Whitwell.* [Kirkstall Abbey ruins] and woods about Leeds; *W. Wood* in *O.B.G.* Esholt; *W. West.* Ardsley; *G. Roberts.* Moortown! Common about Bradford; *J. Willis.*
C. Haw Wood and Langley Wood, Wakefield; *T. W. Gissing.* Halifax; *H. Baines.*
D. Rotherham district; *L. Langley* (1828). Conisborough district; *Y. N. Excurs.* (1873). Old Park Wood, Sheffield; *J. Salt.* Endcliffe Wood; *W. W. Newbould.*
T. Braithwell Woods. Kings Wood, Roche Abbey!

588. Veronica scutellata L.
Area: 63, 64, Top. Bot. Range: 0—1,350 feet.
Native; in ditches, swamps, and by pools, on a peat soil, infrequent. P. June-July.
First record: Salt, 1800.
L. Garsdale. North slope, Bowland Knotts! Cocket Moss; *W. Howson.*
R. Wigglesworth Tarn (!); Foredale Osier-beds, etc., near Settle; *J. Windsor.*
Y. Fountains Hall; Sharow Mires; Pilmoor Car, Sharow (!) etc.; *H. H. Slater.*
N. High Harrogate; *E. Hargrove* (ed. 1809).
W. Rombald's Moor, Askham Bogs (!) *H. Baines.* Near New Wells, Ilkley; *J. Willis.* Ing fields between Carlton and Cononley; *W. Whitwell* (1884).
A. Malham Tarn bog! Ditch near road, Adel Dam! Yeadon Dam; *J. Willis.* Thorp; *G. Roberts.* Bingley; *W. West.* Rawcliffe Rabbit Hills; *H. F. Parsons.*
C. Stream below Gorple water, rare; *A. Stansfield* (1877). Cob Clough, Ripponden; *R. Leyland.*
D. Askern district; *Y. N. Excurs.* (1873). Crooksmoor; *Herb. Salt.* Aldwarke Car; *G. Harrison*; *A. Carr*!

589. Veronica Anagallis L.
Area: 63, 64, Top. Bot. Range: 0—1,250 feet.
Native; watery places, infrequent, with a much wider range in West Yorkshire than in North Yorkshire. P. July-August.
R. Giggleswick and Rathmell; *J. Windsor.*
Y. By brook entering Yore at Hutton Conyers; *H. H. Slater.*
W. Wetherby! Askham Bogs (!) *G. Webster.* Acomb; *W. Whitwell.*
A. Below Attermire (Victoria) Cave; *J. Windsor.* Malham (!) *W. West.* Yeadon; *Id.* Airton and Skipton; *J. Nowell.* Burton Salmon.

C. Ryhill and Hiendley; *G. Roberts.*
D. Askern! Rotherham district; *L. Langley* (1828). Brook at Robroyd, Barnsley; *H. Payne.* Goole Fields; *H. F. Parsons.* Conisborough; *Herb. Salt.* Doncaster Carrs. Rossington; *G. Harrison.*

590. Veronica Beccabunga L. *'Brooklime.'*
Area: 63, 64, 65, Top. Bot. Range: 0—1,500 feet.
Native; in rills and ditches, very common. P. June-August.
First record: Bolton, 1775.
Distribution general; in every district, ascending to the plashy spots about the springs high up Ingleborough and Widdale Fells.

591. Euphrasia officinalis L. *Eyebright.*
Area: 63, 64, 65, Top. Bot. Range: 0—2,100 feet.
Native; heaths, dry pastures, lane-sides, very common. A. June-September.
First records; Bolton, 1775; Salt, 1800.
Distributed generally, throughout every district, although more abundant on the hill pastures of Craven and in the turf of old quarries on the Permian limestone. It varies little in West Yorkshire; the glandulous *E. montana* Jord. I have not seen in the Riding. Our common form is *E. nemorosa* H. Mart., but we have also the forms *E. ericetorum* Jord., and *E. rigidula* Jord., which range under it according to the Townsendian arrangement of this 'polymorphic species' in *Journ. Bot.*, vol. xxii, p. 161 (1884).

Var. **gracilis** Fr.
This slender, rigid, slightly branched form with small flowers occurs on the turfy mountain pastures at the head of Ribblesdale and Garsdale! It seems to have a decided preference for wet, deep peat. It stands on record also from Malham and Otley Chevin (*Miall's Flora*)—the former of which is probably correct, but the plant of the latter spot is certainly only *E. ericetorum.*

592. Bartsia Odontites Huds. *Red Eyebright. 'Cock's-comb.'*
Area: 63, 64, 65, Top. Bot. Range: 0—1,250 feet.
Native; cornfields, damp pastures, and waysides, common. A. July-August.
First record: Salt, 1800.
Distribution general; occurs in every district in the type-form *serotina* Reich.

Forma **O. verna** Reich. I only know as a West Riding plant from once gathering it (in 1879) on a mountain pasture near Linn Gill, Ribblehead, but it will doubtless be found to occur more frequently if specially looked out for. I have gathered it at Countersett, near Semerwater, only some ten miles outside our limits in some plenty.

593. Bartsia alpina L.
Area: 64, Top. Bot. Range: 900—1,250 feet.
Native; peaty mountain pastures, very rare. P. July.
First record: W. Wilson and W. Borrer, 1842.

A. Malham Moor; spn. in *Herb. Borrer*, collected by *W. B.*, May 1842; *W. Howson's Flora* (1850). Marshy ground near Malham Tarn; *W. Wilson* and *J. Tatham* (*Baker's Supp.* 1854). In 1867 still in moderate plenty between Gordale and Malham Tarn; *J. Windsor*. Moor above Gordale Scar going towards the Tarn, odd plants over a mile of ground (1870)!

Obs.—I do not know who first actually detected this plant on Malham Moor, and have been unable to find any record telling me whether Wilson first found it and showed it to Borrer, or whether first seen by the two in company.

594. Pedicularis palustris L.
Area: 63, 64, Top. Bot. Range: 0—1,500 feet.
Native; in swamps and peat bogs, common. P. July-August.
First record: Salt, 1800.
General in its distribution through the river basins, from such low levels as the Thorne and Doncaster Carrs, Askern and Askham bogs, up to the peat bogs of Bleamoor, Bowland Knotts, and Malham Tarn.

595. Pedicularis sylvatica L. *Lousewort.*
Area: 63, 64, 65, Top. Bot. Range: 0—1,500 feet.
Native; damp pastures, peaty moorlands, etc., very common. P. May-July.
First record: Salt, 1800.
Distribution quite general; in every district, but growing rare, like the preceding, near the larger towns, owing to drainage, etc. I can remember it on Hunslet Moor, Leeds, as late as 1860 or thereabout.

596. Rhinanthus Crista-galli L. *'Henpenny.'*
Area: 63, 64, 65, Top. Bot. Range: 0—1,250 feet.
Native; meadows and pastures, commonest on barren land (or the *cause* of the barrenness in grass-production, it being a root-parasite?) and said to be killed off by liming and fostered by manuring; very common, although erratic, a field full of it one year will the next have little or none; and of two adjacent pastures similar in character one may be largely infested and the other free of it. A. Mid May-July.
First record: Salt, 1800.
Distribution general—every river-basin, qualified by what is said above.

VAR. **major** Ehrh.
Pedic. major angustifolia ramosissima, fl. luteo, labello purpureo; Ray.
Area: 63, 64, Top. Bot.
An impermanent parasitic Colonist, occurring *for a few years in fields newly reclaimed* from the moors, chiefly potato or turnips (? because mostly first planted with those crops). Commoner formerly than now, when so little reclaimable arable land remains. A. Late July to end of August.

(Y. Pilmoor Carr near Sharow, boggy fields near Galphay, etc. I fear an error here.)

[N. Within a mile of Boroughbridge amongst the corn, in the wayside from Knaresborough thither; afterwards . . . in great plenty in the large cornfields betwixt Wetherby and Cattall; *Ray, Syn.*, iii*, 284 (1724).]

W. Bramham Moor; *H. Baines.*

[A. Potato fields, Adel Moor, for some years after 1868, as the boys of the *Reformatory* there brought them under cultivation.]

D. & T. Thorne Waste, borders of, 1870!

Obs.—Mr. Jas. Backhouse (senr.) in *Smith's English Flora* (vol. iii, 121, 1825) remarked the occurrence of this species on Stockton Common near York, and in Durham 'in cornfields approaching to peat,' but neither he nor Sir J. E. Smith appear to have connected the occurrence with the bringing of land (mostly moor in the north) under the plough. Rev. A. Ley found it abundantly in Nairn and Sutherland a few years ago on moory land just reclaimed. In 1879 and 1880 it was equally plentiful in potato fields around Barlby, near Selby, in land that very shortly before formed the fringe of Riccall and Skipwith Commons. The early records are, of course, mainly as a *corn*-field plant, because oats and barley were the crops then tried first, almost universally, on newly-stubbed and drained land. It is not so to the same extent now.

597. Melampyrum pratense L. *Cow-wheat.*
M. latifolium luteum; B. Pin.
Area: 63, 64, Top. Bot. Range: 50—2,000 feet.
Native; woods, bushy banks, and amid heather and ling on the high moors; common. An Annual Root-parasite. July-September.
First record: Ray, Syn. (Dillenius), 1724.

M. Woods of Mersey district; *J. Whitehead.*

L. Bowland Knotts Moors. Helk's Wood, Ingleton!

R. Amid ling on Widdale Fell and Cam Fell Ends. Settle Woods; *J. Windsor.* Bolland.

Y. Kirby Malzeard. Hackfall (!) *H. H. Slater.*

N. Pateley Bridge. Scotton Banks, Knaresborough. Brimham.

W. Grassington Woods. Bolton; *W. Whitwell.* Burley. Addingham. Lindley Woods, Washburn Valley. Woodhall. Thorparch Woods; *H. Baines.*

A. Airedale woods, on grits, common. Meanwood Wood! Esholt.

C. In the wood by Dr. Richardson's house at North Bierley (near Bradford); *Ray, Syn.* iii (1724); *Martyn* (1763); *Gough's Cam.*; etc. Hebden Bridge Valley. Kirklees. Halifax. Dungeon Wood, Huddersfield! *P. Inchbald.* Chevet Woods. Bullcliff Wood near Wakefield; *T. W. Gissing.*

D. Woods by the Loxley, Greenside, and Wortley; *A. Carr.* Wharncliffe! Sheffield; *Herb. Salt.* Goole Moor, in the drier spots about trees; *H. F. Parsons.*

VAR. **montanum** Johnst.
Only a state, I think. It occurs mostly where *M. pratense* grows in open spots amid Bilberry, Heath, etc. I have seen it at Wharncliffe Crags; in Helk's Wood, Ingleton; and on the Bowland Knotts Moors.

VAR. **ericetorum** Oliv.
Another *state* rather than a permanent variety, but I have no proof to offer that the seeds if sown in rich soil in shade would produce the larger ordinary *pratense*. This variety is dwarf, almost white-flowered and hairy, but has the toothed bracteas of the type—it grows amid the Ling and Cloudberry on Cam Fell, and about the ridge dividing Langstrothdale from Ribblesdale; and on the top of Sharpah, Flasby Fell (*W. Whitwell*).

598. Melampyrum sylvaticum L.
Area: no record in Top. Bot. Range: 450—900 feet.
Native; bushy thickets on limestone among the hills, very rare. A. Late June-July.
First record: Dixon, 1791.

L. Helk's Wood, Ingleton; *H. Baines.* Feizor Wood; *R. Clapham.*

W. Wood by road near Kettlewell; *W. Horsefield* (1864). Buckden! (Thorparch Woods; *H. Baines.* Error—only *M. pratense* var. *montanum* occurs.)

C. In a lane between Chevet Gates and Chevet Moor Gates, 1791; *D. Dixon.* (Dungeon Wood, Huddersfield; *Hudd. Nat. Hist.* 'An error for *pratense*'; *P. Inchbald.*)

[**Melampyrum cristatum** L. Extinct Casual.
Wakefield, in Walton Field, among the corn, *R. Richardson*; *Ray, Syn.*, iii*, 286 (1724); *Martyn* (1763). Doubtless an error; *Baker, Supp.* I almost think more likely to have occurred as a Casual—casual introductions surely took place then as now?—than to have been a misnomer, but against this, certainly, is the fact that the record is not one by 'the careful Ray' at all, but an addition by Dillenius in the 3rd edition of Ray's Synopsis.

LII. OROBANCHACEÆ.

599. Lathræa squamaria L. *'Corpse-flower.'*
Aublatum Cordi sive Aphyllon; Ray.
Area: 63, 64, Top. Bot. Pelophilous. Range: 0—750 feet.
Native; in stiff, damp soil, in shady places, infrequent, but locally plentiful. P. April-June.
First record: Dillenius in Ray, 1724.

L. Clapdale! Cavehole Wood; *J. Windsor.* Crow Nest Wood, Feizor; *W. Howson.*

R. Occasionally, but rarely, near Settle; *J. Windsor.* Bolton-by-Bolland!

Y. Studley and Mackershaw Woods; *W. Brunton.* Hazel-clad bank near the Yore near Ripon Parks. Sleningford! Hackfall.

N. Wood half a mile from Ripley; *E. Hargrove* (Ed. 1782). Dropping Well Wood, Knaresborough; *Hargrove* (*Hist.*, ed. 1809). Plumpton; *J. Dalton.* Near Ripley; *H. Baines.*

W. Wetherby Grange Woods; *Gough's Camd.* (1789). Thereabouts still in 1880! Bolton Woods; *G. Roberts* (1883).

A. Malham Cove! Wood, Saltaire and near Seven Arches; *J. Willis.* Gilstead; *W. West.*

C. *Prope* Heptonstall; *R. Richardson* in *Ray, Syn.* iii*, 288 (1724). Hebden Valley (1879)! Ripponden, Sowerby, Warley, etc.; *R. Leyland.* Mirfield; *C. P. Hobkirk.*

D. Stainborough; *H. Payne*; *H. Johnson*! Rotherham and Warmsworth; *H. Baines.* Doncaster neighbourhood; *Dr. White* (1804)?= Edlington Wood, 1872; *G. Roberts.* Levitt Hagg! Nether Green, Sheffield, 1801, at roots of Sycamore; *Herb. Salt.* Hampole Wood; *P. Inchbald.*

T. Roche Abbey, Elm roots, abundant; *G. E. Smith, MS.*

600. Orobanche major L. *Broomrape.*
Area: 63, 64, Top. Bot. Range: 0—900 feet.
Native; at the roots of Whin or Broom, rare. P. June-July.
First records: Hargrove, 1782; Dixon, 1792.

L. Near Ingleton; *H. Baines.* [I should expect this to be *O. rubra*—the Thyme-turfed Kingsdale Scars are every way similar to those of Wensleydale and Malham.] Furzy ground near Hackfall!

N. Amongst the Broom in a dry pasture between Spofforth and North Deighton; *E. Hargrove, Hist.* (ed. 1782).

A. Near Gateforth; *H. Baines.* Roundhay limehills in 1867 (!) *J. Abbott.* Perhaps now gone. Baildon Common; *W. West*! On Whins, Lotherton Moor; *G. Webster*!

C. Lawe Hill, 1792; *D. Dixon.* [The Binns, Southowram, Halifax, now eradicated.]

D. Bank at Low Cudworth; *H. Payne.* Thorne Moor. 'In Old Park, Sheffield, plentifull'; *J. Salt* (1800).

[Orobanche ramosa L. *Branched Broomrape.*
Casually on hemp, on a manure heap in a flax-field, between Wighill and Healaugh, several plants, August 1880 ! Alien of course, but not more surprising than some other occurrences.]

601. Orobanche rubra Sm.
Area : '64 Windsor,' Top. Bot.
Native ; on exposed scar-limestone rock at an elevation of 1,500 feet, two stations, and both very scarce and uncertain there : parasitic upon Thyme, it occurs under precisely similar conditions in North Yorkshire at a level rising from 750 feet (about Leyburn) to 1,600 feet on Nappa Scars. P. Late July-August.
R. A single specimen under Attermire Cove (1811) ; *J. Windsor, Flora.* (Sparingly near Settle, Mr. Hardy, of Manchester ; *J. Windsor, Flora.*)
A. A single specimen under the rocks on the east side of Malham Tarn, in 1801 ; *J. Windsor* in *Flora Crav.*; remained unnamed until 1855—see below.

Obs.—A curiously suggestive fact may be mentioned here. In Mr. Townsend's (Woods') copy of O.B.G., against *Lathræa squamaria*, is written in Dr. Windsor's own neat hand : 'In the farther east rocks in the middle fissure above Malham Tarn, J. Winsor.' This entry does *not re-appear* in identical words, in Dr. Windsor's *Flora Cravoniensis* ; showing that the pallid or red withered Broom-rape he gathered in 1801 was only discovered to be not *Lathræa* subsequent to 1837 or 1838, about which time the copy was annotated by him. Mr. Baker kindly annotates the foregoing by giving the following further information : 'I believe I first pointed out to him it was *rubra*, about 1855 ; he showed me his only specimen one evening I spent at his house. He was one of the most excellent of men, but his only time for much field work was when he was an apprentice at Settle ; and all through middle life his time was quite absorbed by his large practice as a medical man in the heart of Manchester ; and when he retired, at about 70, he turned to botany again.'

[Orobanche Hederæ Duby. Incognit.
Dr. Windsor appears to have at one time conjectured that his Malham Scar and Settle plants might be *O. Hederæ*. In Miall and Carrington's Flora the entry runs '*O. Hederæ*. This or *O. rubra* was found by Dr. Windsor . . . in 1806 and . . . 1811,' with a footnote by Dr. Windsor himself : 'It is not improbable that the rocks under which these plants were found had been formerly covered with Ivy, and that these solitary specimens were only the remains of what were once more abundant. But this, I admit, is mere conjecture.' But Ivy reaches its highest limit at 1,200 or 1,250 feet in Yorkshire (nowhere reaching 1,500 as stated by Hooker in the Students' Flora, ed. iii), and the mural scars in question were never ivy-clad.]

602. Orobanche minor Sutt.
Area : 64, Top. Bot. Range : 0—300 feet.
Native ; clover fields, and in open bushy places on *Trifolium* and *Lotus*, very rare. A. July-August.
First record : Dalton or Brunton, 1805.
Y. Dunkirk Wood, Sleningford ; *J. Dalton.* Near Nunwick ; *W. Brunton* ; *O.B.G.* Field near Bishopton ; *H. H. Slater.*
W. Langwith Wood, Collingham ; *J. S. Wesley* ! 'On low plants in open place.'
A. Malham ; *W. Howson's Flora* (1850).

LIII. VERBENACEÆ.

603. Verbena officinalis L. *Vervain.*
Area : 63, 64, Top. Bot. Range : 0—450 feet.
Native ; dry waysides near villages, hedge-banks, and stony fields, rare ; nothing like the troublesome weed in fields, with us, that it is in the south. P. June-August.
First records : Hargrove, 1782 ; Dixon, 1791.
L. Near Wennington ; *W. Howson, Flora* (1850).
Y. Yore banks, not uncommon (!) *H. H. Slater.* Bishop Monckton !
N. Rufforth ; *H. Baines.* Wayside nigh the Abbey at Knaresborough ; *E. Hargrove* (eds. 1782 and 1809). Scriven, near Knaresborough ; *W. Todd* (1876).
W. Lane near Wharfe, beyond Wetherby Union House near the stile entrance to the Ings racecourse !
A. Ferrybridge district ; *Y. N. U. Excurs.* (1873). Knottingley ; *T. W. Gissing.* Wayside near Peckfield bar (!) *Joseph Blackburn.*
C. On the Stennard, Wakefield, in 1791 ; *D. Dixon.*
D. Rotherham district ; *L. Langley* (1828). Conisborough (1798) ; *Herb. Salt* ; and *Y. N. U. Excurs.*, 1879 ! Near Newhill Mill, Barnsley district ; *H. Payne.* Hexthorpe ; *H. F. Parsons.*

LIV. LABIATÆ.

604. Lycopus europæus L. *Gipsy-wort.*
Area : 63, 64, Top. Bot. Range : 0—300 feet.
Native ; wet ditch- and pool-sides, canal-banks etc., rather rare, although locally plentiful. P. July-August.
First record : Salt, 1800.
L. Near Austwick bridge ; *W. Howson.*
R. Near Giggleswick Mill and Town's beck ; *J. Windsor.* Near Ribble, Wigglesworth !
Y. Sharow Mires near Ripon (!) *H. H. Slater.*
W. Askham Bogs ! Moss Carrs and Bardsey, near Wetherby !

A. Leeds ; *H. Denny* in *N.B.G.* By canal between Hunslet and Methley ! Gargrave to Leeds, by canal ! By canal between Shipley and Bradford ; *J. Willis.* Gilstead ; *W. West.* Black-hill dam, Adel, near Leeds !
C. Kirkthorpe and Stanley ; *T. W. Gissing* ; *G. Roberts.*
D. Sheffield, frequent ; *J. Salt.* Barnsley ! Don banks, Holmes ; *G. Harrison.* Goole ; *H. F. Parsons.* By canal, Thorne !

605. Mentha rotundifolia L.
Area : '63 Inchbald, 64 Tatham,' Top. Bot.
Denizen ; river-banks and pool-sides, rare and always an Escape within our area. P. August-September.
First record : Dalton, 1815.
R. Ribble banks near the old Snuff Mill (1832) ; *J. Howson* in *Windsor* ; rep. *H. Baines*, etc.
Y. Studley Woods ; *Rev. J. Dalton* (*O.B.G.*).
D. Adwick-le-Street ; *P. Inchbald* (1856).

VAR. **alopecuroides** Hull.
A. By pond near Pollington, Snaith, 1882-4 ; *P. F. Lee, spn.* ! Later Mr. Lee learnt it was the product of roots out-thrown from the Vicarage garden. Nearly all records such as this are reported at first as wildings. Inquiry at once seldom fails to elicit facts that place the plant-occurrence in its true category, but with the records of by-gone years this is impossible, although one may fairly infer the bushel from the sample. Although allowed, unquestioned, a place as a Native by Mr. Baker in his Supplement to Baines, etc., I have no hesitation whatever in pronouncing *M. rotundifolia*, *M. alopecuroides*, *M. sylvestris*, and *M. viridis* as all alike non-indigenous in West Yorkshire. Mr. J. G. Baker kindly adds (in proof) '*M. sylvestris* is, I think, indigenous in the North Riding, e.g., at Wass. It grows there under similar conditions to those in which *aquatica* and *sativa* commonly occur. I know of no thoroughly good Yorkshire station for the other three ; but *rotundifolia* is truly wild in the Lakes, on Ullswater shores.'

606. Mentha sylvestris L.
Area : 63, 64 sine pers. authority, Top. Bot.
Denizen or Alien—relic of cultivation ; wet places, very rare, almost apocryphal ; if ever indigenous within our limits, clearly now extinct. I am inclined to doubt if Salt meant anything but a form of what we now call *M. aquatica*. That name does not appear in the Flora Sheffieldiensis, and for *M. hirsuta* is given a single station, 'Doncaster Car,' as if a rare plant, yet it is quite common about Sheffield now. I believe Salt meant what we now call *M. rubra* by his *M. hirsuta*, and it grows in Potteric Carrs. P. August.

Y. Mickley Barrows near Ripon ; *T. Simpson* ; *Baker's Suppl.* (1854). [I have thrice failed to find this.]
D. Banks of the river below Sheffield ; *J. Salt, Flora* (1800). At Thorne ; *Mr. Robson* ; *O.B.G.* (1805) = Ditches on the side of Thorne Moor ; *H. Baines.*
(T. Roche Abbey ; *J. H. Aveling*, add. in *Hist.*, name only, 'Horsemint,' but as no other species of the genus are named, it is open to doubt whether *M. hirsuta* was not intended.)

607. Mentha viridis L. *Spear-mint.*
Area : '63, 64 truly wild,' Top. Bot. Range : misleading.
Denizen ; ditch- and stream-sides, infrequent. P. August.
First record : Dalton and Brunton, 1805.
R. Side of Giggleswick beck, Settle. By road near Rathmell ; *J. Windsor.* Near the further Penyghent House ; *Id.*
Y. Ure banks, Ripon ; *W. Brunton* ; and near Newby (with a variety) ; *J. Dalton* ; *O.B.G.* Abundant (but an Escape) by Yore, below Ripon (!) *H. H. Slater.*
W. Wharfe side, Ilkley ; *J. G. Baker.*
[D. No record in *Salt's Flora*, and not found by *A. Carr.*]

Obs.—In 1870, with a narrower experience than I now possess, I thought this to be indigenous. I can, however, no longer consider it so. The Smithian vars., *crispa*, etc., only known in gardens, are suspiciously frequent in the older records.

608. Mentha piperita Huds. *Peppermint.*
Area : 63, 64, Top. Bot. Range : 0—450 feet.
Native (although an Escape in some stations) ; pool-sides and river-banks, rare. P. July-August.
First record : Blackstone, 1746.
R. Road-side near Beggar-wife's Bridge, Settle ; *J. Windsor.*
Y. Ditches at Little Nunwick ; *W. Brunton.* By a dubb, on the opposite side of the river to Nunwick !
W. Brick-ponds, Hob Moor (!) *H. Baines.*
A. [By the beck-side in Water Lane, Holbeck, Leeds ; *Blackst., Spec.* (1746) ; *Martyn* (1763) ; *Gough's Cam.* ; etc. Interesting now only in a historical sense, the 'beck' long a covered-in goit amid mills and streets.] Road-side, Airton ; *J. Windsor.* South Milford, pond-sides !
C. Heath and Purston ; *T. W. Gissing.* Near Halifax *J. Bolton.*
D. In a narrow lane a mile from Pontefract in the road from Snydale ; *Sir T. G. Cullum* (*O.B.G.*). Pond-side in Carlton Lane, near Barnsley ; *H. Payne.* Potteric Carrs ; *E. Miller* (1804). Road-side between Wadsley Bridge and Ecclesfield ; *A. Carr* ! Wardsend Weir, Mark Lane, Ringinglow, near Greenside, and by stream in Storr Valley, near Rowell Bridge, all near Sheffield ; *Id.*

609. Mentha hirsuta L. (*M. aquatica* L. excl. *pubescens*). '*Horsemint.*'
M. aquatica sive Sisymbrium ; Ger. Em.
Area : 63, 64, 65, Top. Bot. Range : 0—1,250 feet.
Native ; swamps, ditches, stream-sides, very common. P. August-September.
First records : Merrett, 1666 ; Bolton, 1775 ; Salt, 1800.
Distribution general—the prevailing form in Districts L., R., Y., N., W., A., C., D., and T.
M. Mersey tribs. district ; *J. Whitehead.* The *only* Mint seen in this area by him.
Obs.—Forma 'tota nigra . . on Ingleborough hill-side towards Settle'; *Merr., Pin.*, 76.

VAR. **citrata** Ehrh.
Ingleton vicinity ; *S. Thompson* and *J. Backhouse* in *Baker's Supp.*

610. Mentha sativa L. (*M. eu-sativa* Hooker's Student's Flora.)
Area : 63, 64, Top. Bot. Range : 0—450 feet.
Native ; marshes, ditch- and stream-sides, common. P. July-August
First record : Salt, 1800.
R. Settle, waste ground by Ribble below the Long stream ; *J. Windsor.*
Y. By the Yore about Newby, and ditches at Boroughbridge !
N. Bilton and Cowthorpe !
W. Langwith Wood by the Wharfe, Wetherby, and Stutton Carrs. Askham Bogs !
A. Skipton ! Baildon ; *W. West.* Belcombe brook, Bradford ; *J. Dalton*, sub nom. '*M. pratensis* Sole.'
C. Coxley Dam and Valley ; *P. F. Lee* !
D. Newbridge, near Goole ; *H. F. Parsons.* About Conisborough ; *J. Salt.*
T. Near Kiveton, by canal ; *A. Carr.* Roche Abbey Valley !

A VAR. **'gracilis'** 'under E. Pollard's court wall, Giggleswick'; *J. Windsor, MS.* in *Towns. O.B.G.*, appears as '*Cardiaca*,' 'by the side of Giggleswick beck,' in *Flora Cravoniensis.* Very likely it was an Escape, as the Basil Mint used to be extensively grown in gardens.

611. Mentha rubra Sm.
Range : 0—450 feet.
Native ; river-banks and ditch-sides, rare. P. August.
First record : Dalton, 1805.
R. Waste ground by the Ribble at Settle below the Long stream, and by Giggleswick beck ; *J. Windsor.*
Y. Common near Ripon ; *J. Dalton, O.B.G.*
W. By the Cock at Stutton, and Wharfe side below Tadcaster !
D. Potteric Carr, Doncaster ; *H. Baines.* Road-side near Ecclesfield ; *A. Carr* ! By the Rivelin ; and in ditch by footpath from Seeman's nursery to Ecclesfield Road ; *Id.*

612. Mentha gentilis L.
Area : no record in Top. Bot. Range : uncertain.
Native ; wet hedge-banks and by stream-sides, rare. P. August.
First record : Salt, 1800.
R. (In the wet ditch between Giggleswick and Beggar's Wife Bridge, Settle ; *H. Baines, Flora.* This station is allocated to *M. piperita* by *J. Windsor* in *Flora Cravoniensis*). Waste ground below Birkbeck's Weir, and also below the Long Stream, Settle ; *J. Windsor.*
D. About Conisborough ; *J. Salt, MS. Flo. Sheff.* By river Don, Thorne ; *H. F. Parsons* : with some little doubt expressed as to correctness of designation.
Obs.—I have not been able to verify the names of the Mints assigned to the above and some other stations. Of the *sativa* forms I know only *eu-sativa* and *rubra* in the Riding.

613. Mentha arvensis L. *Field Mint.*
Area : 63, 64, Top. Bot. Range : 0—1,250 feet.
Native ; arable fields and waste ground, common up to 900 feet, but adventive in two or three places at a considerably higher level. P. August-September.
First record : Salt, 1800.
Distribution general—recorded for every district.
Near Capon House, Malham Tarn ; *J. Windsor* (1865).

614. Mentha Pulegium L. *Pennyroyal.* '*Water Horehound.*'
Area : '64 Leefe spn.,' Top. Bot. ; but Mr. Watson informed me, upon my asking him for the station upon Mr. Leefe's label, that it was near Great Ouseburn, but upon the *east* side of the river and so mis-localised in '64.'
Native ; pool-sides and marshes, very rare. P. July-August.
N. Moor Monckton (in the Ainsty) ; *J. G. Baker.* (Was Mr. Watson mistaken ? The village here named is on the west side of the Ouse.)
D. Near Beighton ; *J. Salt, Flora Sheffieldiensis* (1800). Potteric Carrs ! Doncaster, or York neighbourhood ; *Dr. White* in *Miller* (1804).

615. Thymus Serpyllum Fr. *Wild Thyme.*
Serpyllum hirsutum ; Ger. 571.
Area : 63, 64, 65, Top. Bot. Range : 0—2,000 feet.
Native ; dry banks and stony, turfy ground, generally common in limestone districts abundant, over sandstone areas rather scarce. P. July.
First records : Merrett, 1666 : Salt, 1800 ; Bolton, 1775.
Distribution almost general ; every district except one.
At Bradford, in Yorkshire'; *Merrett*, 112.
No record for Mersey tributaries district.
Obs.—This species and *Origanum* (Calamints also ; *J. G. B.*) are often dimorphic, producing small wholly female and larger bisexual flowers in the same way as *Nepeta Glechoma* does.

[**'Serpyllum magnum latifolium,** a mile off Bradford in the way to Haworth, and at Saresden betwixt the house and the pond, in Yorkshire'; *Merr., Pin.*, 112 (1666). I know not what to make of this.]

616. Thymus Chamædrys Fr.
Area : no record in Top. Bot. Range : 0—200 feet.
Native ; dry banks, calcareous mostly, at a low level, rare. P. July.
First record : the Author.
W. Dry sunny bank (limestone) between Collingham and Wetherby !
T. Warm rocks near Maltby !

617. Origanum vulgare L. *Wild Marjoram.*
Area : 63, 64, Top. Bot. Sub-xerophilous. Range : 0—900 feet.
Native ; dry sandy or calcareous banks and thickets, common. P. July.
First records : Bolton, 1775 ; Salt, 1800.
L. & R. Many places about Ingleton, Feizor, Giggleswick, and Stackhouse ; *J. Windsor.*
Y. & N. Unusually abundant (*H. H. Slater*) on the Ripon and Knaresborough limestone tract.
W. Grassington ! Bolton ; Ilkley ; *W. West.* Collingham, Thorparch, Tadcaster, etc., on the Permian tract.
A. Abundant about Kippax, Garforth, Brotherton, and Pontefract. Malham (!) and Gargrave ; *W. West.* Skipton Castle ; *H. Baines.*
C. Exley, Halifax ; *J. Bolton.* Wakefield, to the east ; *T. W. Gissing.*
D. Plentiful about Went-vale, Warmsworth, etc. Rotherham district ; *L. Langley* (1828). Conisborough, 1798 ; *Herb. Salt.*
T. Roche Abbey ! Kiveton ; *A. Carr.* Firbeck.

618. Calamintha Clinopodium Spenn. '*Basil.*'
Area : 63, 64, Top. Bot. Range : 0—900 feet.
Native ; hedge-banks and thickets, common, and not confined to limestone. P. July-August.
First record : Salt, 1800.
L. Kelcowe, etc., about Settle ; *J. Windsor.*
R. Ingleton. Dentdale !
Y. Copt Hewick, etc., near Ripon ; *H. H. Slater.* Bishop Monckton. Farnham !
N. Cowthorpe. Green Hammerton. Scriven and Bilton Banks !
W. Grassington ! Bolton Woods ! Pool ; *W. West.* Abundant about Collingham, Thorparch, etc. Aberford to Garforth !
A. Skipton !
C. Wakefield, common ; *T. W. Gissing.*
D. Hedges between Sheffield and Brightside ; *J. Salt.* Womersley ; *W. West.* Melton ; *G. Harrison.* Rossington !
T. Plentiful in this district.

619. Calamintha Acinos Clairv.
Area : 63, 64, Top. Bot. Range : 0—900 feet.
Native ; dry-soiled fields and sandy road-side banks, infrequent. A. July.
First record : Johnson's edition of Gerarde, 1633.
Y. Ripon, not uncommon ; *H. H. Slater.* Burton Leonard quarries.
N. St. Helen's Quarry, Crimple Bridge between Spofforth and Deighton !
W. Bramham. Tadcaster and Thorparch ; *H. Baines.*
A. Gordale, 1885, alt. 850 feet ; *H. T. Soppitt.* I first observed it in 1626, a little on this side Pomfret ; *Johns., Ger.* (1633). Ferrybridge district ; *Y. N. U. Excurs.* (1873). Pontefract. Quarry at Fairburn ; and Ledsham ; *G. Roberts.*
D. Rossington (!) *W. Fowler.* Rotherham district ; *L. Langley* (1828). Conisborough ; *A. Carr.* Went-vale (!) *H. Baines.*
T. Firbeck ! Hatfield ; *G. Roberts* !

620. Calamintha Nepeta Savi.
Thymus Nepeta of old Floras.
Area : 64, Top. Bot. Range : 300 feet.
Native ; dry grassy banks, very rare. Treated as if commoner than the next in Baines' Flora, but it is probable Baines did not know the real thing at all. P. August.
Y. Lane-hedge, near Bishop Monckton, Ripon ! The habit of the plant found here was different to the ordinary *C. menthæfolia*—main stem procumbent with short up-curving branches and small leaves.
[A. At Cookridge (near Leeds) ; *Rev. W. Wood* in *O.B.G.* (1805). The plant was probably the next species. Near Ledsham ; *G. Roberts, MS.* By correspondence I learnt that Mr. Gissing named this : probably meaning the next species.].

621. Calamintha menthæfolia Host. *Calamint.*
C. officinalis Benth. in DC. Prod. et omn. Angl. Auct. *Melissa Calamentha* ; Hargr.
Area : 63, 64, Top. Bot. Xerophilous. Range : 50—300 feet.
Native ; dry roadside banks, stony woods, etc., unknown off the Permian limestone, infrequent upon it from south to north of the Riding. P. July-August.
First records : Bolton, 1775 ; Hargrove, 1782.
Y. Newby ; Bishop Monckton ! Hutton-Conyers ; *H. H. Slater.*
N. Abbey rocks, Knaresborough ; *E. Hargrove*, 1782.
W. Wayside, Wetherby to Collingham ! *J. Jackson.* Rigton Bank, near Bardsey (!)—a small form, formerly taken for *C. Nepeta.* Linton ; *J. S. Wesley.*
A. Sherburn ! (Ledsham ; *G. Roberts*—probably this).
C. About Halifax, very scarce, only on Clifton Common (near Brighouse) ; *J. Bolton.* Possibly non-native there even in Bolton's time.

D. Rotherham district; *L. Langley.* By road, Loversall to Wadsworth: *H. Ibbotson.* Between Doncaster and Brodsworth Wood!
T. Near Roche Abbey; *Herb. Salt.* Rocks in Laughton Valley!

Obs.—We have no true *C. officinalis* Mœnch.—the same thing according to Nyman's Conspectus as the large-flowered *sylvatica* of the South—in Yorkshire at all.

[**Melissa officinalis** L. *Balm.* Alien.

A cottager's simple, introduced into England in 1573—a native of South Europe and West Asia,—and occasionally to be found in waste places where garden rubbish has been shot, or (more permanently) upon the sites of destroyed gardens, etc., but hardly naturalised in the Riding.

N. On the cliff rocks, Knaresborough, just above the cottage gardens!
W. Hedge at Rigton (near cottage garden) with *Malva rotundifolia*!]

622. Nepeta Cataria L. *Cat-mint.* '*Nep-in-a-hedge.*'

Area: 63, 64, Top. Bot. Range: 0—300 feet.

Native; sandy or gravelly hedge-banks, by road-sides, etc., infrequent. P. July-August.

First records: Bolton, 1775; Hargrove, 1782.

Y. About Newby and Boroughbridge!
N. Knaresborough, lane near Grimbald Crag, on the York road (!) *E. Hargrove* (1782 et eds. seq.). Near Green Hammerton; *H. Baines.* Road between Spofforth and North Deighton, and on road to Braham Hall Farm!
W. Sweep Hill to Moor End. Wetherby to Thorparch! Wetherby to Tadcaster; *O. A. Moore.* Lane, Wingate Hill, above Jackdaw Crag Quarry, Stutton (!) *H. Ibbotson.* Between Tadcaster and York; *H. Baines.* By the Bishopthorpe road; *R. Whitwell* (1865)!
A. Kippax; *G. Roberts.* Between Pontefract and Purston; *J. Willis.*
C. Millthorp Green, 1790; *D. Dixon.* Halifax (Warley and Ovenden) and Hepton Bridge; *J. Bolton.* Longwood; *Hudd. Nat. Hist.*
D. Near mill in lane leading from Car House to the Car, Doncaster; *H. Baines.* Doncaster neighbourhood; *Dr. White* (1804). Rotherham district; *L. Langley.* Conisborough, plentiful; *J. Salt.* Askern; *A. Walker.* Road-side hedge near toll-gate, Lower Whiston; *A. Carr.* Adwick; *P. Inchbald.* Hampole; *A. F. Gissing.*
T. Finningley (!) *W. Fowler.*

623. Nepeta Glechoma Benth. *Ground Ivy.*

Hedera terrestris montana; Park., 677.

Area: 63, 64, 65, Top. Bot. Range: 0—900 feet.

Native; hedgebanks, woods, and shady rocks, very common, but not ascending high. P. April-June.

First notice: Merrett, 1666.

'At Wiggles-worth near Setle in York-shire, in hedges'; *Merr. Pin.*, 60; rep. *Ray, Syn.*, iii, 243 (1724).

Distribution general—occurs in every district.

A very hirsute form occurs by roadside near Tinsley Toll-bar; *A. Carr*!

FORMA **parviflora**: often taken for a variety when (as usually happens) none of larger corollas are present. The flowers are dimorphic, the larger bi-sexual, the smaller female, and the tendency towards uni-sexuality is so pronounced that where uni-sexual blooms are produced at all there are no bi-sexual flowers on the same plant.

624. Salvia Verbenaca L. *Clary.*

Area: 63, 64, Top. Bot. Range: 0—300 feet.

Native; rough dry banks, rocky pastures, rare. P. June-August.

First record: Hargrove, 1782.

Y. Fountains Abbey! Nunwick; *Miss Morton.* Ellershaw (Hill); *H. Baines.*
N. Among rocks near Priory gate; *E. Hargrove* (1782) = Abbey Gates, Knaresborough; *Id.* (eds. of 1809 and 1832). Pastures about St. Robert's Chapel, Knaresborough (1868)!
W. Lane, entrance to the racecourse, beyond Union House, Wetherby (!) *J. S. Wesley.* Victoria Bar ramparts, York; *J. Ibbotson*, etc.!
A. Near Ferrybridge; *G. Harrison*; *G. Roberts* (1872). Milford (1876); *G. Roberts.*
D. Balby to Doncaster; *J. Salt.* Brightside; *W. W. Newbould.* Smeaton Crags, Went-vale (!) *T. W. Gissing.*
T. Lindrick; *A. Carr.*

[**Salvia pratensis** L. Incognit.

Near Yore, Hutton Conyers; *Dr. Wasse*; *Baker's Suppl.* Probably *S. Verbenaca*, which is at Nunwick, close by, still.]

625. Brunella vulgaris L. *Self-heal.*

Area: 63, 64, 65, Top. Bot. Range: 0—1,800 feet.

Native; road-sides, pastures, and rocky banks, very common. P. July-August.

First record for W.R.; Bolton, 1775.

Distribution general; occurs in every river-basin.

Obs.—The etymology of the generic name is said by Hooker to be doubtful. Surely it is from the German *bräune*—the Quinsy, which it healed, hence Braunella (Dodonæus, Rivinus) the substitution of P. for B. being made, later, from considerations of euphony.

626. Scutellaria galericulata L. *Skull-cap.*

Area: 63, 64, Top. Bot. Range: 0—900 feet.

Native; marshes, pond-sides, ditches, etc., infrequent. P. July-September.

First record: Bolton, 1775.

L. & R. Ditch near Eldroth; *J. Windsor.* Rathmell; *W. Howson.*
Y. Sharow Mires, Ripon; *H. H. Slater.*
N. Pannal! Marsh and pool at Plumpton rocks! Bilton!
W. Askham Bogs (!) *B. B. Le Tall.* Healaugh Priory, near Tadcaster; *J. Jackson.* By mill-pool, Bardsey! Ilkley, Pool, and Washburn Valley (*J. Willis*).
A. Skipton, by canal! Adel Dam, Eccup, Weetwood, near Leeds! Esholt; *W. West.* Selby!
C. Todmorden; *A. Stansfield.* Linley Moor, Halifax; *J. Bolton.* Barnsley canal, Wakefield; *T. W. Gissing.* Stanley and Lofthouse; *G. Roberts.*
D. Near Jordan Dam; *J. Salt* (1800). Don banks, Holmes; *G. Harrison.* Rivelin and Rother banks; *A. Carr.* Doncaster Carrs! Goole fields; *H. F. Parsons.*
T. By stream, Roche Abbey Valley.

627. Scutellaria minor L.

Area: 63, 64, Top. Bot. Range: 0—900 feet.

Native; watery places, spongy bogs, rare. P. July-August.

First record: Bolton, 1775.

M. Gulliver's swamp, Greenfield Valley (!) *J. Whitehead.*
(W. Thorparch; *T. Scatcherd.* Not confirmed.)
A. Adel Bog, north-west of Leeds, by stream on south side! *T. Hick.*
C. Hebden Valley; *Y. N. U. Excurs.* Linley Moor, Halifax; *J. Bolton.* [Bradley Lane, near Huddersfield; *H. Baines.* Probably now gone; a most unlikely place to produce it now; *C. P. Hobkirk.*] Sheard's Dam, Huddersfield; *Hudd. Nat. Hist.* By rivulet at Haigh; *H. Payne.*
D. On wet clayey soil near Doncaster neighbourhood; *S. Appleby* (1831); rep. *H. Baines.* Walkley Bank. By the bottom dam at Redmires; *A. Carr*! Near Cliff Rocher; *J. Salt* (1800).

628. Marrubium vulgare L. *Sweet Horehound.*

Area: 63, 64, Top. Bot. Range: 0—300 feet.

Denizen; waste ground, hedge-banks, rare, and I cannot think indigenous; it was formerly much cultivated. P. July-August.

First record: Dixon, 1790.

Y. Mackershaw; *H. H. Slater.* Near Boroughbridge; *H. Baines.*
N. Green Hammerton, by the York road, a mile to the east (!) *H. Baines.* Ripley, on the way to Harrogate; *L. C. Miall.* Knaresborough; *Mrs. Berkeley.*

W. Thorparch; *J. Emmet* = Bramham Moor. Acomb; *J. G. Baker North Yorkshire.*
A. Near Pontefract; *O. A. Moore.*
C. In abundance on a small common near Horncastle Wood, 1790; *D. Dixon.*
D. Rotherham district; *L. Langley* (1828). Hedges near Parkgate near Swinton; and near Greenside; *A. Carr.* On Cadeby Common near Conisborough, plentiful; *Herb. Salt* (1798). In lane from Grange station to Ecclesfield; *A. Carr.*

629. Ballota nigra L. *Stinking Horehound.*

Area: 63, 64, Top. Bot. Sub-xerophilous. Range: 0—450 feet.

Native; hedges and bushy waste places; common on the Permian limestone, rare off it. P. July-August.

First record: Salt, 1800.

Y. Fountains, Ripon; *H. H. Slater.* Copgrove and Boroughbridge!
N. Knaresborough, common under hedges; *E. Hargrove* (1832).
W. Tadcaster road, York; *R. Whitwell* (!) and about Acomb and Copmanthorpe, on gravel! Common on the Permian belt.
A. Abundant on the magnesian limestone, from Garforth and Sherburn to Pontefract and Hillam. Methley, on sand; *G. Roberts.*
C. Wakefield district, common; *T. W. Gissing.*
D. Went-vale. Kirk Smeaton; *W. West.* Warmsworth. Brightside; *J. Salt.* Conisborough; *Herb. Salt* (1798). Hedge at Tinsley; *A. Carr.*
T. Common from Tickhill (on sand) to Anston (*A. Carr*). Roche Abbey.

630. Stachys Betonica Benth. *Sweet Betony.*

Betonica flore carneo saturo; Merr.

Area: 63, 64, 65, Top. Bot. Range: 0—1,250 feet.

Native; woods, rocky banks, and pastures, very common. P. July-August.

Earliest notices: How, 1650; Merrett, 1666.

'*Betonica flore alba.* In Broadsworth Wood in Yorkshire, plentifully'; *How, Phyt.*, 15.

'In the Cow-pastures at Garforth, Yorkshire; *Mr. Witham*; *Merr.*, 15.

Distribution general—occurs in every district, ascending to the pastures of 'Bordley Mastiles' and above Buckden.

631. Stachys palustris L.

Area: 63, 64, Top. Bot. Range: 0—600 feet.

Native; ditch-banks, badly-drained cornfields, and marshes, common. P. July-August.

First record: Salt, 1800.

Distribution general; occurs in numerous stations in every river basin, but not in any great quantity.

632. Stachys sylvatica L. *Woundwort.*
Area: 63, 64, 65, Top. Bot. Range: 0—900 feet.
Native; hedges and thickets, very common. P. July-August.
First record: Salt, 1800.
Generally distributed—in every district, and in much greater abundance than the last.

633. Stachys ambigua Sm. (*S. sylvatici-palustris* Wirt.)
Area: 63, 64, Top. Bot. Range: Low.
Native; a sterile hybrid; cultivated ground and damp hedge-banks, rare. Varying in characters—some nearer *palustris* than *sylvatica*, never present without both parents near, but long enduring unchanged in any station, because perpetuating its like through its creeping subterraneous stolons.
First record: the Author, 1879.
N. Hedge-bank on left beyond Kirk Deighton towards North Deighton.
D. Hedge-bottom, Keresforth, near Barnsley; *H. Johnson*!

634. Stachys arvensis L.
Area: 63, 64, Top. Bot. Range: 0—600 feet.
Native; in sandy or gravelly arable fields and cultivated garden ground, common. A. July-August.
First record: Bolton, 1775.
Distribution general; occurs in every district, but no actual record for that of Mersey.
Obs.—Is this a species which, Colonist-like, is growing commoner, because now more frequently introduced with foreign seed-corn? The older writers, Bolton, and Baines in his Flora so late as 1840, enumerate special localities; to do so now would fill a page of small type. On the other hand Baines gives no stations for *Galeopsis Ladanum*, considering it too common and general; whilst, to-day, the *Stachys* occurs half-a-dozen times to the *Galeopsis* once.

635. Galeopsis Ladanum L.
Area: 63, 64, Top. Bot. Range: 0—600 feet.
Native; in dry cornfields, gravelly or calcareous, infrequent. A. July-August.
First record: Salt, 1800.
R. About Giggleswick Scar lime-kilns; *J. Windsor.*
Y. Ripon, not uncommon; *H. H. Slater.* Farnham and Flaxby!
W. Arthington! Whitwell farm fields, Linton! Thorparch; *J. Emmet.* Wetherby.
A. Garforth and Micklefield.
C. Wakefield, frequent; *T. W. Gissing.* Kirkheaton, Huddersfield; *Hud. Nat. Hist.*
D. Between Sheffield and Ecclesfield; *J. Salt.* Goole and Thorne.
T. Firbeck.

636. Galeopsis dubia Leers. (*G. ochroleuca* Lamk. *G. villosa* Huds.). *Sideritis arvensis latifolia hirsuta lutea*; Ray.
Area: '63† Leyland ms., [64 requiring verification],' Top. Bot.
Colonist; but more like a Casual in being uncertain; sandy grain fields, rare and fugitive. A. August.
First record: Ray, 1670.
In the West Riding of York, about Wakefield, Darfield, Sheffield, etc., among the corn, plentifully; *Ray, Syn.*, iii, 242 (1724); *Ray, Cat.*, 283 (1670); *Gough's Cam.* (1789); etc.
A. Gateforth, near Selby, comm. by Mr. Backhouse; *C. H. Middleton.*
C. Near Castle Hill and Hall Bower, Huddersfield; *Hud. Nat. Hist.* 'A stray introduced with seed-corn from Holland or Belgium'; *P. Inchbald, MS.* (1876). Fixby; Rastrick; Beacon Hill, Halifax, etc.; *J. Bolton* (1775).
D. Barmby Dun, near Doncaster, in a sandy field, 1885; *H. Johnson*! Cornfields, Cantley; *Gerard E. Smith.*
T. In a cornfield near Bawtry, July 1803; *Herb. Salt.*
Obs.—Ray in a letter to Dr. Richardson, p. 273-4 of Ray's correspondence, seems to imply that about 1700 this plant was deemed peculiar to Yorkshire. Yet then it could have been no more than it is now—viz., an impermanent (if recurrent) Colonist.

637. Galeopsis speciosa Mill. (*G. versicolor* Curt.). *Hemp Nettle.* *Lamium cannabino folio, fl. amplo luteo, labio purpureo*; Ray.
Area: 63, 64, Top. Bot.
Native; river-bank wastes, and potato or turnip fields, seldomer amid grain crops; intermittent, some years abundant, others infrequent, but although uncertain in its localities yet generally distributed. A. August-October.
First record: Richardson, 1724.
R. Cornfields near Settle; *Gough's Cam.* (1789). Rathmell, etc.; *J. Windsor.* Tosside Chapel.
Y. Ripon; *W. Brunton*; and Copgrove; *J. Dalton*; *O.B.G.*
N. Green Hammerton; *R. Teesdale, Linn. Trans.* (1798).
W. Acomb and Copmanthorpe. Wetherby!
A. Per totum Cravoniæ tractum; *Richardson* in *Ray, Syn.* iii, 241. In the hedges near Kighley; *Ray, Syn.* (1724); *Martyn* (1763); etc. Eccup! Pontefract; *J. Willis.* Birkin; *G. Roberts.*
C. Heath, Wakefield; *T. W. Gissing.* Brighouse; *G. L. Lister.*
D. Rotherham district; *L. Langley* (1828). Doncaster; *W. W. Newbould.* Brierley near Barnsley; *H. Johnson.* Sheffield, near the Manor; *J. Salt* (1800), confirmed 'near the Manor Coalpits, and near Handsworth Wood'; *A. Carr* (1877). Woodhouse; *G. Harrison.* Ackworth; *J. Willis.* Warplands about Thorne.
T. Hatfield Moor. Bawtry to Tickhill.

638. Galeopsis Tetrahit L. '*Dea-nettle.*' *Cannabis spuria*; Ger., 709.
Area: 63, 64, 65, Top. Bot. Range: 0—1,250 feet.
Native; corn-fields, potato-fields, and waste ground, very common. A. July-August.
First records: Merrett, 1666; Bolton, 1775; Salt, 1800.
'In Kighly, Yorkshire, plentifully'; *Merr.*, 19 (1666). Referred in *Ray's Syn.* to *G. versicolor.*
Distribution general; occurs in every district.

VAR. **bifida** Bœnn.
A slender plant with paler leaves and small variegated corolla with a notched lower lip—'has been found near Sheffield' (*W. W. Newbould*) and by myself in a sandy field near Thorne.

639. Leonurus Cardiaca L. *Motherwort.*
Area: no record in Top. Bot. Range: 0—450 feet.
Denizen; road-sides and on waste ground where rubbish has been thrown; very rare now, once commoner. P. July-August.
First record: Tofield in Hudson, 1778.
Y. Lane near Melmerby; *W. Brunton.* Not far from Grewelthorpe; *H. Baines* = Hackfall.
[N. About the village of Scotton, near Knaresborough; *J. Dalton* (1805).]
[A. About Leeds; *Rev. W. Wood* (*O.B.G.*).]
C. Plentiful near Crofton Church, 1791; *D. Dixon.* [Dungeon Wood, Huddersfield, evidently only a stray; *P. Inchbald.*]
D. Doncaster neighbourhood; *Dr. White* (1804). Limestone tract west of Doncaster; *S. Appleby* (1831) = Lane leading from the Balby road to Car House, Doncaster; *H. Baines.* (Another of Tofield's introductions, like *Marrubium*?)
T. On rubbish and hedges between Tickhill and Worksop; *Huds.* (1778); *Gough's Cam.* (1789). Between Bawtry and Tickhill: *G. E. Smith* (1876), *MS.*
Obs.—Formerly much used by herbalists and dames as an expectorant; obsolete now, seldom grown and so seldom escaping. In most of the old stations it has become extinct.

640. Lamium amplexicaule L. *Henbit.*
Area: 63, 64, Top. Bot. Range: 0—600 feet.
Native; sandy or loamy arable fields, waste ground, etc., infrequent. A. May-August.
First record: White, 1804.
R. Upper Settle Green; *J. Windsor.*
Y. Ripon, not uncommon; *H. H. Slater.*
W. York (and Doncaster) district; *Dr. White* (1804). Ainsty Fields; *G. Webster.* Healaugh, near Catterton!
A. Methley. Pontefract; *G. Roberts.*

C. Sandal; *G. Roberts*!
D. Rotherham district; *L. Langley* (1828). Doncaster; *J. Willis.* Road-side near Hillsborough, Sheffield; *A. Carr.*
T. Anston; *P. F. Lee*! Roche Abbey; *J. H. Aveling, Hist.*

[**Lamium bifidum** Cyr. (*L. longiflorum*?) Alien.
A pretty, eastern Casual, annual Dead-nettle, with a large corolla throat and a tuft of silky hairs on the helmet, found rather plentifully in 1886, in the field at Three Houses, by G. Roberts!]

[**Lamium intermedium** Fr. Incognit.
Said, in *Yorkshire Naturalists' Recorder*, p. 128 (1873), to have been collected by T. H. Bartlam on Almondbury Bank, Huddersfield. Perhaps not a misnomer, but the record requires confirmation, as there is no other record, either for West Yorkshire, or for any other division of the county.]

641. Lamium incisum Willd.
Area: 63, 64, Top. Bot. Range: 0—600 feet.
Native; in waste ground and in arable land, infrequent. A. May-Sept.
First record: Baines, 1840.
R. By Ribble at Cammock, and Langcliffe, Settle; *J. Windsor.*
N. & W. Ainsty fields; *G. Webster.* Wetherby—once seen in fallow stubble field near Woodhill!
A. Sandy fields about Hambleton!
T. About Crowle; *H. F. Parsons*!

642. Lamium purpureum L. *Red Dead Nettle.*
Area: 63, 64, 65, Top. Bot. Range: 0—1,350 feet.
Native; in fallow and other cultivated fields, lane-sides, on rubbish heaps, as a garden weed, etc., very common. A. April-September.
First West Riding record: Salt, 1800.
Generally distributed; in every district.

[**Lamium maculatum** L. Alien.
Pretty well established on shady banks, in a few places. P. June-Aug.
R. An escape at Settle; *H. F. Parsons.*
A. Stream-banks below Malham village.
D. Opposite a farm-house, in Ecclesfield road from Wadsley Bridge; *A. Carr*!]

643. Lamium album L. *White Archangel.*
Area: 63, 64, Top. Bot. Range: 0—600 feet.
Native; in hedges, foot of walls by waysides, very common at low levels, growing rapidly scarce above 450 feet. P. March-October.
First record: Salt, 1800. 'About Sheffield Manor, plentiful.'
Generally distributed; occurs in every district.
A variety with an unfamiliar facies, due to unusually deeply serrate leaves, and lax flower-whorls, was gathered by P. F. Lee in 1886, in an old quarry at Sandal; perhaps a Casual.

644. Lamium Galeobdolon Crantz. *Weasel-snout. Archangel.*
Area: 63, 64, Top. Bot. Range: 0—750 feet.
Native; damp woods, on clay or calcareous marl, common. P. Early June.
First record: Salt, 1800.

Y. Very common about Ripon; *H. H. Slater.*
N. Abbey rocks, Knaresborough; *E. Hargrove* (1809 and 1832). Pateley Bridge; *W. West.*
W. Grassington! Ilkley; *W. West.* Collingham, Wetherby, Thorparch, Bramham, etc.
A. Belbank, Bingley; *H. Baines.* Esholt; *W. West.* [Batty Wood, Woodhouse Ridge, Leeds; almost gone in 1870]. Gilstead; *J. Willis.* Kellington; *H. F. Parsons.*
C. Kirklees Woods; *Y. N. Excurs.* (1873). Wakefield; *T. W. Gissing.*
D. Doncaster neighbourhood; *Dr. White* (1804). Common near Sheffield; *J. Salt.* Rivelin Valley! Parker Spring; *T. Birks.* Cliff Rocher; *A. Gatty.* Conisborough and Sprotborough! Beeley Wood, Whiteley Wood, Greenside, and Wincobank; *A. Carr.*
T. Maltby Woods.

645. Ajuga reptans L. *Bugle.*
Area: 63, 64, 65, Top. Bot. Range: 0—1,350 feet.
Native; in woods and wet pastures, common. P. May-July.
Earliest notices: How, 1650; Merrett, 1666.
'In the Wood-close belonging to my father's house at Methley, *Mr. Witham*'; *Merr.* 17 (1666).
'Bugula flore alba. In a close of Sir Edward Rodes at Haughton in Darfield parish'; *How, Phyt.* 17.
Distribution general; occurs in every district.

[**Ajuga Iva** Schreb. Alien.
An inconspicuous South European weed sent from the field at Sandal which has yielded so many other unfamiliar things, by Mr. P. F. Lee, in 1886!]

646. Teucrium Scorodonia L. *Wild Sage.*
Area: 63, 64, 65, Top. Bot. Range: 50 to 1,500 feet.
Native; stony and bushy hedge-banks, heaths, etc., common and generally distributed, though rather scarce in some of the more highly cultivated parts of District T.
Seen in every district by myself, but distinctly less prevalent in the calcareous tracts.

[**Teucrium Scordium** L. '*Water Calamint.*'
Potteric Carrs; *E. Miller's List* (1804). A likely place, but no reputable observer has confirmed the record.]

[**Teucrium Chamædrys** L. Incognit.
This name appears in the list of York and Doncaster district plants by *Dr. White*, in *Miller's Hist. Donc.* (1804). I know no more of it. Probably growing on some old wall at York in White's time.]

[HYDROPHYLLACEÆ.

Phacelia tanacetifolia Bth. Alien.
This alien annual, a native of California (occasionally grown in gardens) has occurred as a Casual in a field at Methley (*G. Roberts*), and amid cinders on the railway track between Wetherby and Spofforth in 1881.]

LV. BORAGINACEÆ.

647. Echium vulgare L. *Viper's-bugloss.*
Area: 63, 64, Top. Bot. Range: 0—450 feet.
Native; banks, rough waste ground, quarries and fields, general on limestone, but also on gravel and drift, in the eastern half of the Riding, locally common, on the coal measures and in the hill valleys almost unknown, and a fugitive Casual where it does occur (in cultivated fields) in such districts. B. July-September.
First record: Bolton, 1775.

Y. Below Hewick Bridge (!) and Thieves' Gill, Ripon; *H. H. Slater.*
N. Grimbald's Crag, Knaresborough! St. Helen's Quarry, North Deighton.
W. Collingham; *J. Emmet.* Jackdaw Crag, Stutton! Thorparch; *H. Baines.* Bolton Abbey; *Id.* In a timber yard by railway inside York City Walls*, 1880; *W. Whitwell.*
A. Huddlestone; Sherburn; Brotherton Quarries; Pontefract. Ledstone; *T. W. Gissing.* Clover field, Methley *G. Roberts.*
C. Growing on the top of the ruins of Sandal Castle in 1790; *D. Dixon.* [Warley, Halifax; *J. Bolton* (1775). Gone.] Crofton, Featherstone, and Horbury; *T. W. Gissing.* [Fixby*; *C. P. Hobkirk.*]
D. Near Doncaster; *E. Miller* (1804). Still in plenty on the sandy railway embankments and gravel rises overlooking the lower-lying 'Carr' grounds, between Doncaster and Bawtry and Finningley! Rotherham district; *L. Langley* (1828). Wadsley Forge, Sheffield, July, 1800; *Herb. Salt.* 'Not present near Goole'; *H. F. Parsons.*
T. Roche Abbey (!) *G. Harrison.* Hatfield; *G. Roberts*!

[**Pulmonaria officinalis** L. Alien.
Subspontaneous or escaped from Lodge 'borders' in the wooded gill at Wike Fields, leading to the 'Temple' Woods of Harewood Park, 1869.]

[**Pulmonaria angustifolia** L. Alien. Ambiguity.
Area: '[63],' Top. Bot.
Haw Park, near Wakefield (1857); *T. W. Gissing* in *Flora*, but no specimen seen. Of course planted, whether misnomer or not.]

[**Lithospermum purpureo-cæruleum** L. Alien.
Reported (probably in error, as it is a plant of woods and thickets), as found with exotics, on waste ground by Hirst's Corn Mill, Huddersfield (vide *Yorks. Nat. Recorder*, 1873, p. 128). Two annual species of *Anchusa* (*A. undulata* et *A. hybrida*) are occasionally found introduced with Russian or South European grain, and have somewhat the facies of this purple *Lithospermum.* The plant found is likely to have been one of these.]

648. Lithospermum officinale L. *Gromwell.* '*Grey Millet.*'
Area: 63, 64, Top. Bot. Sub-xerophilous. Range: 0—600 feet.
Native; rocky woods and stony thickets, on calcareous soil, rare, but abundant locally. P. Late May-June.
First record; Hargrove, 1782.

L. Cavehole Wood, rock fissures above, and near the cave (!) *J. Windsor* = Crow-nest Wood; *W. Howson.*
Y. Thickets, Marfield; *H. H. Slater.* Hackfall and Mackershaw!
N. Near Abbey Gates, Knaresborough; *E. Hargrove* (1782 et eds. seq.). Birkham Wood!
W. Collingham, top of Bank Wood, in open bushy places near road (!) *J. Jackson.* Thorparch; *J. Emmet.* Bramham Park, Norwood Bottoms, Thorner, and Jackdaw Crag Quarry, near Stutton!
A. [Near Kirkstall Abbey; *Rev. W. Wood.* Between Headingley and Kirkstall in 1854, since destroyed; *L. C. Miall.* Surely planted there?] Garforth and Kippax (!) *T. W. Gissing.* Ledsham (!) 1868; *G. Roberts.* Lane from Castleford to Ferrybridge; *Id.*
D. Conisborough, 1798; *Herb. Salt.* Bentley, Askern; *G. Harrison.* Brodsworth Wood, near Doncaster; *H. F. Parsons.*
T. Kings Wood, Roche Abbey!

649. Lithospermum arvense L. '*Stone-hard.*'
Area: 63, 64, Top. Bot. Range: 0—500 feet.
Native; cornfields and hedges of fields and lanes, on dry soil, infrequent. A. June-July.
First record: Salt, 1800.

R. Road-side near Settle Mills, south end of Mill Island; *J. Windsor* Adventive with grain?
N. Cornfields, Hay Park, Knaresborough; *E. Hargrove* (1832, not in 1809 ed.). Bilton Park.
W. Fields and hedge-bank by railway at Ingbarrow, Wetherby! Thorparch and Bramham; *H. Baines.* Boston Spa; *W. Whitwell*!

A. Saltaire; *J. Willis.*
C. Near old maltkiln at Horbury Bridge; *P. F. Lee*! Field and old quarry at Sandal; *G. Roberts*, etc.!
D. Askern; *Y. N. Excurs.* (1873). Doncaster! Between Sheffield and Rotherham; *J. Salt.* Rotherham district; *L. Langley* (1828). About Barnsley; *H. Payne.* Eastoft; *H. Johnson.*
T. Thorne, in cornfields bordering moor to south! Anston; *P. F. Lee*, spn.!

650. Myosotis cæspitosa Schultz.
Area: 63, 64, Top. Bot. Range: 0—1,650 feet.
Native; boggy ditches and pool-sides, common. P. June-July.
First record: Baines, 1840.

R. Wigglesworth, by the Ribble; and Slaidburn. Pot' pool in ravine on the west side of Penyghent; *W. Whitwell*!
Y. Ripon, not common; *H. H. Slater.*
W. Rombald's Moor rills; *W. West.* Stutton and Scawy Carrs. Holgate and Dringhouses; *H. Baines.* Askham Bogs!
A. By stream from Adel Dam down through Meanwood Valley! Carleton and Cononley Ings; *W. Whitwell*!
C. Wakefield, frequent; *T. W. Gissing.* [Ditch under Primrose Hill, Huddersfield; *Hobkirk* in *Miall.* Now built upon.]
D. Potteric Carrs. Aldwarke; *G. Harrison.* Dyke-sides about Thorne, plentiful (!) *W. Todd.*
T. Roche Abbey! Bawtry!

651. Myosotis palustris With. '*Water Forget-me-not.*'
Area: 63, 64, Top. Bot. Range; 0—750 feet.
Native; ditches, pool-sides, and marshy places, common. P. June-September.
First record; Salt, 1800.

M. Marked for this district by *J. Whitehead.*
L. Rare in this district, chiefly var. *strigulosa.* Clapdale. Sedbergh!
R. Settle; *J. Windsor.* Wigglesworth. Below Slaidburn.
N. Knaresborough; *E. Hargrove.* North Deighton. Plumpton. Cowthorpe; *J. S. Wesley.* Pannal, near the church; *W. D. Roebuck.*
W. Pool. Brandon, near Shadwell. Wetherby, Tadcaster, etc.
A. Esholt; *W. West.* Skipton! Bingley.
C. Coxley Valley. Colne-vale?
D. Rotherham district; *L. Langley* (1828). Sheffield; *J. Salt.* Oughtibridge; *G. Harrison.* Redmires; and by the Rivelin; *A. Carr.*
T. Bawtry!

VAR. **strigulosa** Reich. is frequent; away from the lowland marshes it is the commoner form, and ascends to 900 feet.

652. Myosotis repens Don.

Area : 63, 64, Top. Bot. Range : 0—2,200 feet.

Native ; peaty places, shady stream-sides in bogs, infrequent, but until late years often overlooked perhaps. P. June-September.

First record : Howson, 1850.

M. Boggy stream-side, Denshaw !

R. Penyghent ; *W. Howson, Flora.*

N. Field between Low Harrogate and Bilton ; *T. J. Foggitt.*

W. Askham Bogs ! Boggy rills of Barden Moor, Bolton, and Ilkley !

A. Rumbald's Moor ; *W. Whitwell.* Keighley ; *J. Backhouse.* Adel Bog, by stream (!) *T. Hick.* Lower down same stream, in Scotland Mills Wood, Meanwood !

D. Potteric Carr (!) Doncaster ; *Y. N. U. Excurs.* ! Bogs near the Lower Dam, Redmires ; *A. Carr* !

653. Myosotis sylvatica Ehrh. '*Wood Forget-me-not.*'

Area : 63, 64, 65, Top. Bot. Range : 0—1,250 feet.

Native ; woods and shady bushy places, generally common, but absent over some rather large tracts. P. May-July.

First record : Jos. Woods, 1838.

L. & R. Very fine about Ingleton, Clapdale, Settle, Catterick, Dent, and other mountain-woods of the gills.

Y. & N. Common about Knaresborough and Ripon respectively. Copgrove ; *H. F. Parsons.*

W. Grassington and Buckden ! Bolton Woods ! Ilkley ! Denton ; *H. F. Parsons.* Thorparch, Woodhall, Collingham Woods, plentiful !

A. Skipton. Bingley Woods ; *L. C. Miall.* Esholt Springs ; *A. Walker.*

C. Hebden Bridge, etc. ; *A. Stansfield.* Shibden Dale and Elland ; *H. Baines.* Fixby, Huddersfield ; *C. P. Hobkirk.*

D. Rocks, Cliff Rocher ; *A. Carr.*

T. Roche Abbey ; *Jos. Woods* (1838) ; *G. E. Smith.*

654. Myosotis arvensis Hoffm. '*Mouse-ear.*' '*Field Forget-me-not.*'

Area : 63, 64, 65, Top. Bot. Range : 0—1,250 feet.

Native ; fields, banks, wall-tops, etc., very common. A. May-August.

First record : Bolton, 1775.

Generally distributed—in all the districts.

VAR. **umbrosa** Bab.

On banks in shady places, not very uncommon, but often by-passed as 'poor' *sylvatica.* I have gathered it in the Bolton-in-Bowland gorge of the Ribble ; in Bramhope 'Staircase' Wood, and at Knaresborough, Wetherby, and Boston Spa.

655. Myosotis collina Hoffm.

Area : 63, 64, Top. Bot. Range : 0—600 feet.

Native ; dry hedge-banks, wall-tops, etc., on both sandstone and limestone, infrequent. A. April-June.

First record ; G. E. Smith, 1838.

L. Clapham to Austwick !

R. Settle ; *J. Windsor.*

Y. Ripon, common ; *H. H. Slater.*

W. Near Aberford ; *G. Roberts.* Near York ; *H. Baines.*

A. Skipton, Haw Bank.

C. Sandal ; *H. Baines* [omitted from *Gissing's Flora*]. Crosland Moor, Birkby ; *Hudd. Nat. Hist.* Coxley Valley ; *P. F. Lee* !

D. Ackworth ; *L. C. Miall.* Armthorpe ; *G. E. Smith.*

656. Myosotis versicolor Reich.

Area : 63, 64, Top. Bot. Range : 0—900 feet.

Native ; dry banks, wall-tops, sandy fields, common. A. May-June.

First record : Gissing, 1864 ; Salt omits the name in his Flora Sheffieldiensis.

Generally distributed—records from every district except Mersey, wherein it may be a real Absentee.

657. Anchusa arvensis Bieb. (*Lycopsis arvensis* L.)

Area : 63, 64, Top. Bot. Range : 0—450 feet.

Colonist, often fugitive and Casual-like ; sandy fields, infrequent. A. July-August.

First record : Salt, 1800.

Y. Ripon, local ; *H. H. Slater.* Givendale and Boroughbridge !

N. Healaugh to Marston. Acomb ; *W. Whitwell* !

W. Tadcaster ; *T. Scatcherd.* Thorparch to Collingham ; *J. Emmet.*

A. Hambleton. Selby. Field at Saltaire ; *J. Willis*='Shipley' ; *L. C. Miall.* Milford.

C. Wakefield ; *T. W. Gissing.* Sandal ; *G. Roberts.*

D. Near Barnsley ; *Salt, MS. Flora.* Rossington and Doncaster. Goole Fields ; *H. F. Parsons.* Thorne !

T. Bawtry and Tickhill !

[**Anchusa officinalis** L. *Alkanet.* Alien. Error.

Recorded in *Y.N. Recorder*, p. 203, as having been collected in the Mixenden or Ogden Valley. The plant was *Anchusa sempervirens*, as I afterwards ascertained by correspondence.]

[**Anchusa undulata** L. (*A. hybrida* Ten.?) Casual.

Introduced with grain from the east. Found in a field at Three Houses, Sandal, by G. Roberts in 1886 !]

658. Anchusa sempervirens L.

Area : 64, Top. Bot. Range : 0—600 feet.

Denizen ; waysides and lane hedges, generally near habitations, rare. P. June.

First record ; Gough's Camden, 1789 ; Teesdale, 1798.

L. & R. Sedbergh* ; *Thos. Cooper* (1870) ! Helk's Wood, Ingleton ; *J. Tatham.* Giggleswick, and in the road between Settle and Ingleborough ; *Gough's Camden* (1789) = By the road between Settle and Ingleton ; *R. Teesdale, Linn. Trans.* (1798) ; rep. *O.B.G.* = Lawkland, Austwick, and Wherf ; *J. Windsor.* Little Bank, two miles west of Settle ; *H. Baines* = Giggleswick ! Horton and Long Preston ; *L. C. Miall.* Stackhouse ; *H. F. Parsons.*

W. In a by-lane in Scarcroft village. Boston Spa ; *W. Whitwell.*

A. Malham Moor ; *L. C. Miall.*

C. Halifax, several places ; *H. Baines.* Ogden and Warley Valleys ; *R. Earnshaw.* Gomersal, by road-side ; *J. Emmet.* Sandal ; *G. Roberts.*

D. Near Conisborough.

[**Borago officinalis** L. *Borage.* Alien.

A Casual only (with us) ; in cornfields (rarely), and more frequently on rubbish heaps in back lanes, out-thrown from gardens, in which it is much cultivated in certain districts as a bees' flower.

R. Bowland ; *J. Windsor.* Giggleswick ; *W. Howson, Flora* (1850) = Roadside, left-hand from Giggleswick to Tarn Barn, 1859 ; *L. C. Miall, Flora.*

N. Hunsingore and Knaresborough, wayside heaps ; and near York ; *H. Baines.*

A. Calcareous cornfield, Brotherton !]

659. Symphytum officinale L. *Comfrey.*

Area : 63, 64, Top. Bot. Range : 0—450 feet.

Native ; stream-banks, pool-sides, bushy wet places, etc., rare, and in many of its stations a Denizen only. P. End of May-July.

First record : J. Salt, 1800.

R. Near Settle, several places ; *J. Windsor.* In two places nearly every year ; *Misses Thompson.* 'In one spot for nearly 60 years to my recollection' ; *Mrs. Thompson.* By the Ribble, Wiggles worth !

N. By stream (plentiful) running from St. Helen's Quarry, North Deighton, into the river Crimple below Newsome Bridge (!) *J. S. Wesley.* Ouse bank nearly opposite Moreby Hall ; *H. Ibbotson.*

W. Ouse ditches, York ! *H. Baines.* Hedge, Stutton, opposite Seavy-Carr Wood ! Arthington. The Green lane, Holgate* ; *W. Whitwell.*

A. Skipton Woods* ; *W. Whitwell.* Bingley ; *W. West.* Brook at Heaton, Bradford ; and by Aire, Esholt ; *L. C. Miall.* By stream below Queen Inn, Chapeltown, going to Moortown (!) *J. Abbott* ; alien there. Roundhay ; *A. Shipley.* Near Knottingley ; *H. Baines.* Ferrybridge, osier-beds !

C. Near Wakefield, doubtfully wild ; *T. W. Gissing.* [Wakefield Road, Huddersfield, with *Bidens* ; *C. P. Hobkirk.* Place now built over.]

D. Rotherham district ; *L. Langley* (1828). Near Doncaster, by Don ; *J. Willis.* Near Brightside ; *J. Salt.* Don-side, below Aldwarke ; *Geo. Harrison* ! By stream near Grange Lane ; *A. Carr.* Dikesides, Goole Fields ; *H. F. Parsons.* Thorne !

T. By Laughton Pool and other places in Roche Abbey Valley, seemingly native !

660. Symphytum tuberosum L.

Area : '64 Tatham spn.,' Top. Bot. Range : 600—300 feet.

Native ; river-banks in wet stony bushy places, rare. P. July.

First record : Tatham, ? date uncertain.

R. By Ribble in several places from Stainforth down to Sawley ! Near Settle Bridge ; *J. Tatham* ; Stainforth ; *W. Howson.* Road between Settle and Anley House ; *H. Baines.*

Y. Hackfall, and lower down Yore, in willow-beds (adventive) !

(W. Cam Fell ; *L. C. Miall.* By Wharfe between Arthington and Pool ; *T. Hartley* in *Miall's Flora.* The occurrence of this species in Wharfedale at all needs confirmation.)

[**Symphytum asperrimum** Willd. Alien.

Occasionally cultivated as a fodder-plant for cattle.

Reported from river-side in Hackfall, and road-sides near Sharow, by *H. H. Slater*, and sent to me from a field-border, Milford, in 1886, by *G. Roberts*, and from Fairburn, by *W. Whitwell.* It will probably establish itself. It has a strong general resemblance to *S. officinale*, but the leaves are not so decurrent, nor is the stem winged as in that plant.]

661. Cynoglossum officinale L. *Hound's-tongue.*

Area : 63, 64, Top. Bot. Range : 0—300 feet.

Native ; dry waste ground by road-sides, mostly on limestone, but also on gravel ; very rare within our area. B. July-August.

First records : Hargrove, 1782 ; Dixon, 1791.

Y. Road-side between Boroughbridge and Ripon ; *H. Baines.* ['Common near Bishopton' ; *Miss Plues.* Once, perhaps, but not now.]

N. Amongst the Abbey Rocks, Knaresborough ; *E. Hargrove* (1782 et eds. seq.). Gone in 1868. About Plumpton ; *H. Baines.* Between Ripley and the river Nidd ; *L. C. Miall.*]

W. Thorparch ; *T. Scatcherd* in *Miall's Flora.* Not seen thereabouts for years ; *J. Emmet.*

C. Sharleston, July 1791; *D. Dixon.* Wakefield, frequent; *Gissing's Flora*—a lapsus calami?

D. Went Vale, 1871; *G. Roberts.* Re-seen in 1878 by *H. F. Parsons.* Near Doncaster; *Dr. White* in *Miller* (1804); *E. Miller*; regathered by *Geo. Harrison* about 1862. Conisborough, near the Castle, 1798; *Herb. Salt.*

[**Asperugo procumbens** L. Alien.

A rare Casual which has occurred twice only.

R. Settle, slope to the river on the Stainforth road just above the Settle Bridge, 1878, in some quantity, 'climbing and spreading about like *Galium Aparine*'; *Miss R. F. Thompson.* Bank cut up soon after, and no re-appearance—a silk mill not far away.

C. Disused quarry, Sandal, 1886, in some plenty; *G. Roberts*!

Obs.—Miss Thompson is a daughter of the late Silvanus Thompson, of York formerly, whose name appears occasionally in the foregoing pages, and a grand-daughter of the late John Tatham.

[**Echinospermum Lappula** Lehm.

A Casual (introduced with wool-waste, its seed-envelope being burr-like) which has occurred on waste ground at Huddersfield, and by Boothroyd's Mill, Meanwood Valley, near Leeds.]

[**Amsinckia lycopsoides** Lehm. Californian Casual.

An increasing, fugitive Casual; on waste ground by rivers where the siftings of grain have been out-thrown (in fields more rarely). Of quite recent appearance.

W. Tadcaster!

A. Gilstead, Bingley; *W. West.* Mirfield. Methley; *G. Roberts*!

D. Conisborough; *W. Hasselby*; *P. F. Lee*! Near Old Goole Mill (for some years); *T. Birks*; etc.!]

LVI. PINGUICULACEÆ.

662. Pinguicula vulgaris L. *Butterwort.* '*Sanicle.*' '*Bog Violet.*'

Area: 63, 64, 65, Top. Bot. Range: 0—1,800 feet.

Native; boggy pastures and spongy rill-sides, common, especially among the hills, growing rare (doubtless gradually disappearing) in the cultivated districts. P. June.

First record: Gerarde, 1597.

In the common fields and sundry bogs, on the hils in Yorkeshire chiefly; *Park., Theat.*, 534.

M. Bogs of Tame stream district; *J. Whitehead.*

L. 'In . . Cragge close . . . upon Ingleborough fels'; *Gerarde* (1597).

R. On every bog near Settle; *W. Curtis* (1782). Common above Settle, and about Slaidburn!

Y. Rare in this district (about Kirby Malzeard and Sawley only), as in West Yorkshire we have only the lower fourth of the whole river-basin.

N. In a low and squallid meadow near Knaresborough; *How, Phyt.*, 98 (1650). Hookstone Crags; *E. Hargrove* (1809). Bogs on Gates Hill, Knaresborough; *E. Hargrove* (1832). Scotton-banks bogs! Aketon, near Spofforth!

W. Common on the moors, down to Collingham !) and Bramham (*J. Emmet*).

A. Rombalds Moor, Skipton (!) *W. Whitwell.* Bog near Addle Dam, three miles from Leeds; *Blackst., Sp.* (1746); *Martyn* (1763). Still thereabout up to 1872, at least. Malham Moor! Boggy field, Skipton Rock!

C. On the Stennard, Wakefield, in 1791; *D. Dixon.* Norland Moor, Sowerby, etc.; *J. Bolton* (1775). Still on all the Calder Vale Moors. Meltham Moss.

D. Near Doncaster; *Dr. White* in *Miller* (1804). Rotherham district; *L. Langley* (1828). Sheffield Moors; *J. Salt.* Went-vale (1871).

663. Utricularia vulgaris L. *Bladder-Wort.*

Area: 63, 64, Top. Bot. Range: Low.

Native; stagnant pools, or slow moving ditches, rare. P. July.

First record: Blackstone, 1746.

Y. Pond near Bull Close, Ripon; *Miss Morton* ?= 'Dub' by the Yore, a mile or more above Ripon Bridge! Ditches near Littlethorpe; *H. H. Slater.*

W. Askham Bog, Dringhouses ponds! Hobmoor brick-ponds; *H. Baines.*

A. On Blackmoor [near Leeds] in a great bog; *Mr. Dawson*; *Blackst., Spec.* (1746); *Martyn* (1763).

D. 'Pottery' Car, near Doncaster; *Herb. Salt.* Black dike, Great Eaa Plantation, near Rossington!

T. Drain by Torne Dike, between Brampton and Auckley.

Obs.—It has been suggested that the Askham Bog plant is *U. neglecta* Lehm.; and in the size of its parts it is certainly smaller than typical *U. vulgaris*, but it very rarely flowers (I have never seen it do so), and is probably rather in a depauperate or dying-out condition.

664. Utricularia minor L.

Lentibularia minor; Pet., Hb.; Ray. *Millefolium palustre, galericulatum minus*; Pluk.

Area: 63, 64, 65, Top. Bot. Hygrophilous. Range: 0—1,800 feet.

Native; in dikes and pools on a peat bottom, often amid *Sphagnum*, rare, but very liable to be overlooked unless in flower. P. August.

First record: Plukenet, 1696.

In agro Eboracensi; *D. Dodsworth*; *Pluk., Almagest.*, 251 (1696).

In (the west of) Yorkshire, *Mr. Dodsworth*; *Ray, Syn.* iii (1724).

R. Peaty pits on Widdale Fell Moss, near the Tarns, at fully 1,800 feet (1885).

N. [Staveley Carrs; *J. Dalton* in *O.B.G.* Now gone from there.] = Between Staveley and Boroughbridge; *Barker* in *O.B.G.* = Wet ditches near the Nidd; *Archdeacon Pierson* (in *With.* i, 1787). [In a ditch in the grounds of Francis Trapps, Esq., near Ripley lime-kilns; *E. Hargrove* (ed. of 1782).] Pool on Long flat, Knaresborough. Hutton Moor (*Pierson*); *H. H. Slater.*

D. & T. Thorne Waste (*Jos. Woods* in *Towns. O.B.G.*), in the dikes and the 'Wells'!

Obs.—'Near Selby'; *H. Baines*, refers to east side the Ouse!

LVII. PRIMULACEÆ.

665. Hottonia palustris L. *Water Violet.* '*Water Yarrow.*' *Featherfoil.*

Area: 63, 64, Top. Bot. Range: 0—300 feet.

Native: dikes and ditches of stagnant or slowly-moving water, common. P. June-July.

First records; Hargrove, 1782; Salt, 1800.

Y. Ripon, in many ponds and 'dubs' about Ripon, and towards Boroughbridge! [Staveley Cars, in most of the ditches; *E. Hargrove* (Ed. of 1782 and down to 1809).]

N. Wet field adjoining Stockwell Lane, Knaresborough; *E. Hargrove* (1832). Dikes in pastures near Scalliber on road to Wetherby, fine and abundant! *J. Jackson* (1881). Ponds between Marston and Wetherby; *H. Baines.* Near Ripley; *L. C. Miall.*

W. Seavy Carr, Stutton! Near Tadcaster; *J. Abbott*! Wetherby, near Spring Wood! Aberford; *L. C. Miall.* Newton Kyme; *J. Emmet.* Church Fenton; *W. Whitwell.* Askham Bogs (!) *B. B. Le Tall*; etc. Middlethorpe, York; *Richd. Whitwell*!

A. Fryston Hall; *G. Roberts* (1871). Osier-bed dikes at Ferrybridge! Dikes between Church Fenton and Hambleton! Carlton by Snaith; *H. F. Parsons.* Oulton; *G. Roberts.*

C. Heath, etc., Wakefield district, frequent; *T. W. Gissing.* Stanley; *G. Roberts.*

D. Rotherham district; *L. Langley* (1828). Potteric Carrs; *G. E. Smith.* Doncaster; *J. Willis.* Near Jordan Dam, Sheffield; *J. Salt* (1800). Ditch by Don near Parkgate Station; *A. Carr.* Wet meadow below Aldwarke Hall; *Geo. Harrison.* 'Marshland' district ditches, frequent; *H. F. Parsons.* Askern (!) *J. Willis.*

T. Rossington and Auckley!

666. Primula vulgaris Huds. *Primrose.* '*Simaruns.*'

Area: 63, 64, 65, Top. Bot. Range: 0—1,350 feet.

Native; shady banks, woods, etc., mostly on calcareous or argillaceous soils, common; rare, owing to uprooting, in the districts near the larger towns, and on the cold coal-measure strata.

First record: Gerarde, 1597; next, Bolton, 1775.

Generally distributed—reported from all the districts.

Obs.—The vulgar name—'Simaruns'—for this plant in the northern part of the county, appears to be an ellipsis of St. Martin's ones, or St. Mary's ones, i.e. St. Martin's or St. Mary's flower, blooming at about the time of the Saint's day (now 'Feast') as appointed. We have instances of a similar kind in the reduction of St. John, St. Maur and St. Leger, as family names, in common speech, to Sinjin, Seymour, and Silleger. The probability of this supposition is borne out by the fact that in Upper Wharfedale about Buckden, and about *Semer*-water (St. Martin's Mere) in Wensleydale—where the name obtains (as well as in the two other Seamers in Yorkshire), the churches are dedicated to St. Martin or St. Mary; although, of course, the name 'Simarun' for Primrose has now spread, just as Heggberry for *Prunus Padus* has done, beyond the centre that first gave rise to it.

Hyb. **Primula variabilis** Goupil. *Cowslip-Oxlip.*

This may be found occasionally in almost all the localities where the Primrose and the Cowslip occur in any plenty; but (as was to be expected, being a hybrid) rarely in any quantity, mostly single roots. Records are for Settle; Barden (*G. Roberts*); Rotherham district (*L. Langley*, 1828); Maltby; Conisborough; Thorner and Wetherby; Thorparch; Owston and Hampole (*A. F. Gissing*); Copgrove (*Trans. Leeds Nat. Club*); Ripon, etc.

Obs.—It was known to *Gerarde*, who in his *Herball* (1597), p. 637, describes a sort of hose-in-hose variety, 'found by Master Hesketh in a wood called Clapdale, three miles from a towne in Yorkeshire called Settle'; and the figure of this '*Primula veris Heskethi*' shows the plant to be *P. vulgaris* β of Bab. Man., Ed. viii, 290, but with sessile flowers (*W. W. Newbould*). In *Johnson's* edition of *Gerarde*, forty years later, a criticism on this sport, communicated by a Mr. Roger Bradshaghe, is inserted by Johnson. Part of it is worth quoting: 'Hesket's Primrose groweth in Clapdale. If Mr. Hesket found it there it was some extraordinary luxurious floure, for now I am well assured no such is there to be seene, but it is only cherished in our gardens. Gerard saith many of these Northerne plants do grow in Cragge Close [in Craven]. In the North, every town and village neere any craggie ground, both with us and in Westmorland, have closes so called, whereby Gerard's Cragge Close is kept close from our knowledge.' *How* (*Phyt. Brit.* p. 98), still later (1650) repeats the record, as does *Merrett*, 98 (1666).

667. Primula officinalis Scop. *Cowslip.* '*Cowstriplings.*'

Area: 63, 64, 65, Top. Bot. Range: 0—900 feet.

Native; meadows, pastures, and banks, very common in most districts, but rather rare in some parts over coal-measures and light sands. P. End of April-May.

First record: uncertain; not in Bolton's Halifax list (1775), nor in Salt's MS. Flora (1800).

Distribution nearly general; not marked as seen in Mersey district by *J. Whitehead.*

668. Primula farinosa L. *Birds'-eyes*, or *ēēn.*

Sanicula alpina minor; Lob. *P. veris flore rubro*; Ger. *Verbasculum umb. alp.*; Ray.

Area: 63?, 64, 65, Top. Bot. Range: 100—1,500 feet.

Native; wet pastures and boggy fields, mostly over limestones, confined to the N.W. half of the Riding, but within that section common. P. June-July.

First record: Stonehouse in How, 1650.

In the mountainous meadows in the West Riding of Yorkshire; *Ray, Cat.*, p. 252 (1670). Provenit in monte Englborrow in Com. Ebor: *Petiver, MS.* in his copy of *Ray, Cat.*, ed. ii, p. 242 (*Brit. Mus., Press Mark* 968, f. 8); *Ray* in *Gibs. Cam.*

L. About Ingleton, plentiful; *D. Turner* in *O.B.G.* Dentdale. Garsdale.

R. Near Settle, very common (!) *J. Windsor.* Selside to Linn Gill; *W. Marshall Watts.* Hellifield.

Y. Near Copgrove, very common; *J. Dalton* in *O.B.G.* Near Ripon (!) *W. Brunton.* Clotherholme (!) *W. Fowler.* Grantley; *H. H. Slater.* Near Kirby Malzeard.

N. Most plentifully, mixt with *Pinguicula* in a very low and squallid meadow near Knaresborough, *Mr. Stonehouse*; *How, Phyt.*, 98; rep. *Merr.*, 98 = Marshy field near Cold Cale, Farnham; *E. Hargrove* = Farnham 'Mires,' still there in 1886; *Wm. Grainge*! Near Lofthouse, Nidderdale.

W. Upper parts of Wharfedale; *W. Wood* in *O.B.G.* Buckden, Arncliffe, Kettlewell, and Grassington!! Conistone Moor; *G. Caley.*

A. Upper parts of Airedale; *W. Wood* in *O.B.G.* Malham and Gordale (!) *J. Windsor.* Malham Tarn. Haw bank, Skipton rock.

669. Trientalis europæa L.

Alsinanthemos; Ray. *Pyrola Alsines flore minor*; Merr.

Area: 63 Leyland spn., 64, Top. Bot. Range: 1,350—750 feet.

Native; stony-peaty heaths, and heathy sub-alpine woods, rare. P. Early June.

First record: Merrett, 1666.

L. Heathy Larch plantation, Garsdale!

Y. Sawley Moor, near Ripon; *W. Brunton* = Laver district; *F. A. Lees* in *West Yorkshire* = Dallow Gill; *J. Dalton* (1805).

N. Near Brimham rocks. Rocky ledges, Ravensgill; *G. E. Smith* (1845) = Moors near Harrogate; *Archdeacon Pierson.* On the right bank slopes opposite Pateley Bridge, in some abundance; *J. E. Clark* (1882). Ripley; *L. C. Miall.*

W. Brackenbrow Woods, Buckden! Blubberhouses Moor; *H. Baines.* Otley Chevin; Almias Cliff; *T. Hartley.* Harper Wood, Sutton (1860); *Miall's Flora.* Hardcastle Moor, Grassington; *Id.*

A. On the east end of Rumbles Meer, near Helwick, in the bogs amongst the rushes, above a great Brook; *C.B.* 100; *P.* 509; *Merr.* 99 (1666); 'formerly,' etc.; *Ray, Cat.*, ii, 1677; rep. cum add.: 'Invenit T. Willisel'; *Ray, Cat.* 256 (1670); *Hist.* 1110 (1688); etc. Bingley; *Martyn* (1763). Eldwick, east end Rumbles Meer; *Id.* In plenty towards the top of Rumbles Moor; *Bolton, Cat.* (1775). Prope Bingley; *Huds.*; *Gough*; etc. Still on Rumbalds Moor in 1877; *H. F. Parsons.*

C. On a moor in Bradshaw, about the pipe-clay pits; *J. Bolton* (1775). On a piece of bog land at the foot of Soil [or Swill] Hill, between Halifax and Haworth; *R. Earnshaw* (1874). Woods near Halifax; *Rev. W. Wood* [another of his loose localisations].

Obs.—It is rather singular that this plant should never have been found in the South Lune, nor yet in any part of the great Ribble with Hodder district. It is one of the few species that seem to have spread to our area after the Glacial Epoch by an 'east coast route' exclusively.

670. Lysimachia thyrsiflora L.

Area: 63? [64 insufficiently vouched], Top. Bot. Range: at 100 feet.

Native; boggy pool-sides and ditches, very rare. P. June-July.

First record: uncertain.

W. By a pool and in a watercourse in a field at Wetherby, between the cemetery and the railway, in small quantity.

(A. Willow-holt marsh, 'between Methley and Castleford, near river Aire (Fairburn side)'; *Joseph Blackburn.* This man, who was a herbalist, and had none of his plants labelled, gave me the specimen out of his herbarium in 1867, declaring he had gathered it in the above place. I failed to find it *in situ.* Still, he was a veracious working-man botanist, had not 'exchanged' at all—indeed, had no plants but rather showy ones of his own gathering, including, however, a few medical aliens (herbalists' remedies), gathered from gardens, such as Boneset, Elecampane, Hyssop, etc.—and he had, so far as I could learn, not collected at a greater distance from Leeds than a day's foot-ramble.)

D. A flowerless Loosestrife with narrower leaves than ordinary *L. vulgaris*, sent to me in a scraggy state by Mr. G. Roberts in 1886 from Barnby-Dun, is *perhaps* this.

671. Lysimachia vulgaris L. *Yellow Loosestrife.*

Area: 63, 64, Top. Bot. Range: 0—450 feet.

Native; marshes, wet thickets, pool- and stream-sides, infrequent, but abundant locally. P. July-August.

First record: How, 1650.

Y. Yore banks, near Newby! Sharow Mires; *H. H. Slater.* Boroughbridge!

N. [Knaresborough; *E. Hargrove.* Lost.] About Bilton.

W. Pool, near Arthington (in a marsh with *Typha*) (!) *J. Abbott.* Sparingly by a pool in fields on north side of road from Wetherby to Walton. Askham Bogs. Bishopthorpe Ings; *W. Whitwell.*

A. By Aire, near Castleford! Birkin; *G. Roberts.* Ferry-Fryston. Harden beck, near Bingley; *L. C. Miall.* Carlton, Snaith; *H. F. Parsons.*

C. On Thornes Common, 1790; *D. Dixon.* Near Bradley Station; *Hudd. Nat. Hist.*

D. By Shirley pool neere Rushie meer; *Heaton* in *How's Phyt.* [Shirley is two miles south of Askern]. Doncaster; *Dr. White* in *Miller* (1804); *J. Willis.* Potteric Carrs! Border of Thorne Moor. By river above Bowbridge, near Rotherham; *J. Salt.* Askern; *A. Walker.*

T. Loversall and Rossington Carrs, by river Idle! cum *W. Fowler.*

Obs.—An excellent plant for the area 'garden' of a smoky town, much better even than Golden Rod; it will defy 'blacks' and cats, and in autumn it will put forth of its golden bloom galore, and (as I think Mr. Grindon has somewhere phrased it) grin cheerfully through the iron railings when all else in its neighbourhood is looking dingy and bedraggled.

672. Lysimachia Nummularia L. *Creeping Jenny. Herb-Tuppence.*

Area: 63, 64, Top. Bot. Range: 0—600 feet.

Native; watery places, osier-holts, ditches, etc., infrequent. P. June-July.

First record: Salt, 1800.

R. Giggleswick; *W. Howson.*

N. Goldsborough; *J. Willis.* Pateley Bridge; *Y.N.U. Excurs.* (1879).

W. Hob Moor, etc.; *W. Whitwell.* Askham Bogs; *B. B. Le Tall.* Clifford and Thorparch; *J. Emmet.* Within three-quarters of a mile of Wetherby! Threshfield; *L. C. Miall.*

A. Osier-beds, Brotherton (!) *G. Roberts.* Bingley Woods: *L. C. Miall.*

C. [Warley Clough; *J. Bolton* (1775)—now gone. Marsh Wood, Huddersfield: *C. P. Hobkirk*—now gone.] Batley*; *G. Roberts.*

D. Askern; *Y.N. Excurs.* (1873). Ackworth; *Mrs. Watson.* Doncaster; *Y.N.U. Excurs.* (1879). Potteric Carrs; *E. Miller* (1804). Brodsworth! Goole; *H. F. Parsons.* Marsh at bottom of Cadeby Common, Conisborough, 1798; *Herb. Salt.* Rother banks, near Woodhouse; *G. Harrison*; *A. Carr*! Margin of Little London Dam, Sheffield; *W. W. Newbould.*

T. Thorne; *J. E. Kenyon, Bot. Chron.* (1864).

673. Lysimachia nemorum L. *Yellow Pimpernel.*

Area: 63, 64, Top. Bot. Range: 0—1,300 feet.

Native; damp woods and thickets, common. P. Late May-August.

First record: Hargrove, 1782.

M. Copses of Mersey tribs. district; *J. Whitehead.*

L. Near Ingleton, Bentham, Clapdale Woods, etc.

R. Kelcowe, Stackhouse, etc., near Settle; *J. Windsor.* Slaidburn!

Y. Kirkby Malzeard; Hackfall!

N. Low end of the Long Walk, Knaresborough; *E. Hargrove* (1782 et eds. seq.) = Wood by river, opposite side to Abbey House; *Id.* (ed. of 1832). South side river!

W. Common in woods from Buckden down through Bolton to Bramham Park Woods. West Woods. Woodhall. Stutton.

A. Skipton Woods. Bingley. By stream, Adel Moor! Bishop Wood, near Hambleton.

C. Wakefield, frequent; *T. W. Gissing.*

D. Rotherham district; *L. Langley* (1828). Edlington Wood, near Conisborough. Near Wentworth; *H. Payne.* Old Park Wood, Sheffield, 1798; *Herb. Salt.* By the Rivelin, Lower Bradfield, and Wharncliffe Wood (!) *A. Carr.*

T. Maltby Wood.

674. Anagallis arvensis L. *Pimpernel.*

Area: 63, 64, Top. Bot.

Colonist; arable land, common, but probably an increasing species. Its stations seem to have been thought noteworthy in 1775—now they are so numerous, and vary so much from year to year, that they are no longer so. A. May-September.

First records: Salt, 1800: Hargrove, 1832.

Distribution general; reported from every district.

VAR. **cærulea** Sm. Very occasional. The only records are:

C. Lofthouse; *G. Roberts.* Sandal; *G. Roberts*!

D. Rotherham district; *L. Langley* (1828). Adwick-le-Street; *P. Inchbald.* Sheffield; *A. Carr.* Conisborough; *L. C. Miall.*

VAR. **carnea** Schrank. Occasional. It stands on record for Bilton and Brodsworth; and I have seen it at Meanwood and Knaresborough.

675. Anagallis tenella L.

Area: 63, 64, Top. Bot. Range: 0—750 feet.

Native; in spongy, peaty-sandy bogs and marshy fields, infrequent. P. July.

First notice: Johnson, 1633.

In two or three places of Yorkshire; *Johnson's Ger.* (1633). Yorks., on bogs; *Gough's Cam.*

L. Bog near Clapham, going up to Bowland Knotts! Conisberwell, Clapham.
R. Near Giggleswick. Rathmell (!) *W. Marshall Watts.*
Y. Clotherholme; field where *Primula farinosa* grows! By the lake in Studley Park. Hutton Moor; *H. H. Slater.*
N. Pateley. Boggy field, right hand of Forest Lane, Knaresborough; *E. Hargrove* (1832). Scotton banks, Knaresborough. Druggen Hill; *H. Baines.* Plumpton Park, by the lake. Below Hookstone Crags; *E. Hargrove* (1809).
W. Addingham, Silsden, and Blubberhouses (lower) moor slopes; *L. C. Miall.* White Car bog, Bramham Moor! Wetherby; *L. C. Miall.*
A. Greenhill, Bingley, and Baildon Moors (*J. Willis*) and Hawksworth Wood (!) (*L. C. Miall*). Adel Bog, near Leeds!
C. Stainland Dean; *J. Bolton* (1775). [Between Ossett and Wakefield; *Miall's Flora*; *T. W. Gissing* and others could never confirm.] Hebden Bridge; *A. Stansfield.*
D. Askern! Cliff Rocher; *J. Salt* (1800). Stannington; *W. W. Newbould.* Near Bell Hagg; *A. Carr.*
T. Rossington: *L. C. Miall.*

676. Centunculus minimus L.

Area: '63, 64 Twining spn.,' Top. Bot. [I fear a mislocalisation—64 for 62—in the case of Twining's specimen]. Arenophilous. Range: 0—100 feet.

Native; in bare muddy-sandy places, where water has stood in winter, on open sandy heaths, or in shade, often in cart-tracks, very rare, but being inconspicuous and ephemeral may be often overlooked. A. June.

W. Wet stony moory lane leading from Compton bank-top to West Woods (Lady Lund's), Bramham, sparingly, but discovered when nearly over! (July, 1880).
T. Open places in Maltby Woods (!) *G. E. Smith, MS.*

677. Glaux maritima L.

Area: unrecorded in Top. Bot. Range: 0—15 feet.

Native; muddy shore and grassy or clayey bank of our main tidal river, locally abundant, growing finer and more succulent as one goes down the river. P. June-July.

First record; H. F. Parsons, 1875.

A. By river Aire, at its junction with the Ouse near Armin.
D. By the Humber and Ouse from Adlingfleet, by Ousefleet, Goole, Hook, upwards.

678. Samolus Valerandi L. *Brookweed.*

Area: 63, 64, Top. Bot. Range: 0—900 feet.

Native; sandy-bottomed ditches and pool-sides, rare. A. & P. July.

First records: Gough, 1789; White, 1804.

'Yorkshire, on bogs'; *Gough's Cam.* (1789).

Y. Hutton Moor; *H. H. Slater.* Boroughbridge; *Y.N.U. Excurs.* (1876).
N. Ditches about Bilton.
W. Askham Bogs; *B. B. Le Tall*, etc. Lane leading to the Kennels, Acomb; *H. Ibbotson.* Dike, amid *Typhæ*, by railway, between Church Fenton and Grimston Bridge! Pool-side by Wharfe below Thorparch Woods; *J. Abbott.*
A. Hamilton [or Hambleton] Dike, near Gateforth.
D. Askern (!) *J. Willis.* Boggy places, Hollow-Meadows, above Sheffield; *A. Carr*! Near Doncaster; *Dr. White* (1804); *E. Miller.* Potteric Carrs! Ousefleet; *H. F. Parsons.*
T. Rossington to Auckley!

LVIII. PLUMBAGINACEÆ.

679. Armeria maritima Willd. *Thrift.*

Caryophyllus montanus minimus; C.B.; Ray.

Area: 64, Top. Bot. Range: 900—1,400 feet.

Native; mountain pastures, very rare, although locally abundant, and having the same stational connexion with old or still-worked lead mines that *Thlaspi occitanum* and (in a less degree) *Arenaria verna* present. P. May-August.

First record: Ray, 1690.

R. Prope Hincklehaugh, and in Bleaberry-gill, at the head of Stockdale Fields, Settle; *Ray, Syn.* i, 149 (1690); *Gibs. Cam.* (1695); *Martyn* (1763); et rep. In Blaeberry Gill *under* Hinklehaugh at the head of . . . etc.; *Derh. litt.*, 214 (1718). Still there 1870! (Near Ingleborough, *Mr. Wood*; *With.* 7th ed., 405 (1830). I fancy Wood, who only copied, here, the earlier record, thought Hincklehaugh meant Ingleborough. It does not—it is the hill now called Rye-loaf, east of Settle).
W. In an elevated moist pasture a little above Stockdale, on the road to Malham from Settle, in 1865; *Windsor's Flora Cravoniensis.* Grassington Moor, about the lead-mines above Cold Conistone.

LIX. PLANTAGINACEÆ.

680. Plantago major L. *Plantain.*

Area: 63, 64, 65, Top. Bot. Range: 0—1,350 feet.

Native; way-sides and pastures, very common. P. June-August.

First record: Salt, 1800.

Generally distributed—in every division.

VAR. **intermedia** Gilib.

Not unfrequent on sandy soil. I have noted it about (W.) Copmanthorpe, (N.) Spofforth, and (D.) Thorne.

681. Plantago media L.

Area: 63, 64, Top. Bot. Range: 0—1,100 feet.

Native; dry banks and pastures, commonest on calcareous soils, but in many places on sandstone also. A. June-August.

First record: Salt, 1800. 'About Conisborough, frequent.'

Distribution nearly general—every district except Mersey.

Stations off the limestones are: Gargrave! Ilkley (adventive in alluvium by river); Hawksworth (*W. West*); Copmanthorpe, Ouse bank sands near Selby, warp-fields about Goole (*H. F. Parsons*); road-side from Sheffield to Ecclesfield (*A. Carr*); etc.

682. Plantago lanceolata L. *Ribwort.*

Area: 63, 64, 65, Top. Bot. Range: 0—1,800 feet.

Native; meadows, pastures, and sandy fields, very common. P. June.

First record: Salt, Flora Sheffieldiensis, 1800.

Distribution general—in every district.

Obs.—A curious *monstrosity* of this—a plant with many heads on a single stalk, was found by A. Carr by a footpath from Ecclesfield to Wadsley Bridge, Sheffield.

683. Plantago maritima L.

Area: 64, Top. Bot. Montane and Maritime. Range: 0—1,250 ft.

Native; stony ground among the hills, and river-banks within influence of the tide, rare, but locally abundant. P. July-September.

First record: Hudson, 1778.

R. & W. Prope Settle; *Huds., Flora Anglica*, Ed. ii, 64 (1778). In the road leading from Kilnsay to Arncliffe, and elsewhere on the mountain sides, near Settle; *W. Curtis* (1782); *Gough's Cam.* (1789). Still thereabout; *W. West* (1882). In the pastures after leaving Bordley Mastiles and beginning to descend towards Kilnsey.
D. About Goole, in many spots on the muddy and grassy bank of the Humber and Ouse, from Ousefleet upwards to above Goole (!) *H. F. Parsons.*

684. Plantago Coronopus L. *Buck's-horn Plantain.*

Area: unrecorded in Top. Bot.? Range: 0—300 feet.

Native; dry sandy ground by road-sides, etc., rare. A. July-August.

First record: Salt, 1800.

W. Aberford; *L. C. Miall.*
C. Heath, Woolley and Outwood; *T. W. Gissing.*
D. Brightside; *J. Salt* and *Herb. Salt.* About Goole and Thorne *H. F. Parsons.* Wentbridge; *G. Roberts.* Armthorpe (!) *H. Baines.*
T. Road-side at junction of two roads, north-east of Shooter's Hill, between Austerfield and Rossington! cum *W. Fowler.*

[**Plantago Lagopus** L. Casual.

C. Found in waste ground at Sandal (with other Aliens) in 1885 by *George Roberts*!]

685. Littorella lacustris L.

Area: 63, 64, Top. Bot. Range: 0—1,250 feet.

Native; shallow margins of pools, dams, etc., upon a gravelly bottom, rare. P. July.

R. Wigglesworth Tarn.
Y. Hutton Moor, Ripon; *W. Brunton* in *O.B.G.* Not seen by *H. H. Slater* (1879-83).
W. Bardsey Pool; *L. C. Miall.*
A. Shore of Malham Tarn! Grassy margin of Adel Dam! *J. Abbott* (1869). Adel Blackmoor near Leeds; *W. Wood.* Yeadon Dam; *J. Willis.* Eccup Reservoir; *L. C. Miall.*

New-Miller Dam, Wakefield! Cold Hiendley Reservoir! *T. W. Gissing* (1867). Plentiful there in 1885; *G. Roberts.*

Grassy shingly edge of reservoir at Dunford Bridge.

[AMARANTHACEÆ.]

[**Amaranthus prostratus** Balb. Alien.

Found casually a few times.

N. By pathway below Flax Mill, Knaresborough.
A. On rubbish (with exotics) by Boothroyd's Mills in the Meanwood Valley!]

[**Amaranthus retroflexus** L. Alien.

Casually on 'wool-waste,' etc., like the preceding.

C. Shoddy heap, Leeds Road, Huddersfield; *C. P. Hobkirk* and *Ingle* (1857-58). Lofthouse (1868); *G. Roberts.*]

LX. CHENOPODIACEÆ.

[**Blitum virgatum** L. *Strawberry Blite.* Casual.

Found by a Mr. James Hamilton in 'an old quarry near Wetherby' (doubtless an Escape) in September 1870, and brought to *James Abbott*, who showed it to me.]

[**Chenopodium polyspermum** L. Casual.

Area: [64 needs confirmation], Top. Bot.

A. Rubbish heap by Aire at Hunslet; *Joseph Blackburn*! Probably pertaining to wool-waste out-thrown where found.
(D. Doncaster fields; *O. A. Moore* in *Baines*, 145. Error; vide *Cyb. Brit.*, ii, 315.)]

686. Chenopodium olidum Curt.
Area: '(63), 64 'Colonist,'' Top. Bot.
A Casual or Colonist in waste ground, rare. A. July-August.
A. Weed in garden at Leeds; *W. Todd*! Leeds; *H. Denny, MS.*
C. Wakefield; *T. W. Gissing.* Sandal; *G. Roberts*!

687. Chenopodium album L. *Goosefoot. 'Fat-hen.'*
Area: 63, 64, 65, Top. Bot. Range: 0—900 feet.
Native; waste ground, arable land, etc., very common.
First record: Salt, MS. Flora Sheffieldiensis, 1800.
Distribution general; occurs in every division.

VAR. **viride** L.
W. Below flour mill, Wetherby!
A. Boothroyd's Mills, Meanwood Valley!
C. Dewsbury; *P. F. Lee*!
D. Attercliffe; *A. Carr*!

[**Chenopodium ficifolium** Sm.
Casual; on waste ground, very rare.
(W. Thorner; *L. C. Miall.* I believe this was *C. rubrum*).
C. Shoddy waste, Leeds Road, Huddersfield; *Hobkirk, Ingle,* and *Stocks,* teste *J. G. Baker.*]

688. Chenopodium murale L.
Area: 64, Top. Bot. Range: immaterial.
Colonist; on manure and in seed-fields, very rare. A. July.
W. Walls of York. Bridge near Hobmoor, on manure!
C. Canal side, below Castleford; *G. Roberts.*
D. Near Ecclesball Wood, Sheffield; *A. Carr.*

[**Chenopodium glaucum** L. Casual.
Rarely; on rich waste ground, or ballast.
W. Near York; *J. Backhouse*; *Baker's Suppl.*
A. Manure heap near Methley Railway Station; *G. Roberts*!
C. Near Pontefract; *G. Roberts.*]

689. Chenopodium urbicum var. **intermedium** M. & K.
Area: '64 'Colonist,'' Top. Bot. Range: immaterial.
Colonist; on rubbish heaps, waste ground, rare. A. July-August.
Y. Ripon (1872).
A. Garden at Meanwood, but possibly the result of seeds from herbarium swept out.
D. By the Don between Tinsley and Rotherham; *A. Carr*!

690. Chenopodium rubrum L. *Red Goosefoot.*
Area: 63, 64, Top. Bot.
Colonist; waste ground, rubbish-heaps in lanes, etc., frequent. A. August.
First record: Salt, 1800.

R. Near Settle—unconfirmed report; *J. Windsor, Fl. Crav.*
Y. Clotherholme near Ripon! Sharow; *H. H. Slater.*
W. Wetherby and Compton! Scarcroft Hill top, in lane on manure! Between Oxton and Catterton, 1880; *W. Whitwell.* Ledsham and Milford; *G. Roberts*! Selby (1876)!
C. Alverthorpe; *T. W. Gissing.* Near Oulton; *G. Roberts.*
D. Near Tinsley, Wardsend Weir, and Parkgate Station; *A. Carr*! Goole; *H. F. Parsons.*
T. Cart-track in wood at Rossington Carr! cum *W. Fowler.*

[**Chenopodium ambrosioides** L. Casual.
On waste ground by Boothroyd's Woollen Mill, Meanwood Valley (1877)!]

691. Chenopodium Bonus-Henricus L. *Good-King-Henry* [the VIth]. *'Mercury.'*
Area: 63, 64, 65, Top. Bot. Range: 0—600 feet.
Denizen; by road-sides and field-borders, common. P. June-August.
First records: Dixon, 1791; Salt, 1800.
R. Settle (!) *J. Windsor.*
Y. Ripon neighbourhood, common; *H. H. Slater*!
N. Not uncommon about Knaresborough.
W. Common about Wetherby, Thorparch, and Tadcaster. Ilkley!
A. Malham! Kildwick and Steeton; *L. C. Miall.* Horton near Bradford; *W. West.* Baildon; *J. Willis.*
C. Stanley; *T. W. Gissing.* Wibsey and Low Moor! *W. West.* Hartshead! Skircoat; *J. Bates.* Chevet Moor gate, near to Joseph Matthewman's house in 1791; *D. Dixon.*
D. Crookes and Rose Wood, Sheffield; *A. Carr.* Fulwood, plentiful; *J. Salt* (1800). Darton; *T. W. Gissing.* Marshland district, near Goole, cultivated as a pot-herb; *H. F. Parsons.*
T. Maltby and Lindrick; *A. Carr.*

692. Atriplex littoralis L.
Area: 63*, Top. Bot.
A Casual; or Denizen since established and constantly reinforced, within tidal influence. A. July-August.
D. On waste ground about the water-side at Goole; *H. F. Parsons*!

693. Atriplex angustifolia Sm.
Area: 63, 64, Top. Bot. Range: 0—1,250 feet.
Native; arable and waste land, very common. A. July.
First record: uncertain.
Distribution general; in every district, ascending as high as there are barns and patches of arable ground.

694. Atriplex erecta Huds. (*A. patula* L.).
Colonist; arable land, common, but not so abundant as the last and liking a lighter soil. A. July-August.
First record: uncertain.
Distribution general; in every district, but much less abundant in some than in others.

695. Atriplex hastata L. (incl. *deltoidea* Bab. and *Smithii*).
Area: 64, Top. Bot. Range: uncertain; but not high.
Colonist; fields and waste ground, on rubbish, infrequent. A. July.
First record: Salt, 1800, sub nom. *hastata.*
(R. Settle; *J. Windsor.* I believe *A. patula,* only, was meant.)
A. Fields at Meanwood, rare! Skipton; *L. C. Miall.*
C. Field and quarry at Sandal; *P. F. Lee*!
D. Pitsmoor, and Nunnery Coal-pits; *A. Carr.* Goole; *H. F. Parsons.*

[**Beta maritima** L. Casual.
A. Clover-field, Methley; *Geo. Roberts*! Hauling-path by river, below Castleford; *G. Roberts* (1885).
C. Shoddy heaps, Leeds Road, Huddersfield; *C. P. Hobkirk* and *Ingle* (1857).]

[**Beta vulgaris** L. *Beet.*
Occasional plants are to be met with in the fields of the lowlands, strays from, or relics of cultivation. Wetherby, Sandal and Ryhill, Methley, Sheffield, etc.]

LXI. POLYGONACEÆ.

696. Rumex conglomeratus Murr.
Area: 63, 64, Top. Bot. Range: 0—900 feet.
Native; waste ground by road-sides, ditches, etc., common. P. July.
First record; Gibson in Baines' Flora, 1840—sub nom. *R. acutus.*
Distribution general—occurs in every district.

697. Rumex nemorosus Schrad. (*R. viridis* Sibth.). *Wood Dock.*
Area: 63, 64, Top. Bot. Range: 0—1,250 feet.
Native; moist woods and shady hedge-banks, etc., common. P. July.
First record: Brunton, 1805—the var.
Distribution general; occurs in every district.

FORMA **sanguineus** L. (*Bloody-veined Dock*) is much rarer than the green-veined type.
Y. Hackfall; *W. Brunton* in *O.B.G.* (1805).
W. Barden Tower!
C. Todmorden; *J. Nowell.*
D. Scraith Wood, Sheffield; *A. Carr.*

698. Rumex maritimus L.
Area: 63, 64, Top. Bot. Range: Low—under 100 feet.
Native; in swampy places, by pools and deep ditches, rare. P. July.
A. Pool near Selby!
D. Marshland, about Goole; *H. F. Parsons*! Dyke by King's Causeway; *Id.* Adlingfleet Common ditches; *T. Birks, junr.*!

699. Rumex palustris Sm.
Area: 64, Top. Bot. Range: 0—300 feet.
Native; marshes and dike-sides, very rare.
First record: Salt, 1803.
A. Thorner Moor; *Miall's Flora.* Seacroft; *A. Shipley* (misprinted Scarcroft in *Miall's Flora*).
D. Ditch-sides (with *R. maritimus*), Adlingfleet Common; *T. Birks, junr.*! In the extreme S.E. corner of the Riding.
(T. Everton Common, August 1st, 1803; *Herb. Salt.* Rightly named? In Riding?)

[**Rumex pulcher** L. Incognit. Alien.
Given in *Dr. White's list* of York and Doncaster district plants, in *Miller's Hist. and Antiq. Donc.* (1804). Once noticed (as a Casual) on rubbish under a garden wall at Hartshead, near Kirklees.]

700. Rumex obtusifolius L. (*R. Friesii*). *Docken.*
Area: 63, 64, 65. Top. Bot. Range: 0—1,300 feet.
Native; ditch-sides and waste ground, very common. P. August.
First record: Salt, 1800.
Distribution general—in every district.
Our plant is all the type, *R. Friesii. R. sylvestris* I have not seen in the Riding.

701. Rumex pratensis M. & K.
Area: 64, Top. Bot. Range: 0—450 feet?
Native; damp ground by water, etc., very rare. P. August.
First record: uncertain.
Y. Sharow Mires, Ripon; *H. H. Slater, Flora* (1873).
(C. Kirklees district; *Y. N. Excurs.* List, *Yorks. Nat. Recorder* (1873), p. 190. Perhaps a misnomer.)
(D. Denby Dale district; *Y. N. Excurs.* List, *Y. N. Rec.* (1872), p. 51. Also erroneous probably.)
Obs.—I have seen no West Yorkshire specimen.

702. Rumex crispus L. *Dock* or *Docken.*
Area: 63, 64, Top. Bot. Range: 0—900 feet.
Native; pastures, ditch-banks, etc., very common. P. July-August
First record: Salt, 1800.
Distribution general; occurs in every district.

703. Rumex aquaticus L. (=**domesticus** Hartm.).
Area: 64, Top. Bot. Range: 100—1,000 feet.
Native; wastes by way-sides, pastures, and river-banks amongst the hills, rare. P. July-August.
First record: Joseph Woods, 1838.
L. Thornton Bridge; *J. W.* (Jos. Woods), *MS.* in *Towns. O.B.G.* Garsdale!
R. By the river Ribble at Settle; *Jos. Woods* (1838). Helwith Moss; *W. Howson.* Near Settle; *J. Backhouse.* About Wigglesworth! Near Paythorne; *L. C. Miall.*
Y. By the river Yore near Boroughbridge, and by the Swale below Myton (washed down from Upper Wensleydale, where it is common).
W. Langstrothdale and down by Wharfe to Grassington (!) probably lower.

704. Rumex Hydrolapathum Huds. *Water Dock.*
Area: 63, 64, Top. Bot. Range: 0—200 feet.
Native; dike- and pool-sides in the low country. P.
First record: Dixon, 1792.
Y. About Ripon, by pools! Sharow Mires and Littlethorpe; *H. H. Slater.* Boroughbridge; *Y. N. U. Excurs.* (1876).
W. Askham Bog dikes, very fine! Stutton. Church Fenton.
A. About Sherburn. Selby. Woodlesford; *G. Roberts.*
D. Plentiful near Smeaton, 1792; *D. Dixon.* Askern (!) *L. C. Miall.* By Don below Sheffield! Rossington Carrs. Potteric Carrs; *E. Miller* (1804)! Goole and Thorne Moor dikes; *H. F. Parsons.* Marshland.
T. Everton Common (on road to Gainsborough), August 1803; *Herb. Salt.* Canal-side, Shireoaks; *A. Carr.*

705. Rumex alpinus L.
Range: 450—100 feet.
Denizen; in waste ground, rare, but well established. P. July-August.
First record: Pinder, 1852.
W. Near Menston; *Rev. G. Pinder* (1852).
C. Springs, Harley Wood, Todmorden; *A. Stansfield* (planted).
D. Road-side, Greenland near Cowick; *H. F. Parsons* (1877).

706. Rumex Acetosa L. *Sour Dock. Sorrel.*
Area: 63, 64, 65, Top. Bot. Range: 0—1,500 feet.
Native; meadows, pastures, and banks of ditches, very common. P. June.
First record: Salt, 1800.
Distribution general—in every district; but I have not seen it at nearly so high an elevation in the West Riding as Baker gives (2,600 feet) for the North Riding.

707. Rumex Acetosella L. *Sheep's Sorrel.*
Area: 63, 64, 65, Top. Bot. Range: 0—2,400 feet.
Native; dry pastures, banks, stony moors, etc. very common, and ascending to the highest summits of the West Riding. P. June-August.
First record: Salt, 1800.
Generally distributed—occurs in every district.

[**Rumex scutatus** L. Alien.
Established in a few places among the hills.
R. About the barns at Gearstones! Dale Head (i.e., Silverdale Head, three miles N. and E. of Great Stainforth—going up to Blishmire House and the Giant's Grave); *J. Tatham.* About the barn there still! In *Herb. Tatham* are two specimens, his own collecting, one labelled 'Top of Silverdale, 1840,' the other Peter's Castle, Penyghent, undated; *W. Whitwell.* Sawley; *R. Brown.*
A. By wall on which *Draba muralis* grows at the end of Malham village; *W. West!*]

[**Oxyria reniformis** Hook. ('*R. digynus*'; O.B.G.). Incognit.
'Welsh Sorrell . . . in Westmoreland, where the curious Dr. R. Richardson tells me he has also found, as well as in Wales, but NOT as yet in Yorkshire, where he lives' (p. 378, Pet. in *Phil. Trans.*, No. 332, 1711). Compare *Pluk.*, *Almagest.*, p. 8.
Mountains in the . . . West Riding; *R. Teesdale* in *O.B.G.* (1805).
'I have a memorandum of this plant having been found on Ingleborough; but although I have ascended this fine mountain thirteen times, I have not met with it'; *J. Windsor, Flora.* Mr. Tatham's collection contains no Yorkshire specimens; *W. Whitwell.*]

[**Polygonum Fagopyrum** L. *Buckwheat.* Alien.
Planted for Pheasant-feeding, and so straying. Occasionally an odd plant occurs in corn or clover.]

708. Polygonum Convolvulus L. '*Black Bindweed.*' *Cornbind.*
Area: 63, 64, Top. Bot. Range: 0—900 feet.
Colonist; corn-fields, and sometimes in clover and on waste ground below flour-mills, common. A. June-August.
Distributed generally; in every district as high as corn-fields are found.

[**Polygonum dumetorum** L. Incognit.
Fixby, but queried; *G. L. Lister* in *Trans. Y. N. U.*, Series E, p. 34. This was the var. *pseudo-dumetorum* of the preceding.]

709. Polygonum aviculare L. *Pigweed. Knotgrass.*
Area: 63, 64, 65, Top. Bot. Range: 0—1,350 feet.
Native; waste ground, arable land, road-sides, railway-banks, etc., very common. A. April (end) to October.
Distribution general; in every district.

VARS. **rurivagum** Jord. and **agrestinum** Jord.
Common!

VAR. **arenastrum** Bor.
Rare. In sandy ground. Near York! About the collieries at Barnsley! and railway-bank at Crookes near the Nunnery Coal-pit; *A. Carr.* By river Yore, Ripon; *H. H. Slater.*

VAR. **microspermum** Jord.
In dry fields and sandy banks, rare. About Thorne Quay! Near Nunnery Coal-pit, Sheffield; *A. Carr*; and marked for Ripon by *H. H. Slater.*

[**Polygonum arenarium** W.K. Casual.
By the river Calder near the bridge at Wakefield; *T. W. Gissing* (in 1860) nom. teste *W. W. Newbould.* A large elegant plant—but only a Casual, probably brought with Hungarian grain.]

710. Polygonum Hydropiper L. *Water Pepper.* '*Arsmart.*'
Area: 63, 64, Top. Bot. Range: 0—600 feet.
Native; ditches and road-ways where water has stood, common. A. August.
First record: Salt, 1800.
Distribution general—occurs in every district, but does not ascend very high.

711. Polygonum minus Huds.
Area: 64, Top. Bot. Range: Low.
Native; marshy grassy turf, goose-greens and pond-sides, rare, and with quite peculiar preferences. A. August-September.
First record: Brunton, 1805.
Y. Wet places near Ripon; *W. Brunton* in *O.B.G.* (1805) = near Ripon; *H. Baines.*
W. Dike-side (near cart-tract across bog), Dringhouses end of Askham Bogs!
A. Gravelly margin of Black Hill Dam, Adel!

712. Polygonum mite Schrank.
Area: 64, Top. Bot. Range: 0—300 feet.
Native; marshes and ditches, rare. A. July-September.
Y. Ditch-sides and Ure bank near Newby. Osier-bed by Yore above Ripon Bridge.
W. Askham Bogs!
A. The shady marsh at the top end of Adel Lower Dam (!) *J. Abbott.*
D. Dike near river at Thorne!

713. Polygonum Persicaria L. *Knotgrass.*
Area: 63, 64, 65, Top. Bot. Range: 0—900 feet.
Native waste ground, arable land, lane-sides, etc., very common. A. July-September.
First record: Salt, 1800.
Distribution general; found in every district.

714. Polygonum lapathifolium L.
Area: 63, 64, Top. Bot. Range: 0—600 feet.
Native; waste ground, rubbish-heaps, arable land, common. A. July-August.
First record: Salt, 1800. '*P. pallidum*, corn-fields.'
Distribution general—occurs in every district.

[**Polygonum maculatum** Trim. & Dyer. Casual.
A. With other aliens on wool-refuse heaps, by Boothroyd's Mill, Meanwood Valley.]

715. Polygonum amphibium L.
Area: 63, 64, Top. Bot. Range: 0—750 feet.
Native; in still waters and wet land adjoining, common. P. July-August.
First record: Salt, 1798.
Distribution nearly general—records wanted for districts M. and L.
R. [Giggleswick Tarn; *J. Windsor.* Gone.] Wigglesworth! Settle; *W. Marshall Watts.* Runley Bridge; *W. Howson.*
Y. Studley. Boroughbridge.
N. Lakes at Plumpton Park! Cowthorpe.
W. Barden reservoirs. Thorparch! Askham. Hobmoor. Middlethorpe Ings.
A. Adel Dam! Skipton. Methley (!) *G. Roberts.*
C. Kirkheaton; *C. P. Hobkirk.* Wakefield, frequent; *T. W. Gissing.*
D. Rotherham district; *L. Langley* (1828). Askern Lake and vicinage! River Don, Conisborough Ferry, 1798; *Herb. Salt.* Doncaster! Goole, a troublesome corn-field weed; *H. F. Parsons.*
T. Roche Abbey pools!

716. Polygonum Bistorta L. *Snakeweed.* '*Eastermore Giants.*'
Bistorta major; Ger.; Ray.
Area: 63, 64, 65, Top. Bot. Range: 0—750 feet.
Denizen, but very well established; in damp pastures, orchards, and stream-sides, common. P. June-July.
First record: Ray, 1670.
M. Stream-side fields in this district; *J. Whitehead.* Denshaw!
L. Sedbergh! Below Clapham.
R. Settle, not uncommon; *J. Windsor.* Sawley.
Y. Near North Leys, Ripon; *J. Dalton.* Mackershaw, etc., near Ripon; *W. Brunton.* Meadow adjoining Fountains Abbey; *D. Turner* in *O.B.G.* Near the Training College, Ripon; *H. H. Slater.*
N. Many places from Pateley Bridge downwards to Cattall!
W. Frequent from Ilkley and Bolton down to Stutton.
A. Bradford; *Martyn* (1763). Shipley; *W. West.* About Leeds, abundantly; *Rev. W. Wood* in *O.B.G.* Wistow; *H. F. Parsons.* Meanwood. Methley (!). Fryston; *G. Roberts.* Cononley Ings; *W. Whitwell.*

C. Hallifax; *Martyn.* Lofthouse; *G. Roberts.* Huddersfield district.
D. Sheffield and several other places of that countrey; *Ray, Cat.*, i, 44; ii, 42 (1670); *Ray, Hist.*; *Martyn*; etc. Doncaster neighbourhood; *Dr. White* (1804). Rotherham; *L. Langley.*
T. Roche Abbey (!) *H. Baines.*

Obs.—This has been accounted a Native in previous Floras of Yorkshire, and may be so in some of its stations; but it is certainly *adventive* in most, if the surrounding circumstances be fairly considered without bias in favour of indigenity.

717. Polygonum viviparum L.
Bistorta minor; Park.; Ger.; Ray.
Area: 64, Top. Bot. Montane. Range: 1,000—600 feet.
Native; damp hilly pastures, very rare. P. June-July.
First records: Parkinson, 1640; Merrett, 1666.
Especially in the North as in . . Yorkeshire; *Park.*, *Theat.*, 392 (1640).

L. Two miles from Settle, . . in a place called Crommack; *Merr.*, 16. [Crummack—a cove-like pasture under Moughton Scars, the site of a lake as shown by the lacustrine deposits. The plant grows there still.] In a mountainous pasture about a mile and a half from Wharf, not far from the foot-way leading thence to Settle, plentifully; *Ray, Cat.*, ii, 42 (1677); rep. *Martyn*=Foot-path between Feizor and Wharf, second field from the lane; and in another field above Bark-house; T. Willisell showed it to Ray; *J. Windsor, MS.* Pasture south of Arco Wood. Spns. in *Herb. Tatham* from 'Moughton' and 'Feizor.'
R. Pastures near Horton in Ribblesdale (!) *W. Brunton* in *O.B.G.*

LXII. THYMELÆACEÆ.

718. Daphne Mezereum L. '*Redberry Laurel.*'
Area: (63), 64, Top. Bot. Range: 100—1,000 feet.
Native, or bird-sown; rocky limestone woods, very rare, in no station in plenty. Shrub. March.
First record: Brunton, 1805.

L. In a wood called Cow-close, near Cantsfield; *J. Howson, MS.* in *Townsend's O.B.G.* Feizor, in a shrubby natural wood; *L. C. Miall*; 'certainly wild, not near a house'; *J. Tatham, MS. note* in *Baines' Flora*; 'planted'; *W. M. Watts' Flora.*
R. Linn Gill, Ribblehead; *H. T. Soppitt*! [A perfectly natural wood in a gorge, where if not Native it must have been sown by birds.]
Y. Oldfield Wood, near Ripon, doubtful if wild; *W. Brunton* in *O.B.G.* (1805).
W. Thorparch Woods; probably planted here.

719. Daphne Laureola L. *Spurge Laurel.*
Area: 63, 64, Top. Bot. Xero-Pelophilous. Range: 0—600 feet.
Native; thickets and hedge-banks, on limestone mainly, not rare, yet hardly to be styled common. P. April-May.
First records: Hargrove, 1782; Dixon, 1791.

L. Natural wood, Feizor; *J. Windsor* (1815); *W. Howson*; 'certainly wild'; *J. Tatham, MS.*; 'planted'; *W. M. Watts.*
Y. Studley! 'Often doubtfully indigenous'; *H. H. Slater.*
N. Abbey Rocks, Knaresborough! Wood near Priory gate; *E. Hargrove* (1782 and 1832); *H. Baines*; etc. Plumpton! Spofforth Haggs (1886); *W. Grainge.*
W. The Wilderness, Wetherby; other spots about Wetherby, Collingham, Thorparch, Tadcaster, and Bramham! Hazlewood; *H. Ibbotson.* Steeton Hall farm plantations; *W. Whitwell.*
A. (Horton near Bradford; *H. Baines*). Kippax! Ferry-bridge and Pontefract; *H. Baines.* Ledstone (1875); *G. Roberts.* (Morwick Hall, near Scholes; *L. C. Miall*). Lumby Lane, Ledsham; *W. Whitwell.*
C. Near Silcoates and near Oakenshaw Wood in 1791; *D. Dixon.*
D. Ackworth; *J. Willis.* Near Doncaster; *Dr. White* (1804). Adwick and Hampole; *P. Inchbald.* Rotherham district; *L. Langley.* Levitt Hagg, Conisborough (!) *Y. N. Excurs.* (1873). Rossington and Warmsworth Woods. Edlington Wood; *Y. N. U. Excurs.* South of Pontefract! Wentbridge.
T. Roche Abbey; *G. E. Smith.* Maltby; *J. Salt* (1800). Tickhill; *A. F. Gissing.* Firbeck!

[ASARACEÆ.]

[Asarum europæum L. Extinct Alien.
Area: '63 Leyland spn., 64 Tatham spn.,' Top. Bot.
Denizen or Alien, I believe originally planted; in woods, always very rare, now extinct.
First record: Dixon, 1791.

R. Formerly near Settle. 'In an old hedge and deep dry ditch between Grainhouse and Fieldgate, two miles west of Settle'; *John Tatham*, spn.! Station now destroyed; vide *H. Baines*, p. 142.
A. In Fryer Wood at Pontefract in 1791; *D. Dixon.*
C. Broad Bottom Wood near Mytholmroyd; *H. Baines, Flora* (1840). 'Stated to be now eradicated'; *Baker's Supp.* 'Two or three small plants in 1858'; *C. Eastwood* in *Miall's Flora* (1862). Now quite gone!
T. Tickhill—twice reported; but not wild in any sense—occurring only in the shrubberies of Tickhill Castle grounds, remains of monastic cultivation.

Obs.—The botanical world has not yet, I think, had the facts as to how the *Asarum* originated at Mytholmroyd.]

[Aristolochia Clematitis L. *Birthwort.* Alien.
Subspontaneous or (more likely) planted, in one or two bushy places about the rocks near the Abbey at Fountains. There were only three or four plants to be seen in 1870; but I noticed it in a *garden*-border also.]

LXIII. EMPETRACEÆ.

720. Empetrum nigrum L. *Crow-* or *Crake-berry.*
Erica baccifera; Matth.; Ray; Ger.; Park.
Area: 63, 64, 65, Top. Bot. Range: 0—2,000 feet.
Native; peaty moors, common on all the hill-ranges where peat has been deposited, and occurring down to sea-level nearly in West Yorkshire (actually to sea-level at Whitby, outside our limits).
First record: Johnson's Gerarde, 1633, 'in great plenty in Yorkshire on the tops of the hills.'
'On Yorkeshire moores'; *How, Phyt.* (1650). Plentifully . . one in Yorkshire; *Merr.*, 36 (1666). Yorkshire, etc.; *Ray, Cat.*; *Hist.*, p. 1631 (1688).

M. Greenfield district moor-slopes!
L. On the sides of Ingleborough; *Gough's Cam.* (1789).
R. Settle Moors, common (!) *J. Windsor, MS.*
Y. & N. Brimham Rocks and Sawley Moor; *W. Brunton* (1805). Moors near Dalton Lodge by Copgrove; *J. Dalton* in *O.B.G.* 'Near' must be elastically interpreted here as in so many of these old records. Nidderdale Moors!
W. On all the moors in Wharfedale down to Otley Chevin.
A. Adel Black Moor near Leeds; *Rev. W. Wood.* Bingley and Baildon Moors; *W. West.*
C. On nearly all the Calder and Colne Valley moorlands!
D. Wharncliffe, 1838; *Jos. Woods, MS.* in *Towns. O.B.G.* Sheffield High Moors; *J. Salt.* About the 'Wells' on Thorne Waste!

LXIV. EUPHORBIACEÆ.

[Buxus sempervirens L. *Box-tree.* Alien.
In thickets and plantations about Parks, occasional, as at Studley, Melton, Parlington, etc., but with no claim to inclusion as even subspontaneous. Always planted.]

721. Euphorbia Helioscopia L. *Sun Spurge.* '*Wart-wort.*'
Area: 63, 64, Top. Bot. Range: 0—900 feet.
Native; cultivated ground, very common. A. June-September.
First records: Bolton, 1775; Salt, 1800.
Distributed generally; in every river-basin.

722. Euphorbia platyphylla L.
Tithymalus verrucosus; Ray, Syn.
Area: (63), 64, Top. Bot. Range: 0—200 feet.
Colonist; corn-fields on a light soil, rare, but perhaps not seldom overlooked. A. July-August.
First record: Robinson in Dillenius, 1724.

W. Near York, *Dr. Robinson*; *Ray, Syn.*, iii, 312 (1724). Must have been this species, *E. platyphylla*, which still occurs in corn-fields on the Triassic sandstones, although Sir J. E. Smith referred it (with some doubt expressed) to *E. stricta.* '*E. stricta* of Sm. was a small state of *platyphylla*'; *H. C. Watson* in *Top. Bot.*, 2, 364. Fields betwixt Acomb and Dringhouses; *G. Webster*=Clayey corn-fields by the foot-road from Acomb to Askham Bryan; *H. Ibbotson.*
A. Hambleton; *T. Birks* (1878). Field between Adel and Headingley *A. Shipley.*
D. One large branching plant in a wheat-field near Adwick-le-Street, 1845; *P. Inchbald.* Fields near Rawcliffe; *O. A. Moore.* Field between Thorpe Healey and Grange station; *A. Carr.*

[Euphorbia amygdaloides L. Extinct Denizen or Alien.
English Type; but the Riding is quite beyond its areal range as an indigenous plant.
Area: 64, Top. Bot.

W. In a copse [Bilton Spring Hagg] between Bilton and Wighill Park; *Oswald A. Moore* (1841).

Obs.—Admitted upon this authority and sole station into 'North Yorkshire' flora by Mr. J. G. Baker, as a 'Native,' without himself visiting the locality, I think; as else I cannot understand so careful an observer classing it as indigenous. The copses and thickets about Wighill, including Bilton Spring Hagg (now cleared of large timber), used as fox coverts, are full of introduced plants, such as *Cornus alba* and *Viburnum Lantana*; and as seed in soil at the roots of the latter, the Spurge might well have been introduced, if it had not a more artificial origin. It is probably extinct now; Mr. William Todd, late of Leeds, and I have both sought for it twice most carefully, without success—and the 'scrub' is only small in extent.]

[Euphorbia Cyparissias L. '*Welcome-Husband-Home.*' Alien.
Planted or subspontaneous in one spot, in a rocky, bushy place on the cliff top at Knaresborough, above the sloping strips of garden. Leeds; *H. Denny, MS.*; *N. B. Guide*, 657.]

723. Euphorbia Peplus L. *Garden Spurge.* '*Wartwort.*'
Area: 63, 64, 65, Top. Bot. Range: 0—1,350 feet.
Native; cultivated garden-and field-land, very common. A. June-Aug.
First record: Salt, 1800.
Distribution general; in every district, as high as there are any garden-patches of potatoes, etc. I once saw it at Middle Houses on Malham Moor.

724. Euphorbia exigua L.
Area: 63, 64, Top. Bot. Range: 0—600 feet.
Colonist; in corn-fields, common. A. July-September.
First records: Bolton, 1775; Salt, 1800.
Distribution nearly general; every district except Mersey, in very numerous stations—not worth while quoting because they change with rotation of crops.

[**Euphorbia Lathyris** L. Alien.
This has occurred two or three times, but has no claim to be called a Denizen even, within our area.
W. A weed (for some years) in broken bushy ground near one of the Lodges of Harewood Park.
C. Luddenden Foot, 1841; *S. King*; *Herb. Tatham.*
D. Campsall; *E. Lankester* (1840). Cultivated ground near Sheffield; *Miall's Flora.*]

725. Mercurialis perennis L. *Dog's Mercury.*
Cynocrambe; Ray.
Area: 63, 64, 65, Top. Bot. Range: 0—900 feet.
Native; dry hedge-banks and woods, very common. P. April-May.
First record: Salt, 1800.
Distribution general; occurs in every district, but is not known to me at an elevation at all approaching 1,500 feet—the limit assigned to it for North Yorkshire. It is *more abundant* upon calcareous soil in West Yorkshire than upon gritstone; but *not* more common—i.e., more frequently recurring—upon lime than sand.

[**Mercurialis annua** L. Casual.
This has occurred on rubbish and in cultivated ground twice only.
A. Garden weed, Rider's Nursery, Moortown—seeds no doubt brought with Rhododendrons or other roots from the south of England.
C. Whitley near Mirfield; *H. F. Parsons* (1878).]

LXV. CERATOPHYLLACEÆ.

726. Ceratophyllum aquaticum Syme. '*Hornwort.*'
Area: '63,' 64 sine personal authority, Top. Bot.
Native; pools and slow-moving waters, rare. P. July-August.
First West Riding record: Gissing, 1865.
A. Drax; *H. F. Parsons*! Canal near Methley; *G. Roberts*!
C. Reservoir at Batley; *P. F. Lee*! Canal, by old Lock, Dewsbury; *Id.* New Miller Dam; *T. W. Gissing* (1865).
D. Ditch near railway, south side Thorne Waste! Canal at Barnsley! In the Cars near Doncaster; *J. Salt.*

LXVI. URTICACEÆ.

727. Parietaria diffusa Koch. *Wall Pellitory.*
Parietaria; Ger.; Ray.
Area: 63, 64, Top. Bot. Range: 0—600 feet.
Native; old damp walls, ruins, rock-faces more rarely; common, yet in some of the more thickly populated manufacturing districts quite rare.
First record: Ray, 1668.
R. Long Preston; *J. Windsor.*
Y. Fountains Abbey (!) *H. H. Slater.*
N. Knaresborough, on the church walls and rocks by the Nidd (!) *E. Hargrove* (1782); etc.
W. Barden and Bolton Abbey. Harewood Castle. Thorparch, etc.
A. On the walls of Kirkstall Abbey, 1800; *D. Dixon.* Malham. Skipton.
C. On the walls of Wragby Church, 1800; *D. Dixon.* Purston and Mirfield; *T. W. Gissing.* Nostell; *G. Roberts.*
D. Brockerdale, Went-vale; *J. Willis.* Near Sheffield; *Ray, Cat.*, 326 (1668). Beauchief Abbey! Rotherham district; *L. Langley* (1828). Conisborough Castle (!) *J. Salt* (1800).
T. Roche Abbey! Tickhill.
Obs.—No record for Mersey district, or for Lune, although I fancy I remember seeing it at Sedbergh.

[**Cannabis sativa** L. *Hemp.* Alien.
This—from being used as bird-seed—occurs occasionally on rich soils, manure-heaps in lanes, near towns; but is tender.]

728. Urtica dioica L. *Stinging Nettle.*
Area: 63, 64, 65, Top. Bot. Range: 0—1,500 feet.
Native; hedges, woods, road-sides, etc. P. June-September.
First records: Bolton, 1775; Salt, 1800.
Distribution general—common in every district, and not becoming infrequent until 900 feet altitude is reached.
Obs.—It is curious how very abundant the Nettle becomes in woods, or warrens, wherever *rabbits* congregate.

729. Urtica urens L. *Little Nettle.*
Area: 63, 64, 65, Top. Bot. Range: 0—900 feet.
Native; waste ground by barns and villages, fields (more rarely), infrequent, and nowhere in great quantity. A. July-September.
First record: Salt, 1800—'near Walkley.'
R. Near Friends' Meeting-House, Settle; *J. Windsor.* Rimmington!
Y. & N. Rather frequent about Ripon (*H. H. Slater*), Knaresborough, Hunsingore.
W. Addingham, Ilkley, Pool, Wetherby, etc., frequent. Hob Moor, York.
A. Silsden! Baildon; *J. Willis.* By farm between Yeadon and Otley Chevin! Eccup!

C. Wakefield, frequent; *T. W. Gissing.*
D. Rotherham district; *L. Langley* (1828). Conisborough! By Rotherham Dam near Whiston; *A. Carr.* I have not noticed it within several miles of Sheffield; *W. W. Newbould.*
T. Firbeck!

730. Humulus Lupulus L. *Hop.*
Area: 63, 64, Top. Bot. Range: 0—450 feet.
Native, but often only a Denizen—formerly an attempt was made to cultivate it, near York; in hedges, infrequent. P. July.
First records: Dixon, 1796; Salt, 1800.
R. Between Giggleswick and Kelcowe, Settle; *J. Windsor.* Between Upper Settle and Runley Bridge; *J. Howson.* Still thereabout, but not truly wild.
Y. Fountains Abbey (on rocks); *H. H. Slater.* Bridge Hewick. Newby to Boroughbridge.
N. Knaresborough. Goldsborough. Bilton. Wighill.
W. Lane at bottom of Race-course Ings, Wetherby, by road to Linton! Walton. Hedges in Wharfedale about Ilkley, Pool, Harewood, etc. Clifford; *J. Emmet.* Church Fenton; *W. Whitwell.*
A. Pontefract; *T. W. Gissing.* Birkin; *G. Roberts.* Haigh and Methley*; *G. Roberts.*
C. Abundant at Kirkthorp, 1796; *D. Dixon.* Kirklees district†; *Y. N. Excurs.* (1873). [Near Todmorden, not truly wild; *A. Stansfield.* Planted all over this district.] Mirfield*.
D. Mexborough and Conisborough! Hedges between Brightside and Rotherham; *J. Salt.* Goole*; *H. F. Parsons.* Wall-side near Endcliffe Wood, Sheffield; *G. Harrison.*
T. Roche Abbey, Tickhill, and Firbeck, ripening fruit well!

731. Ulmus suberosa Ehrh. *English Elm.*
Area: 63, 64, Top. Bot. Range: 0—600 feet.
Denizen; hedge-rows and road-sides, rarely in coppices, and always planted (I think) in West Yorkshire; common in some districts, quite unfrequent in others, as about Bradford (*W. West*). Tree. March-April.
First record: Salt, 1800.
Distribution general; in every district, but qualified by what is said above.

732. Ulmus montana Sm. *Wych-Elm.*
Area: 63, 64, 65, Top. Bot. Range: 0—1,000 feet.
Native; woods and hedges, very common and luxuriant. Tree. March-April.
First record: 'Yorkshire in woods and hedges'; *Gough's Cam.* (1789). Not named by Salt in his Flora Sheffieldiensis.
Distribution general; occurs in every district and in the woods of the gills among the hills, but not ascending nearly so high as the Rowan-tree.

LXVII. AMENTIFERÆ

733. Quercus Robur L. *Oak.* '*Yack.*' '*Ackers,*' '*Yackrans*'—*Acorns.*
Area: 63, 64, Top. Bot. Range: 0—1,200 feet.
Native; in woods and hedges and on stony slopes, but being a tap-rooted tree it is not long-lived in rocky stations, and where there are manufactories it is mostly now in a dwindling state; very common. Tree. May-June.
First notice (and earliest record for a distinctly West Yorkshire plant): W. Turner, 1568.
'A great plentye of galles found upon Oke tues . . . about Hallyfax'; *Turner, Herb.*, 2nd part, fol. 109.
A but little later allusion to West Yorkshire *Oaks* is to be found, with many interesting particulars, in *Evelyn's Sylva*, in connection with Sheffield Park, in which 'in 1646 above 100 trees worth £1,000,' and 'still *two* worth above £20,' were growing. Cf. *Evelyn's Sylva*, 4th (Hunter's) Ed., 1812, pp. 206, 209, and 211, 212 (Vol. ii) for full account of the Worksop Park Shire Oak, and others.
Distribution general—in every district.

VAR. **sessiliflora** Salisb. is much less frequent and less abundant than *pedunculata* Ehrh. The characters assigned in books are found variable in the field, so that it is often impossible to tell to which form a given example belongs. Ripley Park; *W. Brunton* (1805). Anley Crags, 1834; *Herb. Tatham.*

[**Quercus Cerris** L. *Turkey Oak.*
Much planted in the vicinity of the large towns; but no trees are, as yet, above 30 to 40 years old. Often miscalled the 'Maple.']

[**Castanea vulgaris** Lam. *Sweet Chestnut* Alien.
Planted, with increasing frequency, not only in parks and shrubberies but often in wild situations, but it does not, I believe, propagate itself by means of self-sown seeds.]

734. Fagus sylvatica L. *Beech.*
Area: 63, 64†, Top. Bot. Range: 0—900 feet.
Denizen for the most part, but possibly Native on the Permian tract; slopes and way-sides, common in the low country, rare among the hill-valleys. Tree. May.
First record: Salt, 1800.
Distribution general; in every district, but in Lune and Mersey, certainly, only as a planted tree.

735. Corylus Avellana L. *Hazel. Nut-tree.* '*Chats*'—the Catkins.
Area: 63, 64, 65, Top. Bot. Range: 0—1,500 feet.
Native; woods and thickets, very common. Shrub. March-April.
First record: Salt, 1800.
Distribution general; occurs in every district, but rare in Mersey and probably not indigenous therein. The greatest height at present attained in Nidderdale is 1,350 feet, but formerly it grew and ripened its fruit at 1,650 feet; *Joseph Lucas, Studies in Nidderdale*, p. 116. It does not now yield matured nuts anywhere in the West Riding at much above 900 feet.

[**Carpinus Betulus** L. *Hornbeam.* Alien.
Met with occasionally in hedge-rows and plantations, but always as a planted tree; mostly singly. Some old trees at Alder Hill, Meanwood, bordering an old orchard (Whitehead's).]

736. Alnus glutinosa L. *Alder.* '*Eller.*'
Area: 63, 64, 65, Top. Bot. Range: 0—1,000 feet.
Native; damp places by water-courses, very common. Tree. March-April.
First record: Salt, 1800.
Distribution general—in every district.
Obs.—Dr. J. Holland in his 'Tour of the Don' felicitously observes that its fine Alders have *en masse* the colour and effect of Oaks in moist situations; and in Longfellow's 'Hiawatha' we have an equally happy touch of nature, viz.: that one may in ordinary landscapes 'trace the water-courses by them,' sturdily yet not ungracefully fringing the stream-banks as they do.

737. Betula alba L. *White Birch.*
Area: 63, 64, 65, Top. Bot. Range: 0—1,600 feet.
Native; woods, hedges, and stony heaths, common, but showing a distinct preference for sandstone soils over calcareous. Tree. May.
First record: Salt, 1800.
Distribution general; in every district.

Var. **verrucosa** Ehrh. (**pendula** Roth.).
Common in woods and plantations, but at lower levels than the next. The pendulous-twigged form is usually met with singly only, as a planted tree in coppices and parks. No record for this var. in Mersey district.

Var. **glutinosa** Wallr. (**pubescens** Ehrh.).
Very common—much more abundant than either *verrucosa* or its form *pendula*; and is especially *the* form on the stony heaths and moorland cloughs. I have seen this variety in every river-basin.

738. Myrica Gale L. *Sweet Gale.*
Area: 63, 64 Tatham spn., Top. Bot. Range: 0—600 feet, formerly up to 1,500 feet.
Native; boggy heaths, rare, but abundant locally. Shrub. May-June.
First record: Bolton, 1775.
M. Marked for this district by *J. Whitehead.*
L. Lawkland and Austwick Mosses; *J. Tatham*; rep. *H. Baines.*
W. At Askham Bog, plentifully; *Gough's Camd.* (1789). Still in abundance there.
[C. Warley Moor, flat on top towards Haworth parish; *J. Bolton* (1775).]
D. Potteric Carrs; *E. Miller* (1804). In the Car near Doncaster; *J. Salt* (1800). Thorne Waste and Goole Moor, plentiful in parts!
T. New Zealand Moor near Thorne; *J. E. Kenyon* (1864). [Hatfield Moor: nearly gone.]

739. Populus alba L. *White Poplar.*
Area: 63†, 64†, Top. Bot. Range: 0—600 feet.
Denizen, but often in secluded stations appearing indigenous; moist wooded bottoms by streams in the low country, rather common. A short-lived tree, but perpetuating itself successfully by its suckers. April.
First record: uncertain—not in Baines' Flora.
Distribution general, but only thinly spread over the whole of the West Riding.

The Hybrid **canescens** Sm. occurs very sparingly in two or three places, always along with its supposed parents *P. tremula* and *P. alba*.
R. Named (but no locality given) for Settle district by *J. Windsor* in *Flora Cravoniensis.*
W. Damp wooded valley near stream running between Scarcroft and Bramham. One tree by river Cock near Stutton. Sometimes planted by nurserymen?

740. Populus tremula L. *Aspen.*
Area: 63, 64, 65, Top. Bot. Range: 0—1,250 feet.
Native; heathy and moorland thickets, woods, hedges, and bogs on sandy-peaty soil in the lowlands, infrequent, often planted. Tree. April.
First record: Baker's Suppl. to Baines, 1854, apparently—no Poplar mentioned at all by Salt or Baines.
Distribution nearly general, but no record for it (and not seen by writer) in Mersey.
L. Sedbergh. Dentdale. Ingleton (!) *J. Willis.*
R. Settle; *J. Windsor.* Linn Gill and about Rathmell, Sawley, and Tosside.

Y. & N. Pateley and Kirby Malzeard districts. Scotton Banks, Knaresborough!
W. Barden and Washburn Valley! Ilkley. Woodhall. Westwoods, Bramham! Lanes about Wike, Scarcroft, and Brandon!
A. Eccup. Alwoodley. Rare in lower Aire Valley.
C. Hebden Valley! Wakefield, frequent; *T. W. Gissing.*
D. Little Don Dale. Penistone!
T. Noted at Tickhill, near Auckley and Rossington and 'Martin Beck,' Maltby.
Obs.—There are two forms of this, *villosa* and *glabra*, the former with silky leaves and pubescent shoots, the latter with glabrous young leaves and stems.

Var. **glabra** E.B. This is the form mainly used by foresters in laying out plantations. It is that of Westwoods, Bramham (planted woods), and I have noted it, singly, in several places in plantations in the Aire, Don, and Trent districts.

Var. **villosa** E.B. The only form seen by me in aboriginal thickets and on heaths in the West Riding; I consider it the truly native form. It is this which occurs at Ingleton, Ribblesdale, Askham Bogs, Brandon by Shadwell, Hebden Valley, Tickhill, etc.

[**Populus nigra** L. *Black Poplar.* Alien.
Hedgerows, plantations, etc., always planted.]

[**Populus pyramidalis** Roz. *Lombardy Poplar.* Alien.
Planted extensively in the lowland districts. It is perhaps a variety of *P. nigra*.]

[**Populus monilifera** Ait. *Canada Poplar.* Alien.
Shrubberies; always planted of course.]

[**Populus balsamifera** L. *Balsam Poplar.* Alien.
Occasional in shrubberies.
C. A small bush in an old lane at Bretton; *P. F. Lee*!
D. Rivelin Valley, Sheffield; *Amos Carr*!
Obs.—The leaves of this are ovate-elliptical (not cordate at base) and glaucous with reticulated veins beneath. The common Canada Poplar is generally but erroneously styled the 'Balsam Poplar.']

741. Salix pentandra L. *Sweet-Bay Willow.*
S. folio laureo, sive lato glabro odorato; Ray.
Area: 63, 64, Top. Bot. Range: 0—1,250 feet.
Native; damp hedge-banks and stream-sides, common by the hill-streams, rare in the lowlands. Tree. June.
First record: Ray, 1670.
In com. Eboracensis; *Ray, Cat.*, 272 (1670); *Syn.* i; *Gibs. Cam.*; etc.

L. Greta and Wenning banks. Keasdon Beck.
R. Ribble side near Birkbeck's Weir; *J. Windsor, MS.* Ribble from Linn Gill to Sawley. Beggar-wife Bridge; *J. Tatham.*
Y. Ripon, river-bank; *W. Brunton.* Copgrove; *J. Dalton.*
N. Nidd banks from Pateley downwards. Pannal and Spofforth.
W. About Kilnsey; *W. Curtis* (1782); *Gough's Cam.* (1789). Otley; *Rev. W. Wood* (1805). Bolton. Ilkley. Collingham. East Keswick. Askham Bogs; *G. Webster.*
A. By the river, Bradford; *Ray, Syn.* iii; *Martyn* (1763); etc. About Leeds and Bingley; *Rev. W. Wood.* Malham Tarn! Myles Hill stream-side, Meanwood! Cononley and Carlton Ings; *W. Whitwell.*
C. Calder vale, many places to junction with Aire!
D. River-side, Jordan Dam; *J. Salt.* Wharncliffe side, Kimberworth, etc.; *A. Carr.*
T. Roche Abbey Valley!

742. Salix fragilis L. *Crack Willow.* '*Withy.*'
Area: 63, 64, Top. Bot. Range: 0—600 feet.
Native; hedges, river-banks, stream-sides, common. Tree. April.
First record: Salt, 1800.
Distribution general—thinly distributed throughout our area, abundant on the banks of the larger streams in the lower country.

Forma **Russelliana** Sm. (not true *viridis* Fr.).
Range: Low.
First record: Brunton, 1805—'Yore banks.
This may be a hybrid with *S. alba*; but with us in Yorkshire is, apparently, a state of *S. fragilis*, not permitted to attain its full stature, but kept cut down in order that the shoots from the stools may be used as an Osier. In this condition this, the Bedford Willow, is common about Ripon, Ferrybridge, Castleford, Methley, wherever there are Osier-beds.

743. Salix alba L. '*Wilf.*'
Area: 63, 64, Top. Bot. Range: 0—600 feet.
Native; river-banks, damp low-lying woods, etc., common. T. April.
First record: Salt, 1800.
Distribution general—occurs in all the districts, but in no great quantity in any one place.

Var. **vitellina** L. '*Golden Osier.*'
Not uncommon in osier-holts, rarer by stream-sides.
Noted for R., Y., W., A., and D. districts.
First record: Salt's Flora Sheffieldiensis, 1800; 'near waters, common.'
'*Salix cortice intus flavo, folio angusto.* At Rippon and divers other places of Yorkshire. An Salix humilior foliis angustis subcæruleis'; *Merrett, Pin.*, 108 (1666), was most likely the var. *vitellina* of this species; but that, *S. fragilis*, and *S. rubra*, grow in plenty by the Yore side.

744. Salix undulata Ehrh.

Area: '64 (Baker),' Top. Bot. Range: 0—200 feet.

Denizen; osier-beds, rare. Shrub. April-May.

First record: Baines' Flora, 1840, sub nom. *lanceolata*, but 'doubtless erroneous,'—*Baker's Suppl.*, 1854; next record: Baker in Miall, 1862.

Y. Osier-beds and islets in Yore about Nunwick—planted!

W. 'A colony at the mouth of Gill Beck just above Otley . . here truly indigenous'; *J. G. Baker* in *Miall* (1862) [Mr. Baker modifies this opinion now]. Osier-beds at Seavy Carr, Stutton, and Church Fenton!

A. Ferrybridge osier-holts! North side Black Hill Lower Dam, Adel—planted!

D. About Doncaster towards Barnby Dun, by the Don!

Obs.—Hooker considers it a var. of *triandra*, distinguished by its shaggy scales and distinctly developed styles, a non-indigenous basket-osier; whilst Anderson supposes it a hybrid between *S. triandra* and *S. viminalis*.

745. Salix triandra L.

Area: 63, 64, Top. Bot. Range: 0—600 feet.

Native; river-banks, stream-sides, wet hedges, and osier-holts, infrequent. Small Tree. May-June.

First record: Salt, 1800.

L. Greta banks; *W. Howson*. Near Wennington!

R. Giggleswick; *J. Windsor*; confirmed by *W. M. Watts*. Ribble banks, Wigglesworth!

Y. By Yore, Boroughbridge!

N. Pateley Bridge. Crimple side, Spofforth!

W. Frequent about Otley, Harewood, Stutton, Newton Kyme. Askham! Ouse banks, Cawood!

A. By Aire near Methley and lower down.

C. Calder banks and osier-holts about Castleford!

D. By Dearne, Burton Grange, Barnsley. Don banks above Sheffield; *J. Salt*. Conisborough, by Don side; *A. Carr*.

T. Near flour mill below Roche Abbey!

VAR. **amygdalina** L.

Rather rare.

N. Bilton to Wighill!

W. *West* bank of Ouse, opposite Clifton, near York; *G. Webster* Cawood!

A. Selby!

D. Askern!

746. Salix purpurea L. *Rose Willow.* '*Bitter Vilf.*'

Area: 63, 64, Top. Bot. Range: 0—900 feet.

Native; by stream-sides and river-banks, infrequent, but in clearly native stations pretty high up the dales. Shrub. April.

First records: Gough, 1789; Brunton, 1805.

Yorks. in watery places, common; *Gough's Cam.* [sub nom. *Helix*] (1789). Nunwick by Ripon (!) *W. Brunton*.

Distribution general—recorded, or seen by myself, in every division. Two of the varieties are on record.

VAR. **Woolgariana** Borr.

Ripon! Conisborough!

VAR. **Lambertiana** Sm.

Near Settle (!) *J. Windsor*. Scotton banks, by Nidd! Wharfe bank, Harewood and Collingham!

HYBRID **rubra** Huds. (with *S. Helix*). *Red Withy.*

Area: 64, Top. Bot. Range: 0—500 feet.

Denizen; osier-beds and stream-sides, rare. Shrub. April.

First record: Baker, 1854.

R. Settle Bridge, Mill Island, etc.; *J. Windsor*.

Y. Skell banks between Studley and Ripon; *J. G Baker*; etc. Yore side, Ripon!

W. Harewood.

D. Don side, Conisborough! (*S. Forbiana*). Near Brightside, Aldwarke, etc.; *A. Carr*.

Obs.—Mr. Baker remarks that *S. Helix* should go with *S. purpurea* rather than *S. rubra*; but so far as West Riding records go, the stations of each are hardly to be separated.

747. Salix viminalis L. *Common Osier. Basket Osier.*

Area: 63, 64, Top. Bot. Range: 0—750 feet.

Native; damp hedge-banks, river-sides and osier-beds, common.

First record: Salt, 1800.

Distribution general—occurs in every district.

VAR. **stipularis** Sm.—a supposed hybrid with *S. Caprea*—is rare. Baker remarks that it and *S. Smithiana* shade off into *S. viminalis*, but keep quite distinct from *S. cinerea*.

Y. Near Ripon; *W. Brunton* in *O.B.G.*; rep. *With.* (1830); etc. Yes: in osier-beds above Hutton Conyers (1871)!

748. Salix Smithiana Willd.

Area: 64, Top. Bot. Range: 0—600 feet.

Native, but possibly a hybrid and not really distinct; hedges and stream-sides, rare, showing a preference for the north-western dales. Small Tree. May-June.

First record: uncertain; H. Payne, 1860?

L. Wenning side near Bentham.

R. Stainforth, Settle; and Wigglesworth!

Y. By the river Yore near Newby!

W. One tree by the Wharfe, Woodhall! Washburn Valley. Near York; *G. Webster*.

A. Headingley, in shrubbery near Shaw Lane end—planted.

D. Near Oughtibridge; *A. Carr*. By Dearne, Darfield; *H. Payne*.

FORMA **acuminata** Sm.

R. By the Ribble below Wigglesworth! Near Settle; *J. Windsor, Flora Crav.* (1873).

FORMA **ferruginea** And.

W. Reported from Askham Bogs (sub nom. *rugosa*) in *B. B. Le Tall's* account. *G. Webster* marks only *S. Smithiana* for the Ainsty.

749. Salix cinerea L. *Sallow.* '*Sauf*' and '*Saugh.*'

Area: 63, 64, 65, Top. Bot. Range: 0—1,250 feet.

Native; hedges, stream-sides, thickets, moors, and gills among the hills, very common. Small Tree. April-May.

Distribution general—occurs plentifully in every district.

First record: Salt, 1800.

I am unable to distinguish the var. (or perhaps only form) *oleifolia* from the prevailing narrow-leaved type, but

VAR. **aquatica** Sm. (with broader, more obovate leaves, glaucous beneath, and large stipules) is not uncommon in the dale gills by the streams in boggy situations. I have notes of it for districts L., R., Y., N., W., and D.

750. Salix aurita L. *Dwarf Sallow.* '*Moor-palm.*'

Area: 63, 64, 65, Top. Bot. Range: 0—1,500 feet.

Native; damp thickets by streams and on the heaths and mosses, common. Shrub. May.

First record: Salt—'Near Fulwood, plentiful,' 1799.

Generally distributed, but I have no record for the Trent tributaries district.

This has a wide range—it and *S. caprea*, may be found in the gills on the moorlands almost as high as any other woody shrubs.

751. Salix caprea L. *Goat Willow.* '*Great Saugh.*' *Palms*—the catkins.

Area: 63, 64, 65, Top. Bot. Range: 0—1,500 feet.

Native; woods, hedges, thickets on moors and by the streams in the moorland gills, very common. Small Tree. April.

First record: Salt, 1800.

Distribution general—in every district.

Obs.—This varies considerably in the size, shape, breadth, and silkiness of the leaves, some forms shading off into *S. cinerea*, but I have not been able to discriminate any named varieties among our West Riding forms.

752. Salix phylicifolia L.

Area; 64, 65, Top. Bot. Montane. Range: 1,050—350 feet.

Native; mountain gills by the streams, and by the larger rivers somewhat lower, rare. Shrub. May-June.

First record not clear: J. Windsor, 1857, in Phytologist?

L. Rawtheydale, Garsdale and Dentdale!

R. Linn Gill! And by river Ribble in many places from above Horton to Mill Island (*J. Windsor*), and near Rathmell. Langcliffe; *W. M. Watts*.

W. Cowside Beck, Littondale! (*W. West*), Langstrothdale, etc. Hesleden Gill; *J. Tatham*. Langstrothdale and down Wharfe to Grassington!—not seen lower.

HYB. **laurina** Sm. (between *S. phylicifolia* and *S. caprea*), very rare. Small Tree. June.

R. Ribble side near Langcliffe; *J. Windsor*. Ribble bank at Stainforth (one tree). Ribble near Wigglesworth!

W. Below Grass Wood, Grassington, one tree!

753. Salix nigricans Sm.

Area: 64, 65, Top. Bot. Range: 750—350 feet.

Native; rocky mountain-stream banks and gills, very rare. Shrub. May-June.

First record: Howson, 1850.

L. Garsdale, very sparingly! In a moss near Austwick! Near Weathercote; *H. C. Watson*.

R. By the Ribble at Horton; Douk Gill; and at Cammock below Settle; *J. Windsor*; rep. *W. M. Watts*. Ribblesdale; *W. Howson*.

W. Sparingly above Buckden, by the stream in Lower Langstrothdale!

VAR. **Andersoniana** Sm.

Hesleden Gill; *J. Tatham, MS. note* in *Baines*.

[**Salix Myrsinites** L. Incognit. Misnomer.

On the slope of a high hill between Kilnsey and Arncliffe; *W. Curtis* (1782) sub nom. *argentea*; *R. Teesdale* in *Linn. Trans* (1798); *Bot. Guide* (1805). Referred to '*Myrsinites*' in *B.G.* and in *Baker's Suppl.* (117), but Gough's repeated record is placed under the name '*arenaria*,' by which *S. phylicifolia* or (more likely) the large-leaved *argentea* form of *repens* was meant.]

754. Salix repens L. '*Sand-saugh*' (local).

S. arenaria; Gough.

Area: 63, 64, 65, Top. Bot. Range: 0—1,500 feet.

Native; on stony peaty moors, and by rocky streamlets amid the hills, and on sandy heaths and barren places in the lowlands, infrequent. Shrub. May.

First record : Bolton, 1775.

L. Frequent in Garsdale. Austwick and Cocket Mosses. Ingleborough.

R. Blea-moor. Feizor and Wherf ; *J. Windsor*. Linn Gill Moor ! Penyghent slopes. Settle, 'common'; *J. Tatham, MS.*

N. Harrogate Moors. Lanes about Marston ! Ramsgill ; *J. S. Wesley.*

W. Between Kilnsey and Arncliffe ! *Gough's Camd.* (1789) [*S. phylicifolia* likewise grows in this locality in sandy places by the river, and may have been intended]. Langstrothdale ! Barden Moor. Askham Bogs !

A. Malham Moor ! *W. West.* Eldwick ; *W. West.*

C. Halifax Moors ; *J. Bolton.* Heath, Wakefield ; *T. W. Gissing.*

D. South side Thorne Waste, very sparingly ! Stony ground below the quarries near Conisborough station ; cum *W. Todd.* Penistone Moors ; *A. Carr.* Rawcliffe 'Rabbit-hills'; *H. F. Parsons.*

Hyb. **ambigua** Ehrh.

Very rare ; heaths, along with its parent types. I possess specimens, with characters intermediate between *S. aurita* and *S. repens*, gathered at Askham Bogs ; and others from Skipwith Common—a range of forms passing into one or the other—which I consider to be Ehrhart's *ambigua.*

755. Salix herbacea L.

Area : 64, Top. Bot. Range : 2,400—2,300 feet.

Native ; stony mountain summits, very rare. P. June-July.

Earliest record : T. Willisel in Ray, 1670.

On Ingleborough Hill, on the highest rock next to the Beacon (rocks on the north side, Ed. ii), and on a hill called Whernside over against Ingleborough Hill, on the other side the subterraneous river; *Ray, Cat.*, 273 (1670); *Hist.*, 1423; *Syn.*; *Gibs. Cam.* At the top of Ingleborough Hill near Foal-foot, *Mr. Thornbeck*; *Blackst., Spec.*, 87; etc. These stations were at the time recorded under the polynomial designation referred by Sir J. E. Smith and others to *S. reticulata.* There is no doubt, however, that the plant was *S. herbacea*, to which the later recorders rightly refer it. Ingleborough ! *R. Teesdale, Bot. Guide* (1805); *H. Baines* (1840); etc. Whernside ! *R. Clapham* in *Baker's Supp.*, 1854.

W. Penyghent, top of ; *J. Windsor* and *J. Howson.*

[**Salix reticulata** L. Incognit. Misnomer.

S. pumila folio rotundo ; J.B.; Ray.

On Ingleborough, Yorkshire ; *Ray, Cat.* (1670). Mr. Winch affirms that this plant has never been found in the north of England, large specimens of *S. herbacea* having been mistaken for it; *Withering*, vii, 68.

All the repeated later records down to 1789 (*Gough's Camden*) placed above under *S. herbacea* were given as this species, but Ray's interpreters (rather than Ray) were doubtless wrong in putting the English plant down to this species. The plant in Herb. Ray is *S. herbacea* ; *W. W. Newbould.*

J. Bolton, in his *Halifax Catalogue*, says : 'On the top of Norland, Rishworth, and Warley Moors, here and there, not plentiful,' but he doubtless meant some other plant—perhaps dwarfed *S. aurita.*]

LXVIII. CONIFERÆ.

756. Pinus sylvestris L. *Scotch Fir.*

Area : [63, 64], Top. Bot.

Denizen ; in part Native perhaps, if the self-sown trees on Thorne Moor, etc., can be regarded as descendants of the forest once covering the peat area (cones occurring in the peat-earth) in the *sandy* stratum below which the original roots are to be found embedded with upright stumps ; rocky and sandy slopes, common, the largest *existing* trees, however, always planted. Tree. May.

First record : uncertain. 'Alien'—Baker, Suppl., 1854.

Distribution general—as regards planted trees.

[**Pinus austriaca** Höss. (*P. Laricis* Poir.).

Introduced in many parks and plantations, as about Maltby, etc.]

[**Larix europæa** DC. *Larch.* Alien.

Extensively planted for cover in exposed situations, especially amongst the hills, quite up to 1,700 feet in some places.]

[**Abies excelsa** DC. *Spruce.* Alien.

Always planted ; but in many places, as at Studley and Esholt, very fine.]

757. Juniperus communis L. *Juniper.*

Area : 64, Top. Bot.

Native ; limestone scars and precipices, very rare, although locally in plenty. Shrub. May-June.

First record : Blackstone, 1746.

L. Howgill Fells ; *G. Pinder.* In a mountainous meadow called Moughton, two miles from Austwick ; *Mr. Thornbeck* ; *Blackst. Spec.*, 86 (1746). Moughton Fell ; *Martyn* (1763). Above village of Wherf in Craven ; *H. Baines.*

W. A few stunted bushes about the 'swallow-holes' and 'pots' of Heseltine Gill under Fountains Fell.

758. Taxus baccata L. *Yew.* '*Eugh.*'

Area : 63, 64, Top. Bot. Range : 0—1,400 feet.

Native ; limestone scars, and rocky places on the Permian formation, rare, but locally abundant.

In Wm. Turner, A.D. 1548, under *Taxus*, '*Ughe tree,*' is to be found the earliest allusion to a Yorkshire plant—'groweth in diverse parts of Yorkeshyre.'

L. Weathercote Cave. Cavehole Wood ; *J. Windsor.* Ingleton Fells !

R. Giggleswick Scars (!) *W. Curtis* (1782); *Gough*; *J. Tatham*; etc. Linn Gill.

Y. Rocks near Fountains Abbey ; *W. Brunton.* One estimated to be 1,300 years old ; *De Candolle.* Unusually luxuriant there ; *H. H. Slater.*

W. Kilnsey and Grassington Scars—odd bushes ! Oglethorpe Ings (!) near Boston Spa ; *J. Emmet.*

A. Malham Cove ! Gordale ; *W. Brunton* ; *J. Tatham, MS.*

D. Stainborough and Cawthorne Woods ; *H. Payne.* Warmsworth. Conisborough. Hedges near Ecclesfield ; *J. Salt* (1800). Wharncliffe Woods ! Near Bradfield and Dalton ; *A. Carr.*

T. Trent district, all about Maltby, Sandbeck, Letwell, etc., fine and plentiful—a feature in the landscape. Lindrick Common ; *A. Carr.*

MONOCOTYLEDONES.

LXIX. TYPHACEÆ.

759. Typha latifolia L. '*Cat's-tail.*' *Reed-Mace.*

Area : 63, 64, Top. Bot. Range : 0—300 feet.

Native ; pools and stagnant dikes, common, but absent in the higher parts of all the dales. P. July.

Earliest notice : uncertain—not specially recorded by either Bolton or Salt in their early lists for Halifax or Sheffield, near both which places it is quite rare ; whilst Baines (in 1840) well knowing its prevalence about York, etc., gives no stations but registers it as 'frequent' merely.

Y. Sharow Mires (!), ditch, Hewick Bridge, and other dubs (!) near Ripon ; *H. H. Slater.* Staveley to Boroughbridge.

N. Knaresborough. Pools near York !

W. Marsh near Pool (!) *J. Abbott.* Moss-Carrs, Wetherby, and many places in the low-land between Tadcaster and York. Dringhouses, Hobmoor, etc.

A. Adel Dam, near Leeds ! Many places about Methley, Castleford, and Selby. Lower Aire plain, common.

C. Ryhill, Walton, etc., near Wakefield ; *T. W. Gissing.*

D. Rotherham ; *A. Carr.* Many ponds about Doncaster, Thorne, Goole, etc. !

T. Marked for this district by *A. Carr.*

760. Typha angustifolia L.

Area : 63, 64, Top. Bot. Range : 0—100 feet.

Native ; pools and dikes, rare. P. July-August.

First record : Bolton, 1775.

(N. Pond by road between Harrogate and Blubberhouses, three miles from Blubberhouses ; *L. C. Miall.*)

W. [Dringhouses brick-ponds ; *H. Baines.* Now gone.] Dikes by rail, one to two miles from Church Fenton towards Grimston !

A. Selby, July 1835 ; *Herb. Tatham.*

D. Ponds by rail near Doncaster ! Goole ; *H. F. Parsons.*

761. Sparganium ramosum Curtis. '*Burr-reed.*'

Area : 63, 64, Top. Bot. Range : 0—600 feet.

Native ; pools, mill-dams, and slow streams, common. P. August.

First record : uncertain—not named in Salt's Flora.

Distribution general—occurs in every district.

I have seen in situ, or examined, specimens of this, from Hellifield, Methley, Wakefield, Delph, Shipley, Skipton, Frizinghall, Goole, Selby, Askham, Wetherby, Meanwood, Barnsley, Boroughbridge, Bowling, Broughton, and in the Rother, Sheffield.

[**Sparganium neglectum** Beeby, Journ. Bot. Incognit.

Although I have examined upwards of twenty tolerably mature gatherings of *Sparganium*, I have not yet been able to satisfy myself that this new segregate is a West Yorkshire plant.]

762. Sparganium simplex Huds.

Area : 63, 64, Top. Bot.

Native ; marshes and shallow ditches and pools, infrequent. P. August.

First record : Salt, 1800.

R. Rathmell (!) Settle and Giggleswick ; *J. Windsor.*

Y. Yore banks below Hewick Bridge ; *H. H. Slater.*

W. Seavy Carr, Stutton. Askham Bogs !

A. Canal-side, Skipton (the floating form of deep water there also) ; *W. Whitwell* ! Selby !

C. Stanley Canal ; *T. W. Gissing.* Coxley Dam ; *P. F. Lee* !

D. Askern, 1845 ; *P. Inchbald.* Pool near Balby, Doncaster ! Ditch at Park Furnace, near Sheffield, 1800 ; *Herb. Salt.* In the river Don below Doncaster ; *G. Harrison.* Thorne Waste, border dikes. Goole fields ; *H. F. Parsons.*

763. Sparganium minimum Fr.

S. natans of O.B.G.

Area : 63, 64, Top. Bot. Range : 0—450 feet.

Native ; pits, pools, and deep stagnant dikes, rare. P. August.

First record : Salt, 1800.

L. Lawkland Moss; *J. Windsor*; rep. *W. M. Watts.*
[R. Giggleswick Tarn; *J. Windsor.* Gone.]
Y. Staveley Carrs; *J. Dalton.* Ponds near Ripon; *W. Brunton* in *O.B.G.* Pits near Copt Hewick; *H. H. Slater.*
W. Hob Moor ponds, York; *H. Baines.* Askham Bogs dikes! Pond at Aberford.
C. Pond at Gomersal (Spen Valley), introduced; *J. Emmet* in *Miall's Fl.*
D. Dikes at Potteric Carrs, Doncaster; *Herb. Salt* (1800).

LXX. ARACEÆ.

764. Acorus Calamus L. *Sweet Flag. Mace-root.*
Area: 63*, Top. Bot. Range: 0—100 feet.
Denizen; river-banks and canal-sides, rare. P. July-August.
First record: White, 1804.
A. By the river Aire and canal from below Hunslet to beyond Wakefield! Offset canal at Fairburn, in flower, 1885; *G. Roberts*!
C. Sandal, Walton, Winterset Reservoir, near Wakefield; *T. W. Gissing.* Kirkheaton, Milnsbridge, and mill-dam, Luddendenfoot—all planted; i.e., in none of such like situations is it spontaneous.
D. By canal between Thorne and Goole. By the Barnsley and Doncaster Canal (!) at Stairfoot; *H. Payne*; *H. Johnson*; etc. Doncaster; *Dr. White* (1804)—by the Don. Potteric Carrs; *S. Appleby* (1831). Norton Fish-ponds near Sheffield; *H. Baines.*

765. Arum maculatum L. *Cuckow Pint. Lords and Ladies.* '*Wake Robin.*'
Area: 63, 64, Top. Bot. Range: 0—750 feet.
Native; warm hedge-banks, thickets, and woods, common. P. Late April-May.
First record: Salt, 1800.
Distribution nearly general (no record for Mersey district), and ascending in the dale woods to Weathercote, Malham, and Kettlewell.

LXXI. LEMNACEÆ.

766. Lemna trisulca L.
Area: 63, 64, Top. Bot. Range: 0—300 feet.
Native; ponds and stagnant ditches, common. A. June.
First record: Salt, 1800.
Distribution nearly general—no record for Lune district. ['Probably, occasionally, near Settle'; *J. Windsor, Flora.*] Occurs in every other division, although somewhat local.
Sandholes, Swinefleet (in flower, 1878); *T. Birks.*

767. Lemna minor L. *Duckweed.*
Area: 63, 64, 65, Top. Bot. Range: 0—600 feet.
Native; ponds and ditches, very common in the lower lands, but I have no note of seeing it at above the altitude stated. A. June.
Distribution general; occurs in every district.

768. Lemna gibba L.
Area: '63,' Top. Bot. Range: 0—300 feet.
Native; stagnant ponds and dikes, rare. A. July.
First record; Baines, 1840.
Y. Dub by Yore, Ripon!
W. Askham Bog dikes (all four species) (!) *Hy. Ibbotson.*
(A. The Walks, Cottingley; *L. C. Miall.*)
D. Dikes, Potteric and Great Eaa Carrs, Doncaster

769. Lemna polyrhiza L.
Area: 63, 64, Top. Bot. Range: 0—300 feet.
Native; ponds and slow-moving dikes, rare. A. July?
First record: Baines, 1840.
Y. Sharow Mires, Ripon; *H. H. Slater*!
N. Plumpton Lakes! Pool on Knaresborough 'stray.'
W. Pool near Wetherby! Askham Bogs; *H. Ibbotson.*
A. Methley; *L. C. Miall.* Ferrybridge! Drax; *J. F. Parsons.*
C. Pond, Alverthorpe; *T. W. Gissing.* Normanton; *L. C. Miall.*
D. Doncaster and Loversall Carrs!

LXXII. NAIADACEÆ.

770. Potamogeton eu-natans Syme, E.B. *Pondweed.*
Area: 63, 64, Top. Bot. Range: 0—450 feet.
Native; clay-bottomed ponds. in the low country, not much if at all in rivers or canals, locally common. P. August.
First record: Salt, 1800.
R. [Giggleswick Tarn; *J. Windsor.* Gone.] Pond near Hellifield.
Y. Dubs about Ripon and Boroughbridge, common
N. Ponds, Starbeck, Plumpton and Nun Monckton!
W. Arthington! Washburn Valley. Wetherby! Stutton! Askham Bogs!
A. Adel Dam. South Milford! Selby! Drax and Snaith; *H. F. Parsons.*
C. Wakefield district, common; *T. W. Gissing.* Perhaps *oblongus* in part was meant). Nostel! Bretton; *P. F. Lee.*
D. Ponds, Attercliffe Common; *Herb. Salt.* 'A small specimen, but the true plant'; *W. W. Newbould.* By the Rother; *A. Carr*; and *Geo. Harrison.* 'True'; *W. W. Newbould.*
T. Rossington Carrs and pits about Bawtry!

771. Potamogeton polygonifolius Pour. (incl. *oblongus* Viv.).
Area: 63, 64, Top. Bot. Range: 0—1,700 feet.
Native; slow rivers, tarns, moorland ditches, and pools, very common; the most frequent West Yorkshire pond-weed, chiefly in the form *ericetorum* on the heaths of the uplands. P. July-August.
First record: Salt, 1800.
M. Bogs and streams in this district (! var. *ericetorum*); *J. Whitehead.* Gulliver's Swamp.
L. & R. The prevailing form in these river-basins; very common!
Y. Sawley Moor. River Yore at Newby and Boroughbridge!
N. Pateley and Blubberhouses Moors, and in the low country at Aketon bog-field near Spofforth. In the river Nidd below Knaresborough.
W. Langstrothdale, Stake Fell, Barden, and Burnsall Moors! Rombald's Moor. Bramham Moor! Askham Bogs! In the river Wharfe, many places!
A. Malham Tarn! River Aire and Canal, Skipton! Dikes at Bingley; *W. West.* River Aire and Canal from Gargrave to Selby!
C. Todmorden; *A. Stansfield.* Hebden Valley! Norland Moor. Coxley Valley and Batley; *P. F. Lee*! Barnsley Canal!
D. Loxley Valley, Sheffield; *J. Salt.* The Rivelin Valley and dams; *A. Carr.* Bradfield Moors; *Id.*
T. Roche Abbey pools.
Obs.—The small plantain-like leaved form of the Ingleborough, Gilstead (Rombalds Moor), and other stations, recorded in a former Flora as *P. plantagineus*, was this species in its var. or state *ericetorum*; the same thing has been named *pseudo-fluitans.*

772. Potamogeton plantagineus Ducros.
Area: '[64] error,' Top. Bot. (but that dictum of 1873 referred to the high moorland stations reported in Miall and Carrington's Flora). Range: 0—100 feet.
Native; lowland dikes, very rare. P. July-August.
First record: Author.
W. Dike on north side of railway (branch to Tadcaster), one and a half miles north-west of Church Fenton towards Towton!
(A. Gilstead, Rombalds Moor, etc.; *B. Carrington* (1862) in *Miall's Flora.* 'The Peat Pond-weed.' Error. It was *ericetorum*, not the true small-fruited *plantagineus* that Dr. Carrington noticed.)
T. Dikes on the south side of Thorne Moor!

773. Potamogeton rufescens Schrad. (*fluitans* Sm. and O.B.G. list.)
Area: 64, Top. Bot. Range: 0—300 feet.
Native; slow-moving or stagnant water, rare. P. August.
First record: Jos. Woods, MS., about 1838.
R. Ditch by foot-road to Rathmell near Birkbeck's Weir (!) *J. Windsor.*
Y. Ditch near Rippon, *Mr. Brunton*; *MS. note* by *Jos. Woods* in *Towns. O.B.G.*

A. Malham Tarn; *J. Tatham, MS.*; *W. M. Watts.* Canal (Leeds and Liverpool) near Gilstead (!) *W. West.*
C. In the Barnsley and Doncaster Canal!
D. Potteric Carr ditches, August 1845; *J. Hardy* in *Hb. Bailey.*
Obs.—Our plant is all the state *homophyllus* E.B.; and I believe some of the records in Miall's Flora for *P. lucens* are really stations for this species.

774. Potamogeton heterophyllus Schreb.
Area: '63 O. Moore spn.,' 64, Top. Bot. Range: 0—300 feet.
Native; clay-bottomed ponds or slow-moving water, rare.
First record: O. A. Moore, 1850.
A. (Malham Tarn; *H. Baines.* An error, some 'various-leaved' Pond-weed—probably *P. oblongus*—meant). (Adel; *M. and C., Flora.* Error; the Adel Dam plant is *P. polygonifolius*). In a dike near Sherburn!
D. Potteric Carr dikes; *O. A. Moore*!
T. Dikes near Thorne and Auckley!

775. Potamogeton lucens L.
Area: 63, 64 Baker spn., Top. Bot. Range: 0—1,250 feet.
Native; lakes and canals, rare. P. August-September.
First record: Salt, 1801.
Y. In the Yore, not common; *H. H. Slater.*
A. In Malham Tarn (!) *J. Backhouse.* Sparingly in canal, Kirkstall (!) *J. Abbott.*
C. Nostel, etc.; *T. W. Gissing.* Cold Hiendley Reservoir; *G. Roberts*! In the Calder (and Aire); *L. C. Miall.*
D. Near Conisborough Ferry, 1801; *J. Salt.* Dearne Valley Canal; *H. Payne*! Near Doncaster!

776. Potamogeton prælongus Wulf.
Area: 63?, [64 error?], Top. Bot. Range: 1,250 feet.
Native; mountain lakes and tarns, very rare. P. August.
First record: Tatham, 1844.
A. In Malham Tarn; *Spn.* in *Herb. Tatham*, dated June 11th, 1844, *coll. J. T.* Spn. from same locality ('detached and washed up on to shore'; *Fl. Crav.*, 86), July 1865; coll. *J. Windsor* in *Herb. Bailey.*

777. Potamogeton perfoliatus L.
Area: 63, 64, Top. Bot. Range: 0—1,250 feet.
Native; canals, rivers, and (more rarely) pools, common. P. August.
First record: W. Middleton, about 1830.
R. In the Ribble from Rathmell downwards. Sawley; *R. Brown.*
Y. In the river Ure above the mill, Boroughbridge; *Herb. Middleton* in *York Museum.*
N. In the Nidd from Scotton downwards.
W. Common in the Wharfe from Ilkley downwards. In the Ouse from Boroughbridge to Goole.

A. Malham Tarn. Leeds and Liverpool Canal from Skipton. In the river Aire from below Leeds to Snaith.
C. Ryhill and Hiendley Reservoirs; *G. Roberts*! In the canal at Stanley.
D. Dearne Valley Canal (!) *H. Payne.* Worsborough; *G. Harrison.* Barnby-Dun.
T. Canal, Kiveton; *A. Carr.*

Obs.—Many of the localities given in Baines' Flora (without authorities attached) from the Malton and Boroughbridge districts are based upon specimens in the York Phil. Soc. Museum, collected by W. Middleton of Poppleton Lodge—a botanical contemporary of Dalton's later days.

VAR. **Jacksoni** Lees. (Vide Bot. Rec. Club Rep., 1880, p. 150.)
In an isolated pond at Linton near Wetherby; *J. Jackson.* A peculiar form with remarkably abbreviated peduncles.

778. Potamogeton crispus L. *Curled Pondweed.*
Area: 63, 64, Top. Bot. Range: 0—1,250 feet.
Native; ponds, rivers, and canals, common. P. July-August.
First record: Salt, 1800.
R. Near Beggar-Wife Bridge, Settle; *J. Windsor.* In the Ribble (!) *W. Howson*; *J. Tatham*; etc.
Y. Ripon, in the canal, etc.; *H. H. Slater.*
W. In the river Wharfe and Washburn Valley. Bolton Bridge; *J. Willis.*
A. Adel Dam! Leeds and Liverpool Canal!
C. Wakefield, common; *T. W. Gissing.* River Calder; *L. C. Miall.*
D. Near Rotherham; *L. Langley* (1828). Conisborough; *J. Salt.* Canal near Sheffield; *A. Carr.* Barnsley Canal!
T. Kiveton! *A. Carr.*

VAR. **serratus** Huds.
Rather rare.
A. Osier-bed dikes, Brotherton!
D. Ditches near Ecclesfield; *A. Carr*!

779. Potamogeton densus L.
Area: 63, 64, Top. Bot. Range: 0—1,250 feet.
Native; ditches, clay-ponds, and streams, infrequent. P. July.
First record: Salt, 1800.
R. [Giggleswick Tarn and rivulet flowing from it; *J. Windsor.*] In the Ribble; *W. Howson*; *J. Tatham, MS.*
Y. Common in the river Yore; *H. H. Slater.*
W. Acomb; *H. Baines.*
A. Stream running from Malham Tarn! In the river Aire, Skipton; *W. Whitwell.* Canal at Saltaire and Apperley; *W. West.* Bingley, in dikes; *Id.* Sherburn, brick-ponds near station and dikes!
D. Askern; *G. Harrison.* Conisborough; *J. Salt.* Doncaster; *J. Willis.*
T. In the old Torne dikes near Auckley.

780. Potamogeton obtusifolius M. & K.
P. gramineum; Teesdale et aliis.
Area: 63, 64, Top. Bot. Range: 0—100 feet.
Native; pits, deep ditches, and canals, rare (or overlooked).
First record: Langley, 1828.
W. Chandler's Whin ponds, Hob Moor, York, 1881; *G. Webster*!
A. Selby Cut (a canal), 1875! Kildwick (in canal) and Yeadon mill-pools; *B. Carrington.*
C. Barnsley Canal, Wakefield; *T. W. Gissing* in *Wakefield Flora* (sub nom. *gramineus*, as distinct from *pusillus*, but possibly the plant may be *mucronatus*). Halifax Canal; *L. C. Miall.*
D. Rotherham district; *L. Langley* (1828).
T. Canal, Kiveton; *Amos Carr*!

Obs.—I cannot answer for the correctness of any of the above records except the three followed by mark of verification! I have referred the records of former Floras under '*gramineus* L.' to this.

781. Potamogeton mucronatus Schrad.
P. compressus Sm. *P. Friesii* Hooker's Student's Flora.
Area: 64, Top. Bot. Range: 0—100 feet.
Native; rivers and canals, rare (or overlooked). P. August.
First record: the Author.
Y. In one of the ditches by Yore between the mouth of the Ripon Canal and Newby Ferry; *H. H. Slater.*
N. In the river Nidd below Knaresborough, 1869!
A. Ferrybridge; *Y. N. U. Excurs.* (1873). 'Selby Cut,' Selby, 1875!
(C. Stagnant ditch, Coxley Valley; *P. F. Lee*! A rather robust but flowerless plant sent me by Mr. Lee may be this.)
T. Canal, Kiveton; *A. Carr*!

[**Potamogeton acutifolius** Link. Incognit.
Ditch near Newby Ferry; *H. H. Slater, Flora of Ripon* (1873). 'Should be looked for.' Very doubtful.]

782. Potamogeton pusillus L.
Area: 63, 64, Top. Bot. Range 0—1,250 feet.
Native; in ponds and slow-moving or stagnant water-courses, common in one or other of its forms. P. August.
First record: Salt, 1800.
Distribution nearly general; no record for Mersey district.
This species varies considerably, and the preceding, *P. mucronatus*, is probably not distinct; Mr. Bennett treats it as a sub-species.

VAR. **tenuissimus** M. & K.
Not rare.
Y. Dub at Ripon! Sharow Mires; *H. H. Slater*!
W. Ditches and brick-ponds at Hob Moor; and Moss Carrs, Wetherby!
A. Malham Tarn; *W. Todd*! Canal at Shipley; *W. West.*
D. Dikes and pools about Doncaster!

VAR. **rutilus** Wolfg.
In clear clay-bottomed pools at Dringhouses grows a plant I can only refer to this. It is more rigid and tufted in its habit than the other forms, and has a peculiar *facies* from the closely two-ranked regularly alternate leaves, very straight, acute, and gradually attenuated.

783. Potamogeton pectinatus L.
Area: '63, 64 Tatham,' Top. Bot. Range: 0—300 feet.
Native; dikes and canals, rare. P. July-August.
First record: Herb. Harrison, 1862.
R. Ditch near the Willow Island, Settle; *J. Tatham, MS.* (In the river Ribble; *W. Howson.* Possibly *P. flabellatus.*)
N. & W. In the Ouse at York (!) *W. W. Newbould, MS.*, above and below the city.
W. In the Wharfe from Ilkley.
A. In the Selby Cut canal, Selby! In a mill reservoir at Leeds (1867); *J. Abbott*, spn.!
C. Canal, Dewsbury: *P. F. Lee*!
D. Denby Dale; *Y. N. Excurs.* (1872). Don at Worsborough (1862); *G. Harrison*; sub nom. *P. pusillus*; *W. W. Newbould.*
T. Canal, Kiveton; *A. Carr*!

784. Potamogeton flabellatus Bab.
Area: 64, Top. Bot. Range: 0—450 feet.
Native; in rivers and canals, mill-dams (rarely), seldom flowering, but common in a barren state, being one of the species that will grow luxuriantly in swiftly-running water. P. July-August.
First record: the Author, 1873.
In the districts R., Y., N., W., and D., in the main river of each area, in masses among the stones, chiefly in the middle reaches, as about Rathmell and Wigglesworth, Ripon to Boroughbridge, Ripley to Cowthorpe, Bolton Bridge to Newton Kyme, and Wortley to Sheffield.

[**Potamogeton filiformis** Nolte. Incognit.
Barnsley Canal; *Gissing's Flora of Wakefield.* An error.]

785. Zannichellia palustris L.
Area: 63, 64, Top. Bot. Range: 0—150 feet (formerly at 600 feet).
Native; clay-bottomed streams and reservoirs, rare.
First record: Windsor, 1858.

a. macrostemon Gay.
W. In the muddy dike running from North Milford Hall to Coldwell Head Plantation, near where railway crosses it, near Towton! cum *H. T. Soppitt* (1880).

b. brachystemon Gay.
[R. Giggleswick Tarn, east side; *J. Windsor.*]
Y. Sharow Mires, Ripon; and in the canal; *H. H. Slater.*
A. Stream between Monk Fryston and Hambleton; *H. F. Parsons*!
C. Pond near the Union, Wakefield; *T. W. Gissing.* Pond at Batley; *P. F. Lee*!
D. Barnsley Canal (1876 !); *H. Payne.* Askern Pool; *J. Willis.*

c. pedicellata Fries.
In brackish water.
Area: 63, Top. Bot. Range: Low.
(Y. Occasionally (with long pedicels) at Sharow Mires and other places at Ripon, where the other form grows; *H. H. Slater.*) I doubt this being the true var. of Fries.
D. In a pond by the Ouse, in the Hook Road, Goole; *H. F. Parsons*!

LXXIII. ALISMACEÆ.

786. Triglochin palustre L.
Area: 63, 64, 65, Top. Bot. Range: 0—1,250 feet.
Native; marshes and boggy fields, and heaths, common. P. July.
First records: Gough, 1789, 'Yorkshire, common'; Salt, 1799.
L. & R. Common on the lower moorlands. Garsdale! Below Clapham. Cocket Moss; *W. Howson.* Settle, common; *J. Windsor.* Wigglesworth!
Y. Near Sharow Mires; *H. H. Slater.* Boggy pastures, Sawley and Fingerfield!
N. Harrogate, and marsh by the river Crimple at Pannal! Aketon Bleach Works bog-pasture!
W. Ilkley, Rombalds Moor! Wetherby. Askham Bogs; *B. B. Le Tall.*
A. Skipton Moors! Yeadon Ponds; *W. West.* Adel Bog!
D. Rotherham district; *L. Langley.* Askern Marsh! Wharncliffe!
T. Roche Abbey, July 1799; *Herb. Salt.*

787. Triglochin maritimum L.
Area: unrecorded in Top. Bot. Range: Low.
Native; tidal river-banks, very rare, one locality only. P. July.
First record: Birks, 1883.
T. On the muddy-grassy bank of the river Trent, Yorkshire side, at its mouth near Adlingfleet; *T. Birks, junr.*! Perhaps adventive here, sought in vain, many times, by Dr. H. F. Parsons, whilst resident at Goole.

788. Scheuchzeria palustris L.
Area: '63 Harrison spn.,' Top. Bot. Range: Low.
Native; spongy peat bogs, very rare, and perhaps extinct. P. July.
First record: Appleby, 1832.

(A. Hawksworth Moor and other parts of Rombalds Moor; *Miall* and *Carrington's Flora* (1862). Some error.)

T. Thorne Moor, in great plenty; *S. Appleby* (1832) in *Mag. Nat. Hist.*, v, 558. Sought for in vain by Dr. Parsons in 1877-8. I found one flowerless example, by one of the 'Wells,' in company with Mr. W. Todd in 1870; recognisable as *Scheuchzeria* by the pore at the tip of the leaf.

Obs.—First found in Britain in Leckby Carr in 1807 by Rev. J. Dalton.

789. Sagittaria sagittæfolia L. *Arrow-Head.*
Area: 63, 64, Top. Bot. Range: 0—200 feet.
Native; canal- and river-sides, common in the low country. P. July.
First record: Salt, 1795 (Herb.).

W. Dike near Ulleskelf! Marked as seen in Ainsty; *G. Webster.*

A. Canal at Gargrave; *Horsfield* and *Percival* (1866). Skipton! Gilstead and Apperley; *W. West.* Kirkstall, in canal (!) *J. Abbott.* Methley; *G. Roberts.* In Aire, Ferrybridge! Selby.

C. Hiendley and Walton; *T. W. Gissing.* Pond, Milnsbridge; *H. Baines' Flora* (1840), p. 99. Halifax Canal; *L. C. Miall.*

D. Rotherham; *L. Langley* (1828). Dearne Valley Canal; *H. Payne.* In the river Don below Doncaster; *G. Harrison.* Canal at Dovecliffe; *W. W. Newbould.* Near Conisborough Ferry, 1795; *Herb. Salt.* Doncaster and Barnby Dun! Goole; *H. F. Parsons.* Darfield; *T. W. Gissing.*

790. Alisma Plantago L. *Water Plantain.*
Area: 63, 64, Top. Bot. Range: 0—1,250 feet.
Native; ditches, dike- and pond-sides, common. P. July.
First record: Salt, 1800.
Distribution general—in every district, ascending to the margin of Malham Tarn (*Windsor's Flora*) at 1,250 feet.

VAR. **lanceolata** With.

W. Brick-ponds, Hob Moor, York!

A. Pond-sides, Sherburn.

D. Clay-ponds by railway near Doncaster!

791. Alisma ranunculoides L.
Area: 63, 64, Top. Bot. Range: 0—450 feet (formerly at 600 feet).
Native; pools and ditches with a peat or clay bottom, infrequent. P. August-September.
Earliest notice: Blackstone, 1746.

L. In Wennington Moss, near Ingleton, Yorkshire, *Mr. Thornbeck*; *Blackst.*, *Sp.* 75. Perhaps a little out of the Riding.

[R. In Giggleswick Tarn, plentifully; *W. Curtis* (1782); *With.* (1787); *Gough's Cam.* (1789); *Mart.*; *Mill.*; etc. Long gone—the pool drained.]

Y. Bogs on Ripon Common; *W. Brunton.* Ditches about Copgrove; *J. Dalton* in *O.B.G.* Pilmoor Car near Sharow; *H. H. Slater.*

W. Aberford; *L. C. Miall.*

A. Rail-side drains between Sherburn and Ulleskelf! Milford; *G. Roberts.*

C. Canal at Heath near Wakefield; *L. C. Miall.*

D. 'Pottery' Carrs, Doncaster (!) *Herb. Salt.* Askern; *P. Inchbald* and *J. Willis*! Rossington Carrs! Pool at Balby!

T. Dikes near Auckley! Old quarries below Kiveton; *A. Carr.*

[**Alisma natans** L. Incognit.
The Rev. G. Pinder 'believes that he observed it, from a train, in ditches by the railway-side between York and Sherburn' (*Baker, Suppl.* 181). I have more than once examined the wide ballast-dikes parallel with the railway from Sherburn to Ulleskelf, and found only *A. ranunculoides*, the floating form of which I entertain no doubt at all Mr. Pinder really saw.]

792. Butomus umbellatus L. *Flowering Rush.*
Area: 63, 64, Top. Bot. Range: 0—450 feet.
Native; in dikes, still creeks of rivers and ponds.
First records: Dixon, 1800; Miller, 1804.

R. Below Runley Bridge, Settle; *J. Tatham.*

Y. Pond near Ripon; *H. H. Slater.* In the Yore near Newby; *Miss Morton.*

W. Ainsty, near York; *G. Webster*

A. [Formerly by beck below Batty Wood, Meanwood Valley, Leeds. Not since 1860]. Selby! In canal between Gargrave and Skipton; *H. Baines.* Confirmed, 'plentiful,' in 1866 by *Horsfield* and *Percival.* Formerly in the Leeds and Liverpool Canal under Birch Wood, Idle, and near Windhill; *H. Baines.* Perhaps now gone. Canal at Bingley; *J. Willis.*

C. Fish-pond in Heath Park, 1800; *D. Dixon.* Mill-dam at Gomersal; *W. Fowler* (1870). Introduced there? Hemsworth Dam; *Watson.* [Formerly in stream below Porto Bello, Wakefield; *T. W. Gissing*].

D. Potteric Carrs; *E. Miller* (1804); *S. Appleby* 1831). Near Doncaster, sparingly; *H. Baines.* Bolton Common by the river Dearne, three miles below Barnsley (1886); *F. Johnson.*

T. Canal, Kiveton; *A. Carr.* Dikes near Auckley!

LXXIV. HYDROCHARIDACEÆ.

793. Hydrocharis Morsus-ranæ L. *'Frogbit.'*
Potamogeiton latifol. flore albo; Merr.
Area: 63, 64, Top. Bot. Range: 0—100 feet.
Native; dikes and stagnant ponds, rare. P. July.
First records: Merrett, 1666; Hargrove, 1782.

Y. Pond in the Bull Close, Ripon; *Miss Morton.* Perhaps extinct.

N. Staveley Carrs ditches; *E. Hargrove* (1782) = Ponds, Boroughbridge; *H. Baines.*

W. In the Brick-ponds at Hob Moor, York (!) and Askham Bogs; *B. B. Le Tall*; *H. Ibbotson*; etc. Aberford; *T. Hartley.*

A. Snaith; *H. F. Parsons.*

C. At Nostell Abbey, four miles from Wakefield, Yorkshire; *Merr.* 97. Still there in 1870. *Ray* in his *Syn.* relegates it to the doubtful list at the end, but no other than the Frogbit can have been meant.

D. Arksey. Barnby-Dun; *Y. N. Excurs.* (1873). Strawberry Island, Doncaster; *J. Willis.* Doncaster Carrs; *G. Harrison.* In a stagnant water at the bottom of Cadeby Common, 1798; *Herb. Salt.*

T. Rossington dikes. Thorne Moor dikes, near Medge Hall!

794. Stratiotes aloides L. *Water Aloe. Water Soldier.*
Militaris aizoides; Lob.; Ger. *Aloe palustris*; Bauh.
Area: 63, 64, Top. Bot. Range: Low.
Native, but in most of its stations Denizen only; slow dikes and ponds, rare. P. July-August.
First notice: How, 1650, p. 75, 'New ditches of the Dutch workes of Hatfield, within three or four years after they were made.'

N. Pond at Knaresborough; *Miall's Flora* (planted).

W. Kirk Deighton (planted)!

D. & T. Potteric Carrs, Doncaster; *J. Salt*; *J. Windsor, MS.* in *Towns. O.B.G.*; *Salt's Flora*; *E. Miller* (1804). Arksey; *L. C. Miall.* Abundant in the Black Dike, Great Eaa, near the junction of the G.N. main line with that of the Finningley branch, south-east of Doncaster. In the Gyme near Thorne; *Stephen Robson.* Dikes about Thorne and Hatfield still.

Obs.—With us this species flowers only very sparingly, and not every year.

795. Elodea canadensis Mich. (*Udora* Nutt.). *American Waterweed.*
Area: 63, 64, Top. Bot. Range: 0—900 feet.
Denizen; canals, rivers, isolated pools (introduced there by waterfowl?), now too common, but local. Female flowers only found as yet in West Yorkshire. June-July.
First record: Baker, 1854.

R. The Ribble at Wigglesworth.

Y. Ripon district, abundant for a time, now decreasing materially; *H. H. Slater.*

N. In the Nidd and the Crimple in a few places.

W. Wharfe at Ilkley; *J. Willis.* Washburn Valley, mill-race, etc.! Many places about Wetherby, Stutton, Church Fenton, York, etc.

A. Leeds and Liverpool Canal! Skipton! Bingley; *J. Willis.* Fairburn and Methley, abundant in 1870, less plentiful now!

C. Wakefield, almost everywhere; *T. W. Gissing* (1867). Less common now considerably!

D. In the Don and Rother near Sheffield, not common; *A. Carr.*

T. Near Bawtry! Roche Abbey waters; first appearance there about 1865; *Aveling's Hist.*

Obs.—There seems to be a consensus of opinion to the effect that this American plant, after spreading like a wave over the whole country, has now reached the flood-point in abundance, and has begun to die out. After the fashion of some epidemics, it tends to break out or spread into unoccupied areas in which it exhausts its capacities for successful competition. As bearing upon the date of introduction of this plant into Great Britain (usually stated as A.D. 1836) in Aveling's Hist. of Roche Abbey (1870), there occurs the curious statement (by J Bohler probably—he furnished the main part of the plant list which forms the Addendum): 'We noticed it in several places growing with great freedom about half a century ago, and then it disappeared as mysteriously as it came.' This could hardly have been penned later than 1865 or 1866, which would carry the first observance back to about the date of Waterloo.

LXXV. ORCHIDACEÆ.

[**Aceras anthropophora** Br. Incognit.
Area: '63 Mrs. Broadrick,' Top. Bot.

R. Giggleswick Scar; *Watts' School Flora.* 'Included on the verbal authority of a gentleman resident here, at Settle, who, there is no doubt, was mistaken'; *W. M. Watts in litt. ad F.A.L.*, 24th September, 1878.

D. Hampole Wood, near Doncaster, by a Mrs. Broadrick; report to *H. C. Watson* (Vide *Cyb. Brit.*, ii, 432.) Never confirmed.]

796. Orchis pyramidalis L.
O. floribus purpureis; Merr.
Area: 63, 64, Top. Bot. Xerophilous. Range: 0—600 feet.
Native; calcareous fields, broken ground, quarries, etc., common on the Permian Limestone, almost unknown off it. P. July.
First record: Merrett, 1666.

(R. Near Langcliffe—tradition; *J. Windsor*. Tarn field, Giggleswick; *J. Tatham, MS.* note in his copy of Baines, but no spn. The spn. in *Herb. Tatham* is from Wath, North Riding.)

Y. Field next Mackershaw Wood (!), Thieves' Gill, and near Newby, Ripon; *H. H. Slater*.

N. Pastures about Knaresborough, Goldsborough, and Flaxby!

W. Wetherby, rail-banks; Tadcaster, Jackdaw Crag quarry; Bramham! Thorparch; *J. Emmet*. Collingham and Etchell Crags, Scarcroft. Aberford. Towton; *G. Roberts*. Boston Spa; *W. Whitwell*.

A. Ferrybridge district; *Y. N. Excurs.* (1873). Kippax Limehills, etc.(!) *H. Baines*. About Huddlestone Quarry, Micklefield! Pontefract; *B. Carrington*. Garns brows, near Eshton; *J. Windsor*. Ouse bank, Newland; *H. F. Parsons* (adventive—with *Campanula glomerata*). Ledsham.

D. Near Hampole, by Robin Hood's Well, Yorkshire; *Merr., Pin.*, 88; Still common there! Rotherham district; *L. Langley* (1828). Conisborough, Levitt Hagg! Smeaton Crags, and Went-vale; *T. W. Gissing*. Pontefract.

T. Roche Abbey, plentiful; *J. Salt* (1800). Lindrick Common; *A. Carr*.

797. Orchis ustulata L.

Cynorchis minor Pannonica Clusii; How.

Area: 63, 64, Top. Bot. Range: 0—600 feet.

Native; meadows, pastures, and open banks, usually on a limestone soil, but by no means confined to it, occurring on sand by Ouse as on Holderness gravels; common in many parts, unknown in others. P. May-June.

First record; How, 1650.

L. Thornton Bridge; *Jos. Woods, MS.* in *Towns. O.B.G.*

R. Waste ground (one year plentifully), below Birkbeck's Weir, Settle; *J. Windsor*. (Tarn field, Giggleswick; *J. Tatham, MS.* An unlikely spot.)

Y. Near Ripon, abundantly; *W. Brunton*. Clotherholme, and Bridge Hewick; *H. H. Slater*. Mickley; *G. Roberts*.

N. Grimbald Crag pastures; *E. Hargrove*. Meadows about Knaresborough, in plenty!

W. Ouse-side between York and Bishopthorpe; *A. Aiken*. [Adventive; washed down on these alluvial gravelly tracts, as at Clifton Ings, also from stations higher up, but probably existing where now found since late Tertiary period.] Wetherby! Thorparch. Stutton.

A. About Leeds, not uncommon; *W. Wood* in *O.B.G.* [As usual, Wood's limit was wide.] Sherburn; *G. Roberts*. Barlow, near Drax; *H. F. Parsons*.

[C. Katty's Field, Warley; *J. Bolton*. Gone.]

D. On Scasby-lease [Scawsby Leys, two miles N.W. of Doncaster] *Mr. Stonehouse*; *How, Phyt.*, 33 (1650). Rotherham district; *L. Langley* (1828). Brockerdale, Went-vale; *J. Willis*. Hampole; *P. Inchbald*. Conisborough, right of road, near Toll-gate, 1800; *Herb. Salt*. Smeaton (!) *G. Roberts*.

798. Orchis Morio L. *Gander-gosses* (= goslings). '*Foolstones.*'

Area: 63, 64, Top. Bot. Range: 0—600 feet.

Native; pastures and banks, on calcareous soil or alluvial drift, infrequent. P. June.

First record: Salt, 1800.

L. Lower Dentdale, near Sedbergh. Bentham!

R. Settle; *J. Windsor* (sometimes pure white); rep. 1886, *Miss Thompson*. Hellwith Moss; *W. M. Watts*.

Y. Ripon, not uncommon; often in company of *O. ustulata*; *H. H. Slater*.

N. Near St. Robert's Well, Knaresborough; *E. Hargrove* (1832). Birkham pastures!

W. About York, commoner than *O. mascula*! *B. B. Le Tall*. Hob Moor and side of Askham Bogs; *H. Baines*. Thorparch; *Id.* Fields near Cemetery, Wetherby; and Collingham. Pool. Weston Park, Otley; *L. C. Miall*.

A. Near Castleford; *G. Roberts*. Esholt; *J. Willis*.

C. Field near Birkenshaw; *J. Emmet*. Pontefract! 'Below Wakefield, frequent'; *T. W. Gissing* (not within some miles)!

D. Doncaster and Conisborough! Rotherham district; *L. Langley*. South Elmsall; *G. Roberts*. About Newfield Green; *J. Salt*. Goole Fields and Hook; also (with white flowers) at Thorne; *H. F. Parsons*.

T. Sandbeck Park. Kiveton; *A. Carr*.

Obs.—Generally speaking, this orchid likes moister pastures than *O. ustulata*, and is not so partial to limestone soil. In my experience it has often occurred without *O. ustulata*, so they cannot be called geographical allies.

799. Orchis mascula L. '*Foolstones.*'

Area: 63, 64, 65, Top. Bot. Range: 0—1,450 feet.

Native; woods and pastures, common. P. Late April-May.

First record: Salt, 1800.

L. & R. Common in these divisions, ascending Kingsdale and Dentdale to sheltered spots on the lime-scars to 1,450 feet.

Y. Ripon district, common from Sawley down to Boroughbridge.

N. About Knaresborough; *E. Hargrove*. Common at Scotton, Goldsborough, Cattall, etc.!

W. Very common from Buckden and Grassington downwards to Tadcaster.

A. Malham! Skipton Woods; *W. Whitwell*. Esholt; Bingley; *W. West*. Alwoodley, etc., but rather rare in the middle Aire Valley, in the populous districts, common again on the Permian tract about Ledstone, etc.

C. Below Wakefield, frequent. Occasional only in Calder-vale.

D. Conisborough district (!) *Y. N. Excurs.* (1873). Common on the Permian tract. Pastures on the top side of old Park; *J. Salt*. Field near Fulwood.

T. Very common in this district.

800. Orchis incarnata 'L.,' Hook. Stud. Flora, Ed. iii. '*Crowfeet*' or '*Crakefeet.*'

Area: '64 Webster,' Top. Bot. Range: 0—1,500 feet.

Native; marshy meadows, boggy fields, and hill pastures, infrequent in the lower lands, common in the higher dales. P. Before Mid-summer.

First record: Salt, 1800.

M. Greenfield Valley, swampy ground!

L. & R. Common in the damp upland pastures of these divisions, all about Garsdale, Dentdale, Kingsdale, Horton and Settle in Ribblesdale! 'Common'; *W. Howson*.

Y. Hackfall, 1872; *G. Roberts*. Sharow Mires; *H. H. Slater*. Cow Mires; *Miss Morton*.

N. Upper Nidderdale, common—becoming quite rare below Harrogate.

W. About Askham Bogs; *B. B. Le Tall*; etc. Langstrothdale to Kettlewell. Rare below Bolton.

A. Malham Moor! Airton and Skipton Rock! Rare in Lower Airedale.

C. Below Wakefield, frequent; *T. W. Gissing* (I doubt). Rare in Upper Calderdale.

D. Near Conisborough; *J. Salt* (sub nom. *latifolia*). Marshy ground by Don, Conisborough!

T. Borders of Thorne Waste, and further south towards Hatfield.

Obs.—I have ventured, advisedly, to refer all the old records for '*latifolia*' to this.

[**Orchis flore nigro-splendente.** In the pastures near Settle, Yorkshire; *Merr.*, 89 (1666). Included by *Dillenius* in his *Indic. Plant. Dubiarum, App. to Ray's Synopsis* (1724). *O. incarnata*, fine and with very dark-purple flowers, as well as lighter ones, is the only pasture Orchis, besides *O. maculata*, in the locality, at all applying to the description.]

801. Orchis eu-latifolia E.B. (*O. majalis* Top. Bot.)

Native; mountainous boggy pastures, very rare. P. July.

First West Riding record: the Author's.

L. Garsdale, sparingly; *O. incarnata* long past blooming!

W. Near Outershaw, Langstrothdale, in mid July! Askham Bogs (with the preceding species); *B. B. Le Tall* [but I doubt].

(D. Conisborough; *Geo. Harrison, spn.* as '*maculata.*' 'But the specimen may be true *latifolia*'; *MS. note* by *W. W. Newbould* in his copy of Baines' Flora.)

Obs.—I have not found this above half a dozen times, but in every case it was in perfect bloom two to three weeks *later* than the much commoner *O. incarnata*. Prof. Babington's opinion is, I think, in agreement with this, though not that found in Hooker's Student's Flora (Ed. iii, 390); nor that of Bab., Man., 8th Ed., p. 345. The words, 'mostly quite past flowering when sp. 9 is in perfection in the middle of June,' now appended to *O. latifolia*, were rightly affixed to *O. incarnata* in Bab., Man., Ed. 5. Surely one would think Prof. Babington can hardly have *exactly reversed* his opinion between 1862 and 1881; and a reversal, too, opposed to the facts. Many other localities, besides those above given, exist for '*latifolia*' in its aggregate sense, but I believe most of them refer to the segregate *incarnata*, treated here as a distinct species.

802. Orchis maculata L. '*Crowfoot*' or '*Crowfeet.*'

Area: 63, 64, 65, Top. Bot. Range: 0—1,500 feet.

Native; woods, marshes, and moist meadows, very common. P. June.

First record: Salt, 1800.

Distribution general—in every district, varying considerably in colour of the flowers and maculation of the leaves.

[**Orchis militaris** Huds. Error or Ambiguity?

Howroyd Park, Barkisland; *J. Bolton, Cat.* (1775).

Many other statements, presumably errors, of Bolton's are, like this, untraceable now.]

803. Gymnadenia conopsea Br. *Clove Orchis.*

Palmata rubella, cum longis calcaribus rubellis; Ray.

Area: 63, 64, 65, Top. Bot. Range: 50—1,750 feet.

Native; banks and pastures, mostly on limestone, common, with a wide vertical range.

First record: Dillenius, 1724.

L. Common on the limestone grassy turf in Dentdale, Garsdale, Kingsdale!

R. Mill Island, Settle; *Gough's Cam.* Tarn Brow; *W. M. Watts*. Upper Ribblesdale, common!

Y. Mackershaw, Hackfall, Thieves' Gill, Ripon (*H. H. Slater*), etc. Common about Ripon.

N. Common on the Permian tract about Knaresborough.

W. Arncliffe Clouders! Buckden and Grassington, Bolton. Common on the Permian tract about Wetherby, Thorparch, Bramham, etc.

A. *Flore albo in pratis humidis* juxta N. Bierley [Bradford] *Rich.*; *Ray, Syn.*, iii, 381 (1724). Malham; *J. Nowell*. Skipton Rock! Sherburn, towards Micklefield.

C. Todmorden pastures; *A. Stansfield*. Halifax; *H. Baines*.

D. Went-vale; *T. W. Gissing.* Pontefract; *J. Willis.* Rotherham district; *L. Langley.*
T. Roche Abbey; *J. Salt.*

804. Gymnadenia albida Rich.
Orchis palmata thyrso specioso longo, dense stipato, etc.; Ray.
Area: 63, 64, 65, Top. Bot. Montane. Range: 1,700—700 feet.
Native; stony pastures among the hills, mostly upon limestone, rare. P. Mid June-July.
First record: Dillenius, 1724.
L. Pastures above Helks Wood, Ingleton; *Jos. Woods, MS.* in *Towns. O.B.G.* Between Feizor and Wherf Village; *R. Clapham*; rep. *W. M. Watts.* Brackenbrow; *Herb. Tatham.*
R. Croft Closes, Giggleswick (!) = Hillocks near Giggleswick Tarn; *J. Windsor*; still there, and near Lodge, 1883-6; *Misses Thompson.*
Y. Dallow Gill; *W. Brunton, O.B.G.*
W. Slopes of Clouders above Hawkswick, very sparingly. Outershaw, in Langstrothdale, one plant!
A. Juxta Malham, *Richardson*; *Ray, Syn.* iii, 382 (1724)! *Gough's Cam.* (1789). Malham Tarn Moor, above Gordale!
C. Back Rough, near Portsmouth, Todmorden; *A. Stansfield* (1877). This station is, perhaps, just over the county boundary, and therefore in Lancashire.

805. Habenaria viridis Br.
Area: 63, 64, 65, Top. Bot. Range: 0—1,200 feet.
Native; banks and moist pastures, common, but often overlooked, being inconspicuous. P.
First record: Curtis, 1782.
L. Pastures near Feizor; *R. Clapham.* Cavehole Wood; *J. Windsor*; etc. Garsdale!
R. On the moist hilly pastures about Settle; *W. Curtis* (1782); *Gough* (1789). Many places about Settle (!) *J. Windsor.*
Y. Copgrove; Hackfall; *W. Brunton.* Studley Lime-kilns; *W. Brunton* in *O.B.G.* Ripon.
N. Harrogate Moor.
W. Near Arncliffe; *J. Windsor, MS.* in *O.B.G.* Bolton Bridge! Otley Chevin; *J. Willis.* Stainburn; *W. Todd!* By railway, a mile west of Wetherby; *J. S. Wesley!*
A. Malham! Harden Moor and Allerton, near Bradford; *L. C. Miall.* Eldwick; *W. West.* Roundhay; *G. Roberts.* Adel. Keighley; *Thos. Cooper!*
C. Near Halifax; *H. Baines.* Ogden Clough; *J. Bates.* Lofthouse and Ardsley; *G. Roberts.* Barkisland.
D. Rotherham district; *L. Langley* (1828). Heathy ground near Gawber; *H. Payne.*
T. Near Lindrick; *A. Carr.*

806. Habenaria bifolia Bab.
Orchis sphegodes major et minor; Merr., 88. [*H. chlorantha* = the 'major.']
Area: 64, Top. Bot. Range: 0—1,250 feet.
Native; heathy or rushy pastures, mostly amongst the hills, scarce, not nearly so common as the next species.
First record: Dixon, 1792.
L. Thornton Force, Ingleton; *J. Willis.* Clapham; *G. Roberts.* Helwith Moss; *W. Howson.*
R. Kelcowe, about Giggleswick Tarn, etc.; *J. Windsor.* Horton-in-Ribblesdale! Near Lodge, Settle; *Misses Thompson.*
N. Harlow Heath! *T. Hick.* Brimham Moor! Hookstone Crags; *H. Baines.*
W. Near Bolton Bridge; *J. Willis.*
A. Malham Moor pastures (!) *L. C. Miall.*
C. Park Hills, 1792; *D. Dixon.* Todmorden district, cloughs and pastures; *A. Stansfield.*
D. Langsett, S.W. of Penistone, 1885; *H. Johnson*

807. Habenaria chlorantha Bab. '*Butterfly Orchis.*'
Orchis hermaphroditica elegans; How; Ger.
Area: 63, 64, 65, Top. Bot. Range: 0—1,250 feet.
Native; woods, mainly on limestone, and in mountain pastures, common. P. July.
First record: Curtis, 1782.
L. Helwith Moss; *W. Howson.*
R. Mill Island near Settle; *Gough's Cam.* (1789). Settle; *W. Curtis.*
Y. Hackfall and Kirby Malzeard! Field near Mackershaw; *H. H. Slater.* Mickley; *G. Roberts.*
N. Scotton Banks, Knaresborough, and other places!
W. Arncliffe and Grassington Woods. Barden and Bolton. Bramham!
A. Malham! Hawksworth; *W. West.*
C. Todmorden Valley Cloughs; *A. Stansfield.*
D. By Uden [Ewden, two miles south of Middopstones] Lodge in Pennyston Moore, *Mr. Stonehouse*; *How, Phyt.*, 83; *Merr., Pin.*, 85. This may have been what we now call *bifolia* Bab. (not then discriminated), as in the moory district at the head of the Don river it is much the most frequent species. Rotherham district; *L. Langley* (1828). Edlington Wood, Doncaster; *J. Willis.* Meadows near Aldwark; *J. Salt.*
T. Roche Abbey and Kingswood!

[Herminium Monorchis Br. Incognit.
A report, through Mr. Winch, that Mr. Bicheno once found it (vide *New Bot. Guide*, vol. i and vol. ii). No doubt there is some mistake.]

808. Ophrys apifera Huds. *Bee Orchis.*
Area: 63, 64, Top. Bot. Range: 0—600 feet.
Native; dry calcareous ground, grassy turf or bare and recently disturbed, rare, but generally distributed over the limestone. P. July.
First record: Pierson in Hargrove, 1782.
(R. Hellifield; *W. M. Watts*).
Y. Studley Woods and Laver Banks; *W. Brunton.* Copgrove; *J. Dalton* in *O.B.G.*
N. Goldsborough Wood and Belmont Wood, Knaresborough; *E. Hargrove* (1828). Pastures about the Abbey, Knaresborough; *E. Hargrove, Hist.* (1782).
W. Thorparch (!) *W. Wood* in *O.B.G.* Collingham, Tadcaster, Aberford; *H. Baines.* Rail-bank half-mile west of Wetherby Station (!) *J. S. Wesley.*
A. Kippax; *W. Wood* in *O.B.G.* Near Knottingley Station; *J. Willis* = Ledsham. Plantation near Thorleby, Skipton, in 1885 (*W. Eade*) and 1886; *T. W. Edmondson!* Gargrave; *L. C. Miall.* Huddlestone Quarry near Sherburn; *H. Baines.* Birtwhistle Rocks, Skipton; *H. Baines*, rep. *W. Howson.* Glass Houghton lime quarry; *G. Roberts.* Martin's Meadow near Mr. Wilson's Wood, Eshton Hall; *J. Windsor.*
D. Rotherham district; *L. Langley* (1828). Limestone tract west of Doncaster; *S. Appleby* (1831). Hampole; *P. Inchbald.* Warmsworth Cliffs.
T. Old quarry below Kiveton; *A. Carr.* Roche Abbey; *J. H. Aveling.*

[Ophrys aranifera Huds. Ambiguity.
O. sive Testiculus sphegodes, hirsuto flore; Ray.
Area: [64], Top. Bot.
Never confirmed.
W. Juxta Bramham non procul Tadcastria; *Ray, Syn.*, iii, 380 (1724) = 'Dry pastures about Bramham'; *O.B.G.* (1805).]

809. Ophrys muscifera Huds. *Fly Orchis.*
Orchis myodes elegans; How. *O. myodes purpurascens*; Merr.
Area: 63, 64, Top. Bot. Range: 0—600 feet.
Native; calcareous banks and open woods, infrequent. P. May-June.
Earliest record: How, 1650.
L. Skirreth Wood near Ingleton; *Blackstone.* Thornton Force; *J. Willis* (1873). Wood between Settle and Clapham, 1882-86; *Miss F. Thompson.*
R. On the hilly limestone pastures at Stackhouse near Settle; *Gough's Cam.* (1789). Stackhouse 'Borrins', Kelcowe, Lord's Wood, wood near Langcliffe Place, all near Settle; *J. Windsor*; etc.
Y. Ward's field near Copgrove; *J. Dalton.* Mackershaw Woods, etc.; *W. Brunton.*

N. Goldsborough Wood; *E. Hargrove* (1809 and 1828). Pastures near Knaresborough Abbey; *Archdeacon Pierson* in *Hargrove's Hist.* (1782) and in *O.B.G.* Bilton and Knaresborough (!) *P. Inchbald* (1885).
W. Thorparch (!) and Grassington Woods; *Rev. W. Wood* (1805). Wetherby.
A. Kippax (Bland's Wood); *H. Baines.* Hetchell's Wood, Scarcroft (!) *A. Shipley.*
D. In Brodsworth Wood; *Mr. Stonehouse*; *How, Phyt.*, 83; *Merr.*, 86 (1666); *Ray.* Rotherham district; *L. Langley.* Limestone tract west of Doncaster; *S. Appleby* (1831). Went-vale; *J. Willis.* Conisborough; *Yorks. Nat. Exc.* (1873).
T. Roche Abbey, 1796; *J. Salt.* Confirmed; *J. H. Aveling.*

810. Spiranthes autumnalis Rich. '*Ladies' Tresses.*'
Area: 63, 64, Top. Bot. Sub-xerophilous. Range: 0—300 feet.
Native; stony pastures and old lime-quarries, rare. P. August.
First record: Blackstone, 1746.
Y. Quarry Moor, Ripon, plentifully; *H. H. Slater.* Near Copgrove; *W. Middleton.*
N. Scotton Banks, Knaresborough.
W. Bramham Park lawn; *H. Baines.* Barwick-in-Elmet (!) *T. Hick.* Linton Common, Wetherby; *J. S. Wesley!*
A. In the new field near Ledstone Hall; *Mr. Dawson*; *Blackst., Spn.*, 1746; *Martyn* (1763); etc. Kippax limehills (!) *H. Baines.* Micklefield! Ledsham and Glass Houghton; *G. Roberts.*
D. Went-vale; *T. W. Gissing.* At Sprotborough; *Gough's Cam.* (1789). Warmsworth; *H. Baines.* Rotherham district; *L. Langley.* Lime pastures west of Doncaster; *S. Appleby* (1831). Hampole; *P. Inchbald.*
T. Kiveton Common (!) *Amos Carr.* Roche Abbey; and near Norton in a clayey pasture; *J. Salt.* Spoil-banks of quarry opposite the farm-house from the highroad down to the Abbey (Roche); *J. H. Aveling.*

811. Listera cordata Br.!
Bifolium minimum; J. B. Pin.; Ray, Syn.
Area: 63 Leyland spn., 64, Top. Bot. Montane. Range: 1,850—350 feet.
Native; peat-bogs on the fells, rare, but perhaps often overlooked.
In septentrionale parte com. Eboracensis, et in monte Pendle dicta invenitur; *Bob.*, 489 (1699).
First records: Ray, 1670; Bobart, 1699.
L. On Ingleborough Hill; *Sir J. E. Smith* in *O.B.G.* West ascent to Ingleborough, plentifully; *J. Windsor.*
R. Penyghent; *H. Baines*; rep. *H. T. Soppitt* (1885). Ryeloaf Hill above Settle; *J. Tatham, spn.* (*N.B.G.*).

Y. Near the waterfall at Lord Grantley's Lakes on Sawley Moor; *W. Brunton*; *W. Middleton* in *York Mus. Hb.*

A. A mile from Keighley, *Richardson*; *Ray, Syn.*, iii; rep. *Martyn*; etc. On Blackmoor, Leeds, in a ditch among the heath; *Mr. Dawson*; *Blackst., Spec.* (1746); *Martyn* (1763).

C. Several heaths in Yorkshire, particularly not far from Almondbury; *Ray, Cat.*, 44 (1676), and *Hist.*, 1233; *Martyn* (1763). Ogden Clough! Stansfield Moor, and near the Ridge Inn at the bottom of Widdop, and Robin Hood bed, Blackstone Edge; *A. Stansfield.*

D. On the high moor south-west of Sheffield; *J. Salt.* Moors between Sheffield and Chatsworth; *Sir J. E. Smith* in *O.B.G.* Stainborough Park, Rockley; *H. Payne.*

812. Listera ovata Brown! *Twayblade.* '*Swayblade.*'

Area: 63, 64, 65, Top. Bot. Range: 0—1,200 feet.

Native; woods and shady banks, common. P. June.

First record: Gough in Camden, 1789.

Distribution general—occurs in every district, but quite rare in that of Mersey tributaries.

Obs.—There are sometimes three leaves, an accessory one below the two sessile, opposite ones.

813. Neottia Nidus-avis Rich.

Nidus Avis Lugd.; Ray.

Area: 63, 64, Top. Bot. Range: 0—600 feet.

Native; very shady woods on bare ground amid rotten leaves, generally under Beech trees or Hazel 'Scrub,' rare, and never in any great numbers.

First record: Gerarde, 1597.

L. In the hedges where the *Calceolus Mariæ* grows, near Ingleton, *Dr. Richardson*; *Dill.* in *Ray Syn.*, iii, 382.

R. Kelcowe; *Simmonds* (1798). High-ridge Wood, Settle (1806); *J. Windsor.* = Giggleswick; *W. M. Watts.*

Y. Laver banks, Ripon, Mackershaw, and Studley; *W. Brunton.* Farnham!

N. Neer unto a village called Knaresborough; *Gerarde, Herball* (1597). Gibbet and Goldsborough Woods; *E. Hargrove* (1809 and 1828). Gibbet Wood; *Mrs. S. Meanbrook* (1802); *Herb. Salt.* Belmont Wood, Knaresborough (!) *H. Baines.*

W. Thorparch Woods; *Gough's Cam.* (1789). Crag Woods, Boston Spa (1880!) *J. Emmet*; etc.

A. Malham Cove. Gordale; *W. M. Watts.* Holden Gill near Steeton; *J. Willis.*

C. Below Rag Scout, Harley Wood, Todmorden; *A. Stansfield.* [On semi-calcareous Yoredale shales.]

D. Conisborough Carr, 1841; *J. E. Stocks* in *Herb. Tatham.* Scholes Wood, Rotherham; *G. P. Nicholson* in *Herb. Tatham.* Hampole Wood; *P. Inchbald.*

T. Maltby; *T. Dickin* (1878). King's Wood, Roche Abbey; *J. Bohler* (1844). Roche Abbey Wood; *Herb. Bloxam.*

814. Epipactis latifolia Auct. '*Bastard Hellebore.*'

Helleborine latifolia montana; Ray. *Serapias helleborine*; Harg. Hist.

Area: 63, 64 Tatham spn., Top. Bot. Range: 0—900 feet.

Native; woods, infrequent, and somewhat uncertain in its appearances, but distributed thinly through most of the districts. P. Late July.

First record: Blackstone, 1746.

L. Near Ingleton and Clapham, *Mr. Thornbeck*; *Blackst., Spec.* (1746). Feizor and Austwick Woods; *R. Clapham.*

R. Langcliffe, Highbridge, and other woods near Settle (!) *J. Windsor*; etc. Oustre-bank Wood; *Herb. Tatham.*

Y. Mackershaw, Hackfall (!) and Newby districts; *H. H. Slater*; etc.

N. Near the Long Walk at Knaresborough, scarce; *E. Hargrove* (1782).

W. Arncliffe and Buckden Woods! Lindley Woods, Washburn Valley! Thorparch and Wetherby Woods; *H. Baines.* Bramham West Woods! (Knavesmire Wood, 1866; *Richd. Whitwell*—The spn. young and small but suspiciously like *E. violacea*—the '*media*' of some early observers.)

A. Gordale; *W. West.* Bingley Woods; *H. Baines.* Ledsham! (Huddlestone Quarry; *H. Baines*; this is *E. media*!) Bush Cliff Wood near Carlton, 1883; *G. Roberts.*

C. Halifax; *J. Bolton, Cat.* (1775). Fixby Park; *J. Eastwood.* Haw Wood, Wakefield; *T. W. Gissing.* Woods near Whitley Hall and Storthes Hall; *C. P. Hobkirk.*

D. Rotherham district; *L. Langley* (1828). Limestone woods west of Doncaster; *S. Appleby* (1831). Endcliffe Wood, Sheffield; *W. W. Newbould.* At the bottom of Old Park, August 1799; *Herb. Salt*; 'perhaps not the typical plant, this'; *W. W. Newbould.* Cawthorne and Wentworth Woods; *H. Payne.*

T. Kingswood, Roche Abbey!

815. Epipactis ovalis Bab. '*Redshanks.*'

Helleborine altera atro rubente flore; Ray, Syn., iii. *Serapias latifolia β*; Gough.

Area: 64, Top. Bot. Range: 1,100—500 feet.

Native; exposed stony slopes on limestone, rare, but locally abundant enough some years. P. July.

First record: Ray, 1677.

L. In Friar Wood and Mill Bank near Ingleton; *Mr. Thornbeck*; *Blackst., Spec.* (1746). Mill Bank, Ingleton; *Martyn* (1763); with *Thalictrum montanum* and *Rubus saxatilis*, which still grow there. Friar Wood, Ingleton; *Id.* Cavehole Wood; *J. Windsor.* Feizor Wood; *R. Clapham.* Sykes' Wood, Ingleton; *Gough's Cam.*

R. Loose stones below Giggleswick Scar; *Jos. Woods, MS.* in *Towns. O.B.G.*, and *Windsor's Flora* = Hill slope above the Ebbing and Flowing Well (!) 'plentifully there every year'; *Misses Thompson.* Stackhouse Borrins; *H. Baines.*

(Y. Near Ripon; *O. A. Moore*—an error?)

A. Mountain-sides, Malham; *Ray, Cat.*, ed. ii (1677); *Syn.*, i (1690); *Gibs. Cam.*; *Syn.* iii, 383 (1724); rep. *Martyn.* Clefts of rocks above Gordale; *J. Windsor.* Kirby-Malham; *J. Tatham.*

816. Epipactis violacea Dur. (*E. media* Bab. Man.).

Area: '63 "Newbould," 64 Baker spn.,' Top. Bot. Range: Low? replaced apparently in the hill-woods by the preceding.

Native; in clumps in woods and bushy ground on limestone, rare. P. August.

First record: Newbould, 1840?

Y. Mackershaw Woods; *O. A. Moore* in *Baker's Supp.*, sub nom. *ovalis.* The only species I could find there is the same plant as grows in Thorparch Woods; I consider it *E. violacea.*

W. Thorparch Woods, growing in tufts (intermittently) in broken ground about the old limekilns! Knavesmire Wood, August 1843; *Silv. Thompson*; in 1866; *R. Whitwell*!

A. Bushy places about Huddlestone Quarry! Recorded under the aggregate name *latifolia* in *Baines.*

D. Near Sheffield in Yorkshire; *Ray, Cat.*, i. 162; ed. ii, 157 (1670 and 1677), under the same designation as the Malham and Settle plant which is *not* localised in ed. i, but is added in ed. ii—the Sheffield station being *omitted* in the *Synopsis*, whilst the Malham station is retained. This looks as if Ray had come to consider that the two species were not identical, although different from '*H. latifolia montana,*' i.e., *E. latifolia* Auct. I consider the Sheffield sylvestral plant to be *E. violacea.* Ecclesall Wood; *W. W. Newbould.*

817. Epipactis palustris Crantz.

Area: 63, 64 Tatham spn., Top. Bot. Range: 0—1,200 feet.

Native; marshes and wet rushy, sandy, or boggy fields, rare. P. July.

First record: Curtis, 1782.

L. In the boggy part of Sykes Wood [Ingleton]; *Gough's Cam.* (1789). Clapdale; *R. Clapham* = Near Ingleton; *J. Nowell.*

R. In the further part of the wood—'twenty yards from the bridge, *MS.*—near Langcliffe Place, Settle; *J. Windsor.* Oustre-bank Wood, 1835; *Herb. J. Tatham*!

Y. Boggy grounds near Ripon; *W. Brunton* (1805). Cowmire near Galphay, in some plenty (!) *H. H. Slater.*

N. Bog Carr Field near the Sulphur Well, Knaresborough. Boggy field (Ingbarrow) near the Stockeld Park Gate, and by the road, about half way between Wetherby and Spofforth (!) *J. Jackson.*

W. Field near Mill, Boston Spa, east end of town; *J. Emmet*! Brick-ponds, Hob Moor, York; *H. Baines.* Also near Kilnsey; *W. Curtis* (1782); *Gough*; etc.

A. Above Gordale; *J. Nowell.*

D. On strong clay soil, Clayton, etc., near Doncaster; *S. Appleby* (1831). Tinsley Wood, Sheffield; *A. Carr.*

T. Marsh near Letwell, July 1813; *Herb. Salt.*

818. Cephalanthera ensifolia Rich.

Serapias longifolia γ; Huds. *Helleborinem foliis prælongis acutis*; Ray; Bob.

Area: 64, Top. Bot. Range: 0—600 feet.

Native; rocky woods, upon limestones only in West Yorkshire, very rare. P. End May-June.

First record: Merrett, 1666.

L. In Helkwood . . . not far from Ingleborough; *Merr.*, 61. Eandem etiam D. Newton in Eboracensi comitatu, invenit; *Bob.*, *Bot. Plant.*, 488 (1699). Helk's Wood; *J. Windsor.* Mountain-side woods about Ingleton and Clapham; *Hudson* (1762); *Martyn* (1763). Under Brackenbrow, Ingleton, at the end of a wood; *Ray, Fascic.* (1688); *Syn.* i; rep. *Martyn.* Since found in flower by *Dr. Richardson*; *Ray, Syn.* iii, p. 384. Still there 1869! Woods at Settle and Ingleton; *R. Teesdale* in *O.B.G.*

(Y. Hackfall; *Mrs. Berkeley.* Drawing of plant seen by *H. H. Slater*, who thought it *C. grandiflora*; but I have no doubt, looking to their respective ranges in Britain, that it, 'large flowered' though it was, could but be *C. ensifolia.* Vide *Slater's Flora of Ripon*; *Trans. Y. N. U., Series* E., p. 169).

W. Wood at Jackdaw Crag near Tadcaster; *O. A. Moore* (1840); reappeared (!) 1883; *B. B. Thompson*; and again in 1886 (*P. Inchbald*).

[**Cephalanthera rubra** Rich. Misnomer.

Area: '[64],' Top. Bot.

'*Serapias rubra*: woods on the sides of the mountains about Clapham and Ingleton; *Hudson*'; rep. *With.*, 48; and *O.B.G.* (1805). *Epipactis ovalis* was the plant.]

819. Malaxis paludosa Sw.

Area: 63, Top. Bot. Range: 750 feet.

Native; spongy bogs amid Sphagnum, one station only, and rare in that. P. August.

First (only) record: Buxton.

M. Gulliver's Swamp, above Greenfield (!) *R. Buxton*; *J. Whitehead*. Long known—still there in 1876.

820. Cypripedium Calceolus L. *Lady's Slipper.*
Calceolus Mariæ; Ger. Em., 443; Ray, Syn., 385. *Helleb. flore rotundo*; C.B., Pin.

Area: 64 Tatham spn., Top. Bot. Range: 500—850 feet.

Native; rocky woods upon limestone, very rare, uncertain in its appearances, but still occurs in two places. P. Mid June-early July.

First records: Parkinson, 1640; Ray, 1670.

L. In a wood called the Helkes . . . neere the border of Yorkshire; *Park., Theat.*, 218=Helks Wood, fast by Ingleborough; *Ray, Cat.* i, 51; ii, 49=At the end of Helks Wood; *Ray* in *Gibs. Cam.* (1695)=In sylva quadam prope montem Ingleborough Helks Wood dicta; *Bob.*, 488 (1699). Root from Helks Wood; *Hb. Middleton.* Not to be found when I was there in 1796; *Dawson Turner* in *O.B.G.* I searched for it in vain in Helks Wood, a gardener of Ingleton having eradicated every plant for sale; *Mr. Woodward*; *Withering*, ed. 2, vol. ii, p. 1002 (1798). In Friar Wood, Chapel-le-dale, at the foot of Ingleborough, found by Mr. Thornbeck, surgeon at Ingleton, who likewise informs me that it has been lost in Helks Wood for some years; *Blackstone, Spec.* (1746). Sparsely in woods about Clapham; *Hudson* (1762)=In a wood between Clapham and Ingleborough; *Martyn* (1763)=In Friar Wood; *Martyn* (a second entry under town heading of Ingleton as well as under Clapham, but same locality really). Between Ingleton and Chapel-in-the-dale, in 1800; *W. Brunton* in *O.B.G.* Now hardly ever met with; *R. Chambers* (1838). (Deepdale (Dentdale), leaves only (1854); *Miall's Flora.* Cannot be told with certainty from the leaves only—the locality therefore needs confirmation).

W. Woods and hilly pastures about Kilnsay; *W. Curtis* (1782); rep. *Gough's Cam.* (1789). About Kilnsey Crag; *W. Wood* in *With.*, iii. Specimen from 'Kilnsey Wood' in *Hb. Tatham* [as well as that from Arncliffe alluded to below]. About Arncliffe, Litten, and Kettlewell; *T. Knowlton* in *O.B.G.* (1805). In the narrow wood called East Field Wood, and the adjoining ones . . . in 1804 or 5 a few specimens . . . were pointed out to me by Miss Petty; *J. Windsor.* Heseltine Gill; *Alex. Irvine* (*H.B.P.*, 325). Arncliffe, below a tree of *Pyrus Aria*; *J. Tatham* (1833).

Woods about Kettlewell, two specimens were found in 1835; *Baines' Flora.* 'I saw the Cypripedium . . . in Yorkshire (one plant) . . . in 1844'; *W. Borrer* in *Bot. Gaz.*, ii, p. 93. Arncliffe, 1847, really wild; *Archdeacon Boyd.* Specimen, originally sent to J. Tatham, now in *Herb. Whitwell.* [Specimens with two blooms, sent in a fresh state, in 1878, to Mr. Whitwell as 'from Arncliffe,' came out of a garden.] 'Root from Grassington Wood, 1797'; *J. Dalton, Hb.* East Field Wood, 1860; *James Ward*, spn. in *Hb. Lees*; Id., spn., 1867; *Herb. G. E. Smith.* Grass Wood, Grassington, 1875-81; *H. Gleadall*! The Kettlewell Wood, one plant found, and carried off, in 1883; *Archd. Boyd* (1887). (About Thorparch—tradition; found near Pontefract by *J. Backhouse*; *J. Emmet* in *Miall and Carrington's Flora.* Some misunderstanding?).

LXXVI. IRIDACEÆ.

[Iris fœtidissima L. Extinction.

Area: 64, Top. Bot.

Denizen? ditches and wet woods, extinct. Always included in former Floras as an undoubted Native, but looking to its ascertained range in England, and the fact that the local circumstances in each of its reported stations differ so widely from the dry thickets and stony slopes of woods, etc., mostly on calcareous soil, which it affects in the Isle of Wight, Devon, Dorset, and other counties in which it is indisputably wild, I am forced to the conclusion that it is not and never has been an *indigenous* integer in the West Yorkshire Flora.

Y. Bank at Little Nunwick by Ripon; *W. Brunton* in *O.B.G.* (1805); addition 'behind Clark's house'; *H. Baines* (1840). In ditches at Little Nunwick! *T. Simpson* (of Hopetown); *Baker's Supp.* (1854)—the '!' seeming to imply that T. Simpson had seen it himself growing there. Miss Morton has sought it for years there in vain; *Slater's Flora* (1884).

T. Near the railway, Rossington, doubtful; *Miall's Flora.* Low wood by railway, Rossington; *J. Hardy, MS.* (about 1842-4). Sought for in vain by Rev. W. Fowler and myself in 1878.]

821. Iris Pseudacorus L. *Flag. Yellow Flag.*

Area: 63, 64, Top. Bot. Range: 0—600 feet.

Native; pool-sides, swamps, and wet pastures, common. P. June.

First records: Bolton, 1775; Salt, 1800.

Distribution general—in every district, but rare in some parts of the Riding, as about Leeds (a ditch at Bullathorpe, Temple-newsam is perhaps the nearest station) and in the upper parts of the dales.

[Crocus vernus All. Alien.

A garden escape; naturalised in the sense that applies peculiarly to bulbs—persisting when planted, but not spreading much. P. April.

Y. Studley Park! Outcast or, more probably, planted.

A. Weetwood—a waif from a garden, doubtless!

C. Bierley near Bradford; *Miall and Carrington's Flora.* Formerly at Manningham, near the Rifle barracks; *John Willis.*

D. 'Naturalised near . . . Sheffield'; *Stokes, Bot. Comp.*, 205.]

822. Crocus nudiflorus Sm. (*C. speciosus* Hook.)

Area: 63†, Top. Bot. Range: 100—300 feet.

Denizen; meadows and pasture-land in valley bottoms, rare. P. October.

First record: Wood, 1805.

C. Pastures near Halifax; *W. Wood* in *O.B.G.* (1805); rep. *Flo. Brit.* i, 41. Fields at Saville Green, Halifax; *Winch, Add.* in *N.B.G.* (1835). 'This is *C. speciosus*, which differs from *C. nudiflorus* in the greater length of ye stigma.' In many fields near Halifax, naturalised, and not distinct from *C. speciosus*; *R. Leyland, MS.* in *N.B.G.* ii (1837). Wellhead fields, Halifax; *C. Eastwood* (1862). Near Ovenden Wood and Hall; *J. Bates.* Birkby and Huddersfield; *John Bartlam* (admittedly a Casual here).

D. Gathered in a meadow between Norton and Sheffield, 'in the same valley with the Sheffield Nursery'; *Stokes, Bot. Comp.*, 206.

[Crocus aureus Sibth.

Near Otley, garden escape; *L. C. Miall.*]

[Crocus sativus L. *Saffron.* Ambiguity.

C. About Halifax; *Rev. Mr. Wood*; *With., Arr.*, ed. ii, p. 38 (1787). In a field near Great House, Eastwood (two miles east of Todmorden), possibly an outcast from a garden, but occurs in some quantity; *A. Stansfield* (1877). No specimen examined, and I suspect that the Calder Valley plant is all *C. nudiflorus*; both are autumn bloomers, and the stigmas vary much in strength of scent.]

LXXVII. AMARYLLIDACEÆ.

Daffodils,
That come before the Swallow dares, and take
The winds of March with beauty.—*A Winter's Tale*, Act iv, Sc. iv.

823. Narcissus Pseudo-Narcissus L. *Daffodil. Lent-lily.*
N. sylvestris pallidus calyce luteo; Ray.

Area: 63, 64, Top. Bot. Range: 0—450 feet.

Native, but in some of its stations a Denizen only; in woods and pastures on limestone and sand, infrequent, but locally abundant, commoner once—a decreasing species. P. End of March-April.

First record: Dillenius, 1724.

R. Feizor; *Herb. Tatham*—but probably not truly wild.

Y. Galphay Moor; *H. Baines.* Woods by the Laver, Galphay!

N. Pastures about the Abbey, Grimbald's Crag, and Long Flat, Knaresborough. Little Ouseburn; *Anne Pratt.*

W. Lindley Woods and by the Washburn stream, in sandy places abundant (1869-75). Thorparch woods; *H. Baines.* Bolton woods, planted; *J. G. Baker.* Old orchard, Church Fenton; *H. F. Parsons.* Steeton Hall moat, Ainsty* (1879); *W. Whitwell.* Meadows by Wharfe, near Newton Kyme and Nun Appleton.

A. Rawdon Crag Wood; *H. Baines.* Esholt Wood; *L. C. Miall.*

C. [Near Hallifax, præcipue Coley-Hall; *Ray, Syn.*, iii, 371 (1724); *Martyn*; etc. Steps, Warley, Ripponden brook thickets, etc.; *J. Bolton* (1775). Gone.] Todmorden; *A. Stansfield.* Woodsome and Storthes Hall; *Hudd. Nat. Hist.* Bullcliffe Wood and Hiendley; *T. W. Gissing, Add.*

D. Hampole Woods (!) *J. Willis*; *A. F. Gissing*, etc. Womersley; *W. West.* Near Doncaster; *H. Baines.* Below Aldwarke; *J. Salt.* Old pastures near Dalton, Sheffield; *A. Carr.*

T. Woods about Roche Abbey (very plentiful!); *Jos. Woods* (1838); rep. *J. H. Aveling*; etc. Wickersley; *Geo. Harrison.*

[Narcissus major Curt. Alien.

Subspontaneous, escaped from cultivation, or outcast with rubbish, occasional by stream-sides and in old orchards. I have seen it by the Nidd, Wharfe, and Don.]

824. Narcissus biflorus Curt. *Primrose Peerless.* '*Sweet Nancy.*'
N. poeticus of Hudson and Gough.

Denizen; woods, thickets, and alluvial pastures, rare. P. April-May.

First record: Bolton, 1775.

Y. Meadows near Ripon; *W. Brunton* in *O.B.G.*

W. Wharfe banks, Thorparch, in plenty; *T. Knowlton* in *O.B.G.*

C. Meadow, Warley Wood, and field near Kebroyd Mill, Sowerby; *J. Bolton* (1775). Near Halifax; *Gough's Cam.* (1789); and *O.B.G.* auct. *Rev. W. Wood.*

D. Rotherham district; *L. Langley* (1828). Grass-field near Adwick-le-Street; *P. Inchbald.* Hampole*; *J. Willis*; rep. *Rev. J. S. Eastmead.* Near Elmsall! Hooton Pagnell; *J. E. Kenyon* (1864).

[Narcissus poeticus L. Alien.

A garden escape, on alluvial soil by rivers. I cannot answer for the correctness of the name in any of the localities given below, but probably about Hampole both *N. biflorus* and the present species occur.

C. Plentiful in Mr. Armitage's field at Outwood-side, Wakefield, 1800; *D. Dixon.*

D. Hampole Woods* (1875); *P. Inchbald.* Bentley, Doncaster, in a field, naturalised; *H. Baines.*]

[**Narcissus incomparabilis** Mill. Alien.
Straggler from cultivation.
Found on the bank of the Ouse above Goole, by *Dr. H. F. Parsons*, in 1877.]

825. Galanthus nivalis L. *Snowdrop.*
Area: no record, Top. Bot. Range: 0—600 feet.
Not Native, but 'wild' now in the sense of being ineradicably established over considerable areas; and, like the Tulip, or even the Hyacinth and Daffodil, holding its own in competition with the other herbage in shady, sandy places (near streams) on alluvial soils, to which it has mostly been carried by floods, albeit thus spread from a few centres in which doubtless it was originally planted. P. March.

R. Maulifield, Upper Settle; *J. Windsor*; *Misses Thompson*; etc. Beggar-wife Bridge; *Herb. Tatham.* Bolton-in-Bolland Woods!
Y. Skell-crooks, Ripon; *A. Pratt*; and Mackershaw Wood; *J. Dalton* in *O.B.G.*
W. Bolton Woods! Thorparch Woods and Newton Kyme. Collingham. Steeton Hall, Ainsty; *W. Whitwell.*
A. Airton. [Pasture near Kirkstall Abbey; *W. Wood* in *O.B.G.* (1805).] Esholt Springs; *J. Willis.* Rider's Wood, Meanwood*!
C. Woodsome near Huddersfield; and Cromwell Wood, Brighouse; *L. C. Miall.*
D. Near Doncaster; *Dr. White* (1804). 'Old Park—I doubt it being a native there'; *J. Salt* (1800).
T. Roche Abbey Valley!
And very numerous other stations, but always suspicious ones.

LXXVIII. DIOSCOREACEÆ.

826. Tamus communis L. '*Black Bindweed.*' *B. Briony.*
Area: 63, 64, Top. Bot. Range: restricted, not above 600 feet, and rare as high as 500.
Native; hedges and thickets, common. P. May-June.
First record: Salt, 1800.

[M. Not marked for this area by *J. Whitehead.*]
L. Bentham. Crow-nest Wood; *Herb. Tatham.*
R. North-west end of Kelcowe, Settle; *J. Windsor.* Wigglesworth! Sawley; *R. Brown.*
Y. Ripon, common; *H. H. Slater.* Boroughbridge!
N. Common from Harrogate eastward.
W. Near Grassington. Common from Ilkley eastward. Arncliffe Wood; *Herb. Tatham.*
A. Skipton (rare)! Shipley, Apperley, Roundhay; *W. West.* Adel.
C. Common in lower parts of Calder and Colne Valleys.
D. Doncaster; *G. Harrison.* 'Common'; *J. Salt.* Rivelin Valley and Wharncliffe Wood; *A. Carr.*
T. Common herein. Roche Abbey; *J. H. Aveling.*

LXXIX. TRILLIACEÆ.

827. Paris quadrifolia L. '*True-love.*' '*True-love-knot.*' *One-berry. Herba Paris officinalis*; Ger.; Blackst.
Area: 63, 64, 65, Top. Bot. Range: 0—900 feet.
Native; moist woods and rocky mountain gills, infrequent. P. June.
First record: Blackstone, 1746.

L. Dentdale! Wethercote; *G. Roberts.* Clapdale. Feizor Woods; *H. F. Parsons.*
R. Kelkoe road near Settle; *Gough's Camd.* (1789). Kelcowe, Major Springs, etc.; *J. Windsor.* Bolton-by-Bolland!
Y. Studley. Hackfall; *W. Brunton.* Morcar Wood by Copgrove; *J. Dalton* in *O.B.G.*
N. Near Stockeld; *Gough's Cam.* Plumpton Woods. Long Walk, Knaresborough; *E. Hargrove* (1782). Birkham Woods, Knaresborough (!) *E. Hargrove* (1832).
W. Grassington Woods and Buckden! Bolton Woods; *H. F. Parsons*; etc.! Stutton; *G. Roberts.* Thorparch! Near Sir W. Vavasor's Park, Haselwood; *Blackst., Spec.* (1746); *Martyn* (1763). Woodhall and Collingham.
A. In Weet-wood, three miles from Leeds; *Blackst., Sp.*; *Martyn*= Near Leeds; *Rev. W. Wood* in *O.B.G.* Malham Cove; *W. West.* Skipton Woods; *W. Whitwell.*
C. Lupset Woods, 1792; *D. Dixon.* Honley and Storthes Hall Woods; *Hudd. Nat. Hist.*
D. Rotherham district; *L. Langley.* Rockley and wood near Gawber; *H. Payne.* Parker Spring Wood, near Woodthorpe, Sheffield, 1797; *Herb. Salt.*
T. Roche Abbey (!) *J. H. Aveling.* Maltby; *J. Bohler*!

LXXX. LILIACEÆ.

828. Polygonatum multiflorum All.
Area: 64, Top. Bot. Range: 0—450 feet.
Denizen; woods and thickets, rare and doubtfully indigenous. P. June.
First record: Dalton, 1805.

L. Bentham*, probably not wild; *W. Whitwell.*
R. Bluch, Settle*, 1883; *Misses Thompson.*
Y. Studley Woods, abundantly; *W. Brunton* in *O.B.G.* Thickets about the rocks not far from the ruins (1870), questionably wild. 'It may yet be found in plenty—in garden beds at the Lodges'; *H. H. Slater's Flora.*
A. Ditch-bottom at Winterburn between Skipton and Malham; *Rev. W. Wood* in *O.B.G.*=Near Calton; *J. Tatham*; *J. Windsor*=Between Calton and Ribstone; *H. Baines* (misprint, Rylstone was meant.)

829. Polygonatum officinale All. '*Solomon's Seal*' (i.e., Dan. sēgl). *P. floribus ex singularibus pediculis* J.B.; Ray.
Area: 64, Top. Bot. Range: 600—1,200 feet.
Native; rocky thickets on the mountain-limestone scars, rare. P. June.
First notice: Gerarde, 1597.

L. Helks and Sykes' Wood near Ingleton, in the rocky part; *W. Curtis* (1782); *Gough's Camden* (1789); *Dawson Turner* in *O.B.G.* On Ingleborough mountain; *Ray, Cat.*, 248 (1670). About Austwick and Clapham near Ingleton; *Blackst., Spec.* (1746). Two miles off Austwick; *Martyn* (1763). Wharfe; *Id.* [prob. same station as last]. Clapham; *Martyn.* Clapdale Woods; *Gerarde, Hb.*, 758; rep. *Merr.*, 96.
R. On the ledges of the cliffs near Wharfe and Settle, plentifully; *Ray, Cat.*, ii, 238 (1677); *Huds.* (1778); etc. Kelcove, projecting rocks east of the Cave. Rocks just above the broken soil in Cavehole Wood; a quarter of a mile above Major Springs; Benscar, near Birkbeck's Nursery, etc., near Settle; *J. Windsor, MS.* Crag below Springs Wood; *Silv. Thompson* (1877)!
(W. 'Grassington,' 1878—a large specimen given to *W. Whitwell* as from above place, but wildness cannot be vouched for.)
A. Malham Cove! Various scars about Malham; *L. C. Miall.*

830. Convallaria majalis L.! *Lily of the Valley. Lilium convallium*; Ray.
Area: 63, 64, Top. Bot. Range: 0—750 feet.
Native; rocky or loamy woods, on limestone, infrequent. P.
Earliest notice: Ray, 1690.

L. Thornton Force; *J. Willis.* Near Bentham! In woods by the side of Ingleborough Hill; *Ray, Syn.* i, p. 96 (1690); *Gibs. Cam.*; *Martyn* (1763); etc. Feizor Wood; *R. Clapham.*
R. Stainforth Woods, Lord's Wood, Kelcowe, Cavehole Wood, near Settle; *J. Windsor.* Giggleswick Scar, 1884; *W. Whitwell.*
Y. Mackershaw Woods; *J. Dalton.* Studley (!) *Anne Pratt.*
N. Hill on the north side of the Abbey Mill, Knaresborough; *E. Hargrove* (1782).
W. Thorparch and Boston Spa Woods! *J. Emmet*! Becca banks, Aberford. Arncliffe Woods; *H. Baines.*
A. Malham! Gordale; *W. West.*
C. Catholis clough, Todmorden, and Heptonstall 'Eaves' wood (!) truly wild; *S. Hailstone*; rep. *A. Stansfield.* [North Dean; Wood at Little Even in Barkisland, etc., near Halifax; *J. Bolton* (1775).]
D. Rotherham district; *L. Langley.* Conisborough (!) *Y. N. Excurs.* (1873).
T. Roche Abbey woods; *J. Salt*; *G. E. Smith*; *J. H. Aveling*; etc.

[**Maianthemum bifolium** DC. Extinction or Error?
Monophyllon seu unifolium, One-blade; Ger. 409. *Monophyllum*; P. 505. . . . and at Howgill, Yorkshire, in a field next Todberry, in a moist ground belonging to Thomas Oddy; *Merr., Pin.*, 78, 79 (1666). This record is not so unlikely as appears at first sight, but I have twice searched the district indicated in vain. There are, however, holly woods and sheltered thickets on the Lune bank suitable for it. Area: '(64),' *Top. Bot.*]

[**Ruscus aculeatus** L. *Butcher's Broom.*
Alien; only where planted.
Area: '(64),' Top. Bot.

R. Stackhouse, near Settle; *H. F. Parsons* (1878).
Y. 'Near Ripon; *Winch, Add.*'; *N.B.G.* 294 (1832).
C. Woods at Fixby; *Hudd. Nat. Hist.*
T. Plantations about Worksop; *J. Bohler*—introduced, as well as the *Anemone ranunculoides*. 'Woods near the southern border (of the Riding)'; *Miall and Carrington's Flora.*]

[**Asparagus hortensis** W. (*A. officinalis* L.?) Alien.
A garden escape. In a marl-quarry at junction of roads, Hook Moor, near Aberford, 1885; *Geo. Roberts*!]

[**Lilium Martagon** L. '*Turn-cap Lily.*'
Alien; woods, rare, and, although persisting, always only where planted. P.
W. Thorparch Woods (near the hall and the old lime-kilns!); *S. Hailstone.*]

831. Fritillaria Meleagris L.! *Snake's-Head.*
Area: '(64),' Top. Bot. Range: 0—100 feet.
Denizen; hardly a native? old pastures and park land, rare. P. May.
First record: J. Backhouse, 1830.

W. In the Ings pastures by the Wharfe, below Tadcaster, solitarily and very sparingly, but with *Gagea*! Between Tadcaster and Newton Kyme; *H. Baines.*
T. In more than one place in the open turf of Sandbeck Park; abundant in places (1884); known to *Rev. G. E. Smith* in 1845.

832. Tulipa sylvestris L.! *Wild Tulip.*
Area: counties not enumerated, Top. Bot. Range: 0—100 feet.
Denizen; old alluvial pastures by streams, rare. P. May.
First record: Appleby, 1831.

N. Sparingly in the vicinity of Knaresborough; *Anne Pratt* (1864). Wide fields by the river Nidd at Cattall, somewhat sparingly, but over a wide area (1881 and since); *W. Whitwell*!
(C. Wood near Heath, Wakefield; *Rev. G. Pinder*—A garden escape.)

D. On low warp lands liable to be overflowed, immediately adjoining the river Don; *S. Appleby* (1831) = Don banks, Doncaster; *John Tatham*, 1846, in *Herb. W. Whitwell*. There still in 1869-71, but not flowering every year. The exact station is an alluvial meadow (or meadows) between Hexthorpe and Sprotborough broats; *H. Baines*.

833. Gagea lutea Ker.
Ornithogalum luteum; Hudson.
Area: 63, 64, Top. Bot. Range: 0—600 feet.
Native; moist banks by streams, in shade, rare. P. April.
First record: Tofield in Hudson, 1778.

Y. Dunkirk Wood near Sleningford (!) abundant; *J. Dalton* in *O.B.G.* = Near North Stainley; *R. A. Summerfield* in *Slater's Flora*. Red Bank, by river Laver at Bishopton; *W. Fowler* (1875)! Near Sharow Mires; *H. H. Slater*. Hungry Hill, Robin Hood's Bank, and by the river Skell (opposite Littlethorpe); *W. Brunton*. Fields off Wicliffe Lane, Studley; *Anne Pratt*.

N. Knaresborough, corner of second field below St. Robert's Well (!) *E. Hargrove* (1832) = Long Flat at Knaresborough; *Archdeacon Pierson* (1805).

W. In small wood beyond the low end of Boston Spa, on road to Paper Mill Bridge; *J. Emmet*. By Wharfe below Tadcaster!

A. Malham Cove; *Rev. W. Wood* in *O.B.G.*; rep. *L. C. Miall*.

C. Plentiful in Patterdale near Kirkthorpe, 1791; *D. Dixon*.

D. In a pasture near the Don, on the skirts of the wood beyond Hexthorpe, with *Tulipa*! About Doncaster; *Hudson* (1778); *Gough's Cam.* (1789). Alluvial pastures west of Doncaster; *S. Appleby* (1831).

T. About the brook-side, Maltby; *G. E. Smith*. Roche Abbey, wood bank, south side water below the ruins (!) *Y. N. U. Excurs.* At Beighton, six miles east of Sheffield; *J. Salt, MS.* in *Towns. O.B.G.*

[**Ornithogalum nutans** L. Alien.
Banks and thickets, hardly naturalised, very rare. P. May.

W. On a bank by the road-side half way between Tadcaster and York, a dozen roots or so, only; *H. Ibbotson*! Near Aberford; *T. Ward* in *Miall's Flora*.]

834. Ornithogalum umbellatum L. '*Bethlem Star*.'
Area: no record, Top. Bot. Range: 0—300 feet.
Denizen; meadows and pastures, rare. P. May-June.
First record: Blackstone, 1746.

R. Field opposite J. Ellinson's in Upper Settle; *J. Windsor, MS.* in *Towns. O.B.G.* A mile south-east of Settle (but a garden nigh); *Miss F. Thompson* (1885).

Y. Meadows near Ripon, on foot-road to a pasture called Red Bank; *W. Brunton* (1805).

N. Low end of the 'Long Flat, by footpath to Grimbald Bridge'; *E. Hargrove, Hist.* (1782) = Bottom of Long Flat, Knaresborough; *E. Hargrove* (1832); rep. *W. Brunton* in *O.B.G.*

A. In the new field by Ledstone Hall; *Dawson*; *Blackst., Sp.* (1746); *Martyn* (1763); etc. In the field known as Parson's Furze, near Kippax; *H. Baines*.

C. Plentiful in Patterdale, near Kirkthorpe, 1791; *D. Dixon*.

D. Field near Ridge Hall, Adwick-le-Street, several patches, far from cultivation; *P. Inchbald* (1844). Rotherham district; *L. Langley*. River-bank, Goole; *H. F. Parsons*.

835. Scilla nutans Sm. '*Blue-bells*.' '*Squill*.' *Hyacinth*.
Area: 63, 64, 65, Top. Bot. Range: 0—1,000 feet.
Native; shady woods and banks, very common, preferring a moist gritstone soil, but apparently indifferent whether it be light or stiff. P. April-May.
Earliest notices: Bolton, 1775; Salt, 1800; Miller, 1804.
In all the districts.
The white-flowered form not rare, but never in masses by itself, always sparingly interspersed amid the blue.

Obs.—In some parts, especially in the hangers of the dales over the coal-measures, this species will clothe acres of ground to the exclusion of almost anything else, save *Holcus mollis*; and when in full flower resembles nothing so much as a clear Italian sky towards twilight. Surely, there is no tint more beautiful on the palette of Nature, unless it be a wide stretch of *Lychnis Flos-Cuculi* in its first flush of bloom, overspreading a damp upland meadow. Neither Gorse nor Purple Heather en masse are quite equal to these in combination of delicacy with force.

[**Scilla bifolia** L. Alien or Error?
'Wood near Heath, Wakefield,' 1790; *D. Dixon*.]

836. Allium Scorodoprasum L. *Wild Leek*. '*Sandtines*.'
Ampeloprasum, sive Porrum sylvestre; Merr. *A. arenarium*; Sm.
Area: 63 Hailstone spn., 64, 65, Top. Bot. Range: 0—600 feet.
Native; gravelly and limestone fields, banks, and thickets, infrequent. P. July.
First record: Merrett, 1666.

L. Pasture at back of the 'Dun Horse,' Ingleton; *Gough's Cam.* (1789). Lune side, Sedbergh!

R. Mill Island, and between Settle and Langcliffe; *J. Windsor*. And many other places! Lodge Lane, Settle; *Herb. Tatham*.

Y. Copgrove; *J. Dalton*. Near Ripon; *W. Brunton*. Sharow Bridge, Hewick, Mackershaw, etc.; *H. H. Slater*.

N. Top of the rocks at Knaresborough; *Archdeacon Pierson*; *W. Brunton* in *O.B.G.* Fields at Knaresborough, Ribston, and near Cowthorpe!

W. At Sh[k]irethorn in Craven, going from Mawater Tarn to Perolic Bridge before you come to ye River yt maketh Wharf River; *Merr., MS.*, p. 6 (1666). Littondale. Grassington. Langwith Wood, etc., near Collingham. About Thorparch (!) *Rev. W. Wood* in *O.B.G.* (1805); *S. Hailstone*, spn.; etc. Jackdaw Crag, Boston Spa; *J. Willis*; etc.

A. Malham. Kippax; *H. Baines*. River bank, Selby (!) down to Booth Ferry; *H. F. Parsons*.

837. Allium vineale L. *Crow Garlick*.
Area: 63, 64, Top. Bot. Range: 0—300 feet.
Native; sandy fields, hedge-banks, and river-sides, infrequent, but locally abundant, mostly the form *compactum* Thuill. P. July.
First record: Dixon, 1790.

R. Stony field, Major Springs; *J. Windsor*. From Mr. A. Sutcliffe, of Settle, 1806; *Herb. Salt*. Giggleswick; *W. M. Watts*.

N. Poppleton lanes (! *W. Whitwell*), and Tadcaster road near York; *H. Ibbotson*.

W. Road-side, Flint Mill, Thorparch; *H. Baines*.

A. Baildon and Steeton; *L. C. Miall*. Gateforth to Brayton! Selby (var. *bulbiferum*); *H. F. Parsons*. Kellington, 1871; *G. Roberts*.

C. Abundant in Kirkthorpe Wood, near the river Calder, 1790; *D. Dixon*.

D. Conisborough; *Y. N. Excurs.*, sub nom. *A. pratense* (1873). Doncaster; *H. Baines*. Goole; *H. F. Parsons*.

[**Allium carinatum** L. (non Sm.). Casual.
Sandy banks of river Ouse, below Barlby Mill—adventive clearly, but not certain how brought. *Melilotus* near, and the station a likely one for waifs, but native Garlics grow also on the sandy bank near.]

838. Allium oleraceum L. *Field Garlick*.
Area: 63, 64, 65, Top. Bot. Range: 0—600 feet.
Native; meadows, road-side and river-banks, common. P. July.
First notice: Brunton, 1805.

L. Ingleton and Wethercote Cave! Feizor; *W. M. Watts*.

R. Mill Island, Settle; *J. Windsor*; and other places about! Ribble banks to Bolland!

Y. Copgrove; *J. Dalton*. Near Ripon, very common (!) *W. Brunton* (1805). Sharow, Newby, etc.; *H. H. Slater*.

N. Rocks near Knaresborough; *J. Dalton*.

W. Kilnsey. Wharfe banks, Bolton, Ilkley, Pool, Collingham, and Thorparch! Bramhope, Compton bank near Bramham, etc. Holgate; *H. Baines*.

A. About Leeds; *Rev. W. Wood*. Malham. By Aire, Airton. Shipley to Apperley; *W. West*. Selby; Ouse banks from Cawood downwards in many places; *H. F. Parsons*.

D. Corn-field, Conisborough; *A. Carr*.

T. Thorne; *L. C. Miall*.

Var. **complanatum** Bor. (*carinatum* Sm.).
Allium montanum bicorne; Ray.
Area: not treated apart from preceding in Top. Bot. Montane? Range: 600—1,500 feet.
First notice: Ray, 1670.

R. On the scars above Settle; *Ray, Cat.*, i, 13; ii, 12; *Syn.* i, 164; etc. Near Attermire [now Victoria] Cave; *J. Windsor*.

W. Among rocks at Girling Trough, Conystone, Kilnsey, etc.; *W. Curtis* (1782); rep. *Gough's Cam.* (1789).

A. Near Malham Tarn; *J. Windsor* = Awes Scars. Gordale Scar (!) *W. Borrer, MS.* in *Towns. O.B.G.*

839. Allium ursinum L. '*Ramsons*.' '*Bear's-garlick*.'
Area: 63, 64, 65, Top. Bot. Range: 0—900 feet.
Native; woods, thickets, and hedge-banks, common. P. May.
First records: Dixon, 1790; Salt, 1800.
Distribution general—recorded from every district, and locally plentiful in suitable situations in most of them.
'Plentifully in Kirkthorp Wood, Wakefield, 1790'; *D. Dixon*.

840. Narthecium ossifragum Huds. '*Moor-golds*.'
Area: 63, 64, 65, Top. Bot. Range: 0—2,000 feet.
Native; peaty moors and mountain bogs, very common on the high lands, but decreasing on the lower moors, and rare in the eastern half of the Riding. P. July.
First record: Bolton, 1775; and Gough's Camd., 1789.

M. Mosses and swamps of Diggle, Wragstone, and Greenfield valleys.

L. Widdale, Whernside (at quite 2,000 ft.), Cam, and Ingleborough fells. Blea-moor!

R. Boggy places near Settle; *J. Windsor*; etc. Slaidburn and Bowland Knotts! Rathmell; *W. M. Watts*.

Y. Kirby Malzeard moors!

[N. On Harrogate stray, near the old Spaw; *E. Hargrove* (1809 and 1832). Wet land, Longland's Farm, Knaresborough. Gone.]

W. Langstrothdale and Fountains Fell moors! Rombalds Moor, disappearing; *J. Willis*. White Car Bog, south-west corner of Bramham Park boundary wall!

A. Keighley and Bingley moors! Eldwick; *W. West*. Malham Tarn Bog. Adel Bog, etc.

C. Todmorden moors! *A. Stansfield*. Warley, Sowerby, and Norland moors; *J. Bolton* (1775).

D. Doncaster neighbourhood; *Dr. White* (1804). Potteric Carrs; *S. Appleby* (1831). Crooks Moor; *Herb. Salt* (1796). Cliff Rocher; *J. Salt*. Bell Hagg and bogs near Malin Bridge; *A. Carr*. Goole Moors (*H. F. Parsons*) and Thorne (!) under 50 feet.

[Tofieldia palustris Huds. Incognit. Error.

Silsden Moor—a 'report'; *Miall's Flora* (1862). I have no further information. Probably some small starved specimen of *Narthecium* gave rise to the report.]

841. Colchicum autumnale L. '*Autumn Crocus*.' '*Saffron*.'

Area: 63, 64, Top. Bot. Range: 0—600 feet.

Native; meadows and pastures, on calcareous and alluvial soil common, locally abundant, rare elsewhere. P. Leaves, Spring. Fl. August-October.

First record: How, 1650.

L. Between Clapham and Ingleton; *H. Baines*.

R. Pasture near Catterall Hall, Giggleswick, opposite Kelcowe; near Swobeck, etc., about Settle; *J. Windsor, MS.* Paythorne!

Y. Near Ripon too abundant; *W. Brunton* (1805). Copgrove; *J. Dalton*.

N. Most of the meadow grounds near Knaresborough; *E. Hargrove, Hist.* (1782) = 'Long flat' pastures (!) *R. Teesdale* in *O.B.G.*; etc. By Nidd, Cowthorpe and Cattall.

W. In a pasture close by Milsear Lush, near Kilnsey; *Gough's Cam.* (1789). Grassington; *H. T. Soppitt* (1885). Pool and Arthington. By East Keswick Beck, near Rigton. Collingham. Wetherby Ings. Barwick, Sherburn, Beal; *G. Roberts*.

A. In Mr. Wm. Pickering his close neere his Parsonage house at Swillington, *Mr. Stonehouse*; *How, Phyt.*, 29. On clay in Airedale about Apperley Bridge; *Rev. W. Wood* in *O.B.G.* River-side at Fryston; *A. Aiken* in *O.B.G.* About Kippax and Ledstone; *Rev. W. Wood* in *With.* Selby! Garforth. Pontefract; *J. Willis*.

(C. Meadows about Saville Green near Halifax, in plenty; *J. Bolton* (1775). Same station named in *Winch, Add.* (1835), for *Crocus nudiflorus*! Can Bolton have made a mistake? both do not grow there now, and the *Colchicum* appears to be absent from the south-western part of the Riding. Leyland's specimens were really the *Crocus*.)

D. Doncaster neighbourhood; *Dr. White* (1804). Rotherham district; *L. Langley*. Low warp lands by the Don; *S. Appleby* (1831). Askern; *Y. N. Excurs.* (1873).

T. Roche Abbey; *J. Salt*; *J. H. Aveling*; etc.

LXXXI. JUNCACEÆ.

842. Luzula pilosa Willd.

Area: 63, 64, Top. Bot. Range: 0—1,250 feet.

Native; woods and thickets, common. P. May-June.

First record: Salt, 1800.

L. Helks Wood, Ingleton; *W. Howson*. Clapdale! Dentdale!

R. Peart's Crags, Little Bank, etc., Settle; *J. Windsor*; *J. Tatham*; etc. Linn Gill and Stainforth thickets!

Y. Hackfall! Ripon, common; *H. H. Slater*.

N. Bilton banks and woods by Nidd at Knaresborough.

W. Common from Buckden downwards.

A. Airton and Skipton! Occasional in Airedale woods to Fairburn.

C. Kirklees; *Y. N. Excurs.* (1873). Halifax; *H. Baines*. Gawthorpe; *T. W. Gissing*. (The abortive state *Borreri* not seen in W. Riding.)

D. Doncaster; *H. Baines*. Old Park Wood; *J. Salt*. Beeley Wood; *Herb. Salt*. Wharncliffe Woods, and Cliff Rocher; *A. Carr*.

T. Roche Abbey Woods (!) *W. Todd*; *Aveling's Hist.*; etc.

[Luzula Forsteri DC. Incognit.

Westwood; *A. Shipley* in *Miall's Flora*. Whether this was a misprint for *Weetwood*, near Leeds, or meant Westwoods at Bramham, I do not know.

Obs.—It may be told at once from *L. pilosa* by the secund arrangement of the drooping fruiting pedicels.]

843. Luzula sylvatica Gaud. (*L. maxima* DC.) *Wood Rush*.

Area: 63, 64, 65, Top. Bot. Eugeogenous. Range: 0—1,800 feet.

Native; woods and mountain moor-slopes (in open ground but rarely), common, yet absent from the woods of many districts. P. June.

M. Mersey district copses; *J. Whitehead*.

L. Helks Wood, and slopes of Ingleborough! Lawkland Woods; *J. Windsor*.

R. At the foot of the Ryeloaf (hill), near Settle; *W. Curtis* (1782); *Withering* (1787). Peart's Crags, Giggleswick; *J. Windsor*. Linn Gill! Open fields about Little Bank, Settle; *Herb. Tatham*.

Y. Hackfall (!) *H. H. Slater*. Local in this river basin.

N. Bilton banks; not very general in this district. Plumpton.

W. Mountain pastures, Buckden; Bolton Woods; and very abundant in the grit-stone woods about Ilkley, Burley, Menston, and Washburn valley.

A. Mid Airedale, common; rare on the limestones.

C. In every wood in Calder Valley district; *A. Stansfield*. Kirklees; *Y. N. Excurs.* (1873). Dungeon Wood, Huddersfield; *C. P. Hobkirk*. Gawthorpe; *T. W. Gissing*.

D. Old Park Wood; *Herb. Salt*. Endcliffe Wood; *W. W. Newbould*. Cliff Rocher and Handsworth Wood; *A. Carr*.

844. Luzula campestris DC. '*Crow-feet*.' '*Crowfoot'-Rush*.

Area: 63, 64, 65, Top. Bot. Range: 0—1,800 feet.

Native; pastures, both upland and lowland, very common. P. April.

First record: Salt, 1800.

Distribution general; common in every river-basin.

845. Luzula multiflora Koch.

Area: 63, 64, 65, Top. Bot. Range: 0—2,100 feet.

Native; heaths, pastures, and open thickets, common. P. June.

First record: Salt, 1800.

Distribution general; occurs in every district, but is naturally local in the cultivated areas. It has a wide range, extending from Hatfield Moor in the south-east to the summit of Baugh Fell in the north-west.

846. Juncus conglomeratus L. '*Seave*.'

Area: 63, 64, 65, Top. Bot. Range: 0—1,300 feet.

Native; ill-drained pastures, road-side wastes, and damp heaths, common. P. July.

First record: Salt, 1800.

Distribution general; in every district, but not quite so common as *J. effusus*, and often in drier places than that species affects.

847. Juncus effusus L. '*Seave*.' '*Watch-light' Rush*.

Area: 63, 64, 65, Top. Bot. Range: 0—1,800 feet.

Native; wet pastures and marshy places generally, very common. P. July.

First record: Salt, 1800.

Distribution general; abundant enough in every district outside the meadows and arable land.

848. Juncus diffusus Hoppe.

Area: '63? 'Ilkley,'' Top. Bot. [Ilkley is in v.c. 64.] Range: 200—450 feet, but likely occurs both higher and lower.

Native; damp turfy pastures, with *J. effusus* and *J. glaucus*, rare. P. July.

N. Pannal, in valley near River Crimple!

W. Beamsley; and near the bridge, Ilkley; *J. G. Baker*.

Obs.—This hybrid ought scarcely to rank as a species.

849. Juncus glaucus Sibth. '*Seave*.'

Area: 63, 64, 65, Top. Bot. Range: 0—1,250 feet.

Native; heathy or moist pastures, moors, very common. P. July.

First record: Baines, 1840, 'Hobmoor.'

Distribution general. I have seen it in every district.

[Juncus triglumis L. Incognit.

Whernside; *Miall and Carrington's Flora* (1862), p. 45; rep. *Flora Cravoniensis*, and not given in list of doubtful records on p. 93 of former work. No further information obtainable; but I believe some small erect form of *J. supinus* to have been taken for it.]

850. Juncus obtusiflorus Ehrh.

Area: 64 'Moore,' Top. Bot. Range: 0—100 feet.

Native; marshy-peaty ditch- and pool-sides, very rare. P. July-August.

First record: O. A. Moore, 1850.

Y. Sharow Mires, Ripon, sparingly (!) *H. H. Slater*.

W. Askham Bogs; *O. A. Moore*. By the deep ditches surrounding the 'Far Wood' and Jungle!

D. Great Eaa Plantation 'Carr,' south of Doncaster!

851. Juncus acutiflorus Ehrh.

Area: 63, 64, 65, Top. Bot. Range: 0—1,250 feet.

Native; peaty-marshy places on moors, common. P. June-July.

First record: Salt, 1800.

Distributed generally in all the districts.

852. Juncus lamprocarpus Ehrh. *Black-Rush*. *Bog-seave*.

Area: 63, 64, 65, Top. Bot. Range: 0—1,800 feet.

Native; bogs, peaty moors and ditch-sides, very common. P. July.

First record: uncertain, since not discriminated by Salt from '*articulatus*,' and not mentioned by Bolton in Halifax Cat. 'Roche Abbey,' Geo. Harrison Herb. (1862), is perhaps the first specifically West Riding record.

Distribution general; common in every district.

853. Juncus supinus Mœnch.

Area: 63, 64, 65, Top. Bot. Range: 0—2,000 feet.

Native; watery places, sandy peat-bogs and swamps, very common. P. July-August.

First record: Salt, 1800.

Distribution general; occurs in every district.

As near Leeds as Woodhouse Ridge; *W. West*.

There are two or more forms or varieties—the erect-growing, small cæspitose *uliginosus* Sibth., which is not uncommon on the higher peat-bog plateaux, and the supine, floating or decumbent *subverticillatus* Wulf., which is almost as common as the type, being only a state dependent on varying degrees of wetness.

854. Juncus bufonius L.

Area: 63, 64, 65, Top. Bot. Range: 0—1,350 feet.

Native; wet ground where water has stood on moors, by way-sides, and in fields, very common. A. July-August.

First record: Salt, 1800.

Distributed generally; over every district.

855. Juncus Gerardi Lois.

Area: 63*, Top. Bot. Submaritime. Range: Low.

Native; on tidal-river mud, very rare. P. August.

Discovered by Dr. H. F. Parsons in 1877.

D. Growing in and by stagnant dikes parallel with the river Ouse by the Hook road, Goole (*H. F. Parsons*), along with *Carex divisa* and *Scirpus maritimus*!

856. Juncus compressus Jacq.
Area : 63, Top. Bot. Range : 50—500 feet.
Native ; wet pastures, very rare. P. August.
First record : G. E. Smith, 1844.

W. Kilnsey ; *H. T. Soppitt* (1885)!
A. Wet pasture by Aire below Fairburn, a few plants only, 1869 !
(C. Wakefield ; *Gissing's Flora*—' Frequent '—but in error.)
T. Marshy meadow beyond Roche Abbey to the left of the path to the wood ; *Gerard E. Smith* !

857. Juncus squarrosus L. '*Goose-Corn.*' '*Moor-Rush.*' '*Bent.*' *J. acutus Cambro-Britannicus* ; Park.; Ray, Cat. *J. montanus palustris* ; Ray, Syn.
Area : 63, 64, 65, Top. Bot. Range : 0—2,400 feet.
Native ; turfy moor-slopes, mountain pastures, and lowland heaths ; common at high levels, rarer (now) at low ones. P. July-August.
First record : Ray, 1670.
'Common in mountainous morrish . . . . in Yorkshire' ; *Ray, Cat.* (1670).
'Turfy parts of high moors south-west of Sheffield, plentiful' ; *J. Salt* (1800).
Distribution general ; in every district from the summits of Whernside and Ingleborough, down to parts of Hatfield Moor, etc.

LXXXII. CYPERACEÆ.

858. Schœnus nigricans L. *Clock-seave.*
Area : 64, Top. Bot. Range : 200—450 feet.
Native ; peaty-rushy pastures, very rare. P. June-July.
First record : Gough in Camden, 1789.

L. On a bog in Skirreth Wood, Ingleton ; *Gough's Cam.* (1789) ; specimen coll. *J. Tatham*—no date [it would be about 1848-50]—in *Herb. Whitwell.*
R. Boggy place in a field on the left-hand side halfway between Long Preston and Swinden ; *J. Windsor, MS.* in *Towns. O.B.G.* This would be by the brook a little west of Hellifield, near Snell Holme Bridge. Still there 1870.
Y. Sharow Mires ; *H. H. Slater.* Cow-mires, Galphay ; *Miss Morton.*
N. In the boggy marshy pasture behind Aketon Bleach Works, and between them and the Crimple stream—in some plenty (1880) !

859. Cladium Mariscus Br. (*C. germanicum* Schrad.) '*Cut-sedge.*'
Area : 63, 64, Top. Bot. Range : 0—300 feet.
Native ; very watery places in peat-bogs at low levels, rare, but locally abundant. P. August.
First record ; Dalton, 1805.

Y. Sharow Mires (!) *H. H. Slater.*
N. [By Knaresborough, in wet places near the brook, towards Newby ; *J. Dalton.* This would be Staveley Carrs, by the Minskip beck. Now gone from drainage. Mr. Duncombe's low grounds, abundant ; *Id.* in *O.B.G.* (1805). Not visible in Scriven Park, 1868.] Very sparingly near Ripley towards Harrogate, 1860 ; *L. C. Miall.*
W. Askham Bogs (!) *H. Baines*=In the dike on east side of 'Gilson's Bog,' etc. !
D. Askern (!) *P. Inchbald.* In a ditch on the road from Doncaster to the Decoy ; *H. Baines.*

860. Rhynchospora alba Vahl.
Area : 63, 64, Top. Bot. Range : 0—1,500 feet.
Native ; spongy peat-moors, now getting very rare—a decreasing species. P. August.
First record : Bolton, 1775.

Y. Small marshy place in a field, south side of River Yore, near the North Bridge at Ripon ; *H. H. Slater's Flora.*
[A. Black Moor, between Leeds and Harrowgate ; *Rev. W. Wood* in *O.B.G.* = Adel Black-moor ; *H. Denny* (1830). Gone in 1865.]
[C. Sowerby, Rishworth, and Warley Moors ; *J. Bolton* (1775). Not seen for many years.]
D. Thorne Waste (!) *J. Backhouse* ; rep. *H. Baines* etc. Goole Moor ; *H. F. Parsons* !
T. Hatfield Moor ; *H. F. Parsons* (1878).

861. Blysmus compressus Panz.
Area ; 63, 64, Top. Bot. Range : 0—1,250 feet.
Native ; oozy, sandy-peaty ground, about springs, etc., mainly on the limestone. P. July-August.
First record : Curtis, 1782.

L. About Ingleton ; *Gough's Cam.* (1789). Oozy places, Helk's Wood !
R. Ditches near Giggleswick Tarn ; *J. Windsor* in *Towns. O.B.G.*; specimen in *Herb. Tatham*, coll. 1843. Kelbowe Wood, at the lower corner ; *J. Windsor.*
Y. Boggy places near Ripon, frequent ; *W. Brunton* in *O.B.G.*
N. Farnham Mires and elsewhere near Knaresborough ; *J. Dalton* in *O.B.G.*
W. Cowside Beck ; *W. Curtis.* Arncliffe (!) *B. Carrington.* Marshy meadow near the river at Linton ; *L. C. Miall.*
A. Brook from Malham Tarn ; *H. Baines.* Shore of the Tarn ! The Cove (!) *J. Tatham.* Rivulet opposite Gordale House ; *J. Windsor* in *Towns. O.B.G.* Gennett's Cave ; *W. West.*
D. Conisborough district ; *O. A. Moore* ; confirmed, *Y. N. Excurs.* (1873).

862. Scirpus acicularis L.
Area : 63, 64, Top. Bot. Range : 0—600 feet.
Native ; sandy or muddy margins of pools and canals, rare. A. July-August.
First record : Curtis, 1782.

R. On the edge of a rivulet which runs from the Ebbing and Flowing Well, near Giggleswick ; *W. Curtis* (1782) ; *Gough's Camd.* (1789). Side of Giggleswick Tarn at the mouth of the east beck ; *J. Windsor* in *Towns. O.B.G.*
Y. Pilmoor Carr, Sharow (very rare) ; *H. H. Slater's Flora.*
W. Clifford Moor, near Wetherby ; *W. Wood* in *O.B.G.* Ditch-side on Wetherby Race-course Ings.
A. Yeadon Mill dyke ; *B. Carrington.*
C. Border of Cold Hiendley Reservoir ; *H. Johnson* ! In the Calder and Hebble Navigation Canal, near Halifax, 1844 ; *S. Gibson*, spn. in *Herb. Watson* et *Herb. Tatham* ; both sub. nom. 'var. *longicaulis.*' Merely an elongated water form (vide *Bot. Rec. Club. Rep.*, 1880, p. 152). Margin of old Lock by canal at Dewsbury ; *P. F. Lee* !
D. Barnsley, by canal-side under Hymen Wood, abundant (!) *H. Johnson.*

863. Scirpus palustris L. '*Water-seave.*' '*Pond-rush.*'
Area : 63, 64, Top. Bot. Range : 0—1,250 feet.
Native ; sides of ditches, dikes, canals, and ponds, common. P. July.
First record : Salt, 1800.
Distribution general ; occurs frequently in every district.

864. Scirpus multicaulis Sm.
Area : 64, Top. Bot. Range : 0—750 feet.
Native ; peat-bogs and peaty ditches (rarely), rare. P. July-August.
First record ; Brunton, 1803.

L. Blea-moor ! Swarth Moor !
R. Near Settle ; *W. Howson.*
Y. Sawley Moors ; *W. Brunton* in *O.B.G.* = Upon the moors near Brimham Rocks ; *J. Dalton* in *O.B.G.* Near the road from Harrogate to Boroughbridge, thirteen miles from the former place ; *L. C. Miall.*
W. In a wet grassy place by a drain, under the beeches, at the west end of Black Fen, Bramham Park !
D. About the top dams, Redmires, and bogs, Ringinglow ! *Amos Carr* !

865. Scirpus pauciflorus Lightf.
Area : 63, 64, Top. Bot. Range : 0—1,500 feet.
Native ; marshy moors and wet pastures, frequent. P. July.
First record : Salt, 1801.

R. Near Giggleswick Tarn ; *J. Windsor* in *Towns. O.B.G.* Between Long Preston and Swinden, with *Schœnus nigricans* ; *J. Windsor.*

Y. Near Ripon ; *W. Brunton* in *O.B.G.* In a streamlet called Lightwater [North Leys] between Ripon and N. Stainley, plentiful ; *J. Dalton* in *O.B.G.* (1805).
N. Brimham Rocks ; *Miss Morton.*
W. Springs below limestone about Arncliffe and Grassington ! Norwood ; *W. West.*
D. Marsh on the high moors near the Smelting Mill, not far from the toll-bar, about five miles south-west of Sheffield, 1801 ; *Herb. Salt.* In *Herb. Middl.* from Dr. Murray, of Harrogate.

866. Scirpus cæspitosus L. '*Moss-crop.*'
Area : 63, 64, 65, Top. Bot. Range : 0—2,100 feet.
Native ; heaths and moors, common, although growing rarer year by year in the lowland cultivated districts. P. June-July.
First record : Salt, 1800—' Dry heath, Fulwood.'
Distribution general ; in every district, but in this case, as elsewhere, it must not be forgotten that turfy heaths and other suitable stations occur only here and there, often many miles apart, in the south-eastern third of the West Riding.

867. Scirpus fluitans L.
Area : 63, 64, Top. Bot. Range : 0—300 feet.
Native ; in dikes and pools, on sand or peat, rare. P. July.
First record : Herb. Salt, 1800.

W. Askham Bogs ditches ! Dringhouses Moor ponds ; *H. Baines.*
A. Mill-dam, Yeadon ; *B. Carrington.*
D. Potteric Carrs, Doncaster ; *Herb. Salt.* Dike, Great Eaa Plantation ! Rawcliffe Rabbit-hills ; *H. F. Parsons.*
T. Drain near Auckley !

868. Scirpus setaceus L.
Area : 63, 64, Top. Bot. Range : 0—1,500 feet.
Native ; watery sandy-peaty places, by rills on moors, etc., infrequent. A. July.
First record : Salt, 1800.

M. Mersey tributaries district ; *J. Whitehead.*
L. Springs, Kingsdale and near Wethercote Cave ! Dentdale ! Garsdale !
R. Swarth Moor, Settle. Douk Gill, and by Ribble, Horton !
Y. Hutton Moor, Ripon ; *H. H. Slater.* Kirkby Malzeard !
W. Langstrothdale ! Ilkley. Skerry Grange fields, Sicklinghall !
A. Margin of Malham Tarn ! Skipton Rock ! Gordale, and Rawden Crag ; *J. Willis.* Eldwick to Baildon ; *W. West.* Eccup and Adel.
C. Gorse Hill, etc., near Wakefield ; *T. W. Gissing.*
D. Near Sheffield, frequent ; *J. Salt.* Margin of Dunford Bridge reservoirs ! Bell Hagg, Loxley, and Hollow Meadows ; *A. Carr.*
T. By side of boggy rill from a spring, Maltby Valley !

869. Scirpus lacustris L. *Bulrush.* '*Chair-platt Rush.*'
Juncus sive Scirpus medius, B.P. ; Ray.
Area : 63, 64, Top. Bot. Range : 0—300 feet.
Native ; in slow rivers, and in ponds (rarely), infrequent. P. August.
First record : Merrett, 1666.

[R. Giggleswick Tarn ; *J. Windsor.* Gone.]
Y. & C. In the Mill [race] at Cobgrove, not far from Rippon ; and at Nostal Abbey ! *Merrett, Pin.*, 67 (1666). Queen Mary's Dubb, by Yore above Ripon Bridge (!) *H. H. Slater.* Near Boroughbridge ! = Mill-dam there ; *W. Middleton.*
W. River Cock, Stutton !
A. Eshton Tarn, Gargrave ; *S. Hailstone* (1805).
C. New Miller Dam ; *T. W. Gissing.*
D. Doncaster Carrs, July, 1800 ; *J. Salt.* In the River Don, Doncaster and Goole.
T. In the Rivers Idle and Torne about Rossington and Auckley.

870. Scirpus Tabernæmontani Gmel. (*S. glaucus* Sm.)
Area : 63, Top. Bot. Halophilous. Range : Low.
Native ; in brackish dikes or ponds, very rare. P. July-August.
First (and only) record ; Parsons, 1875.

D. Ponds by the railway, Hook Grange, Hook Moor, near Goole ; *H. F. Parsons* !

871. Scirpus maritimus L.
Area : 63, 64, Top. Bot. Range : Low.
Native ; muddy tidal banks and semi-brackish stills, local. P. July-August.
First record : Parsons, 1875.

A. Up the Ouse and Aire (from Goole) as far as Langrick and Newland ; *H. F. Parsons.*
D. Don Banks at Goole to above New Bridge, and Marshland ; *H. F. Parsons.*

872. Scirpus sylvaticus L. *Wood Rush.*
Area : 63, 64, Top. Bot. Range : 0—600 feet.
Native ; sandy swampy ground by rivers and ditches, infrequent. P. July.
First record : Salt, 1800.

L. Helks Wood, Ingleton.
R. Ditches below Ranegill in Bolland ; *J. Windsor* in *Towns. O.B.G.* 'Near Settle' ; *J. Tatham* in *Herb. Whitwell.*
Y. Ure banks, Studley (!) *W. Brunton* in *O.B.G.* Hackfall ! Copgrove, Newby, etc. ; *J. Dalton* in *O.B.G.*
N. Scotton banks, Knaresborough ! By the River Crimple, Aketon ; *J. S. Wesley.*

W. Bolton Woods ; *B. Carrington.* Bog near Ilkley ; *J. G. Baker.*
A. Near Keighley ; *H. T. Soppitt.* Near Leeds, not uncommon ; *Rev. W. Wood* in *O.B.G.* (1805) [Here, as elsewhere, Wood's 'near Leeds' must be read as meaning, possibly, a dozen or sixteen miles away]. Calverley, formerly, now gone ; *J. Willis.* Adel ; *H. Baines.*
C. North Dean ; *H. Baines.* Coxley dam, Dewsbury ; *P. F. Lee* ! Woolley ; *T. W. Gissing.*
D. Ditches and ings by Dearne, Wombwell (*H. Johnson*) and near Burton Grange ! Side of Cliff Wood, Barnsley ; *H. Payne.* Sides of the River Don a little above Sheffield ; *J. Salt.*

873. Eriophorum vaginatum L. '*Hare's-foot.*' '*Moss-crop.*'
Area : 63, 64, 65, Top. Bot. Range : 0—1,800 feet.
Native ; turfy moors and drier heaths, common. P. Late May.
First record : Bolton, 1775.

M. On the moors, abundantly (!) *J. Whitehead.*
L. In the ascent to Ingleborough ; *Gough's Cam.* (1789). Common in Dentdale and on Baugh Fell.
R. Common in Ribblesdale on all the mosses.
Y. Bog on the Ox Close, Ripon ; *W. Brunton* in *O.B.G.* Sawley Moor !
N. Ramsgill and Pateley moors, in plenty.
W. Pastures near Cowside ; *J. Windsor, MS.* in *Towns. O.B.G.* Barden Moors. Rombalds Moor.
A. Malham Tarn bog. Black Moor, near Leeds ; *Rev. W. Wood* in *O.B.G.* (1805). [On dry moory ground at Alwoodley and towards Shadwell (*J. Abbott*) as late as 1871. Now gone.] Bingley and Keighley Moors.
C. Sowerby, Rishworth, and Warley Moors ; *J. Bolton* (1775).
D. Sheffield Moors near Grindleford Bridge, 1799 ; *Herb. Salt.* Ringinglow Moors and Redmires ; *A. Carr.* Goole Moor, abundant.

874. Eriophorum angustifolium Roth. '*Cotton Grass.*' '*Cotton Rush.*' '*Ghost-grass.*'
Area : 63, 64, 65, Top. Bot. Range : 0—2,100 feet.
Native ; peat bogs and moorland mosses, common, rare in the lowlands. P. June.
First record : Bolton, 1775.

M. In this district ; *J. Whitehead.*
L. Very common on the peat moor slopes of Garsdale, Dentdale, and Blea-moor !
R. Common about Settle ; *J. Windsor.* Cocket Moss ; *Hb. Tatham.* Hodder Valley mosses !
Y. Bogs near Ripon ; *W. Brunton* in *O.B.G.* Whitemere, Hutton Moor ; *H. H. Slater.* [Copgrove, Newby, by brook ; *J. Dalton* in *O.B.G.*].

N. Harrogate and Pateley Moors. Common.
W. On all the moors from Buckden to the Washburn ! Far jungle, Askham Bogs !
A. Malham Tarn Bog. Adel Black-moor. Bingley Moor.
C. Sowerby, Warley, and Rishworth Moors ; *J. Bolton* (1775). Ryhill and Woolley Edge ; *T. W. Gissing.* Holmfirth Moors ; *G. Roberts.*
D. Potteric Carrs ; *E. Miller* (1805). Ringinglow and Redmires Moors ; *A. Carr.* Goole Moor, abundant ; *H. F. Parsons.*
T. Roche Abbey ; *Herb. Salt.*

Obs.—En masse upon many of the higher moors in autumn, its red stems give an unique madder-purple hue to the scene in the near distance.

VAR. **elatius** Koch.
This broad-leaved form is rare, so far as my experience goes, within the Riding. I have seen it from

R. Bowland Knotts Moors ! Cocket Moss ; *J. Tatham.* Kingsdale ; *L. C. Miall.*
A. Malham Tarn Bog.
C. Wessenden Valley Moors ; *W. Fowler, spn.* !

It is occasionally misnamed *E. latifolium.*

875. Eriophorum latifolium Hoppe. (*E. pubescens* Auct.)
Area : '64 Tatham spn.,' Top. Bot. Range : 450—1,250 feet.
Native ; oozy, marshy places on moors, but usually upon calcareous soil, appearing to prefer a stiff soil to a light peaty one, rare. P. July-August.
First record : W. Howson's Flora, 1850.

L. Springs by the river Greta, Ingleton !
R. Pasture, Giggleswick ['Tarn field'; *Hb. Tatham*] ; and between Long Preston and Swinden ; *J. Windsor.*
Y. Copgrove ; *H. Baines.*
A. Airton near Malham ; *W. West* (1877) !

876. Carex dioica L.
Area : 63, 64, Top. Bot. Range : 0—1,250 feet.
Native ; bogs and marshy pastures, infrequent. P. June-July.
First record : Bolton, 1775.

L. Bog near Bentham !
R. Boggy pastures at lower end of Kelcowe, etc., about Settle ; *J. Windsor.* Giggleswick ; *Herb. Tatham.*
Y. Near Ripon, plentifully ; *W. Brunton* [not so now from drainage]. Marshy pasture by Laver, near Galphay ! (? Cow Mires). Small marsh near Wormald Green Station ; *H. H. Slater.*

N. Moors near Harrogate, and Hookstone Crags ! Aketon Bleach-works Marsh !
W. Rombalds Moor, and Ilkley ! Bramham Moor. Askham Bogs ; *B. B. Le Tall.*
A. Malham Tarn Bog (!) *J. Nowell.* Malham to Airton ; *W. West.* Near Leeds ; *Rev. W. Wood.* Adel Bog !
C. Sharlestone, Woolley, and Ryhill ; *T. W. Gissing.* Halifax ; *J. Bolton.* Todmorden ; *A. Stansfield.*
D. Stainborough ; *H. Payne.* Crookes, Cliff Rocher, and Wadsley Common ; *A. Carr.* Pottery Carr, Doncaster ; *Herb. Salt.*

[**Carex Davalliana** Sm. Incognit.
This was reported to me from the Sheffield Moors by the late Mr. John Hardy (in litt. 1876). Asking for fuller information I received the statements that follow—they embody all I know of the plant, except that I instructed the late Mr. Amos Carr to gather all the *C. pulicaris* and *C. dioica* forms he could from the high moors to the north-west of Sheffield, keeping a special look-out for a diœcious non-creeping sedge—a tufted plant like a small slender *C. pulicaris.* He made many dozens of gatherings for me, but none of them was *C. Davalliana.* Mr. Hardy (in 1876) wrote : 'My herbarium up to the year 1845 would have settled the question, but unfortunately it passed out of my possession in that year, and the mounted plants were for some time in the Lodge of the Botanical Gardens [at Sheffield]. On inquiry about them some years ago, I learnt that they had been allowed to go to destruction, the young children of a former 'Curator' having been allowed to use them as playthings ; amongst them was *Carex Davalliana*, which I am certain I collected somewhere in the neighbourhood, in a young state, but distinctly recognisable, in the early part of June 1845—a time with me of ill-health and great trouble, and I cannot since remember where I got the plant. I have never ceased to expect hearing of Mr. Newbould having turned it up.']

877. Carex pulicaris L. *Flea Sedge.*
Area : 63, 64, 65, Top. Bot. Range : 0—1,800 feet.
Native ; peat-bogs and turfy heaths, very common. P. June.
First records : Bolton, 1775 ; Gough in Camd., 1789 ; Salt, 1800.
Ogden Clough ; *J. Bolton.* On Ingleborough ; *Gough's Camd.* Crooks Moor ; *J. Salt.*
Distribution general ; occurs in every district, in suitable situations, from the boggy plateau of Baugh Fell to the turf moors of south-west Yorkshire, but in the cultivated eastern third of the Riding it is, of course, restricted to such spots as Askham Bogs, White Carr Bramham, and Morton Bog near Hambleton.

878. Carex divisa Huds.

Area : 63, [64 likely error], Top. Bot. Range : Low.

Native ; brackish marshes and dike-sides, very rare. P. July.

First record : Parsons, 1877.

(Y. Near Copgrove, *B. G.*; *H. Baines, Flora.* Some error—*C. disticha* intended?)

(N. Spongy bog, moor edge, above Ramsgill, 1846 ; *T. Simpson* ! *J. G. Baker* in *Miall's Supp.* Said to have been *C. schœnoides* Host, but in flower only).

(A. Wet places below Malham Cove ; *J. Nowell.* Both *C. disticha* and *Blysmus compressus* occur here).

D. Marshy dike-side, parallel with the river Ouse, Hook Road, Goole, with *Juncus Gerardi*, etc. (!) *H. F. Parsons.*

Obs.—I have not been able to learn whether the inland (Ramsgill) *C. schœnoides* has ever been re-found and verified. I have once only spent half a summer's day searching (in vain) the alleged locality.

879. Carex disticha Huds. (*C. intermedia* Good.).

Area : 63, 64, Top. Bot. Range : 0—1,250 feet.

Native ; marshy pastures and swamps, common. P. June-July.

First record : Dalton, 1803—'canal banks, Ripon.'

Distribution general ; occurs in every district, from Malham Tarn side (*J. Windsor*) down to Askham Bogs, pastures by river Aire at Fairburn, and ditch-sides on Thorne Waste.

Obs.—The peculiar facies which the spike of this species usually has is due to monstrosity—a corky enlargement of the perigynium, which then contains no nutlet. Specimens without this deformity are rare, and wear such a dissimilar aspect as to be frequently unrecognised and misnamed even by fairly good botanists familiar in the field with the majority of Sedges.

880. Carex teretiuscula Good.

Area : '63, 64 Leyland spn.,' Top. Bot. Range : 0—1,250 feet.

Native ; bogs and marshes, very rare. P. June.

First record : Brunton, 1805.

L. Bogs west of Ingleborough ; *B. Carrington.*

[Y. Near Ripon ; *W. Brunton* in *O.B.G.*= Fairfield Woods, Ripon ; *H. Baines.* Near Copgrove ; *J. Dalton.*]

W. Askham Bogs ; *H. Baines* ; an error—*C. paradoxa* being meant, vide *Phyt.* i, 1121—but the difference between the two is so slight that their separation as distinct species is scarcely merited).

A. By Malham Tarn ; *S. Gibson, Phyt.* (1848)—the var. or form *Ehrhartiana*, with short, close spike.

D. Askern, in a boggy meadow, plentiful ; *P. Inchbald* in *Phyt.* (1843). Conisborough district ; *Y. N. Recorder*, p. 205, but the A., signifying Askern, was omitted accidentally. Sandholes, Swinefleet Common ; *T. Birks.*

881. Carex paradoxa Willd.

Area : 64, Top. Bot. Range : Low.

Native ; boggy, bushy thickets and dike-sides, two stations, locally abundant. P. Middle to end of June.

First record : Spruce, 1842—but doubtless gathered previously, though not discriminated from the preceding, by S. Gibson (the blacksmith), H. Baines, and others.

W. Askham Bogs, in the 'Far Wood' principally, but also along with *C. stricta* by the dike-sides ! *R. Spruce* (1842) ; *L. C. Miall* ; *Baker's Suppl.* ; etc. In a Carr between Healaugh and Askham Richard ; *J. G. Baker.*

882. Carex paniculata L. '*Hummock Grass.*' '*Seat-segg.*'

Area : 63, 64, Top. Bot. Range : 0—600 feet.

Native ; boggy, or watery thickets, frequent, locally plentiful. P. July.

First notice : Merrett, 1666—'Gramen Cyperoides . . . Hujus radicibus utantur pro sedilibus in agro Eboracensi' ; *Merr.*, *Pin.*, 51.

L. Near Sedbergh ; *H. Baines.* Near Clapham, towards Black Burton !

R. Hodder Valley !

Y. Hackfall ! Sharow Mires ; *H. H. Slater.*

N. Bilton and Scotton banks, Knaresborough ! Plumpton Rocks, below pools !

W. Washburn valley swamps ! By stream in marshy thicket below Scarcroft Hall ! Bardsey Pool ! Moss Carrs, Wetherby. Stutton Carrs. Collingham.

A. Hawksworth ; *H. T. Soppitt.* Skipton, near the canal towards Gargrave ! Adel Bog, near Seven Arch Aqueduct ; and beyond Adel Bridge in great quantity in a low field by stream, some of the hummocks five feet high, and locally known to children, farm-lads, etc., as '*Seat-seggs*'—an appellation derived from the use made of them, known to Merrett as far back as 1666, and exemplifying very strikingly the persistency of a local custom and term.

C. Widdop, Heptonstall ; *J. Nowell.* Hebden Bridge ; *S. Gibson* ; *Herb. Tatham.* New Miller Dam, Woolley ; *T. W. Gissing.*

D. Sandall Beat Carr, near Doncaster. Great Ea[illegible] Plantation, Potteric Carr. Marsh by the river Rother ; *A. Carr.*

[**Carex brizoides** L. Incognit.

'Error'? Top. Bot.

Studley Woods ; *W. MacIvor*, teste *C. C. Babington, Man.*, ed. v, 368. 'Supposed to have been introduced, but deserves inquiry from resting on good authority.' 'Was probably an escape' ; *Bab., Man.*, ed. viii, 398. If really found, probably planted ; *Engl. Bot.*]

883. Carex vulpina L.

Area : 63, 64, Top. Bot. Range : 0—500 feet.

Native ; moist meadows, marshes, and ditch-banks, common. P. July.

First record : Bolton, 1775.

Distribution nearly general ; recorded from every district (mainly in the lowland areas), except Mersey.

884. Carex muricata L.

Area : 63, 64, Top. Bot. Range : 0—900 feet.

Native ; dry ditch-banks, thickets, etc., unfrequent. P. June-July.

First record : Windsor, 1840.

L. Ingleton !

R. Near Settle ; *J. Windsor* in *Baines' Flora.*

Y. Mackershaw and other places about Ripon !

N. Plumpton Rocks ; Knaresborough cliffs ; and dry ditch-bank by road near the rocky field, Braham Hall, Spofforth.

W. Grassington ! Collingham, Wetherby, Thorparch, etc. !

A. Frequent on the Permian tract about Ledstone and Brotherton.

C. Wakefield, frequent ; *T. W. Gissing.*

D. Lane near Cantley ; *A. Carr.* Conisborough ! Crookes Moor, Sheffield ; *A. Carr* !

T. Maltby !

885. Carex divulsa Good.

Area : 63, 64, Top. Bot. Range : 0—450 feet.

Native ; dry thickets and hedge-banks, rare. P. June.

First record : Carrington, 1862.

L. Helks Wood, Ingleton ; *B. Carrington.*

Y. Mackershaw Wood (!) and Hackfall ; *H. H. Slater.*

N. Cliff slope, Knaresborough !

(A. Near Seven Arches, Bingley ; *Miall's Flora.*)

D. Scraith Lane, near Sheffield ; *A. Carr* !

886. Carex stellulata Good. (*C. echinata* Murr.)

Area : 63, 64, 65, Top. Bot. Range : 0—1,800 feet.

Native ; peat moors and boggy woods, common. P. June.

First record : Salt, 1800—'Crooks Moor.'

'Bogs near Doncaster' ; *Baines' Flora.* 'Near Kiveton' ; *A. Carr.*

Distribution : recorded from every district, abundant in the western two-thirds of the Riding on the moors and heaths, descending to Askham Bog and the stations mentioned above.

887. Carex remota L.

Area : 63, 64, Top. Bot. Range : 0—900 feet.

Native ; pool-sides, ditch-banks, and damp woods, common. P. June.

First record : Salt, 1800—'Parker Spring.'

Distribution general ; in suitable places in every district there are numerous stations, although it rarely occurs in any quantity in any one spot, from Dentdale, Clapdale, and Settle Crags downwards to Thorne Moor.

888. Carex axillaris Good.

Area : 63, 64, Top. Bot. Range : 0—450 feet.

Native ; watery grassy places and pool-sides (on clay mostly), rare. P. July.

First record : Dalton, 1803.

Y. Pits near Copgrove ; *J. Dalton.* Near Ripon ; *W. Brunton.* [Sharow mires, gone ; *H. H. Slater.*]

N. Bog, a mile from Low Harrogate on the road to Ripon ; *L. C. Miall.*

W. Grassy bushy places by the Hobmoor ponds (!) near York.

D. Canal-side, Swinton (*H. Johnson*), and Barnsley (!) by the canal. Near Silkstone (with a doubt) ; *Gissing's Flora.*

889. Carex elongata L.

Area : 63, 64 ?, Top. Bot. Range : Low.

Native ; marshes and swampy fields, very rare. P. Late June.

First record : Salt, 1803.

D. Marshy place at Aldwarke on the river Don below Sheffield ; *J. Salt* (1803). Aldwarke, marshy field near Don river, between the river and the rise of the land towards Aldwarke, a small erection like a bath near to it. There in 1845 ; *J. Hardy.* I sought it in vain in 1874-6, and am told the marsh is much contracted from what it used to be ; *A. Carr* (1876).

890. Carex curta Good.

Area : 63, 64, 65, Top. Bot. Range : 0—1,800 feet.

Native ; peat-bogs and peaty ditch-sides, infrequent. P. June-July.

First record : Brunton, 1805.

L. Garsdale, Baugh Fell slope ! Widdale Fell, near tarns !

R. Blea-moor. Cockitt Moss ; *Herb. Tatham.*

Y. Sawley Moor near Grantley lakes ; *W. Brunton.* Ox-Close, Ripon (!) *O.B.G.*

N. Low Harrogate, one mile along Ripon road ; *H. Baines.*

W. Rombalds Moor (!).

A. Malham Tarn ! Cottingley Moor ; *W. Ainley.* Sharpah ; *W. Whitwell.* Hawksworth. Peaty ditch by plantation through which a foot-road passes, between Heath Nurseries and Alwoodley (!) *H. Ibbotson.*

C. Stansfield Moor, etc., rare in Todmorden district ; *A. Stansfield.* Denholme Bog ; *S. King.*

D. Near Aldwarke by river Don ; *J. Hardy.* Lane near Oughtibridge ; bog near Redmires Dams ; *A. Carr.*

T. Thorne Waste (!) *W. Todd.*

891. Carex ovalis Good.

Area: 63, 64, Top. Bot. Range: 0—1,250 feet.

Native; wet clayey or undrained pastures, common. P. July.

First record: Salt, 1800—'Crooks Moor.'

Distribution general; thinly distributed through every district, from Sedbergh, Swobeck (*J. Windsor*), and Malham Moor (!) down to Aldwarke (*W. W. Newbould*), and a dry furzy lane near Austerfield.

892. Carex stricta Good.

Area: 64, Top. Bot. Range: 0—300 feet.

Native; peat-bottomed pools, rare, although locally abundant. P. June.

First record: Baker, 1854.

Y. By Queen Mary's and other 'dubs' by Yore, a mile to three miles above Ripon Bridge! Sharow mires and pond near Ripon North Bridge; *H. H. Slater*.

W. Askham Bogs, dike-sides bordering the 'Far Wood' chiefly (!) *R. Spruce*; *J. G. Baker*; etc.

D. In wood south-east of Askern! Doncaster Carrs! Near Thorne! ('Marshes near Treeton'; *J. Salt*.) Pool near the Rother; *A. Carr*.

893. Carex acuta L.

Area: 63, 64, Top. Bot. Range: 0—450 feet.

Native; marshes, ditches, and pool-sides, infrequent, but generally in abundance when it does occur. P. June.

First record: Windsor, date uncertain.

[R. About Giggleswick Tarn; *J. Windsor*. Gone—and perhaps doubtful. No specimen in *Herb. Tatham*, nor mentioned in his notes.]

Y. Sharow Mires (!) and Yore side dubs; *H. H. Slater*. Near Ripon and Copgrove; *H. Baines*.

W. Marsh in north corner of Wetherby Racecourse ings! Askham Bogs; *B. B. Le Tall*.

A. By Adel Dam near Leeds—Lower Dam! Marshes by Aire, between Skipton and Gargrave (*W. Whitwell*); about Fairburn (!) etc.

C. Stanley; *T. W. Gissing*.

D. Conisborough district; *Y. N. Excurs.* (1873). By the Rother; *A. Carr*! Potteric Carrs! Salt does not name this, but perhaps his plant of the 'Treeton marshes' under the name of *stricta* refers in part, if not altogether, to this. Goole; *T. Dickin*.

Obs.—Curious varieties (or rather monstrosities) of this Carex are not uncommon; two are mentioned in *O.B.G.*; and Mr. Whitwell has one with the pedicel of lowest catkin 13 inches in length and given off near base of stem, which is often so swollen as to be almost bulbous, possibly (as Mr. Beeby suggests) from attacks of internal-feeding larvæ. It will hybridise with *C. ampullacea*, *C. paludosa*, and *C. vesicaria* when it grows plentifully intermingled (as it does at Adel Dam near Leeds, on Wetherby Ings, and in Hawes 'backwater marsh') with one or the other.

894. Carex rigida Good.

Area: ['64' insufficiently vouched], Top. Bot. Montane. Range: 2,400—1,800 feet.

Native; stony mountain pastures, very rare. P. July.

First record: Dalton, 1797.

L. Ingleborough; *Herb. Dalton* (1797). 'Wet places about Ingleborough; *J. Windsor*,' *MS.* in *Towns. O.B.G.*; but *not* in *Flora Cravoniensis*.

R. Fountains Fell; *Herb. Tatham*. Penyghent; *H. T. Soppitt* (1885)!

895. Carex vulgaris Fr. (*C. Goodenovii* Gay.) '*Seggs*.'

Area: 63, 64, 65, Top. Bot. Range: 0—2,100 feet.

Native; bogs, heaths, moorland pastures, and pool-sides, very common. P. July.

First record: Salt, 1800, sub nom. *cæspitosa*.

Distribution general; occurs in every district.

Obs.—There are at least two forms of this in our area with a distinct enough habit and facies; one, by far the commoner, occurs scattered on the moors, creeping extensively, and having a more or less arcuate fruit stem; the other is more erect, and grows in dense tufts in or by pools, but lacks the membranaceous reticulated leaf-sheaths of true *stricta*.

VAR. **Gibsoni** Bab.

Woodhay, near Hebden Bridge; *S. Gibson*. Now lost; fruit said to have been longer and more gradually acuminate than in the type. I believe a state with corky, abortive perigynia, perhaps galled, is not very unusual. In Mr. Tatham's collection is a specimen labelled *C. vulgaris* var. *Idiogynæa*, near Hebden Bridge, May 1853; *S. Gibson*.

VAR. **juncella** Fr. (*uliginosa*.)

This variety is not uncommon in tufts in very wet peaty places on the moorlands. It has been recorded as '*stricta*' by *Mr. A. Stansfield* from Moss Hall Slades, west of Blackshaw Head, near Todmorden; and I have it from pool-side near Cemetery, Wetherby, from Widdale Tarns, and the Hodder Valley mosses.

896. Carex glauca Scop. '*Seggs*.'

Area: 63, 64, 65, Top. Bot. Range: 0—1,800 feet.

Native; moist meadows, pastures, and moors, very common. P. May.

First record: Salt, 1800.

Distribution general; in every district—the very commonest of the smaller species, and varying in stature and facies considerably.

VAR. **Micheliana** Sm.

This sub-alpine variety is on record from 'Near Ripon,' *W. Brunton* in *O.B.G.*, but I have not seen it there. Mr. Tatham's collection contains a 'var. *recurva*,' Hebden Bridge; *S. Gibson*.

897. Carex limosa L.

Area: 63, 64, Top. Bot. Range: 0—1,250 feet?

Native; spongy bogs, very rare. P. June-July.

First record: Gough in Camd., 1789.

Yorkshire, on turfy bogs; *Gough's Camd.* (1789).

(R. Kendall's Gill, Giggleswick, and in the wood near Langcliffe Place; *J. Windsor*. I believe these stations erroneous—only the small bog forms of *C. vulgaris* and *C. fulva* occur there now. I believe Windsor mistook a form of *fulva* for it.)

[Y. Tanfield Hall Car; *W. Brunton*. Now gone.]

W. Askham Bog, very wet, open turfy spot (1876)!

(A. Malham Tarn Bog; *B. Carrington*. Only *fulva*? Highmoor, Steeton; *Miall's Flora*.)

D. Thorne Moor (near the 'Spectacles' Wells! *H. F. Parsons*. 'Runlet pits,' Thorne Waste! (Bog near Wortley; *Payne's MS. Dearne Flora*. Perhaps an error.)

898. Carex digitata L.

Area: 63, 64, Top. Bot. Range: 0—450 feet.

Native; dry wooded banks on limestones, rare. P. Late May-June.

First record: Frankland or Brunton, about 1803.

Y. Hackfall! *H. H. Slater*. Mackershaw Woods, near Ripon; *W. Middleton*; *W. Brunton*; rep. *H. Baines*.

W. Thorparch Woods; *Frankland* in *O.B.G.* (1805)! Boston Spa, wood near Jackdaw Crag!

D. Levitt Hagg (!) *O. A. Moore*, 1844. Woods between Conisborough and Sprotborough; *A. Carr*!

T. Roche Abbey (!) *Herb. Tofield*; *Aveling's List*; etc.

Obs.—Salt found this 'May 7th, 1801, in Monsall Dale' (*Herb. Salt*), but that locality is in Derbyshire.

899. Carex ornithopoda Willd.

Area: '64 Herb. Borrer,' Top. Bot. Range: 300 feet.

Native; woods on calcareous soil, very rare. P. July.

Y. Mackershaw Wood, Ripon; *Herb. Borrer*, teste *W. W. Newbould*.

900. Carex pilulifera L.

C. montana Linn., Sp. Pl., 1385, et Hudson—teste J. E. Smith.

Area: 63, 64, Top. Bot. Range: 0—1,500 feet.

Native; dry sandy heaths, stony moor slopes, etc., common, but does not like limestone. P. June-July.

First record: Hudson, 1778.

M. Greenfield and Delph Valleys; *J. Whitehead*.

L. Prope Sedberg in comitatu Eboracensi; *W. Hudson*, *Flora Anglica*, ii, 408 (1778).

R. Above Stackhouse Borrins, High Ridge Wood, etc., near Settle; *J. Windsor*. Hodder Valley!

Y. Sawley Moor! Hutton Moor; *H. H. Slater*.

N. Scotton Banks, Knaresborough! Harrogate Moors; *H. Baines*. Plumpton!

W. Barden Tower! Rombalds Moor; *W. West*. Ilkley. Collingham and Etchell Crags. Bramham Moor.

A. Rawcliffe; *T. Birks*.

C. Wakefield, frequent; *T. W. Gissing*. Hebden Valley!

D. Wharncliffe, Redmires, Crookes, etc.; *A. Carr*! Common about Sheffield; *J. Salt*.

VAR. **saxumbra** Lees, Trans. Y. N. U., p. 121. (*Leesii* Ridley).

N. On rocks in shade, Plumpton (1880)! Lowest bract two to three times length of spike, nut fusiform, gradually acuminate into a long rostrate bifid beak, glume with serræ on mid-rib, and a rough, long awn. Dr. Boswell thinks it may prove to be a distinct species; I do not, and I have received intermediate forms from the Aberdeenshire mountains.

901. Carex præcox Jacq.

Area: 63, 64, Top. Bot. Range: 0—1,700 feet.

Native; rather dry pastures, heaths, and rocky slopes, common. P. May.

First record: Salt, 1800.

Distribution general; occurs in every district, ascending to 1,700 feet on the Yoredale-limestone crags of Whernside and Cam Fell, but most distinctively a plant of the down-like pastures over the Permian limestone.

[**Carex tomentosa** L. Incognit.

Leeds: *Denny's Cat.* in *New Bot. Guide*. Another of Mr. Denny's errors—the name probably ticked off through some misapprehension or mental fog, for some other hairy Carex. Such mistakes easily arise where names are not written out and a locality cited to give pause to the understanding.]

[**Carex humilis** Leyss. Error.

Reported in *Yorkshire Naturalists' Recorder* (1873, p. 205) as having been found in the Conisborough district; but I have reason to believe from correspondence that *young C. digitata* was really the species gathered.]

[**Carex vaginata** Tausch. Error.

In same list as the preceding, and like it, an error. The specimens gathered and exhibited at the meeting referred to were not, on that occasion, named by a critical botanist.]

902. Carex pallescens L.

Area : 63, 64, Top. Bot. Range : 0—1,250 feet.

Native ; moist or clayey soil in woods and thickets (less commonly in open ground), infrequent. P. June.

First record : Salt, 1800.

L. Lawkland Hall Wood ; *J. Windsor.* Helks Wood ! Dentdale !

R. Foot of Linn Gill ! High-ridge Wood and Crags, near Settle ; *J. Windsor.* Lodge Gill ; *Herb. Tatham.*

Y. Hackfall ! Pilmoor Carr, Sharow, and field by Mackershaw Wood ; *H. H. Slater.*

N. Scotton banks, Knaresborough.

W. Bolton Woods ! Buckden Woods ! West-woods, Bramham, abundant by the 'rides' ! Fields bordering Askham Bogs !

A. Gordale Scar (!) *L. C. Miall.* Malham. Hambleton !

D. Between Sheffield and Norton ; *J. Salt.* Marshy fields, Silkstone, common ; *H. Johnson* !

T. Maltby Woods !

903. Carex panicea L.

Area : 63, 64, 65, Top. Bot. Range : 0—2,100 feet.

Native ; boggy ground, and heathy or turfy pastures, common above 300 feet elevation, less frequent below. P. June.

First record : Salt, 1800—'Jordan Dam.'

Distribution general ; in every district, but much less common in cultivated places than *C. glauca*, and now mainly (below 100 feet) confined to such swampy localities as Askham and Askern Carrs, Ledsham 'Dale' marsh, Thorne Moor, etc.

904. Carex capillaris L.

Area : '64,' Top. Bot. Range : 800—1,000 feet.

Native ; moorland turf, one station. P. June-July.

A. Plentiful in a limited area on a grassy slope, and in less quantity on the terraced limestone rocks, in Gordale, on the left-hand as the ascent is made, above the large mass of débris ! Discovered by William West in June, 1878. Previously unknown south of Teesdale. Its companions are *Sesleria, Primula farinosa, Gentiana Amarella,* and *Carlina.*

905. Carex pendula Huds.

Area : 63, 64, Top. Bot. Range : 0—600 feet.

Native ; wet clayey woods and thickets, rare. P. July.

First record : Dalton, 1805.

R. Ribble banks, near Settle ; *H. Baines* ; *W. Howson.*

Y. Hackfall (!) *J. Dalton* in *O.B.G.* (1805). Wood near Newby !

N. Dropping Well Wood, Knaresborough ! Wood below Knaresborough ; *Geo. Harrison.*

C. Turner Clough, Ripponden (!) *R. Leyland* in *Baines' Flora.*

D. Roscoe Wood bottom, Sheffield, some examples over eight feet in length ; *A. Carr.*

T. In the canal between Kiveton and Shireoaks ; *Amos Carr.*

906. Carex strigosa Huds.

Area : 64, Top. Bot. Range : 0—600 feet.

Native ; damp shady woods or thickets, very rare. P. June-July.

First record : Dalton, 1802.

(R. Ribble banks near Settle ; *J. Windsor.* Some error.)

Y. Magdalen banks, Hackfall ; *J. Dalton* in *O.B.G.* Studley Woods ; *W. Brunton.* A small patch in Hackfall, south side, and one in Studley (!) *H. H. Slater.*

(W. Arncliffe Wood ; *L. C. Miall,* copied from *H. Baines,* but Arnecliffe Wood in Goathland Dale is the wood referred to—see *Baker's North Yorkshire,'* p. 299.)

(C. Turner Clough ; *C. Eastwood* in *Miall's Flora.* The species there is *C. pendula* !)

(D. Conisborough district ; *Y. N. Excurs.* (1873) ; in mistake for *C. sylvatica.* 'In a wood called Parker Spring, Sheffield' ; *J. Salt.* Perhaps an error—a specimen of *C. sylvatica* is in *Herb. Salt,* with above locality, but none of *C. strigosa.*)

907. Carex sylvatica Huds.

Area : 63, 64, Top. Bot. Range : 0—1,250 feet.

Native ; dry woods, mainly on calcareous soil common. P. June-July.

First record : Salt, 1797—'Wood called Parker Spring' ; Herb. Salt.

Distribution nearly general ; common in every district, except that of Mersey, and infrequent only in Middle Airedale and Calderdale.

D. Lund Wood, Cudworth ; *H. Payne*—off limestone ! Scraith Wood, Sheffield ; *A. Carr*—also on flagstone, as is 'Parker Spring.'

908. Carex lævigata Sm.

Area : 63, 64, Top. Bot. Range : 0—600 feet.

Native ; woods, thickets, and (rarely) open banks by streams, infrequent. P. July.

First record ; Dalton, 1805.

R. Near Settle ; *W. Howson.*

Y. Dallow Gill ; *J. Dalton* in *O.B.G.* Laver banks, above Ripon !

W. Bolton Woods ! Wood near Swinsty, Washburn Valley ; *J. Abbott* ! Harewood Woods, near Temple, and by stream towards Alwoodley !

A. Malham ; *W. West.* Bingley ; *L. C. Miall.* Saltaire Wood ; *W. West*=Trench Wood, Saltaire ; *L. C. Miall.* By Adel Stream, beyond the lower bridge, before the Flour Mills are reached !

C. Keb Clough, Stansfield, Todmorden ; *A. Stansfield.* Hebden Bridge, 1843 ; *S. Gibson.* New-Miller dam ; *T. W. Gissing.* About Luddenden ; *C. Eastwood.*

D. North Royds Wood, Barnsley ; *H. Johnson.* Thickets by moors, Redmires ; *A. Carr.*

909. Carex binervis Sm. '*Seggs.*' '*Moor Palms*'—the polliniferous catkins in bloom.

Area : 63, 64, 65, Top. Bot. Range : 0—2,100 feet.

Native ; dry stony, furzy, or heathery ground, on the moors common and locally plentiful, but infrequent on lower ground. P. August.

First record : Teesdale, 1792.

Yorkshire, not uncommon ; *Gough's Camden* (1789), sub nom. *C. distans.*

M. Drier places on the moor slopes ! Marked by *J. Whitehead,* too, for this district.

L. Common on the drier moors of Dentdale, and on the Keadon Moor !

R. Bowland Knotts. Ribblehead ! Near Giggleswick ; *J. Windsor.* Tosside !

Y. Sawley Moor (!) *W. Brunton.* Copgrove ; *J. Dalton* in *O.B.G.*

N. Driest moors near Harrowgate ; *R. Teesdale* (1792). Dry moors near Brimham rocks ; *J. Dalton* (1805).

W. Otley Chevin ! Barden Moor ! Bramham Moor.

A. Skipton, moor towards Sharpah ; *W. Whitwell.* Common on Keighley and Bingley Moors. Eldwick ; *W. West.* Adel Moor !

C. Turner Clough, Ripponden ; and Ogden Clough, Halifax ; *H. Baines.* Stanley ; *T. W. Gissing*—sub nom. '*fulva.*'

D. Askern ; *Hb. Geo. Harrison* (perhaps *C. distans* ; *W. W. Newbould*). High moors near Fullwood ; *J. Salt,* sub nom. *C. distans.* Near Goole !

T. Near Rossington, towards Austerfield. Maltby !

Obs.—This is really one of the commonest of the Sedges, but (from some confusion with *C. distans*) it seems to have been accounted rare, and was, at least, little known to such early observers as Windsor and Salt.

910. Carex distans L.

Area : '63 G. E. Smith,' Top. Bot. Range : Low.

Native ; marshes in the Triassic plain of the West Riding—that portion of our area which was latest an arm of the sea—very rare. P. July.

First record : G. E. Smith, about 1844.

(R. Settle Crags ; *J. Windsor's Flora*—the plant is *C. binervis.* Edge of Giggleswick Tarn ; *H. Baines*—the plant is either *C. fulva* or *C. binervis* ; both grow thereabout on the moor where the tarn was.)

D. Askern, lake border ! *Hb. Geo. Harrison,* collected 1862, contains a specimen labelled *C. binervis,* which Mr. Newbould thinks may belong here. I have little doubt it does. (Salt-marsh near Goole ; *H. Johnson* (1885)—seems, rather, small *C. binervis.*)

T. Marsh near Maltby ; *G. E. Smith,* spn. ! Roche Abbey district ; *Aveling's List* (1870).

911. Carex fulva Good. (*C. Hornschuchiana* Hoppé.)

Area : 63, 64, 65, Top. Bot. Range : 0—1,800 feet.

Native ; turfy heaths and peat bogs, common on the moors, now rare in the lowlands. P. July.

First record ; Salt, 1800 ?

M. Gulliver's Swamp, above Greenfield (!) *J. Whitehead* (1876).

L. Very common on the peat moors of Dentdale, Garsdale, Whernside, Blea-moor, Keasdon, etc. !

R. Boggy places in the further wood near Langcliffe Place ; *J. Windsor.* Ribble-head.

Y. Near Ripon ; *W. Brunton.* Near Copgrove ; *J. Dalton.* Hackfall ; *H. H. Slater.* Dallow Gill Moor !

N. Ramsgill and Angram Moors ; *J. S. Wesley* !

W. Common on the Langstrothdale and Beamsley Moors, descending to White Car Bog, Bramham ! and Askham Bog !

A. Malham Tarn Bog, and Moor ! Bog near Kirby Malham !

C. & D. Apparently very rare on the moors of these river-basins, perhaps not really so. Thorne Waste, *H. R. Moiser,* 1878 ; *Herb. Whitwell.* I have not seen *C. fulva* there myself.

T. About Roche Abbey ; *J. Salt* ; rep. *Aveling's List.* The plant G. E. Smith identified as *C. distans* ?

[**Carex extensa** Good. Incognit.

Leeds ; *Denny's Cat.* in *New. Bot. Guide.* Another of Denny's mistakes ; I suppose *Carex flava* var. *minor* was intended. A specimen in *Herb. Wesley,* now in *Herb. Whitwell* (the true thing), said to have been found near Wetherby, probably came from Sussex. It was sent to G. Webster as illustrative of what Wesley thought he knew to grow near Wetherby, but mistakenly of course ; the plant he had seen in the field being, doubtless, small *C. flava.*]

912. Carex flava L.

Area : 63, 64, 65, Top. Bot. (aggr. sp.) Range : 0—2,100 feet.

Native ; wet clayey pastures, sandy-peaty or boggy ground, and in woods (more rarely), very common in one or other of its forms. P. June.

First record : Salt, 1800.

Distribution general ; in every district.

The large type is equally general with the smaller variety, but is restricted more to peaty soil over sand. It grows very fine at Ingleton by Thornton Force, by Malham Tarn, and in Aketon Bleach-works marsh.

Var. **argillacea** Towns. Rare.
N. & C. Scotton banks, Knaresborough! Coxley Valley; *P. F. Lee*!

Var. **minor** Towns.
This is the *commoner* form, and the one which stands in the old lists as *C. Œderi*.
M. Frequent in the shaly-boggy ravines (!) and *J. Whitehead*.
R. Neighbourhood of Settle; *J. Windsor*.
Y. Near Ripon, plentifully (!) *W. Brunton*. Cow-mires; *Miss Morton* (sub nom. *C. Œderi*).
N. Common on the moors. Marshy field at St. Francis' Chapel, Stockeld!
W. & A. Common throughout these districts!
C. Heath, Woolley, Ryhill, etc.; *T. W. Gissing*. Common in Upper Calderdale!
D. Goole Moors; *H. F. Parsons*!
T. Bog near Maltby!

[Var. **lepidocarpa** Tausch.
I have not yet seen the true plant within our area. That recorded in *Bot. Rec. Club Rep.*, 1877, p. 244, from Harlow Hill, Harrogate, by *T. J. Foggitt*, was *C. flava* var. *minor*.]

913. Carex Œderi Ehrh.
Area: '63 Carr,' Top. Bot. Range: 500—750 feet.
Native; wet sandy places, very rare. P. June.
First (only) record: Carr, 1875.
(R. Y. A. C. Frequent. Settle and Malham; *L. C. Miall, Flora*. Near Ripon and Hebden Bridge (*S. Gibson*); *H. Baines*. The specimens under this name in *Herb. Tatham* from Settle and Hebden Bridge are really var. *minor* of *C. flava*, and presumably the Malham and Ripon ones were likewise that.)
D. By the Loxley, near a dam, Lower Bradfield; *A. Carr, spn.*! Prof. C. C. Babington identified this, and certifies to the name. I have only seen it inland in Yorkshire in one other place—in the East Riding; but it occurs also I believe on Strensall Common.

914. Carex filiformis L.
Area: 64, Top. Bot. Range: Low.
Native; marshes and pool-sides, very rare. P. July.
First record: Brunton, 1805.
Y. Tanfield Hall Carr, and bog near Hutton Moor; *W. Brunton* (1805) = Whitemere on Hutton Moor (!) *H. H. Slater*.
W. With *C. ampullacea*, *Equisetum*, and *Typha*, in watery carrs a little north of Askham Bog proper, rarely fruiting.

915. Carex hirta L.
Area: 63, 64, Top. Bot. Range: 0—1,250 feet.
Native; marshes, swampy pastures, and pool-sides, common. P. July.
First record: Salt, 1799—'Roche Abbey.'

Distribution general; occurs in every district, in abundance locally, from Malham Moor down to Thorne!
Obs.—Curious forms, the result of hybridisation, are not uncommon. Dalton mentioned one having a compound spike growing at Copgrove, 'constant for three seasons,' as agreeing with one figured by Schkuhr; others are very stunted, and some almost glabrous.

916. Carex Pseudo-cyperus L.
Area: 63, 64, Top. Bot. Range: Low (one record for 900 feet).
Native; pool-sides and stagnant dikes (in woods rarely), rare. P. July.
A decreasing species, probably; always local with us.
First record: Salt, 1800—'Common'; not so now near Sheffield, at any rate.
(R. Betwixt Penyghent House and Penyghent Hill; *J. Windsor*.)
Y. Bog on south side Yore at Ripon; *H. Baines*. Not seen of late years; *H. H. Slater*.
W. Askham Bog (border of Gilson's patch!) *H. Baines*. Hob Moor brick-ponds; *J. G. Baker*.
C. Near Halifax, *S. Gibson*; *Herb. Tatham*.
D. Askern, 1843; *P. Inchbald*. Perhaps now extinct. In plenty on the side of Black Dike, Potteric Carrs, Doncaster, near the signal-box at junction of the main G.N. line and the Finningley branch! *Not* seen near Sheffield, 1874-78, by *A. Carr*.
T. Rossington Carrs!

917. Carex paludosa Good.
Area: 63, 64, Top. Bot. Range: 0—600 feet.
Native; marshes, ditch- and river-sides, but chiefly by canals, very common, the most abundant large Sedge we have in West Yorkshire. P. July.
First record: Windsor, 1855.
M. By the canal, Diggle Valley (!) *J. Whitehead*.
R. Ribble banks from below Settle! Near Settle; *J. Windsor*. No note by *J. Tatham*—who probably did not know it as distinct from *C. riparia*.
Y. Yore banks, Ripon to Boroughbridge! Hackfall; *H. H. Slater*. By canal, Ripon!
N. Knaresborough downwards, by the river Nidd, in many places.
W. Ilkley eastwards, by the river and pool-sides. Wetherby Ings! Askham!
A. Leeds and Liverpool Canal-side, Aire banks, etc., common from Gargrave eastwards to Selby. Adel Lower Dam!
C. Common by the Calder Canal and river from Todmorden to Stanley!
D. Askern (!) *Herb. Geo. Harrison*. Don banks, Conisborough! Barnsley, abundant!
T. By the Kiveton Canal; *A. Carr*!

918. Carex riparia Curt.
Area: 63, 64, Top. Bot. Range: 0—300 feet.
Native; dike-sides, ditches, and by canals, rare, locally abundant. P. June.
First record: Salt, 1800—but preceding intended in part?
(R. Settle Ings, sides of the ditches; *J. Windsor*. 'Common'; *J. Tatham, MS. note*. *C. paludosa* meant.)
Y. Sparingly by the river Yore at and below Ripon! Abundant; *H. H. Slater*.
W. Dikes near Church Fenton. Askham Bogs.
A. Selby. Sherburn! cum *W. Todd*.
C. Wakefield, common; *T. W. Gissing*.
D. Askern; *Geo. Harrison*. 'Banks of rivers'; *J. Salt*. I suspect he included *C. paludosa*, which, although much commoner, he does not name in *Flora Sheffieldiensis*. Potteric Carrs, by the dikes, common! Thorne Moor, bordering dikes! Goole; *H. F. Parsons*! Barnby Dun.

919. Carex ampullacea Good. (*C. rostrata* Stokes). *Bottle Sedge*. *C. inflata*; Hudson and Gough's Camden.
Area: 63, 64, 65, Top. Bot. Range: 0—1,700 feet.
Native; marshes, dam- and pool-sides, and also in peat on the moors. P. July.
First record: Curtis, 1782. Bogs, Yorkshire; Gough's Camd. (1789).
L. Peaty pools on Widdale Fell. Garsdale, Blea-moor, Keasdon, to Bowland Knotts, frequent.
R. Ditch in field near Attermire Cave; *J. Windsor*. [Giggleswick Tarn; *W. Curtis* (1782).] Wigglesworth. Rathmell.
Y. Sharow Mires, Ripon (!) Hackfall, etc.; *H. H. Slater*. Kirby Malzeard!
N. Scotton. Bilton.
W. Barden Moors. Stake Fell Moss! Rombalds Moor! Askham Bogs! Stutton!
A. Malham Tarn Bog. Skipton Moors; *W. Whitwell*. Pools, Yeadon Moor; *B. Carrington*. Adel Dam!
C. Hebden Bridge, Rishworth Moors, and Dunford Moors.
D. Thorne Waste; *H. F. Parsons*.
Obs.—This species will hybridise with *C. vesicaria* and *C. acuta*.

920. Carex vesicaria L. *Bladder Sedge*.
Area: 63, 64, Top. Bot. Range: 0—750 feet.
Native; marshes and pool-sides, infrequent. P. July-August.
First record: Bolton, 1775. Yorkshire, not uncommon; Gough's Camd., 1789.
L. Peat bogs, and by tarns on the moors of Baugh Fell, Widdale.
[R. Near Giggleswick Tarn; *J. Windsor*. Probably gone.]

Y. Copgrove; *J. Dalton*. Pond near the Low Mills, Ripon; *J. Dalton*. Ripon (!) *W. Brunton*. Pilmoor Carr, Sharow (!) *H. H. Slater*.
W. Dike near Stutton. Askham Bogs!
A. Adel Dam, lower end! Water-meadows below Fairburn!
C. Old House Mill, Norland; *J. Bolton*. Near Halifax; *S. Gibson*, 1843; *Herb. Tatham*. New Miller Dam; *T. W. Gissing*.
D. Askern! About Jordan Dam; *J. Salt*. By the river Rother; *A. Carr*!

Var. **pendulina** Blytt. Rare.
W. By a pool in Cundy's Fields, Wetherby!
D. Marsh by the Rother; *A. Carr, spn.*!

LXXXIII. GRAMINEÆ.

[**Lappago racemosa** Willd. Casual.
A shoddy-waste introduction.
C. Leeds Road, Huddersfield; *J. B. Ingle*, etc., in *Hudd. Nat. Hist.* (1857-58).]

[**Digitaria sanguinalis** Scop. Casual.
An introduction with foreign wool.
C. Leeds Road, Huddersfield, on a shoddy heap; *C. P. Hobkirk* and *J. B. Ingle* (1857).]

[**Echinochloa Crus-galli** Beauv. Casual.
A. Boothroyd's Mill, Meanwood!
C. On Mungo-waste, Leeds Road, Huddersfield (1858); coll. *J. B. Ingle*; *Herb. W. Kirkby*!]

[**Setaria viridis** Beauv. Casual.
On wool-refuse, fields of Italian Rye-grass, and waste ground, rare.
W. Amid refuse from a flour-mill, Tadcaster!
A. Boothroyd's Mill, Meanwood!
C. Near Wakefield (1859); *T. W. Gissing*.
D. Thorne Quay!]

[**Setaria verticillata** Beauv. Casual.
A shoddy-refuse waif.
C. Leeds Road, Huddersfield; *J. B. Ingle*, etc. (1857-8).]

921. Anthoxanthum odoratum L. *Sweet Vernal Grass*.
Area: 63, 64, 65, Top. Bot. Range: 0—1,250 feet.
Native; meadows and banks, very common. P. June.
First record: uncertain.
Distribution general; in every district.

[**Anthoxanthum Puelii** Lec. et Lam. Casual.

Seen once only in a sandy 'seed-field' of sown grass, clovers, etc., near Spofforth, in 1880—a Casual merely, not likely to naturalise with us, being an annual hardly of vigorous combative habit, but likely to turn up anywhere for a season or two. German seed-growers palm it off as a substitute for the perennial species, from which it may be easily known by its having stems branched at the nodes from the leaf-sheath axis as in No. 923.]

922. Digraphis arundinacea Trin. *Ribbon Grass. Reed Grass.*

Area : 63, 64, 65, Top. Bot. Range : 0—900 feet.

Native ; ditches, river-banks, and swamps, common. P. July.

First record : Salt, 1800.

Distribution general ; in every district.

[**Phalaris canariensis** L. *Canary Grass.* Alien.

A waif often occurring (through its use as Canary-seed) on manure heaps and road-sides about towns and villages.

W. Lane near Park Hill, Wetherby !

A. Frizinghall and Woodhouse Ridge ; *W. West.*

C. Castle Hill, Huddersfield, occasionally ; *C. P. Hobkirk.*

D. Corn-field, Sharrow ; *Geo. Harrison, MS.* Various places about Sheffield ; *J. Salt.* Field near Brindsworth and quarry near Crookes ; *Amos Carr.*]

[**Phalaris paradoxa** L. Alien.

A casual introduced with grain or wool.

C. Fields, Whitley-Willows, 1858 ; *Hudd. Nat. Hist.*]

923. Alopecurus agrestis L. '*Black Grass.*'

Area : 63, 64, Top. Bot. Range : 0—600 feet.

Colonist ; corn- and clover-fields, rather rare (fortunately). A. July.

First record : Salt, 1800.

R. Near Settle and Rathmell ; *J. Windsor.*

W. Field at Moor-end, Boston Spa ! Compton bank top ! Ainsty ; *G. Webster.*

A. Frizinghall ; *W. West.* Eccup !

C. Wakefield, frequent ; *T. W. Gissing.*

D. Road-sides about Sheffield ; *J. Salt.*

924. Alopecurus fulvus Sm.

Range : Low.

Native ; watery places, pool-sides, very rare. P. June.

Y. Pilmoor Car, Sharow ; *H. H. Slater.* I have not seen any specimen of this, but Mr. Slater knew *A. geniculatus* too well, I think, to take it for *A. fulvus.*

A. Near Selby (1836) ; *S. Gibson, Baines' Flora*, 142. Not seen since ?

D. Wet boggy field near Woodhouse, Sheffield ; *Amos Carr.*

925. Alopecurus geniculatus L.

Area : 63, 64, Top. Bot. Range : 0—1,250 feet.

Native ; watery plashes in meadows and by pool-sides, common. P. June.

First record : Salt, 1800.

Distribution general ; occurs in every district, and ascends to Malham Moor !

926. Alopecurus pratensis L. *Meadow Foxtail.*

Area : 63, 64, 65, Top. Bot. Range : 0—1,250 feet.

Native ; meadows and banks, very common. L. End of May.

First record : Salt, 1800.

Distribution general ; in every district.

927. Phleum pratense L. '*Timothy.*'

Area : 63, 64, 65, Top. Bot. Range : 0—1,250 feet.

Native ; pastures and hedge-banks, very common, sometimes viviparous. P. June-July.

First record : Salt, 1800.

Distribution general ; in every district.

Var. **nodosum** L. Rather rare, in swampy places.

R. Below Birkbeck's Weir, Settle ; *J. Windsor.* New Hall Meadow ; *Herb. Tatham.*

Y. Sharow Mires, Ripon, plentifully ; *H. H. Slater.*

W. Wetherby Ings !

928. Sesleria cærulea Scop. *Blue Rock Grass.* '*Blue Bent.*'

Gramen spicatum mont. asperum ; Ray, Cat. *G. parv. mont. spica crassiore purp.-cærulea brevi* ; Ray, Syn.

Area : 64, 65, Top. Bot. Montane. Range : 1,300 down to 300 feet at Bolton.

Native ; limestone crags and banks, rare, but locally plentiful on stony river-banks low down, adventive. P. June.

First record : Ray, 1670.

L. In fissures of the rocks on Ingleborough Hill in dense tufts ; *Ray, Cat.*, i, 155 (1670). Ingleton ; *R. Teesdale.* Dentdale Crags !

R. Rocks near Settle ; *W. Curtis* (1782) ; *Gough's Camd.* (1789). Kelcove ; *J. Windsor.* Giggleswick.

[Y. Mackershaw Wood ; *J. Dalton.* Extinct, I fear ; no one seems to have seen it since.]

W. Bolton Abbey ; *Rev. W. Wood* in *O.B.G.* Kilnsey Crag ; *J. Willis.*

A. Malham Cove ; *J. Windsor, MS.* in *Towns. O.B.G.* Still there (1877) ; *T. Gough.* Gordale and Awes Scar !

Obs.—I believe the 'Gramen spicatum Ingleburgense asperum, *D. Newton*,' of *Plukenet's Almagest*, p. 173 (1696), to have been this.

[**Polypogon monspeliensis** Desf. Casual.

An introduction with wool, found on shoddy-waste.

A. Boothroyd's Mill, Meanwood Valley ; *James Abbott* !

C. Whitley-Willows, Huddersfield (1858) ; *Hudd. Nat. Hist.*]

929. Agrostis Spica-venti L.

Area : 63, (64), Top. Bot.

Colonist : in sandy fields, rare, abundant locally. A. June-July.

First record : Davies, 1805.

Y. Copgrove ; *H. Baines.* Stump Cross Field, near Boroughbridge ; *W. Middleton.*

A. Bradford district; as an Alien, 1885 ; *H. T. Soppitt.* On hair-refuse, Boothroyd's Mill, Meanwood, as an Alien !

D. & T. Near Doncaster ; *Rev. H. Davies* in *O.B.G.* (1805). About Cantley, Bessacarr (!) and Rossington, in arable ground, too common ; *G. E. Smith.*

930. Agrostis canina L.

Area : 63, 64, Top. Bot. Range : 0—1,500 feet.

Native ; heaths and sandy-moory pastures, locally but not universally common. P. June-July.

First record : uncertain.

R. Probably near Settle ; *J. Windsor.* Helwith Moss, 1844 ; *Herb. Tatham.*

Y. Hutton Moor ; *H. H. Slater.*

N. About Harrogate, Hookstone Crags, and Plumpton !

W. Ilkley and Otley Chevin ! Etchell Crags, Scarcroft, and lanes about Wike and Brandon !

A. Baildon Moor ; *W. West.* Adel Black-moor ! Moortown to Shadwell !

C. Wakefield, common ; *T. W. Gissing.* Woolley Edge !

D. Wharncliffe ! Wadsley Common ; near Crooks and near Crosspool ; *Amos Carr.*

T. Furzy ground about Rossington !

Obs.—I do not consider this a common grass ; it is rather local, and certainly rare on the limestones.

931. Agrostis alba L. '*Twitch.*' '*Wicks.*' *Fiorin.*

Area : 63, 64, Top. Bot. Range : 0—1,250 feet.

Native ; moist hedges and ill-drained fields, common. P. July.

First record : uncertain.

Distribution general ; occurs in every district.

The *stoloniferous* state stands on record for ditches near Settle (*J. Windsor*) ; brook-side, Bradford ; and Doncaster Carrs (*H. Baines*). I have not noticed it in the Riding myself.

932. Agrostis vulgaris With. '*Bents.*' '*Twitch.*'

Area : 63, 64, 65, Top. Bot. Range : 0—2,000 feet.

Native ; ill-drained fields, pastures, heaths, and road-sides, very common and with a wide range, in a variety of forms, of which the dwarfed montane are 'pumila' (*Lightf.*) ; and others (creeping extensively in arable land) are styled '*Twitch*' and '*Wicks*' indifferently with *A. alba.* A. or P. July.

First record : Salt, 1800.

Distributed generally ; in every district.

933. Calamagrostis Epigejos Roth. *Wood-reed.*

Area : 63, 64, Top. Bot. Range : 0—450 feet.

Native ; damp woods and shady swamps, mostly on limestone soil, rare. P. July.

First record : Pierson, 1801.

(R. Near Langcliffe, Settle—uncertain ; *J. Windsor.* No specimen or note in *Herb. Tatham.*)

Y. Near Monckton Common Field : *W. Brunton* = Ripon Field, near Hollin Hall ! Aldfield ; *H. H. Slater.*

N. Wood near Boroughbridge ; *Archdeacon Pierson* in *O.B.G.* Wood near Ripley and Scotton !

W. Lane between Askham Bryan and Rufforth ! *H. Ibbotson.* Woods at Bramham, especially in Black Fen, and on the Kidhall Lane (south-east) side, abundant !

T. Roche Abbey, 'Dicy Car Wood, July 31, 1803' ; *Herb. Salt.*

934. Calamagrostis lanceolata Roth.

Area : 63, 64, Top. Bot. Range : 0—300 feet.

Native ; swamps and boggy thickets, rare, locally plentiful. P. Late June.

First record : Dalton, 1805.

Y. What appears to be an old pond near the brook, half-way between Copgrove and Newby, is full of it ; *J. Dalton* in *O.B.G.* (1805). Ripon ; *H. Baines.* Not recently seen ; *H. H. Slater.*

(N. Near Harrogate ; *Miall, Flora.* This is the preceding species.)

W. Askham Bog ; (Gilson's patch !) *H. Baines.*

(D. Sheffield ; *L. C. Miall.* The preceding meant ?) Great Eaa plantation, near junction of main G. N. R. and Finningley branch line !)

T. Thorne, very abundant ; *Jos. Woods, MS.* in *Towns. O.B.G.* Dike borders by the Waste !

[**Calamagrostis sylvatica** Schrad. Casual.

Twice near Hebden Bridge ; *S. Gibson* (vide *Baker, Suppl.* 135). It may have occurred amid sown grass or on wool- or cotton-waste, quite possibly ; but I know of no specimens ; Mr. S. Gibson, however, reported many other strange things.]

935. Phragmites communis Trin. *Reed.*
Area: 63, 64, Top. Bot. Range: 0—1,250 feet.
Native; swamps, ditches, and wet hedgerows, common in the low country, rare in the dales, but ascending to Malham Tarn in our district. I have seen it at a much higher altitude in the North Riding than the 750 feet Mr. Baker mentions, viz.: by Locker Tarn, above Carperby, at 1,400 feet. P. July.
First record: Salt, 1800.
Distribution general; every district except Mersey, but only a *prevailing* species in the vale east of the belt of Permian limestone.

936. Milium effusum L. '*Wood Millet.*'
Area: 63, 64, Top. Bot.
Native; moist woods, infrequent, local, abundant enough in most woods on grit, rarer on limestone. P. June.
First record: Salt, 1800.
R. Lund's Wood and Lord's Wood, Settle; *J. Windsor.*
Y. Ripon, not common. Hutton Moor, Swinton Burn; *H. H. Slater.*
N. Bilton Banks, Knaresborough! (No records for M., L., W., or T.)
A. Heaton, Saltaire, and Esholt Woods; *W. West.*
C. Oakwell Wood, Birstall! Wakefield; *T. W. Gissing.* Hebden Valley! Frequent in almost every wood in this district; *C. P. Hobkirk.*
D. Hampole; *P. Inchbald.* About Sheffield, frequent; *J. Salt.* Barnsley! Wharncliffe Woods; *Amos Carr.*

937. Aira cæspitosa L. '*Bull's-foreheads.*'
Area: 63, 64, 65, Top. Bot. Range: 0—1,800 feet.
Native; woods, moory ground, ditch-banks, etc., very common. P. July.
First record: Salt, 1800.
Distribution general; in every district, one of the commonest of the larger grasses, especially in the less highly-cultivated districts. On the exposed moors the form with abbreviated leaves is more usual than the type.

938. Aira flexuosa L. *Silver Heath Grass.*
A. montana Huds.
Area: 63, 64, 65, Top. Bot. Range: 0—2,100 feet.
Native; sandy-heathy places, turf moors, etc., common. P. July.
First record: Gough, 1789.
Sandy (Yorkshire) heaths; *Gough's Camd.* (1789).
M. Greenfield and Denshaw Valley slopes (!) *J. Whitehead.*
L. & R. Common on the moors; *J. Windsor et aliis!*
W. Askham Bog, rare (but this, like *Pinguicula* and *Narthecium*, is dying or has died out there). On all the drier moors from Buckden by Bolton and Ilkley eastwards.
A. Skipton Fells, Bingley Moors, Adel Black-moor, Meanwood Wood, etc.!
C. D. & T. Frequent in suitable places in all these areas.

VAR. **setacea** Huds. (*uliginosa* Weihe).
Native; in spongy peat-bogs, very rare.
D. Sparingly (along with *Carex limosa*) in very wet peat by the wells in the southern part of Thorne Moor, near where *Scheuchzeria* used to occur!

939. Aira caryophyllea L. *Hair-grass.*
Area: 63, 64, Top. Bot. Range: 0—900 feet.
Native; sandy or calcareous dry pastures, unfrequent. A. June.
First record: Salt, 1800.
R. Long Preston Moor, about Giggleswick limekilns, etc.; *J. Windsor.* Tosside!
Y. Ripon, common; *H. H. Slater.*
N. Frequent about Grimbald's Crag, Knaresborough! Spofforth! On the Triassic sand, common.
W. Rombalds Moor and Otley Chevin. Healaugh! and in the Ainsty.
A. Skipton Rock! Baildon Moor; *W. West.*
C. Wakefield; *T. W. Gissing.*
D. Crookes Moor, Sheffield; *Herb. Salt.*

940. Aira præcox L.
Area: 63, 64, Top. Bot. Range: 0—1,500 feet.
Native; dry bare ground, wall-tops, stony banks, common. A. May
First record: Salt, 1800.
Distribution general; in every district, and often in abundance, but being fugitive it is easily overlooked. A record for the Y. river-basin is lacking, as Slater does not give it in his Ripon Flora at all, and I have no note of having seen it therein myself, but there can be no doubt it occurs.

[**Chloris compressa** DC. Casual.
An introduction with shoddy.
C. Leeds Road, Huddersfield; *J. B. Ingle*, etc., in *Hudd. Nat. Hist.* (1857-8).]

941. Trisetum pratense Pers. (*Avena flavescens* L.) '*Oat-grass.*'
Area: 63, 64, 65, Top. Bot. Range: 0—1,500 feet.
Native; meadows, pastures, road-sides, very common. P. July.
First record: Salt, 1800.
Distribution general; occurs in every district and on all sorts of soils, although finest and most abundant on a calcareous soil or calcareous gravel such as the drift in many parts supplies. It ascends to quite 1,500 feet on the warm lime-scars of Whernside.

942. Avena pubescens L.
Area: 63, 64, Top. Bot. Range: 0—1,250 feet.
Native; meadows, pastures, and banks, common, although somewhat local. P. Mid-June.
First record: Brunton, 1805.
L. & R. Common in the Dale pastures on limestone; *J. Windsor et aliis.*

Y. Near Ripon, plentifully; *W. Brunton* in *O.B.G.* (1805). Copgrove; *J. Dalton.*
N. About Knaresborough, abundant!
W. Ainsty; *G. Webster.* Common about Wetherby, Collingham, Tadcaster, etc., on the Permian tract.
A. Gordale, Malham Moor! Common on the limestone about Garforth, Kippax, Brotherton, etc.
D. Conisborough, Went-vale, and Warmsworth! Askern. Barnsley*; *H. Johnson.*
T. Roche Abbey!

943. Avena pratensis L.
Area: 63, 64, Top. Bot. Range: 0—1,500 feet.
Native; dry pastures and stony river and road-side banks, on limestone, unfrequent. P. July.
First record: Baines, 1840.
L. Ingleton, Wethercote Cave, Dentdale!
R. Settle (!), Giggleswick Scar, High Hill, Castleberg (*Silv. Thompson*, spn.), etc.; *J. Windsor.*
W. Arncliffe to Grassington, by the river!
A. Malham Moor, Gordale (*J. Windsor*), and by Aire to Gargrave. Ouse bank, near Selby!
D. Worsborough Dale; *Herb. Geo. Harrison* (as *A. fatua*). Goole; *H. F. Parsons.* Near Thorne; *H. Baines.*
T. Roche Abbey; *J. H. Aveling* in *Hist.* (1870).
Obs.—The stunted state '*alpina*' occurs on the limestone crags about Malham, Settle, and Ingleton.

944. Avena fatua L. '*Haver.*' *Wild Oat.*
Range: 0—600 feet.
Colonist; fields, rare. A. July-August.
Y. Fields near Ripon; *W. Brunton* in *O.B.G.* 'A fair test of farming; most abundant on the worst-managed farms'; *H. H. Slater's Flora* (1884). Aldborough; *J. Dalton.*
N. North Deighton!
A. Kippax; *H. Baines.*
D. Near Brindsworth; *Amos Carr.* Goole, on warped lands; *H. F. Parsons.* Thorne; *H. Baines.*

[**Avena cærulea**, Blew Haver. It is sown at Keighly (Keighley) in Yorkshire'; *Merrett, Pin.*, 13.]

[**Avena strigosa** Schreb. Casual, hardly Colonist because impermanent.
In Yorkshire; *Dr. Smith* in *O.B.G.*
A. Sandy field at Eccup!—a few plants in a 'seed field.'
C. Hebden Bridge, 1840; *S. Gibson* in *Herb. Tatham.*

945. Avena elatior L. (*Arrhenatherum avenaceum* Beauv.). '*Wick*' or '*Wicken.*'
Area: 63, 64, 65, Top. Bot. Range: 0—1,250 feet.
Native; pastures, hedge-banks, and river-banks, very common. P. July.
First record: Salt, 1800.
Distribution general; plentiful in every district.
Obs.—The form which is bulbous about the root (*A. bulbosum* Presl.) is vulgarly known as *Knot-grass*, and is commoner than the type in stiff soil.

[**Stipa pennata** L. Extinct Casual.
W. Rombalds Moor (1839); *S. Gibson* in *Baines*, p. 142. 'Near the pathway between Ilkley and Keighley; *J. Tatham*, vide *Comp. Cyb. Brit.*, p. 593.
C. Leeds Road, Huddersfield, on shoddy-waste; *J. B. Ingle*, etc., vide *Hudd. Nat. Hist.* (1857).]

946. Holcus mollis L. '*Quick*' or '*Pig-wick.*'
Area: 63, 64, 65, Top. Bot. Range: 0—1,200 feet.
Native; woods and shady banks, common, locally abundant. P. July.
First record: Salt, 1800.
Distribution general; in every district, in great sheets on the slopes of woods upon the flagstones and gritstones, less abundant in the woods on limestone. A great nuisance when it spreads into light soil.

947. Holcus lanatus L. '*Yorkshire Fog.*'
Area: 63, 64, Top. Bot. Range: 0—1,500 feet.
Native; meadows, pastures, hedge-rows, etc., very common. P. June.
First record: Salt, 1800.
Distribution general—in every district; commoner than *H. mollis*, in that it is more frequently recurring, although it is rarely found in such exclusive local abundance as that species.

948. Triodia decumbens Beauv.
Area: 63, 64, 65, Top. Bot. Range: 0—1,800 feet.
Native; wet moory pasture-ground, and on heaths, common, although rarely in obtrusive abundance. P. July-August.
First West Riding record: uncertain, not mentioned by Salt.
M. Greenfield! *J. Whitehead.*
L. Whernside slopes! Dentdale. Blea-moor.
R. Kelcowe, Major Springs, etc., near Settle; *J. Windsor.*
Y. Between Grewelthorpe and Swinton; *H. H. Slater.*
N. Kirkby-Overblow and Spofforth. Brimham Moor!
W. Ilkley! Bramham Moor! Askham Bogs!
A. Baildon, Eldwick, Harden, etc.; *W. West.* Adel Moor! Ledstone Park!
C. Todmorden; *A. Stansfield.* Halifax, 1837, *J. Sutcliffe*; *Hb. Tatham.*
D. Wharncliffe Crags! Ringinglow; *Amos Carr.*

949. Koeleria cristata Pers.

Area : 63, 64, 65, Top. Bot. Range : 0—1,500 feet.

Native ; dry pastures, banks, and stony hill-slopes, mainly upon the limestones, common. P. Late June-July.

First record : Windsor, 1855.

L. Ingleton and Dentdale, on the dry limestone banks and rocks, common !

R. High Hills, etc., near Settle ; *J. Windsor et aliis.* Selside Crags !

Y. Mackershaw Wood, Ripon ! Burton Leonard quarries. Not mentioned as a Ripon species by *H. H. Slater.*

N. Knaresborough !

W. Common about Kilnsey and Grassington. Wetherby, Thorparch !

A. Malham ; *J. Willis.* Skipton Rock ! Ledstone, Kippax, etc. Ouse bank, Selby !

D. Conisborough. Went-vale !

T. Maltby and Firbeck. Kiveton ; *Amos Carr.*

950. Molinia cærulea Mœnch. *Blue Moor-grass.*

Area : 63, 64, 65, Top. Bot. Range : 0—1,800 feet.

Native ; heathy or moory ground, common, abundant locally. P. July.

First record : Salt, 'meadows about Riveline,' 1800.

Distribution general ; in every district, and appearing to be unfrequent only in the highly-cultivated lowland districts. Its stations are far too numerous to be worth recording, from Widdale moors and Baugh Fell, through Malham Moor, Adel Dam, and Eldwick, to Askham and Morton Bogs, Thorne and Hatfield Moors in the south-east.

951. Melica nutans L.

Area : 63, 64, Top. Bot. Montane. Range : 1,000—300 feet.

Native ; rocky woods, lanes, and shady scars on limestone, unfrequent. P. June.

First record : Gough, 1789.

L. In Skirreth Wood, Ingleton ; *Gough's Camd.* Helks Wood ; *Mr. Woodward* in *O.B.G.* (1805).

R. Lund's Wood, Cavehole Wood, etc., near Settle ; *J. Windsor.* Stainforth ; *Miss F. P. Thompson* (1886).

Y. Mackershaw (!) and Studley Woods ; *W. Brunton* in *O.B.G.* Hackfall ! Clotherholme ; *H. H. Slater.*

W. Between Thorparch and Wetherby (!) *Rev. W. Wood* in *O.B.G.* Dalton Lane, between Bramham West Woods and Beilby Grange, abundant !

C. Hebden Valley (1879). Eaves' Wood, Heptonstall ! *A. Stansfield.*

D. Rotherham district ; *L. Langley* (1828). Conisborough. Warmsworth Cliffs !

T. Roche Abbey Woods ; *G. E. Smith.*

952. Melica uniflora Retz. *Melic-grass.*

Area : 63, 64, 65, Top. Bot. Range : 0—1,200 feet.

Native ; dry woods, and thickets on flag, grit and lime, common. P. June.

First record : Salt, 1800.

Distribution general ; occurs in every district, although not an abundant grass, counting individuals.

953. Catabrosa aquatica Beauv.

Area : 63, 64, Top. Bot. Range : 0—600 feet.

Native ; shallow ditches, pool-sides, etc., unfrequent. P. June-July.

First record : Salt, 1800.

L. Near Austwick Bridge ; *J. Windsor.*

Y. By Ripon Canal (!) and Sharow Mires ; *H. H. Slater.*

N. By a pool on the Long Flat, Knaresborough ! Pool on Knavesmire, York ; *W. Whitwell* !

A. Sherburn Dikes ! Beck at Newton near Castleford ; *G. Roberts.*

C. Wakefield, frequent ; *T. W. Gissing.*

D. Pottery Carr, Doncaster, July, 1800 ; *Herb. Sal* ! Near Ecclesfield and Wadsley Bridge ; *A. Carr.*

T. Rossington ; *G. Harrison.*

954. Glyceria fluitans Br. *Flote-grass.*

Area : 63, 64, 65, Top. Bot. Range : 0—1,300 feet.

Native ; pools and slow-moving ditches, very common. P. June-July.

First record : Salt, 1800 ?

Distribution general ; in every district, very frequently ergoted.

Var. **pedicellata** Towns.

Marshy clayey ground by canals, generally by running water, rare, or perhaps not yet sufficiently discriminated by field observers. It, like the type, is often ergoted, as well as smutted.

N. By the river Crimple near Spofforth (!) *J. S. Wesley.*

W. By rivulet in Wetherby Race-course Ings, and by pool in Linton Lane !

D. By the Dearne, near the Sewage Works below Barnsley ! cum *H. Johnson.*

955. Glyceria plicata Fries.

Area : 63, 64, Top. Bot. Range : 0—450 feet.

Native ; marshy places in peaty or sandy soil, by pools or dikes, rare. P.

First record : the Author, 1876.

W. Marshy ground under Etchell Crags, Scarcroft, by the stream running north ! Field-corner by dike on that side of the 'Far Wood' at Askham Bogs nearest the highroad ! cum *Y. Todd.*

A. Ellerbeck, Skipton woods ; *W. Whitwell* ! Adel Blackmoor, by stream below Reformatory !

Obs.—I have tried, in vain, to satisfy myself of the distinctness of this species from the preceding ; the two seem connected through the variety I have empirically assigned to *G. fluitans.*

956. Glyceria aquatica Sm. '*Cut-grass.*' '*Water Sweet-grass.*'

Area : 63, 64, Top. Bot. Range : 0—450 feet.

Native ; borders of ponds, canals, dams, and in ditches, common. P. July.

M. By Leeds and Liverpool Canal, Diggle Valley.

R. Rathmell ; *J. Windsor.*

Y. Ripon Canal ! Sharow Mires ; *H. H. Slater.*

N. Ditches by the Crimple Beck and river Nidd. Cowthorpe, etc. !

W. Common from Ilkley eastwards. Askham Bogs.

A. By Leeds and Liverpool Canal from Gargrave by Saltaire (*W. West*) to the river Aire, Selby.

C. Todmorden district, by canal, etc. ; *A. Stansfield.* Stanley, etc., by canal ! Wakefield ; *T. W. Gissing.*

D. Potteric Carrs ; *E. Miller* (1804). Barnsley Canal (!) *H. Johnson.* Aldwarke and Tinsley ; *Amos Carr.* Goole ; *H. F. Parsons.*

T. Roche Abbey Mill ! Bawtry and Rossington dikes ! Kiveton Canal ; *Amos Carr.*

Obs.—All the *Glyceriæ* are more or less frequently ergoted ; and this species by the Barnsley Canal and other places is infested with the Elongated Smut (*Ustilago longissima*), giving the leaves a black-striped and wrinkled appearance which is most singular. The smutted spikes of these *Glyceriæ* and Wheatears are vulgarly spoken of as Sleëan (i.e. slain).

957. Sclerochloa maritima Lindl.

Area : 63, Top. Bot. Halophilous. Range : Low.

Native, or adventive ? tidal mud and 'warped' soil about tidal rivers, local. P. June-July.

First record : Parsons, 1875.

D. About Goole (*H. F. Parsons*) !—in many muddy spots by the Humber and Ouse from Adlingfleet by Ousefleet to Goole.

958. Sclerochloa distans Bab. *Sand-grass.*

Aira aquatica β distans Huds.

Area : 63, 64 Hailstone spn., Top. Bot. Range : 0—150 feet.

Native ; sandy fields and waste ground, rare or unknown over the western two-thirds of the Riding, common (locally abundant) on gravelly soil over coal-measures and trias. A. & P. July-August.

Earliest records : Hudson, 1778 ; Gough, 1789 ; Withering, 1796. 'In sandy places in Yorkshire.'

W. Thorparch ; *S. Hailstone*, spn. York ; *H. Baines.* Ainsty ; *G. Webster.*

A. Selby ! Knottingley ! By Gas Works near Selby ; *H. F. Parsons.*

C. Featherstone to Knottingley !

D. Waste ground by canal above Barnsley (!) *F. Johnson.* About London Road Station, Sheffield ; *G. E. Smith A. Carr.* Between Sheffield and Rotherham ; *J. Salt.* And Doncaster ; *O. A. Moore.* Thorne Quay ; *H. Johnson* ! Rossington. Cantley ; *G. E. Smith.* Rawcliffe Bridge and Swinefleet ; *H. F. Parsons.*

T. Maltby ; *G. E. Smith.* Hooton Roberts (Aug. 1838 ; *Joseph Woods, MS.* in *Towns. O.B.G.*

Obs.—Traceable all along the low valley from Gainsborough to Selby, and Rotherham to Wakefield. A relic of the time when the Triassic tract was an inlet of the sea ?

959. Sclerochloa rigida Link.

Area : 63, 64, Top. Bot. Range : 0—600 feet.

Native ; wall-tops, dry banks, quarried wastes, mostly in limestone districts, unfrequent. A. June-July.

First record : Salt, 1800 (! J. Tatham).

R. Kelcowe and Giggleswick Limekilns, Settle ; *J. Windsor.*

Y. Hackfall and Fountains Abbey ; *H. H. Slater.*

N. Knaresborough !

W. Wetherby, Collingham, Thorparch, Tadcaster—common ! Huddlestone Quarry ; *W. Whitwell.*

A. Skipton Rock ! Baildon ; *H. Baines.* Brotherton and Ledstone ! Monk Fryston ; *H. F. Parsons.*

C. Heath, Woolley, etc. ; *T. W. Gissing.*

D. Askern ! Levitt Hagg, Conisborough ; *G. Harrison.* Rotherham ; *L. Langley* (1828). Near the Rivelin Paper Mills ; *A. Carr.*

T. Near Maltby ; *J. Salt.* Firbeck !

960. Poa annua L.

Area : 63, 64, 65, Top. Bot. Range : 0—1,700 feet.

Native ; waste ground, way-sides, gardens, etc., very common. A. May-June.

First record : Salt, 1800.

Distribution general ; in every district, but I cannot say I have myself ever seen it at a higher level than the summit road over Blackstone Edge. It is given by Baker as ascending to 2,400 feet, but in West Yorkshire it is rare (it has to be keenly looked for) at any elevation over 1,250 feet.

961. Poa alpina L.

Area : 64 Gibson spn., Top. Bot. Montane. Range : 2,300—1,850 feet.

Native ; sub-alpine grit crags, rare, and sparingly. P. July.

First record : J. Windsor (1802) cum T. W. Simmonds, but not published until 1830.

L. On the crags of Whernside, about the spring on north-east shoulder! In several nearly inaccessible places on the north-west side of Ingleborough; and two or three places on the right-hand going down Foalfoot, on the north-east rocks; *J. Windsor, MS.* in *Towns. O.B.G.* North-east rocks of Ingleborough, near the *Saxifraga aizoides* (!) *J. Windsor, Flora Cravoniensis.* Steep rocks on the west side of Ingleborough; *Silvanus Thompson*! Id., specimen, July 1844; *Herb. Tatham.* Not on 'limestone' rocks, as is stated in Baines.

[**Poa glauca** Sm. Ambiguity.

Ingleborough Hill and other mountains; *R. Teesdale* in *O.B.G.* (1805). What we now call *Poa Balfourii* was intended. In Townsend's copy of O.B.G. Mr. Jos. Woods has appended an obelisk to the printed record, referring to a foot-note in his hand-writing: 'Nearly allied to *P. nemoralis*, but differs in the form and density of the panicle and shape of the glumes. I can find no difference in the nerves. *Mr. Brunton.*' On Ingleborough, on soil covering schistose rocks; *J. Windsor* in *Engl. Flora*, i, 128. '*Poa glauca* from Ingleborough, *W. Middleton* apparently, in *Herb. York, W. W. Newbould.* It is apparently the same as the Ben Lawers plant; *W. W. Newbould.* Given in *Flora Cravoniensis* as a var. of *P. nemoralis*, with admission that it may be referred to *P. Balfourii.*

I consider the plant to which the above records refer to be Parnell's *P. montana* (perhaps *P. dissitiflora* R. & S.), i.e., an abbreviated state of *P. Balfourii*, which varies in size, number of spikelets, length of pedicels, etc.]

962. Poa Balfourii Bab.

Area: 64, Baker spn., Top. Bot. Range: 1,200—2,100 feet.
Native; sub-alpine limestone rocks, rare. P. July.
First record: Windsor, 1813.

L. Ingleborough Hill, rocks descending by Foalfoot (!) *J. Windsor* in 1813; *B. Carrington*; *S. Thompson*; *J. G. Baker*; etc. Recorded at first as a glaucous var. of *P. nemoralis.*

R. Limestone scars above Linn Gill, Upper Ribblesdale! Larger in stature than the Foalfoot plant, because at a lower elevation.

VAR. **montana** Parn.

Ingleborough; *B. Carrington.* In the same spot in Lune district as the type occurs in. I imagine it to be a state only.

[**Poa cæsia** Sm.

Ingleborough; *Miall and Carrington's Flora.* The plant referred to is essentially the same as the Ben Lawers *P. dissitiflora* and the *P. montana* referred to above.]

963. Poa nemoralis L.

Area: 63, 64, Top. Bot. Range: 0—1,400 feet.
Native; moist shady banks, woods, and rocks, infrequent. P. June.
First record: Salt, 1800.

L. Several places near Ingleborough; *H. Baines.* Dentdale! Clapdale! Cavehole Wood; *J. Windsor.*

R. Kendall's Gill, Major Springs, etc., Settle; *J. Windsor.* Douk Gill, Horton!

Y. Hackfall!

W. Bolton Woods, not common. Woods near Buckden!

A. Bishop Wood, Hambleton. Esholt Springs; *J. Willis.*

D. Between Brightside and Blackburn Forge; *Herb. Salt.* Parker Spring Wood, Sheffield; *Amos Carr*!

T. Roche Abbey; *J. H. Aveling* in *Hist.*

964. Poa compressa L.

Area: 63, 64, Top. Bot. Range: 0—600 feet.
Native; on wall-tops, dry ground, etc., unfrequent. P. June-July.
First record: Salt, 1800.

R. Probably near Settle; *J. Windsor.*

N. Walls and rocks, Knaresborough Castle!

W. Collingham! Moor End, Boston Spa, near cottages! Towton!

A. Skipton! Rimmington! Shipley; *W. West.* Allerton Bywater!

C. Near a coal-pit at Lofthouse; *L. C. Miall.*

D. Stone-quarry near the London Road at Adwick-le-Street; *S. Gibson.* Near the reservoirs on Crooks Moor, Sheffield; *J. Salt.* Old quarries near Barnsley; *H. Payne.*

Obs.—This varies much, and many forms have been discriminated. In my experience it is not a common plant in the West Riding.

965. Poa pratensis L.

Area: 63, 64, Top. Bot. Range: 0—1,800 feet.
Native; meadows, dry banks, and rocky slopes, wall-tops, etc., in a variety of forms, very common. P. June-July.
First record: Salt, 1800.
Distribution general; very common in all the districts.

The VAR. **subcærulea** Sm. (the '*P. humilis*' of Old Bot. Guide), is, as *W. Brunton* remarked in 1805, very common on barren soils and rocks in the hilly portion of the Riding. It stands recorded for Crooks Moor, Sheffield, Halifax, Heptonstall, Settle, etc.

966. Poa trivialis L.

Area: 63, 64, Top. Bot. Range: 0—1,500 feet.
Native; moist meadows, woods, and shady hedgerows, common. P. June-August.
First record: Salt, 1800.
Distribution general; in every district.

VAR. **parviflora** Parn. A slender state (occasionally mistaken for *P. nemoralis*). I have noticed it only in damp woods at Woodhall by the Wharfe, under rocks in shade, and *Mr. H. Johnson* sent it me from Hymen Wood, Barnsley.

967. Briza media L. *Trembling-grass.* '*Doddering Dick.*'

Area: 63, 64, 65, Top. Bot. Range: 0—1,500 feet.
Native; meadows and pastures, common. P. June.
First record: Salt, 1800.
Distribution general; in every district, although running out so rapidly as to make it distinctly rare above 900 feet altitude.

[**Briza minor** L. Incognit.

Gramen Tremulum glumis albis. In a meadow at Garforth; *Mr. Witham*; *Merr., Pin.*, 59. I am not sure what this was.

Recorded in *Yorks. Nat. Recorder* (p. 18, 1873) as having been found near Chevet Hall; but I learnt by correspondence that a pale unbronzed specimen of *B. media* was so named.]

968. Cynosurus cristatus L. '*Windlestraw.*' '*Dog's-tail.*'

Area: 63, 64, 65, Top. Bot. Range: 0—1,500 feet.
Native; meadows, pastures, and grassy way-sides, very common. P. June-July.
First record: Salt, 1800.
Distributed, plentifully, through every river-basin.

[**Cynosurus echinatus** L. Casual.

In fields of rye-grass, and in fields manured with wool-waste, rare, and fugitive.

W. Near Thorparch; *S. Hailstone* (*Ann. Nat. Hist.*, second series, vol. 5, p. 147 (1850) = In a field of rye-grass between Thorparch and Clifford, 1849; *Baker's Suppl.*

C. Ashgrove, Calderdale; *S. Gibson* (*Phyt.*, i, 773). Corn-field near Brighouse, *T. Barnes*; *Miall's Flora.*]

969. Dactylis glomerata L. '*Cock's-foot.*'

Area: 63, 64, 65, Top. Bot. Range: 0—1,250 feet.
Native; fields and hedge-banks, very common. P. June.
First record: Salt, 1800.
Distribution general; in every district.

970. Festuca Pseudo-myurus Soy.

Colonist-Casual, perhaps not permanent; among sown grasses or clover in sandy fields, rare. A. July.
First record: G. E. Smith, 1846.

A. Clover field near Smithy Mills, Meanwood!

C. Broad Bottoms, Mytholmroyd, near Eastwood; *L. C. Miall.*

D. Between Cantley and Doncaster (in seed-fields); *G. E. Smith.* Thorne Quay!

971. Festuca sciuroides Roth. (*F. bromoides* Flora).

Area: 63, 64, Top. Bot. Range: 0—600 feet.
Native; sandy fields, banks, wall-tops, etc., unfrequent. A. June.
First record: Salt, 1800.

Y. Sandy lanes (! and fields on the Red Sand) to the east of Ripon; *J. Dalton.* Ellershaw Hill; *H. Baines.* Fields about Boroughbridge!

N. Plumpton, and Braham Hall Rocks, Spofforth!

W. Common on the Red Sandstone of the Ainsty.

A. Baildon Green; *W. West*! Eccup Lane (!) and Adel Blackmoor; *J. Abbott*! Hambleton.

D. Near Doncaster! Barnsley; *H. Payne.* Woolley Edge to Cudworth! Near Sheffield; *J. Salt.* Dovecliffe; *G. Harrison.*

972. Festuca ovina L. '*Sheep's-bent.*'

Area: 63, 64, 65, Top. Bot. Range: 0—2,200 feet.
Native; pastures, banks, and hill-slopes, very common, and polymorphic. P. June-July.
First record: Herb. Salt, 1799—'Crooksmoor.'
Distribution general; in every district, varying in size, length of leaves, facies, etc. I have not kept notes of varieties, but a glaucous form (*glaucantha*?), liable to be mistaken at first sight for *Poa Balfourii*, occurs on the limestone terraces of Penyghent! whilst the var. *tenuifolia* is frequent on the hill-slopes.

FORMA **vivipara.**

Not uncommon on the higher hills, in the basins of L., R., and W. Whernside, and Widdale Fell! 'Near the summit of Ingleborough'; *Gough's Camd.* (1789). Fountains Fell (*J. Tatham*) and Penyghent (*J. Windsor*).

973. Festuca rubra L. var. **duriuscula** Sm.

Area: 63, 64, Top. Bot. Range: 0—1,500 feet.
Native; pastures, sandy waste ground, etc., common, but most frequent in the low country. P. July.
First record: Salt, 1800.
Distribution general; in every district.

Obs.—The viviparous state has occurred on Ingleborough (*W. Brunton*).

974. Festuca sylvatica Vill.

Area; '63 Leyland spn., 64 Baker spn.,' Top. Bot. Range: 0—600 feet.
Native; rocky woods, rare. P. July.
First record: Leyland, 1832.

R. Linn Gill!

W. Posforth Gill, Wharfedale; *J. G. Baker.* Thorparch; *H. Baines* (1840).

A. Wood near Airton, Bell Busk, 1870 ! Bingley Woods ; *H. Baines.*
C. Stanelly Clough, near Todmorden ; *A. Stansfield.* Heptonstall Eaves ; *R. Leyland* (1832). Near Halifax, 'frequent'; *L. C. Miall*, but doubtful if it is really so.
(T. Roche Abbey ; *Aveling's List*—but *F. elatior* or *B. giganteus* probably meant.)

975. Festuca elatior L.
Area : 63, 64, Top. Bot. Range : 0—1,250 feet.
Native ; moist hedge-banks, river-banks, and swamps, common. P. July.
First record : Curtis, 1782 ; Gough's Camd., 1789—Ribble sides, near Settle.
Distribution nearly general ; unrecorded only for the Mersey district.

976. Festuca pratensis Huds.
Area : 63, 64, Top. Bot. Range : 0—1,250 feet.
Native ; meadows, pastures, and hedge-banks, common.
First record : Salt, 1800.
Distribution general ; occurs in every area.

HYB. VAR. **loliacea** Huds., Fl. Ang., Ed. ii, 47—with racemose inflorescence and apparently distichous spikelets, a variable debased form, the result of hybridisation with *Lolium* pollen—is frequent. *W. Brunton* recorded it for Ripon meadows ; *J. Dalton* from Copgrove ; and *Jos. Woods* has the MS. note 'Leeds' (where it still occurs) in *Townsend's copy* of *O.B.G.* (1805).

977. Bromus giganteus L. *Tall Wood-brome.*
Area : 63, 64, 65, Top. Bot. Range : 0—900 feet.
Native ; woods and hedge-rows, common. P. July.
First record : Gough in Camden, 1789.
L. Bentham and Lower Dentdale !
R. Oustre-bank Wood ; *Herb. Tatham*, spn. [Under the stone walls in the road from Settle to Giggleswick ; *Gough's Camd.* (1789). Could this have been the var. *triflorus* of dry open ground ? Only *B. sterilis* to be found there in 1870.]
Y. Ripon, common ; *H. H. Slater.*
N. & W. Common throughout these divisions.
A. Frequent, but not universal in the manufacturing districts.
C. Oakwell Wood, Birstall ! Hebden Bridge !
D. Sheffield, frequent ; *J. Salt.* Rose Wood, Old Park Wood, etc. ; *Amos Carr.*
T. Common about Sandbeck Park and Tickhill !
Obs.—The depraved form *triflorus* is not rare, being the usual state in open ground. First recorded from Helks Wood, Ingleton.

978. Bromus asper Murr. *Rough Hedge-brome.*
B. nemoralis Huds.
Area : 63, 64, Top. Bot. Range : 0—1,250 feet.
Native ; hedge-rows, moist woods, and thickets, common. A. or P. July.
First records : Gough in Camden, 1789 ; Salt, 1800.
Distribution general ; in every district, but rather uncommon about Leeds and Bradford, at least not so obtrusively abundant as it is in many of the rural districts upon limestone soil.

979. Bromus erectus Huds.
Area : 63, 64, Top. Bot. Xerophilous. Range : 0—300 feet.
Native ; dry-soiled meadows and banks, mainly on limestone, unfrequent, but abundant locally. P. June.
First record : Salt, 1799.
Y. Pastures at Burton Leonard and Copgrove !
N. About Knaresborough, common ! Wood below Knaresborough ; *Geo. Harrison, Hb.*
W. Common on the Permian tract at Wetherby, Bramham, and Tadcaster !
A. Allerton Bywater and Ledsham ! Kippax ; *G. Roberts.* Pontefract Castle, *T. Gray* ; *Miall* and *Carrington's Flora.*
C. Hebden Bridge*, 1836 ; *J. Sutcliffe* in *Herb. Tatham* ; but as a Casual there, doubtless.
D. Loversall, Doncaster ; *H. Baines.* Went-vale !
T. Roach Abbey, 1799 ; *Herb. Salt* ; 1838 ; *Jos. Woods, MS.* in *Towns. O.B.G.* Maltby (!) *O. A. Moore.*

980. Bromus sterilis L.
Area : 63, 64, 65, Top. Bot. Range : 0—600 feet.
Native ; way-sides, old walls, dry fields, etc., common, but with a restricted range. A. June-July.
First record : Salt, 1800.
Distribution general ; at low levels in every district, ascending to flag-quarries, etc., on Otley Chevin.

[**Bromus diandrus** Curt.
Reported as found at Conisborough at a meeting of the West Riding Consolidated Naturalists' Society in 1873, in error for *B. erectus*; *Yorks. Nat. Recorder* (1873), p. 205. By the *Bromus maximus* of same list *B. giganteus* was meant.]

981. Bromus secalinus L.
Range : 0—450 feet.
Colonist-Casual, fugitive ; grain- and seed-fields, rare. A. July-August.
First record : Knapp, 1802, in Herb. Salt.
Y. Corn-fields about Clotherholme, Ripon !

W. Barley-field, Kirk Deighton ! 'Thorparch, 1802, ***Mr. Knapp*** ;' *Herb. Salt.*
C. Todmorden, 1842, *S. G.* ; *Herb. Tatham.*
A. Saltaire ; *W. West.*
D. Near Barnsley ; *H. Payne.*

FORMA **velutinus.**
This is on record for Todmorden (*S. Gibson*), Saltaire (*Miall* and *Carrington's Flora*), and doubtless occurs in other localities.

982. Bromus racemosus L.
Range : 0—600 feet.
Colonist ; in fields, often introduced with grass-seed, unfrequent. A. or B. July.
R. Near Settle ; *J. Tatham.*
Y. Near Ripon ; *H. Baines.* Boroughbridge !
W. About Wetherby, Tadcaster, Towton, etc., frequent.
A. Fields about Leeds (*J. Abbott* !), Roundhay, etc.
D. Ackworth ; *Mrs. Watson* in *Gissing's Flora.* Thorne ! Goole fields ! Rossington !

983. Bromus commutatus Schrad.
Area : 63, 64, Top. Bot. Range : 0—900 feet.
Native ; meadows and cultivated fields, common. B. June-July.
First West Riding record : uncertain—Salt did not, I believe, discriminate it from *B. mollis.*
Distribution general ; in every district.

984. Bromus mollis L. '*Mouse-ear Grass.*' '*Goose-grass.*'
Area : 63, 64, 65, Top. Bot. Range : 0—1,300 feet.
Native ; meadows, pastures, grassy way-sides, and barren waste ground, very common. B. June-July.
First record : Salt, 1800.
Distribution general ; abundant in every district.

[**Bromus arvensis** L. Casual.
Occasionally occurring among grain, but more commonly on the heaps of shoddy-waste so often to be found about the woollen mills of the clothing districts.
M. About Greenfield and other places nearer Staleybridge ; *J. Whitehead.*
A. Boothroyd's Mill, Meanwood Valley ! Yeadon ; *B. Carrington.*
C. Leeds Road, Huddersfield ; *Hudd. Nat. Hist.* Brighouse, Sowerby, Hebden Bridge ; *S. Gibson* and *S. King.*
D. Thorne Quay (1870) ! Goole, waste ground !]

[**Bromus patulus** M. & K. Casual.
Occasional on waste ground—a shoddy-waste or corn-field alien.
A. Wool-waste by Boothroyd's Mill, Meanwood !
C. Near Hebden Bridge ; *S. Gibson.*]

[**Bromus tectorum** L. Casual.
Occasional on waste ground.
A. Hair-refuse heaps, Boothroyd's Mill, Meanwood Valley !
D. Newly-made railway embankment, Cudworth !]

985. Brachypodium sylvaticum R. & S.
Area : 63, 64, Top. Bot. Range : 0—1,250 feet.
Native ; woods, shady rocks in glens, etc., common. P. June-July.
First record : Salt, 1800.
R. Langcliffe, Hollin Hall, and other woods near Settle ; *J. Windsor.* Sawley ; *R. Brown.*
Y. Hackfall. Ripon, not uncommon ; *H. H. Slater*
N. Knaresborough ; *G. Harrison.* Scotton Banks !
W. Grassington and Bolton Woods ! Washburn Valley and Bramham !
A. Wood below bridge, Kirby Malham ; *J. Windsor.*
C. Haw Wood and New Miller Dam, Wakefield ; *T. W. Gissing.* Hebden Bridge, 1842, *S. Gibson* ; *Herb. Tatham.*
D. Old Park, Sheffield ; *J. Salt.* Wentworth, Ecclesfield, Handsworth, etc. ; *Amos Carr.*
T. Roche Abbey ; *Herb. G. Nicholson* and *J. H. Aveling.*

986. Brachypodium pinnatum Beauv.
Gr. spica Brizæ majus ; Ray.
Area : 63, 64, Top. Bot. Xerophilous. Range : 0—600 feet.
Native ; pastures and thickets, rare, but locally abundant ; follows the Permian tract very closely. P. July.
First record : Dillenius, 1724.
R. Cavehole Wood, amongst the stones, and under a wall on the side of Giggleswick Scar, next Stackhouse ; *J. Windsor.* Settle ; *H. F. Parsons.*
Y. Walkingham Warren, near Farnham ; *W. Brunton* in *O.B.G.* There yet. South Stainley !
N. On all the commons for twenty miles on this side York, *Sherard* ; *Ray, Syn.*, iii, 392 (1724). Still there. 'On this side' must, however, be interpreted upon the limestone tract, which lies west of York. Goldsborough, Bilton Haggs, etc. !
W. About Wetherby, plentifully ; *Rev. W. Wood* (1805). Norwood Bottoms, Thorner ! Near Barkston ; *W. Whitwell* !
A. Roundhay limehills ; *W. West.* Ledstone Park ! Near Hambleton ; *H. F. Parsons.*
D. Road-sides between Doncaster and Ferrybridge ; *Archdeacon Pierson* in *O.B.G.* (1805). Went-vale !
T. Maltby ; *G. E. Smith.* Firbeck. Anston. On the magnesian limestone east of Sheffield, very abundant (7.8.38) ; *Jos. Woods, MS.*

987. Triticum caninum Huds. *Dog-wheat.*
Gramen caninum var. cum spica aristata; Ray, Cat. *Agropyron*; Gærtn.
Area: 63, 64, Top. Bot. Range: 0—900 feet.
Native; hedge-rows and open woods, common. P. July.
First record: Ray, 1677.
L. Lower Dentdale! Ingleton to Bentham!
R. Circa Settle; *Ray, Cat.*, ii. 139 (1677). Below Settle Bridge and below Birkbeck's Weir; *J. Windsor.*
Y. Ripon, common; *H. H. Slater.*
N. Common.
W. Very common from Grassington eastwards to Tadcaster!
A. Gilstead and Hawksworth; *W. West.* Skipton!
C. Wakefield, common; *T. W. Gissing.*
D. Near Sheffield; *G. Harrison*; *A. Carr*; etc.
T. Roche Abbey, 1799; *Herb. Salt.*

988. Triticum repens L. (*Agropyron* Gærtn.). *Couch-grass. 'Quitch.' 'Dog-wheat.'*
Gramen repens, officinorum forte spicæ Triticeæ; Ray.
Area: 63, 64, 65, Top. Bot. Range: 0—1,200 feet.
Native; hedge-rows and waste and cultivated ground, very common. P. June.
First record: 'Circa Settle, in agro Ebor.'; Ray, Hist., 1255 (1688).
Distribution general; in every district.

[**Triticum pungens** VAR. **pycnanthum** G. & G. Casual.
I can only refer to this a glaucous plant with tufted stems and involute leaves, gathered by *P. F. Lee* in July 1886, on waste ground by a road-side at Hay-beck, West Ardsley—possibly brought with ballast from some sea-coast.]

989. Lolium perenne L. *Rye-grass.*
Area: 63, 64, 65, Top. Bot. Range: 0—1,500 feet.
Native; pastures, way-sides, very common. P. June-July.
First records: D. Dixon, 'on Woodhouse Common,' 1791; Salt, 1800.
Distribution general; in every district.
Obs.—Abnormal forms, with the internodes of the rachis abbreviated, forming curious broad and flat spikes, are, as noticed by Baines, not uncommon. I have known such to be reported as *Sclerochloa procumbens.* The form *multiflora* Lamk. and the aristate awned variety are occasionally seen in, or as escapes from, cultivation.

[**Lolium italicum** Braun. *Italian Rye-grass.* Casual.
This occurs occasionally in corn-fields and amid clover, sometimes as a sown crop by itself; and also now and then on shoddy-waste heaps about Huddersfield, Leeds, Batley, etc. First cultivated in England about 1825.]

990. Lolium temulentum L. (including awnless state *arvense*). *'Droke'* (pronounced Drōōack). *Darnel.*
Area: 63, 64, Top. Bot. Range: 0—450 feet.
Colonist; corn-fields, often fugitive (less common than formerly?), unfrequent.
First record: Dixon, 1791.
Y. Ripon, frequent; *W. Brunton* in *O.B.G.* Too common around Copgrove; *J. Dalton* in *O.B.G.* Near Blois Hall, Sharow—one plant, 1881; *H. H. Slater.*
A. Clayey field near Oil Mills, Meanwood Valley, abundant in 1870-71. Now gone.
C. Greetland; *W. West.* On Woodhouse Common near Normanton, 1791; *D. Dixon, MS.*
D. Near Barnsley; *H. Payne.*

991. Lepturus filiformis Trin.
Area: 63*, Top. Bot.
Native; maritime grass-flats, very rare. P. August.
D. Muddy tidal bank of river Ouse, Hook Road, Goole, with *Glaux*, etc.! *T. Hick* (*W. West*, etc.) in 1877, sparingly.
Obs.—Native, in the sense of having been introduced by natural causes, i.e., the flow of tide from Spurn and Grimsby shores, where *Lepturus* is common on grassy flats; and yet, since the station is not a usual one, in sense only Adventive. The same remark applies to *Carex divisa, Glaux, Triglochin maritimum*, etc.

992. Hordeum sylvaticum Huds. *Wood Barley.*
Elymus europæus (L.), Smith, and O.B.G.
Area: 63, 64, Top. Bot. Range: 0—600 feet.
Native; woods and thickets on calcareous soil, rare, but often abundant locally. P. July.
First record: Salt, 1799.
R. Lord's Wood, between Settle and Stackhouse, Kelcowe Cove, and Mill Island; *J. Windsor.* Cave Ha' Wood; *Herb. Tatham.*
Y. Hackfall; Studley Park; *W. Brunton* in *O.B.G.*
W. Very common about Bilton Spring and Wighill Park! Bramham Park Woods. Thorparch (!) *P. Inchbald.*
A. Wood at Malham Cove; *J. Windsor.*
D. In Old Park, at the top corner, next Cooke Wood, Sheffield, 1799; *Herb. Salt.*
T. Roche Abbey; *J. H. Aveling.*

993. Hordeum pratense Huds. *Wild Barley-grass.*
Area: 63, 64, Top. Bot. Range: 0—300 feet.
Native; dry meadows, pastures, and road-side banks, unfrequent and local. P. June-July.

First record: Salt, 1796.
Confined to the Permian tract and sand-lands east of it.
Y. Ripon Park, and field by Yore at Bridge Hewick; *H. H. Slater.*
W. Fields and lanes about Flint Mill, between Wetherby and Thorparch! Fields near Grimston, between Stutton and Towton, plentifully! In the Ainsty, near York; *G. Webster.* Low meadows by the Ouse, York; *H. Ibbotson.*
A. Near Pontefract; *T. W. Gissing.*
D. By the road-side between Rotherham and Maltby, 1796; *Herb. Salt.*
T. Pastures near Roche Abbey; *J. Salt*; etc. Tickhill!

994. Hordeum murinum L. *'Way-bent.'*
Area: 63, 64, Top. Bot. Range: 0—300 feet.
Native; waste dry ground, rare and local. A. July-August.
First record: Salt, 1800.
A. Way-side banks at Knostrop; *James Abbott*! Near Pontefract and Knottingley! Ferrybridge district; *Y. N. Excurs.* (1873).
C. A Casual on shoddy-waste, Leeds Road, Huddersfield, 1857, *H. Ingle*; *Hudd. Nat. Hist.* Wakefield, 'common'; *T. W. Gissing.* Sandal; *P. F. Lee*!
D. Doncaster; *J. Willis.* Goole; *H. F. Parsons.* Thorne! Near Conisborough Castle; *J. Salt.* Masborough; *W. W. Newbould.* Wath; *G. P. Nicholson* in *Herb. Tatham.*

[**Hordeum maritimum** With. Incognit.
'Near the confluence of the Trent and Ouse'; *Rev. W. Wood* in *O.B.G.* Not yet observed about Bosom Cross, Adlingfleet Clough, or Ousefleet; but the locality *may* have been in the East Riding, on the north bank of the Ouse, about Blacktoft.]

995. Nardus stricta L. *Mat-grass.*
Area: 63, 64, 65, Top. Bot. Range: 0—2,000 feet.
Native; heaths, moors, and pastures, common. P. July.
First record: Salt, 1800.
Distribution general; occurs in every district, but, of course, less frequently in the highly cultivated lowlands than in the upland tracts, and *not* recorded for the Roche Abbey district in the *Aveling-Bohler* list.
Obs.—I have heard this called *'Sheep's Bent'* by undiscerning dale farmers, but the grass rightly so named is *Festuca ovina.* The moorland sheep *avoid* the *Nardus*, on account of its wiry, innutritious blade, quite as unmistakeably as do cattle the *Colchicum* of alluvial pasture-land.

ACOTYLEDONES.

LXXXIV. FILICES.

996. Hymenophyllum unilaterale Willd. (*H. Wilsoni* Hook.)
Area: '63 Leyland spn., 64 Tatham, 65,' Top. Bot. Range: 500—2,000 feet.
Native; wet shady rocks by cascades, rare. P. July-August.
First records: Dillenius, 1724; Bolton, 1785.
M. Seal Bark Rocks, head of Greenfield Valley! *J. Whitehead.* Greenfield; *W. Wilson* (*Phyt.*, i, 453, 1855).
L. Howgill Fells; *Rev. G. Pinder* (1855); rep. *Miall and Carrington's Flora* (1862). The station is Cautley Spout, on rocks in spray, upper part of falls! *W. West* (1878). On the rock called Foalfoot, Ingleborough (!) *J. Bolton* (1785); rep. *R. Teesdale* (1798); *O.B.G.* (1805); rep. *Baker's Suppl.*; *Miall and Carrington's Flora* (1862); etc.
N. Ravensgill, near Pateley Bridge! *F. T. Walker*; etc.
C. Turner Clough, Rishworth Valley; *S. Gibson* (1838). 'Near Halifax'; *W. Wilson* (no doubt Turner Clough was meant here).
Obs.—The Ingleborough, Greenfield, and other stations referred by me, from later information, to *H. Wilsoni*, were first recorded, it must be remembered, under the name of *H. tunbridgense*, the early authors not distinguishing between the two.

997. Hymenophyllum tunbridgense Sm.
Adiant. petræum perpusillum anglicum fol. bifidis vel trif.; Ray.
Area: '63, [64 needing verification],' Top. Bot. Range: 300—750 ft.
Native; wet rocks, very rare. July-August.
First record: Dillenius in Ray, 1724.
[L. or R. Near Settle; *Richardson*; *Ray, Syn.*, iii (1724); rep. *Martyn* (1763). 'I have not found it in this neighbourhood'; *J. Windsor* (1858). Probably the Foal-foot station was meant, Ingleborough being 'near Settle' in Ray's sense. No spec. in *Herb. Tatham.*]
N. [Low Harrogate, near the Sulphur Springs; *L. C. Miall* (1862). Rocks near Lower Harrogate; *J. Backhouse* (1854). Reported as *H. Wilsoni* in *Baker's Suppl.*, but 'the *Hymenophyllum Wilsoni* found at Harrogate turns out to be *H. tunbridgense*, as I was recently informed by Mr. Backhouse'; *J. Emmet* in *Miall and Carrington's Flora*, p. 61 (1862).] 'Reynard Crags, Wilson's Wood, Birstwith!' *J. G. Baker, H. B. Brady*, and *J. Backhouse* in *Baker's Suppl.* (1854); rep. *F. T. Walker* (1884).
M. & C. Near Halifax and at Greenfield; *W. Wilson* (1855); *Newman's Ferns* (1844).
Obs.—There has been much confusion between this and *H. Wilsoni* from collectors (even within the last ten years) reporting what they have gathered as this or that species, indifferently.

998. Trichomanes radicans Swartz. *Killarney Fern.*
Filix humilis repens, fol. pellucidis et splendentibus, caule alato; Ray, Syn. *Hymenophyllum alatum*; Smith. *T. pyxidiferum*; Bolton.

Area : '64 extinct ?,' Top. Bot. Range : 300—900 feet.

Native ; moist shady rocks, very rare indeed and in its first-discovered station extinct. P. August.

First record : Dillenius in Ray, 1724.

A. Found by Dr. Richardson at Belbank, scarce half a mile from Bingley, at the head of a remarkable spring, and nowhere else that he knows of ; *Ray, Syn.*, iii, 128 (1724). Two figures are given in Ray, the larger one unmistakably *T. radicans*; the smaller, fig. 4, is in a very immature state, and in this last condition it was found by Dickson and Teesdale at Belbank long after Richardson's day (vide *Smith's Eng. Flora*, iv, 325). 'First discovered by Dr. Richardson in a little dark cavern, under a dripping rock, below the spring of Elm Crag well in Bell Bank . . . In this place I saw it in plenty in the year 1758 : afterwards some alterations being made about the well . . . the cavern was destroyed, the plant perished, and was lost to Great Britain till the year 1782, at which time, ... after several researches, I found a root under a dripping rock to the left side of the current, and about fifteen yards above the cistern. From this root I have sent specimens to one or two of my friends, and have in my possession the best of them, from which this figure and description was taken'; *J. Bolton, Filices Brit.*, p. 57 (1785). Under *Trichomanes tunbridgense*, at p. 59 of the same work, Bolton remarks that he suspects the *T. pyxidiferum* of Bell Bank, to be a luxuriant variety of *T. tunbridgense*; but Mr. Newbould has a MS. note to the effect that Dr. Richardson's specimen, now in Herb. Sloane in the British Museum, shows that idea to be untenable, and he further remarks that the *Trichomanes* must have been native—'it can be traced from its discovery till early in this century, and they did not cultivate ferns then'; *W. W. Newbould* (1878). *R. Teesdale* in *Linn. Trans.* (1800), v, p. 75, records having found it in September, 1782, 'in Belbank Wood near Bingley,' the place mentioned in *Ray's Syn.* The station is repeated in *Old Bot. Guide* (1805), and by *Sir J. E. Smith* in *English Flora* (1828), but I do not know if anyone gathered it after Dickson. It has long been extinct at Bingley.

D. or T. *Still* in one station in the West Riding ; *James Backhouse* in litt. I cannot more precisely indicate the locality.

999. Pteris aquilina L. *Bracken.* '*Brakes.*'

Area : 63, 64, 65, Top. Bot. Range : 0—1,800 feet.

Native ; woods, moors, and borders of fields, very common. July-September.

First record : Salt, Flora Sheffieldiensis, 1800.

Distribution general ; in every district, but in some of the more thickly populated and cultivated parts, as about Birkenshaw, it is growing scarce, and what there is seems stunted and sickly.

1000. Cryptogramme crispa Br. *Parsley Fern.*
Adiantum album crispum alpinum; Ray.

Area : '63 'extinct,' 64,' Top. Bot. Range ; 2,000—1,600 feet.

Native ; amid grit- or slate-rock debris on mountains, rare. P. Aug.

First record ; Dillenius, 1724.

[M. In Knotty Lane, Saddleworth ; *Ray, Syn.*, iii, 126 (1724); *Martyn* (1763).]

L. Howgill Fells (!) *G. Pinder.* On Ingleborough ; *T. Knowlton* in *O.B.G.* On Ingleborough little hill (!) plentifully ; *J. Windsor.*

R. Swobeck, near Settle ; wall in Mitchell Lane above Settle ; *J. Windsor.* West slope of south end of Penyghent, plentiful (1884) ; *W. Whitwell* ! Catlow Fell, Hodderdale ; *W. Fowler* !

W. Fountains Fell (!) *J. Tatham* ; *H. Ibbotson.* Bolton and fells above Threshfield ; *Miall's Flora.*

A. Haworth ; *T. Knowlton* in *O.B.G.*

C. Near Todmorden ; *A. Stansfield.* Head of Hebden Valley. Warley ; *J. Bolton.* (Reported to *A. F. Gissing* from 'a little wood at Walton'—probably an error, but if not, certainly planted.)

1001. Blechnum boreale Sw. '*Hard Fern.*'

Area : 63, 64, 65, Top. Bot. Range : 0—1,800 feet.

Native ; stony heaths and boggy woods and moors, common. P. July-September.

First record : Bolton, 1785.

Distribution general ; in every district, but on the limestone rare, and in the cultivated districts confined to the bits of unreclaimed wood, heath, or carr.

In Old Park, Sheffield, frequent ; *J. Salt* (1800).

Heath lands east of Doncaster ; *S. Appleby* (1831).

Obs.—Localities for upwards of a dozen sports or Mooreian 'varieties' are (uselessly) given in a former Flora, but I think no good purpose would be served by repeating them here.

1002. Asplenium Ruta-muraria L. '*Wall Rue.*'
Ad. album sive Ruta-muraria; Merr.; Ray.

Area : 63, 64, 65, Top. Bot. Range : 0—1,800 feet.

Native ; on rocks and walls, in the limestone districts very common, rare off them. P. July-September.

First record : Merrett, 1666.

Distribution general ; recorded from every river-basin.

M. Mersey district : *J. Whitehead.*

N. On most of the rocks about Knaresborough ; *E. Hargrove* (1782). Hunsingore !

W. On a wall by Mr. Wade's house at Kilnsey in Craven ; *Merr.*, 2 (1666). Compare *Ray, Syn.*, iii, 128 ; Merrett's plant was a variegated variety, not uncommon on dry sunburnt walls. Something (in the rain-drops, or is it due to larvæ ?) blights portions of the cuticle, and produces a silvery variegation of the fronds. Grit-wall half a mile from Harewood village on the high-road to Leeds !

A. Skipton Castle walls (!) *W. Whitwell.* Skipton Rock. Drax near Selby ; *H. F. Parsons.*

C. Kirklees Hall ; *Y. N. Excurs.* (1873). [Sowerby Bridge, in plenty ; *J. Bolton.*]

D. Limestone tract west of Doncaster ; *S. Appleby* (1831). Common in the Sheffield district ; *J. Hardy.* Rockley Abbey ; *H. Payne.* Old wall, Darnall ; *Amos Carr.* Adlingfleet ; *H. F. Parsons.*

T. Roche Abbey (!) *Aveling's List.*

Obs.—Only unusual stations, or localities in districts where it is now rare, are given above.

[**Asplenium septentrionale** Hull. Incognit.
Acrostichum septentrionale Hudson.

Area : '64' sine personal authority, Top. Bot.

Schistose mountain rocks, extinct ?

First record : Hudson, 1762.

L. In monte Ingleborough ; *Huds.*, 451 (1762) ; *Gough* in *Camden* (1789). On Ingleborough, Yorkshire ; *Mr. Tofield, O.B.G.* (1805) ; *Smith, Eng. Flora*, iv, 308 (1828). 'It does not appear to have been found by any subsequent botanist'; *Dawson Turner* in *O.B.G.* 'A specimen, labelled Ingleborough, was in the herbarium of my brother, the late Thos. Backhouse ; but I do not think any living botanist has seen it there'; *J. Backhouse* (1854) in *Baker's Suppl.*]

1003. Asplenium Trichomanes L. *Black Spleenwort.*

Area : 63, 64, 65, Top. Bot. Range : 0—1,800 feet.

Native ; old rocks and walls, formerly commoner than now ; common in the retired dales, rare in the populous districts. P. July-Sept.

First record : Bolton, 1775.

Distribution nearly general ; every district except Mersey.

L. & R. Common in these districts.

Y. Fountains Abbey (!) *H. H. Slater.*

N. Ramsgill and Pateley Bridge ; *J. S. Wesley.*

W. Common from Buckden to Bolton Bridge, rare to the eastward. Jackdaw Crag, Stutton. Still there, 1880 !

A. Malham plateau to Skipton, common. Bingley, Eldwick, and Haworth ; *L. C. Miall.* Nearly exterminated. Swillington Church walls ; *G. Roberts.*

C. Soyland, Ogden Clough, Hebden Bridge ; *J. Bolton.* Still, high up Hebden Valley ! Mytholmroyd. Storthes Hall, Huddersfield. New Miller Dam ; *T. W. Gissing.* [Sandal Church walls, 1790 ; *D. Dixon.*]

D. Womersley and Went-vale (!) *T. W. Gissing.* Lime-rock tract west of Doncaster ; *S. Appleby* (1831). Conisborough ! Cliff Rocher ; *J. Salt.* Walls, Aldwarke and Eccleshall ; *W. W. Newbould.*

T. Roche Abbey ! Anston ; *J. Hardy.*

Obs.—The pinnules are very variable as to incision—sometimes prettily and deeply toothed, but the forms have no permanence ; and to repeat the record of the localities in which they happen to have been observed would be a useless waste of space.

1004. Asplenium viride Huds. *Green Spleenwort.*
Trichomanes ramosum; J. B. ; Ray.

Area : 63 ? 64, 65, Top. Bot. Range : 0—1,800 feet.

Native ; on shady rocks (or in fissures) of mountain or Yoredale limestone, common, on high grit rocks, rare. P. July-August.

First record : Dillenius, 1724.

On wet rocks, Yorkshire ; *Gough's Camd.* On calcareous rocks, Craven ; *Ray, Syn.*, iii, 119 (1724).

M. Reported for sheltered rocks at Seal Bark, head of Greenfield Valley, but not seen by *J. Whitehead.*

L. Very common in Dentdale, Garsdale, and Kingsdale ! Rocks between Stackhouse and Feizor ; *J. Windsor.* Ingleborough ; *J. Bolton.* Between Settle and Ingleton ; *R. Teesdale.*

R. Common from the Crags of Cam Fell through Penyghent to below Settle !

W. Between Darnbrook and Arncliffe ; *J. Windsor.*—Cowside-beck dale ! Kilnsey and Grassington !

A. Malham ; *Rev. W. Wood.* Gordale (!) and other places on the limestone.

C. Hardcastle Crags. Below the great rock called Ogden Kirk, Ogden Clough, four miles from Halifax ; *J. Bolton.* Rocks at head of Wessenden Valley ; *W. Fowler.* Woodsome, Huddersfield—planted.

Obs.—Occasionally the rachis in this fern becomes forked by division in the upper third, or even lower down. Dr. Richardson observed it about 1690. It has occurred in several places.

1005. Asplenium lanceolatum Huds.

Area : '63, [64 at best doubtful],' Top. Bot. Range : 600 feet.

Native ; stony and sandy banks, very rare. P. July.

First record : Bolton, 1785.

[L. On a wall in the village of Wherf ; *J. Bolton* (1785) ; *O.B.G.* (1805). Now gone ! Perhaps this was *A. Adiantum-nigrum*; vide fig. and description, *Bolton's Fil. Brit.*, p. 31.]

D. On an old wall and dry road-side bank, with a southern aspect, near Agden reservoir, Upper Bradfield (!) *Rev. R. A. Gatty.*

1006. Asplenium Adiantum-nigrum L.
Adiantum nigrum officinarum; Ray.
Area: 63, 64, Top. Bot. Range: 0—600 feet.
Native; on dry rocks (mostly sandstone) and hedge-banks, rare; a decreasing species; if ever common, now nearly exterminated. P. July-August.
First records: Hargrove, 1782; Bolton, 1785.

L. Wall, right-hand, Ingleton Brow; *J. Windsor.* Wherf; *J. Bolton, Fil. Brit.*
R. Kendall's Gill, Giggleswick; *J. Windsor.*
N. [Upon most of the rocks near Knaresborough; *E. Hargrove* (1809). Gone or nearly so in 1868.] Spofforth Castle, south-east wall (!) *J. Emmet.* [Amongst the rocks at Plumpton; *E. Hargrove, Hist.* (ed. 1782).
A. Garden-wall at Baildon; *J. Willis.*
C. [Near Luddenden; *C. Eastwood* in *Miall's Flora.* Lofthouse; *C. Forrest.* A mistake this; *G. Roberts.* Dungeon Wood, Huddersfield; *C. P. Hobkirk.*] Near Mount Skip, Hebden Bridge, and Scout, Hareley Wood; *T. Stansfield.*
D. Limestone tract west of Doncaster; *S. Appleby* (1831). Rather uncommon in the Sheffield district; *J. Hardy.* Rockley Abbey; *H. Payne.*
T. Roche Abbey; *J. Salt*; *J. H. Aveling*; etc. Anston Rocks; *P. F. Lee*, spn.!

[**Asplenium fontanum** L. Error—Imposition?
It cannot be doubted that, *if* this species was really that gathered in the localities reported, it was *planted*; but it is only fair to science to remark that Mr. S. Gibson was singularly unfortunate in finding so many alien species in stations very unlikely to be native. The grades of alienity *v.* indigenity were, however, little understood half a century back.

A. Near Skipton Castle, 1835; *S. Gibson* (*Phyt.*, i, 452).
D. Wharncliffe Wood, 1838; *R. M. Redhead* (*Phyt.*, i, 1084).]

1007. Athyrium Filix-fœmina Bernh. *Lady Fern.*
Area: 63, 64, 65, Top. Bot. Range: 0—1,350 feet.
Native; woods, lane-banks, hedges, and boggy moors, common. P. July-September.
First record: Salt, 1800.
Generally distributed; in all the districts.

a. rhæticum Roth. (*convexum* Newm.)
Exposed peaty banks and dike-sides, not common. The rachis is often of a fine purple.

L. Ingleborough; *B. Carrington.*

R. Linn Gill!
Y. Mickley Barrows, Ripon; *L. C. Miall.*
W. Barden Tower! Askham Bogs; *H. Ibbotson.*
C. Todmorden! *A. Stansfield.* Hebden Bridge; *L. C. Miall.* Halifax; *Id.* Coxley Wood; *P. F. Lee*!
D. Sheffield; *J. Hardy.* Cliff Rocher; *Amos Carr.*

b. molle Roth.
Shady dryish situations, rare.

L. Ingleborough; *L. C. Miall.*
C. Hebden Bridge; *S. Gibson.*

Obs.—I have not attempted to re-record the score or more nurserymen's 'varieties' of this Fern, the Male-Fern, and the Hart's-Tongue, entered without discrimination in a former Flora of the West Riding. I regard the great majority of them as mere sports, and to christen each individual showing a trifling aberration from a type, seems to me a 'reductio ad absurdum.' Furthermore, few of them occur twice in the same spot (especially if the root bearing the variable frond be religiously hauled away to a greenhouse-pot prison for safety, if not for sale!), and hardly a tithe, even of the type-forms once existing in the Calder Valley cloughs, are now to be found. The tourist or botanist gathering fronds for the herbarium are not the worst enemies ferns have: witness the horticultural columns of 'The Exchange and Mart.'

1008. Athyrium incisum Newm. *Lady Fern.*
Native; wet rocky woods among the hills and on grit-rock chiefly, not rare. A noble, tripinnate plant, its vase-like tufts often attaining a frond-length of three to four feet in sheltered gills.
Noted in districts R., N., W., A., C. (C. Coxley Wood, west of Dewsbury; *P. F. Lee.*)

Obs.—This fern better deserves the rank of a species than either of the two usually classed with it as alike varieties of an ideal 'Lady Fern.' It is not merely unlike the varieties in facies, but seems essentially distinct in its structural design, and has been proved to retain unchanged its character under cultivation. It is much further from them in its physiognomy than they are from each other.

1009. Ceterach officinarum Willd. '*Rustyback.*'
Area: 63, 64, Top. Bot. Range: 300—1,450 feet.
Native; dry limestone rocks and walls, very rare. P. July.
First record: Ray, 1724.

L. Chapel-le-dale; *R. Clapham.*
(R. Several places near Settle; *L. C. Miall.* Not confirmed by *J. Windsor.* Near Settle, a 'var. *crenatum* Moore'; *T. Holden.* Not in *Tatham's Herb.* nor in his Notes for Settle.)

A. Near Malham Tarne on the chalk rocks; *R. Richardson*; *Ray, Syn.*, iii, 118; *Martyn* (1763). Addition—'towards the east'; *Ray, O.B.G.* Still there at base of cliffs to east of Malham Tarn—there in 1880 and in 1885 (*P. F. Lee*)! Between Malham and Trenhouse; *H. Baines.* Near Apperley Bridge; *Rev. W. Wood* in *O.B.G.*
W. On Pool Bridge! *P. F. Lee* (1883). Probably of recent origin; not there in 1870. There in 1872; *G. Hainsworth.*
C. Old garden walls, Kirklees Park; *H. Baines.* Storthes Hall, Huddersfield; but introduced there; *P. Inchbald.*
[D. Heeley Bridge, near Sheffield, about 1790; *Herb. Salt.*]

1010. Scolopendrium vulgare Sm. *Hart's-Tongue.*
Phyllitis; Camden; Ger.; Ray. *Lingua cervina*; Bauh.
Area: 63, 64, 65, Top. Bot. Range: 0—1,400 feet.
Native; rocky woods, and scars in calcareous districts, common. P. August.
Earliest notice: Gerarde, 1597.

L. Upon Ingleborough hils; *Gerarde, Hb.*, 977 (var. *multifida*); *Merr.*, 93; *Ray, Cat.*, 241.
R. Common in Ribble district from Linn Gill to Settle.
Y. Hackfall and Fountains (!) *H. H. Slater.*
N. Knaresborough, on rocky slopes by river; *E. Hargrove* (1832). On the rocks by the Nidd from Scotton eastwards to Grimbald's Crag, but growing rare.
W. Abundant in Upper Wharfedale down to Bolton Woods, reappearing again on the Permian craggy slopes by the river at Wetherby, Flint Mill, and Thorparch.
A. Malham. Gordale! Camblesforth, now exterminated; *H. F. Parsons.*
C. Ouzlewell Green, in a well, 1869; *G. Roberts.* Nostal and New-Miller Dam; *A. F. Gissing.* Heath Common, in a well, but now extinct on Hell Lane-side; *G. Roberts.* [Lofthouse; *Id.* Gone.]
D. Lime-rock tract west of Doncaster; *S. Appleby* (1831). Conisborough Crags, growing rare! Old pit shaft, Wellthorpe; *H. Payne.* Went-vale.
T. Roche Abbey, *Mr. Salt*; *MS.* by *Jos. Woods* in *Towns. O.B.G.* Near Maltby! Kiveton; *Amos Carr.*

1011. Cystopteris fragilis Bernh. *Bladder Fern.*
Polypodium fragile; Huds., 459. *Filix saxatilis caule tenui fragile*; Ray.
Area: 63, 64, Top. Bot. Range: 0—1,800 feet.
Native; on rocks (mainly limestone) and walls, common in the north-west half of the Riding, rare in the east. P. July-August.
First record: Ray, 1670—Upon old stone walls and rocks in the West Riding of York; *Cat.*, i, 114.

M. Seal Bark Rocks, head of Greenfield Valley; *J. Whitehead* ('*dentata*').
L. & R. Very common, and in variety in both these districts.
Y. Near Cayton Gill, Ripon; *H. H. Slater.* Just surviving.
N. On the dropping rock, Knaresborough; *Martyn* (1763); rep. *E. Hargrove* (Ed. 1782 et seq.). There still, 1868!
W. Common in Upper Wharfedale down to Bolton. Rare on the Permian tract. Cock-beck Bridge between Tadcaster and Stutton (!) *J. Emmet*!
A. Gordale; *Mr. Curtis*; *Smith's English Flora*, iv, 301. Common at Malham! Gilstead; *W. West.* West Haddlesey; *H. F. Parsons.*
C. Shibden-dale head; *H. Baines.* Horsebridge Clough and a bridge top of Hebden Valley; *T. Stansfield.*
T. Anston Rocks; *J. Hardy.*

Obs.—This Fern varies much in the cutting of the frond and disposition of the clusters, and the forms named *anthriscifolia*, *cynapifolia*, and *angustata* all occur—together with the var. *dentata* Hook., which has by far the most distinct facies, and appears to be restricted in its area to the limestone rocks—from Ingleton and Weathercote in the west, through Settle to Arncliffe and Malham on the east.

[**Cystopteris alpina** Desv. Incognit—for the present.
Sheltered rock-niches, not confined (abroad) to limestone.
There is reason to suppose this Fern an inhabitant of the West Riding; *B. Carrington* in *W. R. Flora*, 56.

M. The diminutive tripinnate *Cystopteris* of the sheltered crevices of high rocks at the head of the Greenfield Valley is not improbably this species.
(N. Rocks at the Dropping Well, Knaresborough; *Francis.* The plant there—Martyn's station—is ordinary *C. fragilis.*)]

1012. Aspidium Lonchitis Sw. *Holly Fern.*
Area: 64 Tatham spn., Top. Bot. Range: 1,000—1,600 feet.
Native; calcareous mountain rocks, very rare (now). P. July-August.
First record: Martyn (1763).

R. Linn Gill, Ribblehead, a very few plants, 1879! Sparingly on lime stone rocks, 1,550 feet alt., 1,050 above town of Settle, in fissures in company with *Lastrea rigida*; *J. Tatham, junr.*, in *Newman* on '*L. rigida*' (*Mag. Nat. Hist.*, 1840). [This locality was] Langcliff or Attermire Scars, two miles east of Settle (*Baines*). Moughton Fell; *R. Clapham* (*Baker's Suppl.*); still there about 1867; *J. Ardill.* A spn. from rocks at Settle in *Herb. Tatham* is dated 1844.

(N. 'Collected at Howstean, May 30th'; *W. Storey*—sub nom. 'Holly Fern—in *Trans. Leeds Nat. Club*, vol. i (1886), p. 43, but without doubt a misnomer I think. At any rate, seeing that *Vicia lutea* (nowhere found wild in North England at all) is named in the same paragraph, and that the 'Diary of Observations' in which the record occurs teems with obvious errors both of fact and editorial judgment, not the slightest reliance can be placed on the correctness of the name).

W. Near Malham; *Wilson* in *Phyt.*, i, 451 (1855). Arncliffe Clouders, about eighty plants in 1839; *H. Ibbotson*! The 'near Malham' doubtless referred either to the Attermire station in Ribble basin or the Arncliffe one in W. area.

[A. Bingley; *Martyn* (1763); rep. *Huds.* (1778).]

1013. Aspidium aculeatum Sw. *Prickly Shield Fern.*

Area: 63, 64, 65, Top. Bot. Range: 0—1,800 feet.

Native; woods, thickets, among rocks on the limestone and hedge-banks, common. P. July-August.

First record: Salt, 1800.

L. Dentdale! Ingleton. Limestone pavement of Ingleborough (!) 1849; *Hb. Tatham*, spn. Whernside Lower Scars!

R. Attermire Cave, etc., Settle; *J. Windsor.* Linn Gill! Horton, etc., common!

Y. Studley; *H. Baines.* Hackfall!

N. Rennie Crags, Birstwith, near Ripley, Scotton, Knaresborough!

W. Common in Upper Wharfedale. Bolton Woods! Common about Collingham! 'Hell Hole,' Healaugh; *Ferns of York.* Tadcaster. Nun Appleton.

A. Gordale and Malham! Hawksworth; *W. West.* Bishop Wood, Hambleton!

[C. Ardsley; *T. W. Gissing.* Probably gone from there now.]

D. Wentbridge; *T. W. Gissing.* Lime-rock tract west of Doncaster; *S. Appleby* (1831). Conisborough! Back of a house in Old Park, Sheffield; *J. Salt.* Not common near Sheffield; *J. Hardy.*

T. Roche Abbey; *J. H. Aveling*—sub nom. *lobatum.*

Obs.—The form *lobatum* Sm. is frequent wherever the type occurs, but is not a true variety.

1014. Aspidium angulare Willd.

Area: 63, 64, Top. Bot. Range: 0—600 feet.

Native; dry, shady sandy-soiled woods and banks, very rare; whether a running-out species or one that has decreased from eradication it is difficult to decide. P. August.

First record: *H. Baines*, 1840.

R. Occasionally in similar situations to the preceding; *J. Windsor.* Did he really know it? if so why not have given some locality? Linn Gill (1886); the *Misses Thompson*—but I have not seen the spn.

W. Cuddy Wood, Collingham; *J. Emmet.* Nearly gone in 1880.

C. Shibden; *L. C. Miall.* Near Elland; *H. Baines* = Beestones Wood, Stainland; *T. Stansfield.* Nostell, *Mr. Watson*; *T. W. Gissing.*

D. Edlington Wood near Conisborough; *J. Hardy.*

Obs.—From some accounts this Fern would seem to have once been much commoner than now, but looking to its distribution in Britain and its climatic requirements, one is forced to think that can hardly have been the case; and that the 'Frequent' (without any specified stations) with which the account of the type in a former Flora was dismissed (whilst a dozen localities were expressly cited for the commoner, hardier *P. lobatum*) was but a cloak to hide the fact of a lack of information.

1015. Nephrodium Filix-mas Rich. *Male Fern.*

Area: 63, 64, 65, Top. Bot. Range: 0—2,300 feet.

Native; woods, thickets, banks, and rocks, very common. P. July-August.

First records: Bolton, 1785; Salt, 1800.

Distribution general; in every district plentiful.

VAR. **affine** Bab.

Unfrequent; in dense woods chiefly.

Y. Hackfall!

W. Bolton Woods! Bramham Westwoods (!) and Collingham; *J. Emmet.*

D. Wharncliffe Woods; *A. Carr*!

VAR. **Borreri** Newm. (*paleacea* Borr.).

Unfrequent, but often locally plentiful.

R. Near Settle; *B. Carrington.*

N. Plumpton, and wood near Stockeld Park gates

W. Collingham (!) *J. Emmet.* Lund Wood and West Woods, Bramham! Askham Bogs and Acomb; *Ferns of York.*

C. Todmorden, plentiful; *T. Stansfield.*

D. Wood near Barnsley! Old Park Wood, etc., Sheffield; *A. Carr*!

T. Maltby Woods!

VAR. **abbreviatum** DC.

Rare; in crevices of exposed rocks, 1,500 up to 2,300 feet.

L. On Ingleborough (!) *G. Pinder*, *Newm. Ferns* (1844, 202), in company with *Lastrea rigida* on the pavement of Ingleborough; but more plentifully on the grit rocks on west side and near the summit.

W. Fountains Fell, near the summit, with *Allosorus crispus*, only six or eight inches high, yet with one row of sori!

(A. Wilsden; *T. Stansfield.*)

1016. Nephrodium rigidum Desv.

Area: '64,' Top. Bot. Range: 900—1,550 feet.

Native; dry limestone rocks, rare. P. August.

First record: Bree, 1815.

L. The Seven Forks of Whernside near Ingleton! *S. Thompson* [the limestone plateau about the forks of the Kingsdale beck]. Whernside; *W. Wilson* (1837). Helks Wood, Ingleton; *J. Nowell*; *Herb. Tatham*, spn. coll. 1849. On Ingleborough, near the hill foot towards the village, in 1815; *Bree* in *Francis' B. Ferns* (1837) [the village was Clapham; the locality meant either Clapdale or Moughton Scars]. Moughton Fell, pavement above Crummock; *J. G. Baker.*

R. Fissures of limestone rock (1,550 feet alt.), 1,050 feet above Settle town, in company with *Aspl. viride* and *Polyst. Lonchitis*; *J. Tatham*; *Ed. Newman* in *Mag. Nat. Hist.* (1840); spn. (1854) *Hb. Tatham.* The rocks referred to here were the scars variously called Attermire, Langcliffe, or Settle Scars, two miles to the east of Settle, on the way to Malham Tarn. Attarmire Scars north of the (Victoria) Cave, in 1802, by Simmonds and Windsor; *J. Windsor* in *Fl. Crav.* (1873) and *Phyt.* (1855); *S. S. Burlingham* (1886).

(N. Greenhow Hill; *F. T. Walker* in *Ferns of York.*)

A. Malham Moor; *J. Nowell* = Pavement crevices above the Cove!

1017. Nephrodium cristatum Rich.

Area: '63, 64?,' Top. Bot. Range: Low.

Native; boggy thickets, very rare. P. July-August.

First record: W. Casson, 1856.

[N. Plumpton rocks near Knaresborough; *H. Baines.* This referred to *N. dilatatum* or some form of Male-fern.]

W. Askham Bog, south end! *Thornton* (1873); *H. Ibbotson* (1874). Gathered again in 1875 and 1877 (spn. in *Hb. Whitwell*) by *R. M. Christy* and *J. W. West.* Seen in 1876 by author, two or three roots only.

D. Thorne Waste; *Wm. Casson* (1856), *Herb. Backh.* Still there, 1872 (!) in a boggy bushy place near border of the Waste, to the south-west side. 'Thorne Moor near Doncaster, *H. Baines?*'; *Miall's Flora*, but *Baines' Flora* does *not* give any such entry.

1018. Nephrodium spinulosum Desv.

Area: 63, 64, Top. Bot. Range: 0—600 feet.

Native; boggy woods, bushy marshy places, etc., rare, avoiding limestone, yet locally plentiful. P. August.

First record: S. Appleby, 1831.

(L. Cavehole Wood (near Feizor), lower part; *J. Windsor.*)

R. ('Loose stones near Attermire Cave; *J. Windsor*' (1802), *MS.* in his hand, *Towns. O.B.G.* Dr. Windsor afterwards showed it to Sowerby (fide *Fl. Crav.*), and the plant was *Last. rigida.*) Kendall's Gill, Giggleswick; *J. Windsor, Fl.* Helwith Moss (1844); *Herb. Tatham*, spn.

Y. Hackfall; *H. H. Slater.* Boroughbridge!

N. Ripley and Scotton banks by Nidd! Harrogate and Spofforth! Acomb Grange and Rufforth Woods; *Ferns of York.*

W. Rare in this dale above Ilkley. Jackson's Wood, a mile east of Wetherby Station! Askham Bog!

A. Rare above Leeds. Ditch, Alwoodley; *H. Ibbotson.* Morton Bog, Hambleton!

C. Many woods near Wakefield; *T. W. Gissing.* Coxley Wood, south-east of Dewsbury; *P. F. Lee*!

D. Potteric Carrs; *S. Appleby* (1831)! Limestone tract west of Doncaster and sand-lands on east; *S. Appleby* (1831). Thorne (specimen collected by *W. Wilson*, of Warrington, in *Hb. Whitwell*!) and Goole Moors (borders!) *H. F. Parsons.* Between Loxley and Bradfield; *Amos Carr.*

T. Rossington Carrs. Carrs by the Old Torne Dike near Wroot, Hatfield Moor.

VAR. **uliginosum** Newm.

W. Askham Bog; *R. M. Christy.* An intermediate beween *cristatum* and *spinulosum.*

1019. Nephrodium dilatatum Desv. *Buckler Fern.*

Aspidium vel *Polypodium cristatum* of Old Lists.

Area: 63, 64, 65, Top. Bot. Range: 0—2,100 feet.

Native; woods, heaths, lane-banks, and peat-bogs, common and in great variety. P. July-September.

First record: Salt, 1800—sub nom. *Polypodium cristatum.*

Distribution general; occurs in every district.

VAR. **collina** Newm.

In peat-bogs on the moors of the hills in the north-west of our area (*G. Pinder et aliis*), Baugh Fell, Cam Fell, Widdale, etc. Todmorden Moors; *T. Stansfield.*

VAR. **nana** Newm.

Near Settle; *J. Tatham* = Peart's Crags (1886); *Miss R. F. Thompson*—known as 'the Tatham Fern.'

1020. Nephrodium æmulum Baker. (*L. recurva* Newm.)

Area: [64 likely an error], Top. Bot. Range: 0—450 feet.

Native; moist light-soiled sheltered woods, very rare. P. August.

First record; T. Simpson, 1850.

(R. Settle; *J. Tatham* teste *E. Newman*; rep. *B. Carrington*. The fact that there is no West Riding specimen in John Tatham's own collection is suggestive of error.)
Y. Hackfall, sparingly (!) *T. Simpson*; rep. *H. H. Slater*.
A. Harden Woods; *L. C. Miall*. Near Shipley Glen; *J. Emmet*.
C. Luddenden Foot; *C. Eastwood* in *Carrington's Flora*.

Obs.—Reported for Settle in *Cybele Brit.*, iii, 272; yet placed within [ ] for v.c. 64, by Watson, with his riper experience, in 1875. It certainly grows in Hackfall, and I have seen a root, garden-grown, said to have been obtained near Shipley Glen, but I cannot vouch for its occurrence in districts A. or C. myself.

1021. Nephrodium Thelypteris Desv.
Area: 63, 64, Top. Bot. Range: 0—100 feet.
Native; marshes and boggy thickets, rare, locally plentiful. P. July-September.
First record: Bolton, 1775, but the plant he intended under that name was *N. Oreopteris*, I think.
(R. Near Settle; *Baines*; rep. *Carrington*. But never really seen there. No specimen in *Herb. Tatham* or in his Notes.)
W. Askham Bogs, abundantly.
N. Near Copgrove. Staveley Carrs, formerly; now long gone. In Sharow Mires, near Ripon (!) plentifully; *H. H. Slater*.
D. Bog at Askern (!) *G. Harrison, Herb.* (1862). Potteric Carrs, Doncaster; *S. Appleby* (1831); 'rare'; *J. Hardy* (1844).

1022. Nephrodium Oreopteris Desv. '*Hay Fern.*'
Area: 63, 64, 65, Top. Bot. Range: 100—1,500 feet.
Native; rocky moist pastures, woods and heaths among the hills, common. P. August.
First record: Salt, 1800; but Bolton, sub nom. *L. Thelypteris*, gives a Halifax locality in his *Fil. Brit.* (1785).
M. Greenfield Valley (!) *J. Whitehead*.
L. Garsdale! Great Blake Gill, Dentdale head! Cavehole Wood; *J. Windsor*. Bog on Ingleborough! Road between Settle and Clapham (1849); *Herb. Tatham*.
R. Kendall's Gill, Giggleswick; *J. Windsor*. Ribblehead! Hodder Valley!
Y. Head of Lord Grantley's fish-pond on Sawley Moor; *J. Dalton, O.B.G.*
N. Birk Crag Valley; Pateley; Harlow Heath, Harrogate; Plumpton (nearly gone!); etc.
W. Common about Buckden. Fountains Fell Gills! Valley of Desolation, Bolton! Not known in Ainsty; *H. Ibbotson*.
A. Rare in Airedale. Morton Bog, Hambleton! (two plants only seen).
C. Several places near Halifax; *J. Bolton*. Rishworth Valley, opposite quarries, above Darby Inn, plentiful!
D. Rawson Spring; *J. Salt*. Common near Sheffield; *J. Hardy*. By the Rivelin near the Paper Mills; *Amos Carr*.

1023. Polypodium vulgare L. *Polypode.*
Area: 63, 64, 65, Top. Bot. Range: 0—1,900 feet.
Native; old walls, thickets, on mossy tree-trunks and sandy banks, common, but like most other ferns, growing rare in the smoky populous areas. P. August.
First record: Salt, 1800.
Distribution general; numerous stations on record for all the districts, but extinct in some of them. Many named or unnameable varieties on record. Mr. Roberts traverses my statement as to commonness given above, and says it is not known to him in the eastern part of the Wakefield district; and Dr. Parsons remarks that it is 'rare in the Marshland district,' near Goole.

1024. Polypodium Phegopteris L. *Beech Fern.*
Area: 63, 64, 65, Top. Bot. Range: 0—1,500 feet.
Native; shady rocks, rocky woods, sometimes in more exposed situations in the gills, unfrequent. P. June-July.
First records: Bolton, 1775; Gough in Cam., 1789; Salt, 1800.
M. Head of Greenfield and Diggle Valleys; *J. Whitehead*.
L. Opposite the slate quarries in Helks Wood, Ingleton; *W. Howson*. Wood on lower side road opposite Cavehole Wood; *J. Windsor*.
R. Gill near Huntworth House, Giggleswick; *J. Windsor*. Hodder Valley; *W. Fowler*. Cave Ha' Wood (1842); *Hb. Tatham*.
Y. Hackfall; *H. H. Slater*. Old-field Wood, by Ripon; *W. Brunton*. Dallow Gill; *J. Dalton*.
W. Buckden Woods! Bolton Woods! Kettlewell; *L. C. Miall*.
A. Waterford Gill, source of the Skipton Ellerbeck; *W. Whitwell*. [Formerly at Lidget Green, near Bradford.] Eldwick; *L. C. Miall*. Bishop Wood near Selby; *Y. N. U. Excurs.* (1877).
C. Wood in Soyland, just by brook below Kebroyd Mill, and woody ground above Oldhouse Mill, Norland; *J. Bolton*. Todmorden, Harely Wood, and other places; *T. Stansfield*. Shibden and Booth Wood, Luddenden; *C. Eastwood*. Wood near Meltham; *C. P. Hobkirk*.
D. Rivelin; *J. Salt*. Upper Bradfield!
T. Anston Rocks; *J. Hardy*.

1025. Polypodium Dryopteris L. *Oak Fern.*
Area: 63, 64, 65, Top. Bot. Range: 0—1,800 feet.
Native; stony woods and shady rocks; formerly common, now growing scarce, although still locally abundant, most so on gritstone, although by no means absent on the Yoredale and scar limestone. P. July-August.
First record: Martyn, 1763.
M. Marked for this district by *J. Whitehead*.

L. Rawthey Dale! Dentdale! Clapham. Wood, lower side of the road opposite Cavehole Wood, and wood opposite Helks Wood; *J. Windsor*.
R. Sannet Gill, Settle; *S. S. Burlingham*.
Y. Dallow Gill; *J. Dalton*. Hackfall (!) and on Sawley Moor; *W. Brunton* in *O.B.G.*
N. Brimham, Pateley, Rennie Crags, Ravensgill, Guy's Cliff, etc.; *Ferns of York*.
W. Hesletine Gill! About Grassington and Barden! Burley-in-Wharfedale; *L. C. Miall*. Bolton Abbey Woods; *Rev. W. Wood*.
A. [Formerly in several places near Bradford and Thornton; now gone.] Elslack Moor; *W. Whitwell*. Hawksworth; *W. West*. Bingley, frequent; *L. C. Miall*.
C. [North Bierley (near Bradford); *Martyn* (1763).] Hebden Bridge! Birk's Wood, North Dean Wood, etc., near Halifax; *J. Bolton*. Netherton; *T. W. Gissing*. [Lofthouse; *C. Forrest*. Soothill Wood, etc.—Gone.] Wood near Meltham; *C. P. Hobkirk*.
D. Cliff Rocher; *Amos Carr*. Hartley, Denby-dale, *Mr. Hardman* (1869); *A. F. Gissing*.

1026. Polypodium Robertianum Hoffm. (*P. calcareum* Sm.)
P. Dryopteris of Dickson and Bolton.
Area: 63, 64, Top. Bot. Xerophilous. Range: 300—1,800 feet.
Native; on calcareous rocks and stony slopes amid the limestone scars, mostly under and among the talus, or débris, below the mural crags. P. July.
First record: Curtis, 1782.
L. Swilla Bottom, Ingleton; *Misses Thompson*. Kingsdale Scars! Clapdale; *C. Deighton*. Ingleborough Lower Scars! Austwick Scars; *R. Clapham*.
R. Loose stones near Attermire Cave; and Giggleswick Scar; *J. Windsor*; *Hb. Tatham*; etc.
W. On the limestone of Howgate Edge and Cam End overlooking Outershaw! Among the limestones on the right hand side of the Girling Trough near Conistone and near Kilnsey; *W. Curtis* (1782); *Gough's Cam.* (1789). Arncliffe Clouders! Eastfield Wood, Arncliffe; *J. Windsor*. In the old masonry of the terrace wall in front of the ruined Hall in Bramham Park (!) *J. Emmet*. Not native here but well established.
A. Awes Scar, Malham Moor (!) *J. Windsor*. Malham Cove and Gordale (!) *W. Howson*, *G. Pinder*, *T. Stansfield*, *W. West* et aliis.
(C. Woods near Halifax; *Rev. W. Wood* in *O.B.G.* Another of his mistakes.)
T. Anston rocks, plentiful; *J. Hardy*.

1027. Osmunda regalis L. *Royal Fern. Flowering Fern.*
Filix florida sive O. regalis; Ger.; Ray.
Area: 63, 64, Top. Bot. Range: 0—100 feet, formerly to 600 feet.
Native; boggy bushy places, rare, formerly abundant. P. July.
First record: Ray, 1670.
In Yorkshire . . . . in boggy and wet grounds, in the valleys among the mountains, abundantly; *Ray, Cat.*, i, 113 (1670).
Y. One place near Ripon, four or five plants in 1871, but locality I withhold. [Copgrove, 1828; *W. Grainge*. Gone.]
N. Brimham Rocks*; *W. Storey* in *Trans. Leeds Nat. Club*, i, 43. [Knaresborough, ditch near Hay Park Windmill; *E. Hargrove* (1832). Up to 1874 (*W. Grainge*). Now gone.] Close by a farm-house called Susacres, near Ripley; *Hargrove's Hist.*, 1782, et seq.; *Archd. Pierson*. Near Harrogate; *Miss Ann Walker*, spn. in *Hb. Whitwell*, ex. coll. *H. Baines*. [Harrogate (1816); *Loidis and Elmete*.]
W. Askham Bogs—still in some plenty, in the almost inaccessible 'far wood' surrounded by wide ditches.
A. [About Keighley; *T. Knowlton* (1805). Addle Bog, near Leeds; *A. Shipley*. Long gone. Formerly at Birkin, below Castleford, but now gone; *G. Roberts*.] Camblesforth; *H. F. Parsons*.
C. [Formerly in several places near Halifax; *J. Bolton*; but now gone. Birkenshaw; *J. Emmet*; now gone.] Near Horbury*; *W. Rushforth*! Bellmoor, Todmorden—nearly extinct; *T. Stansfield*.
D. Potteric Carrs! *E. Miller* (1804); *S. Appleby* (1831); etc. Still in one spot! Sand and heath lands east of Doncaster, on borders of Hatfield Chase; *S. Appleby* (1831). Goole Moors; *H. F. Parsons*. Thorne Waste! Still there, but very scarce now, having been dug up by nurserymen.

[**Onoclea sensibilis** L. Extinct Alien.
Lane, Moreby, near York; *H. Baines* (*Phyt.*, i, 492). An introduction, on a hedge-bank—'long gone'; *H. Ibbotson*.]

1028. Ophioglossum vulgatum L. *Adder's-tongue.*
Area: 63, 64, Top. Bot. Range: 0—600 feet.
Native; damp pastures, meadows, and (more rarely) in woods, common, easily overlooked.
First records: Bolton, 1775; Salt, 1800.
Yorkshire, not uncommon; *Gough's Cam.* (1789).
L. Bentham. Sedbergh! Clapham; *C. Deighton*.
R. High Ridge Wood, Settle; *J. Windsor*; etc. Tarnfield, Giggleswick (1885); *S. S. Burlingham*.
Y. Copt Hewick; *H. H. Slater*. Markington; *J. S. Tute*.
N. Belmont Wood, Knaresborough; *E. Hargrove*. Birstwith and Birk Crag, Harrogate; *Ferns of York*. Buttersike Bar; *W. D. Roebuck*.

W. Pickhill Rash Wood, Bramham! Preston or Spring Wood pasture, by rail-line a mile east of Wetherby (!) *J. Emmet.* Parlington; *H. F. Parsons.*

A. Baildon; *W. West.* Cookridge; *G. Roberts.* In plenty in the Green Lane, Ledsham; *W. Whitwell* (1878-81). Cononley ings.

C. Luddenden Dene; *Y.N.U.* (1879). Storthes Hall and Farnley; *Hudd. Nat. Hist.* Lofthouse; *G. Roberts.* Todmorden; *T. Stansfield.* Howley. Coxley Valley, near Dewsbury; *P. F. Lee.*

D. On the low 'warp' lands by the Don; *S. Appleby* (1831). Goole, very fine in new fields on the moor-borders; *H. F. Parsons.* Old Park, Sheffield; *J. Salt.* Barnsley (!) *H. Payne.*

T. Roche Abbey; *J. H. Aveling.*

1029. Botrychium Lunaria Sw. *Moonwort.*

Lunaria minor; Ray.

Area: 63, 64, 65, Top. Bot. Range: 0—1,600 feet.

Native; damp stony pastures, mostly amid the hills, unfrequent. P. June-July.

First records: Gerarde, 1597; Ray, 1670; etc.

M. Mersey tributaries district; *J. Whitehead.*

L. On Mear bank by Sykes Wood, Ingleton; *Gough's Camden* (1789). I observed it on the top of Ingleborough Hill and the hills fast by Settle; *Ray, Cat.*, i, 199; ii, 190 (1670).

R. On the mountains near Settle; *Ray, Hist.* (1686); *Syn.* i; *Martyn* (1763); etc. Nowhere so plentiful, rank, and large as on the tops of some mountains near Settle; *Ray* in *Gibs. Cam.* (1695). Three places near Settle; *J. Windsor.* Still common; *Misses Thompson* (1886).

Y. Pastures near Ripon; *W. Brunton.* Near Copgrove; *J. Dalton.* Ingerthorpe, Markington, Mackershaw; *H. H. Slater.*

N. Half's (or Halves) Farm pasture, Knaresborough; *E. Hargrove* (1782 et seq. eds.). Rennie Crags, Greenhow Hill, Birk Crag near Harrogate; *Ferns of York.*

A. Plentiful in John Nun's cow-pasture adjoyning to his house at Methley, *Mr. Witham*; *Merr.*, 74 (1666)—a variety 'foliis dissectis' (cf. *Sm. E. F.* iv, 329). (Bradley Moor, Skipton; *Report.*)

C. Neere unto Wakefield in the close where Sir George Savell [Saville] his house standeth, called the Heath Hall; *Gerarde, Hb.* (1597); rep. *D. Dixon*, 1790—'and again in 1842.' Stanley and Featherstone; *T. W. Gissing.* Mixenden, Sowerby, Northowram, etc., near Halifax; *J. Bolton* (1775). Kidroyd and Fixby; *Hudd. Nat. Hist.* Rake Hay, Royd Hills, etc., Todmorden; *T. Stansfield.* Coxley Valley, near Dewsbury; *W. Rushforth.*

D. On clay soils (as at Clayton, eight miles from) Doncaster; *S. Appleby* (1831). Hook, near Goole; *T. Bunker*! Sheffield; *J. Hardy.*

LXXXV. LYCOPODIACEÆ.

1030. Lycopodium clavatum L. *'Stag's Horn Moss.' 'Wolf's Claw.' 'Fox's-tail'* (Wordsworth).

Muscus clavatus sive lycopodium; Ger.; Park.; Ray.

Area: 63, 64, 65, Top. Bot. Range: 25—2,400 feet.

Native; stony heaths and moors, common amid the gritstone hills, rare and decreasing in the lowland districts. P. July-August.

First records: Ray, 1670; Bolton, 1775; Salt, 1800.

'On the high hills of Yorkshire'; *Ray, Cat.* (1670); *Hist.* (1686); etc.

M. Millstone Edge, Denshaw! (1876).

L. & R. Common still on the moor-slopes and grit-crags in both districts!

N. Ramsgill; *J. S. Wesley.* Rennie Crags, Pateley moors! Hookstone Quarries, Pannal (!) *T. Hick.*

W. Beamsley and Barden Fells! Rombalds Moor and Otley Chevin! (nearly gone there now). Almias Cliff! Ethell Crags, Scarcroft —still there in 1880. Heathy lane near Wike, Shadwell (1871).

A. Bingley and Keighley moors, growing scarce. Formerly in pastures at foot of walls behind Meanwood Hall, going down to Scotland Mills, and on Adel Blackmoor—I fear now gone, but not quite in 1878. Brayton, in the railway-cutting; *W. N. Cheesman*—at 25 feet only above sea level.

C. Heath, Wakefield, up to 1866; *T. W. Gissing.* Hebden and Halifax Moors; *J. Bolton*!

D. High moors, south-west from Sheffield; *J. Salt*; *J. Hardy*; etc. Bradfield, 1877! Woolley Edge; *T. W. Gissing.*

1031. Lycopodium alpinum L.

Muscus clavatus foliis Cypressi; Merr.; Ger. Em.

Area: 63, 64, 65, Top. Bot. Montane. Range: 1,000—2,400 feet.

Native; stony peaty moors and crags, rare, locally plentiful. P. Aug.

First records: Merrett, 1666; Ray, 1670; Bolton, 1775.

L. Howgill Fells (!) *W. West*; *H. F. Parsons*; etc. On Ingleborough hill, abundantly; *Ray, Cat.*, 214; *Hist.*; *Martyn* (1763); *W. Curtis* (1782); etc. Still plentiful! Whernside (!) *B. Carrington.*

Y. Sawley Moor; *W. Brunton.* Moors near Grantley; *J. Dalton.*

A. About Keighley; *T. Knowlton* in *O.B.G.*

C. Near the mile-cross west from Halifax; *Merr.*, *Pin.*, 80 (1666). Sowerby, Wadsworth, and Midgeley moors; *J. Bolton.* Langfield and Stansfield moors, rare; *T. Stansfield.*

D. Sheffield high moors; *J. Salt.* Not common there; *J. Hardy.* Rocks above Abbey brook, very sparingly!

1032. Lycopodium inundatum L.

Muscus terrestris denticulatus repens; Clus. Merr.). *Musc. terr. repens clavis singulis erectis*; Ray.

Area: '63 Leyland spn.,' Top. Bot. Range: 0—900 feet.

Native; on moors where turf has been pared off, or on bare wet peat, very rare (intermittingly?) P. July.

First record: Merrett, 1666.

L. [On Ingleborough Hill; *Ray, Cat.* (1670); *Gibs. Camd.* (1695); etc. No recent confirmation.] (Side of the rivulet that springs from Giggleswick Scar, beyond the first milestone; *J. Windsor, MS.* in *Towns. O.B.G.*, but referred to *L. Selaginoides* in *Flora Cravoniensis.*)

[N. Harrogate; *J. Dalton* in *O.B.G.*]

W. Askham Bog, in the jungle, very sparingly (1876).

[A. On the east side of Rumble-moor, above Helwick, in the deep heath; *Merr.*, 80 (1666). Blackmoor, near Leeds; *Rev. W. Wood.*]

C. Norland Moor; *L. C. Miall.* Halifax and Todmorden Moors; *H. Baines.*

Obs.—Is not this a species uncertain in its localities, and impermanent because its special requirements (bare peat, moisture, and some warmth without active competition) rarely persist long unchanged in any given spot? Mr. Watson (in *Top. Bot.*, 527) asked if it was really very rare in Wales. My experience is that throughout Britain it is rare at high levels on tracts over hard rocks, where there is little peat and little warmth. Normally, it is a low-level peat-on-sand species, and its areal census of 50 vice-counties greatly over-represents its relative commonness.

1033. Lycopodium Selago L. *Fir Moss.*

Chamæpeuce fœmina seu polyspermos; Merr.; Huds. *Muscus erectus abietiformis*; Ray.

Area: 63, 64, 65, Top. Bot. Montane. Range: 25—2,300 feet.

Native; rocky or heathery places, on the hills mostly, rare. P. July.

First records: Merrett, 1666; Ray, 1670; etc.

M. Rocks on Moor-edge, head of Greenfield Valley; *J. Whitehead.* Wicken Clough.

L. Howgill Fells. Baugh Fell and Whernside! Ingleborough Hill; *Ray, Cat.*; *Hist.*; *Petiv.*, *Concord.* (1713); *Martyn* (1763); etc.

R. Settle Hills; *J. Windsor.* Ribble-head and Cam! Bowland Knotts!

N. Great Whernside down to Rennie Crags, Pateley.

W. Rocks of Cam Fell, Penyghent (*J. Tatham*), Fountains Fell, and Barden!

A. Otterburn Head. Above Malham; *W. Kirkby.* Brayton railway cutting, in a delving, a mile west of Selby; *W. N. Cheesman* (1880)! A relic.

C. At Dovestones in the side of Yorkshire, near Widdop; *Merrett*, 25 (1666). Still thereabouts in many of the exposed craggy places on the summit-ridge of the Pennine range dividing Yorkshire from Lancashire, both north and south of Todmorden! Halifax Moors; *J. Bolton.* Heath, Wakefield; *W. Banks*; *T. W. Gissing.*

D. Sheffield High Moors; *J. Salt*; *J. Hardy*; etc.

1034. Selaginella Selaginoides Gray.

Muscus polyspermos; Merr. *Selaginoides foliis spinosis*; Ray.

Area: 63, 64, 65, Top. Bot. Montane. Range: 2,100—400 feet.

Native; oozy and boggy places on mountains, and by rills, unfrequent. P. July.

First record: Merrett, 1666.

L. Howgill Fells; *G. Pinder.* Whernside slopes. Wethercote! Kingsdale; *L. C. Miall.* On the north of Ingleborough, near the under-ground river; *Merr.*, *Pin.*, 79. On Ingleborough; *Ray, Cat.*, 215 (1670). Near Clapham; *C. Deighton.*

R. Blea-moor. Horton, slopes of Penyghent! In a field on the left between Long Preston and Swinden (1813); *J. Windsor.* Rill-side, Giggleswick Scar, beyond the first milestone; *J. Windsor.* Catterick Force; *J. Tatham.*

[Y. Abbey's Marsh, near Copgrove; *J. Dalton* (180?).]

W. Outershaw, Langstrothdale! Arncliffe (!) *W. West.*

A. Above Gordale; *Jos. Woods, MS.* High-close bog on east side Malham Tarn (!) *J. Windsor.*

Obs.—Ray in *Gibson's Camden* (1695) remarks on *all* five sorts being found upon Ingleborough; but *L. inundatum* does not appear to have been gathered there for many years, and is the rarest kind in Yorkshire.

LXXXVI. MARSILEACEÆ.

1035. Pilularia globulifera L. *Pillwort.*

Area: 63, 64 sine pers. auth., Top. Bot. Range: Low.

Native; margins of pools with a gravelly-peaty bottom, very rare. P. July-August.

First record: Parsons, 1876.

D. By the pools on Rawcliffe Common! *H. F. Parsons*!

[**Isoetes lacustris** L. Extinct Introduction.

Subularia repens, foliis convexo-planis . . nascitur sub aquis. The Creeping Subularia with Leafes flat on the inside and round on the other . . *Rich. Richardson primo observavit in lacubus prope Llanberris*; unde in piscinas suis *in comitatu Eboracensi* detulit, etc.; *Dill., Hist. Musc.* (1741), p. 542, 543. I know no more of this, but doubtless long gone from the fish-ponds now. But according to *Newman* (*Hist. Brit. Ferns*, 385) the '*Subularia repens*' of *Dillenius's Ray's Synopsis* was the *Littorella* of modern nomenclature, but the description of the leaf negatives that conclusion.

('I found this upon Rombalds Moor in 1799'; *D. Dixon, MS.* note in his copy of Berkenhout. I suppose Dixon meant *Littorella.*)]

LXXXVII. EQUISETACEÆ.

1036. Equisetum arvense L. '*Toad-pipes*'—the fertile stems.

Area: 63, 64, 65, Top. Bot. Range: 0—1,500 feet.

Native; ill-drained fields, waste ground, rail-banks, etc., very common. P. March-May.

First record: Salt, 1800.

Distribution general; common in every district.

Obs.—A curious monstrous form—the fertile stem bearing a whorl of branches from the upper sheaths somewhat sylvaticum-like—was found by Malham Tarn in 1878 by Dr. H. F. Parsons.

1037. Equisetum maximum Lam. *Giant Horsetail.*

E. primum Math. quo utuntur arenarii ad sagittas poliendas; Merr., 36. (*E. Telmateia* Ehrh.)

Area: 63, 64, 65, Top. Bot. Range: 0—900 feet.

Native; swamps, usually on slopes in shade, common. P. May.

First record: Merrett, 1666.

L. By the Wenning near Bentham and Clapham!

R. Langcliffe and Ribble banks, Cammock; *J. Windsor.* Bolton-in-Bowland!

Y. Hackfall! Boroughbridge; *Y. N. U. Excurs.* (1876).

N. Scotton banks and other swamps in woods near Knaresborough.

W. Bolton Bridge to Gill Beck; *W. West.* Buckden! Washburn Valley! Langwith 'scrub,' Collingham (!) *James Abbott.*

A. At Skippon, and Craven; *Merr., Pin.*, 36 (1666). Still common there!

C. Hebden Bridge Valley. Haw Wood, Gawthorpe; *T. W. Gissing.*

D. Rigby Woods, near Pontefract; *Merrett, Pinax*, 36. Still there! Near Barnsley (!) *H. Payne.* Wharncliffe Woods; *Amos Carr.*

(T. In the Roche at Roche Abbey; *H. Baines.*)

1038. Equisetum sylvaticum L. *Wood Horsetail.*

Area; 63, 64, Top. Bot. Range: 0—1,000 feet.

Native; woods and damp shady slopes, by rivers, common, but not fruiting freely. P. May.

First record: Salt, 1800.

M. Greenfield Valley, swampy banks (!) *J. Whitehead.*

R. Linn Gill! Near Settle; *J. Windsor.* Sawley. Slaidburn (!) *W. Fowler.*

Y. Dallow Gill. Morcar Wood by Copgrove; *J. Dalton.* Oldfield Wood and Laver banks, Ripon; *W. Brunton* in *O.B.G.* Studley and Hackfall; *H. H. Slater.*

N. Pateley Bridge. Near Knaresborough! Green Hammerton; *H. Baines.*

W. Grassington! Bolton. Washburn Valley below Leathley! Wood below Harewood Park gates (!) *James Abbott.*

A. About Leeds; *W. Wood* in *O.B.G.* Bingley; *W. West.* Camblesforth; *H. F. Parsons!*

C. Hebden Valley! Leaventhorpe, Lofthouse; *G. Roberts.* Hiendley Dam; Stanley, etc.; *T. W. Gissing.* Coxley Wood, near Dewsbury; *P. F. Lee.*

D. Old Park; *J. Salt.* Beighton; *Geo. Harrison, Hb.* Tinsley Park Wood and Parker Spring; *Amos Carr.* Near Barnsley; *H. Payne.*

1039. Equisetum palustre L.

Area: 63, 64, 65, Top. Bot. Range: 0—1,900 feet.

Native; marshes, wet moors in peat or sand, common. P. July.

First record: Salt, 1800, 'near Treeton.'

Distribution general—in every district, but most plentiful on the moors.

The State **polystachyum** occurs (rarely) on moors.

Var. **nudum** DC.

Sandy places, rare.

N. Greenhow Hill, Pateley; *Henry Ibbotson.*

W. Askham Bog!

1040. Equisetum limosum L. '*Paddock-*'(O.E. for toad)'*-pipes.*'

E. nudum, lævius nostras; Ray.

Area: 63, 64, Top. Bot. Range: 0—1,500 feet.

Native; in marshes, stagnant pools, dams, and ditches, common. P. July.

First records: Dillenius, 1724; Salt, 1800; E. Miller, 1804, 'Potteric Carr.'

Distribution general; to be found in every district.

At Scippon and Craven; *Ray, Syn.*, iii, 131. *Smith, Eng. Flo.*, 4, 340, refers this to *E. hyemale*, but that does not grow at Skipton; it is Merrett whom Dillenius quotes for the locality, placing it under his '*E. nudum junceum*,' and further, 'in palustribus et aquosis' does not at all well express the characteristic local conditions *E. hyemale* prefers, viz.: wet clayey banks in shade.

Var. **fluviatile** L. Not uncommon.

R. Oustre-bank Wood, Settle; *Herb. Tatham.*

Y. Laver banks, Ripon; *W. Brunton.*

N. Below Plumpton pools!

W. Hobmoor ponds, York; *W. Whitwell.*

A. About Leeds; *Rev. W. Wood* (1805).

T. Roche Abbey!

1041. Equisetum hyemale L. '*Shave Grass.*'

Area: 63, 64, Top. Bot. Pelophilous. Range: 100—450 feet.

Native; wet clayey banks, in hedges or thickets, rare, and rather peculiar as well as irregular in its distribution. Has been oftener misreported than any other West Riding plant; but it is probably a decreasing species. P. August.

Earliest record: Bolton, 1775.

L. Among bushes close by the brook at the lower end of the Helk near Ingleton; *J. Bolton* (1790).

Y. In Hackfall; *Rev. J. Dalton*; *H. H. Slater* (1879). Skell banks, Mackershaw, and Laver banks; *W. Brunton.*

N. Between Knaresborough and Ripley. Hartwith Dam and Spring Wood, Winsley; *H. Ibbotson, Ferns of York.*

W. Hedge of lane near Scholes, half-way between Leeds and Tadcaster, near 7th milestone! *H. Ibbotson* (1875).

A. [At Skipton and other parts of Craven; *Withering*; *S. Hailstone* (1816). These two authors, I fear, merely quoted the mis-referred Dillenian record alluded to above.] Ingrow, Haworth Valley, above Keighley; *T. Hebden.* I am assured the last is correct by Mr. Soppitt. Malham Cove; *L. C. Miall* in 3rd ed. of *Whitaker's Craven* (1878).

C. From Wakefield and Pontefract (*H. Baines*), about Oakenshaw, Normanton, etc. Near Halifax; *H. Baines.* Field on Upper Brier estate, Northowram; *J. Bolton, Filices Brit.* (1790). Near Mixenden; *J. Bolton, Cat.* (1775). North Dean bottom, in Greetland! *Steph. Hartley*; *J. Bolton, Cat.* (1775). Near Lascelles Hall, Mirfield.

T. Clayey bank by brook, Maltby (!) *Gerard E. Smith.*

1042. Equisetum variegatum Schl.

Area: '64 Tatham,' Top. Bot. Range: 800—900 feet.

Native; oozy, sandy-peaty or stony places by mountain streams, very rare. P. July-September.

First (only) record: Howson, 1850.

L. Swarth Moor; *W. Howson's Flora* (1850); rep. *J. Windsor*; rep. *W. Marshall Watts.* Small and uncharacteristic, like the plant of Bain Gill near Semerwater in Wensleydale; and only occurring sparingly, but unquestionably the right thing. The two Wensleydale stations have been doubted as well, but I had no difficulty in verifying all three the first time of search.

Summary of West Yorkshire species of Phanerogams, with Ferns, etc.—1,042.

Phanerogams	995
Filices	34
Lycopods	5
Pillwort	1
Equisetums	7
Total	1,042

CHARACEÆ.

CHARACEÆ.

[The position of this class in the scale of the Vegetable Kingdom has long been debatable. Sachs gives it a place intermediate between Thallophytes and Muscineæ; related as the Charas are to the Algæ in some respects (differing from them in the form of their antherozoids), and to the Muscineæ in other respects, although unlike them in the structure of their reproductive organs; but if these be not forgotten their sequence in a local flora matters little. Their customary place in English text-books, and in the London Catalogue of British Plants, is therefore adhered to.

The arrangement of species followed is that adopted by the Messrs. Groves in their 'Review' (Journal of Botany, 1880), two divisions being recognised (*Charæ* and *Nitellæ*), each divided into two genera, the second of the first division, *Lychnothamnus*, being unrepresented in Yorkshire.

Yorkshire workers among the Characeæ have been few. Beyond some dozen old records, the material at command is the result of the gatherings of some half-dozen botanists—Messrs. H. F. Parsons, P. F. Lee, Amos Carr, W. Todd, W. West, Geo. Webster, and the writer—during the last ten years. All the names have at one time or another been determined or verified either by Prof. C. C. Babington or Messrs. Groves, with the exception of *C. contraria* and *N. capitata*.]

1. Chara fragilis Desv.

'York S.W. and M.W.,' Groves' Review. Range: 0—1,250 feet.

Reservoirs, streams, and ditches, common. July-August.

W. Hob Moor and Chandler's Whins ponds, near York; *G. Webster*. In the river Wharfe, nearly under Linton Bridge, near Wetherby!

A. Malham Tarn (!) *W. Todd*.

C. Stream between canal and Wintersett reservoir near Wakefield; *P. F. Lee*! Coxley Dam, near Dewsbury; *P. F. Lee*!

D. Askern!

Var. **fulcrata** Gant.

C. Old quarry-pool at Sandal; *P. F. Lee*!

2. Chara aspera Willd.

'York S.W.,' Groves' Review, p. 8.

Shallow pools and rivulets on limestone strata. Rare. August.

A. Malham Tarn, and in streams at Kirby Malham; *B. Carrington* in *W. R. Flora* (1862).

T. Langould near Maltby; *G. E. Smith* teste *James Groves*; spn. in *Herb. Brit. Mus.*

3. Chara polyacantha Braun.

Peaty- or sandy-bottomed pools, very rare. June.

D. Pond at Hook Grange, near Goole (1877); *H. F. Parsons*. Not since seen I believe.

4. Chara hispida L.

'York M.W.,' Groves' Review.

In lakes, Yorkshire; *Gough's Camd.* (1789).

Lakes, tarns, canals, and slow-moving dikes, infrequent. June.

First record: uncertain—Wood or Brunton, about 1803.

Y. Ditches near Ripon; *W. Brunton* in *O.B.G.* (1805).

N. Farnham Mires, Knaresborough; *J. Dalton*; rep. *H. Baines*.

W. Scarthingwell Park; *H. F. Parsons* teste *H. and J. Groves*.

A. In the rivulet issuing from Malham Tarn; *Rev. W. Wood* in *O.B.G.* (1805); 'before it falls into the ground'; *S. Hailstone*, in *Whitaker's Craven* (1815). In Malham Tarn (!) *B. Carrington*. Pools south of Malham; *J. Nowell*. Hambleton dike below Monk Frystone; *H. Baines* (1840). Brick garth near Selby; *Id.*

D. In the 'lake' at Askern; *P. F. Lee*.

5. Chara vulgaris L. (*C. fœtida* Braun). '*Stoneweed*.'

'York S.W. and M.W.,' Groves' Review.

Stagnant clayey- or peaty-bottomed ponds, and clayey- or limestone-bottomed rivulets, common. May-June.

In Craven; *S. Hailstone* in *Whitaker's Craven* (1815).

R. Near Birkbeck's Weir, Settle; *J. Windsor*. Beck below the Ebbing and Flowing Well, under Giggleswick Scar; *W. Whitwell* (1884). In the Ribble near Wigglesworth!

W. Hob Moor Brick-ponds; *Geo. Webster*! Askham Bogs, near York; *H. Baines*. Dike at Church Fenton; *G. Webster*. Newly-made water-cut at the eastern end of White Car Bog—which lies at the south-west corner of Bramham Park demesne.

A. Rivulet from (and bog to east of!) Malham Tarn, just before it sinks into the ground (along with another species); *J. Windsor*, *Flora Cravoniensis*. In Malham Tarn; *B. Carrington*. Stream from Malham Cove; *Id.* Gordale beck; *P. F. Lee*! Brick-pond pools near station, Sherburn (!) *W. Todd*.

C. Wakefield, in ponds, varying much in appearance; *T. W. Gissing*. Coxley Dam—var. *longibracteata*; *P. F. Lee*!

T. Pools and rivulet (on limestone), Lindrick Common; *Amos Carr*!

6. Chara contraria Kütz.

Probably not very rare, but I have only found it once.

W. Dike by railway between Stutton and Church Fenton, July 5th, 1881!

7. Tolypella glomerata Leonh.

Ponds and slow-moving dikes, very rare. May.

D. In a drain near Goole; *T. Birks, junr.* (1881). Pool by Warping-syke, 1877! [Broomfleet; *H. F. Parsons*. But the station is in the E. Riding.]

8. Tolypella intricata Leonh.

Pools and ponds, very rare. April-May.

D. Ditch between Ibbetson's and Duckett's farms, Goole fields, March 1882; *T. Birks*!

T. Pools on Lindrick Common, 1878; *Amos Carr*! (Was mixed with the *C. vulgaris* which Mr. Carr gathered at same time in same place).

9. Nitella translucens Ag.

Still clear pools and dikes, very rare. July.

A. In a clear dike (sewage dike), Sherburn! cum *W. Todd*.

C. Wintersett reservoir; *T. W. Gissing's Flora*. [Bowling Brook, near Bradford; *H. Baines* (1840).]

10. Nitella flexilis Ag.

Ponds and slowly-running dikes, rare. June.

First record: Ray, 1724.

(N. In a bog where the stinking Spaw empties itself, nigh Knaresborough; *Ray, Syn.*, iii, 133; rep. *Martyn* (1763); etc.)

W. Pond near Wetherby, 1882; *J. Jackson*.

A. Sandy-bottomed 'cut,' feeding an old silted-up dam, by stream above Scotland Mills, between Meanwood and Adel (!) *James Abbott* = Stream below Seven Arches, Adel; *Bot. Rec. Rep.* 1877, p. 228. [South-east corner of the new dam, Bradford; *H. Baines* (1840).]

C. Bowling Brook, near Bradford; *H. Baines*. Tong Park Mills; *W. West* (1879). Ditch, Howley; *P. F. Lee*.

11. Nitella opaca Ag.

'York M.W.,' Groves' Review.

Shallow clayey or peaty-bottomed pools and slowly-moving 'cuts,' common. May-June.

W. Pool by road-side, Rigton Hill-top, going towards Thorner! York; *Trans. Y. N. U.*, Series E, p. 77. In the Cock Beck near Tadcaster; *J. Jackson*.

A. Adel Upper Dam.

C. Slow-running stream by Coxley Dam, Dewsbury; *P. F. Lee*!

D. Rawcliffe Common; *W. West* (misnamed *syncarpa*; *Trans. Y.N.U.*, Series E, p. 85). Dike at Bessacarr, Doncaster; *T. Birks* (1881).

12. Nitella capitata Ag.

Peaty-bottomed lakes and dikes, very rare. July.

A. (In a peat 'cut,' Malham Tarn Bog!) *Not* in Tarn itself, as is erroneously reported in *Trans. Y. N. U.*, Series E, p. 22. The above record has been doubted, and perhaps needs confirmation, but I believe it to be correct. The fertile heads were profusely mucilaginous when gathered in 1870; but it is not easy to demonstrate it in the only specimen preserved.

BRYOPHYTA.

BRYOPHYTA.

Class I.—MUSCI.

[In the ordinal and generic arrangement of this list the Second Edition of the *London Catalogue* (1880), founded upon Jaeger's system, has been followed: this being done for the sake of convenience of reference, Dr. Braithwaite's magnificent and scholarly *British Moss Flora* being as yet unfinished, the first volume only (down to the end of Tortulaceæ) having been issued. For the sake of consistency, however, the citation of authorities has (by the kind labour of Mr. W. West) been revised by comparison with Schimper's *Synopsis Musc. Europæorum* (2nd Ed., 1876), inasmuch as it has been incontrovertibly pointed out that no fixed principle was followed in the matter of quoting authorities either in the *London Catalogue* or Hobkirk's *Synopsis*, and special thanks are due—and here tendered—to Mr. West for that help, and for assistance which no bryologist in the county was more competent than he to render.

The students and workers in this class are to-day, for the most part, the same as those whose names have already been given as contributing records to the Phanerogamic lists; with the addition of Drs. B. Carrington and R. Spruce, and Messrs. M. B. Slater, J. M. Barnes, G. Stabler, T. Hebden, and J. R. Murdoch, but recognition should be made here of the honourable fact that the solid foundations of our present knowledge in this class were laid by the pioneer *William Wilson*, and the artizan-genius *John Nowell*. Most of their early discoveries have been confirmed by the latest observers.

It only remains to say that certain species included in Dr. Carrington's and Mr. Hobkirk's lists of West Riding Mosses (1862 and 1873) will *not* be found in the following enumeration. My erasures from former lists, with the reason for the exclusion in this, are as follows:—

1. *Rhabdoweisia denticulata*. 'Green's Clough, Todmorden,' the station upon which it was included, is out of the Riding.
2. *Seligeria calcarea*. 'Bishop's Burton,' the station upon which included, is in the East Riding.
3. *Campylostelium saxicola*. 'Ramsden Clough,' the long-known but sole station, is in Lancashire.
5. *Orthotrichum obtusifolium*. Messrs. Wilson and Spruce's 'near York' station is in the East Riding.
6. *Orthotrichum fallax*. 'Hellwith' (properly Hill-Wath), near Ripon, is in the North Riding.
7. *Mielichoferia nitida*. Mudd's 'Ingleby-Greenhow' locality is in the North-east of Yorkshire, near Stokesley.
8. *Hypnum elodes*. 'Stockton Forest,' near York, is in the North Riding.]

I. SPHAGNACEÆ.

1. Sphagnum acutifolium Ehrh. *Bog-moss.* '*Pack-moss.*'
Montane and lowland swamps, frequent, widely distributed, and in considerable variety.
Area general.

VAR. **deflexum** Schpr.
A. Malham Tarn Bog; Baildon Moor; *W. West.*

VAR. **rubellum** Wils.
Moorland mosses, rather rare. Montane.
L. Ingleborough; *B. Carrington.* Whernside; *W. West.*
R. Hobson Moss!
Y. Kirby Malzeard Moor; *W. West.*
A. Rombalds Moor; *B. Carrington.* Beamsley Beacon; *W. West.*
C. Marsden. Stansfield Moor, Todmorden; *J. Nowell.*
T. Martin Beck Bog, near Bawtry; *G. Webster.*

VAR. **tenue** Braithw.
A. Baildon Moor; *W. West.*

VAR. **patulum** Schpr.
A. Baildon Moor; *W. West.*

2. Sphagnum fimbriatum Wils.
Moist turfy ground, rare.
M. Greenfield Valley; *J. Whitehead, MS.*
W. Wharfedale; *C. P. Hobkirk.*
A. Moors about Bingley; *Y. Nat. Rec.*, p. 175. Eldwick, in fruit; *W. West.*
C. Keb Lumb, Stansfield Moor; *T. Stansfield.*
D. Goole Moor; *Nat.*, iii, 32.
T. Martin Beck Bog; *G. Webster.*

3. Sphagnum squarrosum Pers.
Swamps, not very common.
L. Whernside.
R. Settle; *J. Windsor.* Bowland Knotts.
N. Swampy ground by Hookstone Quarries, Pannal.
W. Askham Bog. Rombalds Moor and Beamsley Beacon; *W. West.*
A. Denholme; *W. West.*
C. Calder district moors; *C. P. Hobkirk.*
D. Midhope and Bradfield Moors. Goole Moor; *H. F. Parsons.*

4. Sphagnum teres Angst.
L. In Dent Valley on south-west side of river, in 1872; *G. Stabler* in *Nat.*, iv, p. 113.

5. Sphagnum intermedium Hoffm. (*S. recurvum* P. Beauv.)
Wet moors, common; undetected up to 1871.
M. Greenfield; *J. Whitehead.*
L. Barbon Fell; *J. M. Barnes.* Dent Town's Fell. Whernside slopes!
W. Beamsley. Rombalds Moor and Beamsley Beacon; *W. West.*
A. Baildon; Nab Wood, Bingley; Newsholme Dean, Keighley, and Adel Bog; *W. West.*
C. Marsden Moors, near Redbrook; *C. P. Hobkirk.*
D. Bogs of Rivelin; *Amos Carr*!

6. Sphagnum cuspidatum Ehrh.
Frequent in both lowland and mountain bogs from Goole Moor (*H. F. Parsons*) to Blea-moor (*F. A. Lees*) and Bell Hagg Bogs (*Amos Carr*).
First record: Richardson in Ray's Syn., 1724.
Distribution general; all districts except Trent.

VAR. **plumosum** Schpr.
On all the moors, in peaty pools and drains, commoner than the type.

7. Sphagnum compactum Brid. (*S. rigidum* var. β Schpr.)
Swampy moors, scarce.
L. Whernside; *B. Carrington.*
W. Beamsley Rocks, near Bolton (!) *J. G. Baker.*
A. Baildon Moor, common; *W. West.*
C. Stansfield Moor; *J. Nowell.* Hudson Moor and Widdop; *T. Stansfield.*
D. Goole Moor; *H. F. Parsons.*
T. Martin Beck Bog, near Bawtry; *G. Webster.*

8. Sphagnum subsecundum Nees & Hornsch
Moorland swamps and woodland ditches. The type much rarer than the variety *contortum.*
M. Greenfield; *J. Whitehead.*
W. Devonshire Wood, Stockeld (!) *J. S. Wesley.* Askham Bogs.
A. Baildon Moor and Adel Bog; *W. West.*
C. Marsden; *W. West.*
D. Rivelin and Bell Hagg; *Amos Carr.*

VAR. **contortum** Schultz.
M. Greenfield Valley (!) *J. Whitehead.*
L. Moors, common.
R. Common.
W. Harlow Heath, Harrogate! Beamsley Beacon; *W. West.*
A. Flasby Fell and other moors. Rombalds Moor. Cullingworth; *W. West.*
C. Norland Moor; *H. F. Parsons.*
D. Many places above Sheffield; *Amos Carr.*
T. Martin Beck Bog, near Bawtry; *G. Webster.*

VAR. **auriculatum** Schpr.
A. Adel; Haworth; *W. West.*

9. Sphagnum laricinum Spruce.
L. In the pass between Dent and Barbon before the summit is reached. It occurs hereabouts both in Westmorland and Yorkshire! Discovered by *C. J. Wild.*

10. Sphagnum tenellum Ehrh. (*S. molluscum* Bruch).
Rare.
L. Ingleborough; *B. Carrington.*
A. Moor above Morton, in fruit; *W. West.*
C. Stansfield Moor, but not recently; *T. Stansfield.*

11. Sphagnum papillosum Lindb.
Moorland swamps, common, though chiefly in the VAR. *confertum*, although only first recognised as a W. R. species by myself in 1876.
L. Whernside slopes. Dentdale head.
R. Blea-moor. Knotts between Slaidburn and Clapham Station (1876).
W. Buckden. Stake Fell moss.
A. Baildon Moor; *W. West.*
D. Goole moors.
T. Martin Beck Bog, near Bawtry; *G. Webster.*

12. Sphagnum cymbifolium Ehrh. *Bog-moss.* '*Pack-moss.*'
Boggy and swampy heaths, in every district, though not quite so common as *S. acutifolium.*

VAR. **squarrosulum** N. & H.
A. Bingley; *W. West.*

II. ANDREÆACEÆ.

13. Andreæa petrophila Ehrh. (*A. rupestris* Hedw.)
High gritstone rocks, rare.
M. Greenfield; *J. Whitehead.*
L. Winder Fell, Sedbergh; *H. F. Parsons*; *W. West.* Higher ridges of Whernside. Ingleborough; *B. Carrington.*
N. Brimham rocks; *R. Teesdale* in *Linn. Trans.* (1798)—first record.
W. Penyghent, slope towards Hesletine Gill; *W. West*!
C. Bottom of Hudson Moor, Hareley Wood, nearly extinct; *T. Stansfield.*

14. Andreæa alpina Turn.
Slate rocks, rare.
L. Cautley Spout; *W. West* and *W. Nuttall* (1879 and 1881)! On Ingleborough (L.) and at Brimham rocks (N.), upon *R. Teesdale's* authority, supported by *S. Hailstone*; but not seen in these localities, I believe, by any recent observers.

15. Andreæa Rothii Web. & Mor. (*Jungermannia rupestris* L.)
L. Baugh Fell (possibly the var. δ, but Braithwaite does not give that var. for any West Yorkshire station). Ingleborough; *B. Carrington*; 'western rocky precipices'; *W. West.*
W. Wharfedale; *S. Gibson* (perhaps this referred to var. β).
C. Hudson and Staups Clough, and Hebden Valley rocks, rare; *A. Stansfield.*

VAR. **grimsulana** Bruch. (*A. Rothii β frigida* Hueb.)
(L. Ingleborough; *Hobkirk, Syn.*, Ed. i. An error, as I was informed by Dr. Braithwaite. Record omitted from 2nd Ed. of Synopsis.)
W. Beamsley Fell, Bolton; *J. G. Baker* (1858). (Misprinted 'Brandsley Falls' in *Hobkirk's Syn.*, ed. i.)

VAR. **falcata** Schpr.
L. Ingleborough (1878); *W. West.* Baugh Fell (1879); *W. West*; *H. F. Parsons* and *W. Nuttall.*
W. Penyghent (1880); *W. West*!

16. Andreæa crassinervis Bruch.
W. On the rocks of Penyghent at 2,000 feet; *J. Whitehead* (1868).
C. Hebden Bridge; *G. E. Hunt* (1864).

III. WEISIACEÆ.

17. Gymnostomum tenue Schrad. (*Mollia tenuis* (Schrad.).)
On sandstones in shade, rare.
Y. Walls of Abbot's house, Fountains Abbey; *A. Hutchinson* in *Baker's Suppl.* (1854).
W. Thorparch; *R. Spruce* = Near Wetherby; *J. Dickson.*

[**Gymnostomum squarrosum** (N. & H.) Wils. Incognit.
Erroneously recorded by Mr. Hobkirk in *Journ. Bot.*, N. S., viii, 339, on faith of name in *Nat.*, iv, 159, as having been collected in Ingleton district. W. West's 'Brant Fell' gathering was made on the Westmorland (not the Yorkshire) hill of that name.]

18. Gymnostomum rupestre Schwg. (*Mollia æruginosa* (Sm.).)
Wet shady rocks and banks, common among the hills.
L. Great Blake Gill, Whernside. Howgill Fells; *W. West.* Garsdale!
R. Hodder Valley. Douk Gill, Horton.
N. Upper Nidderdale; *J. S. Wesley.*
W. Bolton Abbey.
A. Malham; *B. Carrington.* Belbank Wood, Bingley; *L. C. Miall.*
T. Shaly rock-edges, by sides of the Abbey Brook tributaries, south-west of Bradfield; *Amos Carr*!

19. Gymnostomum curvirostrum (Ehrh.) Schpr. (*Barbula curvirostris* Lindb.
Very rare.
L. Ingleborough; *W. J. Hooker*, confirmed by *B. Carrington.*
A. Malham district; *W. West.*
W. Arncliffe; *W. West.*

20. Gymnostomum tortile Schwg. (*Mollia tortilis* Braithw.)
A. Malham district; *W. West*, teste *H. Boswell* (*Nat.*, ix, 26)!

21. Gymnostomum microstomum Hedw. (*Mollia microstoma* Lindb.)
L. Ingleborough; *W. West.*
W. Ilkley; *H. F. Parsons.*
A. Malham; *B. Carrington.* Eldwick; *W. West.*
C. Hareley Wood and Heptonstall; *A. Stansfield.*

VAR. **obliqua** N. & Hornsch.
L. Helks Wood, Ingleton; *J. Nowell* (1850).

22. Gymnostomum commutatum Mitt. (*Barbula curvirostris* v. β Lindb.)
L. Discovered to occur plentifully in large masses high up on the Coniston slate rocks at Cautley Spout, Rawtheydale, by *W. West*, April, 1881.

23. Anæctangium compactum (Schl.) Schwaeg.
High damp gritstone rocks, very rare.
L. Whernside; *B. Carrington.*

24. Weisia viridula Brid. (*W. controversa* Hedw.) (*Mollia viridula* Lindb.)
Rocks, walls, and crumbling sandy banks, very common.
Area general; in all the districts.

VAR. **densifolia** Schpr.
A. Malham district; *W. West* (1882).

25. Weisia crispula Hedw.
Alpine rocks, usually on slates or schists, very rare.
L. Howgill fells (slate); *W. West*—amongst a miscellaneous lot of species gathered in that district, the exact locality not known!
N. Marked for Harrogate district in a list furnished to me by *Thos. Hick*; but I fear some error and have seen no specimen.

26. Weisia cirrhata Hedw.
Walls (sandstone), banks, and trees, very frequent.
Distribution general—reported from every district.
Y. Ripon; *W. Brunton.*
N. Pateley Bridge; *B. Carrington.* Spofforth.

W. Nesfield to Bolton; *W. West.* Ilkley; *J. S. Wesley.*
A. Skipton; *W. West.* Shipley; *Id.* Rawden; *B. Carrington.* Hambleton Hough. Kellington; *H. F. Parsons*; etc.
C. Todmorden; *A. Stansfield.* Storthes Hall; *C. P. Hobkirk.*
D. Rose Wood, and other stations on sandstone rocks near Sheffield; *A. Carr.*
T. Tickhill.

27. Rhabdoweisia fugax Hedw.
Arenophilous. Mountain rocks, rare.
L. Howgill Fells; *H. F. Parsons* and *W. West*! On the west side of Ingleborough; *B. Carrington.*
W. Between Bolton Abbey and Barden Tower; *R. Spruce.* Penyghent; *W. Nuttall* and *W. West.*

[**Rhabdoweisia denticulata** Brid. Incognit.
Improperly included in Mr. Hobkirk's 1873 List of West Riding Mosses, on faith of a station (in Lancashire) outside Riding. Also reported for Howgill Fells, but further examination proved the specimens to be *R. fugax.* Inserted by Mr. Hobkirk in a list in *Journ. Bot.*, 1880, p. 19.]

28. Cynodontium Bruntoni (Sm.) B. & S. (*Oncophorus* Braith.)
Montane; rare.
Y. Hackfall, first gathered here in 1801 by W. Brunton, after whom it was named by Bruch and Schimper.
W. Rocks opposite the Strid, Bolton (!) *J. Nowell.*
C. Rag Scouts, near Todmorden, very rare; *A. Stansfield.*

29. Dichodontium pellucidum (L.) Schpr.
Wet rocks and sandy banks, common, except in low grounds; recorded for every district except Trent, from which it may be really absent. On Wharfe found not below Ryther.

30. Dichodontium flavescens Lindb. (*D. pellucidum* var. *serratum.*)
Stony, sandy banks of streams, commoner than the preceding in some districts.
M. Near Dory Castle, Saddleworth district.
R. Slaidburn.
Y. Skell banks; *W. Fowler.*
N. By the Nidd, Cowthorpe.
W. Thorpe Fell, Grassington; *W. West.*
A. Bingley, Denholme, etc.; *W. West.*

31. Dicranella crispa (Ehrh.) Schpr.
Sandy banks, not common.
(W. Sandy bank at Spofforth; *J. S. Wesley*?—a doubtful plant.)
A. Shipley Glen; *W. West* (1879).

32. Dicranella Schreberi (Swartz) Schpr.
C. Only known on Langfield Moor, Todmorden; *J. Nowell* teste *T. Stansfield.*

33. Dicranella squarrosa (Schrad.) Schpr. *Muscus trichoides palustris erectis foliis reflexis*; Ray, Hist., iii, 38 (1704). Confirmed by E. M. Holmes as referring to this species.
Wet stony places and shady shale-banks, frequent.
M. Greenfield district; *J. Whitehead.*
L. Cautley Spout; *W. West.* Garsdale shales, frequent.
R. Hodder Valley.
Y. Dallow Gill; *W. Brunton.*
N. Knaresborough. How Stean beck, Ramsgill, etc.; *J. S. Wesley.* Brimham district; *G. Webster.*
W. Bolton; *S. Gibson.*
A. Baildon; Denholme; *W. West.* Elslack Moor; *W. Whitwell.*
C. Yoredale shale cliff, Hebden Valley (!). Todmorden; *T. Stansfield.* Harden Moss Wood; *C. P. Hobkirk.* North Bierley, *Dr. Richardson* (1702); *Herb. Buddle*, spn. This station and the Rayian description have hitherto been quoted under *Jungermannia trichophylla.*
D. Rivelin Valley, Sheffield; *Amos Carr.*

34. Dicranella cerviculata (Hedw.) Schpr.
Clayey banks, local.
M. Greenfield; *J. Whitehead.*
Y. Cow-mires, Galphay; *Miss Morton.*
W. Bramhope. Rombalds Moor; *B. Carrington.*
A. Thornton; Eldwick; *W. West.*
C. Todmorden; *T. Stansfield.* Marsden; *W. West.*

35. Dicranella varia (Hedw.) Schpr. (*Anisothecium rubrum* Braith.)
Damp fallows and clayey banks, very common. In all the districts.

VAR. **tenuifolium** Bruch. (*fallax* Wils. MS.).
C. Near Meltham; *C. P. Hobkirk.* And probably elsewhere.

36. Dicranella rufescens (Dicks.) Schpr. (*Anisothecium* Braith.)
Clay-banks, fallows, etc.
M. Greenfield; *J. Whitehead.*
N. Spofforth.
W. Bramhope Staircase.
A. Esholt and Idle Woods; *B. Carrington.*
C. Langfield Moor and Pennant Clough, Todmorden; *T. Stansfield.*

37. Dicranella subulata (Hedw.) Schpr.
C. Hareley Wood and Strawbridge near Todmorden; *J. Nowell.*

38. Dicranella heteromalla (Dill.) Schpr.
On sandy and heathy banks in woods, etc., very common.
In all the districts.

VAR. **interrupta** Braith., Mon.
C. Marsden; *W. West* (1880).

39. Dicranum Scottianum Turn.
Only one record, and that possibly erroneous—at any rate never confirmed by present-day search.
A. Idle Wood; *B. Carrington.*

40. Dicranum fuscescens Turn.
Submontane.
Shady rocks, and on rotting bark, rare—a decreasing and ill-understood species, 'duller, more opaque, and more flexuose, in facies' (W.W.).
L. Ingleborough; *W. West.*
N. Pateley Bridge; *G. Webster.*
W. Barden Tower, on trees. Bolton Woods; *S. Gibson.* In rock crevices, Beamsley Beacon; *W. West* (1882)—a robust form.
A. Malham; *B. Carrington.* Bingley; *Id.*
C. Hebden Valley; *A. Stansfield.* Heptonstall; *J. Nowell.*
D. Wharncliffe Woods; *A. Carr*!—perhaps not the type but the var.

FORMA **saxicola** Fergusson—a plant still sub judice as to its affinities, and perhaps 'a rock form' [see *Fergusson* in *Nat.*, vol. iv, p. 116] of *montanum*; has been found in small quantity at Wharncliffe Crags by *H. F. Parsons*, and a similar plant at Pateley Bridge by *J. S. Wesley.*

41. Dicranum scoparium (L.) Hedw. *Fork-moss.*
Woodland and heatherland, generally distributed. In all the districts.

VAR. **paludosum** Schpr.
L. Whernside and Ingleborough; *W. West.*
N. Moors near Ramsgill; *J. S. Wesley.* Wet field, Aketon Bleach Works, near Spofforth!

42. Dicranum majus Turn.
Shady banks in woods, etc., not rare.
L. Helks Wood, Ingleton, and Clapdale; *B. Carrington.*
R. Bolton-by-Bolland; *W. Kirkby.*
Y. Hackfall!
N. Pateley Bridge. Plumpton. Middlesmoor; *W. West.*
W. Bolton! Harewood! Barden; *C. P. Hobkirk.*
A. Bingley; *L. C. Miall.* Shipley Glen; *W. West.*
C. Rag Scouts, Todmorden, and Hebden Valley; *A. Stansfield.*
D. Wharncliffe; *A. Carr*! Bradfield.

43. Dicranum Bonjeanii De Not. (*D. palustre* B. & S.).
Marshy heaths and wet banks at low levels.
M. Greenfield; *J. Whitehead.*
W. Rombalds Moor; *B. Carrington.* Askham Bog! Knavesmire; *W. Whitwell.*
A. Morton Bog near Hambleton; *W. N. Cheesman.* Baildon; *W. West.*
C. Todmorden; *A. Stansfield.* Sharleston Common; *Nat.*, ii, 192.
D. Ackworth; *I. Brown.* Warmsworth; *H. F. Parsons.*
T. Martin Beck Bog; *G. Webster.*

44. Dicranum spurium Hedw.
Damp lowland peat-moors, very rare.
Y. Sawley Moor, near Ripon; *J. G. Baker* (*M.* and *C. Flora*, p. 73). Ripon: *Braith., Mon.*, 152.
D. Hatfield Chase (near Thorne); *J. Bohler.*

45. Dicranodontium longirostre (Web. & Mohr.) Schpr. (*Didymodon denudatus* Brid.)
Sub-alpine turfy banks and on decaying wood, rare.
W. Bolton; *Dr. Wood.*
(A. 'Shipley Glen; *Dr. Parsons*'; *Hobkirk's List.* An error—the plant was another species.)
C. Highgreen Wood, Heptonstall; *J. Nowell*!

46. Campylopus alpinus Schpr. (*Dicranodontium longirostre* var. *alpinum* M.C.)
Very rare.
L. Summit of Ingleborough; *G. E. Hunt* (1867). Not re-found until sent to the compiler by *W. West* amongst other gatherings from Ingleborough in April 1878. The report of its occurrence at Shipley Glen in July 1877, turned out to be an error in nomenclature.

47. Campylopus atro-virens De Not. (*C. longipilus* Brid.)
Damp slaty rocks, very rare.
L. Thornton Force; *W. West.* Cautley Spout, Howgill Fells; *W. West*! Near the villages of Wharfe (*J. Nowell*) and Austwick (*B. Carrington*).

48. Campylopus flexuosus (L.) Brid.
Damp turfy heaths and fir-plantations, rather rare, locally abundant.
N. Ravensgill, Pateley; *J. S. Wesley.*
W. Ilkley; *B. Carrington.*
A. Idle and Rawden Woods; *B. Carrington.* Bingley; *J. Nowell.* Adel fir wood. Bradford district, abundant: *W. West.*
D. Near Sandhill Wood, Goole; *T. Birks, junr.* Bell Hagg and Rivelin Valley; *Amos Carr.*

49. Campylopus paradoxus Wils.
In sheets in the turf, Rombalds Moor, Ilkley; *J. S. Wesley*! Rivelin Valley above Sheffield; *A. Carr*! Horsforth (Moseley Wood); *J. R. Murdoch*! There has been some controversy about this plant, which is possibly only a doubtful species. Mr. Boswell held the Ilkley plant to be *C. flexuosus* only; Dr. Braithwaite, Mr. Fergusson and Mr. Barnes think it certainly Wilson's *C. paradoxus.* It has not yet occurred in fruit. Mr. Carr's plant is identical with the Ilkley one.

50. Campylopus fragilis (Dicks.) Brid.
Sandy and turfy sub-alpine moors and woods, rare.
L. Garsdale, sparingly.
R. Hodder Valley.
A. Otley Chevin; *B. Carrington.* Adel Bog Wood. Brayton; *W. N. Cheesman.* Bingley, frequent; *W. West.*
C. Todmorden cloughs; *T. Stansfield.*

FORMA **densus.**
This form, not admitted as a constant variety by Dr. Braithwaite, is almost as common as the type, occurring with it, from Malham (*J. Nowell*), Todmorden (*T. Stansfield*) and Baildon (*W. West*), etc., to Goole Moor.

51. Campylopus pyriformis (Schultz.) Brid. (= *turfaceus* B. & S.).
Wet turfy banks in woods, ditch-sides, etc., apparently rare.
M. Greenfield; *G. E. Hunt.*
L. Baugh Fell; *W. West.*
Y. Kirby Malzeard Moor; *W. West.*
N. Cranberry Car, Hunsingore!
W. West-wood, Wetherby; *J. S. Wesley.*
A. Moortown, Meanwood! Headingley; *B. Carrington.*
C. Staups Clough, Todmorden; *T. Stansfield.*
D. Rawcliffe Rabbit Hills; *W. West.*

IV. BRUCHIACEÆ.

52. Archidium phascoides Brid. (*Phascum alternifolium* Dicks.)
Very rare.
C. On garden-beds in autumn at Todmorden; *T. Stansfield.* Hareley Wood, Todmorden; *J. Nowell.*

53. Pleuridium nitidum (Hedw.) Rabenh.
Clay-banks and garden-beds, rare.
M. Greenfield; *J. Whitehead.*
L. Bowland Knotts!

R. Near Settle; *J. Windsor.*
A. A fine variety by ponds on Yeadon Moor; *B. Carrington.* Marley, Bingley; *W. West.*
C. Todmorden; *T. Stansfield.*

54. Pleuridium subulatum (Huds.) Rabenh.
Clay-banks, frequent.
R. Settle; *J. Windsor.*
Y. Ripon; *Miss Rose Morton.*
W. Harewood! Collingham! Ilkley; *J. S. Wesley.*
C. Royd Hills and Hareley woods, Todmorden; *T. Stansfield.*
D. Sheffield. Doncaster Race-course; *W. West.*

55. Pleuridium alternifolium (Kaulf.) Rabenh.
W. Wharfe banks, Bolton Park; *B. Carrington.*
C. Wessenden Valley; *C. P. Hobkirk.*

V. LEUCOBRYACEÆ.

56. Leucobryum glaucum (L.) Schpr. *Cuss-moss.* '*Sponges.*'
On mossy heaths, wet heathy pastures, etc., with a wide vertical range, yet not very frequent.
M. Greenfield; *J. Whitehead.*
L. Howgill Fells; *W. West.*
N. Harlow Heath, Harrogate. Great Whernside; *W. West.*
W. Askham Bog! Ilkley, and Otley Chevin (!) *B. Carrington.*
A. Hill-side below Norton Tower; *W. Whitwell.* Shipley Glen; *W. West.* [Adel Black-moor, up to about 1870—now extinct.]
C. Heptonstall. Dungeon Wood, Huddersfield; and Marsden moors; *C. P. Hobkirk.*
D. Goole moors; *H. F. Parsons.* Bradfield Moor!

VI. SELIGERIACEÆ.

57. Seligeria Donii (Sm.) C. Muell. (*Anodus Donianus* B. & S.)
W. Cowside Beck, off Littondale; *W. West.*
C. Highgreen Wood and Mitholme Clough, Heptonstall; *J. Nowell.* Hardcastle Crags; *G. E. Hunt* (1867).

58. Seligeria pusilla (Ehrh.) B. & S.
Shady rocks, mostly on limestones, rare.
L. Howgill and Dent Fells; *G. Pinder.* Dentdale; *W. West* (1879). Gaping Gill Hole, Ingleborough; *B. Carrington.* Clapham and Ingleton; *W. West* and *W. Nuttall.*
W. Litton and Cowside-beck Dale, near Arncliffe; *W. West*; *J. Whitehead* and *Ashton* = Heseltine Gill; *W. West.*
A. Malham Cove (!) *J. Nowell*; *W. West.* Gordale; *G. E. Hunt.*

59. Seligeria acutifolia Lindb. var. **longiseta** Lindb.
Gathered as *S. pusilla* on dry limestone rocks at Arncliffe in 1868 (*Nat.*, 1879, iv, 108) by *J. Whitehead.* At first referred to the type, but Dr. Braithwaite says (*Mon.* 118) *all* the British specimens have the obliquely-beaked lid to the capsule.

60. Seligeria recurvata Hedw. (*S. setacea* Wulf., Lindb.)
Damp shady rock-faces, mostly on sandstone, not rare, but readily overlooked.
M. Greenfield; *J. Whitehead* (1867)—on stones near the stream.
L. Clapham Cave. Dent and Howgill Fells; *W. West.* Great Blake Gill, Whernside! *W. Nuttall* and *W. West.*
Y. Rocks near Copgrove; *J. Dalton* (1803).
N. Shady grit-rocks by the lakes at Plumpton.
W. Littondale, with the next; *J. Whitehead* and *Ashton* (1868). Bolton; *B. Carrington.*
C. On Iron cobs in Hareley Wood; *J. Nowell* (1858). Hardcastle Crags, Hebden Valley; *G. E. Hunt* (1867).

61. Seligeria trifaria (Brid.) Lindb. (*S. tristicha* B. & S., Bry. Eur.)
On dripping limestone rocks, very rare. Montane abroad.
W. Littondale; *J. Whitehead* and *Ashton* (1878).

[**Seligeria paucifolia** Dicks. (*S. subcernua* Schpr.) Incognit.
Only one old record, never confirmed. On ('brick rubbish'; *Braithwaite, Moss Flora*, 120) a brick wall, near Wetherby, 1801; *J. Dickson.*]

62. Brachydontium trichodes (W. & M.) Fürnr. (*Brachyodus* B. & S.)
Wet grit rocks.
M. Greenfield; *Hobson* (1835) = On 'rocks at Ashway Gap'; *J. Whitehead.*
L. Black Force, near Tebay (!) *J. E. Barnes.*
N. Guy's Cliff, Pateley Bridge; *J. Nowell.*
C. Holme Moss, on sandstone rocks; *C. P. Hobkirk.*

63. Blindia acuta (Huds.) B. & S. (and FORMA **trichodes** Braith., olim).
Range: 450—2,000 feet.
Dripping grit rocks, rare.
L. Whernside, at 2,000 feet! Garsdale.
W. In a gill near Outershaw, Langstrothdale head (1885). Bolton; *S. Gibson* (seems a very low level for this, but other things which occur higher up (e.g. *Sesleria*), come as low down on streams rising among the hills, and liable to great spates).
C. Wessenden Valley; *J. Whitehead.*

Obs.—The tall decumbent form leafless at base of stem (*trichodes* of Braith., olim) is included in the above; both occur near together generally, it being a question of situation exposing the growing plants to much or little stress from falling water.

VII. POTTIACEÆ.

64. Sphærangium muticum (Schreb.) Schpr.
W. Between Ilkley and Addingham; *L. C. Miall.*
A. Walls near Esholt; *B. Carrington.* Meanwood, in corn stubble; *J. Abbott* (1871).
C. Stoneyroyd, Todmorden; *J. Nowell.*

65. Phascum cuspidatum Schreb.
In fallow fields and on clayey banks, common and generally distributed. Recorded from every district.

66. Phascum bryoides Dicks.
Apparently rare—few records.
Y. Ripon; *Miss Morton.*
A. Malham; *J. Windsor.*
C. Holmfirth moors; *C. P. Hobkirk.*

67. Phascum curvicollum Hedw.
A. Castleford; *Dr. Wood.*
C. Todmorden; *J. Nowell.*
D. Pontefract; *Dr. Wood.* Ackworth; *I. Brown.*

68. Phascum rectum With.
Fields, and road-side banks bare of vegetation, rare.
(Y. Ripon; *Miss Morton.*)
W. Near York; *R. Spruce.* (Wetherby; *J. S. Wesley.*)
A. Rawden and Esholt; *B. Carrington.*
C. Todmorden; *C. P. Hobkirk.*
D. Pontefract; *G. E. Hunt* (1867).

69. Pottia truncatula (L.) Lindb.
Fields and banks, common.
M. Greenfield; *J. Whitehead.*
W. Wetherby; *J. S. Wesley.* Stockeld.
A. Meanwood. Keighley; *W. West.*
C. Huddersfield; *C. P. Hobkirk.*
D. Ranmoor, Sheffield; *Amos Carr.*

70. Pottia intermedia (Turn.) Fuernr.
W. Wetherby; *J. S. Wesley* (1878) teste *H. Boswell.*
A. Rawden and Guiseley; *B. Carrington.*

71. Pottia Heimii (Hedw.) Fuernr.
Y. Quarry Moor, Ripon; *Miss Morton.*
W. Thorparch Bridge; *R. Spruce*!
D. Goole; *H. F. Parsons.* Doncaster; *Amos Carr.*

72. Pottia lanceolata (Hedw.) C. Muell. (*Anacalypta lanceolata* Röhl.)
W. Wetherby; *J. S. Wesley.* Kirk-Deighton quarry, by the kiln!
A. Skipton; *B. Carrington.*
D. Ackworth; *J. W. Watson.* Pontefract; *B. Carrington.*

73. Pottia Starkei (Hedw.) C. Muell.
Very rare.
W. Dringhouses brickponds; *R. Spruce.*

Var. **Davallii** (Sm.) Lindb. (*P. minutula* Schwaeg.)
Fields, rare, local.
W. Tadcaster; *R. Spruce.* Wetherby; *J. S. Wesley* ('var. *conica*.')
A. Rawden and Apperley; *B. Carrington.* Barlow, Selby; *W. N. Cheesman.* Newland; *H. F. Parsons.*
D. Pontefract; *J. Nowell.*

74. Didymodon rubellus (Roth.) B. & S. (*Barbula rubella* Mitt.)
Rotten rock, crumbling walls and stony banks, with a xerophilous restriction, very common on the limestones, rarer elsewhere.
Distribution general; reported from every area, but rather rare on the grits of Calder district.

75. Didymodon luridus Hornsch. (*Barbula lurida* Lindb.)
L. Ingleborough; *W. West*!
W. Lime-rocks in cutting, west of Wetherby (!) *J. S. Wesley.*
C. Sharleston Common; *Y. N. U. Excurs.* (1877).
D. Wentbridge; *H. F. Parsons.*

76. Didymodon flexifolius (Dicks.) H. & T. (*Leptodontium flexifolium* Hampe.)
Rocky places in turfy districts.
W. On thatch near Addingham; *B. Carrington.*
A. Malham Moor; *B. Carrington.* Bell Bank, Bingley, and Baildon; *W. West.*
C. Eaves Wood, near Heptonstall; *J. Nowell.* Rocks at bottom of Hudson Moor, Todmorden, rare; *T. Stansfield.*
D. Ackworth; *I. Brown* (?).

Var. **gemmascens** Mitt.
L. Ingleborough; *B. Carrington.*

77. Didymodon cylindricus (Bruch) B. & S.
N. Added to the flora of the Riding by *J. S. Wesley*, who discovered it on January 7th, 1878, growing on a 'loamy hedge-bank at Spofforth near Wetherby.' Name certified by Mr. H. Boswell.

78. Didymodon sinuosus (Wils.) Schpr. (*Tortula sinuosa* Mitt.)
Very rare.
(W. Near Wetherby?; *J. S. Wesley.* Doubtfully reported for the Wetherby district in 1877, from 'a wood near Collingham' by *C. P. Hobkirk.*)
D. Endcliffe Valley, Sheffield; *Amos Carr* (1876) *spn.*!

79. Eucladium verticillatum (L.) Br. (*Weissia* Brid.; *Mollia* Braith.)
L. Helks Wood, Ingleton! *B. Carrington*; *W. West.*
N. Knaresborough (!) *J. Nowell*; confirmed (if a Nowellian record can be said to require it) by Messrs. Hobkirk and West as recently as 1885.
W. Rocks by Wharfe, Bolton; *B. Carrington.* Thorparch Cliffs!
A. Gordale; *B. Carrington.*
D. Went-vale; *J. W. Watson*; confirmed by *H. F. Parsons.*
T. Roche Abbey! The '*Weissia calcarea*' (Seligeria) given for this locality in the list at end of Aveling's *History* was this species in all probability. Notice of it as an excluded species was omitted on page 543.

80. Ditrichum homomallum (Hedw.) Braith. (*Trichostomum* B. & S.)
Sandy banks and grit rocks, rare; apparently a decreasing species.
L. Upper Garsdale, 1886!
R. Slaidburn district!
W. Ilkley; *J. S. Wesley.*
A. Idle and Esholt Woods; *B. Carrington.*
C. Todmorden; *J. Nowell*; *T. Stansfield.* Hebden, and Wessenden Valleys; *C. P. Hobkirk.*

81. Ditrichum flexicaule (Schleich.) Braith. (*Leptotrichum* Hampe.)
Range: 100—2,000 feet.
Xerophilous. Dry banks, one of the commonest species upon the limestones, rare off them.
L. Dentdale. Garsdale. Clapham. Whernside crags (!) *W. West.*
R. Linn Gill. Moughton Fell. Settle.
Y. Ripon and Copgrove.
W. Cam Fell; *J. G. Baker.* Grassington; *C. P. Hobkirk.* Kirk Deighton. Wetherby and Tadcaster quarries.
A. Malham. Banks between Skipton and Bolton; etc.
T. About Roche Abbey.

Var. **densum** B. & S.
L. Ingleton Fells. Clapham; *W. West.*
W. Arncliffe; *W. West.*
A. Malham Moor; *C. P. Hobkirk.* Gordale; *W. West.*

82. Trichostomum tophaceum Brid.
L. Ingleborough; *W. West.* Ingleton; *W. Fowler.*
Y. Old quarry, South Stainley; *W. Fowler.*
W. Otley; *B. Carrington.* Pool! Wetherby; *J. S. Wesley.*

A. Gordale; *J. Nowell.* Rylstone Road, Skipton; *W. Whitwell.* Bellbank woods, Bingley; *L. C. Miall.* Baildon; *W. West.* Newthorpe road, Sherburn; *W. West.* Monk Fryston; *H. F. Parsons*; *W. N. Cheesman.*
C. Highgreen Wood, Todmorden; *J. Nowell*; *T. Stansfield.*
D. Went-bridge.

83. Trichostomum mutabile Bruch.
L. Whernside and Clapdale; *B. Carrington.* Common at Ingleton; *W. West.*
R. Linn Gill.
W. Arncliffe; *W. West.* Near the 'Strid,' Bolton; *B. Carrington.*
A. Malham Cove; *J. Nowell*—large form re-found in 1879 by *W. West.*
D. Brockerdale, Went-bridge; *H. F. Parsons*; *Nat.*, iii, 10 (vide *Trans. Y. N. U.*, *Series* E, p. 35).

84. Trichostomum crispulum Bruch.
Rock-fissures, rare.
L. Ingleborough; *B. Carrington*—this was probably var. *nigroviride* Braith.
A. Gordale; *J. Nowell.* Malham; *H. F. Parsons*, *spn.* (1877).

Var. **nigro-viride** Braithw.
L. Near summit of Ingleborough at 2,000 feet; *J. Nowell* (1857). 'Yoredale crags of'; *W. West* (1882).

Var. **elatum** Schpr.
L. Ingleton; *W. West* (1882).
A. Gordale; *W. West* (1880).

85. Barbula rigida Schultz. (*Tortula stellata* Schreb.)
Mud-topped walls.
Y. Near Ripon; *W. Brunton.*
N. Knaresborough; *W. Brunton* (1805).
W. Addingham; *J. G. Baker* (1859). Hob Moor, near York; *H. Ibbotson*; *R. Spruce.*
C. Castleford, abundant; *B. Carrington.*
D. Pontefract; *J. B. Wood.*
T. Maltby. Roche Abbey; *J. H. Aveling* in *Hist.* (1870).

86. Barbula ambigua B. & S. (*T. ericæfolia* [Neck.] Lindb.)
Mud-topped walls and dry banks in limestone districts.
W. Wetherby (!) *J. S. Wesley.* Barkston-Ash; *J. Nowell.* Addingham; teste *H. Boswell.* Menston, and near Barden; *W. West.*
A. Cottingley; *W. West.* Castleford (!) *J. Nowell.*
D. Pontefract; *J. B. Wood.* Knottingley; *J. G. Baker.* Askern!
T. Roche Abbey; *H. F. Parsons.*

87. Barbula aloides Fuernv.
On grit rocks and mud-capped walls.
L. Ingleborough; *Hudson.*
R. Stackhouse, near Settle; *A. O. Black.*
N. Near Ripley; *J. G. Baker.*
A. Monk Fryston; *H. F. Parsons* and *W. N. Cheesman.*
C. Rag Scouts, Todmorden; *T. Stansfield.* Grimescar Wood, Huddersfield; *C. P. Hobkirk.*
D. Wenthill, near Pontefract; *J. Nowell.*

88. Barbula pusilla (Hedw.) Lees. (*Tortula pusilla* Mitt.; *Pottia cavifolia* Ehrh.; and *Gymnostomum ovatum* Hedw.)
Mud-topped walls and banks, rare, though locally abundant.
First record: Wood, 1849.
A. Clay-bank between Leeds and Bramhope; *James Abbott*! Drying-up mud of Adel Lower Dam in 1870, with *Limosella.* Garforth; Monk Fryston; *H. F. Parsons.* Knottingley; *J. G. Baker.*
D. Pontefract and Darrington; *J. B. Wood.* Eastoft, near Goole; *H. F. Parsons.*
T. Roche Abbey; *J. H. Aveling* in *Hist.* (1870).

89. Barbula lamellata (Lindb.) Lond. Cat., 2. (*B. cavifolia* var. δ *gracilis.*)
With the last, and similarly limited in its distribution.
A. Fairburn! Castleford; *J. Nowell.*
D. Ackworth; *I. Brown.* Buffit Farm, Pontefract; *J. B. Wood.*

90. Barbula muralis (L.) Hedw.
Walls, thatches, etc., ubiquitous up to 1,800 feet.
Distribution general; all the districts.

Var. **rupestris** Schultz.
Common in the mountain limestone area.
L. Dentdale, Whernside.
R. Settle crags, etc.
N. Old quarry, Wormald Green (*W. Fowler*), and Kirk Deighton!

91. Barbula unguiculata (Dill.) Hedw.
Walls, rocks, banks, generally distributed, although hardly to be called common everywhere.
Reported from numerous stations, in every district.

92. Barbula fallax Hedw.
Grassy and earthy banks, waste ground by road-sides, etc., very common.
Generally distributed; every district.

Var. γ **brevifolia** Brid.
Wall, Bradley Wood, near Huddersfield, in fruit; *C. P. Hobkirk.*

93. Barbula recurvifolia Schpr. (*B. fallax* var. δ Carrington's Flora.)
Walls and clayey banks, rare.
L. Ingleborough; *J. Nowell.*
R. Lime rocks, Douk Gill, Horton, long drawn-out female form.
W. Bolton Woods; *W. West.* Wetherby, male plant, wall-tops; *J. S. Wesley.* Litton; *W. Nuttall* and *W. West.*
A. Malham Moor (!) *J. Nowell.* This, which like the Horton plant, is a large form, and in facies mimics *Zygodon Nowellii*, is the *Didymodon aquaticus* of Nowell's Suppl. to Baines (p. 182); and in Hobkirk's 1873 List of W. R. Mosses was erroneously assigned to *Geehebia cataractarum* = *Tortula hibernica* Mitt., i.e., Mitten's Irish *recurvifolia.*
D. Dykemarsh near Thorne; *H. F. Parsons.*

[**Tortula gracilis** H. & G. Incognit.
Perhaps the '*Tortula brevifolia*' reported by *Dillenius* was the Ingleborough *recurvifolia.*]

94. Barbula rigidula (Dicks.) Schpr.
Walls and rocks, occasional, and widely distributed.
L. Bridge, Sedbergh; *W. Whitwell*! Whernside; *W. West.*
W. Bolton Abbey; *B. Carrington.*
A. Near Guiseley; *B. Carrington.*
D. Pontefract; *H. F. Parsons.*
T. Roche Abbey; *H. F. Parsons.*

95. Barbula spadicea (Mitt.) L. C., 2. (*Trichostomum rigidulum* Wils.)
On wet stones or rocks. Wesley the first recorder of it.
L. Whernside, to 1,800 feet! Helks Wood, Ingleton; *J. Nowell*!
W. On stones in river above Ilkley, in fine fruit; *W. West.* On large stones by river below Flint Mill, Wetherby (!) *J. S. Wesley.*
A. Gordale; *W. West.*
D. Dykemarsh, Thorne; *H. F. Parsons*!

96. Barbula cylindrica (Tayl.) Schpr. (*T. insulana* De Not.)
Distribution hardly ascertained as yet. Perhaps overlooked as *B. fallax*, which it resembles in facies.
W. Harewood Park, sunk fence wall (1871).
A. Middlehouses; Malham; and Gordale; *W. West.*
C. Grimescar Wood, Huddersfield; *C. P. Hobkirk* (1876).

97. Barbula vinealis Brid.
Common on walls where it occurs at all, but not generally distributed.
L. Ingleborough; *B. Carrington.*
R. Horton; *B. Carrington.*
W. Bolton Bridge; *J. Nowell*; rocks near the Abbey; *W. West.* Ilkley; *B. Carrington.* Tadcaster; *J. S. Wesley.*
D. Upper Bradfield; *Amos Carr*!

98. Barbula Hornschuchiana Schultz.
Banks, fields, wall-tops, rare.
W. Wetherby, wall of Manor House garden; *J. S. Wesley*! Wet fields beyond Hob-moor, York, 1879; *Geo. Webster.*
D. Askern; *M. B. Slater* (1886). Doncaster race-course, both male and female plants; *W. West* (1878).

99. Barbula revoluta Schwg.
Walls, etc., rare.
L. Ingleton. Clapham; *W. West.* Upper Garsdale!
R. Settle; *W. West.*
W. Bramham. Arncliffe; *W. West.*
A. Esholt; *W. West.* Kellington; *H. F. Parsons*!
C. Todmorden; *T. Stansfield.*

100. Barbula convoluta Hedw.
Bare ground in stony places, and walls, common.
L. Ingleton; *B. Carrington.*
R. Tosside Chapel, Wigglesworth.
N. Spofforth!
W. Stutton and Wetherby quarries and rail-banks! Addingham; *W. West.*
A. Monk Fryston and Cawood; *H. F. Parsons.*
C. Norland Moor, North Dean; *H. F. Parsons.* Todmorden; *John Nowell.* Almondbury; *C. P. Hobkirk.*
D. Thorne; *H. F. Parsons.* Penistone!
T. Roche Abbey; *H. F. Parsons.*

101. Barbula tortuosa (L.) W. & M.
Montane and Sub-xerophilous. On rocks and walls, local, but abundant all over the scar-limestone area, and not rare on the calcareous shales.
L. Sedbergh. Dentdale. Whernside. Bentham. Clapham; *W. West.* Ingleborough; *B. Carrington.* Wethercote, in fruit; *W. Fowler.*
R. Linn Gill, Ribblehead. Horton.
W. Buckden. Kilnsey; *R. Teesdale* (1789). Burnsall and Ilkley; *B. Carrington.* Jackdaw Crag, Tadcaster; *J. S. Wesley*!
A. Common on the limestones above Skipton.
C. Keb Clough, Todmorden; *T. Stansfield.*
D. Brockerdale, Went-vale, and Conisborough; *H. F. Parsons.*
T. Maltby.

102. Barbula subulata (L.) P. de Beauv.
Sandy banks and walls, common. Very fine by river Nidd at Cowthorpe, and at Tickhill.
Distribution general; reported from every area.

103. Barbula lævipila (Brid.) B. & S.
Tree-trunks and walls occasionally. Has a rather low vertical range.
R. Walls, Settle; *J. Nowell* (1864)!
Y. Walls, Mickley and Sleningford (!) *H. H. Slater.*
N. Knaresborough; *W. West.*
W. Bolton Bridge (!) *J. Nowell.* Ilkley; *B. Carrington.* Harewood (orchard trees).
A. Malham; *B. Carrington.*
D. Rawcliffe; *H. F. Parsons.*
T. Apple-tree trunks, Roche Abbey.

104. Barbula latifolia (Bruch.) B. & S.
Stream-sides where liable to be overflowed, but fruiting seldom.
W. Banks of Wharfe near Tadcaster; *J. Nowell.* West banks of Ouse near York; *R. Spruce.* Bolton Woods; Burley; *W. West.*
A. Esholt (this is frequent, associated with *Leskea polycarpa* in both Aire and Wharfe valleys); *W. West.*

105. Barbula ruralis (L.) Hedw.
Thatches, sandy wall-tops, banks, rarer than the next.
L. Bentham. Whernside. Dentdale. In fine fruit on plants three or four inches long at Wethercote; *W. West.*
R. Settle. Ribblehead.
N. Harrogate; *T. Hick.*
W. Bolton Abbey; *B. Carrington.* Arncliffe; *W. West.*
A. Skipton; *B. Carrington.* Malham; *C. P. Hobkirk.*
D. Rawcliffe and Reedness; *H. F. Parsons.*

106. Barbula intermedia (Brid.) Milde.
Walls, roofs, thatches, faces of lime rock, etc., common.
L. Ingleton.
R. Slaidburn. Horton-in-Ribblesdale.
N. Nidd banks above Pateley; *C. P. Hobkirk.* Hunsingore.
W. Kilnsey. Bramham; *J. S. Wesley.* Wetherby. Jackdaw Crag Quarry, Stutton. Steeton Hall; *W. Whitwell.*
A. Malham; *W. Wilson*, fide *M. J. Berkeley*, 251. Bell Busk; *W. West.*
D. Wentbridge; *H. F. Parsons.*
T. Roche Abbey (!) *J. H. Aveling* in *Hist.*—sub nom. *ruralis.*

107. Barbula papillosa (Wils.) C. Muell.
On trees.
N. Knaresborough; *W. West* (with *Lecidea canescens*).
W. Between Bolton and Ilkley, and Harewood; *B. Carrington.* Wetherby; *W. West.*

108. Ceratodon purpureus (L.) Brid.
Banks, walls, bare heaths, etc., very common, and with a wide vertical range, descending to Eastoft, and ascending to 2,200 feet.
Generally distributed; in every district.

109. Ceratodon conicus (Hampe) Lindb.
D. Wet exposed rocks, Bell Hagg, near Sheffield; *Amos Carr* (1877 and 1880). The only record as yet, but will be proved to occur in most of the elevated heathery areas, I think.

110. Distichium capillaceum (L.) B. & S.
Calcareous rocks, rather rare.
M. 'Greenfield: *J. Tinker*, 1822'; *C. P. Hobkirk.*
L. Ingleborough; *H. Boswell.* Garsdale!
R. Thorns Gill, Gearstones! Selside-in-Ribblesdale.
W. Bolton Woods; *B. Carrington.* Cowside-beck Dale, Arncliffe; *W. West.* Ilkley; *B. Carrington.*
A. Malham Moor (!) *B. Carrington.*

VIII. CALYMPERACEÆ.

111. Encalypta vulgaris Hedw. *Extinguisher Moss.*
Walls and rocks, frequent, but xerophilous in its restrictions.
L. Clapham; *James Abbott.*
N. Knaresborough!
W. Strid, Bolton; *G. Webster.* Wetherby (!) and Tadcaster; *J. S. Wesley.*
A. Malham; *John Nowell*; and *W. West* (with var. *pilifera*). Skipton; *W. West.* Newthorpe, Sherburn; *W. Whitwell.*
D. Brockerdale, Smeaton, and Conisborough; *H. F. Parsons.*
T. Roche Abbey; *J. Bohler*; and *J. H. Aveling* in *Hist.*

112. Encalypta rhabdocarpa Schwg.
Montane. Very rare.
L. Ingleborough; *W. J. Hooker* (1800). 'On dead moss near the summit of Ingleborough'; *J. Nowell.* Ingleborough, rock crevices at 2,000 feet, where *P. catenulata* grows, but a rather doubtful gathering, since barren, although the leaf-cells, etc., certainly agreed with book description; *W. West.*

113. Encalypta ciliata Hedw.
Rocks, locally common, but Montane and Xerophilous in its restrictions.
First record: Teesdale, 1789.
L. Ingleborough; *R. Teesdale*; *Gough's Camden*; *John Nowell*! Whernside, Kingsdale; *B. Carrington.* Howgill Fells; *G. Pinder.* N.E. end of Whernside. Source of Great Blake Ghyll! cum *W. West.*
R. Settle Crags.
W. Arncliffe. Buckden Scars. Outershaw.
A. Gordale and Malham Cove (!) *J. S. Wesley*; etc.

114. Encalypta streptocarpa Hedw.
Xerophilous. Limestone walls and rocks, very common in the calcareous tracts, rare off them, but reported from Grimescar Wood, Huddersfield, by Mr. Hobkirk. In fruit in old quarry south of Tadcaster, April, 1878; *J. S. Wesley.*
In all the districts except Mersey; localities too numerous to deserve separate mention.

IX. GRIMMIACEÆ.

115. Grimmia apocarpa (L.) Hedw.
On rocks, walls, etc., common, and varying much in facies.
Distribution general; reported from every district except the Mersey and the lower Colne Valley areas.

Var. **rivularis** Brid.
Common among the western hills.
L. Baugh Fell; *W. West.* Clapham, Ingleton; *W. West.*
R. Ribblehead.
W. Outershaw. Arncliffe. The Strid at Bolton; *W. West.*

116. Grimmia conferta Funk.
L. Discovered on hills above Dent, Dentdale, on crumbling Yoredal shales; *J. M. Barnes* (reported by *G. Stabler* in *Nat.*, iv, 113).

117. Grimmia pulvinata (Dill.) Sm.
Common on walls, etc., in all save the populous manufacturing districts. Area general.

[**Grimmia orbicularis** B. & S.
Reported by *Dr. H. F. Parsons* from walls near Hawes Junction railway-station, but there is some little doubt whether the locality was really within the limit of Mid-West Yorkshire.]

118. Grimmia funalis (Schwg.) B. & S. (*G. spiralis* H. & T.)
High grit rocks, west of Ingleborough; *B. Carrington*; but not confirmed recently, I believe.

119. Grimmia trichophylla Grev.
Walls and rocks, very rare, but why so is not easily explicable. Montane and Arenophilous.
R. On slate-rock plates near Horton-in-Ribblesdale.
N. Walls near Pateley Bridge; *J. Nowell.*
W. Old walls near Barden Tower (!) and Bolton Woods; *J. Nowell.*

120. Grimmia Doniana Sm.
Montane. Moorland rocks and walls, not rare at high elevations.
L. Grisedale Beck to Baugh Fell; *H. F. Parsons.* Wall on the summit ridge of Whernside, in plenty! On slates, Ingleborough, very abundantly; *W. Brunton* (1805). Winder Fell, Sedbergh, abundant; *W. West*, 1879.
(A. Walls near Esholt Wood; *B. Carrington.* ? An error.)
C. Stansfield Moor, now seldom seen; *A. Stansfield.*

121. Grimmia ovata W. & M.
R. Wall high up the Hodder Valley above Slaidburn! cum *W. Todd.* Not previously found in Yorkshire, but the species has a very wide vertical range, occurring low down in Leicestershire, and up to 3,500 feet in Clova.

122. Rhacomitrium aciculare (L.) Brid.
Wet rocks, stones in becks, etc., frequent.
M. Greenfield; *J. Whitehead.*
L. Dentdale head, Whernside. Baugh Fell (*W. West*) and Kingsdale. Ingleborough.
Y. Sawley Moor; *W. Brunton* (1805).
N. Oak beck, Harrogate. Ravensgill, near Pateley; *Geo. Webster.*
W. Langstrothdale, Outershaw. Bolton and Ilkley; *H. F. Parsons.* Farnley, near Otley; *H. F. Parsons.*
A. Malham Tarn; *H. F. Parsons.* Shipley Glen; *W. West.*
C. Hudson Clough, Hebden Valley (!) etc.; *A. Stansfield.* Norland Moor; *H. F. Parsons.*
D. Little Don Dale. Bradfield; *A. Carr.*

123. Rhacomitrium protensum A. Braun.
Stones in streams, and wet rocks.
L. Austwick; *J. Nowell.* Ingleborough, in fruit; *W. West.*
W. Bolton Woods; *B. Carrington.*

124. Rhacomitrium sudeticum (Funk.) B. & S.
On high grit scars, very rare, and perhaps some misnomer.
L. Foal-foot, Ingleborough; and Clapdale; *B. Carrington.*

[**Rhacomitrium ellipticum** Turn. Error.
Crimsworth Dean, Todmorden; *S. Gibson*; *Baines* (1840). 'Is erroneous'; *J. Nowell* in *Baker's Supp.*, 1854.]

125. Rhacomitrium heterostichum (Hedw.) Brid.
Damp banks and rocks, rare.
L. Whernside, at 2,400 feet (var. *alopecurum*)! Austwick; *J. Nowell.* Sedbergh; *W. West.*
R. Near Settle and Giggleswick; *J. Windsor.*
W. Ilkley; *H. F. Parsons.* Barden Tower; *J. Nowell.*
A. Near Malham Tarn; *Y. N. U. Excurs.*
C. Red Brook, Marsden; *C. P. Hobkirk.*

126. Rhacomitrium fasciculare (Schrad.) Brid.
Damp mountain rocks and banks, frequent.
M. Greenfield; *J. Whitehead.*
L. Dentdale. Whernside. Garsdale. Howgill Fells; *W. West.*
R. Blea-moor, stream-side. Penyghent.
N. Pateley Bridge; *J. S. Wesley.*
W. Cam and Outershaw Fells, Langstrothdale head!
A. Allerton; *W. West.*
C. Todmorden; *A. Stansfield*; *H. F. Parsons.*
D. Bell Hagg, Sheffield; *Amos Carr.*

127. Rhacomitrium lanuginosum (Hedw.) Brid.
Montane. Range: 2,400 feet down to 20 feet.
Walls and rocks, very common among the hills of the N.W. half of the Riding. Stations too numerous to mention; rarer in the S.W. half.
D. Don-head moors; *A. Carr.* Dykemarsh near Thorne, on grit coping of a wall of an old clough, perhaps introduced with stone from the moors; *H. F. Parsons.*

128. Rhacomitrium canescens (Hedw.) Brid.
Rocks, sandy heaths, etc.
L. Whernside slopes at 1,800 feet. Baugh Fell, in fruit; *W. West.*
W. Buckden. Ilkley.
A. Steeton and Rombalds Moor; *B. Carrington.* Malham; *W. West.*

129. Ptychomitrium polyphyllum (Dicks.) B. & S.
L. Helks Wood, Ingleton; *B. Carrington.* Clapham; *James Abbott.* Cautley Spout; *W. West.*
N. & Y. Brimham rocks and Sawley Moor; *R. Teesdale*; *W. Brunton.*
W. Bolton Fells. Burley-in-Wharfedale; *W. West.* Woodhall, near the Wharfe.
A. Esholt Wood; *B. Carrington.* Malham Tarn; *Y. N. U. Excurs.*
C. Hebden Valley.
D. Penistone Moors; *Amos Carr.* Dykemarsh, with No. 127—adventive; *H. F. Parsons.*

130. Amphoridium Mougeotii (B. & S.) Schpr. (*Zygodon* H. & T.)
Damp rocks, rare. Montane.
M. Greenfield; *J. Whitehead.*
L. Brant Fell, Howgills; *W. West.* Garsdale! Ingleborough; *J. Nowell.*
R. Penyghent; *Henry Ibbotson.*
N. Great Whernside; *W. West.*
W. Near Strid, Bolton Woods; *H. Ibbotson.* Outershaw.
A. Gordale; *J. Nowell.*
C. At Heptonstall, Cliviger and Rattan Clough, near Todmorden; *J. Nowell*; *T. Stansfield.*

131. Zygodon viridissimus (Dicks.) Brid.
On tree-trunks, rare.
N. Cowthorpe !
W. Near the Strid, Bolton; *B. Carrington.* On old hawthorns at Pool ! Aberford !
A. Skipton; *J. Nowell.* Malham; *W. West.*
D. Pontefract district ! Rawcliffe, Fockerby, and Eastoft; *H. F. Parsons.* Askern; *M. B. Slater* (1886).
T. Roche Abbey !

132. Zygodon Nowellii Schpr. (*Z. gracilis* Wils. MS.)
L. On Ingleborough, April, 1878; *W. West* !
W. Arncliffe, 1879; *J. Whitehead.*
A. Walls near Malham Tarn; *W. Wilson,* 1861; *J. Nowell.*

133. Ulota Drummondii (Grev.) Brid.
Montane. On trees, rare.
L. & R. Hacker Gill at the head of Dentdale ! and about Widdale Fell; *J. G. Baker.*
Gibson's 'High-green Wood' moss, given as this in *Baines' Flora,* p. 135 (as well as his *O. Hutchinsiæ* from same place), was a variety of *O. saxatile.*

134. Ulota Bruchii Hornsch.
Tree-trunks, the commonest of the montane bark forms, but rare in the lowlands.
L. Dentdale head ! Rawthey dale; *W. West.* Great Blake Gill, Whernside ! Helks Wood, Ingleton; *B. Carrington.*
R. Ash-trees, Douk Gill, Linn Gill, etc., common.
Y. Fountains Abbey.
W. Bolton Woods and Ilkley; *B. Carrington.*
A. Gordale and Malham Cove; *J. Nowell.* Malham (!) *W. West.*

135. Ulota crispa (Hedw.) Brid.
Trunks of trees; very rare with us. I have not seen the true plant.
L. Clapdale; *B. Carrington.* Dent; *W. West.*
Y. Studley Park; *B. Carrington.*
W. Bolton; *J. Nowell.* Parlington Park; *J. Bohler.* (Wetherby; *J. S. Wesley*).
A. Gordale; *J. Nowell* (this was certainly *U. Bruchii*).

136. Ulota intermedia Schpr.
A recently-separated species, 'frequent on tree-boles,' perhaps the very commonest in this puzzling genus, now somewhat simplified by the separation of this. Hitherto confused with, and recorded indifferently as, *U. Bruchii* and *U. crispa*; and most probably some of the

preceding records for both those species belong to this one. Dr. Carrington, in 1862, giving Bolton and Studley for both, says of *U. Bruchii,* 'more abundant than *U. crispa,*' it is only amongst the hills. Mr. Boswell remarks, 'in *U. crispa* the leaves are very closely and much curled and crisped; in *U. crispula* and *U. intermedia* considerably so; in *U. Bruchii* only slightly.' Yet the latter evidently including *U. intermedia*) is defined in *Hobkirk's Synopsis,* Ed. 1, as having leaves 'contorted when dry, sometimes wavy'; but they are separated and more clearly defined in Ed. 2.
W. West Woods, Bramham. Parlington Park.
A. Stainor Wood, Selby; *W. N. Cheesman.*
T. Roche Abbey Woods and Kings Wood.

[**Ulota Hutchinsiæ** (Sm.) Schpr. Incognit.
The only old record, 'High-green Wood, Heptonstall; *S. Gibson,*' was an error, a form of *O. saxatile* occurring there being intended.]

137. Ulota phyllantha Brid.
On trees, very rare; and perhaps the (barren) plants gathered were mostly young or dwarfed *O. Lyellii.* All the *Ulotæ* seem to me to be *decreasing* species in the West Riding.
L. & W. Clapham, and between Bolton and Ilkley; *B. Carrington.*
R. or A. N.W. side Malham Tarn; *J. Nowell.* Not recently confirmed.

[**Orthotrichum anomalum** Hedw. Incognit.
It is uncertain whether we have this species or the next only.]

138. Orthotrichum saxatile Brid.
Xerophilous. Rocks, walls, and trees occasionally, not rare.
L. Bentham. Dentdale. Baugh Fell; *W. West.*
R. Settle, near Giggleswick; *J. Windsor.*
W. Bolton, on trees; *B. Carrington.* Wetherby (!) and Tadcaster; *J. S. Wesley.*
A. Malham; *C. P. Hobkirk.*
D. Whitgift; *H. F. Parsons.*

139. Orthotrichum cupulatum Hoffm.
On calcareous stone, common in Craven and Lune district.
L. Whernside Crags to 1,800 feet. Dentdale. Clapham; *J. Nowell.*
R. Settle; *J. Windsor.*
W. Between Skipton and Bolton; *B. Carrington.*
A. Malham Moor; *J. Nowell.* Bell Busk; *W. West.*

VAR. **nudum** Dicks. (Var. *riparium* B. & S.)
Common on the higher hills, on stones in streams.
L. Clapdale; *B. Carrington.* Whernside; *W. West.*
W. Outershaw. Bolton; *J. Nowell.* Burley; *W. West.*

140. Orthotrichum rupestre Schl.
L. On high rocks, Whernside, at 2,300—2,400 feet (!) *F. A. Lees* and *W. West* (1878). One tuft near Ingleborough in 1857; *B. Carrington's Flora.*
(W. Near Bolton Bridge; *Baines' Flora,* 1840.)

141. Orthotrichum affine Schrad.
On trees; the commonest corticolous species of the genus, although specifically localised records are few.
L. Clapham; *James Abbott.* Dentdale.
R. Settle. Horton. Wigglesworth.
N. Harrogate.
W. Grassington. Wetherby (!) *J. S. Wesley.* Bolton Bridge; *H. Baines.*
A. Drax; *H. F. Parsons:* Selby; *W. N. Cheesman.* Malham: *J. Nowell.*
C. On a young Ash, Horsebridge Clough, Hebden Valley; *F. A. Lees.*
D. Hook and Rawcliffe; *H. F. Parsons.* Ecclesall and Waleswood; *Amos Carr.*
T. Roche Abbey Valley; *J. H. Aveling* in *Hist.* (1870).

142. Orthotrichum speciosum Nees.
Only one positive record, and that one a little doubtful.
R. Settle; *C. P. Hobkirk.*
Obs.—Gibson's Bolton plant, 'probably only a form of *O. anomalum*'; *B. Carrington* in *W. R. Flora,* p. 76, *saxatile* being meant by *anomalum.*

143. Orthotrichum stramineum Hornsch.
A decreasing species; no recent records?
L. Clapham; *B. Carrington.*
Y. Studley Park; *B. Carrington.*
W. Between Bolton and Ilkley, frequent; *B. Carrington.* (Fairfield Hall, near Bolton; *J. Nowell.*) [This station given for *O. diaphanum* in *Baines' Flora,* 1840.]
A. Kirby Malham; *B. Carrington.*

144. Orthotrichum tenellum Bruch.
W. Ilkley; *B. Carrington.* Trees west side of Ouse below York; *M. B Slater.* By Cock Beck, near Tadcaster (!) *R. Spruce.*
A. Malham; *J. Nowell.*

145. Orthotrichum pallens Bruch.
Y. On trees on south side of Ure near Mickley; *J. G. Baker.*
N. On trees by Ouse (west side) above York; *H. Ibbotson.*
W. A 'var. *obtusifolia* from hedges between Addingham and Bolton is recorded by *B. Carrington* in *West Riding Flora,* p. 75.

146. Orthotrichum diaphanum Schrad.
Trees, rather common.
L. Garsdale; *W. West.*
W. Bolton Bridge and Fairfield Hall; *H. Baines.* Wetherby (!) *J. S. Wesley.* Bilbrough. Dringhouses; *G. Webster.*
A. On elders, Kirk Sike, near Malham; *J. Nowell.*
C. Kirklees Park !
D. Goole and Rawcliffe; *H. F. Parsons.*

147. Orthotrichum pulchellum Sm.
L. Helks Wood, Ingleton; *J. Nowell.*
N. Near Pateley Bridge; *J. Nowell.*
W. Barden Tower. Bolton and Ilkley; *B. Carrington.*
A. Gordale and Malham; *H. Baines.*
D. Smeaton Crags; *H. F. Parsons.*

148. Orthotrichum Lyellii H. & T.
Sub-montane. Rare, but locally common; when young and barren very likely to be taken for *O. phyllantha.*
L. Ingleton; common in Rawtheydale; and Dent, in fruit; *W. West.*
A. Malham; *W. West.* In fruit on trees near Gennest's Cave, Gordale; *J. Nowell.*

149. Orthotrichum leiocarpum B. & S.
N. Harrogate; *T. Hick.*
W. Pool.
A. Near Malham; *J. Nowell.*

150. Orthotrichum Sprucei Mont.
Very local; less plentiful than it used to be !
W. Frequent on trees on bank of Wharfe and Cock Beck, near Tadcaster; *R. Spruce.* Willow stumps near the Ouse, York; *H. Ibbotson.*

151. Orthotrichum rivulare Turn.
L. Ingleton; *W. West.*
Y. Skell banks, near Ripon; *W. Brunton.*
N. Scotton banks, Knaresborough.
A. Stones in or near rills at Malham; *J. Windsor.*

X. SPLACHNACEÆ.

152. Œdipodium Griffithianum (Dicks.) Schw.
Alpine rocky banks, very rare, but, being inconspicuous, may not be really so among the higher hills.
L. Barbon Fell, west side of Deepdale, but just within Yorkshire limits; *Geo. Stabler* ! Ingleborough; *Sir W. J. Hooker* and *Rev. J. Dalton* (1800). Later, on shady rocks at Foal-foot, by *J. Nowell.* Not seen in this station since ?

153. Tetraplodon mnioides (L. fil.) B. & S.
Wet rocky places, very rare. Montane.
(M. Greenfield, near Saddleworth; *H. Baines* (1840). This was immature *Splachnum ampullaceum*.)
L. Cautley Crags, near Sedbergh; *Rev. G. Pinder*. Side of Ingleborough, near Foal-foot and near the summit; *W. J. Hooker* (1800); *J. Windsor*; later by *J. Nowell*!
(A. Malham Moor; *J. S. Wesley* (1878). Malham Tarn; *H. F. Parsons* in *Bot. Rep.* for 1878, *Y. N. U. Trans.* I feel a doubt about this. West does not give it in his Malham list (*Nat.*, Sept. 1873), and not fully developed examples of *Splachnum* have been mistaken for it.)

[**Tetraplodon angustatus** (L. fil.) B. & S. Incognit.
Ingleborough; *Sir W. J. Hooker* in *Townsend's* copy of *O.B.G.* Still unconfirmed.]

154. Splachnum sphæricum L. fil.
M. Warmton Moor, Greenfield; *J. Whitehead*.
L. Whernside. Widdale Fell, by tarns. Howgill Fells; *W. West*. Ingleborough; *W. J. Hooker*; *B. Carrington*.
R. Blea-moor, near Ribblehead.
Y. Dallow Gill Moors; *W. Brunton* (1805).
N. Great Whernside; *W. West*.
A. Gordale; *J. Nowell*. Malham Moor! Skipton Moor; *W. Whitwell*. Rombalds Moor; *W. West*.
C. Holme Moss; *C. P. Hobkirk*. Stansfield Moor, Todmorden; *T. Stansfield*.

155. Splachnum ampullaceum L.
On dung, much rarer than the preceding.
M. Greenfield; *J. Whitehead*!
Y. Dallow Gill; *J. Dalton* (1805)=Near Ripon; *S. Hailstone*.
A. Malham Tarn Bog; *W. West*. Adel Bog (1860); *W. Kirkby*!

XI. FUNARIACEÆ.

156. Discelium nudum (Dicks.) Brid.
Clayey or damp shaly banks, rare. Sub-montane.
M. Ram's Clough, Delph; *J. Whitehead*.
W. Bolton; *S. Gibson* in *Nowell's Suppl.*
C. Shibden, near Halifax; *L. C. Miall* in *Miall and Carrington's Flora*. Heptonstall neighbourhood; *H. Baines* (1840). Pennant Clough and Hebden Valley, Todmorden; *A. Stansfield*.

157. Ephemerum serratum (Schreb.) Hampe. (*Phascum serratum* Wils.)
Y. Field near Mackershaw Wood; *W. Brunton*.
C. Rail-bank, Stoney-royd, Hareley Wood, Todmorden; *J. Nowell*.

158. Physcomitrium pyriforme (L.) Brid.
On bare mud-banks by streams, not rare in the lowlands.
N. Well-field, Harrogate.
W. In great plenty on mud out-thrown from ditch at Scaliber, between Kirk Deighton and Knaresborough. Ditch between Spofforth and Wetherby; *J. S. Wesley*! Bolton Woods; *W. West*. Near Poppleton Station and near Acomb; *W. Whitwell*. Holgate; *G. Webster*.
A. Yeadon Old Mill ponds; *B. Carrington*. Barlow; *W. N. Cheesman*.
C. Valley to south of Thornhill Edge; *C. P. Hobkirk*. By the river at Castleford. Hareley Wood; *A. Stansfield*.
D. Goole; *H. F. Parsons*.

159. Entosthodon ericetorum (Bals.) Schpr.
M. Greenfield Valley (!) *J. Whitehead*.
R. Peaty rill-bank, Ribblehead.
W. Outershaw, Langstrothdale (1885).
A. Newsholme Dean, Keighley; *W. West*.
C. Hudson Clough, Todmorden; *A. Stansfield*. Hareley Royd; *J. Nowell*.

160. Entosthodon Templetoni (Hook.) Schw.
L. Cautley Spout Crags; *W. West*! (1879).

161. Funaria calcarea Wahl. (*F. Muhlenbergii* Schwg.)
L. Ingleborough, near summit; *B. Carrington*. Clapham; *W. West*.
R. Linn Gill.
A. Malham and Gordale (!) *J. Nowell*.

162. Funaria hygrometrica (L.) Hedw.
Waste bare ground, cinders, railway tracks, tree-pots, old brick walls, etc., ubiquitous and generally distributed, especially plentiful on anything that has been charred by fire.

XII. BARTRAMIACEÆ.

163. Amblyodon dealbatus (Dicks.) Pal. de Beauv.
L. On Ingleborough; *H. Baines*.
R. Wet pasture, Stockdale (near the Farm!) on the way from Settle to Grisedale; *R. Teesdale*.
A. On slaty rocks, Gordale; *B. Carrington*=Malham Moor; *J. Nowell*; *J. S. Wesley*!
C. On ground covered with lime, Stansfield Moor; *J. Nowell*.

[**Catascopium nigritum** (Hedw.) Brid. Incognit.
A plant sent to me found on wet exposed rocks, Bell Hagg, near Sheffield, by Mr. Amos Carr of Sheffield, in 1877, and re-sent to me in 1880, which was thought to be this, turned out to be *Ceratodon conicus* Hampe.]

[**Meesia uliginosa** Hedw.
Reported by Mr. Hobkirk to the Author as growing on Goole Moor, at the meeting of Y.N.U., Aug. 6, 1877, but never confirmed.]

[**Paludella squarrosa** (L.) Hedw. Incognit.
Dr. Carrington in his *Flora* wrote, 'I think I have seen a specimen collected by Mr. Nowell, between Malham and Arncliffe?' but the pseudo-record has never yet met with confirmation.]

164. Bartramia ithyphylla Brid.
Very rare. Montane.
L. Near Ingleborough summit; *J. Nowell*. Whernside; *W. West*.
W. Barden Tower (!) *B. Carrington*. Bolton Woods; *W. West*.
A. Malham Moor; *J. Nowell*.

165. Bartramia pomiformis (L.) Hedw.
Arenophilous.
M. Greenfield; *J. Whitehead*.
N. Pateley Bridge. Bank of road between Spofforth and Aketon.
W. Bank of road, Bolton Abbey, just outside grounds towards Barden; *W. West*. Near Strid; *J. Nowell*. Burley; *W. Nuttall*!
A. Otley Chevin (!) Old sandy lane between Hambleton and the 'Hough.'
C. Hebden Valley (!) *W. West*.

166. Bartramia Halleriana Hedw.
Very rare, preferring metamorphic or slaty rocks.
L. Dentdale and Howgill Fells; *G. Pinder*. Cautley Spout; *W. West* (1881). Swarth Fell; *J. W. Watson*. Austwick, slate scars; *J. Nowell*.

167. Bartramia Œderi (Gunn.) Swartz.
Montane. Xerophilous. The commonest species on the damp shady limestones of the hill districts.
L. Garsdale. Deepdale, Dentdale; *W. West*! Dent-head. Whernside (!) *C. P. Hobkirk*. Helks Wood, Ingleton, and in the Swallow-holes below Ingleborough; *B. Carrington*.
R. Linn Gill. Horton Scars.
W. Heseltine Gill. Arncliffe; *W. West*. Buckden. Bolton Woods; *R. Spruce*.
A. Malham Cove (!). Gordale; *John Nowell*.
Obs.—A constant associate of *Amphoridium Mougeotii* and *Leskea subrufa*.

168. Philonotis fontana (L.) Brid.
Swampy places and springs, very common on all the higher hills.
First record: J. Salt, 1800.
Descends to Brayton railway cutting, near Selby, and Rawcliffe, near Goole (*H. F. Parsons*). Reported from every district except Trent, from which it may be a true absentee. Stations too numerous to be worth mentioning.

169. Philonotis calcarea B. & S.
In calcareous turf or wet tufa by springs, rare.
M. Ram's Clough, Delph; *J. Whitehead*.
R. Douk Gill!
Y. Mickley Wood near Ripon; *B. Carrington*.
W. Heseltine Gill in Littondale; *W. Borrer*; *J. Nowell*.
A. Gordale, at both 700 and 1,300 feet! Malham Cove; *B. Carrington*!
C. Todmorden; *J. Nowell*.

170. Breutelia arcuata Dicks.
Wet stony boggy places on mountains, rare, but often abundant locally.
L. Cautley Spout and Howgill; *G. Pinder*; in fruit (1881); *W. West*. Garsdale. Dent head. Great Blake Gill, Whernside, to 1,800 feet; *F. A. Lees* and *W. West*. Thornton Force; *J. Nowell*=Helks Wood *B. Carrington*. Ingleborough foot, and caves; *Jos. Woods*; *J. Nowell*. Baugh Fell; *W. West*.
R. Widdale Fell, south-west end. Cam Fell.
W. North side of Penyghent, and Langstrothdale. Arncliffe; *W. West*.
A. Above Malham; *W. West*.

XIII. BRYACEÆ.

171. Orthodontium gracile (Wils.) Schw.
W. On rocks near the Strid, Bolton; *R. Spruce*. Jumps—in its distribution—from S. England and N. Wales to the Abyssinian Alps.

172. Leptobryum pyriforme (L.) Schpr.
Old walls, mainly calcareous, and in greenhouse pots.
Y. Abbot's House, Fountains Abbey; *J. G. Baker*. Ripon.
N. Very fine on field side of old wall near the N.E. Lodge-gate of Stockeld Park, between Wetherby and Spofforth (!) *J. S. Wesley*.
W. On bridge at Thorparch, and by road there (!) *J. G. Baker*.
A. Micklefield; *G. Webster*. Pots in greenhouse, Meanwood.
C. Stansfield Moor (once) and in greenhouse pots; *A. Stansfield*. Also in greenhouses in Huddersfield; *C. P. Hobkirk*.
D. Limestone quarry near Castleford; *J. Nowell*. Ackworth; *J. W. Watson*.

173. Webera elongata (Dicks.) Schpr.
Very rare—a little doubtful?
L. Sedbergh, Dent, and Howgill Fells; *W. West*. Foal's-foot, Ingleborough, doubtful; *B. Carrington*.
C. Near Halifax; *Dickson*. Seatcliffe Wood, Todmorden; *J. Nowell*.
'This moss may be the *B. acuminatum* B. & S.'—*B. Carrington* in *Flora of West Riding*.

174. Webera nutans (Schreb.) Hedw.

Heaths and peaty moors, very common, although not now obtrusively so in the better cultivated districts.

Distribution general; in every district. Ascends to the summit plateau of Widdale and Ingleborough at 2,300 feet; descends to Goole Moor and Thorne Waste. Stations too numerous for mention; one curious one is worth giving, however—on the tops of the huge isolated blocks of gritstone near the road-side at Stockeld near Spofforth, along with a little heather, although now long gone from the arable land surrounding them.

175. Webera cruda (Schreb.) Schpr.

L. Whernside, by springs, high up N.E. side! Helks Wood; *B. Carrington.*
R. Settle; *C. P. Hobkirk.*
W. Bolton Woods; *J. Nowell.*
A. Gordale; *B. Carrington.*
C. Storthes Hall Woods; *C. P. Hobkirk.*

176. Webera annotina (Hedw.) Schw.

Arenophilous.

W. Sand-bank of quarry south of lime-quarry at Kirk Deighton, Wetherby! [Not *Wesley* auct.]
A. Esholt Wood and Idle; *B. Carrington.* Baildon; *W. West.* Kellington; *H. F. Parsons.*
C. Slaithwaite; *W. West.*

177. Webera carnea (L.) Schpr.

Banks on loamy soil near springs, rare.

M. Greenfield: *J. Whitehead.*
N. Nidd side, Cowthorpe. Knaresborough; *C. P. Hobkirk.*
W. Burley-in-Wharfedale, c. fr.; *W. West.* Wharfe-bank, Bolton; *J. Nowell.*
A. Thorpe Willoughby; *H. F. Parsons.* Brayton Barf. Malham, c.fr.; *C. P. Hobkirk*, 1885.
C. Stanelly, Todmorden; *A. Stansfield.* Ogden Clough, Halifax!

178. Webera albicans (Wahl.) Schpr. (*Bryum Wahlenbergii* Schwg.)

Wet places, frequent.

M. Greenfield; *J. Whitehead.*
L. Howgill Fells; *W. West.* Ingleborough; Austwick; *B. Carrington.*
Y. Hackfall.
N. Upper Nidderdale; *J. S. Wesley.*
W. Wetherby; *J. S. Wesley.* Stream from Ilkley Wells; *B. Carrington.* Nether Poppleton; *W. Whitwell.*
A. Skipton woods; *W. Whitwell.* Brayton and Thorpe; *H. F. Parsons.* Bingley and Baildon; *W. West.* Barlow; *W. N. Cheesman.*
C. Hebden Valley (!) *A. Stansfield.* Hareley Wood; *J. Nowell.* Black Clough, Todmorden; *J. Nowell.* Halifax; *W. West.*
D. Many places to the west of Sheffield; *Amos Carr.*

179. Zieria julacea Schpr. (*Bryum Zierii* Dicks.)

L. Howgill Fells; *W. West.* Whernside Crags to 2,000 feet. Dentdale! Garsdale! Ingleborough; *B. Carrington.*
R. Settle; *J. Windsor.* Widdale Fell. Penyghent Scars.
W. Hesletine Gill. Gills of Langstrothdale.
A. Malham Moor, above the Cove; *J. Nowell.*

180. Bryum pendulum (Hornsch.) Schpr. (*B. cernuum* B. & S.)

L. Near Clapham; *B. Carrington.*
Y. Hackfall.
N. Pateley; *Y. N. U. Excurs.*
W. Bolton Abbey; *B. Carrington.*
A. Embsay Moor; *W. West.* Edge of Malham Tarn; *J. S. Wesley.* Kirby Malham; *J. Nowell.* Yeadon and Apperley Lane; *B. Carrington.* Newthorpe near Sherburn, 1879; *W. Whitwell.*
D. Endcliffe Valley; *Amos Carr.*

181. Bryum rufum Ferguss. (Nat., 1880).

W. Litton, near Arncliffe; *J. Whitehead* (1879). On loose earth amid limestone rocks.

182. Bryum inclinatum (Swartz.) B. & S.

'Very frequent both in the high and low grounds; *R. Spruce.*' I have not found this so common as the quotation implies, and records are few.

N. Greenhow Hill, Pateley; *J. S. Wesley.*
W. Near Bolton Abbey and Bridge; *J. S. Wesley.*
A. Banks near Malham.
D. Walls, Rivelin Valley; *Amos Carr.*

183. Bryum uliginosum (Bruch) B. & S.

Y. By Yore above Hackfall; *J. G. Baker.*
C. White Reaps and Gorple Clough, Hebden Valley head; *A. Stansfield.*
D. Ackworth; *J. W. Watson* ('on a stone in a pond.')

184. Bryum intermedium (W. & M.) B. & S.

Y. Fountains Abbey; *J. G. Baker.*
W. Dringhouses brick-ponds; *R. Spruce.* Walls between Bolton and Skipton; *B. Carrington.*
A. Malham Moor; *B. Carrington.* Yeadon; *Id.*
C. Black Clough, etc., Todmorden; *J. Nowell.*

185. Bryum bimum Schreb. (including FORM **cuspidatum**).

Boggy places, infrequent. Ranges up to 1,250 feet.

Y. Bishopton, Ripon; *Miss Morton.*
A. Rombalds and Cottingley Moors; *B. Carrington.* Adel Bog! Baildon Common and Shipley Glen; *W. West.* Malham; *W. West.*
D. Wintersett, Ackworth; *J. W. Watson.*
T. Maltby, marsh by stream.

186. Bryum pallescens Schl.

R. Wall at Stackhouse, near Settle; *A. O. Black.*
Y. About the Ure on limestone rocks at Tanfield; *J. G. Baker.*
W. 'Arncliffe!' *C. P. Hobkirk.*
A. Rocks on Malham Moor near lead mines; *J. Nowell.*

187. Bryum erythrocarpum Schwg. (*B. sanguineum* Brid.)

Clayey ground, turf-topped walls, etc., rare.

D. Ackworth; *A. O. Brown.*

188. Bryum atropurpureum W. & M.

Sandy heaths, etc.

W. Near Ilkley; *B. Carrington.* By kiln in quarry at Sicklinghall, near Wetherby! Roadside, Bramhope; *James Abbott.* Harewood.
A. Rawden and Apperley Bridge; *B. Carrington.* Pasture below Malham Cove; *J. Nowell.*

189. Bryum alpinum L.

L. On slate rocks above Austwick (!) *J. Nowell*; *J. Windsor.*
C. Hipping's Farm on Staup's Moor, Stansfield, near Todmorden; *J. Nowell.*

190. Bryum cæspiticium L.

Walls, etc., very common, only less so, perhaps, than *B. capillare.*

Generally distributed; every district.

191. Bryum argenteum L.

Roofs, mortared walls, and rocks, very common, ascending to 2,200 feet.

Generally distributed; reported from every district.

VAR. **majus** Wils. (Often taken for *B. filiforme*).

L. On wet rocks by Grisedale Beck, east of Baugh Fell; *W. West*!

192. Bryum capillare L.

On walls, rocks, trees, etc.; the commonest *Bryum.*

Generally distributed; found in every area.

193. Bryum obconicum Hornsch.

Arenophilous. On walls and rocks; only twice recorded.

N. Ripley; *J. Nowell.*
A. Guiseley; *B. Carrington.*

194. Bryum pallens Swartz.

Montane. Moist banks, rocks, etc. In barren state very common; stations almost too numerous to be worth mentioning. Fruits in June.

M. Greenfield; *J. Whitehead.*
L. Dentdale. Ingleton. Clapham, in fruit; *W. West.*
R. Blea-moor. Horton. Settle. Slaidburn.
Y. Fountains Abbey; *J. G. Baker.*
N. Nidderdale, common; *J. S. Wesley.*

W. Barden Moor; *W. Whitwell.* By reservoir, Washburn; *H. F. Parsons.* Thorparch; *R. Spruce.* Quarry on Sicklinghall Road, Wetherby, c. fr.; *J. S. Wesley.* Bolton. Ilkley.
A. Embsay Moor; *H. F. Parsons.* Skipton, in fruit; *W. West.* Esholt Wood; *B. Carrington.* Below Malham Cove; *J. Nowell.*
C. Todmorden; *A. Stansfield.* Hebden Bridge; *S. Gibson.*
D. Ackworth; *J. W. Watson.* Waleswood; *Amos Carr.*

Obs.—An alpine form, growing in compact radiculose tufts on the ridge of Whernside (*F. A. Lees* and *W. West*, 1878), at 2,300 feet, will possibly turn out to be *B. barbatum* or *B. origanum.*

195. Bryum pseudotriquetrum (Hedw.) Schw. (*B. cubitale* Dicks.)

Bogs and marshes, common.

L. Whernside. Ingleborough and Austwick; *B. Carrington.* Howgill Fells; *W. West.*
Y. Sawley Moor; *W. Brunton.*
N. Stream from smelting mill, near Pateley—a small form; *J. S. Wesley*!
W. Ilkley, plentiful; *B. Carrington*; *J. S. Wesley*! Wetherby!
A. Malham Moor (!). Apperley Bridge: *J. Nowell.* Frequent in Aire basin; *W. West.*
C. Slaithwaite; *C. P. Hobkirk.* Luddenden; *W. West.*
D. Wentbridge; and Goole; *H. F. Parsons.*
T. Roche Abbey valley; *J. H. Aveling.*

196. Bryum roseum Schreb.

Montane. Damp woodland banks, rare.

Y. Mackershaw Wood near Studley; *Miss Morton.*
A. Gennett's Cave, Gordale; *W. West*, 1879!
C. Pennant Clough, Hebden Valley; *A. Stansfield.*

197. Bryum filiforme Dicks. (*B. julaceum* Sm.

L. South ascent of Ingleborough; *J. Nowell* (1858); *B. Carrington.* Ingleborough; *W. West*! Whernside and Cautley Spout; *W. West.*
(D. Dunford Bridge, Yorkshire side; *C. P. Hobkirk* in *Journ. Bot.* 1879. There is very considerable doubt about this, the specimen being very small, and mislaid, too, just now; *C. P. H.*, ms. note.)

XIV. MNIACEÆ.

198. Cinclidium stygium Swartz.

Mossy bogs, very rare.

L. Howgill Fells near Tebay, in fruit; *C. H. Binstead* (1886); *Nat.* Mar. 1887. Ingleborough, near summit; *J. Nowell.*
(W. Above Ilkley; *J. S. Wesley*, May 1878.)
A. Moor near 'Maltham' Tarn (1836); *J. Nowell*; *T. Leyland, Rep. Bot. Soc. Edin.*, Jan. 1837. Small bog between the Tarn at Malham and Cowside Beck, not far from the path near the 'Middle-House' Farm; *J. Whitehead* and *J. Percival.* Still there in 1885.

199. Mnium cuspidatum Hedw.
Frequent, yet few records.
L. Wall opposite Ingleborough Cave; *B. Carrington.* Ingleborough, frequent; *W. West.*
W. Bolton Abbey; *B. Carrington.*
A. From Malham down to Esholt.
C. Hareley Wood; *J. Nowell.* Pastures below Rough, Stansfield; *A. Stansfield.*

200. Mnium affine Bland.
Shady banks and marshy places, rare.
A. Malham; *W. West.* Rombalds Moor; *B. Carrington.* Nab Wood, Bingley; *W. West.*
(W. Reported from Boroughbridge Road, York, by *W. Whitwell*; but I fear some mistake, perhaps No. 199 was the species found.)
C. Todmorden; *A. Stansfield.*

VAR. **elatum** Schpr.
A. Bog by Malham Tarn, very fine; *W. West*! Recently this has been split off as a distinct species.

201. Mnium undulatum Hedw. (*Bryum ligulatum* Schreb.)
Damp shady banks by streams, common, but rather rare in fruit.
Reported from every drainage area.
R. & N. In fine fruit on road-side bank at North Deighton, and between Spofforth and Aketon; and in Douk Gill (*W. West*).

202. Mnium rostratum (Schrad.) Schw.
Shady banks and swamps, common. Reported from every district. Up to 2,000 feet on Whernside, and down to 15 feet above sea at Rawcliffe (*H. F. Parsons*) and Lindrick Common (*Amos Carr*).

203. Mnium hornum L.
Stream-sides, banks, woods, stumps, etc., everywhere common, and in every district.

204. Mnium serratum (Schrad.) Brid.
Shady banks and rocks by the larger streams, locally abundant, but absent or very rare in the lowlands near the Ouse. Reported from every district except Mersey and Trent.
Very fine on Nidd banks, Scotton to Cowthorpe; on Wharfe banks, Grassington to Thorparch; Skell banks (*W. Fowler*); Sedbergh (*G. Pinder*); Ribble banks, Stainforth to Bolton and Sawley Abbey; Ingleton; Don banks (*Amos Carr*); etc.

205. Mnium orthorhynchum B. & S.
Shady banks and rock ledges, quite rare.
L. Ingleborough and Ingleton; *B. Carrington.*
W. Arncliffe Clouder, c. fr.; *John Nowell.*
A. Malham Moor; *John Nowell*; *W. West*; etc.

206. Mnium stellare Hedw.
Shady loamy banks on limestone, rare.
L. Ingleborough district; *Nat.*, iv, 159. (Clapham and Dent.)
N. Pateley Bridge.
W. Bolton Woods (!) *John Nowell*; *W. West*; etc.
A. Malham Cove; *J. Nowell.*
C. Todmorden; *A. Stansfield.*
T. Roche Abbey; *H. F. Parsons.*

207. Mnium punctatum Hedw.
Shady damp banks, wet rocks, bogs, common.
Reported from every district.
Earliest record: Ray, 1704, as *Muscus foliis serpylli rotundis erectus major.*

208. Mnium subglobosum B. & S.
Wet peaty ground, local.
M. Greenfield; *Jethro Tinker.*
L. South side of Ingleborough; *B. Carrington.*
W. Ilkley; *Y. N. U. Excurs.*
A. Near Malham Tarn, at 1,300 feet, in fruit (1880); *W. West.* Near Bingley; *Id.*
C. Stansfield and Longfield Moors, Todmorden; *A. Stansfield.* Wessenden Valley; *C. P. Hobkirk.* Luddenden; *W. West.*

209. Aulacomnium androgynum (L.) Schw.
Banks, tree-stumps, walls, etc., on sandy soil, uncommon, yet widely distributed; commoner not only in Yorkshire but all over England in the east than the west.
First notice: S. Gibson, 1838. Range: 20—1,800 feet.
L. Dentdale, Kingsdale, and Clapdale; *B. Carrington.*
R. Linn Gill. Wigglesworth.
Y. Hackfall. Near Staveley.
N. Ramsgill; *J. S. Wesley.* Scotton banks.
W. Near Bolton Strid; *J. Nowell.* Harewood Park. Wetherby; *J. S. Wesley.* Askham; *W. West.* Holgate; *W. Whitwell.*
A. Grit wall by Nanny Bridge, Alwoodley, c. fr.! Gargrave. Brayton Barff; *W. N. Cheesman.* Bishop Wood, Hambleton; *G. Webster.* Carlton, Drax, and Heck; *H. F. Parsons.*
C. Heptonstall; *S. Gibson.* Netherton; *W. West.*
D. Ackworth; *J. W. Watson.* Roscoe Wood, Sheffield; *Amos Carr.* Madam Wood, Goole; *T. Birks.*

210. Aulacomnium palustre (L.) Schwg.
Bogs and swamps, common, though rare in fruit. Known since 1686 (Ray). Range: 20—1,800 feet.
Reported from every district except Trent.
Fruiting abundantly in the Bleachworks-field, Aketon (!) *J. S. Wesley*; Askern (*H. F. Parsons*); etc.

211. Aulacomnium turgidum (Wahl.) Schwg.
L. Discovered in Aug. 1878 on the north-east slope of Whernside by W. West and myself, growing with *Sphagnum papillosum*, but being taken for a pale imbricate var. of *A. palustre* only one tuft was actually brought away.

XV. GEORGIACEÆ.

212. Tetraphis pellucida L. (*Georgia* C. Muell.)
Shady banks, dead wood, on peat, etc., common, but in fruiting condition rare. Every district.
In fruit in Askham Bogs; near 'Strid,' Bolton (*B. Carrington*); Wetherby (*J. S. Wesley*).

213. Tetrodontium Brownianum Schwg. (*G. Brownii* (Dicks.) Muell.)
Montane and Arenophilous.
M. Greenfield (!) *J. Whitehead.*
Y. Sawley; *W. Brunton.* Hackfall and Dallow Gill; *J. Dalton* (1805).
N. Plumpton Rocks (!) *J. Dalton.* Grits by road to Kirby Overblow, just above Spofforth Station.
A. Near Harden (!) *T. Hebden.* Harden Moss; *C. P. Hobkirk.*
C. Marsden Moor; *J. Whitehead.* Keb and Staup's Cloughs, Todmorden; *A. Stansfield.*
D. Upper Bradfield, Harecliff Rocher!

XVI. POLYTRICHACEÆ.

214. Oligotrichum hercynicum (Ehrh.) Lam. & Del. (*O. incurvum* Huds.)
M. Greenfield Valley head; *J. Whitehead.*
L. Whernside Ridge, near the Cairn.
N. Great Whernside peak; *J. G. Baker.*
A. Harden Moss; *C. P. Hobkirk.*
C. Stansfield Moor, rare; *A. Stansfield.* Hebden Valley; *W. West.*

215. Atrichum undulatum (L.) Pal. de Beauv. (*Catharinea* James.)
Shady stream-banks and wet woodland, very common.
Distribution general; in every area.

216. Atrichum crispum Sull. (*A. laxifolium* Wils. et *Cath. crispa* James.)
Local; by sandy stream-sides.
First record: Nowell, 1849.
M. Greenfield (!) *J. Whitehead.* Seal Bark!
A. Newsholme Dean; *W. West.* Hawksworth; *H. F. Parsons.*

C. Hebden Valley (!) *J. Nowell* (1860). Marsden Valley; *J. Whitehead.* Todmorden; *A. Stansfield.*
D. Hollow meadows stream, north-west of Sheffield; *Amos Carr.*

217. Pogonatum nanum (Dill.) Pal. de Beauv.
Bare stony and turfy ground, rare.
N. Abundant on gritty waste ground about the Hookstone quarries, Pannal.
C. Hareley Wood, Todmorden; *A. Stansfield* (first record for Riding).
D. Race-course, Doncaster (1878); *W. West.*

218. Pogonatum aloides (Hedw.) Pal. de Beauv.
Sandy and turfy banks, very common in the heathy and moorland districts, rarer in the more cultivated tracts, but found in all.
Distribution general.

219. Pogonatum urnigerum (L.) Pal. de Beauv.
Stony and turfy banks, rare.
M. Greenfield; *Jethro Tinker.*
L. Howgill Fells; *W. West.* Ribblehead. Whernside (!) *W. Nuttall.* Between Tosside and Slaidburn.
N. Under trees (*Pinus*) border of Stockeld Park (!) *J. S. Wesley.*
C. Stansfield Moor; *A. Stansfield.*
D. High moors near Hallam; *J. Salt.*

220. Pogonatum alpinum (L.) Röhl.
L. Whernside summit. Foal-foot, Ingleborough; *B. Carrington.*
N. Ramsgill Moor; *J. S. Wesley.*
W. Bolton Woods; *S. Gibson.*
A. Malham Moor (!) *J. Nowell*; *W. West*; etc.
C. Stiperden Bank; *T. Stansfield.*

221. Polytrichum gracile Dicks. & Menz.
Turfy ground, under fir-trees in fir-needle mould, etc., rare.
L. Baugh Fell; *W. West.* Ingleborough; *W. J. Hooker.*
N. Swamp two miles west of Pateley Bridge; *J. S. Wesley.* Cranberry Car, north of Hunsingore.
A. Hill-side below Norton Tower (1884); *W. Whitwell.* Side of turf-dike below Rider's Nursery, Moortown, near Leeds (1877).
C. Langfield Moor; *A. Stansfield.* Hebden Bridge; *C. P. Hobkirk.*

222. Polytrichum formosum Hedw. (*P. attenuatum* Menz.)
Turbaries and peaty wet ground in woods, rare.
L. Clapham and Ingleton. Whernside; *W. West.*
Y. Studley Park; *W. Brunton.*
W. Otley Chevin, Ilkley, Bolton; *B. Carrington.*
A. Keighley; Baildon; Hawksworth; *W. West.*
C. High Green Wood, Heptonstall; *J. Nowell.*
D. Goole Moor; *H. F. Parsons*!

223. Polytrichum piliferum Schreb.
Dry stony heaths, common.
Distribution general; reported from every district, from Sedbergh and Garsdale down to Rawcliffe (*B. Saynor*) and Ecclesall (*Amos Carr*).

224. Polytrichum juniperinum Willd.
Heaths, rather common.
L. Bowland Knotts. Ingleborough (a var. *alpestre*); *B. Carrington.* [Was not this the next species? if so, the priority of discovery belongs to *Dr. Carrington.*]
Y. Hackfall.
N. Harrogate; *T. Hick.* About Hookstone quarries, Pannal.
W. Buckden Moor. Almias Cliff and Washburn Valley.
A. Rombalds Moor; *C. P. Hobkirk*; etc. Common; *W. West.* Brayton Railway-cutting; *W. N. Cheesman.*
C. On the moors, but few records. Rishworth Valley.
D. Rawcliffe, etc.; *H. F. Parsons.*

225. Polytrichum strictum Banks.
Montane. Two districts only.
L. Whernside slopes! Baugh Fell; *W. West.* Ingleborough Hill; *W. West*—discovered by him in April 1878, in fruit, and authenticated by *H. Boswell*, but probably Dr. Carrington and Sir W. J. Hooker had gathered it long prior, regarding it as the alpestral form of No. 224.
R. Blea-moor.

226. Polytrichum commune L.
Wooded swamps and peat-bogs, common.
First record: Ray, 1686.
Distribution general; in every district.

227. Diphyscium foliosum (L.) Möhr. (*Webera sessilis* (Schmid) Lindb.)
M. Greenfield Valley; *J. Whitehead.*
L. Brant Fell, Sedbergh; *H. F. Parsons.* Cautley Spout; *W. West.* On rocks on south and west of Ingleborough; *J. Nowell*; *B. Carrington.*
W. Valley of Desolation, ascending to Beamsley Beacon.
C. Staup's Clough, Eastwood, and Gorple Clough, head of the Hebden Valley; *T. Stansfield.*
(T. Roche Abbey; *J. H. Aveling* in *Hist.* (1870), but probably an error.)

[BUXBAUMIACEÆ.]

[**Buxbaumia aphylla** L. Incognit.
Reported twenty-five years ago from Sawley Moor near Ripon by *T. Simpson*, of Bedale, but not confirmed by subsequent investigation.]

XVII. FISSIDENTACEÆ.

228. Fissidens bryoides (L.) Hedw.
Wet ditch-banks and clayey or stony ground, common.
First record: Richardson, 1690.
Generally distributed; in every district.

229. Fissidens exilis Hedw.
N. North Deighton!
W. Wetherby; *J. S. Wesley.* Banks of Walton Lane.

230. Fissidens incurvus Starke.
W. Wetherby; *J. S. Wesley.* Collingham.
C. Todmorden; *A. Stansfield.*

VAR. **tamarindifolius** Don. (sp. Brid.).
C. Hipping's Farm, Stansfield; *J. Nowell*; etc.

231. Fissidens viridulus (Swartz.) Wahl.
Y. Stones in Yore, Swinton; *G. Webster.*
W. Wetherby; *J. S. Wesley.* Arncliffe; *W. West.* Potterton near Aberford; *G. Webster.*
A. On grit boulders, Nab Wood, Bingley; *W. West.*
D. Near Pontefract; *J. Nowell.*

232. Fissidens pusillus Wils.
Y. Stones in lane at Fingerfield near Mickley.
W. Rocks in stream from Ilkley Wells; *B. Carrington.*
A. Shipley Glen; *W. West.*
C. Near Halifax; *W. West*! [Nowell's Todmorden locality is outside the Riding.]
D. Pontefract; *Dr. Wood* (1850). Old quarry at Badsworth; *G. Webster.*

233. Fissidens osmundoides (Swartz.) Hedw.
L. Bogs, Kingsdale; *B. Carrington.*
C. Staup's Clough, etc., Todmorden; *A. Stansfield.*

234. Fissidens crassipes Wils.
W. Arncliffe; and rocks in Wharfe at Bolton; *W. West*!
A. Gordale; *W. West.*

235. Fissidens rufulus B. & S.
W. Bolton; *W. West* (Mar. 1883). 'On rocks in the Wharfe with *Amblystegium fluviatile* and *Barbula spadicea.*'

236. Fissidens adiantoides (L.) Hedw.
Wet banks, limestone quarries, etc., common.
Ascends high. Range: 50—2,000 feet on Cam Fell (*J. G. Baker*) and Whernside. Descends to wet quarries at Kirk Deighton (!) and Anston (*Amos Carr*).

237. Fissidens taxifolius (L.) Hedw.
Shady and wet wood banks, very common.
Distribution general; in all the districts.

XVIII. SCHISTOSTEGACEÆ.

238. Schistostega osmundacea (Dicks.) W. & M.
Sandstone caves, very rare. Typically arenophilous.
N. Guy's Cliff, Pateley; *J. Nowell.*
C. Grit-boulder fissures, head of Hebden Valley; *A. Stansfield.*
D. Quarry hole near Bradfield; *R. A. Gatty.*

XIX. RIPARIACEÆ.

239. Cinclidotus fontinaloides (Hedw.) P. de Beauv.
On wet rocks in streams and rivers, common, from 1,500 feet alt. down to 10 or 15 feet within reach of tide, though not of salt-water.
Abundant on rocks in rivers Lune, Ribble, Yore, Nidd, Wharfe, Aire (to Selby; *H. F. Parsons*), and Don (to Wharncliffe; *J. Holland*).
No record for Trent or Mersey districts.

240. Fontinalis antipyretica L. *Trough-weed.*
On stones, etc., in streams and rivers, also in isolated pools, very common. Range wide; from Rawtheydale to Goole Fields (*T. Birks*).
Reported from every district.

241. Fontinalis squamosa L.
Montane. In streams, replacing the preceding at higher levels.
M. Mersey district.
R. In the Hodder at Slaidburn.
N. Upper Nidderdale; *J. S. Wesley* (1878)!
W. Wharfe at Bolton; *S. Gibson.*
(A. Gordale; *B. Carrington*—Perhaps the next, really.)
C. Gorple Water and Hebden Valley stream (!) *J. Nowell.*

242. Fontinalis gracilis Lindb. (*F. minor* Wahl.)
Only known at Malham Cove, where it was discovered by *Prof. T. Barker*, of Owens College, in 1874. From its facies it is liable to be passed over as *Hypnum cordifolium* or *fluitans.*

XX. CRYPHÆACEÆ.

243. Hedwigia ciliata (Dicks.) Ehrh.
Montane. Rocks and walls, rare.
L. Howgill Fells; *H. F. Parsons* and *W. West.* Above Austwick, and walls in Wharf village; *J. Nowell.*
Y. Sawley; *W. Brunton.*
N. Upper Nidderdale; *J. S. Wesley* (1877).
W. Barden Tower; *B. Carrington.*
A. 'Malham Moor!' *C. P. Hobkirk.*
(T. Roche Abbey valley; *J. H. Aveling* in *Hist.* (1870)—doubtful.)

244. Cryphæa heteromalla (Hedw.) Mohr.
On decaying bark, or wet stones, very rare. Only one old record.
A. 'On an old willow at Broughton Hall, near Skipton'; *J. Nowell.*

XXI. LEUCODONTACEÆ.

245. Leucodon sciuroides (L.) Schw.
On trees mostly, but also on walls, not very common.
L. Ingleton. Dent; *W. West.*
R. Slaidburn. Stainforth.
W. Bolton Woods, cum fructu; *J. Nowell.* Wetherby. Aberford.
A. Malhamdale; *J. Nowell.* Armin; *H. F. Parsons.*
D. Eastoft; *H. F. Parsons.*
T. Sandbeck Park.

246. Antitrichia curtipendula (L.) Brid.
On walls and on bark, rare.
L. Dent; *W. West.*
R. Widdale Fell; *J. G. Baker.* Horton.
N. Old wall at Stockeld (!) *J. S. Wesley.*
W. Kettlewell. Barden Tower and near the 'Strid,' c. fr.; *J. Nowell.*
A. Walls at Middle-house Farm, Malham Moor (!) *J. Nowell.*

XXII. NECKERACEÆ.

247. Neckera pumila Hedw.
On bark, and more rarely on stone, very rare.
L. Dent (!) *W. West.*
Y. Studley Woods; *J. G. Baker.*
W. Bolton Woods, c. fr.; *J. Nowell.*

248. Neckera crispa (L.) Hedw.
Xerophilous. Rocks and trees in calcareous districts, common.
First record: Doody in Ray's Syn., ii, App. 337 (1677).
Localities too numerous to be worth recording, from Whernside limestone crags up to 1,800 feet, down to Yoredale shales in Hebden Valley, rocks in Boston Spa Woods, Went-vale, and Roche Abbey. Occasionally off limestones, as in Moseley Wood, Horsforth (*J. R. Murdoch*), and, in starved condition, in Old Park Wood, Sheffield (1878); *Amos Carr*.
No record for Mersey district.

249. Neckera complanata (L.) Hüben.
On walls and old trees, chiefly in calcareous districts. Recorded from every district except Mersey and Calder, but doubtless occurs in the latter at least; although dwarfed, as well as very rare, in the smoky districts near the large towns.
Localities too numerous for mention.

250. Homalia trichomanoides (Schreb.) B. & S.
On decaying trees or shady old walls.
L. Helks Wood (abundant (!) *B. Carrington*) and other places.
R. Bolton-in-Bolland.
N. Cowthorpe Weir.
W. Bolton Woods (!) *J. Nowell*; etc. Langwith Scrub, above Collingham (!) *J. S. Wesley*. Flint Mill, Wetherby. Stutton.
A. Ledsham. (Bishop Wood; *H. F. Parsons*.) Stainor Wood, near Selby; *W. N. Cheesman*.
C. Near Hudson Mill, Heptonstall, but very rare; *T. Stansfield*.
D. King's Wood, Roche Abbey.

XXIII. HOOKERIACEÆ.

251. Pterygophyllum lucens (L.) Brid. (*Hookeria lucens* Sm.)
Damp shady rocks and banks, rather common, but local and rare in fruit, and absent over considerable areas. Pelophilous—likes clay and shale.
M. Greenfield; *J. Whitehead*.
L. Sedbergh; *G. Pinder*. Howgill Fells and Cautley Spout; *W. West*.
Y. Hackfall!
N. Pateley Bridge. Middlesmoor; *W. West*.
W. Bolton (in fruit!) *B. Carrington*.
A. Bingley; *R. Richardson* (1690). Shipley Glen, &c., but frequent on the grit; *W. West*.
C. Hebden Valley (c. fructu!) *A. Stansfield*.
D. Cliff Rocher and Bell Hagg Crags, Sheffield; *Amos Carr*! Stannington.

XXIV. FABRONIACEÆ.

252. Myrinia pulvinata (Wahl.) Schpr. (*Leskea* Hedw.)
N. On inundated tree-boles near Cowthorpe by Nidd (!).
W. Banks of river Cock, Tadcaster; *R. Spruce*.
D. 'Rawcliffe!' *C. P. Hobkirk*.

253. Habrodon Notarisii Schpr.
Very rare. On sycamore and on ash near the river Dee, S.W. side, near Dent; *G. Stabler* (vide *Nat.*, iv, p. 113).

XXV. LESKEACEÆ.

254. Myurella julacea (Vill.) B. & S. (*Hypnum moniliforme* Wahl.)
L. Rocks near Ingleborough summit; *B. Carrington*=Yoredale scars on west side of Ingleborough; *J. Nowell*; at 2,000 feet!
R. Penyghent scars; *Henry Ibbotson* (N.W. face!).

255. Heterocladium heteropterum (Bruch.) B. & S.
L. Cautley Spout; *W. West*.
R. '? Stackhouse'; *J. Windsor*.
W. Heseltine Gill, shady ledges of shales. Rocks by water, Woodhall.
A. Shipley Grove and Esholt Wood, on boulders; (! W. W.) *B. Carrington*.
C. Stanelly Clough; *A. Stansfield*.
D. Endcliffe Valley; *Amos Carr*.

[**Heterocladium dimorphum** (Brid.) B. & S. Incognit. (*H. squarrulosum* Dicks.)
Recorded in *Mr. Townsend's* interleaved copy of *Old Botanists' Guide* under the name '*squarrulosum*,' as having been found at Wethercote Cave, Ingleborough, by *Sir W. J. Hooker*; 'Mackershaw Wood near Ripon' by *W. Brunton*; and 'Studley' by *J. Dalton*. The first station is not so very unlikely; but the independent observation by two botanists of some moss in the Ripon district supposed to be the same as that found by Sir W. J. Hooker, seems to indicate that the alpine species '*dimorphum* Brid.' was not intended.]

256. Leskea polycarpa Ehrh.
Tree-boles, and roots, by rivers, not rare.
R. Bolton-by-Bolland.
Y. Skell banks; *W. Brunton* (1805). Studley; *B. Carrington*.
N. Knaresborough; *W. West*. On alders by Nidd, Cowthorpe (!) *J. S. Wesley*.
W. Bolton (! W. W.) *J. G. Baker*. Barden Tower; *J. S. Wesley*. Ilkley (1880)! Tadcaster; *J. Nowell*. Cawood (1876). Thorparch.
A. Skipton; *C. J. Wild* (1878). Apperley; *W. West*. Drax; *W. N. Cheesman*.
D. Fishlake; *H. F. Parsons*.

257. Anomodon viticulosus (L.) H. & T.
Xerophilous. Range: 1,800—100 feet.
Calcareous rocks of both formations, tree-roots, springs, etc., common; scarcer on the grits.
Rare off limestone; rare in fruit.
No record for Mersey area.
L. Very common. In fruit at Dent; *W. West*.
R. Very common.
Y. Ripon. Fountains Abbey; *Rev. W. Fowler*.
N. Upper Nidderdale limestones and Knaresborough, etc., on the Permian.
W. Common down to Thorparch.
A. Common down to, rare below, Skipton. In fruit, Gordale; Esholt; *W. West*.
D. Wentbridge; *H. F. Parsons*.
T. Waleswood; *Amos Carr*! Maltby.

258. Pseudoleskea catenulata (Brid.) Schpr. (*Thuidium catenulatum* De Not.)
Montane; very rare.
L. On Yoredale shales, west of Ingleborough, and on rocks near the summit (!) *J. Nowell*. Still there, 2,000 feet (1886).
R. Penyghent; *Henry Ibbotson*.
A. One mile east of Malham Tarn, abundant; *W. West*.

259. Thuidium tamariscinum (Hedw.) B. & S.
Damp shady banks, decaying bark, etc., very common.
Generally distributed; reported from every district, but, oddly, rare over considerable areas, as just around Leeds, etc.

260. Thuidium recognitum (Hedw.) Schpr. (*H. delicatulum* L.)
Xerophilous. Dry shady stony banks, very rare—or overlooked?
L. Helks Wood, Ingleton; *W. Wilson*.
R. Linn Gill (1879), a very little among other mosses off the dry rocks on the north side of the gill, high up.
W. Bolton Woods; *J. Nowell*.
Obs.—The old records are two of the very few resting on unimpeachable authority which are up to the present time unconfirmed.

261. Thuidium abietinum (L.) B. & S.
W. Known only in the neighbourhood of Wetherby, where it was discovered, in Dec. 1877, by *Dr. J. S. Wesley*, in two limestone quarries, one near Kirk Deighton (!), the other near Stutton. In Jackdaw Crag Quarry it grows on the ground along with *Frullania Tamarisci*; and on bare ground at the cross-roads, Stubbing Moor, Rigton, along with *Asperula cynanchica*.

262. Thuidium Blandovii (W. & M.) B. & S.
Very rare.
A. Near Malham Tarn; *J. Whitehead* and *J. Percival* (1868)!

XXVI. HYPNACEÆ.

263. Pterogonium gracile (Dill.) Swartz.
In the mountain region, rare, on trees and stones in ravines.
L. Near Low Gill, Sedbergh! On rocks above Austwick; *J. Nowell*.
Obs.—This species, by affinity and in natural sequence, properly belongs to Leucodontaceæ, and should follow the species numbered 245; but to place it there would be a radical departure from the *London Catalogue* arrangement followed here, and which had the sanction of so authoritative a bryologist as Mr. H. Boswell. The genera Thamnium and Climacium are in a much-the-same debatable position.

264. Cylindrothecium concinnum De Not. (*C. Montagnei* B. & S.)
Montane.
R. Giggleswick Scar slopes.
W. Lane at Linton, near Wetherby (!) *J. S. Wesley*! and Jackdaw Crag Quarry, Stutton.
A. Malham and Gordale; *J. Nowell*. Still there, 1885; *W. West*!

265. Thamnium alopecurum (L.) Schpr.
Shady rocks and banks, common.
Distribution general; recorded, in numerous stations, from every district, but *not* common near the chief manufacturing centres.

266. Climacium dendroides (L.) W. & M.
Swamps, wet moors, etc., common.
First record: Richardson, 1724.
Distributed generally; reported from every district, although rare in fruit, and by far the most frequent amongst the hills.
Range: from north slope of Ingleborough at 2,000 feet down to 50 feet on limestone rocks, South Anston (*Amos Carr*).

267. Pylaisia polyantha (Schreb.) B. & S. (*Leskea polyantha* Hedw.)
R. On old thorn-trees between Broughton Hall and Elslack Station; *J. Nowell*.
W. Bolton Woods; *S. Gibson*; *C. P. Hobkirk*. Askham Bog; *R. Spruce*; *J. G. Baker*.

268. Isothecium myurum (Poll.) Brid. (*Hypnum curvatum* Swartz.)
Tree-roots, stony ground, etc., not common, but widely diffused.
L. Bentham. Dentdale. Clapham; *W. West*.
R. Not uncommon; *J. Windsor*. Slaidburn.
Y. Cayton Gill near Ripley; *Rev. W. Fowler*.
W. Grassington. Bolton; *J. Nowell*; etc. Collingham.
A. Barlow near Selby; *W. Norwood Cheesman*.
D. Wentbridge; *H. F. Parsons*.
T. Kingswood, Firbeck.

269. Orthothecium intricatum (Hartm.) B. & S. (*Leskea subrufa* Wils.)
Montane and Xerophilous. Rare in fruit.
On limestone rock or shale, a constant associate of *Zieria julacea*, common on the limestones of the N.W. portion of the Riding.
L. Clapdale; *B. Carrington*. Whernside. Ingleborough. Dentdale. Howgill.
R. Linn Gill; Douk Gill; Penyghent; etc.
W. Outershaw Crags. Heseltine Gill. Arncliffe; etc.
A. Gordale; *W. West*.
C. On Yoredale shales in Eaves Wood, Heptonstall; *J. Nowell*.

270. Orthothecium rufescens (Dicks.) B. & S. (*Leskea rufescens* Schw.)
Montane. A fine moss with a facies of its own.
R. Silverdale. Penyghent.
W. Heseltine Gill; *J. Whitehead*. Between Malham and Kilnsey Crag; *L. C. Miall*. Cowside-beck Dale; *W. West*.
A. Slaty rocks, Gordale; *J. Nowell*. Still there; *W. West* (1,250 feet).

271. Homalothecium sericeum (L.) B. & S. (*Leskea sericea* Dill.)
On old walls and decaying wood, common, only rare near the large towns.
Distribution general; recorded from every district.
First record: Richardson in Ray, 1690.

272. Camptothecium lutescens (Huds.) B. & S.
Xerophilous. Dry banks, rare off the limestone tracts, but very prevalent on them wherever they cross the drainage areas.
Recorded from every district (on the lime tracts) except Mersey.
Range: 1,500—50 feet.

273. Camptothecium nitens (Schreb.) Schpr.
Swamps. Two localities only.
N. Boggy field behind the Bleach Works at Aketon, near Spofforth, abundantly (!) *J. S. Wesley* (1878).
A. Swampy meadow on north-west side of Malham Tarn; *J. Nowell*; etc.; and recently.

274. Scleropodium cæspitosum (Wils.) Schpr.
N. By Nidd at Cowthorpe Weir.
W. Thorparch. Banks of river Cock near Tadcaster; *R. Spruce*.
A. Shipley; *H. F. Parsons* (1877).

275. Brachythecium glareosum B. & S.
Arenophilous. Sandy moory banks, rare.
L. Helks Wood (!) *J. Nowell* = Wilson's *salebrosum* station?
R. Settle.

N. By Nidd below Knaresborough, among grass.
W. Bolton Woods; *B. Carrington*. Kirk Deighton. Jackdaw Crag Quarry near Stutton, with *Cylindrothecium*.
A. Malham; *W. West*.
C. Todmorden; *A. Stansfield*.

276. Brachythecium salebrosum (Hoffm.) B. & S.
L. Helks Wood, Ingleton; *W. Wilson* (1838); *Baines' Flora*, 139.
(Y. Near Grantley's lake, by Rippon; *W. Brunton*.)
A. On decaying leaves and rotting sticks, with *B. rutabulum*, near stream that runs at the foot of Ledsham Park towards Ledsham village, in 1876! Unrecognised, until I found it again in greater quantity near Market Rasen, North Lincolnshire, in 1877.
Obs.—I believe Hooker and Taylor's *salebrosum* (= *glareosum*) was the plant Brunton found.

277. Brachythecium albicans (Neck.) B. & S.
Arenophilous, rather rare. The leaves have often somewhat *recurved* margins—a character not mentioned in books.
N. On sand rocks at Knaresborough.
W. Kirk Deighton quarry, and Stubbing Moor, Rigton.
A. Baildon Moor, among loose grit-sand; *W. West*
C. Dry banks near Todmorden; *J. Nowell*.
D. Rawcliffe; *H. F. Parsons* (1877).

278. Brachythecium velutinum (L.) B. & S.
Shady banks, rocks, walls, common.
Distribution general; recorded from every district. Numerous localities, from the lead-mines on Malham Moor (1,400 feet) down to Rawcliffe and Carlton (*H. F. Parsons*).

279. Brachythecium rutabulum (L.) B. & S.
On shady banks, bark, stones, etc., very common.
Distribution general; in every district in plenty.

280. Brachythecium rivulare B. & S.
Rocks by streams, frequent. Range very wide
L. Dentdale. Great Blake Gill, Whernside. Slaidburn.
Y. Sawley; *W. Brunton*.
N. By Spofforth (Crimple) Beck (!) *J. S. Wesley*. Cowthorpe.
W. Bolton; *J. Nowell*. Burley; *W. West*. Ilkley. Collingham.
A. Malham Beck; *C. P. Hobkirk*. Airton and Stirton; *W. Whitwell*. Esholt; *B. Carrington*.
D. By Don, Sheffield district; *Amos Carr*!

VAR. **chrysoleucum** Spruce.
The fruitless female plant of this diœcious species, an amphibious form of *rivulare*—a much softer plant than the type, and notable for

the hooked branches and subsecund apical leaves of a pale green with a slight golden lustre—occurs in wet grassy places in quarries about Wetherby, near the Crimple Beck at Spofforth, by the Weir at Boston Spa, and elsewhere. It is often by-passed as *B. rutabulum*.

281. Brachythecium populeum (Hedw.) B. & S.
Hilly districts, rather common on rocks and tree-trunks.
M. Greenfield; *J. Whitehead*.
L. Ingleton; *W. West*.
R. By Mill Island, Settle; *J. Windsor*. Douk Gill, Horton.
Y. Near Ripon; *W. Brunton* (1805).
N. Spofforth; *J. S. Wesley*! Cowthorpe.
W. Wetherby. West Woods, Bramham. Bolton; *W. West*.
A. Adel. Bingley; *W. West*.
C. Huddersfield: *C. P. Hobkirk*. Todmorden; *A. Stansfield*.
D. Oughtibridge; *A. Carr*.
T. Roche Abbey; *H. F. Parsons*.

The VAR. **majus** of Bry. Eur. was found by myself and *M. B. Slater* on a stump on Mickley Bank, near Kirkby Malzeard, in 1880. It was at first erroneously thought to be *Eurynchium striatulum*.

282. Brachythecium plumosum (Swartz.) B. & S.
On water-washed sandstone rocks, frequent.
M. Greenfield; *J. Whitehead*.
L. Dentdale head. Howgill Fells; *W. West*. Clapdale; *B. Carrington*.
R. Slaidburn! Penyghent; *W. West*.
N. Ravensgill, Pateley; *J. S. Wesley*. Spofforth; *J. S. Wesley*.
W. Barden; *J. S. Wesley*. Collingham.
A. Bingley; *Y. N. Rec.*, p. 175. Shipley Glen; Allerton; *W. West*.
C. Todmorden; *A. Stansfield*. Huddersfield; *C. P. Hobkirk*.

283. Eurhynchium myosuroides (L.) Schpr.
Tree-boles and banks, not common.
L. Dentdale. Ingleton. Clapham; *W. West*.
R. Bolton-by-Bolland.
W. Near Wetherby; *J. S. Wesley*. Bramham West Woods. Addingham; *W. West*.
A. Camblesforth; *H. F. Parsons*.
C. Pennant Clough and Rag Scouts, Hebden; *A. Stansfield*.
D. Wentbridge. Scraith Wood, Sheffield; *Amos Carr*!

284. Eurhynchium striatum (Schreb.) B. & S.
Wooded banks, especially in the low country, very common, except near the big towns.
Distribution general; reported from every district except that of Mersey.

[**Eurhynchium striatulum** (Spruce) B. & S. Incognit.
Erroneously recorded in the *Naturalist* for May 1880.—Vide Vol. v, p. 169.]

285. Eurhynchium crassinervium (Taylor) Schpr.
On shady stones, liking old arches.
L. Dentdale. Whernside.
R. Hodder banks, Slaidburn, and nearer Clitheroe; *Dr. Wood*; 'in perfect fruit' in February!
N. Knaresborough; *W. West*.
W. Thorparch Weir. Askham; *W. West*.
A. Skipton; *W. West*.
D. Wentbridge; *H. F. Parsons*. Rivelin Valley; *Amos Carr*!
T. Waleswood; *Amos Carr*.

286. Eurhynchium piliferum (Schreb.) B. & S.
Damp banks, not common.
L. Bentham; *F. A. Lees*. Dent; *W. West*.
R. Barnoldswick; *W. West*.
W. Between Spofforth and Wetherby; *J. S. Wesley*. Bolton Woods; *John Nowell*.
A. Gordale; *H. F. Parsons*. Rombalds Moor; *B. Carrington*. Malham district; *W. West*.

287. Eurhynchium Swartzii (Turn.) L.C. (*E. prælongum* var. *atro-viride* B. & S.)
Damp shady places, rather common.
L. Clapdale; *B. Carrington*. Dentdale.
R. Wigglesworth. Barnoldswick; *W. West*.
N. Hunsingore. Scotton banks, Knaresborough.
W. Wetherby (!) *J. S. Wesley*. Cawood; *H. F. Parsons*. Bolton; *J. Nowell*. Bramham, c. fr.; *J. S. Wesley*. Askham Road, York; *W. Whitwell*.
A. Baildon; Apperley; *W. West*.
C. Todmorden; *A. Stansfield*.
D. Wentbridge; *H. F. Parsons*.
T. Waleswood; *Amos Carr*. Maltby.

288. Eurhynchium prælongum (Dill.) Schpr.
Wet thickets and hedge-banks, very common.
Distribution general; in all the districts.

289. Eurhynchium pumilum (Wils.) Schpr.
On shady sandstone rocks.
L. Swallow-hole, Crina Bottom, Ingleborough; *B. Carrington*.
A. Shipley Glen; *W. West*.
C. Shady rocks near Todmorden; *John Nowell*.

290. Eurhynchium Teesdalii (Sm.) Schpr.
On shady rocks near rivulets.
L. Near Sedbergh; *G. Pinder*=Cautley Spout. Crina Bottom, Ingleborough; *B. Carrington*. Near Thornton Force, in fruit; *W. West.*
W. On Kilnsey Crag; *John Nowell.*
A. Malham district; *W. West.*

291. Hyocomium flagellare (Dicks.) B. & S. (*H. umbratum* Sm., Eng. Flora).
Wet sandstone rocks in and by rushing streams, common.
M. Greenfield; *J. Whitehead.*
L. Howgill Fells. Dentdale.
Y. Dallow Gill; *W. Brunton.* Hackfall.
N. Ravensgill, Pateley (!) *J. S. Wesley.*
W. Ilkley; *J. S. Wesley*; *H. F. Parsons.* Bolton; *S. Gibson.*
C. Hebden Bridge (!) Norland Moor; *H. F. Parsons.* Wessenden Valley; *C. P. Hobkirk.* Marsden; *W. West.*
D. Upper Bradfield. Loxley stream; *Amos Carr.*

292. Rhynchostegium tenellum (Dicks.) B. & S.
Xerophilous; on rocks and old walls.
L. Helks Wood; *B. Carrington.*
R. Giggleswick Scars; *B. Carrington.* Slaidburn.
N. Knaresborough; *J. G. Baker.*
W. Bolton; *C. P. Hobkirk.* Wetherby; *J. S. Wesley.*
A. Gordale, rock at mouth of Gennett's Cave, and Cove; *John Nowell.* Malham; *W. West.*
D. Endcliffe Valley, and near Greystones, Sheffield; *Amos Carr*! Wentbridge; *H. F. Parsons.*
T. Roche Abbey Valley; *J. H. Aveling* in *Hist.* (1870).

293. Rhynchostegium depressum (Bruch) B. & S.
On and among shady rocks.
L. Helks Wood; *B. Carrington.* Grensell Crags, Whernside. Near Thornton Force; *W. West.*
R. Giggleswick Scar; *B. Carrington.*
A. Gordale; *John Nowell.*
C. Hough-stone, Stansfield; *A. Stansfield.*

294. Rhynchostegium confertum (Dicks.) B. & S.
Sandy and shady hedge-banks, walls, etc., common, but few specific stations are on record.
M. Greenfield; *J. Whitehead.*
L. Ingleton; *W. West.*
R. Wigglesworth.
Y. Copgrove; *J. Dalton.*

W. Wetherby (!) *J. S. Wesley.* Ilkley; *W. West.*
A. Skipton; Bingley; Allerton; *W. West.* Selby; *W. N. Cheesman.*
C. Todmorden; *A. Stansfield.* Mirfield; *C. P. Hobkirk.*
D. Wentbridge, Pontefract, Eastoft, and Rawcliffe; *H. F. Parsons.* Kexborough; *C. P. Hobkirk.*

295. Rhynchostegium murale (Hedw.) B. & S.
Xerophilous. On mortar, brick walls, etc., locally common.
L. Clapham; *B. Carrington.*
R. Slaidburn.
Y. Studley; *W. Brunton.*
N. Knaresborough, c. fr.; *C. P. Hobkirk.*
W. Wetherby; *J. S. Wesley.* Bolton Bridge; *John Nowell.* Askham Road, York; *W. Whitwell.*
A. Gordale, and Belbank, Bingley; *John Nowell.* Gordale; Skipton; Allerton; *W. West.*
C. Todmorden; *A. Stansfield.*
D. Wentbridge; Warmsworth; *H. F. Parsons.* Askern; *M. B. Slater.*

VAR. **complanatum** Schpr.
Rare.
A. Chellow Dean; *W. West.*
C. On a wall at Liversedge; *W. Fowler.*

296. Rhynchostegium rusciforme B. & S.
Dripping rocks or submerged stones in streams, very common, and with a very wide vertical range.
Area general; reported from every district.

297. Plagiothecium latebricola B. & S. (*Leskea latebricola* Wils.)
On shady trunks of trees, on both living and putrescent bark.
C. Oak-tree roots, Stanelly Clough, Hareley, rare; *John Nowell*; *A. Stansfield.*
D. Park Wood, Stainforth; *H. F. Parsons* (1878).

298. Plagiothecium pulchellum (Hedw.) B. & S.
Montane. Limestone rock ledges.
L. Head of Great Blake Gill, N.E. shoulder of Whernside!
R. Penyghent, N.W. crags.
W. Heseltine Gill. Outershaw.
A. Malham district; *W. West.*
C. Gibson Wood, Heptonstall; *J. Nowell.*

299. Plagiothecium nitidulum (Wahl.) B. & S.
In hollows of tree-roots, and on putrid wood.
L. Howgill Fells; *W. West* (1879)—teste *H. Boswell.*
W. Heseltine Gill; *J. Whitehead* and *J. Percival* (1868).

300. Plagiothecium denticulatum (L.) B. & S. b. **sulcatum** Spr.
Wood-banks, tree-boles, and rocks, very common.
Area general; in every district.
Obs.—There is a minor form which is often called *elegans.* I have not seen *a. aptychus* Spruce in the Riding.

301. Plagiothecium Borrerianum (Spruce).
Montane. Arenophilous. Very shady rocks, locally common on grit, but absent or nearly so on limestone.
L. Ingleborough; *W. West.*
R. Penyghent; *W. West.*
Y. Near Ripon; *Miss Morton.*
N. Spofforth. Ramsgill; *J. S. Wesley.*
W. Near Bolton 'Strid,' c. fr.; *B. Carrington.* Ilkley (!) *W. West.* Near Collingham, c. fr.; *C. P. Hobkirk.* Wetherby; *J. S. Wesley.* Wistow; *H. F. Parsons.*
A. Meanwood Wood. Idle and Esholt Woods; *B. Carrington.* Bingley; *J. Nowell.* Shipley Glen; *W. West.*
C. Hebden Bridge (!) *C. P. Hobkirk.* Todmorden; *A. Stansfield.*
D. Rivelin Valley; *A. Carr*! Upper Bradfield.
T. Abbey brook shales, above Derwent Chapel!

302. Plagiothecium sylvaticum (L.) B. & S.
In woods, not rare, but large forms of *P. denticulatum* often pass for it. In fruit the rostrate lid will distinguish it, as well as the inflorescence; when barren the best character appears to be afforded by the loss of glossiness and shrinkage of the dark green foliage whilst drying; also by its larger cells and always-entire leaves.
R. Bolton-in-Bolland, woods by Ribble. Slaidburn.
N. Near Pannal; *W. West.*
W. Harewood; and in fruit in wood opposite Woodhall. Below Bolton, c. fr.; *W. West.*
A. Idle and Esholt Woods; *B. Carrington.* Kildwick; Baildon; Allerton; Bingley; *W. West.*
C. Horbury; near Halifax; *W. West.*

303. Plagiothecium undulatum (L.) B. & S.
Damp turfy woods, under fir-trees, etc., common, though often local.
Distribution general; reported from every district.

304. Amblystegium Sprucei (Bruch) B. & S.
Moist stony places, rare.
L. Thornton Force; *J. Nowell.* Helks Wood and Hell Gill, Ingleborough; *B. Carrington.*
A. Gordale; *J. Nowell.*

305. Amblystegium serpens (L.) B. & S.
Old rocks, walls, hollows and roots of trees, etc., very common.
Area general; reported from every district.
There is a large variety or form of this which is liable to be taken for *E. pumilum* or *H. Sommerfelti*, but the former moss has bifarious and the latter broad-based nerveless leaves; on the other hand, *A. radicale*—the next species but two—has been confused with this large state of *A. serpens* by even good authorities.

306. Amblystegium irriguum (Wils.) Schpr.
Chiefly on siliceous rocks in rivulets, and on crumbling wet rocks.
L. Chapel-le-dale; *B. Carrington.*
Y. At Tanfield, in the Yore; *J. G. Baker.*
W. Askham Bog; *R. Spruce*; *H. F. Parsons* (1879).
C. Near Mirfield; *C. P. Hobkirk.*
D. Kexborough; *C. P. Hobkirk.*

307. Amblystegium fluviatile (Swartz) B. & S.
Commoner than the preceding, on stones in streams.
L. Ingleton; *W. West.*
Y. Tanfield, in the Yore; *J. G. Baker.* Rivulets near Ripon; *W. Brunton* (1805).
N. Dripping rocks near Dacre banks; *W. Fowler*!
W. In the Wharfe at Bolton; *J. Nowell*; *J. Whitehead*; *W. West*; etc.
D. In the Don, Little Don Dale; *Amos Carr.*

308. Amblystegium radicale (P. de Beauv.) B. & S.
Moist stones, by weirs, etc., very rare.
R. On a moist stone, Hodder Valley, Slaidburn (1876).
W. Askham Bogs; *R. Spruce.*
D. Askern; *M. B. Slater* (1886).

309. Amblystegium riparium (L.) B. & S.
R. Ribble banks, Settle; *J. Windsor.*
W. Askham Bogs; *R. Spruce.* Wetherby (!) *J. S. Wesley.* Pool in field on Walton Road, Cawood; *W. N. Cheesman.*
A. Airton; *W. Whitwell.* Meanwood Beck. Frizinghall; *W. West.* Monk Fryston; *H. F. Parsons.*
C. Well trough, Liversedge; *W. Fowler.*
D. Wentbridge; *H. F. Parsons.*
T. By Laughton Pool, Roche Abbey.

VAR. **inundatum** Schpr.
A. Seacroft; *W. West.*

310. Hypnum aduncum Hedw. (*H. Kneiffii* Bry. Brit.)
Grassy swamps, rare.
(L. On Clapham Common, rare; *B. Carrington.* I saw only *H. exannulatum* there.)
W. Swampy field to north of Walton Road, Wetherby; teste *H. Boswell.* Askham Bogs; *G. Webster.*
D. Ringinglow Bogs, Sheffield; *Amos Carr*—teste *H. Boswell.*

311. Hypnum exannulatum Gümb. (*H. aduncum* Wils.)
Hill swamps, rather common.
L. Clapham. Bowland Knotts. Garsdale pastures. Ingleborough slope near Gaping Gill Hole; *J. G. Baker.*
R. Blea-moor. Cam Fell slopes.
Y. Cow-mires, Galphay; *Miss Morton.* Wet field, Fingerfield (!) *F. A. Lees* and *H. H. Slater.*
N. Nidderdale Moors about Ramsgill; *J. S. Wesley.*
W. Ilkley Moor; *J. S. Wesley*; *W. West*; etc.
A. Wilsden; *B. Carrington.* Baildon; *H. F. Parsons*; *W. West*; etc.
C. Near Todmorden; *A. Stansfield.*

312. Hypnum vernicosum Lindb.
In turfy places.
N. Ramsgill Beck; *J. S. Wesley*!
W. Bog at Ilkley; *J. S. Wesley* (1878).
(A. Baildon; *H. F. Parsons* teste *C. P. Hobkirk.* This, according to Mr. Boswell's unimpeachable authority, was *H. exannulatum.*)
C. Slaithwaite; *C. P. Hobkirk.*

313. Hypnum Cossoni Schpr.
In swampy places.
L. Brant Fell on Howgill Fells; *W. West* (1879). Dr. Parsons also sent me this from the same district, but I believe it was gathered on same occasion as the above.

314. Hypnum Sendtneri Schpr.
In swampy and turfy places.
L. Rill-side on moor rising from Clapham Junction Station to Bowland Knotts (1876) 'Forma minor' teste *H. Boswell.* Ingleborough; *W. West.*
N. Abundant and fine in boggy field at back of Bleach Works at Aketon, near Spofforth (!) *J. S. Wesley.*
[W. Incorrectly reported for 'Wetherby; *J. S. Wesley*' in *Y. N. U. Trans.*, Series E, p. 89.]

315. Hypnum lycopodioides Schw.
Sandy-peaty moors, rare, not to say almost apocryphal.
(A. Malham Moor bogs; *John Nowell.* I doubt; *W. West* and I myself have only noticed large *H. scorpioides.*)
D. Thorne Waste; *spn. leg. J. Bohler*! I could only find *H. scorpioides.*
Obs.—A 'Var. *falcatum*' is recorded in *Hobkirk's Syn. Brit. Mosses*, p. 169, Ed. i (not in Ed. ii) for 'Copgrove, Yorks.' Some error?

316. Hypnum revolvens Swartz.
Bogs.
L. Near Ingleborough; *B. Carrington.* Bowland Knotts. Whernside.
R. Blea-moor.
W. Ilkley Moor (c.fr. 1878) (!) *J. S. Wesley*; et aliis. By stream from Ilkley Wells; *B. Carrington.* Arncliffe; *W. West.*
A. Near Malham Tarn (!) *J. Nowell*; *W. West*; &c.
C. Langfield Moor; *J. Nowell.*

317. Hypnum fluitans L.
Swamps and rill-heads on moors, frequent.
L. Whernside (!) Ingleborough; *W. West.* Garsdale.
R. Ribblehead.
W. Stake Fell. Ilkley, c. fr.; *H. F. Parsons* (1878). Park Hill, Wetherby; *J. S. Wesley.*
A. Near Malham Tarn; and on Elslack Moor; *W. Whitwell.* Baildon; *H. F. Parsons*; *W. West.* Haddlesey; *W. N. Cheesman.* Thornton Moor; Wibsey Slack; *W. West.*
C. Heptonstall Moor; *A. Stansfield.* Todmorden *J. Nowell.* Marsden Moors; *C. P. Hobkirk.*
D. Goole and Rawcliffe; *H. F. Parsons.*

318. Hypnum uncinatum Hedw.
Wet grassy banks, old walls, and tree-roots, rather rare.
N. Wall near Gouthwaite Hall; *J. S. Wesley.*
W. Barden Tower; *J. S. Wesley.* Bolton; *J. Nowell.* Rombalds Moor; *W. West.*
A. Gordale! Bellbank, Bingley (!) *W. West.* Barden Moor, Aire drainage side; *W. Whitwell* (1882). Hawksworth; *W. West.* Steeton Grove; *B. Carrington.*
C. Hebden Valley, rare; *A. Stansfield.*
D. Went-vale; *C. P. Hobkirk.* On rocks in the Rivelin, c. fr., at Black-brook, and elsewhere among the hills west of Sheffield; *Amos Carr*!

319. Hypnum filicinum L.
Wet calcareous rocks, clayey banks and road-sides, common.
Area general, reported from every district; localities too numerous to deserve mention, although in the flagstone manufacturing area it is not a common moss.

Var. **vallisclausæ** Brid.
W. Ingbarrow marsh-field, Wetherby (!) *J. S. Wesley.* In Wharfe, Bolton; *Id.*
A. Spring at Yeadon; *B. Carrington.*

Var. **elatum** Schpr.
A. Gordale; *W. West.*

320. Hypnum commutatum Hedw.
Wet places, dripping rocks, mostly on calcareous soil, common, and locally abundant.
Area general; recorded from every district, although unfrequent in the lowland sandy area.
First record: Malham; *Ray*, 1690.

321. Hypnum falcatum Brid.
Bogs, etc., not so frequent as the preceding, but often accompanying it. There are innumerable connecting links between them.
L. Garsdale. Whernside. Ingleborough; *W. West.*
N. Aketon Bleach-works marsh (!) *J. S. Wesley.* Ramsgill Beck; *Id.*
W. Bolton Woods; *B. Carrington.*
A. Malham Tarn Bog; *J. Nowell*; *J. S. Wesley*; etc. Runnel by road between Thornton and Elslack; *W. Whitwell.* Shipley Glen.
T. Lindrick Common; *Amos Carr.*

Var. **virescens** Boulay.
L. Clapham; *W. West.*
R. In the beck, Tarn fields, Giggleswick, 1884; *W. Whitwell*!
W. Arncliffe; *W. West.*
A. Malham Cove; *W. West*!

322. Hypnum rugosum Ehrh.
Very local, on exposed elevated stony declivities.
R. Giggleswick Scar; *John Nowell.*
A. Gordale and Malham Cove (!) *J. Nowell*; *W. West*; etc.

323. Hypnum incurvatum Schrad.
Montane. Very rare; not gathered very recently.
L. Walls in the lower part of Helks Wood, Ingleton; *W. Wilson* (1838); regathered in 1861 by J. Nowell, who sent me fresh spns.; *H. Boswell.*
Y. Studley Park; *B. Carrington.*

[**Hypnum hamulosum** B. & S. Incognit.
W. Wilson had a specimen (vide *Carrington's W. R. Flora*, p. 84) said to have come from 'a plantation near Headingley, Leeds,' but the species is strictly montane, the record has gone unconfirmed, and the most probable explanation of it is that some accidental transposition or mingling of specimens has occurred. Specimens gathered as this species near Clapham, by W. West, in 1880, were only a form of No. 324; *H. Boswell.*]

324. Hypnum cupressiforme L.
Tree-boles, old walls, etc., very common indeed.
Distribution general; in every district.

Var. **ericetorum** B. & S. (*compressum*).
L. Ingleborough; Whernside; *W. West.*
W. Rombalds Moor; Thornton Moor; *W. West.*

Var. **tectorum** (Schpr.) B. & S. (*lacunosum* Wils.)
L. On dry limestone rocks near Bentham, and Weathercote Cave.
W. Wetherby, etc.
C. Huddersfield; *C. P. Hobkirk.*

Var. **filiforme** Schpr.
L. Sedbergh. Whernside and Clapham; *W. West.*
W. Bolton Woods; *W. West.*

Var. **resupinatum** Wils.
Common, and recorded for all the districts except Mersey and Nidd. It ascends to 2,200 feet on Whernside, and descends to Roche Abbey. It (*not* the *Pylaisia*) is the *Hypnum polyanthus* of Aveling's List in *Hist. Roche Abbey.*

325. Hypnum patientiæ Lindb. (*H. arcuatum* Angl. Auct., non Sull. and Lindb. *H. Lindbergii* Mitt.)
Grassy woodland banks.
Y. Bank of field above Hackfall. *J. G. Baker* reports it from the north side of river at same place.
W. Ride border, Devonshire Wood, Sicklinghall ( ) *J. S. Wesley.*
C. Hareley Wood; *J. Nowell*; conf. *A. Stansfield.* Thurstonland; *C. P. Hobkirk.*

326. Hypnum molluscum Hedw.
Very common in the limestone districts, scarce off them.
Reported from every district except that of Mersey.
Range: wide—2,200 feet on Cam, Widdale and other fells of the north-west, down to 100 feet in Trent district.

Var. **condensatum** Schpr.
W. Bramham; *W. West.*
A. Hawksworth; *W. West.*

Var. **palustre** = var. **erectum** Schpr.
A large, erect-growing variety, of distinct facies, but not seen in fruit, grows in Aketon Bleach-works swamp-field, on grit stratum.

327. Hypnum Crista-castrensis L.
Very rare; a dying-out species?
L. Dentdale and Sedbergh; *Rev. G. Pinder.*
R. Near the Ebbing and Flowing Well, Stackhouse, Settle; *A. O. Black* (1840).

328. Hypnum palustre L. (*Limnobium palustre* B. & S.)
L. Howgills; *W. West.* Clapham; *Id.* Whernside. Garsdale. Helks Wood: *W. Wilson* (1838).
R. Ribblehead, Settle, etc.
Y. Fountains Abbey, and Sawley; *W. Brunton.*
N. Ramsgill; *J. S. Wesley.* Pateley; *G. Webster.* Scotton banks, Knaresborough.

W. Bolton; *B. Carrington*. Wetherby; *J. S. Wesley*.
A. Malham Cove; *J. Nowell*. Tarn House Lane, Skipton; *W. Whitwell*. Allerton; Baildon; *W. West*.
D. Ribbleden near Sheffield; *Amos Carr*!

VAR. **subsphæricarpon** B. & S.
L. Whernside, in rill of Great Blake Gill, at 1,500—1,800 feet; and Ingleton.

329. Hypnum eugyrium Schpr.
L. Cautley Spout cascade, Rawtheydale; *J. Nowell*, auct. *Schimper* (1842). Re-found there by *W. West* (1879). 'Dent (Yorkshire)'; *C. P. Hobkirk, Syn. Brit. Mosses*, 228. This may refer to same station.

330. Hypnum ochraceum Turn.
M. Diggle; *J. Whitehead*. Saddleworth; *C. P. Hobkirk*.
L. Howgill Fells; *H. F. Parsons*; *W. West*; etc.
[W. Beamsley Rocks; *L. C. Miall* (Vide *Nat.*, 1866, p. 266).]
A. Denholme; Eldwick; *W. West*.
C. Gorple Clough, etc.; *A. Stansfield*.

VAR. **flaccidum** Milde.
Marsden; *W. West* (teste *Fergusson*).

331. Hypnum polymorphum (Hedw.) Br. ms. (*Sommerfelti* Myr.)
On limestone rocks, as yet rare, or overlooked probably because it grows generally in small quantity and intermingled with *H. chrysophyllum*, etc.
L. Wethercote Cave; *W. J. Hooker*.
Y. Studley (Mackershaw Wood); *W. Brunton*.
W. Jackson's Quarry, Kirk Deighton! Jackdaw Crag Quarry, Stutton! Flint-mill Wood, Wetherby! Collingham Bank Wood; *W. West*!

332. Hypnum chrysophyllium Brid.
Xerophilous. On calcareous rocks, rather rare.
L. Clapham; *B. Carrington*.
Y. Burton Leonard Quarries. Mackershaw Wood.
W. Addingham and Ilkley; *B. Carrington*. Rail-cutting and quarries at Wetherby (!) *J. S. Wesley*. Jackdaw Crag, Stutton; *H. Ibbotson* (1859).
A. Malham; *W. West*. Burton Salmon. Newthorpe, Sherburn; *W. Whitwell*. Monk Fryston; *H. F. Parsons*.
D. Wentbridge, and Warmsworth; *H. F. Parsons*.
T. Maltby; *Amos Carr*.

333. Hypnum polygamum (B. & S.) Schpr. (*H. androgynum* Wils.)
Very rare.
W. Askham Bogs near York; *R. Spruce*!

334. Hypnum stellatum Schreb.
Spongy bogs and wet sandy heaths, quarries, etc., not rare.
L. Garsdale. Ingleborough; *W. West*.
Y. Cow-mires, Galphay; *Miss R. Morton*.
N. Aketon Bleach-works marsh (!); *J. S. Wesley*.
W. Wetherby. Jackdaw Crag Quarry, Stutton; *H. Ibbotson*.
A. Malham Moor; *W. West*. Field near Horsforth; *J. R. Murdoch*. Cottingley Moor; *L. C. Miall*. Eldwick; Marley; and Adel Bog; *W. West*. Monk Fryston; *H. F. Parsons*.
C. Todmorden; *A. Stansfield*.
D. Goole Fields; *H. F. Parsons*.
T. Kiveton; *Amos Carr*.

VAR. **protensum** Brid.
W. In limestone quarries, Wetherby; *J. S. Wesley*.
A. Malham; *W. West*.

335. Hypnum cordifolium Hedw.
Bogs and marshes, rather rare.
M. Near Saddleworth; *C. P. Hobkirk*.
L. Whernside.
Y. Bogs near Ripon; *W. Brunton*.
N. Near Poppleton; *W. Whitwell* (1879).
W. Ilkley Moor; *B. Carrington*; *J. S. Wesley*. Wetherby, one place; *J. S. Wesley*. Askham Bogs (!) *H. Ibbotson*. Chandler's Whins, Hob Moor; *G. Webster*.
A. Near Malham Tarn; *W. Whitwell*. Yeadon Dam; *B. Carrington*. Bog above Shipley Glen; *James Abbott*. Baildon; *H. F. Parsons*; *W. West*. South Milford, pool sides.
C. Sharleston Common; *H. F. Parsons*.
D. Potteric Carrs, Doncaster! Thorne Waste; *H. F. Parsons*.
T. Martin Beck bog near Bawtry; *G. Webster*.

336. Hypnum giganteum Schpr.
Very rare.
W. Rills near the Cow and Calf Rocks, Ilkley; *J. S. Wesley* (1878). In dike by railway, near Towton. Chandler's Whins, Dringhouses, York; *G. Webster*; *W. West*.
A. Lower Gordale; *W. West*.

337. Hypnum sarmentosum Wahl.
Very rare.
L. Near the summit of Ingleborough; *B. Carrington*.

338. Hypnum cuspidatum L.
Watery meadows, dike-sides, etc., one of the very commonest of the Hypna.
Area general; reported from every district.

VAR. **pungens** Schpr.
At the base of the Jackdaw-Crag Quarry cliffs, near Stutton; *J. S. Wesley*!

339. Hypnum Schreberi Willd.
Heathy ground, common, although absent over considerable areas where an aggressive agriculture has destroyed the aboriginal character of the surface.
M. Greenfield Valley; *J. Whitehead*.
L. & R. Common.
Y. Sawley; *W. Brunton*. Grewelthorpe; *W. West*.
N. Great Whernside; *W. West*. Harrogate; *C. P. Hobkirk*. Pateley Bridge. Ramsgill; *J. S. Wesley*.
W. Ilkley. Bolton. Wetherby. Bramham.
A. Malham. Skipton Moor; *W. Whitwell*. Bingley Moors. Camblesforth; *W. N. Cheesman*.
C. Todmorden, Hebden Bridge, Luddenden, etc.
D. Rawcliffe; *H. F. Parsons*. Bell Hagg, Rivelin Valley, etc.; *Amos Carr*.
T. Roche Abbey Valley; *J. H. Aveling* in *Hist.* (1870).

340. Hypnum purum L.
Wet meadows, stream-banks, very common, yet in some few districts about the large manufacturing towns by no means obtrusively plentiful.
Area general; reported from every district.

341. Hypnum stramineum Dicks.
Moorland swamps, not rare, but seldom in any quantity.
M. Greenfield; *W. Wilson* (1832); *J. Whitehead*.
L. Baugh Fell, Garsdale slope. Whernside (!) *W. West*.
N. Great Whernside; *W. West*. Ramsgill; *J. S. Wesley* (1878).
W. Rombalds Moor, Ilkley (1858); *H. Ibbotson*! *H. F. Parsons* (1878).
A. Malham Moor. Keighley; Baildon Moor; and Adel Bog; *W. West*.
C. Stansfield and other moors, Todmorden; *A. Stansfield*.

342. Hypnum scorpioides L.
Bogs, not very rare.
L. Ingleborough slopes; *B. Carrington*. Near the Austwick Beck above Wharf; *W. West*. Bowland Forest moors.
[R. Giggleswick Tarn; *B. Carrington*. Tarn now gone; is the moss, too?]
W. Beamsley Moor; *L. C. Miall*.
A. Bog near Malham Tarn; *J. Nowell*. Cullingworth; *W. West*.
C. Stansfield Moor, Todmorden, very rare; *A. Stansfield*.
D. Thorne Waste.

343. Hylocomium splendens (Hedw.) Schpr.
Heaths, grassy banks, etc., not common in the south-east, but the reverse of rare, and often locally abundant in the north-west half of the Riding.

L. Rawtheydale. Dentdale. Garsdale. Ingleton, etc.
R. Linn Gill. Pearts Crags, Settle; *J. Windsor*. Tosside Chapel.
Y. Sawley Moor; *W. Brunton*. Hackfall.
N. Scotton and Bilton banks. Gateswood, Knaresborough; *W. Kirkby*.
W. Bolton; *B. Carrington*. Grassington. Wetherby; *J. S. Wesley*. Bramham. Stubbing-moor, Rigton. Arncliffe; *W. West*.
A. Malham and Bingley; *J. Nowell*.
C. Todmorden; *A. Stansfield*.

[**Hylocomium umbratum** (Schreb.) Schpr. Incognit.
My only record, '*H. umbratum*, Dallow Gill near Ripon; *W. Brunton*,' a *MS.* note in *Joseph Woods*' handwriting in *Townsend's copy of O.B.G.* (1805), referred to the '*umbratum*' of Smith's English Flora, *not* of Schreber, and therefore to what we now know as *H. flagellare*.]

344. Hylocomium brevirostre (Ehrh.) Schpr.
Rocky wooded places, rare; a decreasing species.
L. Dentdale. Helks Wood, Ingleton; *J. Nowell*.
R. Linn Gill, Ribblehead.
N. Pateley Bridge; *Nat.*, iv. 175.
W. Bolton Woods, abundant; *B. Carrington*.
A. Malham Cove; *J. Nowell*.
C. Hebden Valley, very rare; *T. Stansfield*.

345. Hylocomium squarrosum (L.) Schpr.
Wet grassy banks in woods and swamps, and on upland pastures, common.
Distribution general; recorded for numerous localities, from every drainage district, yet decidedly uncommon about Leeds, and 'growing scarcer about Todmorden' (*A. Stansfield*).
Ranges from 1,800 feet down to near sea level at Rawcliffe.

VAR. **calvescens** Wils.
Wetherby; *J. S. Wesley*.

346. Hylocomium loreum (L.) Schpr.
Montane. Woods and heaths, rather rare.
M. Greenfield Valley; *J. Whitehead*.
L. Dentdale. To 1,800 feet on Whernside. In fruit in Kingsdale and Helks Wood; *W. West*.
R. Long Preston. Linn Gill, Ribblehead. Cleatop Park, Settle; *J. Windsor*.
N. Harrogate; *T. Hick*. Middlesmoor; *W. West*.
W. Washburn Valley; *J. W. Davis* and *W. Todd*! Near Wetherby; *J. S. Wesley*. Bolton; *B. Carrington*.
A. Bingley Woods; *L. C. Miall*.
C. Hebden Bridge. Todmorden; *A. Stansfield*. Holmfirth Moors; *C. P. Hobkirk*.
D. Midhope and Bradfield Moors; *J. Holland*. Langsett; *Amos Carr*!
(T. Roche Abbey Valley; *J. H. Aveling* in *Hist.* A little doubtful.)

347. Hylocomium triquetrum (L.) Schpr.

Sub-xerophilous, woodland banks, common in most districts, but scarce about the large towns.

L. Bentham. Wethercote. Slaidburn. Ingleton (fruit); *W. West.*
R. Langcliffe Wood, etc., Settle; *J. Windsor.* Linn Gill.
Y. Studley Woods.
N. Middlesmoor; *W. West.* About Knaresborough.
W. Grassington; Bolton! Wetherby (!) *J. S. Wesley.* Becca banks, Aberford.
A. Malham; *W. West.* Skipton Woods; *W. Whitwell.* Bishop's Wood; *W. N. Cheesman.*
C. Todmorden; *A. Stansfield.* Dungeon Wood, Huddersfield (now extinct); *C. P. Hobkirk.*
D. Conisborough. Went-vale; *H. F. Parsons.*
T. About Maltby. Roche Abbey; *J. H. Aveling.* Kiveton; *Amos Carr.*

Three hundred and forty-seven species are included in the foregoing list, as now to be found in the West Riding, all the forms being clearly Native where found, if we except *Leptobryum pyriforme*, which is *adventive* in its greenhouse-pots habitats.

Seventeen species have been enclosed within brackets as Incogniti, and not numbered, the evidence in favour of their occurrence being too slight to warrant inclusion. One of these only—*Buxbaumia*—is, properly, an Extinction, and comes in the same category as the Belbank *Trichomanes*.

A much larger proportion of our mosses are *Montane*, i.e. boreal, in their distribution throughout Britain than obtains amongst the Flowering Plants and Ferns. The following tabular list specifies these, and shows in which of the four zones they occur. Those restricted to the two highest are usually most abundant in the lower of those two, because of the greatly restricted area of the surface in the fourth zone above 1,800 feet.

MONTANE MOSSES.	Zone 1. 0—300 ft. in Trent dist. only	Zone 2. 300 to 900 feet.	Zone 3. 900 to 1800 feet.	Zone 4. 1800 to 2400 feet.
Sphagnum intermedium		2	3	
,, laricinum			3	
,, rubellum			3	
Andreæa petrophila		2?	3	4
,, alpina			3	4
,, Rothii			3	4
,, crassinervis			3	4
Gymnostomum rupestre		2	3	
,, curvirostrum			3	
,, microstomum			3	4?
,, commutatum				4

MONTANE MOSSES.	Zone 1. 0—300 ft. in Trent dist. only	Zone 2. 300 to 900 feet.	Zone 3. 900 to 1800 feet.	Zone 4. 1800 to 2400 feet.
Anæctangium compactum				4
Weissia crispula			3	
,, cirrhata	1	2	3	
Rhabdoweissia fugax		2?	3	4
Cynodontium Bruntoni		2	3	
Dicranella squarrosa		2	3	
,, Scottianum		2		
,, fuscescens		2	3	4
Campylopus alpinus				4
,, atro-virens		2	3	
Seligeria Donii		2	3	
,, pusilla		2	3	4
,, recurvata		2	3	4
,, trifaria			3	
Brachyodus trichodes		2	3	
Blindia acuta		2?	3	4
Didymodon flexifolius		2	3	4
Eucladium verticillatum		2	3	
Barbula tortuosa	1	2	3	4
Trichostomum crispulum		2	3	
Distichium capillaceum		2	3	4
Encalypta rhabdocarpa				4
,, ciliata		2	3	4
Grimmia funalis				4
,, trichophylla		2	3	
,, Doniana		2?	3	4
,, ovata			3	
Rhacomitrium aciculare		2	3	4
,, protensum		2		
,, sudeticum			3	4
,, heterostichum		2	3	4
,, fasciculare		2	3	4
,, lanuginosum	1*	2	3	4
Ptychomitrium polyphyllum	1*	2	3	
Amphoridium Mougeotii		2	3	4
Ulota Drummondii			3	
,, Bruchii		2	3	
Orthotrichum cupulatum		2	3	
,, rupestre				4
,, Lyellii		2	3	
Œdipodium Griffithianum			3	

* Both these doubtless *Adventive* here—they were growing together.

MONTANE MOSSES.	Zone 1. 0—300 ft. in Trent dist. only	Zone 2. 300 to 900 feet.	Zone 3. 900 to 1800 feet.	Zone 4. 1800 to 2400 feet.
Tetraplodon mnioides			3?	4
Splachnum sphæricum			3	4
Discelium nudum		2		
Entosthodon ericetorum		2	3	
,, Templetoni				4
Bartramia ithyphylla		2	3	4
,, Halleriana		2	3	
,, Œderi		2	3	4
Philonotis calcarea		2	3	
Breutelia arcuata		2	3	4
Webera elongata		2	3	
,, albicans		2	3	4
Zieria julacea		2	3	4
Bryum pallescens		2	3	
,, alpinum		2	3	
,, pallens	1	2	3	4
,, roseum		2		
,, filiforme			3	4
Cinclidium stygium			3	4
Mnium orthorhynchum			3	
,, subglobosum			3	
Tetrodontium Brownianum		2	3	
Oligotrichum hercynicum			3	4
Pogonatum alpinum		2	3	4
Polytrichum strictum			3	4
Diphyscium foliosum	1?	2	3	
Fissidens osmundoides		2		
Schistostega osmundacea		2	3	
Fontinalis squamosa		2	3	
Hedwigia ciliata		2	3	
Antitrichia curtipendula		2	3	
Neckera crispa	1	2	3	
Pterygophyllum lucens		2	3	
Myurella julacea				4
Heterocladium heteropterum		2	3	
Pseudo-leskea catenulata				4
Pterogonium gracile		2		
Cylindrothecium concinnum		2	3	
Orthothecium intricatum		2	3	4
,, rufescens			3	
Brachythecium plumosum		2	3	
Eurhynchium pumilum		2	3	

MONTANE MOSSES.	Zone 1. 0—300 ft. in Trent dist. only	Zone 2. 300 to 900 feet.	Zone 3. 900 to 1800 feet.	Zone 4. 1800 to 2400 feet.
Eurhynchium Teesdalii		2	3	
Hyocomium flagellare		2	3	
Rhynchostegium depressum		2	3	
,, Borrerianum		2	3	
Plagiothecium pulchellum			3	4
,, nitidulum			3	
Amblystegium Sprucei		2	3	
Hypnum exannulatum		2	3	
,, Cossoni			3	
,, vernicosum		2	3	
,, revolvens		2	3	
,, uncinatum		2	3	
,, rugosum		2		
,, incurvatum		2		
,, crista-castrensis		2	3	
,, eugyrium			3	4
,, ochraceum		2	3	
,, sarmentosum				4
,, scorpioides	1	2	3	
,, brevirostrum		2	3	
,, loreum		2	3	
	5	77	96	46

We thus see the presence of 46 species above 1,800 feet, 10 of these confined to this zone; 96 species between 1,800 and 900 feet, 13 of these confined to this zone; 77 species out of the 115 found below 900 feet, but only eight—one doubtful—occurring only below that elevation; whilst only 5 of these Montanes occur at all in Zone 1 in the level lowland district of Trent, and of these 5 two are adventive to the locality in which they occur, whilst the others should perhaps hardly be reckoned as Montane at all.

This proportion of 115 Montanes to 347 species, about 33 per cent., is much higher than what we get amongst the Flowering-plants and Ferns. This is to be accounted for by the constitution of Mosses, as a class, fitting them to thrive at a lower temperature, or higher altitude (much the same thing), in the presence of more constant moisture; in fact, to resist greater vicissitudes of climate than can more highly organised and complex vegetable forms. On the other hand, in consequence of their requirements, the species which are *Xerophilous*—dry-loving—in their distribution, and therefore with us found to restrict themselves to the calcareous rock tracts, are comparatively few, numbering 27 only. Some are distinctly *Pelophilous* (clay-lovers) like *Physcomitrium*, etc., but the great majority are arenophilous in their rock preference; whilst a certain minority are distinctly *lowland* and hygrophilous in their distribution, affecting for the most part the level peat-bogs and marshes, and the often-submerged tree-trunks bordering the larger rivers, where they wind through the Permian and Triassic tracts towards the Ouse, whose bed marks the lowest point of the central Vale of York.

HEPATICÆ.

In this class comparatively little has been done, except by Mr. William West and the writer, since Dr. Carrington ceased to be a resident worker in the Riding; and that little has been almost altogether confined to the multiplication of stational records as apart from critical investigation. Still, the distribution of the species has been pretty fully worked out, and the following list is believed to mirror with tolerable accuracy the present knowledge of a group which presents certain difficulties which the Mosses and Fungi do not present. The lack of any complete recent work on British Hepaticæ precludes otherwise willing workers from investigating them fully. Acknowledgment must here be made for several valuable emendations and additions suggested by Mr. W. H. Pearson after the list was in type, whereby it has been purged of at least one undoubted misconception into which the writer had fallen.

The arrangement adopted is that of the London Catalogue of Hepatics.

Meagre as are the ancient Yorkshire records, Hepatics appear to have attracted the attention of Richardson and others in Ray's time. There are reported stations dating back to 1666, and many of at least a hundred years ago.

Four *types* only can be discriminated with any approach to accuracy—General, Northern, Southern, and Western, according to the British distribution of the species. Most of the species are Hygrophilous, a very few—in number—showing Xerophilous preferences, whilst many are Montane in the same sense of the term as applied to Mosses.

I. MARCHANTIACEÆ.

1. Marchantia polymorpha L. *Liverwort.*

General type.

On stones, rocks, earth, in damp situations usually, very common. July.

Distribution general; in all the districts, male receptacles common, female capitula less frequent.

Range: Wide, to 2,000 feet alt.

Earliest record as a West Yorkshire plant: Richardson, 1677.

2. Preissia commutata Nees.

Montane. Southern type.

On damp ground, on stones, or in rock-fissures in hilly districts, with a preference for such as are calcareous. Sub-xerophilous; widely distributed, but not common, although Mr. Pearson says it is much more frequent than *Asterella*. May.

Range: 1,800—300 feet.

Earliest record: Brunton, 1805.

L. Ingleborough; Helks Wood, Ingleton; *B. Carrington.*
R. Linn Gill, Ribblehead!
Y. Studley Park; *W. Brunton* (1805).
A. Malham Moor; *B. Carrington.*
C. Near Todmorden; *J. Nowell.*
T. Rocks, Roche Abbey (1876)!

3. Conocephalus conicus Neck. (*Fegatella conica* Corda.)

Native. General type.

On moist stones, rocks, and earth, by rills, ditches, etc., very common, and fruiting not unfrequently. February-April. Aromatic when bruised.

Range: 0—1,250 feet.

Earliest record: Bolton, 1775.

Distribution general; all the districts.

4. Asterella hemisphærica (L.) Beauv. (*Reboulia hemisphærica* Raddi.)

Montane. General type.

On soil among exposed rocks and banks in the hill districts, rare; much more so than No. 2, and specimens named *Asterella* are often *Preissia* according to Mr. Pearson. A very frequent companion of this is *Solorina saccata.* June.

Range: 500—2,400 feet.

L. Dentdale; *G. Stabler* (1872). Clapdale (!) and Ingleborough summit; *B. Carrington.* Thornton Force; *W. West*!
R. Settle; *H. F. Parsons.*
W. About Arncliffe and Buckden.
A. Malham Moor (!) *W. West.*

5. Dumortiera hirsuta (Sw.) N. ab E. (*March. irrigua* 'Tayl.' Wils.)

Alien. Western type.

Warm moist rocky places in shade, confined to the south and west, but amenable to cultivation. Intentionally planted—a reprehensible experiment—in 1878 near Dent by *Mr. G. Stabler* (see *Naturalist*, vol. iv, p. 114). Seeing, however, that it occurs in England (near Hastings, Sussex) in a native condition, and that our area does nourish some few Atlantic type species, such as *Wahlenbergia hederacea, Trichomanes radicans*, etc., it is by no means improbable that it may yet be discovered in one of our sheltered glens amid the western hills. The frond has some resemblance to that of *Pellia epiphylla* in habit, is large (2 to 5 inches long, by half to three-quarters in width), membranaceous—fleshy, bilobed, undulate-edged, without pores, and of a deep, bright, velvety green colour.

L. Near Dent, Dentdale; *G. Stabler*—introduced.

6. Lunularia vulgaris Mich. (*Marchantia cruciata* L.)

Native. Southern type. Sub-xerophilous.

On moist calcareous banks, crumbling lime-rocks and walls, rather frequent, but only locally abundant. Its crescentic scyphi, containing gemmæ, are plentifully produced from February to October. Fruit in July and August, very rare, although plants of both sexes grow together not seldom.

Range: Imperfectly ascertained, but not high apparently. Alt. 900 feet—at Douk Gill, and above Clapham (*W. W.*)—the highest I have seen.

Earliest record: Bolton, 1775.

L. Sedbergh. Clapham; *W. West.*
R. Douk Gill, Horton!
Y. Kirkby Malzeard!
N. Nidd banks, Cowthorpe Weir!
W. Barden Tower. Burley; *W. West.* About Wetherby, plentiful!
A. Gordale; Bingley and Manningham; *W. West.* Selby.
C. Halifax; *J. Bolton.* Todmorden (1871); *B. Carrington, spn.*! *Carr. et Pears., Exsicc.*, ii, 148.
D. Sheffield; *Amos Carr*! Rawcliffe; *G. Webster.*
T. Kiveton; *Amos Carr*!

7. Targionia hypophylla L.

Native. Southern type. Xerophilous.

Dry rocks or banks with a warm aspect or south-west exposure, very rare. May.

Range: not clearly ascertained.

Earliest record: Knowlton (Teesdale), Linn. Trans., 1798, v, 86.

L. Bank near Cowgill, Dentdale Head.
A. Near Keighley; *T. Knowlton, O.B.G.* (1805); etc.

8. Riccia glauca L.

Native. General type. Pelophilous.

Stubble and fallow fields, and ditch-banks on heavy soils, rather rare, but perhaps often overlooked. November-March.

Range: Not above 300 feet.

First record: 1775, Bolton, as *Riccia minima*—the var. c of Moss Catalogue, Ed. ii.

Y. Bishop Monckton.
N. Long Flat fields, Knaresborough. Pannal to Spofforth; *W. West.*
W. Priest Hill, Wetherby.
A. Selby; *Y. N. U. Trans.*, Series E, p. 81; *W. N. Cheesman.*
C. Halifax; *J. Bolton.*
D. Goole; *H. F. Parsons.*
T. Waleswood; *Amos Carr*!

9. Ricciella fluitans (L.) Al. Braun.
Southern type. Range : 20—50 feet.
Still ditches and stagnant pools, among Lemnæ, very rare.
Three or four stations, but all in the low-lying district around Goole and Snaith.
First record : Parsons, 1878.
A. Hirst Courtney and Drax ; *H. F. Parsons.*
D. Pond near Pollington ; *T. Birks, junr.* = Near Goole ; *H. F. Parsons* (1878).

10. Ricciocarpus natans (L.) Corda.
Southern type. Range : Low.
Stagnant peaty pools and ditches, very rare.
First record : Birks, 1878.
W. Askham Bog, dike near rail by 'the jungle' (!) *H. F. Parsons* (1879). = Chandler's Whin ponds ; *G. Webster.*
D. In a ditch at Ousefleet ; *Thos. Birks, junr.* (1878). Between Adlingfleet and Fockerby (*H. T. Gardiner*).

II. JUNGERMANIACEÆ.

11. Frullania dilatata (L.) Nees.
General type. Range : 20—1,500 feet.
On banks, rocks, but most frequent on bark.
Earliest notice : Bolton, 1775.
L. Sedbergh ; *H. F. Parsons* ! Rawthey Dale ; *W. West.* Ingleton ; *Id.*
R. Linn Gill ! Giggleswick ; *W. West.*
Y. Banks, Mickley, Sleningford.
N. Rocks, Knaresborough.
W. Bolton ; *B. Carrington.* Wetherby. Flint Mill. Becca, Aberford.
A. Malham ; *W. West.* Bishop's Wood ; *W. N. Cheesman.* Kelfield ; *H. F. Parsons.*
[C. Halifax district ; *J. Bolton* (1775). Now gone ?].
D. Eastoft ; *H. F. Parsons.*
T. Roche Abbey (!) *H. F. Parsons.*

12. Frullania fragilifolia Tayl.
Southern type. Range : Ill ascertained.
On bark or warm dry rock-ledges along with other mosses, very rare.
First record : Stabler, 1876.
L. Dent ; *G. Stabler* (1876).
R. Linn Gill, on rocks with *Neckera* ! (1879).

13. Frullania Tamarisci (L.) Nees.
Montane type. Range : 300—1,500 feet.
On heathy banks, stones, and bark, rather rare.
Earliest notice : Teesdale, 1798.

L. Ingleton ; *W. West.* Brant Fell ; *W. West.* Cowgill, Denthead ; *W. West* !
R. Rocks, Douk Gill, above Horton !
N. Brimham Rocks ; *R. Teesdale.*
W. On ground amidst *Cladoniæ*, on turfy hillocks in Jackdaw Crag Quarry, near Stutton !
A. Malham ; *W. West.*

14. Lejeunea hamatifolia (Hook.) Dum.
Western type.
Trees and rocks, generally the latter, but very rare.
One record only : alt. 600 feet ?
L. Near Dent, Dentdale ; *G. Stabler* (1877).

15. Lejeunia echinata Hook. (*L. calcarea* Lib.).
Southern type. Xerophilous. Range : 600—1,200 feet.
Among mosses, etc., on limestone rocks and vegetable humus.
Earliest notice : Hooker, 1802.
L. Kingsdale, on vertical rock-faces ; *W. West.* Rocks near Clapham Cave ; *B. Carrington.* Rocks at Ingleborough ; *Sir W. J. Hooker.*
R. Linn Gill, amid other gatherings. Penyghent ; *W. West.*
W. Dripping lime-rocks, Litton ; *J. Whitehead.* Hesleden Gill ; *W. West* !
A. Malham Cove, stream-bank rocks ; *B. Carrington* ; conf. *W. West.* Gordale ; *J. Nowell.*

16. Lejeunia ovata (Dicks.) Tayl.
Western type. Very rare.
Reported for the *County* by Teesdale, 1798.
L. Detected among gatherings from damp rocks, Whernside (!) made by *F. A. Lees* and *W. West*, Aug. 1878.

17. Lejeunia ulicina Tayl. (*minutissima* Auct. olim non *Sm.*).
Western type.
On bark or mosses, very rare, but perhaps occasionally overlooked.
L. On trees, Dent ; *G. Stabler.*
[R. On an ash-tree near Clitheroe ; *J. Nowell.* Probably out of the Riding.]
W. On *Neckera crispa*, Bolton Abbey ; *Schimper, Herb.*, detected by *R. Spruce*—vide Journ. Bot., 1881, p. 36, and 1887, p. 74.

18. Lejeunia inconspicua (Raddi) Nees. (*J. minutissima* Sm.).
An extinct form, so far as the one recorded station goes, if the species is distinct from the preceding. Dr. Carrington gives it as an exstipulate variety in his *W. R. Flora* of 1862 ; Dr. Spruce in *Journ. Bot.*, 1881, pp. 36-37, corrects a mistake of early years, and shows that Smith's *minutissima* (in fine fruit in Sowerby's fig.) was not *J. ulicina* Tayl.
[C. 'On an old ash near Shedden Clough, now cut down ; *J. Nowell* !' *B. Carrington.*]

19. Lejeunia serpyllifolia Libert.
General type. Range : 0—1,500 feet.
On trees and vegetable humus overlying rock-ledges, usually much intermingled with mosses, frequent.
First record : Carrington, 1862.
L. Ingleton and Whernside (!) Great Blake Gill ; *W. West.*
Y. Near Ripon !
W. Wetherby and Bramham West Woods! Bolton Woods ; *B. Carrington, spn., Exsicc.*, iii, 195.
A. Malham ; *W. West.*
C. Todmorden district ; *A. Stansfield.*
T. Roche Abbey !

20. Lejeunia Mackaii (Hook.) (*Phragmicoma* Dum.).
A Western type species, hardly to be included.
L. One small tuft among pickings from Ingleborough neighbourhood, requiring confirmation ; *B. Carrington.*

21. Radula complanata (L.) Dum.
General type. Range : 0—1,500 feet.
On bark, with a wide range.
Earliest notice : Bolton, 1775.
L. Ingleton ; *Y. N. U. Excurs.* Deepdale ; *W. West.*
R. Bolton-in-Bolland, Gisburn, etc.
Y. Mickley, etc., near Yore, Ripon.
N. Knaresborough and Scotton Banks.
W. Wetherby (!) *J. S. Wesley.* Bolton Woods (!) *W. West.*
A. Malham ; *W. West.* Kelfield ; *H. F. Parsons.* Stainor Wood near Selby ; *W. N. Cheesman.*
C. Halifax ; *J. Bolton* in *Cat.* (1775).
D. Rawcliffe ; *H. F. Parsons.* Fockerby ; *H. F. Parsons.*
T. Firbeck. Roche Abbey ; *A. Carr.*

22. Porella lævigata (Schrad.) Lindb. (*Madotheca lævigata* Dum.)
General type. Xerophilous. Range : 450—1,500 feet.
On calcareous rocks in hilly districts, rare or local.
First record : Nowell, 1842.
L. Whernside. Ingleborough Scars ; *B. Carrington.* Deepdale ; *Y. N. U. Trans.*, Series E, p. 66.
W. Bolton Woods ; *J. Nowell.*
A. Malham Cove ; *B. Carrington* ; *W. West* !

23. Porella platyphylla (L.) Lindb.
General type. Xerophilous. Range : 50—1,500 feet.
On calcareous rocks and limestone walls.
First record : Bolton, 1775.
L. Ingleton ; Clapham ; and Dent ; *W. West.* Clapham ; *Carr. et Pears., Exsicc.*, i, 71.

R. Linn Gill, Ribblehead. Settle Scars.
Y. Mickley bank, Sleningford !
N. Grimbald's Crag, Knaresborough (!) *J. Jackson.*
W. Bolton Woods (!) in fruit ; *B. Carrington.* Woodhall Quarry !
A. Craven district, frequent ; *B. Carrington.* Malham.
C. Halifax ; *J. Bolton* in *Cat.* (1775).
D. Wentbridge ; *H. F. Parsons.* Endcliffe Valley, Sheffield ; *A. Carr* !
T. Roche Abbey (!) *J. H. Aveling* in *Hist.* (1870) ; *H. F. Parsons.*

24. Porella rivularis (Nees) Lindb.
Northern type. Range : 150—600 feet.
Rocks and tree-boles in or near water, rare.
First record : Nowell, 1842.
L. Clapham district ; *W. West* ! (Given me by gatherer from there (with others) as *P. platyphylla.* It no doubt occurred by the Clapham beck.)
W. Wet lime-rocks by Wharfe at Flint Mill, Wetherby ! On tree-boles by the water, Collingham ! (Ilkley, in a tarn ; *G. Webster.*)
A. Rocks in the stream from Malham Cove (!) *B. Carrington.* Still there in plenty.

[**Porella pinnata** (Dicks.) Lindb. (*Madotheca Porella* Nees.)
Incognit.
The Wharfe-bank plant was at first erroneously considered to be this.]

[**Pleurozia purpurea** (Lightf.) Dum. (*cochleariformis* Weiss.).
Incognit.
The first and last record (for I can find no confirmation of its occurrence) is that by *R. Teesdale* (*Linn. Trans.*, v, p. 83, A.D. 1798), who wrote : 'Moist heaths, rarely found in fructification. I once found it in that state near Harrowgate.' By this some other species is evidently meant ; *W. H. Pearson.*]

25. Lepidozia reptans (L.) Nees.
General type. Range : Not high apparently ; 0—900 feet.
In woods, on mossy turf and stones, frequent.
First record : Bolton, 1775.
M. Seal Bark, Greenfield (very large) ; *W. Wilson* (1832).
L. Wood between Ingleton and Bentham. Clapham ; Kingsdale ; *W. West.*
R. Woods, Giggleswick !
N. Cranberry Carr, Hunsingore !
W. Bolton Woods ; *B. Carrington.* Wetherby.
A. Tarn House Lane, Skipton ; *W. Whitwell.* Bingley ; *W. West.* Idle and Esholt ; *B. Carrington.* Moseley Wood, Horsforth. Meanwood Wood !
C. Near Todmorden ; *J. Nowell.* Hebden Valley. Halifax ; *J. Bolton.* North Dean ; *H. F. Parsons.* Huddersfield district ; *C. P. Hobkirk.*
D. Wharncliffe Woods ! Hollow meadows, Sheffield ; *Amos Carr* !

26. Lepidozia cupressina (Sw.) Lindb. (*tumidula* Tayl.).
Western type.
Damp grit rocks in shade, rare.
One record only (1857); should occur in other places.
A. On gritstone boulders, Idle Woods; *B. Carrington* (1857).

27. Lepidozia setacea (Web.) Mitt. (*Jungermania multiflora* of Huds., teste Spruce. *Blepharostoma setacea* Dum.)
Western type. Range: Ill ascertained.
In shady turfy places, rare.
First record: Nowell, 1842.
W. Collingham Wood! On *Sphagnum*, Rombalds Moor; *B. Carrington*—this was a creeping form, named *sertularioides* in the 1862 *W.R. Flora*.
A. Thornton Moor; *W. West*!
C. Moist woods near Todmorden; *J. Nowell* = Hareley Wood; *T. Stansfield*. Marsden; *W. West*. Stansfield Moor, scarce; *A. Stansfield*.

28. Bazzania trilobata (L.) Gray.
Northern type. Montane.
Rocky woods and thickets, rare.
M. Seal Bark, Greenfield; *W. Wilson* (1832).
L. Helks Wood; *B. Carrington*. Ingleborough; *W. West*. Dent; *G. Stabler*.
R. Fir-wooded lane bank, Hodder Valley (!) *W. Todd*.
Y. Hackfall!
N. Near Pateley Bridge; *J. Nowell*.
W. Sparingly, amid *Hylocomium triquetrum* and *H. splendens*, in a bushy corner of Jackdaw Crag Quarry, Stutton!
A. Idle Woods; *B. Carrington*. Bingley; *H. F. Parsons*. Shipley Glen; Newsholme Dean; *W. West*.
C. Hebden Valley; *A. Stansfield*.
D. Bell Hagg, Sheffield; *Amos Carr*!

29. Bazzania tricrenata (Wahl.) Lindb. (*deflexa* Mart.)
Northern type. One record; altitude unknown to me.
'Rocky places, rare'; *B. Carrington*.
L. On Whernside, and a few stems from Ingleborough; *B. Carrington* (1857).

30. Odontoschisma Sphagni (Dicks.) Dum. (*Cephalozia Sphagni* Spruce. *Sphagnœcetis communis* Nees.)
General type. Range: 50—1,800 feet.
In bogs and swamps, amidst *Sphagnum*, *Leucobryum*, etc., rare.
First record: Carrington, 1857.
L. Moor west of Ingleborough; *B. Carrington*.
R. Moor between Linn Gill and Ribblehead!
N. Harlow Heath, Harrogate!
W. Beamsley Beacon; *W. West*. Askham Bog.
A. Malham Tarn bog; and Harden Moor; *W. West*.
C. Stansfield Moor; *John Nowell*.

[**Cephalozia byssacea** (Roth.) Mart.
Moist banks, tolerably common. The name occurs (without No. 31 being also named) in the Todmorden List of Hepaticæ, kindly supplied by Mr. Stansfield; but as no record for the species occurs in Dr. Carrington's Flora, the next was doubtless the plant intended. The older authors regarded Roth's *J. byssacea* and Smith's Eng. Bot. *J. divaricata* as synonymous.]

31. Cephalozia divaricata (Sm., E.B.). (*Jung. Starkii* Nees.)
General type.
On moors and moist loamy banks, rather common.
First record: Carrington, 1862, vel Spruce.
L. Ingleton; *W. West*.
Y. Studley Park; *B. Carrington*.
W. Bolton Woods; and between Eastby and Barden; *B. Carrington*. Near York; *R. Spruce*.
A. Skipton. Baildon; Thornton Moor; *W. West*.
C. About Todmorden tolerably common; *A. Stansfield*. Old wall under Oscar Wood, opposite Sewage Works Flats, Barnsley!
D. Endcliffe Valley; *Amos Carr*.

32. Cephalozia bicuspidata (L.) Dum.
General type. Range: Low? ill ascertained.
Banks, heathy ground, damp rocks, very common, and rather variable in segmentation of the leaves. Fr., April-May.
First record: Bolton, 1775.
L. Ingleborough; *W. West*.
R. Wigglesworth to Tosside!
N. Harlow Heath!
W. Wetherby (!) *J. S. Wesley*.
A. Shipley Glen; Bingley; and Hawksworth; *W. West*. Adel Moor! Morton Bog near Selby (!) *H. F. Parsons*.
C. North Dean; *H. F. Parsons*. Halifax; *J. Bolton*. Todmorden; *A. Stansfield*.
D. Went-vale; *H. F. Parsons*.

33. Cephalozia Lammersiana (Hübn.). (*C. uliginosa* Spruce?)
General type. Range: Ill ascertained.
Moors and swamps; until recent years considered a var. of *C. bicuspidata*. Probably common.
First record for W. Riding: Lees, 1880.
R. Blea-moor.
N. Harlow Heath, Harrogate, on peat, fine!
W. Beamsley Beacon; *W. West*. Askham Bogs.
A. Baildon; *W. West, spn.*! teste *M. B. Slater*.
D. Thorne Waste.

34. Cephalozia curvifolia (Dicks.) Spruce.
Northern type. 'Alpine woods' (*B. Carrington*), decaying trunks of trees, very rare. Range: not made out.
R. Penyghent; *W. West*—a little doubtful!
N. Brimham rocks; *R. Teesdale* (1798).
C. Near Todmorden; *J. Nowell*, vide *R. Spruce* 'On Cephalozia.'

35. Cephalozia connivens (Dicks.).
General type. Range: 0—1,250 feet.
Wet mosses, trailing on *Sphagna*, rare or overlooked.
First record: Carrington, 1857.
R. Penyghent; *W. West*.
Y. Grewelthorpe Moor; *W. West*.
W. Bolton Woods; *B. Carrington*.
A. (Walls, Esholt; *Carr.*, *W. R. Flora*, 89—a 'forma *minor*,' but doubtless some error.) Malham Moor. Bingley; *W. West*.
C. Near Todmorden; *John Nowell*.

36. Cephalozia multiflora Spruce. (*Jungermania connivens* Mult. Auct. Angl.)
General type. Range: 0—1,250 feet.
Heathy banks in shade, on rotting bark, and more rarely on sand-rock, often fruiting luxuriantly; also on peaty moors, amid *Sphagnum*, *Leucobryum*, etc., usually sterile, common. Only clearly differentiated by recent observers who have profited by the disquisition thereon of *R. Spruce* and *M. B. Slater*. Hudson's *multiflora* was *L. setacea*.
First record: Pearson, 1878.
R. Ribblehead Moor (1879)!
W. Askham Bogs (1880)! Bolton Woods, on earth by the woodland paths; *W. H. Pearson*, spn. (1878), *Exsicc.*, iii, 174.
A. Esholt; *W. West*! teste *W. H. Pearson* (1880). Probably Dr. Carrington's wall plant mentioned under No. 35 was this.

[**Cephalozia heterostipa** Carr. & Spruce.
The specimen of the species provisionally named as above, from Ingleborough, given in Spruce's monograph 'On Cephalozia,' is 'now referred by Dr. Spruce to *Jungermania inflata*'; *W. H. Pearson*.]

37. Cephalozia catenulata (Hübn.).
General type? Range: not known.
On rotting wood, turfy shady banks, and rocks; very rare or else ill-understood as yet by resident West Riding workers.
First record: Carrington, 1857.
A. Shady walls, Esholt Wood; and Idle Wood; *B. Carrington*.
C. Hareley Wood, Todmorden; *John Nowell*.

38. Lophocolea bidentata (L.) Nees.
General type. Range: 0—1,800 feet.
On rotting wood, moist banks amid grass, or on bare clay, one of the very commonest of Hepatics, and very variable in size and facies.
Distribution general; occurs in every area.
First notice: not to be ascertained clearly, as is not unfrequently the case with very common species, observer after observer not thinking it worth while to put the specific observance on record. Bolton in 1775, however, gives it as a Halifax plant in his Catalogue.

39. Lophocolea heterophylla (Schrad.) Nees.
General type. Range: Not well made out.
On rotting stumps and sticks and on mossy grit-rocks in shade, rare.
First record: Carrington, 1862.
R. Slaidburn!
Y. Studley Park; *B. Carrington*.
N. Devonshire Wood, Stockeld!
W. Askham Bog, in the 'Far Wood,' on rotting stumps with fruiting *Tetraphis*. Bolton Woods; *B. Carrington*. Washburn Valley; *W. West*.
A. Bingley and Baildon; *W. West*. Bishops Wood near Selby; *W. N. Cheesman*.

40. Leptoscyphus interruptus (Nees). (*Plagiochila interrupta* M. & N.)
Southern type. Range: 0—1,250 feet.
Dry shady rocks, very rare; 'often confounded with *Chiloscyphus*.' Fr. May.
First record: Carrington, 1857.
W. Near 'the Strid,' Bolton, abundant; and in fruit (1857); *B. Carrington*. (Again in 1878; *Carr. & Pears.*, *Exsicc.*, ii, 86.)
A. Malham Moor; *John Nowell*.

Var. **pyrenaicus** Spruce. (*J. polymorpha* Carrington, W.R. Flora.)
A. Malham Moor; *B. Carrington* (1857).

41. Chiloscyphus polyanthos (L.) Corda.
On wet ground, in or by clear rills, in wells and springs, etc., frequent.
Range: 100—1,500 feet.
First record: Bolton, 1775.
L. Howgill Fells; *W. West*. Ingleborough, and Austwick Scar; *B. Carrington*.
W. Rombalds Moor, Ilkley (!) *J. S. Wesley*! *G. Webster*. Bolton Woods, very fine; *B. Carrington*. Woodhall!

A. Malham; *H. F. Parsons*. Baildon; *W. West*. Runnel between Cononley and Carlton; *W. Whitwell*. Rothwell Wood! Pond, Bingley Wood; *B. Carrington*. Mill dam, Yeadon; *B. Carrington*. Bishops Wood near Selby; *W. N. Cheesman*.

C. Spring in Oakwell Wood, Birstall! Near Huddersfield; *C. P. Hobkirk*. Hareley Wood, Todmorden; *A. Stansfield*. Halifax; *J. Bolton, Cat.* Marsden; *W. West*.

D. Saxon Well, Upper Bradfield (!) *R. A. Gatty*. Hollow Meadows stream, Rivelin, Sheffield; *Amos Carr*!

The VAR. **rivularis** is of more frequent occurrence than the type in our hilly county, so far as my experience goes. I have not separated the records for it from those of the type, but the inflorescence of var. *rivularis* is said to be *autoicous*.

42. Harpanthus scutatus (Web.) Spruce.

Western type. Moist shady, stony, wooded banks, very rare.
Range: 500 feet.
First record: Carrington, 1862.

W. In dense compact tufts on rocks near the Strid; *B. Carrington*. Arncliffe.

(A. Knottingley; *Y. N. U. Trans.*, Series E, p. 24. But I fear error of designation.)

43. Saccogyna viticulosa (L.) Dum.

Northern type. Range: ill known.
On damp rocks and banks, amidst mosses, rare.
First record: Bolton, 1775.

L. Dent; *G. Stabler*.

A. Malham Cove; *R. Teesdale* in *Linn. Trans.* (1798).

C. Dulesgate, Todmorden; *John Nowell*. Halifax; *J. Bolton, Cat.* (1775).

44. Kantia Trichomanis (L.) Gray. (*Calypogeia trichomanis* (L.) Corda.)

Native. General type. Range: 0—1,200 feet.
Wet clayey and mossy banks in shade, common. Known to Richardson in 1724; *Ray, Syn.*, iii.

R. Slaidburn = Hodder Valley; *Rev. W. Fowler*.

Y. Ripon.

N. Scotton Bank.

W. Bolton. Ilkley Woods. Lindley Woods. Wetherby. Devonshire Wood, Sicklinghall! Quaker's Wood, Acomb (var. *fissa*); *Geo. Webster*.

A. Esholt Wood, c. fr.; *B. Carrington*. Bingley. Malham; *R. Teesdale* (1798). Kildwick, Adel, etc.; *W. West*.

C. North Dean; *H. F. Parsons*. Hebden Valley (!) Todmorden; *John Nowell* (and var. *fissa* Radd.). Near Woolley; *W. West*.

D. Wharncliffe Woods! Many places about Sheffield; *Amos Carr*!

45. Kantia Sprengelii Mart.

Native. General type. Range: Ill known.
Habitat somewhat different from the preceding, more partial to mossy turf in woods and open bogs and heaths; frequent probably, but not discriminated by observers until 1879.
First record: Parsons, *spn.*—here!

Y. Wood near Mickley.

A. Brayton Barf; *H. F. Parsons, spn.*! Bishop Wood, Hambleton!

46. Kantia arguta (Nees & Mont.).

Diœcious. Leaves bidentate at apex, divergent, smaller and more remote on the elongated stem than in *K. trichomanis*, often tipped with gonidiferous gemmæ. Amphigastria deeply bifid. In facies to the unaided vision rather like small *Saccogyna*. Uniformly sterile.
General type. Range: Not known.
On wet sandy ground, or on moss or other hepatics, rare so far as present knowledge goes. Not previously recorded for the West Riding.

W. Wood bank by river Cock, Stutton!

47. Trichocolea tomentella (Ehrh.) Nees.

Northern type. Range: 100—1,100 feet.
Shady, rocky, or mossy banks, rather rare.
First record: Richardson in Ray's Syn., iii (1724). 'Plentifully in Yorkshire'; teste spn. in *Herb. Buddle*.

L. Helks Wood, Ingleton, and Clapdale Scars; *B. Carrington*.

R. Linn Gill, sparingly.

N. Pateley Bridge; *T. Hick* = Ravensgill.

W. Bolton Woods, c. fr.; *John Nowell*; *B. Carrington*.

A. Shipley Glen and Holden Gill; *H. T. Soppitt* and *W. West* (1885).

C. Stansfield Moor, not common; *A. Stansfield*.

48. Blepharozia ciliaris (L.) Dum. (*Ptilidium ciliare* Nees.)

1. TYPE. **B. pulcherrima** Hoffm. (*P. ciliare* var. *ericetorum* Syn. Hep.)

General type. On heathery or grassy moory ground, frequent.
Range: 0—1,500 feet.
Earliest notice; Bolton, 1775.

L. Ingleborough; *B. Carrington*; rep. *W. West*!

N. Harrogate Heath; *R. Teesdale* in *Linn. Trans.* (1798).

W. Otley Chevin and Rombalds Moor; *B. Carrington*. In Jackdaw Crag Quarry, Stutton, sparingly, and on heathy ground at the Cross roads, Stubbing Moor, Rigton. Beamsley Beacon; *W. West*.

A. Baildon Moor; *W. West*. Haddlesey; *W. N. Cheesman*.

C. Stansfield Moor, Todmorden; *A. Stansfield*. Halifax; *J. Bolton*.

D. Bell Hagg and Sandygate near Sheffield; *Amos Carr*!

2. VAR. **Wallrothiana** Nees.

Native. Northern type. Range: About 800 feet.
On rocks and bark of trees, very rare, or (?) overlooked.
First notice: F. A. Lees, 1879. Previously known only in the Castle Howard district, on bark (R. Spruce).

R. On mossy limestone rock, a spur of Penyghent, above Douk Gill, Horton-in-Ribblesdale, growing in compact circinate patches, very sparingly, with a distinct facies, like a stunted brown-velvety *Frullania*.

49. Anthelia julacea (L.) Lightf.

High exposed moist grit or metamorphic rocks, very rare.
Northern type. Montane. Range: 2,000—2,200 feet.
First record: Teesdale, 1798.

L. Howgill Fells; *H. F. Parsons* and *W. West*.

[N. Brimham Rocks; *R. Teesdale* in *Linn. Trans.* (1798). Can Teesdale have made some mistake, or is the florula of Brimham Rocks less alpine now than formerly? Alpines exist still, at low levels in Yorkshire, and unless Teesdale's specimens—I have not access to them—prove to be something else, I must believe that, although now gone, it once existed there.]

50. Blepharostoma trichophylla (L.) Mitt. (*Jungermania* Hook.)

On peaty banks, amid other mosses, and among *Sphagnum* in swamps, rare—or overlooked.
General type. Range: 0—1,800 feet.
Earliest notice: Ray, 1696.

L. Whernside. Clapdale; *B. Carrington*. Ingleborough; *B. Carrington*.

W. Bolton Woods; *B. Carrington*; rep. *W. West*.

A. Near Keighley; *W. West*.

C. North Bierley; *Dr. Richardson* in *Ray's Syn.*, ed. ii, p. 338, fide *S. Doody*; *Martyn*, 1746. Marsden; *W. N. Cheesman* and *W. West*! Near Todmorden; *A. Stansfield*.

D. Among gatherings of *Sphagnum* from Red Brook, Sheffield; *A. Carr*.

51. Scapania compacta (Roth.) Lindb. (*S. resupinata* Tayl.).

Southern type. Range: 700 feet (approx.).
Heathy or turfy banks, in rather dry stony places; xerophilous?—in its only West Yorkshire station, very rare. Has a resemblance in physiognomy to embrowned *Radula* or the stunted cæspitose *Scapania nemorosa* of exposed places. Said in 'the books' to be common, but for 25 years no additional locality has been discovered.
First record: Baker in Miall and Carrington's Flora, 1862.

A. Near Malham Cove; *J. G. Baker*; *B. Carrington* = Left-hand side going up to Cove top, with *Helianthemum*, *Thymus Serpyllum*, etc. (c. fr. June 1880). West says *S. æquiloba* occurs here, and he never found *compacta*.

52. Scapania undulata (Dum.).

General type. Range: very wide, 50—2,200 feet.
Wet rocks in streams, by cascades, etc., common among the hills.
First record: Bolton, 1775.

M. Head of Greenfield Valley! Denshaw Clough!

L. Dentdale, Whernside rills, Ingleborough, common!

R. Trough between Slaidburn and Clitheroe; *W. Fowler*!

N. Nidderdale head streams; *J. S. Wesley*, spn.! Pateley Bridge; *R. Teesdale* in *Linn. Trans.* (1798).

W. Bolton Woods; *B. Carrington*. Ilkley Moor-slope; *J. S. Wesley*! Blubberhouses Moor.

A. Top of Tarn House Lane, Skipton, and near foot of Norton Tower; *W. Whitwell*. Yeadon Moor; *B. Carrington*. Idle Wood; *B. Carrington* = (?) Eccleshill; *W. Nuttall*. Near Druids' Altar, Bingley, and Baildon; *W. West*. Moseley Wood, Horsforth; *J. R. Murdoch*! Burn, near Selby; *W. N. Cheesman*.

C. North Dean, Halifax; *J. Bolton*; *H. F. Parsons*. Todmorden, Gorple; *John Nowell*. Rishworth Valley stream, abundant!

D. Wharncliffe; *Amos Carr*! Eccleshall Wood; *A. Carr*!

53. Scapania uliginosa (M. & N.) Nees.

Northern type. Range: 1,600—1,900 feet?
Marshy heathery places on mountains; only one record.

L. South slope of Ingleborough, very rare; *B. Carrington* (1857).

54. Scapania nemorosa (L.) Dum.

General type. Range: 100—1,800 feet.
Wet shady banks, old walls, quarry slopes, rocks, etc., next to *S. undulata* the commonest of the genus. Fr. May-June.
First record: Carrington, 1862.

L. Clapdale; *G. Stabler*. Dentdale Head. Whernside Crags!

R. Ribblehead.

W. Arncliffe; *W. West*. Bolton and Ilkley; *B. Carrington*; *spn., Carr. & Pears.*, ii, 92. Farnley; *H. F. Parsons*. Wet rocks in old lime quarries at Kirk Deighton; near Stutton; Stubbing Moor, etc., near Wetherby!

A. Shipley Grove; *B. Carrington*.

C. Todmorden; *A. Stansfield*.

D. Old quarry ledges near Conisborough; *Amos Carr*.

T. Roche Abbey; *J. H. Aveling* in *Hist.*

55. Scapania resupinata Dum. (*Jungermania nemorosa* var. Tayl.)

Western type. Range: Ill known.
Open heathy and stony banks, very rare, I think, in the West Riding. Dr. Moore writes of the peculiar (and constant) fawn-colour of its growing patches—the only time I gathered it I did not notice this.

First record : Carrington, 1862.
R. On dry slaty rocks near Horton-in-Ribblesdale (1879) !
W. Ilkley Wells, c. fr.; *G. Stabler*; *B. Carrington*; *G. Webster*.
A. Malham Moor; *B. Carrington*.

56. Scapania purpurea (Dill.) Carr.
General type. Montane? Range : 700—1,500 feet.
Moorland swamps, rare.
First record : Nowell in Carrington's Flora, 1862, but station not in Riding.
N. Up Ramsgill Beck; *J. S. Wesley* (1878) !
C. (Ramsden Clough, Todmorden; *John Nowell*. This station not in West Riding.) Plentiful in bogs by stream, amid grass and *Sphagnum*, above the Derby Inn, Rishworth Valley (!)
D. Wet rocks, Rivelin Valley, near Sheffield; *Amos Carr* !

57. Scapania æquiloba (Schwæg.) Nees.
Northern type. Montane. Range : 450—900 feet.
Rocky places, rare.
First record : Carrington, 1862.
L. Crina Bottom, etc., near Ingleborough; *B. Carrington*. Clapham; *W. West*.
R. Giggleswick Scar; *J. G. Baker*.
W. Arncliffe; *W. West*. Bolton Woods; *G. Stabler*.
A. Gordale; *B. Carrington*; *W. West* (?) = Malham Cove; *G. Stabler*.

58. Scapania Bartlingii Nees. (*S. rupestris* Dum.)
Type?
Moist rocks, one station.
(Y. R. Teesdale's *Jungermania 'rupestris'* (1798) was an *Andreæa*.)
W. Rocks near the Strid, April 1858; *B. Carrington*; *spn.*, *Carr. & Pears.*, *Exsicc.*, i, 18.

59. Scapania umbrosa (Schrad.) Nees.
Western type.
Only one record—'moist rocks and banks'—but I never found the species, which is a very small one, with serrate, ovate leaves.
W. Bolton; *B. Carrington, Flora* (1862); *Carr. & Pears.*, *Exsicc.*, i, 20. Confirmed August 1883, being found in Bolton Woods by *G. A. Holt*, teste *W. H. Pearson*.

60. Scapania curta (Mart.) Nees.
General type.
Moist stony banks, rare. Much smaller and more compact than *S. nemorosa*, and with the leaves not so uniformly toothed.
First record : Carrington, 1860.
W. Rigton Hill top, on broken ground between 'Spring' and 'Lund's' Wood.
A. Idle Woods; *B. Carrington*. Yeadon; *B. Carrington*. Shipley Glen and Newsholme Dean; *W. West*.

VAR. **spinulosa** Nees.
A. On walls, Esholt Woods, c. fr., 1860; *B. Carrington*.

61. Diplophyllum albicans (L.) Dum.
General type. Range : 50—2,400 feet.
Damp shady banks and rocks, very common and variable in facies. Fr. April-May.
Distribution general; seen or recorded from every district.
First record : Teesdale, 1798, but there may be an earlier; it is difficult to assign with certainty the many-worded descriptions of the ante-Linnean writers to their proper place, in all cases.

62. Plagiochila asplenioides (L.) M. & N.
General type. Range : 50—1,200 feet.
Loamy or rocky woodland banks, common, yet absent over considerable areas in its large form.
First record : Bolton, 1775.
L. Dent; *G. Stabler*. Austwick. Clapham, etc.; *W. West*.
R. Douk Gill, Horton, fruiting; *W. West*.
Y. Hackfall.
W. Bolton Woods. Woods about Wetherby, common; and in lanes on open banks (the small form). Washburn Valley.
A. Malham; *J. G. Baker*. Bingley; *W. West*. Bishop's Wood near Selby; *W. N. Cheesman*.
C. Todmorden district; *A. Stansfield*. Halifax; *J. Bolton, Cat.*
T. Roche Abbey ! *J. H. Aveling*, and *H. F. Parsons*.
D. Went-vale; *H. F. Parsons*—perhaps the variety.

VAR. **Dillenii** Tayl.—a small form—is commoner than the type, especially in the lowland districts on the crumbling lime-rock of the Permian formation, but few distinctive notices are on record. I have seen it in districts R., N., W., A., D., and T., and it probably occurs in the whole ten.

63. Plagiochila spinulosa (Dicks.) Nees.
General type. Range : 450—900 feet.
Moist shady banks, rare?
First record : Carrington, 1858.
L. Dent; *G. Stabler* ! Ingleborough; *J. G. Baker*.
W. Bolton Woods, near the Strid; *John Nowell*; *B. Carrington*.
C. Stiperden Clough, Todmorden; *A. Stansfield*.

64. Mylia Taylori (Hook.) Gray.
Northern type. Montane. Range : 1,500—2,400 feet.
On grit rocks by streams and where subject to mist, mostly at high levels, frequent there, absent elsewhere.
First record : Carrington, 1862.
M. Seal Bark, Greenfield; *W. Wilson* (1832), *Nat.*, p. 182 (1887).

L. Whernside. Near Ingleborough summit (!) *B. Carrington* (1856), confirmed by *W. West*.
R. Penyghent; *W. West*—away from any stream.
(N. 'Middlesmoor,' Masham Exc., Y. N. U. Was an error, the plant being *Jungermania cordifolia*.)
W. In fruit (extremely rare) in the Upper Wharfe valley; *G. A. Holt* (1883).
C. Todmorden; *John Nowell*; *A. Stansfield*.

65. Mylia anomala (Hook.) Gray.
Southern type. Range : Not well ascertained.
On mosses and moors, mainly at *low* levels—as opposed to the preceding—rare.
First record : Carrington, 1862.
N. Harlow Heath !
W. Ilkley Moor (!) *B. Carrington*. Askham Bog !
A. Thornton Moor; *W. West*.
C. Stansfield Moor; *John Nowell*. Really the preceding?

66. Eucalyx obovata (Nees.) Lindb. (*Nardia* Carr., *Brit. Hep.*) (*J. tersa* N.)
General type. Range : Not clearly ascertained.
On wet banks and rocks.
First record : Carrington, 1862.
L. Clapham; *W. West*.
W. By the Strid, Bolton; *B. Carrington*.
A. Shipley Grove; *B. Carrington*.

67. Eucalyx hyalina (Lyell) Lindb.
General type. Range : 450—1,500 feet.
Wet rocks amid moss, etc., rather common.
First record : Carrington, 1862.
L. Howgill Fells; *W. West*. Ingleborough; *G. Stabler*.
N. Near Pateley Bridge; *B. Carrington*. Moist rocks in railway-cutting west of Spofforth station; *J. Jackson* !
W. Bolton Woods; *B. Carrington*.
C. Hebden Bridge; *G. E. Hunt*.

68. Jungermania lanceolata Nees. (*Liochlæna* Nees.)
General type.
Wet rocks, very rare, but perhaps often overlooked from being an insignificant species, seldom growing massed together, and with no striking habit.
First record : West, junr., 1880.
(L. 'Ingleborough'; *W. West* (specimens lost, doubtful).)
A. Shipley Glen; *W. West, junr.* ! (1880). Alt., 350 feet.

69. Jungermania Schraderi Mart.
Northern type. Montane.
I have this, testified to by authority, from the station mentioned below, but know nothing more about it. No other Yorkshire record exists, but the species occurs in North Wales and Westmorland.
L. Whernside (at about 2,000 feet ?); *F. A. Lees* cum *W. West* (1879).

70. Jungermania crenulata Sm.
Moist moory banks, frequent in its smaller gracile form at least, but records are scanty, it being an inconspicuous species.
Range : Ill ascertained.
First record : Carrington, 1862.
L. Ingleborough and Clapham; *W. West* !
A. Yeadon Moor; *B. Carrington*.
C. Todmorden; *John Nowell*; 'common'; *A. Stansfield*.
D. Rose Wood, Sheffield; *A. Carr*.

VAR. **gracillima** Sm. (*J. Genthiana* Hübn.)
L. Dent; *G. Stabler*.
A. Shipley Glen; and sand quarry, Esholt Wood; *B. Carrington*. Nab Wood, Bingley; *W. West*.

71. Jungermania pumila With.
General type. Range : 0—1,200 feet.
Moist rocks, frequent.
First record : Hooker, 1802.
L. Howgill; *W. West*. Dent; *G. Stabler*. Near Clapham Cave; *B. Carrington*.
Y. Studley; *W. J. Hooker* (1802).
W. Bolton Woods; *B. Carrington*.
C. Todmorden; *A. Stansfield*.

VAR. **nigricans** Hook.
A. Swallow-hole above Malham Cove; *B. Carrington*.

VAR. **Zeyheri** Hübn.
W. Boulder sides, near 'Strid,' Bolton Woods; *B. Carrington*.

72. Jungermania sphærocarpa Hook.
General type. Range : Uncertain—to 1,800 feet?
Damp stony places, amid moss; few records, but, perhaps, not rare.
First notice : Nowell, 1842.
R. Penyghent; *W. West*.
N. Railway-bank shales, Spofforth end of Pannal tunnel !
A. Bingley; *W. West*.
C. Hareley Wood, on shales; *John Nowell* ! Todmorden; *G. A. Holt* (1880).

73. Jungermania lurida Dumort. (*J. nana* Nees.)

W. On shale in a wet shady gully, Roundhay, June 1882; *W. West, spn.*! The plant here was first said to be *E. obovata* by Mr. W. H. Pearson, but he afterwards diagnosed it as '*J. lurida* approaching *sphærocarpa*—paroicous.' Perhaps Dumort's plant is a bad species; my idea written before I received the corroborative opinion from Mr. Pearson, viz., that 'there is no satisfactory character by which to separate *lurida* and *sphærocarpa*.' The Spofforth gathering put under No. 72 seems identical with it. Suggestive of easy error as well as affinity, however, is the fact that the perichætium or colesule in both *lurida* and Nos. 69 and 70 is trigonal and crested—points that in part characterise Lindberg's *Eucalyx*, under which both *obovata* and *hyalina* are now usually placed, but Mr. Pearson points out that this Roundhay plant cannot be *Eucalyx* because the perianth is free to the base.

74. Jungermania cordifolia Hook.

Northern type. Montane. Range: 600—2,400 feet.

On wet rocks, and on stones in alpine streams, rare, but locally abundant.

First record: Nowell, 1840.

L. Howgill Fells. Whernside, N.E. shoulder! Foal-foot, Ingleborough; *John Nowell*. Helks Wood, Ingleton; *B. Carrington*. Clapham; *W. West*.

R. Penyghent; *W. West*!—'Spring between Hull Pot and summit, at 1,600 feet.'

N. Middlesmoor; *W. West*; *Y. N. U. Excurs.*!

W. Buckden Gavel end of Great Whernside, by spring (1885).

75. Jungermania riparia Tayl.

General type. Sub-xerophilous. Range: 450—2,000 feet.

On banks and rock by streams in 'limestone ravines' (Carrington) chiefly, rare.

First record: Carrington, 1857.

Taylor thought this was frequently by-passed as *J. pumila*; Dr. Moore remarks that 'when strong it is more likely to be overlooked at first sight for *Chiloscyphus polyanthos*.'

L. Dent. Clapdale; *B. Carrington* (1857). Helks Wood; *W. West* (var. '*minor*') teste *W. H. Pearson*.

W. Rocks by Wharfe, Bolton; *B. Carrington*.

A. Gordale; *B. Carrington*; at 1,250 feet; *W. West*. Baildon; *W. West*.

(C. Shedden Clough, Todmorden; *John Nowell*. Out of Riding?)

Var. **minor.**

On rocks in bed of stream near the Strid; *B. Carrington, spn., Exsicc.*, iii, 169.

Var. **cæspitosa** Spruce *in litt.*

L. On Yoredale crags, Ingleborough, at 2,000 feet, with *Distichium capillaceum*; *W. West* teste *Spruce*.

76. Jungermania Bantriensis Hook.

Western type.

On mossy rocks, one station only?

'Common in the West Riding'; *R. Spruce*.

W. On rocks between Barden Tower and Bolton Abbey; *B. Carrington*.

77. Jungermania Mülleri Nees.

Northern type.

On rocks. I know nothing of this beyond the two Carringtonian records which follow.

L. On stones in the stream from Clapham Cave; *B. Carrington*.

A. Swallow-hole above Malham Cove; *B. Carrington*.

78. Jungermania Schreberi Nees. (*J. barbata* Schreb., type, excl. vars.)

General type. Range: 0—600 feet.

Stony, heathy, and mossy banks, frequent.

First specific record: Wesley, 1878.

R. Moory bank, Tosside Chapel, Slaidburn district.

W. Ilkley; *J. S. Wesley*! (1878).

A. Malham; *W. West*. Baildon; *W. West*.

C. Hebden Bridge; *Y. N. U. Excurs.*, 1879!

D. Wharncliffe Woods! Bell Hagg, Rivelin; *Amos Carr*!

79. Jungermania attenuata Lindbg.

General type. Range: Low?—a middle zone hepatic—of woods and heath-shaded banks.

Stony, heathy, and bushy banks, perhaps the commonest West Yorkshire form of the puzzling *barbata* group, of which it looks like a slender variety with leaves only bi-tridentate at the apices. In facies *J. intermedia* is somewhat like it, but that species has capitate apices, whereas in this (No. 79) they are attenuate.

First record: Carrington, 1862.

L. Bowland Knotts to Clapham! On shales about Burton!

N. Scotton banks, Knaresborough!

W. Farnley near Ilkley; *H. F. Parsons*. Rombalds Moor (1,000 feet); *W. West*. Ilkley (!) *G. Webster*.

A. Idle Wood, and Belbank, Bingley; *B. Carrington*. Baildon (and Nab Wood, Bingley, 300 feet); *W. West*.

C. Ripponden to Rishworth (to 1,200 feet about).

D. Upper Bradfield, Sheffield; *Amos Carr*.

80. Jungermania Flœrkii Web. & Mohr.

Rocks, rare? 'Leaves connivent.' Range: Ill known yet.

First record: Carrington, 1862.

L. Whernside! Ingleborough; *B. Carrington*.

W. Ilkley; *G. Webster*. Beamsley Beacon; *W. West*.

A. Baildon; *W. West* teste *W. H. Pearson*.

C. Near Todmorden; *W. Wilson*; *Carrington's Flora*; *G. A. Holt*, spn.! (1880).

81. Jungermania quinquedentata Nees. (*J. Lyoni* Tayl.)

Northern type? Range: uncertain.

Rocky banks, rare, or confused with *J. Schreberi*?

First record: Bolton, 1775—sub nom. '*quinquedentata*,' but probably *eu-barbata* was meant.

L. Ingleborough; *B. Carrington*.

C. Near Todmorden; *John Nowell*. Old walls, common; *A. Stansfield*.

82. Jungermania lycopodioides Wallr.

Type; northern?

On limestone rocks, rare? I know little of it; no records but Dr. Carrington's, in his Flora.

L. Clapdale and Austwick Scars; *B. Carrington*.

R. Giggleswick Scars; *B. Carrington*.

A. Malham Moor; *B. Carrington*. Gordale; *W. West*.

83. Jungermania exsecta Schmid.

On banks in damp woods, etc., very rare? Is said to bear a likeness to *J. ventricosa*, but 'the stems are usually arranged in a stellate manner,' and to bear upon the leaves *reddish* gemmæ in late autumn.

First record: West, 1880.

A. Baildon; *W. West*!

84. Jungermania porphyroleuca Nees.

Shady banks. One record.

W. Bolton; *B. Carrington* (1876).

85. Jungermania ventricosa Dicks.

General type. Montane. Range: 300—1,200 feet.

Shady, rocky turfy banks, rather rare. The leaves bear conspicuous yellow gemmæ.

First notice: Nowell, 1842.

R. Linn Gill, Ribblehead!

Y. Brimham rocks.

N. Moorland near Harrogate, by Oak-beck; *W. Grainge*!

W. Wharfedale; *R. Spruce*. Rombalds Moor; *B. Carrington*. Moor, Eastby to Barden; *B. Carrington*.

A. Esholt Wood; *B. Carrington*. Nab Wood, Bingley, Baildon, and Thornton Moor; *W. West*.

C. Heptonstall; *John Nowell*.

86. Jungermania bicrenata Lindb. (*J. excisa* Dicks., Sm., E.B.)

Northern type? Range: Uncertain.

Mossy, turfy banks, rare? If this is a northern species (as its provincial distribution would seem to indicate) it is curious it should hitherto have turned up only in the warmest, southernmost drainage-area of the county.

First record: Carr, 1877.

T. Roche Abbey, c. fr.; *H. F. Parsons* (1878). Lindrick Common; *Amos Carr*!

87. Jungermania capitata Hook. (*J. intermedia* Lindb.)

General type. Range: 50—1,000 feet.

On stones and banks in shady situations, less frequent on limestone than on grit.

First record: Wesley, 1878.

R. Ribblehead; *W. West*.

N. Between Harrogate and Fewston; *W. Grainge, spn.*! Ramsgill; *J. S. Wesley*!

A. On grit boulders in Moseley Wood (Horsforth); *J. R. Murdoch* (1879).

C. Ripponden Woods.

D. Wharncliffe Woods (!) *Amos Carr*.

T. On stones in woods near Roche Abbey.

88. Jungermania incisa Schrad.

General type. Range: 300—1,800 feet.

Peaty, turfy banks, under heath, etc., rare apparently.

First record: Hooker, about 1802.

L. Rocky parts of Ingleborough; *W. J. Hooker*; still there (1887); *W. West*. Whernside; *B. Carrington*.

N. Between Harrogate and Fewston; *W. Grainge, spn.*! Harlow Heath!

A. Thornton Moor; *W. West*—this the *granuliferous* 'variety' of Nees.

(C. Rocks, Cliviger near Todmorden; *John Nowell*. Not in Riding.)

89. Jungermania inflata Huds.

General type.

On heaths and moors, very common.

Distribution general; I have it from every district, from 2,000 feet alt. on Ingleborough, to 15 feet on Goole Moor (*H. F. Parsons*!).

Earliest notice: Teesdale, 1798.

90. Jungermania turbinata Raddi. (*Wilsoniana* Nees; *affinis* Wils.)

General type. Xerophilous. Range: 50—1,200 feet.

On the face of moist crumbling rocks, chiefly limestone, frequent.

First record: Carrington, 1862.

L. Crina Bottom and other places near Clapham; *B. Carrington*.

Y. Old quarries near Bishop Monkton.

N. Plumpton rocks! Knaresborough!

W. By the Wharfe near Bolton; *B. Carrington.* Thorparch (!) *J. G. Baker.* Collingham, Langwith Scrub.
A. Gordale; *John Nowell.*
D. Wentbridge; *H. F. Parsons.*

91. Jungermania minuta Crantz.
Northern type. Range: uncertain.
Heathy stony banks, very rare.
First record: Wm. Wilson, 1832.
M. Near Greenfield; *W. Wilson* (1832), *Nat.*, 1887, 183.
R. On stones in bog by stream, Blea-moor, Ribblehead!
C. Haworth; *B. Carrington* in *Flora.* Near Todmorden; *John Nowell.*

92. Nardia emarginata (Ehrh.) Gray. (*Sarcoscyphus Ehrharti* Nees.)
Northern type. Montane. Range: 600—1,800 feet.
Wet sub-alpine rocks, rare.
First record: Carrington, 1862.
L. Foal-foot, Ingleborough; *B. Carrington.*
R. Penyghent; *W. West.*
W. Near Barden Tower; *B. Carrington.*
C. Beaumont Clough, Hebden Valley; *A. Stansfield.*

93. Nardia compressa (Hook.) Gray.
Northern type. Montane. Range: 1,400—2,000 feet.
Wet rocks by cascades, sandy banks by mountain springs, rare.
First record: Nowell, 1842.
L. West side Ingleborough; *John Nowell*; *B. Carrington.*
R. Langdon Valley, Hodder Valley; *W. Fowler* (1881), *spn.*! teste *W. H. Pearson.*
C. Todmorden; *J. Nowell*; *Holt*, *spn.* (1880). Rishworth Valley (1881)!

94. Nardia scalaris (Schrad.) Gray. (*Alicularia scalaris* Corda.)
General type. Range: Wide; 300—1,800 feet.
On clayey banks, or on bare turf over peat on moors, common, and variable. The various forms have not been exhaustively discriminated as to distribution.
First record: Teesdale, 1798.
L. Whernside.
R. Blea-moor.
Y. Fountains Abbey; *B. Carrington.* Brimham Rocks; *R. Teesdale.*
W. Rombalds Moor; Ilkley (!) *Henry Ibbotson.* Wetherby.
A. Yeadon; *B. Carrington.* Baildon; *W. West.*
C. Todmorden; *John Nowell.* Norland Moor; *H. F. Parsons.* Rishworth.
D. Hollow meadows, Redmires Moors; *A. Carr*!
T. Roche Abbey; *J. H. Aveling* in *Hist.*

Var. **repanda** Hübn.
N. Rail-bank a mile west of Spofforth station!

Var. **geoscypha** De Not. (*hæmatostictum* N.).
Perhaps the commonest form on the moors, and one to which, doubtless, several of the localities given for the type or aggregate really refer.
W. Penyghent; *W. West*!
A. Baildon Moor; *W. West*!

Var. **rivularis** Lindb. (A beautiful, giant, aquatic form.)
R. Langdon Valley; *W. Fowler* teste *G. Stabler.*
C. Rishworth Valley, in spray!

95. Gymnomitrium concinnatum Corda.
Northern type. Montane. Range: 900—2,400 feet.
High alpine rocks, very rare, and perhaps extinct on Ingleborough, but there seems to be no doubt it has occurred.
First record: Hooker, 1802.
L. Summit of Ingleborough; *W. J. Hooker*; '!' *B. Carrington.* (Whernside; *W. West* in *Y. N. U. Trans.* But in error, vide p. 50.) Howgill Fells; *W. West* (1879) spn.!
N. Brimham Rocks; *R. Teesdale* (1798)—sub nom. '*Jungermania concinnata* Lightf.'

96. Fossombronia pusilla Nees.
Banks of ditches, on clay, etc., very rare.
Western type. Range: Low.
First record: Nowell, 1842.
A. Keighley; *T. Hebden.* Hawksworth; *W. West.*
C. Henshaw, near Todmorden; *John Nowell* = Royd Hills; *A. Stansfield.*

97. Blasia pusilla L.
Wet clayey banks in woods, by pools or where water has stood, rare.
General type. Pelophilous. Range: Low.
First record: Bolton, 1775.
L. Near Clapham; *B. Carrington.*
A. Clay pool-side near Sherburn (1877)!
C. Halifax; *J. Bolton.* Todmorden; *John Nowell.* Pennant Clough, Hebden Valley; *A. Stansfield.*

98. Pellia epiphylla (L.) Raddi.
General type. Range: 0—1,800 feet.
Damp clayey ditch-banks, wet shaly rock-slopes in woods and open ground, common.
Distribution general; reported from every district, and seen in most by myself.
First record: Bolton, 1775.

99. Pellia calycina (Tayl.) Nees.
General type. Range: 100—1,000 feet.
Wet clayey banks in shade, and in swamps, one form in water, rare.
First record: Carr, 1877.
L. Dent; *G. Stabler.* Howgill Fells; *W. West*! (named *Blyttia* in *Y. N. U. Trans.*, 81).
R. Douk Gill, Horton. Bolton-by-Bolland.
Y. Hackfall.
W. Rombalds Moor; *J. S. Wesley.* Ingbarrow Marsh-field, half way between Wetherby and Spofforth, *in water*, with *Hypnum vallis-clausæ*! Langwith Scrub, Collingham!
A. Baildon Moor and Hawksworth; *W. West.*
D. Bell Hagg rocks, etc., Sheffield; *Amos Carr*, *spn.*, 1877!

100. Aneura pinguis (L.) Dum. (*Riccardia* Gray.)
General type. Range: 0—1,500 feet.
Wet grassy or mossy banks, amidst *Sphagnum* on heaths, etc., frequent.
First record: Bolton, 1775.
L. Clapham; *B. Carrington.*
N. Scotton Banks, Knaresborough!
W. Rombalds Moor; *B. Carrington.* By Wharfe; *J. Bolton*! Moor between Barden Bridge and Eastby; *B. Carrington.* Rigton, Wetherby!
A. By Adel Dam near Leeds, amongst grass!
C. Halifax; *J. Bolton* (1775). Todmorden; *John Nowell.*
D. Rose Wood, Sheffield; *Amos Carr.*

101. Aneura palmata (Hedw.) Nees.
Northern type?
On rich rotten mould, or bark, very rare. Only one record.
T. Waleswood, near Sheffield; *Amos Carr* (1878)!

102. Aneura sinuata (Dicks.) L.C. (*A. pinnatifida* Nees).
Western type. Range: 500—900 feet.
In cool clear water of springs and wells, in detached, floating cushion-like masses, rare.
First record: Dillenius, 1724.
R. Douk Gill, Horton!
W. Ilkley, c. fr.; *H. F. Parsons* (1878).
A. Elm Crag Well, Bellbank Wood, Bingley; *Dillenius*; *B. Carrington.*
C. Near Huddersfield; *C. P. Hobkirk.*
D. Old Saxon Well at Upper Bradfield (!) very fine; *R. A. Gatty.* Rivelin Valley, etc.; *Amos Carr*!

103. Aneura multifida (Dill.) Gray.
General type. Range: 0—1,000 feet.
Among wet grass, moss, heath, etc., on banks where water has stood, rather common.
First record: Nowell, 1842.
L. Lower Garsdale. Clapham; *W. West.*
Y. Studley Park; *B. Carrington.*
W. Many places about Wetherby! Rigton. Compton bank. Washburn reservoirs; *H. F. Parsons.*
A. Near Yeadon Old Mill-dam; *B. Carrington.* Malham district; *W. West.* Baildon; *W. West.*
C. Mytholm Clough, Todmorden, etc.; *J. Nowell*; *A. Stansfield.*
D. Goole; *H. F. Parsons.* Wincobank Wood, Sheffield; *Amos Carr*! Eccleshall churchyard; *Amos Carr*!

Var. **ambrosioides** Nees.
I have generally found this intermixed with the type; surely it is hardly worthy of varietal rank? Once only unmingled with the type—on Compton Bank Wood near Collingham (1880).

104. Metzgeria furcata (L.) Dum.
General type. Range: Wide; 100—2,000 feet.
On rotting bark, and damp rocks and mossy soil, frequent.
First record: Bolton, 1775.
L. Ingleton. Silurian slate rocks, Cautley Spout; *W. West.* Dentdale.
R. Hodder Valley! Settle.
N. Hackfall, near Masham; *G. Webster.*
W. Buckden! Bolton Woods; *W. West.* Outershaw!
A. Malham; *W. West*! (var. *nuda*). Little Fenton; *H. F. Parsons.*
C. Staups Clough, etc., near Todmorden; *A. Stansfield.* Halifax; *J. Bolton*, *Cat.* Rishworth Valley.
D. Eastoft; *H. F. Parsons.* Many places west of Sheffield; *Amos Carr*!
T. Waleswood; *Amos Carr*!
Obs.—Is not the large form of *furcata* at Malham *hamata* Lindb.?

105. Metzgeria conjugata Lindb.
Western type. Range: 900—1,800 feet.
Damp mossy rocks, sometimes, but rarely, on bark.
First record: West, 1880.
L. On trees, Howgill Fells, with *Zieria julacea*; *W. West.* Dentdale; *G. Stabler.* Ingleton; *W. West*!
Obs.—Readily distinguished from *M. furcata* by the more robust, paler yellower-green colour of the matted fronds, and the profusion of flowers. Coarser in facies, with fewer hairs and more horn-like to the touch.

106. Metzgeria pubescens Schrank.

Northern type. Montane. Range: 400—1,800 feet.

On damp limestone rocks, and on rotting trunks occasionally. Xerophilous, and uniformly sterile. Common on the scar-limestone and Yoredale shales, unknown off them.

First record: Nowell, 1840.

L. Clapdale; *B. Carrington.* Dentdale! *W. West!* Ingleton.

R. Linn Gill and Douk Gill, Ribblesdale.

W. Outershaw! Bolton Woods; *B. Carrington.* Outershaw and Arncliffe, Litton; *John Whitehead, spn.*; *Carr. & Pears., Exsicc.* ii, 149.

A. Malham; *John Nowell.* Gordale; *W. West.*

III. ANTHOCEROTACEÆ.

107. Anthoceros punctatus L.

Western type. Range: 300—600 feet.

Wet sandy or gravelly banks of ditches or lanes, very rare. In the barren state the flabby semi-translucent fronds may be known from the common Marchantia by the absence of pores.

First record: Merrett, 1666.

R. Wet bank, Douk Gill near Horton-in-Ribblesdale!

C. In a lane at Greenwood Lee, a mile from Heptonstall; *Martyn's App.*, p. 107 (1763); repeated from *Ray, Syn.*, iii, 110 (1724); rep. *Gough's Camden.*

108. Anthoceros lævis L.

Southern type. Pelophilous. Alt. 300 feet.

In clayey stubble-fields or on clayey banks in woods, very rare with us. One record only.

D. Rose Wood, near Sheffield; *Amos Carr* (1876).

LICHENES.

LICHENES.

[The arrangement adopted in the following list of West Riding species is that of the Rev. W. A. Leighton, as given in his 'Lichen Flora of Great Britain,' for not only is it now, but must long continue, the standard work on the subject for English students. To save space the more generally distributed forms have had no localities specially assigned to them. Collectors will soon discover, however—and should remember—that even the very commonest species are not to be found everywhere. Lichens especially love a pure air uncontaminated by smoke; being, also, of slow growth, they cannot attain maturity under the conditions of frequent interference from the surface alterations that are inseparable from thickly-populated manufacturing districts. In the vicinity of the larger towns they will be sought in vain, save in a rudimentary state; and over comparatively large areas upon the coal-measures the varieties are few, the individuals not abundant, and even those depauperate. Some species seem dying out; not a few, formerly of frequent occurrence, are now rarely seen. A much greater proportion of the forms are Xerophilous in type and Montane in their rôle of distribution than is the case with the Flowering Plants, or even the Mosses, and with the exception of the wooded Roche Abbey, Studley and Aberford districts, saxicolous species are throughout the Riding incomparably more abundant than those of corticolous habit. It is amongst the slate and limestone hill-regions that they flourish in greatest variety and luxuriance. Sedbergh, Slaidburn, Settle, Arncliffe, and Bolton in the north-west, Studley, Aberford, and Roche Abbey on the limestones of the lower levels, are the best centres for their convenient study.]

FAM. I.—COLLEMACEÆ.

I. COLLEMEI.

1. Synalissa symphorea (DC.).

Xerophilous.

Rocks, generally maritime, rare.

L. & A. Kingsdale and Gordale; *B. Carrington.*

2. Collema pulposum (Bernh.).

Walls and dry earthy banks, not rare.

L. Clapdale; *B. Carrington.*

W. & A. Not uncommon in Upper Wharfedale and Airedale. Kirk Deighton!

VAR. **tenax** Ach.

A. On moss in Gordale; *B. Carrington.*

3. Collema melænum Ach.
Xerophilous. Calcareous rocks, frequent.
L. Kingsdale !
R. Near Settle ; *Sir J. E. Smith.*
W. Jackdaw Crag Quarry, Tadcaster !
W. & A. Bolton Bridge and Malham ; *S. Hailstone.*
A. Malham Cove ; *R. Richardson* in *Ray's Syn.*, iii (1724).
D. Went-vale ; *H. F. Parsons.*

VAR. **complicatum** Schær.
L. Ingleborough ; *J. G. Baker.*
A. Malham.

4. Collema cheileum Ach.
Walls and rocks, rare.
W. Ilkley ; *B. Carrington.*

5. Collema limosum Ach.
Wet clay soil, only recorded from
W. Near the Strid at Bolton ; *B. Carrington.*

[**Collema plicatile** Ach. Incognit.
A strictly maritime rock species, recorded from Malham by *S. Hailstone*, sixty years ago. Seen since ?]

6. Collema crispum (Huds.) Ach.
W. On gravelly earth in Sweep Lane, Wetherby !

7. Collema ceranoides Borrer.
Xerophilous, rare. Only reported by *Dr. B. Carrington* from near Clapham Railway Station (L.) and neighbourhood of Bell Busk (A.).

8. Collema fluviatile (Huds.) Ach.
On stones in streams in hilly districts.
A. Known from 'under Malham Cove' since 1704 ; *Dr. Richardson* in *Ray's Hist. Plant.*, vol. iii, p. 30. Skipton Springs ; *Samuel Hailstone.*

9. Collema furvum Ach.
On sub-alpine rocks, not unfrequent.
L. Whernside. Ingleborough ; *B. Carrington.*
Y. Walls about Ripon ; *W. Brunton.*
N. Kirk-Deighton, on rocks in quarry, with other sub-alpines !
W. Bolton woods !

10. Collema nigrescens (Huds.) Ach.
On old trees. Only known 'near Ripon ;' *W. Brunton.*

11. Collema flaccidum Ach.
Rocks and walls, rare.
Giggleswick (R.), Bolton Woods (W.), and Gordale Scars (A.) ; *B. Carrington.* (W.) Cowside Beck, Arncliffe ; *W. West.* Jackdaw Crag, Tadcaster !

12. Collema multipartitum Sm.
Sub-alpine calcareous rocks.
L. About Wethercote Cove, first recorded by *W. Borrer.*

13. Leptogium turgidum (Ach.).
On limestone rock, rare.
A. Malham Cove ; *B. Carrington.*

14. Leptogium dermatinum Borr.
A. Only known on the limestone rocks about Malham, whence first reported by *W. Borrer* himself.

15. Leptogium saturninum (Dicks.).
Old tree-trunks in hill districts.
A. Gordale ; *B. Carrington.* The only known locality.

16. Leptogium lacerum (Ach.).
On old walls, rare, in Craven (*B. Carrington*), about Settle (R.), Clapham (L.), Arncliffe (W.), Malham (*W. West*) and Skipton (A.).

VAR. **pulvinatum** Hoffm.
In Helks Wood (L.), and near the bridge at Bolton (W.); *B. Carrington.*

[**Leptogium tenuissimum** (Dicks.). Incognit.
Near Copgrove, in plenty ; *J. Dalton.* An error ?]

FAM. II.—LICHENACEI.

II. CALICIEI.

17. Sphinctrina turbinata Pers.
Parasitic on *Pertusariæ*, not rare.
In Bolton Woods (W.) on *P. fallax* ; *B. Carrington* ; and elsewhere in hilly districts on *P. communis.*

18. Calicium chrysocephalum Ach.
On trees and palings, very rare.
W. On ash-trees at Bolton Abbey ; *B. Carrington.*

19. Calicium hyperellum Ach.
Common upon various trees, but chiefly alder and oak.
L. Sedbergh.
R. Near Slaidburn.
Y. Ripon.
W. Bolton Bridge. Bramham Park !
T. Laughton Valley, Roche Abbey ; *J. Bohler.*

20. Calicium quercinum Pers.
Y. On oaks in Studley Park ; *W. Brunton.*

21. Calicium citrinum (Leight.).
Parasitic on thallus of *Lecidea lucida.*
L. Dent, near Sedbergh, in tolerable plenty ; *J. A. Martindale.* Between Cautley Spout and Sedbergh, where its host is abundant ; *W. West.* Very rare throughout England.

22. Coniocybe furfuracea Ach.
L. & W. On tree-roots in Helks Wood and Bolton Woods ; *B. Carrington.*

23. Trachylia tympanella Fries.
On posts, palings, decaying tree-trunks, common in the districts where lichens are common at all.
W. Bramhope staircase.

III. SPHÆROPHOREI.

24. Sphærophoron compressum Ach.
On rocks, 'rare.'
R. Slaidburn.
W. Bolton and Ilkley ; *B. Carrington.*
C. High-green Wood, Heptonstall ; *B. Carrington.*

25. Sphærophoron coralloides Pers. '*Rock-coral.*'
On rocks, common.
L. Baugh Fell and Whernside ! *W. West.* Ingleborough and Ingleton.
R. Peart's Crags, Stockdale Gill, and Giggleswick Farm, all near Settle ; *J. Windsor.*
C. Hebden Bridge Valley ! Near Todmorden, formerly common, but now rare ; *T. Stansfield.*
D. Rocks on high moors above Sheffield ; *Amos Carr* !

26. Sphærophoron fragile Pers.
All my records are old ones.
Y. Copgrove ; *J. Dalton.*
N. Harrogate ; *R. Teesdale.* Brimham ; *W. Brunton.*

IV. BÆOMYCEI.

27. Bæomyces rufus DC.
On stones and earth, frequent.
Y. Sawley Moor ; *W. Brunton.*
W. Bolton and Otley Chevin ; *B. Carrington.* Near Ilkley ; *W. West.* Jackson's Quarry, Deighton !
A. Esholt Woods ; *B. Carrington.* Kildwick ; *W. West.* Baildon Common ; *J. G. Baker.*
C. Nab Wood, Todmorden ; *T. Stansfield.*
T. Roche Abbey and near Shireoaks ; *J. Bohler.*

28. Bæomyces roseus Pers.
On earth, rare.
L. Park House, near Ingleton ; *J. Windsor.*
N. Crumbling rocks in lane above Spofforth Station !
C. Lanes in Stansfield ; *T. Stansfield.*

29. Bæomyces icmadophilus (Ehrh.).
On the earth, rare.
L. Moughton Fell ; *J. Windsor.*
A. Malham Moor ; *B. Carrington.*
C. Todmorden Moors ; *T. Stansfield.*

V. CLADONIEI.

30. Cladonia papillaria Duf.
N. Brimham ; *W. Brunton.*

31. Cladonia pungens Flk.
L. Helks Wood. Ingleton ; *B. Carrington.*
D. Whitley Thorpe ; *H. F. Parsons.*

32. Cladonia cervicornis Schær.
Hilly districts, on high exposed rocks, not rare.
L. Ingleborough ; *B. Carrington.*
N. & W. Riggs Moor ; and Great Whernside ; *W. West.*
A. Malham Moor and Rombalds Moor ; *B. Carrington.* Newsholme Dean, Baildon ; *W. West.*

33. Cladonia delicata Flk.
On dead wood, rare.
T. Near Shireoaks ; *J. Bohler.*

34. Cladonia alcicornis Flk.
A. On Malham Moor ; *B. Carrington.*

35. Cladonia pyxidata Fr. *Common Fairy-cup Lichen.*
On heathy banks, moors, in woods, etc., very common and in great variety of form. Generally distributed in all the districts.

36. Cladonia gracilis Hoffm.
Moors and heathy hill slopes, not rare in its VAR. *chordalis.*
L. Ingleborough ! Whernside ; *B. Carrington.*
A. Bingley ; *S. Hailstone.*

37. Cladonia degenerans Flk.
A. Near Malham Tarn ; *B. Carrington.*

38. Cladonia furcata Hoffm.

Heaths, both mountain and lowland, frequent, from Craven district, Apperley, Adel, Wharncliffe, etc., to Askham and Thorne.

39. Cladonia squamosa Hoffm.

Mossy woods.

N. Harrogate Heath; *R. Teesdale.*

W. Bolton, Rombalds Moor, and Otley Chevin; *B. Carrington.*

D. Goole Moor; *H. F. Parsons.*

40. Cladonia cornucopioides Fr. *Red-Horn-of-Plenty Lichen.*

On earthy banks, often confused with *C. pyxidata.*

Common from Howgill and Baugh Fells (*W. West*), Ingleborough, Settle (*J. Windsor*), and Todmorden (*T. Stansfield*) to the upper Coxley Valley (*W. West*), Askham Bogs, Bishop's Wood (*W. N. Cheesman*), and Thorne Waste (*H. F. Parsons*).

41. Cladonia deformis Hoffm. '*Scarlet-cup Lichen.*'

With the preceding species, but rarer.

42. Cladonia digitata Hoffm.

Mossy woods, frequent, in endless variety.

L. Whernside and Ingleborough; *W. West.*

R. Slaidburn!

C. Todmorden; *T. Stansfield.*

VAR. **macilenta** Hoffm.

C. Norland Moor; *H. F. Parsons.*

43. Cladonia Florkeana Fr.

A. Malham Moor; *John Nowell.*

44. Cladonia bellidiflora Schær.

L. On Dent Fells and Whernside, but very rare.

45. Cladonia sylvatica Hoffm.

Occasional, with the next species.

46. Cladonia rangiferina Hoffm. '*Reindeer Moss.*'

On mountains and heaths, an alpine species, but with a singularly wide vertical range. Found from the summit of Whernside (2,414 feet) down to the gritstone moors, and the peat mosses about Goole and Thorne and York, but in the well-cultivated districts rare, and in the populous ones of the dales fast disappearing.

47. Cladonia uncialis Hoffm.

Heaths, probably in every district except Trent, common in the hill districts about Baugh Fell, Whernside, Ingleborough, Malham, Rombalds Moor, and Halifax.

C. About Todmorden formerly common, but has now become very scarce; *T. Stansfield.*

VI. STEREOCAULEI.

48. Stereocaulon paschale Ach.

On rocks, rare.

L. Fine on slate rocks in Dale Beck Dale; *W. West.* Ingleborough; *B. Carrington.*

C. Todmorden; *T. Stansfield.*

VII. USNEEI.

49. Usnea barbata (L.) Fr. '*Tree-beard*' *Lichen.*

On bark, palings, etc., frequent and variable.

Generally distributed; in all the districts.

VAR. **florida** (L.) Fr.

N. Scotton banks, Knaresborough!

C. Hebden Valley; *T. Stansfield.*

VAR. **plicata** (L.) Fr.

Y., D., & T. Studley Woods, Wharncliffe, Roche Abbey, etc.!

VIII. RAMALINEI.

50. Alectoria bicolor (Ehrh.).

On alpine rocks.

C. Stansfield Moor, Todmorden; *J. G. Baker.*

51. Alectoria jubata (L.) Nyl. '*Rock-hair.*'

Trees and rocks, frequent amongst the hills, and more sparingly in the low country.

R. Peart's Crags, Settle; *J. Windsor.*

N. Brimham Rocks; *W. Brunton.*

W. Ilkley and Otley Chevin; *B. Carrington.*

A. Baildon; *J. G. Baker.*

C. Todmorden; *T. Stansfield.*

T. Anston Woods; *R. Deakin.*

52. Evernia furfuracea Mann.

On trees, palings, wall-copings among the hills, etc., not rare.

L. Garsdale, Dentdale, Ingleborough; *W. West.*

R. Giggleswick; *J. Windsor.*

Y. Copgrove; *J. Dalton.*

N. Brimham; *R. Teesdale.*

W. Penyghent; *W. West.* Bolton and Ilkley. Beamsley Beacon; *W. West.*

A. Newsholme Dean, Shipley Glen; *W. West.*

D. Wharncliffe Woods!

53. Evernia prunastri L. '*Rags.*' '*Bark-rag.*'

On trees, frequent.

L. Clapdale; *B. Carrington.* Lower Garsdale. Dent, Ingleton, and Austwick; *W. West.*

R. Hollin Hall, near Settle; *J. Windsor.* Slaidburn!

Y. Near Copgrove; *J. Dalton.* Studley!

W. Bolton and Arncliffe; *W. West.*

A. Malham. Rawcliffe; *H. F. Parsons.* Brayton; *W. N. Cheesman.*

C. Hebden Valley; *T. Stansfield.*

[**Ramalina scopulorum** (Dicks.) Ach. Incognit.

On rocks, exclusively a littoral species, and I believe Dr. Windsor's record 'Near Settle'—the sole one—to have been the next species, which is not so servilely maritime. Mr. Leighton quoted it under this in his first edition, but it must be classed amongst the Incognits, since in *Dr. Windsor's Flora Cravoniensis* (1873) the station is given for '*polymorphum*' only, the latter being originally considered a variety of *scopulorum.*]

54. Ramalina polymorpha Ach.

On rocks, a maritime species usually.

L. On Moughton Scars, near station of *Silene maritima*!

R. Near Settle; *J. Windsor.*

55. Ramalina calicaris (Hoffm.) Fr. '*Gray-beard*' and '*Tree-beard*' *Lichen.*

On trees, frequent, its VAR. *fraxinea* much the commoner.

R. Slaidburn!

Y. Studley.

A. Malham; *W. West.*

VAR. **farinacea** (L.).

L. Dent, Ingleton, Clapham; *W. West.*

R. Stackhouse Borrins, near Settle; *J. Windsor.*

T. Roche Abbey.

VAR. **fraxinea** (L.).

L. Lower Dentdale and Garsdale. Ingleton and Clapham; *W. West.*

R. Douk Gill, Horton! Settle; *W. West.*

A. Malham; *W. West.* Rawcliffe; *H. F. Parsons.*

C. Todmorden; *T. Stansfield.*

D. Hook; *H. F. Parsons.*

T. Roche Abbey Valley (!) *J. H. Aveling* in *Hist.* (1870).

56. Ramalina pollinaria Ach.

On trees, rare.

Y. Near Fountains Abbey.

[Luddington; *H. F. Parsons.* In Lincolnshire, but near the border.]

57. Ramalina fastigiata (Pers.) Ach.

R. Giggleswick; *W. West.*

A. Malham; *W. West.* Burn; *W. N. Cheesman.*

T. Roche Abbey Valley; *J. H. Aveling* in *Hist.* (1870).

IX. CETRARIEI.

58. Cetraria islandica L. *Iceland Moss.*

Sub-alpine heaths, aud sometimes lowland fir-woods.

L. Baugh Fell; *H. F. Parsons.* Ingleborough; *J. Windsor.* Whernside and Gragreth!

R. & W. Penyghent; *W. West.*

59. Cetraria aculeata Fr.

On heaths, moors, etc., not rare, from the higher ridges to the lowland mosses.

L. Whernside! Ingleborough; *W. West.*

R. Settle; *J. Windsor.*

N. Brimham Rocks! Middlesmoor; *W. West.*

W. Otley Chevin! Askham! Barden Moors; *W. West.*

A. Malham Moor; *B. Carrington.* Baildon, etc.; *W. West.* Adel Bog!

D. Thorne Moor; *H. F. Parsons*! Wharncliffe.

60. Platysma glaucum (L.).

On rocks, stones, tree-boles, etc., common among the hills.

L. Howgill, Baugh Fell, Whernside! *W. West.*

R. Peart's Crags, Settle; *J. Windsor.*

W. Beamsley Beacon; *W. West.*

A. Walls near Malham Tarn; *B. Carrington.* Rombalds Moor.

C. Todmorden; *T. Stansfield.* Hebden Bridge Valley!

D. Thorne Moor.

X. PELTIGEREI.

61. Nephromium lævigatum Ach.

Rocky woods, on stones or bark, rare.

L. Helks Wood, Ingleton; *B. Carrington.* Wethercote Cave.

R. Linn Gill!

W. Old woods near Buckden! Langstrothdale!

62. Nephromium lusitanicum Schær.

Mossy walls, etc., rare.

L. Wall, Dentdale Head!

63. Peltigera aphthosa (L.).

Shady alpine rocks.

L. Thornton Force; *W. West.* Ingleborough; *Hudson.* Kingsdale; *B. Carrington.*

A. Gordale; *B. Carrington.*

64. Peltigera canina (L.). '*Dog's Liverwort.*'
On heathy banks, stones, etc., common and generally distributed.
L. Bentham! Dentdale! Cavehole Wood, Ingleborough, and Kendall's Gill; *J. Windsor.*
R. Hodder Valley!
N. Brimham! Aketon and Kirk Deighton!
W. Bolton Woods; *W. West.* Otley Chevin! Bramham West Woods.
A. Bingley; *W. West.* Rawcliffe; *H. F. Parsons.* Selby; *W. N. Cheesman.*
D. Wharncliffe! and Rivelin Valley; *A. Carr.* Goole Moor; *H. F. Parsons.*
T. Roche Abbey Valley; *J. H. Aveling* in *Hist.* (1870).

65. Peltigera rufescens (Hoffm.).
On mossy earth, very rare.
C. Todmorden, rare; *T. Stansfield.*

66. Peltigera polydactyla Hoffm.
Mossy walls, banks, etc., rare.
L. Whernside and Dentdale Head!
Y. Studley Woods; *W. Brunton.*
N. Jackson's Quarry, Kirk Deighton!
W. Bolton Woods; *B. Carrington.*
A. Malham; *W. West.* Yeadon Moor; *B. Carrington.*
D. Wharncliffe.

67. Peltigera horizontalis (L.).
Walls and trees, rare.
L. Thornton Force. Ingleborough; *S. Hailstone.*
R. Linn Gill!
Y. Studley Woods; *W. Brunton.*
W. Ilkley; *Sam. Hailstone.*

68. Solorina saccata (L.).
On rocky ground in shady sub-alpine situations, frequent.
L. Dent and Clapham; *W. West.* N.W. side of Ingleborough; *J. G. Baker.* Helks Wood; *H. F. Parsons.* Yordas Cave; *Dr. Smith.*
Y. Studley Park; *W. Brunton.*
W. Outershaw. Penyghent and Heseltine Gill; *J. Windsor.* Kettlewell; *W. West.* By the Strid at Bolton; *B. Carrington.* Stubbing Moor near Rigton!
A. Malham; *B. Carrington.* Gordale, common; *W. West.*
C. On calcareous shales, the Eaves, Heptonstall; *T. Stansfield.*
D. Warmsworth; *H. F. Parsons.*

69. Solorina limbata Sommf.
On damp earth amongst rocks or under spray, very rare.
L. Source of Great Blake Gill, Whernside, profusely; *W. West.* Ingleborough; *B. Carrington.*
A. Malham; *W. West.*

XI. PARMELIEI.

70. Stictina sylvatica (L.).
Mossy rocks, rare.
L. Helks Wood, Ingleton; *B. Carrington.*
W. Bolton Woods; *S. Gibson.*

71. Stictina scrobiculata Scop.
'On trees, rare.'
L. Foal-foot, Ingleborough; *B. Carrington.*
Y. Rocks at Grantley near Ripon; *W. Brunton.*

72. Stictina fuliginosa Dicks.
L. Thornton Force; *W. West.*

73. Sticta pulmonacea Ach.
On the bark of trees, local, and by no means so luxuriant as in the S.W. of England.
L. Garsdale! Helks Wood, Ingleton; *B. Carrington.*
R. Slaidburn!
Y. Studley Royal, Ripon! *W. Brunton.*
W. Bolton Woods; *J. Nowell*!
C. Hebden Valley; *T. Stansfield.*
T. Roche Abbey!

74. Ricasolia amplissima (Scop.).
On trees, rare.
L. Sedbergh; *S. Slinger.* Kingsdale; *B. Carrington.*

75. Ricasolia lætevirens Lightf.
On rocks as well as bark, not rare amongst the hills.
L. Sedbergh; *S. Slinger.* Dent; *W. West.* Dentdale Head! Helks Wood, Ingleton; *B. Carrington.*
Y. Studley! Hackfall near Masham; *W. Brunton.* Copgrove; *J. Dalton.*
W. Bolton Woods; *B. Carrington.*

76. Parmelia caperata (L.).
On trees and rocks, not rare at lower levels.
L. Sedbergh, Lunedale; *S. Slinger.* Dentdale! Helks Wood, Ingleton; *W. West.*
R. Slaidburn! Settle; *J. Windsor.*
W. Arncliffe; *W. West.* Bolton! Aberford.
A. Malham! Bingley and Idle; *Benj. Carrington.*
D. Wharncliffe; *Amos Carr.*
T. Maltby!

77. Parmelia olivacea (L.).
On trees, palings, and rocks, widely distributed but rather rare.
L. Walls, Clapham, common, and in fruit; *B. Carrington.* Dent, etc.; *W. West.*

R. Settle; *W. West.*
W. Outershaw. Arncliffe, Barden, etc.; *W. West.*
A. Rombalds Moor; *H. F. Parsons.*
C. Todmorden, occasional; *T. Stansfield.*
D. Hook near Goole; *H. F. Parsons.*
T. Roche Abbey Valley; *J. H. Aveling* in *Hist.* (1870).

78. Parmelia stygia (L.).
L. & A. On the limestone scars of Ingleborough and Malham; *B. Carrington.*

[**Parmelia lanata** (L.). Incognit.
N. 'Harrogate; *R. Teesdale*,' but probably some other *Parmelia* was intended.]

79. Parmelia physodes (L.). '*Heather-rags.*'
On rocks, trees, heath-stems, etc., not rare, from high elevations to very low ones, on the heaths.
L. Dentdale! Sedbergh; Ingleton; Clapham; *W. West.*
R. Settle; *J. Windsor.* Slaidburn! Austwick; *W. West.*
Y. Ripon; *W. Brunton.*
N. Middlesmoor; *W. West.*
W. Askham!
A. Near Malham Tarn, in fruit; *B. Carrington.*. Adel Moor! Rawcliffe, near Goole; *H. F. Parsons.* Camblesforth; *W. N. Cheesman.*
C. Very common about Todmorden; *T. Stansfield.* Halifax; *W. West.*
D. Wharncliffe!

VAR. **labrosa** Ach.
L. Sedbergh; Clapham; *W. West.*
W. Beamsley Beacon; *W. West.*
A. Malham; *W. West.*

80. Parmelia perlata (L.).
On trees.
L. Helks Wood, Ingleton; *B. Carrington.* Clapham; Dent; *W. West.*
R. Hodder Valley, near Slaidburn! Giggleswick; *W. West.*
A. Malham; *W. West.*

VAR. **ciliata** DC.
L. Sedbergh; *S. Slinger.* Dent; *W. West.*

81. Parmelia aleurites Ach. [*Platysma diffusum* (Web.)]
On trees, etc.
L. Clapham Common; *B. Carrington.*

82. Parmelia tiliacea (Ach.) VAR. **scortea** Ach.
On rocks and wall-stones, rather common.
L. Dentdale; Clapham; *W. West.* Ingleton and Wethercote.
R. Gearstones. Settle; *J. Windsor.*

83. Parmelia conspersa (Ehrh.).
On grit and slate rock, local.
L. Ingleton (on slate) and Austwick Scar; *B. Carrington.* Moughton Fell (on scar limestone); *J. Nowell.* Sedbergh; Dent; Ingleborough (on slate, and frequent); *W. West.*
C. Near Cross-stone Church (on grit), Todmorden! *T. Stansfield.*

VAR. **Mougeotii** Schær.
W. Grit walls, Otley Chevin; *B. Carrington.*
A. Near Malham Tarn; *B. Carrington.*

84. Parmelia sinuosa (Sm.).
L. Limestone walls near Clapham! and Helks Wood; *B. Carrington.*

85. Parmelia acetabulum (Neck.).
Y. On trees, Studley near Ripon; *W. Brunton.*

86. Parmelia saxatilis (L.).
On rocks, trees, etc., common, and generally distributed, from the highest peaks to the walls of the lowland districts. It does not fruit freely, however. We have two, if not three, of the varieties, of which the most obvious,

VAR. **omphalodes** (L.) occurs abundantly on the calcareous rocks of (L.) Whernside! (R.) Penyghent! etc., on the grit rocks of (L. & R.) Bowland Knotts! (N.) Brimham rocks (*W. Brunton*), (W.) Ilkley and Otley Chevin (*B. Carrington*), (A.) Keighley Moor and Baildon Moor (*W. West*), (C.) Todmorden (*T. Stansfield*), (D.) Wharncliffe Crags near Sheffield! and (T.) Roche Abbey valley (*J. H. Aveling* in *Hist.*, 1870), but is hardly ever fertile.

VAR. **furfuracea** (Schær.) is commonest on trees, and fruits in Linn Gill (R.), about Slaidburn, and Barden (W.).

[**Physcia flavicans** (Sw.). Incognit.
Reported from Settle by *Dr. J. Windsor* to Mr. Leighton, but *Mr. J. M. Crombie* (*Lichen Flora*, p. 479, ed. ii) wisely expresses great doubt. No specimens have been seen and no one has confirmed the discovery. This and the next are submaritime species, showing a preference for coast-rocks or trees within influence of sea-breezes. All the localities given by Leighton in Ed. 3 are southern. Still it is just possible that one or both may once have grown on the mural scars of Craven, and survived to Dr. Windsor's time. Some maritime species of plants even linger there still, attestations to the truth of the geological evidence as to the Craven cliffs having once been washed by sea-waves.]

[**Physcia chrysophthalma** (L.). Incognit.
'Near Settle; *J. Windsor.*' Somewhat less likely than even the last species to have really occurred.]

87. Physcia parietina (L.).
On trees and rocks, very common and both generally as well as widely (vertically) distributed throughout the Riding, from Ribblesdale and Upper Wharfedale ! Todmorden (*T. Stansfield*) to Pontefract, Rawcliffe and Hook near Goole (*H. F. Parsons*), Barlow near Selby (*W. N. Cheesman*), and Roche Abbey Valley (*J. H. Aveling* in *Hist.*, 1870).

VAR. **laciniosa** (Duf.).
W. Beamsley, on a mortared wall; *W. West*.

88. Physcia ciliaris (L.). (*Borrera* Ach., Sm. E. F.)
On trees, stone walls, and rocks, not rare.
L. Chapel-le-dale ! Garsdale ! Dentdale.
R. Settle; *J. Windsor*. Austwick; *W. West*.
C. Todmorden; *T. Stansfield*.

89. Physcia pulverulenta (Schreb.), FORMA **pityrea** (Ach.).
Y. Studley Park !
W. Aberford. Baildon; Allerton; *W. West*.
W. & A. On trees, Bolton Abbey and Gordale; *B. Carrington*.
T. Roche Abbey Valley; *J. H. Aveling* in *Hist.* (1870).

90. Physcia obscura (Ehrh.).
A. On apple-trees, Rawden in Airedale; *B. Carrington*. The only record.

91. Physcia stellaris (L.).
On trees, apparently frequent, especially the variety *tenella*.
L. Ingleton !
W. Bolton; *B. Carrington*. Hawkswick; *W. West*.
A. Malham; *J. Nowell*. Skipton to Earby; *W. West*.

VAR. **tenella** (Scop.). (*Borrera* Ach.)
L. Clapham; Ingleton; Dent; *W. West*.
R. Cavehole Wood and near the old corn mill, Settle; *J. Windsor*.
C. On ferruginous rocks, Hareley Wood, Todmorden; *T. Stansfield*.
A. & D. Rawcliffe and Hook; *H. F. Parsons*. Malham; *W. West*.
T. Roche Abbey; *J. H. Aveling* in *Hist.* (1870).

92. Physcia astroidea (Clem.).
On trees, 'rare.'
R. Near Settle; *J. Windsor*.
W. Bolton; *B. Carrington*.

XII. GYROPHOREI.

93. Umbilicaria pustulata (Hoffm.).
On rocks and stones in hill districts.
C. Near Halifax; *James Bolton*. A very old record; perhaps the next species was intended.

94. Umbilicaria polyphylla (L.).
On grit-rocks amongst the hills, chiefly the solitary monophylline form.
L. Baugh Fell. Summit of Ingleborough; *B. Carrington*. Austwick; *W. West*.
R. Widdale and Cam Fells !
N. Harrogate and Brimham; *W. Brunton*.
W. On the grits of Langstrothdale and Fountains Fell !
A. Harden Moor; Baildon Moor; *W. West*.
C. Whirlaw near Todmorden; *T. Stansfield*. Stansfield Moor, Todmorden; *J. Nowell*. The two last localities possibly refer to one station.
D. Head of Dondale !

95. Umbilicaria flocculosa (Wulf.).
On mountain rocks.
A. Malham Moor; *B. Carrington*.

96. Umbilicaria proboscidea (L.).
L. Baugh Fell; *H. F. Parsons* and *W. West*.

97. Umbilicaria erosa (Web.).
On alpine rocks.
C. 'Stansfield Moor near Halifax'; *W. Brunton*.

98. Umbilicaria polyrhiza (L.).
Sub-alpine grit rocks.
L. Ingleborough; *R. Teesdale*.
R. Settle; *Martyn* (1763).
N. Brimham; *W. Brunton*. Plumpton grit !
C. Todmorden; *T. Stansfield*.

XIII. LECANOREI.

99. Psoroma hypnorum (Vahl.).
On mosses, Malham Moor; *B. Carrington*.

100. Pannaria rubiginosa (Thunb.).
On moist shady rocks and trees.
L. Thornton Force, Ingleton; *J. Nowell*.
R. Cavehole Wood, Settle; *J. Windsor*.
W. Upper Wharfedale near Buckden !

101. Pannaria pezizoides (Web.).
Y. On grit rock, Sawley Moor; *W. Brunton*.
C. Stansfield Moor near Todmorden; *J. Nowell*.

102. Pannaria nigra (Huds.).
On limestone rocks, not rare, nor confined to the hills.
L. Dentdale and Garsdale. Helks Wood; Ingleborough; *B. Carrington*
N. Kirk Deighton !
A. Malham Moor; *B. Carrington*.
D. Went-vale; *H. F. Parsons*.
T. Roche Abbey valley; *J. H. Aveling* in *Hist.* (1870) as *Collema nigrum*. Roche Abbey; *J. Bohler*.

[**Pannaria plumbea** (Lightf.). Incognit.
'Trees near Bradford; *Hudson*' in *Bot. Guide*; but perhaps not this species.]

103. Amphiloma lanuginosum (Ach.).
L. Ingleton and Clapham; *W. West*.
W. On mossy rocks by the Strid, Bolton Woods; *B. Carrington*.

104. Squamaria crassa (Huds.).
On calcareous earth, not rare in the Craven district.
L. Ingleton !
R. Kelcowe, near Settle; *J. Windsor*.
N. Knaresborough. Near Harrogate; *R. Teesdale*.
A. Gordale and Malham; *B. Carrington*.

105. Squamaria saxicola (Poll.).
On rocks, flints, and brick walls.
L. Garsdale, Dentdale !
R. Settle; *J. Windsor*.
W. Ilkley; *B. Carrington*. Old quarry near Tadcaster !
A. Malham; *B. Carrington*.
D. Went-vale; *H. F. Parsons*. Brayton Barf; *W. N. Cheesman*.

106. Placodium murorum (Hoffm.).
On walls and rocks, one of the more generally distributed species on lime-rock from Dentdale Head ! to Went-vale (*H. F. Parsons*). One of the showier saxicolous species, but less common in the elevated tracts than the next.

VAR. **miniatum** Hoffm. [sp. in Leight. III].
Y. Tower of Fountains Abbey !

107. Placodium callopismum (Ach.).
On walls and rocks, chiefly in Craven.
L. Ingleborough pavement !
R. Cam Fell. Giggleswick Scar; *B. Carrington*.
A. Malham Moor; *B. Carrington*.

108. Placodium citrinum (Ach.).
On the mortar of walls.
Craven, Yorkshire; *Sam. Hailstone*.

109. Placodium candicans (Dicks.).
On limestone rocks.
L. Thornton Force; *W. West*. Ingleborough; *B. Carrington*.
A. Gordale Scars; *B. Carrington*.
D. Went-vale; *H. F. Parsons*.

110. Lecanora vitellina (Ach.).
On walls, rocks, decaying posts, etc., not rare and widely distributed, from the Craven district and (C.) Todmorden *T. Stansfield*) to (D.) Sheffield (*Amos Carr*) and (T.) Roche Abbey (!) *J. H. Aveling* in *Hist.* (1870). (W.) Ilkley; *H. F. Parsons*.

111. Lecanora candelaria (Ach.).
On old mossy trees, rare.
L. Ingleton; *W. West*.
R. Cavehole Wood, Settle; *J. Windsor* !
W. Bolton Woods; *B. Carrington*.
A. Near Gennet's Cave, Gordale; *B. Carrington*.

112. Lecanora squamulosa (Schrad.).
On mountain rocks, rare.
L. or R. Bowland Knotts !

113. Lecanora fuscata Schrad.
A. On gritstone walls on Rombalds Moor, and at Rawden; *B. Carrington*.

114. Lecanora tartarea (L.). '*Cudbear.*'
On rocks, rather local than rare.
N. Brimham Rocks; *W. Brunton*.
A. Malham Moor; *B. Carrington*.
C. Todmorden Common; *T. Stansfield*.
D. Upper Bradfield; *Amos Carr*.

115. Lecanora parella L.
On rocks and trees, not rare, but chiefly the pallescent form.
L. Dent and Garsdale. Ingleton, Clapham; *W. West*.
R. Moughton Fell and Giggleswick; *J. Windsor*.
W. Ilkley; *H. F. Parsons*.
W. & A. Airedale and Wharfedale, frequent; *B. Carrington*.
C. Hareley Wood, Todmorden; *T. Stansfield*.

VAR. **Turneri** Sm.
L. Giggleswick to Austwick (on trees); *W. West*.
Y. On oaks at Studley; *W. Brunton*.

116. Lecanora rupestris (Scop.).
One of the commoner of the xerophilous species, confining itself to dry limestone rocks and walls. Common in Craven, about Clapham (L.), Malham and Skipton (A.), and Bolton (W.), chiefly as the form *calva* of Dickson.

Y. Mackershaw Woods ; *W. Brunton.*
W. Arncliffe ; *W. West.* Wetherby district !
T. Maltby ; *J. Bohler.* Roche Abbey valley ; *J. H. Aveling* in *Hist.* (1870) as *Lecidea rupestris.*

117. Lecanora varia (Ehrh.).
On tree-boles, old palings, etc., frequent.
W. Harewood Park !
T. Roche Abbey !

118. Lecanora atra (Huds.).
Stone walls, rocks, and also on trees, frequent.
L. Ingleton. Clapham ; *B. Carrington.*
R. High-hill, Settle ; *J. Windsor.*
N. Hookstone Quarries, Harrogate !
W. Ilkley ; *B. Carrington.*
A. Apperley and Yeadon ; *B. Carrington.*
C. Todmorden ; *T. Stansfield.*
D. Grit walls, Bradfield, near Sheffield !

119. Lecanora circinata (Pers.).
On calcareous rocks.
L. Deepdale near Kirkby Lonsdale ; *Sir J. E. Smith.* Mossdale Moor, Hawes Junction !
D. Near Conisborough ; *J. Bohler.*

VAR. **cinerea** (L.).
L. Rocks on Ingleborough ; *B. Carrington.*

120. Lecanora polytropa (Ehrh.).
On sub-alpine grit-rock, not rare.
L. Clapham ; *B. Carrington.*
A. Malham Moor and Rombalds Moor ; *B. Carrington.*

121. Lecanora sulphurea (Hoffm.).
L. On slate rocks, Ingleton ; *B. Carrington.*
R. Walls near High-hill, Settle ; *J. Windsor.*
A. Skipton.

122. Lecanora subfusca (L.).
On trees and old wooden rails, very generally distributed, sometimes on rocks, from Dentdale ! Garsdale, Malham Moor (*B. Carrington*), Todmorden (*T. Stansfield*), and Hackfall ! to Selby (*W. N. Cheesman*), Swinefleet and Hook near Goole (*H. F. Parsons*). There are many forms, of which *argentata* is the commonest, and *albella* is reported by *Dr. Carrington* from limestone rocks above Malham Cove, and by *J. H. Aveling* (in *Hist.*, 1870) from Roche Abbey Valley.

123. Lecanora umbrina (Ehrh.) VAR. **crenulata** (Dicks.).
Y. Stones near Ripon ; *W. Brunton.*
A. On slate rocks, Gordale ; *B. Carrington.*

124. Lecanora calcarea (L.).
On limestone-rocks, one of the more frequent xerophilous species.
L. Clapham ; Ingleton ; *W. West.*
R. Settle ; *B. Carrington.*
Y. About Ripon (FORMA *Hoffmanni* Ach.) ; *W. Brunton.*
W. By the Strid at Bolton ; *B. Carrington.*
A. In the Craven district about Malham ; *B. Carrington.*

125. Lecanora gibbosa (Ach.).
N. Rocks near Harrogate ; *R. Teesdale.*

126. Lecanora Dicksonii (Ach.).
L. On rocks near the summit of Ingleborough ; *B. Carrington.*

127. Lecanora badia Ach.
On mountain rocks.
L. Baugh Fell ; *W. West.* Ingleborough ; *B. Carrington.*
W. Otley Chevin and Rombalds Moor ; *B. Carrington.* Washburn Valley ; *H. F. Parsons.*

128. Lecanora glaucoma (Hoffm.).
L. Ingleton ; *W. West.*
A. On limestone rocks, Malham Moor ; *B. Carrington.*

129. Lecanora aurantiaca (Lightf.).
On trees and rocks.
L., W. & A. Ingleton, Clapdale, Bolton, and Malham ; *B. Carrington.*

130. Lecanora ochracea Schær.
A. On calcareous 'rocks near Malham Tarn' ; *Sam. Hailstone.*

131. Lecanora ferruginea (Huds.).
Not rare in its rough crustaceous form on rocks at (L.) Clapham ; (R.) Newby Head, Settle, and (W.) Outershaw and Ilkley ; *B. Carrington.* In its smooth corticolous form rarer ; (T.) on trees, Roche Abbey ; *J. Bohler* ; rep. *J. H. Aveling* in *Hist.* (1870).

132. Lecanora cerina (Ehrh.).
Y. On ashes near Ripon ; *W. Brunton.*

VAR. **stillicidiorum** (Hornem.).
A. On dead moss, Gordale ; *B. Carrington.*

133. Lecanora pyracea (Ach.).
On trees.
Y. Near Masham ; *W. Mudd* in *Leighton's Lichen-Flora.*

VAR. **ulmicola** DC. (*Lecid. luteo-alba* Ach.)
T. Roche Abbey ; *J. H. Aveling* in *Hist.* (1870).

134. Lecanora sophodes VAR. **exigua** (Ach.).
Y. On trees near Ripon ; *W. Brunton.*

135. Lecanora hæmatomma (Ehrh.). *Blood-spot Lecanora.*
On rocks, widely distributed, but not common.
L. Deepdale. Ingleton ; *B. Carrington.*
R. Dod and Cam Fell grits.
Y. Studley Park ! *W. Brunton.*
W. Outershaw.
A. Fountains Fell.
C. Shaded stones, Todmorden ; *T. Stansfield.*
D. Bradfield near Sheffield !

136. Lecanora ventosa (L.).
L. Austwick (on slate) ; *W. West.* Howgill Fells (slate). South-east of Wethercote Cave (on grits).
N. Brimham Rocks (grit) ; *W. Brunton.*
C. High grit rocks, Todmorden ; *T. Stansfield.*

137. Urceolaria scruposa (L.).
On rocks, walls, bark, moss, etc., tolerably common.
L. Crina Bottom (bryophilous form) ; *B. Carrington.* Ingleton.
R. Settle ; *J. Windsor.* Ribblehead and Horton.
A. Malham Moor ; *B. Carrington.*
C. Todmorden ; *T. Stansfield.*
T. Near Anston ; *J. Bohler.*

138. Pertusaria multipunctata (Turn.).
A. On trees, Malham ; *B. Carrington.*

139. Pertusaria dealbata (Ach.).
On rocks and trees, widely distributed throughout the lowlands.
N. Brimham Rocks ; *W. Brunton.*
W. Harewood Park ! Wall near Otley Chevin ; *B. Carrington.*
A. Selby ; *R. Teesdale.*
C. Todmorden, not unfrequent ; *T. Stansfield.*

140. Pertusaria communis DC.
On trees, common almost everywhere in the country away from town smoke, from (R.) Slaidburn ! to (T.) Roche Abbey ! and (W.) Cawood (*W. N. Cheesman*).

141. Pertusaria fallax (Pers.).
On trees and rocks, frequent.
L. Ingleton ; *W. West.*
R. Giggleswick ; *W. West.*
Y. Oaks about Ripon ; *W. Brunton.*
W. Bolton Woods ; *B. Carrington.*
A. Malham ! Bell Busk ; *B. Carrington.*
T. Shireoaks and Roche Abbey ; *J. Bohler.*

142. Pertusaria faginea (L.).
On trees, not rare.
W. Parlington Park !
C. Mostly on oaks near Todmorden ; *T. Stansfield.*
T. Roche Abbey Valley ; *J. H. Aveling* in *Hist.* (1870) as *Variolaria faginea.*

143. Pertusaria globulifera (Turn.).
L. & A. On trees, Clapdale and Malham ; *B. Carrington.*
T. Roche Abbey valley ; *J. H. Aveling* in *Hist.* (1870) as *Variolaria discoidea.*

144. Thelotrema lepadinum Ach.
On trees, widely distributed.
L., W. & A. Ingleborough, Bolton, and Malham ; *B. Carrington.*
R. Giggleswick ; *W. West.*
Y. On hollies and oaks in Studley Woods ; *W. Brunton.*
T. Roche Abbey ; *J. Bohler* ; and *J. H. Aveling* in *Hist.* (1870).

XIV. LECIDEINEI.

145. Lecidea lurida (Swartz).
On calcareous earth or rock, rare.
Y. Mackershaw Wood ; *W. Brunton.*
A. Gordale and Malham Cove ; *B. Carrington.*

146. Lecidea ostreata (Hoffm.).
On trees, rare.
Y. Aisenby near Boroughbridge ; *J. G. Baker.*
A. Idle Wood ; *B. Carrington.*

147. Lecidea crustulata (Ach.).
On grit rock, rare.
L. Summit of Gragreth ; *J. Windsor.*
A. Yeadon and Otley Chevin ; *B. Carrington.*

148. Lecidea lucida Ach.
On walls, rocks, etc.
L. Dent, Dentdale ; *J. A. Martindale.* Sedbergh ; *W. West.*

149. Lecidea decolorans Flk.
On the earth.
W. Ilkley ; *B. Carrington.* Jackdaw Crag Quarry !
A. Yeadon Moor ; *B. Carrington.*
C. North Dean ; *H. F. Parsons.*

150. Lecidea quernea (Dicks.).
L. On aged oaks, Helks Wood, Ingleton ; *B. Carrington.*
T. Roche Abbey valley ; *J. H. Aveling* in *Hist.* (1870).

151. Lecidea sanguinaria L.
On rocks.
N. Brimham ; *W. Brunton.*
W. Ilkley, and Rombalds Moor ; *B. Carrington.*
A. Newsholme Dean ; Baildon Moor ; *W. West.*
C. Tolerably common near Todmorden ; *T. Stansfield.*

152. Lecidea parasema (Ach.).
On trees and palings.
W. Beamsley, on railings ; *W. West.*
D. Swinefleet and Hook ; *H. F. Parsons.*
T. Roche Abbey valley ; *J. H. Aveling* in *Hist.* (1870). Maltby ; *J. Bohler.*

153. Lecidea uliginosa (Schrad.).
On turfy or peaty earth-banks.
W. Langstrothdale peat-haggs.
A. Malham Moor and Rombalds Moor ; *B. Carrington.* Thornton Moor ; *W. West.*
T. Goole Moor ; *H. F. Parsons* ; *W. West.*

154. Lecidea coarctata (Sm.).
On walls and stones, not rare, but variable.
R. Hodder Valley, Slaidburn !
Y. Hutton Moor, Ripon ; *W. Brunton.*
W. Wall near Otley Chevin ; *B. Carrington.*

155. Lecidea mollis (Wahl.).
W. Farnley, Ilkley ; *H. F. Parsons* in *Y. N. U. Trans.*

156. Lecidea polycarpa Flk.
W. & A. Sub-alpine rocks, Bolton, Ilkley, and Yeadon ; *B. Carrington.*

157. Lecidea rivulosa Ach.
Mountain rocks, rare.
L. & A. Clapham and Malham Moor ; *B. Carrington.*

158. Lecidea fusco-atra Ach. FORMA **fumosa** Ach.
On gritstone rocks.
N. Near Harrogate ; *R. Teesdale.*

159. Lecidea contigua Fr.
On rocks, grit, slate, and shales, frequent, and in many varieties.
L. Howgill and Baugh Fells ; *H. F. Parsons.* Garsdale. Whernside !
Y. Hutton Moor, Ripon ; *W. Brunton.*
N. Middlesmoor ; *W. West.*
W. Outershaw ! Arncliffe ; *W. West.* Farnley, Ilkley ; *H. F. Parsons.* Bolton Woods ; *W. West.*
A. Keighley and Baildon ; *W. West.*
C. Hebden Bridge ; Luddenden Dene ; *W. West.*

160. Lecidea albo-cærulescens (Wulf.).
Mountain rocks, rare.
A. Malham and Skipton ; *B. Carrington.*

161. Lecidea confluens (Web.).
Mountain rocks, not common.
Y. Sawley Moor ; *W. Brunton.*
C. Near Todmorden, not unfrequent ; *T. Stansfield.*

162. Lecidea calcivora (Ehrh.).
On calcareous rocks. Frequent in Craven !
L. Pavement on Ingleborough ; *B. Carrington.*
R. Newby Head. Ribblehead. Moughton ; *W. West.*
Y. Near Masham ; *W. Mudd.*

163. Lecidea canescens (Dicks.).
On walls and trees, frequent in a barren state.
Common in Craven ; *B. Carrington.*
L. Dent ; *W. West.*
N. Knaresborough ; *W. West.*
A. Rawcliffe ; *H. F. Parsons.* Wistow near Selby ; *W. N. Cheesman.*
C. Todmorden ; *T. Stansfield.*
D. Hook near Goole ; *H. F. Parsons.*

164. Lecidea saxatilis (Schær.).
L. & A. On sub-alpine rocks, Ingleborough, and Malham Moor ; *B. Carrington.*

165. Lecidea myriocarpa (DC.).
W. On trees, Bolton Abbey woods ; *B. Carrington* ; and in the wood below Otley Chevin !

166. Lecidea grossa (Pers.).
L. & W. On bark, Helks Wood, Clapdale, and Bolton Abbey ; *B. Carrington.*

167. Lecidea chalybeia Borr.
L. On wall near Ingleton ; *B. Carrington.*
R. On slate rocks near Horton-in-Ribblesdale !

[**Lecidea candida** (Web.). Incognit.
The Malham plant is the next named species.]

168. Lecidea vesicularis Hoffm. (*L. cæruleo-nigricans* (Lightf.).)
On earth or mosses among calcareous rock.
L. & R. Kingsdale, Ingleborough pavement ! Clapham to Horton ; *W. West.* Giggleswick ; *Id.*
Y. Mackershaw Wood near Ripon ; *W. Brunton.*
W. Outershaw ! Jackdaw Crag quarry, Tadcaster
A. Malham Moor ; *B. Carrington.* Gordale ; *W. West.*

169. Lecidea tricolor (With.).
On the bark of oaks and firs.
D. Wharncliffe Wood ; *Amos Carr.*

170. Lecidea alboatra (Hoffm.).
On trees and palings.
Y. Masham ; *W. Mudd.* On ashes near Ripon ; *W. Brunton.*
A. Bell Busk ; *B. Carrington.*

VAR. **epipolia** (Ach.).
Y. On rocks in Mackershaw Woods ; *W. Brunton.*

171. Lecidea abietina Ach.
W. On trees near Bolton 'Strid' ; *B. Carrington.*

172. Lecidea exanthematica (Sm.).
Xerophilous.
L. & A. On scar limestone in Kingsdale, on Ingleborough, and in Gordale ; *B. Carrington.*

173. Lecidea sphæroides (Dicks.).
L. Whernside, on moss !
Y. On trees near Ripon ; *W. Brunton.*

174. Lecidea milliaria Fries.
On banks and dying wood.
L., W. & A. Ingleborough, Otley Chevin, and Yeadon ; *B. Carrington.*

175. Lecidea carneola Ach.
Y. On trees, Studley Wood ; *W. Brunton.*
W. Bolton ; *B. Carrington.*
A. Near Gennet's Cave, Gordale !

176. Lecidea endoleuca Nyl.
On trees.
L. Helks Wood ; *B. Carrington.*
Y. Hackfall !
W. Wharfedale ; *B. Carrington.*

177. Lecidea rosella (Pers.).
Y. On trees near Ripon ; *W. Brunton.*
Obs.—This, not the next, is the *L. porriginosus* of *Bot. Guide.*

178. Lecidea rubella (Ehrh.).
On trees.
W. Bolton Woods ; *B. Carrington.*
A. Gordale, and Esholt near Apperley Bridge ; *B. Carrington.*

179. Lecidea muscorum (Sw.).
A. On mossy earth, Malham Moor ; *B. Carrington.*

180. Lecidea effusa (Sm.).
Y. On beeches, Studley Park ; *W. Brunton.*

181. Lecidea pelidna Leight.
L. On walls, Clapham Common ; *B. Carrington.*

182. Lecidea geographica (L.). *Map Lichen.*
On sub-alpine rocks, but chiefly those of the older formations.
L. On slate-rock, Howgill Fells, not rare ! *H. F. Parsons* and *W. West.* Park House, Ingleton ; *J. Windsor.* Walls near Clapham ; *B. Carrington.* Upper Garsdale.
R. Grits of Dod and Cam Fells. At the foot of Penyghent in Ribblesdale!
Y. Grantley and Sawley Moor (on grit) ; *W. Brunton.* On grit near Markington ; *W. Fowler.*
W. Grit débris, Langstrothdale.
C. On exposed rocks, near Todmorden, common ; *T. Stansfield.*
Obs.—Oftener noticed than many other *Lecideæ* because of its striking and distinct appearance.

VAR. **atrovirens** (L.).
N. Near Harrogate ; *R. Teesdale.*

183. Lecidea petræa Wulf.
L. & A. Clapham, and Malham Moor ; *B. Carrington.* But Leighton says only 'on granitic and schistose rocks.'

184. Lecidea concentrica Dav.
On calcareous rocks and walls, not rare in Craven, and especially common on walls at Horton-in-Ribblesdale ; yet not noticed apparently by previous collectors in the Riding.

185. Lecidea rimosa (Dicks.).
A. On slaty rocks, Gordale ; *B. Carrington.*

186. Lecidea fuscolutea Dicks.
A. On mosses—'one specimen in Gordale' ; *B. Carrington.*

187. Lecidea cupularis (Ehrh.).
On limestone rock, rather common when looked out for, being inconspicuous.
L. On the mortar of the Lune Bridge, joining Yorkshire to Westmorland, at foot of Howgill Fells ; *W. West.* Whernside, Grensell Crags ! Kingsdale ; *B. Carrington.* Dent Clapham, Ingleton ; *W. West.*
R. Newby Head.
Y. Mackershaw Wood ; *W. Brunton.*
W. Outershaw. Kettlewell ; *W. West.* Pannal to Wetherby ; *W. West.*
A. Upper Gordale ; *W. West.* Malham ; *B. Carrington.*
D. Went-vale crags.

188. Lecidea morio (Ram.).
On alpine rocks, rare.
L. Ingleborough ; *B. Carrington.*

189. Lecidea Parmeliarum Smrf.
Parasitic on the thallus of *Parmelia omphalodes*, etc.
L. Whernside ! Clapham ; *B. Carrington.*
R. Settle ; *J. Windsor.*

XV. GRAPHIDEI.

190. Lithographa cerebrina Schær.
L. & A. On calcareous rocks, Ingleborough and Gordale ; *B. Carrington.*

191. Graphis elegans (Sm.).
On holly and other trees, not rare.
Y. Studley Woods ; *W. Brunton.*
W. In Craven, and Bolton Woods ; *B. Carrington.* Barden ; *W. West.*
A. Holly Grove, Cookridge !
T. Roche Abbey (!) *J. H. Aveling* in *Hist.* (1870).

192. Graphis scripta Ach.
In shady woods, not rare in many places.
L. Helks Wood. Dentdale. Sedbergh ; *W. West.*
Y. Ripon and Studley.
N. Pateley Bridge.
W. Bolton Woods ; *W. West.* Hazlewood !
A. Haddlesey : *W. N. Cheesman.* Gennet's Cave, Gordale ; *W. West.*
T. Roche Abbey Valley ; *J. H. Aveling* in *Hist.* (1870). Sandbeck Park !

193. Graphis sophistica Nyl.
On trees, rare.
L. A. & W. Helks Wood, Malham, and Bolton ; *B. Carrington.*

194. Opegrapha herpetica Ach. (incl. *rufescens* Pers.).
On trees.
L. & W. Helks Wood, Clapdale, and Bolton ; *B. Carrington.* Dent ; *W. West.*
T. Roche Abbey Valley ; *J. H. Aveling* in *Hist.* (1870). Near Worksop ; *J. Bohler.*

195. Opegrapha atra Pers.
On trees, chiefly hazel and ash, generally distributed, and to be found in all parts of the Riding.

196. Opegrapha saxicola Ach.
On rocks (chiefly limestone), frequent, and variable.
L. Kingsdale ; *B. Carrington.* Howgill Fells ; *W. West.*
R. Ribblesdale !
W. By Wharfe, Bolton ; *B. Carrington.* Bramham, Collingham, etc.
A. Malham Cove ! Between Bell Busk and Kirkby Malham ; *B. Carrington.*
D. Brockerdale and Smeaton. Conisborough !

197. Opegrapha varia Pers.
On oak and other trees, frequent.
L. Helks Wood.
R. Giggleswick ; *W. West.*
W. Wharfedale, at Bolton, and elsewhere. Thorparch !
A. In Gordale ; *B. Carrington.*

198. Opegrapha vulgata Ach.
On trees, common, in most parts of the Riding, from Dentdale and Helks Wood, Studley, Bolton, and Ilkley to Wharncliffe and Roche Abbey.

199. Opegrapha lyncea (Sm.).
On old oaks, rare.
W. Parlington Park.

200. Opegrapha viridis Pers.
W. & A. On young ashes, Bolton Woods and Gordale ; *B. Carrington.*

201. Stigmatidium crassum Duby.
T. Old trees near Roche Abbey ; *J. Bohler.*

202. Arthonia epipasta (Ach.).
T. On trees, Shireoaks near Worksop ; *J. Bohler.*

203. Arthonia Swartziana Ach.
W. & A. On trees, Bolton Park and Malham Cove ; *B. Carrington.*

204. Arthonia cinnabarina (Wallr.).
W. On old trees, Bolton Woods ; *B. Carrington.* Bramham !

205. Arthonia pruinosa Ach.
N. & Y. On old trees, Harrogate, and on oaks at Studley ; *W. Brunton.*

XVI. PYRENOCARPEI.

206. Normandina lætevirens Turn. & Borr.
L. Giggleswick to Austwick ; *W. West.*
A. On rocks above Malham Cove ; *T. Stansfield.*

207. Endocarpon miniatum L.
Sub-montane. On calcareous rocks, rather rare.
L. Helks Wood, Ingleton ! *B. Carrington.* Clapham ; *W. West.* Wethercote Cave !
R. North end of High-hill, opposite Attermire Scars, Settle ; *J. Windsor.* Douk Gill, near Horton-in-Ribblesdale !
W. Arncliffe Clouders ! Jackdaw Crag Quarry near Stutton, in plenty !
A. Malham ! Gordale ; *W. West.*

VAR. **complicatum,** prefers wet rock to dry, and is found in Cautley Spout, in stream (*W. West*), Gordale, Malham Cove, rocks in stream, and on Ingleborough according to *Dr. Carrington.*

208. Endocarpon fluviatile DC.
L. On moist rocks in Kingsdale and Wethercote ! West of Ingleborough ; *B. Carrington.*
R. On wet rocks in Linn Gill !
T. Waleswood near Maltby, on wet rock in quarry ; *Amos Carr* !

209. Endocarpon rufescens Ach.
On earth in sub-alpine regions.
L. Kingsdale ; Clapham ; *W. West.*
R. Linn Gill, Ribblehead !
W. Outershaw Gill, Langstrothdale !
A. Gordale and Malham Cove ; *B. Carrington.*

210. Endocarpon hepaticum Ach.
D. Wentbridge ; *H. F. Parsons.*

211. Verrucaria Dufourei DC.
L. & A. On limestone rocks, Kingsdale, Helks Wood, and Malham ; *B. Carrington.*

212. Verrucaria margacea Wahlnb.
On wet rocks.
A. & W. On stones in stream running from Malham Cove ; and at Bolton Bridge. The species in these stations has been called *aquatilis* Mudd, but that is a maritime species.

213. Verrucaria nigrescens (Pers.).
On limestone rocks, not rare.
L. Clapham ; *B. Carrington.*
N. Knaresborough !
W. Jackdaw Crag Quarry, near Stutton !
D. Went-vale ; *H. F. Parsons.*
T. Roche Abbey Valley ; *J. H. Aveling* in *Hist.* (1870).

214. Verrucaria macrostoma (Duf.).
D. Smeaton Crags ; *H. F. Parsons.*

215. Verrucaria fuscella Turn.
W. Walls in Washburn Valley ; *H. F. Parsons.*
VAR. **plumbea** Ach.
L. & A. On limestone, Ingleborough and Gordale ; *B. Carrington.*

216. Verrucaria viridula (Schrad.).
W. On old walls at Bolton Bridge ; *B. Carrington.*

217. Verrucaria murina Leight.
On calcareous rocks.
T. Anston near Shireoaks ; *R. Deakin.*

218. Verrucaria rupestris Schrad.
On flints, sandstone, etc.
W. Craven district and Bolton ; *B. Carrington.* Boston !
D. Went-vale ; *H. F. Parsons.*

219. Verrucaria calciseda DC.
R., L. & A. Frequent on the Craven scar-limestone, as about Giggleswick Scar (*B. Carrington*), Ingleborough pavement, Malham 'streets,' etc.

220. Verrucaria tephroides (Ach.).
A. On the pavement of limestone at Malham (above the Cove !) ; *B. Carrington.*

221. Verrucaria conoidea Fries.
L. On limestone, Ingleborough ; *B. Carrington* in *Leighton's Lichen Flora* (ed. ii), p. 430.

222. Verrucaria gemmata Ach.
On trees, 'common.'
R. Giggleswick ; *W. West.*
W. Bolton Woods !
T. Roche Abbey Woods ; *J. H. Aveling* in *Hist.* (1870).

223. Verrucaria epidermidis Ach.
On the smooth bark of beech and birch, rare.
W. Aberford ! Thorparch !
T. Near Shireoaks ; *J. Bohler.* Roche Abbey Valley ; *J. H. Aveling* in *Hist.* (1870).

224. Verrucaria cinerea (Pers.).
T. Roche Abbey Valley ; *J. H. Aveling* in *Hist.* (1870).

225. Verrucaria immersa Leight.
Limestone rocks, frequent in Craven district.
L. Ingleborough ; *B. Carrington.*

226. Verrucaria capnodes Nyl.
A. On trees about Malham Cove ; *B. Carrington.*

227. Verrucaria biformis Borr.
On trees.
W. & A. Bolton and Malham !

228. Verrucaria rhyponta Ach.
A. On bark, Gennett's Cave, Gordale ; *B. Carrington.*

229. Verrucaria incavata Nyl.
On calcareous rock.
L. Near Ingleborough ; *B. Carrington.*

230. Verrucaria nitida (Weig.).
On trees, not rare, but perhaps its distribution is better known than that of species of more inconspicuous facies.
L. Helks Wood, Ingleton (!) *B. Carrington.*
R. Giggleswick ; *W. West.*
Y. & W. Studley and Bolton Woods !
T. Roche Abbey Valley ; *J. Bohler.*

231. Verrucaria glabrata Ach. VAR. **dermatodes** Borr.
A. On smooth bark, Gordale ; *B. Carrington.*

232. Verrucaria umbrina Wahlnb.
On wet calcareous rocks in Craven district, but not common.
L. Dentdale ; *W. West.*

233. Verrucaria rimosicola Leight.
A. Parasitic on *Lecanora calcarea* on Malham Moor ; *B. Carrington.*

234. Strigula Babingtonii Berk.
Parasitic on evergreen leaves.
A. Old laurels bordering Meanwood Park, 1879 !

FUNGI.

FUNGI.

[In the compilation of this portion of the Flora the author has been assisted by the knowledge and experience of Mr. H. T. Soppitt.

For convenience of reference, these plants are arranged conformably to Dr. M. C. Cooke's '*Handbook*' (1871), there being as yet no other complete descriptive enumeration of the British forms. In so doing, the authors do not wish to be understood as acquiescing in any views the arrangement may express or imply. They would have preferred to adopt the nomenclature of the *Uredinei* as revised by Dr. Winter, and that of Saccardo in the *Pyrenomycetes*, as more in accord with present knowledge, but so little has been done in these groups within our area, that it seemed hardly worth while to depart from a published arrangement for the sake of a few names.

Although tolerably representative for certain districts, this list must not be regarded as exhaustive for the Riding as a whole. Owing to lack of resident workers, who can always investigate most satisfactorily, several of the river-basins have been imperfectly explored. Until the last seven years, since the time of James Bolton—a hundred years ago!—the records published have been very few. The earliest date back to 1677, the pre-Linnean time of Dr. Martin Lister, but down to 1775 the species described as growing within our limits were less than a score. Bolton's *History of Fungusses*, therefore, furnishes the kernel of this fungus-florula, albeit some of his species are hard nuts to crack.

Recent years have found workers in the Rev. W. Fowler, Dr. H. F. Parsons, Messrs. James Abbott, Thomas Birks, the late J. Bohler, W. N. Cheesman, A. Clarke, T. Hebden, Thomas Hick, G. E. Massee, W. West, and the writers.

The first Fungus Foray ever held in Yorkshire took place in 1881, and gave a decided impetus to the study of Fungi, having the great advantage of the services of the Rev. J. E. Vize and Messrs. W. Phillips and C. B. Plowright. The districts then investigated were Harrogate, Studley, and Selby, many species new to the Riding being found.

Care has been taken to mark as 'common' only those forms which are truly and indisputably so within the Riding itself; but, as many species are remarkably local, the stations of many even of these common ones have been cited.

Nine hundred and eighty-seven 'species' names are herein enumerated, but some of them refer to conditions or states of other species. Substantially, however, their number makes the proportion of West Riding to British forms 24·5 per cent. This is a smaller percentage than obtains in the Mosses, Lichens, etc., but this paucity of fungoid forms is not real. In the falling off indicated, set against those lists dealing with the higher forms of vegetable life, we have simply a botanical milestone showing us what a little way we have gone, how far on the road we have yet to go, before a point is reached at which we may rest even tolerably satisfied.

The reference to 'Hudd. Bot. Soc.' (sometimes contracted to the initials H. B. S.) after some records stands for Huddersfield Botanical Society, being the results of the investigation of members of that association who do not wish their names to be separately indicated. Localities marked '!,' as in other lists, have been verified by the author of the great bulk of the Flora; those unmarked and without authority are (*in this list only*) vouched for by H. T. Soppitt.]

HYMENOMYCETES.

I. AGARICINI.

1. Agaricus (Amanita) vaginatus Bull.
In woods, common.
W. Ryther; *W. N. Cheesman.*
A. Eccup! Bingley, etc.
C. Huddersfield district; *Hudd. Bot. Soc.*
D. Wharncliffe; *H. F. Parsons.*
T. Roche Abbey Valley; *J. H. Aveling* in *Hist.* (1870).

2. Agaricus (Amanita) Ceciliæ B. & Br. (*A. strangulatus* Fr.)
A. Shipley Glen, 1886, rare.
T. Roche Abbey Valley; *J. H. Aveling* in *Hist.* (1870).

3. Agaricus (Amanita) vernus Bull.
C. In woods about Huddersfield; *Hudd. Bot. Soc.*

4. Agaricus (Amanita) phalloides Fr.
Woods, not common. Poisonous.
A. Haddlesey; *W. N. Cheesman.* Bingley Woods, Shipley Glen, etc.
C. Halifax; *James Bolton.* The earliest notice.

5. Agaricus (Amanita) mappa Batsch.
A. Bingley Woods, not common.

6. Agaricus (Amanita) muscarius L. '*The Fly Agaric.*'
Woods, not common.
N. Beckwithshaw.
W. Bolton Woods!
A. Horsforth Wood; *J. Abbott.* Hawksworth; *B. Illingworth.* Stainor Wood; *W. N. Cheesman.* Hainworth; *T. Hebden.* Nab Wood Saltaire. Scotland Mills and Adel!
D. Wharncliffe Woods!

7. Agaricus (Amanita) excelsus Fr.
C. Meltham Woods; *Hudd. Bot. Soc.*
T. Roche Abbey Valley; *J. H. Aveling* in *Hist.* (1870).

8. Agaricus (Amanita) pantherinus DC.
Under trees.
N. Harrogate; *Exc.*, 1881.
C. North Dean; *H. F. Parsons.* Huddersfield district; *Hudd. Bot. Soc.*

9. Agaricus (Amanita) rubescens P. '*The Blusher.*'
In woods, common. Esculent.
R. Slaidburn!
N. Beckwithshaw.
W. Bolton Woods! Buckden Woods!
A. Meanwood! Hambleton and Camblesforth; *H. F. Parsons.* Bingley. Hawksworth. Horsforth.
C. Huddersfield.

10. Agaricus (Amanita) asper Fr.
C. Woods, rare. First noticed in 1790 at Ovenden near Halifax by *James Bolton.*

11. Agaricus (Lepiota) procerus Scop. '*Parasol Mushroom.*'
Pastures, not common. Esculent.
N. Harrogate; *T. Hick.*
A. Headingley, etc.; *J. Abbott.* Brayton; *H. F. Parsons.* Stainor Wood, Selby; *W. N. Cheesman.* Hawksworth; *W. Loxam.*
C. Thornes Park, Wakefield; *H. Sims.*

12. Agaricus (Lepiota) rachodes Vitt.
A. A fine specimen made its appearance in Dr. Parkinson's greenhouse, Bradford, 1882; *W. West.*
T. Roche Abbey Valley; *J. H. Aveling* in *Hist.* (1870).

13. Agaricus (Lepiota) Badhami B. and Br.
T. Under yews, Roche Abbey Wood, October 1876!

14. Agaricus (Lepiota) cristatus Fr.
Amongst grass, not common.
Y. Studley; *Exc.*, 1881.
W. Ilkley. Harewood, 1886.
A. Bingley.

15. Agaricus (Lepiota) hispidus Lasch.
C. Plantation near Longley Hall; *Hudd. Bot. Soc.*

16. Agaricus (Lepiota) licmophorus B. & Br.
A. In a fernery at Bradford; *J. Eastwood.*

17. Agaricus (Lepiota) granulosus Batsch.
Woods and heaths, common.
C. Burks Wood, Halifax, 1788; *J. Bolton.* The first record.

18. Agaricus (Lepiota) polystictus Berk.
C. Old pastures, Fixby, Lepton, and Bradley; *Hudd. Bot. Soc.*

19. Agaricus (Lepiota) Bucknalli B. & Br.
Y. Studley; *Exc.*, 1881. Very rare.

20. Agaricus (Lepiota) seminudus Lasch.
Y. Studley; *Exc.*, 1881. Rare.

21. Agaricus (Lepiota) carcharias Pers.
N. Harrogate; *Exc.*, 1881.
A. Hilly pastures, not common. Baildon. Bingley.

22. Agaricus (Armillaria) melleus Vahl. '*Honey Agaric.*'
On decaying stumps, etc., very common.
R. Slaidburn!
W. Harewood; *J. Abbott.*
A. Meanwood; *J. Abbott.* Saltaire. Allerton; *W. West.* Selby; *W. N. Cheesman.* Bingley.
C. Birstall; *W. West.*
D. Goole; *H. F. Parsons.*
T. Roche Abbey Valley; *J. H. Aveling* in *Hist.* (1870).

23. Agaricus (Tricholoma) sejunctus Sow.
A. Stainor Wood; *W. N. Cheesman.*

24. Agaricus (Tricholoma) spermaticus Fr.
A. In fir woods. Adel Dam Wood!

25. Agaricus (Tricholoma) fulvellus Fr.
In woods, not common.
A. Bingley.

26. Agaricus (Tricholoma) flavo-brunneus Fr.
A. Barlow Woods; *W. N. Cheesman.* Bingley Woods.

27. Agaricus (Tricholoma) rutilans Schæff.
Stumps of trees in fir and pine plantations, frequent.
N. Harrogate; *Exc.*, 1881.
W. Plantation between Kidhall Lane and Bramham Park, 1786; *James Bolton.* Earliest record.
A. Adel! Heaton. Bingley. Barlow; *W. N. Cheesman.*
C. Meltham Woods; *Hudd. Bot. Soc.* North Dean; *H. F. Parsons.*
D. Wharncliffe; *H. F. Parsons.*

28. Agaricus (Tricholoma) luridus Schæff.
Woods, not common.
A. Camblesforth; *W. N. Cheesman.*

29. Agaricus (Tricholoma) imbricatus Fr.
T. Roche Abbey Valley; *J. H. Aveling* in *Hist.* (1870).

30. Agaricus (Tricholoma) terreus Schæff.
Fir woods.
Y. Studley; *Exc.*, 1881.
N. Harrogate; *Exc.*, 1881.
A. Adel, 1876!

31. Agaricus (Tricholoma) saponaceus Fr.
In woods.
Y. Studley; *Exc.*, 1881.
D. Thorne Waste; *H. F. Parsons.*

32. Agaricus (Tricholoma) sulphureus Bull.
In woods, not common.
A. Nab Wood, Saltaire.

33. Agaricus (Tricholoma) lascivus Fr.
A. Bishops Wood; *W. N. Cheesman.*

34. Agaricus (Tricholoma) bufonius P.
Local.
C. Huddersfield district; *Hudd. Bot. Soc.*

35. Agaricus (Tricholoma) immundus.
In pastures, rare.
C. Huddersfield district; *Hudd. Bot. Soc.*

36. Agaricus (Tricholoma) carneus Bull.
In fields.
A. Hambleton, near Selby; *H. F. Parsons.*

37. Agaricus (Tricholoma) gambosus Fr. *St. George's Mushroom.*
A true vernal and edible species, not uncommon.
A. Bramhope; *J. Abbott.* Bingley. Hawksworth.
D. Adlingfleet; *H. F. Parsons.*

38. Agaricus (Tricholoma) grammopodius Bull.
In pastures, in rings.
A. Hawksworth.
C. Halifax, 1788; *James Bolton.* Huddersfield district; *Hudd. Bot. Soc.*
D. Goole Moor; *H. F. Parsons.*

39. Agaricus (Tricholoma) melaleucus P.
Rare.
A. Temple Hirst; *H. F. Parsons.*

40. Agaricus (Tricholoma) brevipes Bull.
Not common. A good esculent species.
C. Near Huddersfield, June 1885; *J. A. Farrand*; *Hudd. Bot. Soc.*

41. Agaricus (Tricholoma) murinaceus Bull.
A. Heaton Woods, 1884.
C. Huddersfield district; *Hudd. Bot. Soc.*

42. Agaricus (Tricholoma) nudus Bull.
Y. Studley; *Exc.*, 1881.
T. Roche Abbey Valley; *J. H. Aveling* in *Hist.* (1870).

43. Agaricus (Tricholoma) cinerascens Bull.
A. Barlow Woods; *W. N. Cheesman.*

44. Agaricus (Tricholoma) panæolus Fr.
On dung, in pastures, common.
Y. Studley; *Exc.*, 1881.
A. Cottingley. Baildon.

45. Agaricus (Tricholoma) subpulverulentus Pers.
C. In pastures, Huddersfield district; *Hudd. Bot. Soc.*

46. Agaricus (Clitocybe) nebularis Batsch.
In woods, not common.
A. Meanwood Wood; *C. B. Plowright.* Wistow; *W. N. Cheesman.*

47. Agaricus (Clitocybe) odorus Bull.
In woods. Esculent. Remarkable for its strong smell of aniseed.
A. Heaton Woods, 1884.

48. Agaricus (Clitocybe) phyllophilus Fr.
A. Meanwood Wood; Bingley; *W. West.* Shipley Glen. Carlton Woods; *W. N. Cheesman.*
C. Huddersfield district; *Hudd. Bot. Soc.*

49. Agaricus (Clitocybe) candicans Fr.
In woods amongst leaves.
A. Allerton and Horsforth; *W. West.* Heaton Woods.
C. Larch plantation, Fixby Hall; *James Bolton.* Birstal; *W. West.*

50. Agaricus (Clitocybe) dealbatus P.
A. Brayton; *W. N. Cheesman.*
C. Thornes Park, Wakefield, 1875; *H. Sims.*

51. Agaricus (Clitocybe) gallinaceus Scop.
Pastures, not uncommon!

52. Agaricus (Clitocybe) fumosus P.
Y. Studley; *Exc.*, 1881.
A. Waste ground, Bradford, 1880.
C. Huddersfield district; *Hudd. Bot. Soc.*

53. Agaricus (Clitocybe) maximus Fr.
Meadows and woods.
A. Marked as a species of the Leeds district by *James Abbott.*

54. Agaricus (Clitocybe) infundibuliformis Schæff. '*Funnel Agaric.*'
Y. Studley; *Exc.*, 1881.
A. Allerton near Bradford; *W. West.* Saltaire.
C. Huddersfield district, rare; *Hudd. Bot. Soc.*
T. Roche Abbey Valley; *J. H. Aveling* in *Hist.* (1870).

55. Agaricus (Clitocybe) geotropus Bull.
Woods, etc., not rare.
R. Slaidburn!
W. Barwick-in-Elmet!
A. Gateforth; *H. F. Parsons.* Baildon.
C. Fixby; *Hudd. Bot. Soc.*

56. Agaricus (Clitocybe) cyathiformis Fr.
Meadows, etc.
A. Meanwood! Brayton; *H. F. Parsons.* Bishops Wood; *W. N. Cheesman.*
C. Spa near Halifax, 1790; *James Bolton.*

57. Agaricus (Clitocybe) brumalis Fr.
In woods.
A. Steeton.
C. Huddersfield district; *Hudd. Bot. Soc.*

58. Agaricus (Clitocybe) metachrous Fr.
N. Harrogate; *Exc.*, 1881.
A. Nab Wood, Saltaire.

59. Agaricus (Clitocybe) fragrans Sow.
Y. Studley; *Exc.*, 1881.
A. Heaton Wood. Bishops Wood; *W. N. Cheesman.*
D. Conisborough; *H. F. Parsons.*

60. Agaricus (Clitocybe) bellus Fr.
A. In woods. Shipley Glen, 1885.

61. Agaricus (Clitocybe) laccatus Scop.
A. Meanwood; *J. Abbott.* Brayton; *H. F. Parsons.* Saltaire Bingley; *W. West.* Selby; *W. N. Cheesman.*
C. Very common in woods, etc., near Halifax and Fixby Hall; *J. Bolton.*

62. Agaricus (Clitocybe) difformis Pers.
C. In fir plantations near Halifax (towards the end of last century); *James Bolton, Hist. of Fungusses*, pl. 17.

63. Agaricus (Clitocybe) ericetorum Bull.
A. Bingley Woods. Shipley Glen.

64. Agaricus (Pleurotus) dryinus P.
On tree-trunks.
Y. Studley; *Exc.*, 1881.
C. Wyke near Lightcliffe.

65. Agaricus (Pleurotus) spongiosus Fr.
C. On dead poplars, Odsal Wood near Wibsey, 18[illegible].

66. Agaricus (Pleurotus) ulmarius Bull.
On tree-trunks, chiefly elm.
A. Esholt. In a timber-yard, Selby; *W. N. Cheesman.*
C. Odsal Wood, near Wibsey.

67. Agaricus (Pleurotus) fimbriatus Bolt.
C. On tree-trunks, Northowram, 1787; *J. Bolton.* The only record.

68. Agaricus (Pleurotus) salignus Fr.
A. On tree trunks, Idle, January 1887.

69. Agaricus (Pleurotus) ostreatus Jacq.
Not common.
A. Harewood; *J. Abbott.* Haddlesey; *W. N. Cheesman.*
C. Huddersfield district, occasionally; *Hudd. Bot. Soc.*

70. Agaricus (Pleurotus) applicatus Batsch.
On dead twigs, not unfrequent.

71. Agaricus (Pleurotus) mitis P.
A. On larch in a wood-yard at Steeton.
T. Roche Abbey Valley; *J. H. Aveling* in *Hist.* (1870).

72. Agaricus (Pleurotus) hypnophilus P.
A. On a moss-covered wall, Malham; *W. West.*

73. Agaricus (Pleurotus) chioneus P.
A. On a decayed tree, Saltaire.

74. Agaricus (Collybia) radicatus Relh.
On old stumps, frequent.
A. Meanwood; *J. Abbott.* Shipley Glen. Bingley. Horsforth; *W. West.*
C. Battye-Ford; *H. F. Parsons.* Vicarage shrubbery, Liversedge!
T. Roche Abbey Valley; *J. H. Aveling* in *Hist.* (1870).

75. Agaricus (Collybia) longipes Bull.
On old stumps, not common.
A. Apperley. Bingley.

76. Agaricus (Collybia) platyphyllus Fr.
In woods, rare.
A. Bingley, 1884.

77. Agaricus (Collybia) semitalis Fr.
A. In pastures, Wilsden.

78. Agaricus (Collybia) fusipes Bull. *Spindleshank.*
On stumps, frequent.
N. Harrogate; *Exc.*, 1881.
A. Bingley. Shipley Glen.
C. Birstal; *W. West.*
T. Roche Abbey Valley; *J. H. Aveling* in *Hist.* (1870).

79. Agaricus (Collybia) maculatus A. & S.
In woods, locally common.
N. Harrogate; *Exc.*, 1881.
A. Shipley. Bingley. Carlton; *W. N. Cheesman.*

80. Agaricus (Collybia) butyraceus Bull.
N. Harrogate; *Exc.*, 1881.
A. In woods, frequent. Shipley Glen. Carlton; *W. N. Cheesman.*

81. Agaricus (Collybia) velutipes Curt.
On tree-trunks, very common, especially in winter.
L. Sedbergh and Ingleton; *W. West.*
A. Bingley. Esholt. Saltaire. Steeton. Horton; *W. West.*
C. Southowram, 1789; *J. Bolton.*
D. Goole; *H. F. Parsons.*

82. Agaricus (Collybia) confluens P.
Frequent in woods, etc.
A. Adel! Bishops Wood; *W. N. Cheesman.* Thackley. Hawksworth.

83. Agaricus (Collybia) tuberosus Bull.
Amongst leaves, etc., in woods, and usually attached to a *Sclerotium.*
W. Bolton Woods.
A. Horsforth; *W. West.* Shipley Glen.

84. Agaricus (Collybia) dryophilus Bull.
Among leaves in woods, frequent.
A. Meanwood. Shipley Glen. Heaton. Wistow; *W. N. Cheesman.*
T. Roche Abbey Valley; *J. H. Aveling* in *Hist.* (1870).

85. Agaricus (Collybia) stipitarius Fr.
A. On the ground, Newsholme Dean near Keighley.
T. Roche Abbey Valley; *J. H. Aveling* in *Hist.* (1870).

86. Agaricus (Collybia) cirrhatus Schum.
Amongst leaves, etc., not common.
A. Meanwood, 1881. Church Fenton; *G. Massee*

87. Agaricus (Collybia) clavus Bull; Vaill. (*A. esculentus* 'Wulf.,' Stev.)
C. A minute and doubtful species, found by Bolton 'about Halifax,' and figured by him (plate 39). Not found since, and Berkeley says the plant may have been *Mycena acicula* Schæff.; whilst Stevenson refers it to *C. esculentus.*

88. Agaricus (Mycena) pelianthinus Fr.
C. Northowram, 1787; *James Bolton.*

89. Agaricus (Mycena) rosellus Fr.
Rare.
N. Beckwithshaw; *Exc.*, 1881.

90. Agaricus (Mycena) purus P.
Amongst woodland leaves, frequent.
Y. Studley; *Exc.*, 1881.
A. Shipley Glen. Allerton; *W. West.*

91. Agaricus (Mycena) Iris Berk.
T. Roche Abbey Valley ; *J. H. Aveling* in *Hist.* (1870).

92. Agaricus (Mycena) flavo-albus Fr.
Y. Studley ; *Exc.*, 1881.
A. On mossy tree-trunks, Nab Wood, Saltaire.

93. Agaricus (Mycena) lacteus P.
In fir-woods, etc., not common.
Y. Studley ; *Exc.*, 1881.
A. Barlow Woods ; *W. N. Cheesman.*

94. Agaricus (Mycena) rugosus Fr.
A. On fallen trees, Nab Wood, Saltaire.

95. Agaricus (Mycena) sudorus Fr.
W. On stumps, Denton near Ilkley.

96. Agaricus (Mycena) galericulatus Scop.
On tree-trunks, etc., very common.
A. Temple-hurst; *H. F. Parsons.* Bingley. Allerton; Meanwood; *W. West.*

97. Agaricus (Mycena) niveus Quel.
Rare.
N. Harrogate ; *Exc.*, 1881.

98. Agaricus (Mycena) polygrammus Bull.
On tree-trunks, generally distributed.
Y. Studley ; *Exc.*, 1881.
A. Shipley Glen.

99. Agaricus (Mycena) atro-albus Bolt. (*Hist. Fung.*, t. 137).
Amongst moss at tree-roots.
A. Cullingworth.
C. About Halifax ; *J. Bolton.* Battye Ford ; *H. F. Parsons.*

100. Agaricus (Mycena) dissiliens Fr.
C. Lee Bridge, Halifax, 1790 ; *James Bolton.*

101. Agaricus (Mycena) leptocephalus Pers.
On lawns, etc. Remarkable for its strong nitrous odour.
W. Ilkley.
A. Hawksworth. Peel Park, Bradford. Baildon.

102. Agaricus (Mycena) alcalinus Fr.
On tree-trunks, frequent.
A. Shipley Glen. Bingley ; *W. West.*

103. Agaricus (Mycena) vitreus Fr.
Not common.
A. Hawksworth, 1885.

104. Agaricus (Mycena) vitilis Fr.
A. Amongst leaves in woods, etc., Hawksworth.

105. Agaricus (Mycena) speireus Fr.
A. On a tree-trunk, Thackley, 1886.

106. Agaricus (Mycena) tenuis Bolt.
C. North Dean Wood, 1783 ; *J. Bolton.*

107. Agaricus (Mycena) filopes Bull.
A. Church Fenton ; *G. Massee.* Hambleton ; *H. F. Parsons.* Bishops Wood ; *W. N. Cheesman.*
C. North Dean ; *H. F. Parsons.*

108. Agaricus (Mycena) sanguinolentus A. & S.
Amongst leaves in woods.
A. Cullingworth. Meanwood. Camblesforth ; *W. N. Cheesman.*

109. Agaricus (Mycena) galopus Schrad.
Y. Common in woods. Studley ; *Exc.*, 1881.
A. Shipley Glen. Bingley ; *W. West.*
C. Liversedge ; *W. Fowler.*
T. Roche Abbey Valley ; *J. H. Aveling* in *Hist.* (1870).

110. Agaricus (Mycena) epipterygius Scop.
On heaths and in woods, frequent.
N. Harrogate ; *Exc.*, 1881.
W. Rombalds Moor ; *W. West.*
A. Heaton Wood. Baildon. Temple Hurst ; *W. N. Cheesman.*
C. Wibsey ; *W. West.*

111. Agaricus (Mycena) pelliculosus Fr.
A. On heaths, etc., not common. Baildon Moor.

112. Agaricus (Mycena) vulgaris P.
In larch woods, not common.
T. Roche Abbey Valley ; *J. H. Aveling* in *Hist.* (1870).

113. Agaricus (Mycena) stylobates P.
On dead fern, rushes, etc., common.
A. Baildon. Hawksworth.
T. Roche Abbey Valley ; *J. H. Aveling* in *Hist.* (1870).

114. Agaricus (Mycena) tenerrimus Berk.
On sticks, etc., in moist situations.
A. Shipley Glen. Baildon.

115. Agaricus (Mycena) corticola Schum.
L. On tree-trunks, Ingleton ; *W. West.*
Y. Studley ; *Exc.*, 1881.

116. Agaricus (Mycena) hiemalis Osbeck.
A. On trunks of trees, Nab Wood, Saltaire.

117. Agaricus (Mycena) capillaris Schum.
R. Amongst moss, Linn Gill, 1886 ; *F. Rhodes.*

118. Agaricus (Mycena) juncicola Fr.
A. On dead rushes, not common. Baildon Moor. Rombalds Moor; *W. West.*

119. Agaricus (Omphalia) pyxidatus Bull.
In pastures.
R. Hodder Valley, near Slaidburn, 1876 !
A. Church Fenton, 1881 ; *G. Massee.*

120. Agaricus (Omphalia) oniscus Fr.
C. On Halifax peat moors ; *J. Bolton.*

121. Agaricus (Omphalia) muralis Sow.
W. Ilkley ; *H. F. Parsons.*
A. Seven Arches, Saltaire. Eldwick.

122. Agaricus (Omphalia) umbelliferus L.
Common on swampy heaths, &c.
W. Ilkley.
A. Baildon. Cullingworth. Barlow ; *W. N. Cheesman.*

123. Agaricus (Omphalia) fibula Bull.
Amongst moss in winter, common.

124. Agaricus (Omphalia) integrellus P.
On dead twigs, rare.
C. Odsal Wood, 1885.

125. Agaricus (Omphalia) campanella Batsch.
A. Church Fenton, 1881 ; *G. Massee.*

126. Agaricus (Volvaria) gloiocephalus Fr.
A. Apperley, 1886.
C. On a waste-heap, Huddersfield district ; *Hudd. Bot. Soc.*

127. Agaricus (Volvaria) speciosus Fr.
On dung-hills and rubbish-heaps.
A. Camblesforth ; *H. F. Parsons.* Carlton ; *W. N. Cheesman.*
T. Roche Abbey Valley ; *J. H. Aveling* in *Hist.* (1870).

128. Agaricus (Chamæota) cretaceus Fr. '*False Mushroom.*' '*Cockstool.*'
In meadows, occasional. Bears a resemblance to the common edible, *A. campestris*, but the gills are white. Eaten in Wensleydale, but is unwholesome, producing vertigo *without* vomiting !
W. Wetherby, 1879 !
A. Adel, 1876 !
T. Roche Abbey Valley ; *J. H. Aveling* in *Hist.* (1870).

129. Agaricus (Pluteus) cervinus Schæff.
On tree-trunks, not common.
A. Bingley. Esholt.

130. Agaricus (Pluteus) umbrosus Pers.
A. On old stumps, etc. Nab Wood, Saltaire, 1884.

131. Agaricus (Entoloma) sinuatus Fr.
Rare.
Y. Studley ; *Exc.*, 1881.
A. Camblesforth ; *W. N. Cheesman.*

132. Agaricus (Entoloma) placenta Batsch.
A. Shipley Glen, 1885.

133. Agaricus (Entoloma) repandus Bull.
Rare.
A. Cullingworth, 1885 ; *T. Hebden.*

134. Agaricus (Entoloma) jubatus Fr.
W. Ilkley.
A. In pastures. Bingley ; *W. West.* Hawksworth. Baildon.
C. Huddersfield district ; *Hudd. Bot. Soc.*

135. Agaricus (Entoloma) clypeatus L.
A. Bingley Wood, 1885.
C. Bracken Bed Wood, Halifax ; *J. Bolton.* Huddersfield district ; *H. B. S.*

136. Agaricus (Entoloma) rhodopolius Fr.
A. Shipley Glen, 1884.

137. Agaricus (Entoloma) costatus Fr.
In meadows and pastures, frequent.
A. Near Bradford. Baildon. Hawksworth.
C. Liversedge ; *Rev. W. Fowler.*

138. Agaricus (Entoloma) sericellus Fr.
C. Huddersfield district ; *Hudd. Bot. Soc.*

139. Agaricus (Entoloma) sericeus Bull.
In pastures, very common.
N. Harrogate ; *J. Abbott.*
A. Bingley ; *W. West.* Baildon.
C. Battye-ford.
D. Goole Moor ; *H. F. Parsons.*

140. Agaricus (Clitopilus) prunulus Scop.
In woods and shady places. A good edible species.
A. Near Malham.
C. Thornes Park, Wakefield ; *H. Sims.*

141. Agaricus (Claudopus) variabilis. '*Ears.*'
On dead sticks, etc., frequent.
R. Slaidburn !
N. Harrogate ; *Exc.*, 1881.
A. Meanwood Wood. Saltaire. Ingrow.

142. Agaricus (Claudopus) euosmus Berk.
C. Huddersfield district, rare; *Hudd. Bot. Soc.*

143. Agaricus (Leptonia) æthiops Fr.
A. In pastures near Bingley Woods.

144. Agaricus (Leptonia) incanus Fr.
R. Wet pastures in Hodder Valley above Slaidburn, 1876!

145. Agaricus (Nolanea) pascuus P.
Y. Studley; *Exc.*, 1881.
N. Harrogate; *Exc.*, 1881.
A. Whitley; *H. F. Parsons.* Baildon.
C. In pastures near Halifax, 1787; *J. Bolton.*

146. Agaricus (Nolanea) pisciodorus Ces.
Rare.
Y. Studley; *Exc.*, 1881.

147. Agaricus (Pholiota) durus Bolt.
C. In gardens about Halifax; *J. Bolton.* Huddersfield district; *H.B.S.*

148. Agaricus (Pholiota) præcox P.
Common in spring on waste ground, etc.
A. Bradford. Baildon.
C. Huddersfield; *J. A. Farrand.*

149. Agaricus (Pholiota) Leveillianus D. & M.
C. Rare. Toothill near Rastrick; *Hudd. Bot. Soc.*

150. Agaricus (Pholiota) comosus Fr.
C. Ramsdene, Halifax, 1786; *J. Bolton.*

151. Agaricus (Pholiota) squarrosus Müll.
Tree-trunks, frequent.
Y. Studley; *Exc.*, 1881.
W. Bolton Woods.
A. Meanwood; *J. Abbott.* Camblesforth; *H. F. Parsons.* Bradford.
D. Wharncliffe! Goole.

VAR. **Mulleri.**
C. Fixby; *Hudd. Bot. Soc.*

152. Agaricus Pholiota) mutabilis Schæff.
On trunks, etc.
Y. Studley; *Exc.*, 1881.
A. Hawksworth. Barlow Woods; *W. N. Cheesman.*
T. Roche Abbey Valley; *J. H. Aveling* in *Hist.* (1870).

153. Agaricus (Pholiota) spectabilis Fr.
On tree-stumps.
A. & C. Horsforth Woods and Storthes Hall; *Hudd. Bot. Soc.*

154. Agaricus (Pholiota) marginatus Batsch.
Y. Studley; *Exc.*, 1881.

155. Agaricus (Hebeloma) crustuliniformis Bull.
D. Goole Moor; *H. F. Parsons.*
T. Roche Abbey Valley; *J. H. Aveling* in *Hist.* (1870).

156. Agaricus (Hebeloma) fastibilis Fr.
In woods in summer, frequent.

157. Agaricus (Hebeloma) longicaudus P.
In woods, not common.
A. Nab Wood, Saltaire.
T. Roche Abbey Valley; *J. H. Aveling* in *Hist.* (1870).

158. Agaricus (Hebeloma) relicinus Fr.
In wet fir-woods amongst *Sphagnum.*
A. Scotland Mill Wood, Adel, above the Aqueduct!

159. Agaricus (Hebeloma) scaber Müll.
N. Harrogate; *Exc.*, 1881.
A. Nab Wood, Bingley.

160. Agaricus (Hebeloma) plumosus Bolt.
C. Ramsdene, Ovenden, 1787; *J. Bolton.*

161. Agaricus (Hebeloma) pyriodorus Pers.
Y. Studley; *Exc.*, 1881.
A. Haddlesey; *W. N. Cheesman.*

162. Agaricus (Hebeloma) obscurus P.
A. Heaton Wood, 1884.

163. Agaricus (Hebeloma) fastigiatus Fr.
A. Church Fenton, 1881; *G. Massee.*
T. Roche Abbey Valley; *J. H. Aveling* in *Hist.* (1870).

164. Agaricus (Hebeloma) fibrosus Sow.
A. Heaton Wood, 1884.

165. Agaricus (Hebeloma) rimosus Bull.
Y. Studley; *Exc.*, 1881.
N. Harrogate; *Exc.*, 1881.
A. Heaton. Steeton. Wistow; *W. N. Cheesman.*
C. 'North Dean'; *H. F. Parsons.* Huddersfield district; *Hudd. Bot. Soc.*

166. Agaricus (Hebeloma) trechisporus Berk.
A. Airedale; *J. Mackenzie.* Heaton Wood.
T. Roche Abbey Valley; *J. H. Aveling* in *Hist.* (1870).

167. Agaricus (Hebeloma) geophyllus Sow.
On bare ground in woods, frequent.
Y. Studley; *Exc.*, 1881.
N. Harrogate; *Exc.*, 1881.
A. Meanwood.

168. Agaricus (Flammula) spumosus Fr.
A. Heaton Wood.

169. Agaricus (Flammula) carbonarius Fr.
A. On burnt earth in Shipley Glen, 1884.

170. Agaricus (Flammula) inopus Fr.
C. On fir-tree trunk, Fixby Hall; *J. Bolton.* Not seen since.

171. Agaricus (Crepidotus) alveolus Lasch.
A. On dead trees, not common, Malham, 1882.

172. Agaricus (Crepidotus) mollis Schæff.
On old stumps in early autumn, frequent.
Y. Studley; *Exc.*, 1881.
W. Bolton Woods. Grassington.
A. Malham. Steeton.

173. Agaricus (Crepidotus) rubi Berk.
On dead bramble, rare.
A. Airedale; *J. Mackenzie.*
T. Roche Abbey Valley; *J. H. Aveling* in *Hist.* (1870).

174. Agaricus (Crepidotus) chimonophilus B. & Br.
A. On logs in a wood-yard at Steeton, 1885.

175. Agaricus (Naucoria) melinoides Fr.
A. On lawns in Peel Park, Bradford, common, 1885.
T. Roche Abbey Valley! *J. H. Aveling* in *Hist.* (1870).

176. Agaricus (Naucoria) semiorbicularis Bull.
In pastures, common.
A. Baildon. Hawksworth.
T. Roche Abbey Valley; *J. H. Aveling* in *Hist.* (1870).

177. Agaricus (Naucoria) escharoides Fr.
A. Moist ground, Esholt.

178. Agaricus (Naucoria) nuceus Bolt.
C. Mount Pellon, Halifax, 1787; *J. Bolton.*

179. Agaricus (Naucoria) scorpioides Fr.
A. Meanwood Wood, 1881.

180. Agaricus (Galera) tener Schæff.
Rich pastures, common.
A. Saltaire. Bingley. Eldwick; *W. West.*
C. Near Halifax; *J. Bolton.* Sharlston; *H. F. Parsons.* Huddersfield: *Hudd. Bot. Soc.*

181. Agaricus (Galera) confertus Bolt.
C. 'In a pine-apple stove at Halifax, with *A. luteus*, in 1785'; *J. Bolton.*

182. Agaricus (Galera) sparteus Fr.
C. Amongst moss, in meadows, Halifax, 1788; *J. Bolton.*

183. Agaricus (Galera) hypnorum Batsch.
Common amongst moss.
L. Ingleton; *W. West.*
Y. Studley; *Exc.*, 1881.
W. Bolton Woods; *W. West.*
A. Bingley. Esholt; Horsforth; *W. West.*
C. North Dean; *H. F. Parsons.* Hebden Bridge; *W. West.*
D. & T. Wharncliffe and Roche woods!

184. Agaricus (Galera) sphagnorum Pers.
A. In *Sphagnum* swamps, Rombalds Moor.

185. Agaricus (Galera) lateritius Fr.
A. Church Fenton; *G. Massee.*

186. Agaricus (Galera) embolus Fr.
A. On heathy ground. Baildon Moor.

187. Agaricus (Galera) paludosus Fr.
A. Swampy ground in Nab Wood, Saltaire.

188. Agaricus (Tubaria) furfuraceus P.
On wood-yard rubbish, etc., frequent, and rather pretty.
W. York.
A. Leeds. Rawcliffe. Esholt, etc. Carlton; *W. N. Cheesman.*

189. Agaricus (Psalliota) arvensis Schæff. *Horse Mushroom.*
In meadows, common, and generally distributed. Often attaining a large size. Esculent, and frequently eaten for the next species, of which it is perhaps only a variety.

190. Agaricus (Psalliota) campestris L. *True Mushroom.*
In rich old pastures, common, and generally distributed, but more abundant in some seasons than in others. In dry seasons and on elevated limestone pastures the variety *rufescens*, with a scaly red-brown pileus, is as common as the type.

191. Agaricus (Psalliota) sylvaticus Schæff.
A. Abundant in Nab Wood, Saltaire, 1880; not seen since.

192. Agaricus (Stropharia) æruginosus Curt.
In woods, meadows, and pastures, common.
N. Harrogate; *Exc.*, 1881.
W. Eccup!
A. Brayton; *H. F. Parsons.* Saltaire. Bingley. Eldwick; *W. West.*
C. Birstall; *W. West.*

193. Agaricus (Stropharia) albo-cyaneus Desm.
A. In fields, but not common. Baildon.
C. Huddersfield district; *Hudd. Bot. Soc.*

194. Agaricus (Stropharia) inunctus Fr.
W. In pastures. Common near Ilkley, 1886.

195. Agaricus (Stropharia) squamosus Fr
N. Harrogate; *Exc.*, 1881.
W. Ryther; *W. N. Cheesman.*
A. Nab Wood, Saltaire.

196. Agaricus (Stropharia) stercorarius Fr.
On dung, frequent.
W. Bolton.
A. Heaton. Eldwick; *W. West.*

197. Agaricus (Stropharia) semiglobatus Batsch.
On dung, very common.
N. Harrogate.
A. Rawcliffe. Eldwick; Esholt; *W. West.*
C. Birstall, etc. Wibsey; *W. West.*
D. Goole.
Obs.—Sometimes gathered and eaten (as well as the preceding species) for 'young mushrooms,' with serious effects.

198. Agaricus (Hypholoma) sublateritius Fr.
On old stumps, common.
A. Hambleton; *H. F. Parsons.* Temple Hurst; *W. N. Cheesman.* Shipley Glen. Heaton. Bingley; Horsforth; *W. West.*

199. Agaricus (Hypholoma) fascicularis Huds. *Yellow Tufted Agaric.*
On old tree-stumps, very common.
N. Beckwithshaw.
A. Meanwood. Baildon; Bingley; Headingley; *W. West.*
C. Halifax; *J. Bolton.* Huddersfield.
D. Goole.
T. Roche Abbey Valley; *J. H. Aveling* in *Hist.* (1870).

200. Agaricus (Hypholoma) epixanthus Fr.
A. On fir stumps, Heaton.

201. Agaricus (Hypholoma) dispersus Fr.
A. On stumps in fir-woods. Adel Dam Wood!

202. Agaricus (Hypholoma) lachrymabundus Fr.
In woods.
N. Harrogate; *Exc.*, 1881.
A. Nab Wood, Saltaire. Bingley. Haddlesey; *W. N. Cheesman.* Horsforth; *W. West.*
C. Huddersfield district; *Hudd. Bot. Soc.*

203. Agaricus (Hypholoma) velutinus P.
On stumps, frequent and generally distributed.
A. Bingley. Heaton.
T. Roche Abbey Valley; *J. H. Aveling* in *Hist.* (1870).

204. Agaricus (Hypholoma) Candolleanus Fr.
C. On fallen trees, rare. Odsal Wood, near Wibsey, 1885.

205. Agaricus (Hypholoma) lanaripes Cooke.
A. On soil in a conservatory at Bradford, 1880.

206. Agaricus (Hypholoma) appendiculatus Bull.
A. On old stumps, frequent. Heaton. Saltaire.

207. Agaricus (Hypholoma) hydrophilus Bull.
In woods, etc., locally common.
A. Abundant in Bingley Woods, 1884.

208. Agaricus (Psilocybe) ericæus P.
On wet heaths, etc.
A. Common on Baildon Moor. Wistow Common; *W. N. Cheesman.*

209. Agaricus (Psilocybe) areolatus Klot.
C. Huddersfield district; *Hudd. Bot. Soc.*

210. Agaricus (Psilocybe) semilanceatus Fr. *'Liberty-cap.'*
In hilly pastures, extremely common. Poisonous.
A. Adel, 1876! Baildon, Bingley, Saltaire, etc.
C. Skircoat Moor; *W. West.* Wibsey.
D. Wharncliffe.

211. Agaricus (Psilocybe) spadiceus Schæff.
A. In woods on logs and stumps. Shipley Glen.

212. Agaricus (Psilocybe) fœnisecii P.
Amongst grass, frequent.
A. Adel! Airedale; *J. Mackenzie.* Saltaire. Church Fenton; *G. Massee.*

213. Agaricus (Psathyra) obtusatus Fr.
A. On logs. Goitstock, near Bingley, 1884.

214. Agaricus (Psathyra) mastiger B. & Br.
A. Amongst grass. Nab Wood, Saltaire.

215. Agaricus (Psathyra) corrugis P.
A. In meadows and pastures, not common. Baildon, 1884.

216. Agaricus (Psathyra) fibrillosus P.
A. In woods. Nab Wood, Saltaire.

217. Agaricus (Psathyra) gossypinus Fr.
In woods, on the ground, rare.
A. Bell Bank, Bingley, 1784 and 1788; *J. Bolton.* Bingley Woods, 1884.

218. Agaricus (Psathyra) semivestitus B. & Br.
A. Amongst grass on road-sides. Saltaire.

219. Agaricus (Deconica) bullaceus Bull.
A. On horse-dung. Heaton.

220. Agaricus (Deconica) coprophilus Bull.
A. Church Fenton, 1881; *G. Massee.*

221. Agaricus (Panæolus) separatus L.
On dung; common.
Y. Studley; *Exc.*, 1881.
A. Esholt, Saltaire, etc. Brayton and Rawcliffe; *H. F. Parsons.* Cookridge; *W. West.*
C. Whalley near Mirfield. Birstall; Greetland; *W. West.*

222. Agaricus (Panæolus) fimiputris Bull.
On dung, frequent.
Y. Studley; *Exc.*, 1881.
A. Heaton. Saltaire.
C. Huddersfield district; *Hudd. Bot. Soc.*
T. Roche Abbey Valley; *J. H. Aveling* in *Hist.* (1870).

223. Agaricus (Panæolus) campanulatus L.
A. In meadows. Saltaire. Baildon Moor.

224. Agaricus (Panæolus) papilionaceus Bull.
On dung, frequent.
W. Cawood; *W. N. Cheesman.*
A. Bingley.

225. Agaricus (Panæolus) cinctulus Bolt.
C. On dung-hills. Halifax, 1789; *J. Bolton.* Not seen since.

226. Agaricus (Panæolus) fimicola Fr.
C. Amongst grass in pastures at Halifax; *J. Bolton.* Not seen since.

227. Agaricus (Psathyrella) gracilis Fr.
A. Amongst grass. Hawksworth. Ryther; *W. N. Cheesman.*

228. Agaricus (Psathyrella) atomatus Fr.
In grassy places.
Y. Studley; *Exc.*, 1881.
A. Baildon Moor, Saltaire. Stainor Wood; *W. N. Cheesman.*

229. Agaricus (Psathyrella) disseminatus P.
About fallen trees, etc., frequent, and very pretty.
W. Bolton Woods.
A. Shipley Glen. Esholt.

230. Coprinus comatus Fr. *'Maned'* or *'Bell' Mushroom.* *'Wigstool.'*
Rich waste ground, road-sides, pastures, lawns, etc., common and generally distributed. One of the most delicious of edible fungi.
Y. Studley; *Exc.*, 1881.
A. Selby; *W. N. Cheesman.* Esholt, Frizinghall, Hawksworth, and Bradford, where many stones in weight were gathered and eaten during the autumns of 1885-6.
C. Ravensthorpe; *W. Fowler.*

231. Coprinus ovatus Fr.
In similar situations, and bearing a resemblance to the last-named.
A. Frizinghall near Bradford.

232. Coprinus atramentarius Fr. *Inky Mushroom.*
On the ground, sometimes on stumps, frequent. A good edible species.
Y. Studley; *Exc.*, 1881.
A. Selby; *W. N. Cheesman.* Meanwood; *J. Abbott.* Esholt. Hawksworth. Horton; *W. West.*
C. Odsal near Wibsey.
D. Goole; *H. F. Parsons.*
T. Roche Abbey Valley; *J. H. Aveling* in *Hist.* (1870).

[**Coprinus luridus** Fr. Incognit.
Halifax; *J. Bolton.* Not found since, and, according to Dr. Cooke's Handbook of British Fungi, p. 162, probably not a distinct species.]

233. Coprinus aphthosus Fr.
C. Halifax; *J. Bolton.*

234. Coprinus extinctorius Fr.
C. On the ground, rare. Halifax; *J. Bolton.*

235. Coprinus fimetarius Fr.
A. On dung-hills in spring, common. Bradford, Heaton, and Esholt.

236. Coprinus tomentosus Fr.
A. Heaton.
C. Halifax; *J. Bolton.*

237. Coprinus niveus Fr. *Snowy Coprinus.*
On horse-dung, frequent, and generally distributed.
W. Bramham!
A. Bradford, Heaton, and Bingley.

238. Coprinus micaceus Fr.
On old logs and stumps, very common, and generally distributed.

239. Coprinus lagopus Fr.
A. On dung. Saltaire.

240. Coprinus ephemerus Fr.
Y. Studley; *Exc.*, 1881.
N. Harrogate; *Exc.*, 1881.
A. On dung-hills, etc., frequent. Bradford. Heaton.

241. Coprinus plicatilis Fr.
In pastures, common, but very fugacious.
Y. Studley; *Exc.*, 1881.
A. Heaton. Brayton; *W. N. Cheesman.*
C. Huddersfield district; *Hudd. Bot. Soc.*

242. Coprinus hemerobius Fr.
A. Esholt, 1885.
C. By a road-side, Halifax; *J. Bolton.*

243. Coprinus radiatus Fr.
On dung, common. Very small and delicate.
A. Saltaire, Bradford, Heaton. Horton; *W. West.*

244. Bolbitius Boltoni Fr.
C. Halifax, on dung; *J. Bolton.*

245. Bolbitius fragilis Fr.
On dung.
A. Selby; *W. N. Cheesman.*
C. Halifax; *J. Bolton.*

246. Bolbitius titubans Fr.
A. Amongst grass, common. Meanwood.

247. Bolbitius tener B.
A. Pastures at Baildon.
C. Huddersfield district, common; *Hudd. Bot. Soc.*
T. Roche Abbey Valley; *J. H. Aveling* in *Hist.* (1870).

248. Cortinarius (Phlegmacium) cyanipes Fr. (*C. cyanopus* Berk.)
L. In woods, rare. Helks Wood, Ingleton.

249. Cortinarius (Phlegmacium) cærulescens Fr.
Y. Studley; *Exc.*, 1881.

250. Cortinarius (Phlegmacium) purpurascens Fr.
In woods, especially fir and larch, frequent.
A. Shipley Glen.

251. Cortinarius (Phlegmacium) callochrous Fr.
T. Roche Abbey Valley; *J. H. Aveling* in *Hist.* (1870).

252. Cortinarius (Myxacium) collinitus Fr.
In woods.
Y. Studley; *Exc.*, 1881.
A. Bingley and Shipley Glen, common.

253. Cortinarius (Myxacium) elatior Fr.
In woods, not unfrequent.
R. Slaidburn!
N. Harrogate; *Exc.*, 1881.
A. Shipley Glen.

254. Cortinarius (Dermocybe) tabularis Fr.
A. In woods, frequent. Heaton and Bingley.

255. Cortinarius (Dermocybe) anomalus Fr.
In woods, frequent, and generally distributed.

256. Cortinarius (Dermocybe) cinnabarinus Fr.
W. Cawood Wood; *W. N. Cheesman.*

257. Cortinarius (Dermocybe) sanguineus Fr.
In woods, rare.
R. Slaidburn, 1876!
C. Halifax; *J. Bolton.*

258. Cortinarius (Dermocybe) cinnamomeus Fr.
In woods, a very pretty species.
A. Selby; *W. N. Cheesman.* Bingley.
C. Ogden Kirk, Halifax, 1782; *J. Bolton.* Honley Wood; *Hudd. Bot. Soc.*

259. Cortinarius (Telamonia) limonius Fr.
In fir-woods, rare.
A. Plantation near Cullingworth, 1884.

260. Cortinarius (Telamonia) hinnuleus Fr.
N. Generally distributed in woods. Harrogate; *Exc.*, 1881.

261. Cortinarius (Telamonia) iliopodius Fr.
A. In woods, frequent in early summer. Bingley, 1884.

262. Cortinarius (Hygrocybe) castaneus Fr.
Y. Studley; *Exc.*, 1881.
A. In gardens, Meanwood!
C. Odsal Wood near Wibsey, 1885.

[**Cortinarius (Hygrocybe) dilutus** Fr. Incognit. Halifax; *J. Bolton.* Not found since.]

263. Cortinarius (Hygrocybe) acutus Fr.
A. Wet places in woods. Hawksworth.

264. Lepista nuda Bull. (*Tricholoma nudus* Berk.) *Wood Blewitts.*
In fir-plantations, frequent.
W. Bramham!
A. Adel fir-wood!
C. Chevin-edge Wood, Halifax! Near Huddersfield; *Hudd. Bot. Soc.*
D. Wharncliffe! Cantley!

265. Lepista personata (Fr.). (*Tricholoma personatus* Bull.) '*Blue-bottoms.*' '*Blewitts.*'
In pastures, common, often growing in large rings. A fairly good edible species, when young, or in dry weather.
N. Harrogate; *T. Hick.*

A. Meanwood; *J. Abbott.* Eccleshill, Hawksworth, and Heaton.
C. Lepton; *Hudd. Bot. Soc.*
D. Near Goole; *H. F. Parsons.*
T. Roche Abbey Valley; *J. H. Aveling* in *Hist.* (1870).

266. Paxillus involutus Fr.
One of our commonest fungi, abundant in most woods throughout the Riding. Distribution general.

267. Hygrophorus cossus Fr.
Y. In woods, rare. Studley; *Exc.*, 1881.

268. Hygrophorus hypothejus Fr.
N. In fir-woods, Beckwithshaw near Harrogate; *Exc.*, 1881.

269. Hygrophorus pratensis Fr.
Upland pastures, frequent. A good edible species.
Y. Studley; *Exc.*, 1881.
W. Ilkley.
A. Common in Airedale. Burn; *W. N. Cheesman.*
C. Halifax; *J. Bolton.* Huddersfield; *Hudd. Bot. Soc.*

270. Hygrophorus virgineus Fr. *Ivory Caps.*
Pastures, common. Esculent.
A. Brayton; *H. F. Parsons.* Meanwood; *J. Abbott.* Saltaire, Baildon.
C. Huddersfield; *Hudd. Bot. Soc.*

271. Hygrophorus niveus Fr.
Often growing in company with the last-named.
A. Bingley, Heaton, and Saltaire.

272. Hygrophorus distans Berk.
A. In woods, rare. Shipley Glen, 1884.
T. Roche Abbey Valley; *J. H. Aveling* in *Hist.* (1870).

273. Hygrophorus ovinus Fr.
A. In pastures, not common. Hawksworth, 1885.

274. Hygrophorus fornicatus Fr.
Y. Studley; *Exc.*, 1881.

275. Hygrophorus ceraceus Fr.
In pastures, occasional.
R. Slaidburn!
A. Bingley.
C. Huddersfield; *Hudd. Bot. Soc.*

276. Hygrophorus coccineus Fr.
A common species, and a pretty one.
Y. Studley; *Exc.*, 1881.
A. Brayton; *H. F. Parsons.* Selby; *W. N. Cheesman.* Moor Allerton! Saltaire and Bingley.
C. Huddersfield; *Hudd. Bot. Soc.*

277. Hygrophorus puniceus Fr.
Y. Studley; *Exc.*, 1881.
A. Bingley.
C. Halifax; *J. Bolton.* Huddersfield; *Hudd. Bot. Soc.*

278. Hygrophorus miniatus Fr.
A. On moist heaths. Baildon Moor.

279. Hygrophorus eburneus Fr.
T. Roche Abbey Valley; *J. H. Aveling* in *Hist.* (1870).

280. Hygrophorus conicus Fr.
In pastures, variable in facies, but always attractive in appearance.
Y. Studley; *Exc.*, 1881.
A. Brayton; *H. F. Parsons.* Heaton.
C. Huddersfield; *Hudd. Bot. Soc.*
T. Roche Abbey!

281. Hygrophorus psittacinus Fr.
In fields, common.
Y. Studley; *Exc.*, 1881.
A. Saltaire. Bingley.
C. Battye-Ford; *H. F. Parsons.* Huddersfield; *Hudd. Bot. Soc.*
D. Wharncliffe; *H. F. Parsons.*

282. Hygrophorus unguinosus Fr.
In woods, etc., rare.
Y. Studley; *Exc.*, 1881.
A. Meanwood Wood; *Exc.*, 1881.

283. Hygrophorus murinaceus Fr. (*H. nitratus* Fr., Hym. Eur.)
A. Pastures. Baildon, 1885.

284. Gomphidius glutinosus Fr.
A. In fir-woods, not common. Adel!

285. Gomphidius viscidus Fr.
A. Bishops Wood; *W. N. Cheesman.*

286. Lactarius torminosus Fr.
N. In woods, not common. Harrogate; *Exc.*, 1881

287. Lactarius turpis Fr.
In woods, common. A very dingy species.
W. Otley.
A. Shipley Glen. Heaton. Meanwood and Horsforth; *W. West.*

288. Lactarius controversus Pers.
A. Under trees, rare. Hawksworth.

289. Lactarius insulsus Fr.
In woods and under trees.
C. Huddersfield; *Hudd. Bot. Soc.*
T. Roche Abbey Valley; *J. H. Aveling* in *Hist.* (1870).

290. Lactarius blennius Fr.
A. Heaton. Meanwood.

291. Lactarius trivialis Fr.
A. Heaton Wood, 1884.

292. Lactarius uvidus Fr.
N. Harrogate; *Exc.*, 1881.

293. Lactarius pyrogalus Fr.
A. Heaton Wood.
C. Lepton Wood, etc.; *Hudd. Bot. Soc.*

294. Lactarius acris Fr.
In woods, rare.
L. Ingleton; *W. West.*
C. Woodhouse Wood, Halifax; *J. Bolton.*

295. Lactarius chrysorrheus Fr.
In woods, not common.
N. Harrogate; *Exc.*, 1881.
C. Halifax; *J. Bolton.*

296. Lactarius piperatus Fr.
In woods, not common. Poisonous. The earliest noticed West Riding Fungus, recorded in 1677 by Ray in *Cat. Plant. Angl.*, ed. ii, as found by one Martin Lister, in Marton Woods under Pinno Moor, Craven. Well described by Dr. Lister in a letter to Ray, October 12th, 1672, in Derham's Letters.
A. Brayton; *W. N. Cheesman.*
C. Halifax.
T. Roche Abbey Valley; *J. H. Aveling* in *Hist.* (1870).

297. Lactarius vellereus Fr.
A. Shipley Glen Wood, Heaton, Bingley, etc.

Var. **exsuccus** Otto.
A. Shipley Glen.

298. Lactarius deliciosus Fr. '*Red-milk' Mushroom.*
In fir-woods, very local.
N. Harrogate district; *T. Hick.*
A. Fir-wood near Adel Dam! Bishops Wood; *W. N. Cheesman.*

299. Lactarius pallidus Fr.
A. Heaton Woods.
C. Huddersfield district; *Hudd. Bot. Soc.*

300. Lactarius quietus Fr.
In woods, common.
A. Shipley Glen; *W. West.* Bingley, Heaton, etc.
C. North Dean; *H. F. Parsons.*
D. Wharncliffe!

301. Lactarius theiogalus Fr.
In woods, rather rare.
N. Harrogate; *Exc.*, 1881.
A. Shipley Glen; *W. West.* Bingley.
C. Ramsden Wood, Halifax; *J. Bolton.* Near Huddersfield; *H.B.S.*

302. Lactarius rufus Fr.
In fir-woods, frequent. Poisonous.
A. Meanwood! Bingley, etc.
D. Thorne Waste; *H. F. Parsons.*

303. Lactarius glyciosmus Fr.
N. Harrogate; *Exc.*, 1881.
A. Saltaire, Bingley, etc.
C. Near Wakefield.

304. Lactarius serifluus Fr.
In woods, probably common.
Y. Studley; *Exc.*, 1881.
N. Harrogate; *Exc.*, 1881.
A. Horsforth; Bingley; Allerton; *W. West.* Shipley Glen.
C. Near Huddersfield; *Hudd. Bot. Soc.*
T. Roche Abbey Valley; *J. H. Aveling* in *Hist.* (1870).

305. Lactarius volemus Fr.
A. Only seen once, at Shipley Glen.

306. Lactarius mitissimus Fr.
Y. Studley; *Exc.*, 1881.
A. Heaton Wood near Bradford.

307. Lactarius subdulcis Fr.
In woods, locally abundant.
Y. Studley; *Exc.*, 1881.
N. Harrogate; *Exc.*, 1881.
A. Bingley. Meanwood Wood.
C. Woods about Huddersfield; *Hudd. Bot. Soc.*

308. Lactarius mammosus Fr.
A. Meanwood Wood near Leeds; *Exc.*, 1881.

309. Russula nigricans Fr.
In woods, locally common.
Y. Studley; *Exc.*, 1881.

N. Harrogate; *Exc.*, 1881.
A. Near Wilsden. Heaton.
C. Huddersfield district; *Hudd. Bot. Soc.*
D. Wharncliffe Woods!

310. Russula adusta Fr.
A. Currer Wood, Steeton.

311. Russula furcata Fr.
C. Huddersfield district, not common; *Hudd. Bot. Soc.*

312. Russula sanguinea Fr.
A. Newlay Woods.

313. Russula sardonia Fr.
A. Near Currer Wood, Steeton, 1884.

314. Russula depallens Fr.
A. Hawksworth Spring.

315. Russula lactea Fr.
A. Under trees near Cottingley Bridge.

316. Russula virescens Fr.
A. Heaton and Bingley Woods.
D. Wharncliffe; *H. F. Parsons.*
T. Roche Abbey Valley; *J. H. Aveling* in *Hist.* (1870).

317. Russula rubra Fr.
N. Harrogate; *Exc.*, 1881.
W. Harewood; *J. Abbott.*
A. Micklefield; *J. Abbott.* Heaton Woods.

318. Russula vesca Fr.
A. Bingley.
C. Armitage Bridge Wood, 1883; *Hudd. Bot. Soc.*

319. Russula cyanoxantha Fr.
A. Shipley Glen Wood. Under trees, Cottingley.

320. Russula heterophylla Fr.
C. Frequent, woods near Huddersfield; *Hudd. Bot. Soc.*
T. Roche Abbey Valley; *J. H. Aveling* in *Hist.* (1870).

321. Russula fœtens Fr.
In woods, not common.
N. Harrogate; *Exc.*, 1881.
A. Meanwood Wood, 1881.

322. Russula emetica Fr.
In woods, common.
Y. Studley; *Exc.*, 1881.

A. Hambleton; *H. F. Parsons.* Saltaire, Horsforth. Barlow Woods; *W. N. Cheesman.* Harden; *W. West.*
C. Birstal; North Dean; *W. West.*
D. Goole; *H. F. Parsons.*

323. Russula ochroleuca Fr.
Common in woods.
N. Harrogate; *Exc.*, 1881.
A. Horsforth; *W. West.* Shipley Glen and Bingley.

324. Russula fragilis Fr.
In woods.
Y. Studley; *Exc.*, 1881.
N. Harrogate; *Exc.*, 1881.
A. Nab Wood, Saltaire.

325. Russula aurata Fr.
A. Bingley Woods, 1884, rare.

326. Russula nitida Fr.
A. Bingley Woods, 1884.
T. Roche Abbey Valley; *J. H. Aveling* in *Hist.* (1870).

327. Russula lutea Fr.
N. Harrogate; *Exc.*, 1881.
A. Goitstock Wood near Bingley, 1885. Saltaire; *W. West.*

328. Russula alutacea Fr.
In woods, not common. Esculent.
Y. Studley; *Exc.*, 1881.
A. Hawksworth. Barlow Woods; *W. N. Cheesman.*
C. Armitage Bridge Wood, 1883; *Hudd. Bot. Soc.*
T. Roche Abbey Valley; *J. H. Aveling* in *Hist.* (1870).

329. Russula fellea Fr.
Y. Studley; *Exc.*, 1881.

330. Cantharellus cibarius Fr. *Edible Chantarelle.*
In woods. A first-class edible species, but apparently very rare in the West Riding, since a fungus so striking in appearance, from its yolk-yellow colour and trumpet-like shape, is little likely to have been overlooked.
A. Near Leeds; *J. Abbott.*

331. Cantharellus aurantiacus. *False Chantarelle.*
On heaths and in woods, frequent.
N. Harrogate; *Exc.*, 1881.
A. Airedale; *J. Mackenzie.* Fir-wood on Adel Blackmoor! Bingley, Shipley Glen, etc. Haddlesey; *W. N. Cheesman.*
C. Odsal near Wibsey.

332. Cantharellus tubæformis Fr.
A. Bingley Woods, 1881. Not seen since.

333. Cantharellus infundibuliformis Fr.
Y. In woods. Studley; *Exc.*, 1881.

334. Cantharellus radicosus B. & Br. (*C. carbonarius* A. & S.)
D. Rare. On ashes where charcoal had been made, Wharncliffe Woods!

335. Cantharellus cinereus Fr.
C. In woods, Halifax; *J. Bolton.*

336. Cantharellus muscigenus Fr.
R. or A. On mosses between Malham and Settle; *H. F. Parsons.*

337. Cantharellus lobatus Fr.
C. On mosses in swamps, Halifax; *J. Bolton.*

338. Nyctalis parasitica Fr.
Parasitic on various species of *Russula*, rather rare.
W. Devonshire Wood near Sicklinghall, and Aberford, 1879!
A. Barlow Woods; *W. N. Cheesman.*
T. Roche Abbey Valley; *J. H. Aveling* in *Hist.* (1870).

339. Marasmius urens Fr.
In woods, etc., rather common.
A. Goit-stock Wood near Bingley, 1885.
T. Roche Abbey Valley; *J. H. Aveling* in *Hist.* (1870).

VAR. **peronatus** Fr.
Y. Studley; *Exc.*, 1881.
A. Meanwood Wood.
C. Northowram; *J. Bolton.*
T. Roche Abbey Valley; *J. H. Aveling* in *Hist.* (1870).

340. Marasmius oreades Fr. *Fairy-ring Mushroom.*
Forming rings in open fields, not nearly so common in West Yorkshire as it is said to be in many other districts. A most delicious edible species.
A. Bingley. Selby; *W. N. Cheesman.*
C. Huddersfield district; *Hudd. Bot. Soc.*
T. Roche Abbey Valley; *J. H. Aveling* in *Hist.* (1870).

341. Marasmius languidus Fr.
A. On withered grass in a fir-plantation near Cullingworth.

342. Marasmius amadelphus Fr.
C. On dead twigs, Odsal Wood, 1885.

343. Marasmius ramealis Fr.
Y. Studley; *Exc.*, 1881.
A. Common in the shade, on dead twigs and branches. Shipley Glen. Bingley; *W. West.*

344. Marasmius rotula Fr.
On fallen twigs, etc., common and generally distributed.
A. Shipley Glen.
T. Roche Abbey Valley; *J. H. Aveling* in *Hist.* (1870).

345. Marasmius graminum B. & Br.
A. On leaves of grass. Heaton Wood.
T. Roche Abbey Valley; *J. H. Aveling* in *Hist.* (1870).

346. Marasmius epiphyllus Fr.
Y. On dead leaves, twigs, etc. Studley; *Exc.*, 188 .

347. Marasmius insititius Fr.
T. Roche Abbey Valley; *J. H. Aveling* in *Hist.* (1870).

348. Marasmius erythropus Fr.
Y. Amongst leaves, etc., Studley; *Exc.*, 1881.

349. Marasmius androsaceus Fr.
On dead leaves, etc., in woods.
A. Saltaire, Rombalds Moor (*W. West*), Bingley, etc.
C. 'Causey Foot, near Halifax, near where *Trientalis* and *L. cordata* grow'; *J. Bolton.* The fungus may possibly be found there still, but what about the two flowering plants? The district has seen a few changes during the last eighty years!

350. Marasmius epichloe Fr.
W. On dead grass near Ilkley.

351. Lentinus lepideus Fr.
On fir-stumps and logs.
A. On wood in a cellar at Bradford; *W. West.*
C. & D. Sowerby Bridge and Darton; *Hudd. Bot. Soc.*

352. Lentinus cochleatus Fr.
On the ground in woods, etc. Has a powerful odour of aniseed.
A. Heaton Wood.
C. Bradley Gate Wood near Huddersfield; *H.B.S.* Odsal Wood.

[**Lentinus flabelliformis** Fr. Incognit.
C. On stumps, Halifax; *J. Bolton.* Not seen since, and, according to *Berkeley*, Bolton's figure (on plate 157 of Hist. Fung.) may be only *Agaricus salignus.*]

353. Panus torulosus Fr.
On old stumps, not common.
A. Hawksworth.
C. Halifax; *J. Bolton.* Huddersfield district; *Hudd. Bot. Soc.*
D. Hook near Goole; *H. F. Parsons.*

354. Panus stypticus Fr.
On dead logs and stumps, not common. Formerly dried and used for arresting hæmorrhage.
A. In a wood-yard, Steeton, 1885.

355. Panus conchatus Fr.
A. Selby; *W. N. Cheesman.*

356. Schizophyllum commune Fr.
On dead wood, very rare.
A. In a wood-yard at Steeton, 1884. Selby; *W. N. Cheesman.*

357. Lenzites betulina Fr.
W. Bolton Woods.
A. On old stumps, etc., frequent. Steeton.
D. Near Sheffield; *Amos Carr.*

358. Lenzites flaccida Fr.
C. On stumps, Shibden Hall near Halifax; *J. Bolton.*

II. POLYPOREI.

359. Boletus luteus L.
A. In fir-woods, occasional. Hainworth near Keighley.

360. Boletus elegans Schum.
N. In woods, Harrogate; *Exc.*, 1881.
A. Esholt Woods, 1886.

361. Boletus flavus With.
W. Grassington. Near Stutton; *G. Massee!*
A. Frequent in woods. Bingley. Esholt. Hawksworth.
C. Burks Hall; *J. Bolton.* Meltham; *Hudd. Bot. Soc.*

362. Boletus laricinus Berk.
M. Warmfield larch-wood, Greenfield!
W. Sweep Wood, Wetherby!
A. In larch plantations, occasional. Church Fenton, 1881; *G. Massee.*

363. Boletus granulatus L.
A. Fir-plantation near Cullingworth. Bingley Woods. Bishops Wood; *W. N. Cheesman.*

364. Boletus bovinus Fr.
A. Nab Wood, Saltaire. Bingley Wood. Brayton; *W. N. Cheesman.*

365. Boletus badius Fr.
N. Beckwithshaw Wood near Harrogate; *Exc.*, 1881.
A. Bingley Woods. Camblesforth; *W. N. Cheesman.*

366. Boletus piperatus Bull.
Not common.
A. Nab Wood, Saltaire. Near Cullingworth.
C. Shelley near Huddersfield.

367. Boletus rubinus Smith.
W. Under trees, rare. Field near Bolton Woods, 1885.

368. Boletus parasiticus Bull.
T. Roche Abbey Valley; *J. H. Aveling* in *Hist.* (1870).

369. Boletus chrysenteron Fr.
In fields, woods, etc., the commonest species of the genus.
A. Camblesforth; *H. F. Parsons.* Bingley, Shipley Glen, etc.
C. Huddersfield district; *Hudd. Bot. Soc.*

370. Boletus subtomentosus L.
In woods, frequent.
N. Harrogate; *Exc.*, 1881.
A. Heaton. Bingley; *W. West.*
C. North Dean; *H. F. Parsons.* Huddersfield district *Hudd. Bot. Soc.*

371. Boletus calopus Fr.
A. Bingley, August 1886.
C. Halifax; *J. Bolton.*

372. Boletus pachypus Fr.
Local.
A. Frequent in Bingley Woods, along with the var. *amarus* Saund. & Sm.

373. Boletus olivaceus Schæff.
C. In woods, rare. Lillands Wood, Raistrick; *Hudd. Bot. Soc.*

374. Boletus edulis Bull.
Dry woods, locally common. A very good edible species.
N. Harrogate; *T. Hick.*
A. Bramhope; *J. Abbott.* Bingley. Selby; *W. N. Cheesman.*
T. Roche Abbey Valley; *J. H. Aveling* in *Hist.* (870).

375. Boletus luridus Fr.
In woods, not common. Poisonous.
Y. Studley!
W. Grassington.
A. Malham. Bingley.
D. Wharncliffe; *H. F. Parsons.*

376. Boletus erythropus P.
Under trees in fields and woods, not common. A very pretty species.
A. Hawksworth. Shipley Glen.
C. Huddersfield; *Hudd. Bot. Soc.*

377. Boletus purpureus Fr.
A. Heaton Woods, rare.

378. Boletus versipellis Fr.
A. Bingley Woods, not common.

379. Boletus scaber Fr.
In woods, frequent. Edible.
R. Linn Gill, 1,000 feet.
N. Harrogate; *Exc.*, 1881.
W. Ryther; *W. N. Cheesman.*
A. Saltaire; *W. West.* Bingley, etc.
C. Huddersfield district; *Hudd. Bot. Soc.*
D. Goole Moor; *H. F. Parsons.*

380. Boletus felleus Bull.
A. In woods, a rare and interesting species. Bingley, 1884.

381. Polyporus lentus Berk.
T. Roche Abbey Valley; *J. H. Aveling* in *Hist.* (1870).

382. Polyporus fuscidulus Fr.
C. North Dean; *J. Bolton.* Not seen since.

383. Polyporus Schweinizii Fr.
W. Farnley Wood near Otley, 1885; *R. Dewhirst.*

384. Polyporus perennis Fr.
Rare.
A. Near Selby; *W. N. Cheesman.* Shipley Glen; *J. Beanland.*
C. Southowram, 1784; *J. Bolton.*

385. Polyporus squamosus Fr.
A common species on tree-trunks.
R. Bolland Gorge, common!
W. Bramham, abundant; *W. West.*
A. Leeds district! Shipley; *W. West.* Bishops Wood; *W. N. Cheesman.* Apperley, etc.
C. Huddersfield; *Hudd. Bot. Soc.* Whitley. Wakefield; *H. Sims.*
D. Goole. Askern; *H. F. Parsons.*
T. Roche Abbey Valley; *J. H. Aveling* in *Hist.* (1870).

386. Polyporus varius Fr.
W. Bolton Woods.
C. On tree-trunks, Shibden Hall; *J. Bolton.*

387. Polyporus elegans Fr.
C. Woodhouse, near the Calder; *J. Bolton.*

VAR. **nummularius** Fr.
Y. Studley; *Exc.*, 1881.

388. Polyporus giganteus Fr.
On stumps, not common.
A. Nab Wood, Saltaire. Horsforth, 1886.
C. Halifax, 1786; *J. Bolton.* Huddersfield; *Hudd. Bot. Soc.*

389. Polyporus sulphureus Fr.
On tree-trunks. Very handsome.
W. Bilton Haggs near Thorparch!
A. Leeds; *J. Abbott.* Sherburn! Stainor Wood; *W. N. Cheesman.*
C. Thornes Park, Wakefield; *H. Sims.* Huddersfield district; *H.B.S.*

390. Polyporus intybaceus Fr.
A. On tree-trunks, rare. Shipley Glen; *W. West.*

391. Polyporus heteroclitus Fr.
C. 'On the ground under oaks, Fixby Hall'; *J. Bolton.*

392. Polyporus salignus Fr.
C. On willows, Copley; *J. Bolton.*

393. Polyporus destructor Fr.
C. Huddersfield district, 1883; *Hudd. Bot. Soc.*

394. Polyporus cæsius Fr.
Y. Studley; *Exc.*, 1881.

395. Polyporus adustus Fr.
On stumps, not common.
Y. Studley; *Exc.*, 1881.
W. Bolton Woods. Boston Spa; *J. Emmet.*

396. Polyporus hispidus Fr.
On living tree-boles, frequent.
A. Hambleton; *H. F. Parsons.* Leeds district; *J. Abbott.* Saltaire, Hawksworth, etc. Wistow; *W. N. Cheesman.*
C. Halifax; *J. Bolton.*

397. Polyporus cuticularis Fr.
A. On trunks, rare. Hawksworth Springs, 1885.

398. Polyporus spumeus Fr.
On trunks of trees.
A. Malham, 1883.
T. Roche Abbey Valley; *J. H. Aveling* in *Hist.* (1870).

399. Polyporus dryadeus Fr.
On decaying trees, not always oaks.
N. Oak stump, North Deighton!
W. On an old elm, near Bolton Bridge, 1878; *W. West.*
A. Leeds district; *J. Abbott.*
C. Huddersfield district, 1883; *Hudd. Bot. Soc.*

400. Polyporus betulinus Fr.
On birch-tree boles, common and generally distributed.
T. Roche Abbey Valley; *J. H. Aveling* in *Hist.* (1870).

401. Polyporus applanatus Fr.
On tree-stumps, etc.
A. Seven Arches near Saltaire.

402. Polyporus fomentarius Fr.
A. On trunks, occasional. Meanwood. Rawcliffe; *H. F. Parsons.*

403. Polyporus igniarius Fr.
A. On various trees, not common. Keighley.

404. Polyporus salicinus Fr.
On willow-trees by streams.
A. By Myles Hill beck, Meanwood! Church Fenton, 1881; *G. Massee.*

405. Polyporus ulmarius Fr.
On old elms, frequent.
C. Huddersfield district; *Hudd. Bot. Soc.*
T. Roche Abbey Valley; *J. H. Aveling* in *Hist.* (1870).

406. Polyporus fraxineus Fr.
D. On ash trees, rare. Wentbridge; *Nat.*, vol. iii, p. 10.

407. Polyporus cervinus Pers. (*P. mollis* Fr.)
R. On beech, not common. Horton-in-Ribblesdale.

408. Polyporus annosus Fr.
Frequent on larch stumps, etc.
Y. Studley; *Exc.*, 1881.
A. Esholt. Hawksworth. Meanwood.

409. Polyporus velutinus Fr.
W. On twigs, Church Fenton, 1881; *G. Massee.* Sweep Wood, Wetherby!
T. Roche Abbey Valley; *J. H. Aveling* in *Hist.* (1870).

410. Polyporus versicolor Fr.
On stumps, etc., the commonest of the Polyporei. Distributed generally throughout the Riding, and ascending high. The earliest book record, however, is
T. Roche Abbey Valley; *J. H. Aveling* in *Hist.* (1870).

411. Polyporus abietinus Fr.
A. On fir-trunks, not common. Barden, 1885.

412. Polyporus ferruginosus Fr.
L. On fallen trees, etc., not common, Ingleton, 1884.

413. Polyporus medulla-panis Fr.
C. Halifax; *J. Bolton.*

414. Polyporus vulgaris Fr.
On dead wood.
Y. & N. Studley and Harrogate; *Excs.*, 1881.
T. Roche Abbey Valley; *J. H. Aveling* in *Hist.* (1870).

415. Polyporus sanguinolentus Fr.
W. Uncommon, covering the under-side of railway sleepers at Church Fenton; *G. Massee.* Sweep Wood, Wetherby, on larch logs!

416. Polyporus molluscus Fr.
A. On decaying wood, Steeton, 1885.

417. Polyporus Gordoniensis B. & Br.
D. On fir, rare, Doncaster; *H. F. Parsons.*

418. Polyporus vaporarius Fr.
On fallen sticks, common and generally distributed.

419. Trametes suaveolens Fr.
On willows, limes, etc., rare.
W. Cawood; *H. F. Parsons.*
A. Selby; *W. N. Cheesman.*

420. Trametes odora Fr.
C. Shibden Hall; *J. Bolton.* Not found since.

421. Dædalea quercina P.
On oak stumps and rails, not common.
R. Linn Gill near Ribblehead.
A. Selby; *W. N. Cheesman.* Leeds district; *J. Abbott.* Steeton.
C. Halifax; *J. Bolton.* Huddersfield district; *Hudd. Bot. Soc.*

422. Dædalea confragosa P.
Rare.
W. On willows in Askham Bog, June, 1879; *W. West.*
A. Church Fenton; *G. Massee.*
C. Fixby Hall; *J. Bolton.*

423. Dædalea unicolor Fr.
A. On stumps, etc., very frequent. Leeds; *J. Abbott.* Saltaire.

424. Merulius corium Fr.
On dead trunks, not common.
W. Bolton Woods; *W. West.*
A. Leeds; *J. Abbott.* Esholt.
C. Horbury; and North Dean Woods; *W. West.*

425. Merulius lachrymans Fr. *'Dry-Rot.'*
Only too common, both in cellars, roofs, and hollows in trees.

426. Fistulina hepatica Fr. *The Liver Fungus.*
On old oaks, by no means common.
Y. Studley Woods!
W. Near Bolton Bridge; *W. West.* Ilkley.
A. Leeds district; *J. Abbott.* Hambleton; *H. F. Parsons.*
C. Huddersfield district, only once; *Hudd. Bot. Soc.*
T. Roche Abbey!

III. HYDNEI.

427. Hydnum imbricatum L.
'In pine woods, on the ground.'
C. 'I found this in North Dean, Greetland, in pretty great plenty, 1786'; *J. Bolton* (fig. 88, *Hist. of Fungusses*). But Cooke in his Handbook does not quote the figure under this species.

428. Hydnum repandum L.
In woods, on the ground, occasional. Edible.
Y. Studley; *Exc.*, 1881.
A. Shipley Glen. Nab Wood, Saltaire. Bingley.
C. Meltham; *Hudd. Bot. Soc.*
D. Askern; *Hudd. Bot. Soc.*

429. Hydnum squalinum Fr.
C. Halifax, 1788; *J. Bolton.* Not seen since.

430. Hydnum ferruginosum Fr.
D. Doncaster; *H. F. Parsons.*

431. Hydnum farinaceum Pers.
A. On dead wood, Bingley.

432. Hydnum argutum Fr.
A. Hambleton; *H. F. Parsons.*

433. Hydnum alutaceum Fr.
W. On a prostrate stump in Horn-bank Wood, near Flint Mills, Wetherby; *G. Massee.*

434. Irpex obliquus Fr.
C. On the surface of decaying wood, Halifax; *J. Bolton.*

435. Radulum orbiculare Fr.
A. On decaying birch, Cullingworth, 1885.

436. Phlebia merismoides Fr.
A. Holden Wood near Keighley, 1885.

437. Phlebia vaga Fr.
On decayed wood, occasional.

438. Grandinia granulosa Fr.
A. On twigs, Church Fenton, 1881; *G. Massee.*

439. Odontia fimbriata Pers.
Y. Studley; *Exc.*, 1881.

IV. AURICULARINI.

440. Craterellus lutescens Fr.
C. Ramsden Wood, Halifax; *J. Bolton.*

441. Thelephora laciniata Pers.
On branches and on heaths, etc., frequent.
Y. Studley; *Exc.*, 1881.
A. Hambleton; *H. F. Parsons.* Hainworth; *T. Febden.*
C. Lee Bridge, Halifax; *J. Bolton.*

442. Thelephora terrestris Ehr.
A. On the ground at Horsforth; *W. West.*
T. Roche Abbey Valley; *J. H. Aveling* in *Hist.* (1870).

443. Thelephora mollissima Pers.
A. Airedale; *J. Mackenzie* in *Nat.*, 1878.

444. Thelephora sebacea Pers.
A. On the ground, Bishops Wood; *W. N. Cheesman.* Apperley.

445. Thelephora anthochroa Pers. VAR. **versicolor** Berk.
On sycamore twigs, rare.
T. Roche Abbey Valley; *J. H. Aveling* in *Hist.* (1870).

446. Stereum purpureum Pers.
On fallen logs, etc., not rare. Handsome.
R. Slaidburn!
A. Selby; *H. F. Parsons.* Heaton, etc.

447. Stereum hirsutum Fr.
On tree-stumps, etc., ubiquitous. Distribution general.

448. Stereum spadiceum Fr.
On ash-sticks, etc.
A. Carlton and Rawcliffe; *H. F. Parsons.* Camblesforth; *W. N. Cheesman.*

449. Stereum rugosum Fr.
On hazel-stumps, frequent.

450. Stereum acerinum Fr.
W. On boles of *Acer campestre*, Thorparch.

451. Stereum sanguinolentum Fr.
On fir-wood, etc., frequent.
W. Church Fenton (!) *W. West*, etc. Wetherby!
A. Horsforth; *W. West.* Bingley, Esholt, etc.
C. Mirfield.

452. Hymenochæte rubiginosa Lev. (*Stereum rubiginosum* Fr.)
On stoops, gate-posts, etc., not common.
L. Ingleton.
A. Selby, etc.

453. Hymenochæte tabacina Lev. (*Stereum tabacinum* Fr.)
C. Halifax; *J. Bolton.*

454. Hymenochæte corrugata Berk. (*Corticium corrugatum* Fr.)
On fallen sticks in woods, not common.
L. Helks Wood, Ingleton.

455. Auricularia mesenterica Fr.
C. Halifax; *J. Bolton.*

456. Corticium giganteum Fr.
A. Church Fenton, 1881; *G. Massee.*

457. Corticium lacteum Fr.
A. On a fallen tree, Hawksworth.

458. Corticium sanguineum Fr.
A. On larch-sticks, rare. Scotland Mill Wood, Adel!

459. Corticium cæruleum Fr.
On palings, dead wood, etc., rare.
A. Church Fenton, 1881; *G. Massee.*
T. Swinefleet; *H. F. Parsons.*

460. Corticium quercinum Pers.
A. Common on oak twigs and branches, Saltaire, etc.

461. Corticium lividum Pers.
A. Church Fenton, 1881; *G. Massee.*

462. Corticium cinereum Fr.
A. On dead twigs, etc., Nab Wood, Saltaire.

463. Corticium incarnatum Fr.
L. Ingleton; *W. West.*
A. On decaying wood. Hawksworth. Near Selby.

464. Corticium sambuci Pers.
A. On elder twigs, etc. Saltaire, Bingley, etc.

465. Cyphella muscigena Fr.
A. On a mossy wall, Steeton, 1884-5.

466. Cyphella Curreyi B. & Br.
A. On dead furze, etc., Baildon Moor.

467. Solenia ochracea Hoffm.
A. On dead holly, etc., not common, Holden Wood near Keighley, 1885.

V. CLAVARIEI.

468. Clavaria umbrina Berk.
T. Roche Abbey Valley; *J. H. Aveling* in *Hist.* (1870).

469. Clavaria amethystina Bull.
A. Amongst grass, rare.
T. Roche Abbey Valley; *J. H. Aveling* in *Hist.* (1870). Lindrick Common; *Amos Carr*!

VAR. **lilacina** Fr.
A. Saltaire; *J. W. Carter.*

470. Clavaria fastigiata L.
In pastures, frequent, and generally distributed.

471. Clavaria muscoides L.
Banks, etc., pretty, but rather rare.
R. Near Settle.
W. Bolton Woods; *W. West.* West-woods, Bramham!

472. Clavaria cinerea Bull.
Y. In woods, frequent. Studley; *Exc.*, 1881.
A. Meanwood. Heaton. Bingley.

473. Clavaria coralloides L.
A. In woods, Micklefield, 1881.

474. Clavaria rugosa Bull.
N. Harrogate; *Exc.*, 1881.
W. West Woods, Bramham.
A. Bingley Woods.
T. Roche Abbey Valley; *J. H. Aveling* in *Hist.* (1870).

475. Clavaria cristata Pers.
Y. Studley; *Exc.*, 1881.
A. Newsholme Dean, 1884.

476. Clavaria abietina Schum.
A. Church Fenton, 1881; *G. Massee.*

477. Clavaria purpurea Müll.
W. Bank near Bolton Woods; *W. West*!
A. Meanwood Wood!

478. Clavaria fusiformis Sow.
A. Micklefield, 1881. Ingrow. Calverley.

479. Clavaria inæqualis Müll.
Amongst grass in woods, common.

480. Clavaria argillacea Fr.
A. On heathy soil. Horsforth; *W. West.* Shipley Glen.

481. Clavaria tenuipes B. & Br.
A. Railway embankment, Shipley.

482. Clavaria fragilis Holmsk.
C. Halifax; *J. Bolton.* Not found since.

483. Clavaria vermicularis Scop.
On lawns and in pastures, not common.
A. Apperley.
C. Fixby; Lepton; and Toothill; *Hudd. Bot. Soc.*

484. Clavaria fumosa P.
C. In fields occasionally about Huddersfield; *Hudd. Bot. Soc.*

485. Calocera viscosa Fr.
On fir-tree stumps, pretty, and not rare in fir-woods.
W. Wetherby! Wood on Otley Chevin.
A. Currer Wood, Steeton. Eccup! Shipley Glen.
C. Huddersfield district; *Hudd. Bot. Soc.*
D. Wharncliffe; *H. F. Parsons.*
T. Lindrick and Anston; *Amos Carr!*

486. Calocera cornea Fr.
N. On stumps, Harrogate; *Exc.*, 1881.
A. Esholt.
D. Goole; *H. F. Parsons.*

487. Typhula erythropus Desm.
On dead herbaceous plants, frequent.
W. Ilkley, etc.
A. Horsforth; *W. West.* Shipley Glen.
C. Halifax; *J. Bolton*, t. 112.

488. Pistillaria quisquiliaris Fr.
On dead brackens, etc., frequent.

VI. TREMELLINI.

489. Tremella mesenterica Retz.
On dead sticks, common. Generally distributed.

490. Tremella fimbriata Pers.
A. On dead wood, Steeton, 1885.

491. Tremella foliacea Pers.
C. Decayed wood in Lepton Wood; *Hudd. Bot. Soc.*

492. Tremella albida Huds.
N. On dead branches, Harrogate; *T. Hick.*

493. Tremella viscosa Pers.
N. On dead wood, Harrogate district; *T. Hick.*

494. Exidia glandulosa Fr. '*Witches' Butter.*'
Y. On dead oak-twigs, Studley; *Exc.*, 1881.
W. Bolton Woods; *W. West.* Woodhall!

495. Hirneola Auricula-Judæ Berk. '*Jew's Ears.*'
On elder, local.
W. Harewood! Flint-mill Wood, Wetherby!
A. Airedale; *J. Mackenzie.* Leeds district; *J. Abbott.* Heaton.
C. Halifax; *J. Bolton.*
D. Sprotborough, etc.; *H. F. Parsons.*
T. Roche Abbey! Roche Abbey Valley; *J. H. Aveling* in *Hist.* (1870).

496. Næmatelia encephala Fr.
W. On pine rails, rare. Border of West Woods, Bramham!

497. Dacrymyces deliquescens Duby.
Y. & N. Studley and Harrogate; *Excs.*, 1881.

498. Dacrymyces stillatus Nees.
W. On pine rails, Wetherby; *G. E. Massee.*
A. Saltaire, etc.
D. Goole; *H. F. Parsons.*

499. Dacrymyces chrysocomus Tul.
On branches of fir.
W. Wetherby! Beaumont's Wood, Boston Spa!
C. Battye-ford; *H. F. Parsons.*
D. Wentbridge; *H. F. Parsons* (*Nat.*, iii, 10).

GASTEROMYCETES.

VII. HYPOGÆI.

500. Melanogaster variegatus Tul. *Red Truffle.*
W. Rare. Near Ilkley, 1883; *R. Lishman.*

VIII. PHALLOIDEI.

501. Phallus impudicus L. *Stink-horn.*
Frequent in woods and generally distributed.
First record: Rotherham Woods; *Gough's Camden*, 1789.

502. Cynophallus caninus Fr.
In woods, local.
A. Nab Wood, Saltaire. Bingley Woods. New Wortley near Leeds; *Dr. Hick* (vide *Nat.*, 1886, p. 332).
D. Wharncliffe Crag Woods!
T. Roche Abbey!

IX. TRICHOGASTRES.

503. Geaster hygrometricus Pers.
C. Near Halifax; *J. Bolton.*

504. Geaster fornicatus Fr.
D. Near Doncaster; *D. Tofield*; *Huds.*, Ed. ii (1778), 644; rep. *Gough's Camden* (1789).

505. Geaster fimbriatus Fr.
A. Hambleton Haugh!
Y. Studley; *Exc.*, 1881.

506. Geaster lageniformis Vilt.
A. Beckfoot Lane, Bingley, 1882-3; not seen since.

507. Bovista nigrescens Pers.
A. Goitstock near Bingley, 1886.
C. Halifax; *J. Bolton.*
T. Roche Abbey Valley; *J. H. Aveling* in *Hist.* (1870).

508. Bovista plumbea Pers.
Frequent in pastures.
A. Leeds district; *J. Abbott.* Near Cullingworth.
C. Thornes Park, Wakefield; *H. Sims.*
D. Endcliffe Valley; *Amos Carr!*
T. Roche Abbey Valley; *J. H. Aveling* in *Hist.* (1870).

509. Bovista ammophila Lev.
A. In a pasture at Cullingworth, 1885.

510. Lycoperdon giganteum Batsch. *Giant Puff-ball.*
In pastures, rather rare. Edible when young.
W. Arthington; *Matthew Taylor*, 1881.
A. Near Leeds; *J. Abbott.* Saltaire. Cullingworth; *T. Hebden.*
C. Wakefield; *H. Sims.*
T. Roche Abbey Valley; *J. H. Aveling* in *Hist.* (1870).

511. Lycoperdon cælatum Fr.
In pastures, frequent.
W. Bramham Park!
A. Heaton. Cullingworth.
T. Roche Abbey Valley; *J. H. Aveling* in *Hist.* (1870).

512. Lycoperdon pusillum Fr.
A. Bolton Woods; *W. West.*
C. Halifax; *J. Bolton.*

513. Lycoperdon gemmatum Fr.
C. Frequent in meadows, etc. Halifax; *J. Bolton.*

514. Lycoperdon pyriforme Schæff.
On dead stumps.
Y. Studley; *Exc.*, 1881.
A. Shipley Glen, etc.
C. Battye-ford; *H. F. Parsons.* Huddersfield district; *Hudd. Bot. Soc.*
D. Conisborough; *H. F. Parsons.*
T. Roche Abbey Valley; *J. H. Aveling* in *Hist.* (1870).

515. Lycoperdon saccatum Vahl.
W. Bolton Woods, 1878; *W. West.*
A. Rare, Heaton Woods, 1884. Bingley.

516. Scleroderma vulgare Fr.
Woods, etc., frequent and generally distributed.

517. Scleroderma Bovista Fr.
On the ground in woods.
Y. Studley; *Exc.*, 1881.
A. Heaton Wood.

X. MYXOGASTRES.

518. Lycogala epidendrum Fr.
On stumps.
N. Harrogate; *T. Hick.*
W. Wetherby! Askham Bog; *W. West.*
A. Bramhope! Bingley. Holden Wood near Keighley! Selby; *W. N. Cheesman.* Horsforth; *W. West.*
C. Halifax; *J. Bolton.* Hebden Bridge!

519. Reticularia atra Fr.
A. On a fallen tree, Esholt.

520. Reticularia umbrina Fr.
On stumps.
N. Harrogate district; *T. Hick.*
W. Westwoods, Bramham!
A. Selby; *W. N. Cheesman.* Fågley near Bradford.
C. Halifax; *J. Bolton.* Sharleston; *H. F. Parsons.* Huddersfield district; *Hudd. Bot. Soc.*

521. Æthalium vaporarium Fr.
A. Bramley; *W. West.*
C. Halifax; *J. Bolton.*

522. Æthalium septicum Fr. '*Yolk-of-egg*' *Fungus.*
On mosses, grass, twigs, etc., in woods.
W. Boston Spa (!) *J. Emmet.*
A. Heaton Wood. Nab Wood, Saltaire. Shipley Glen!

523. Spumaria alba DC. *Cuckoo-spit Fungus.*
On living herbaceous stems. Generally distributed.
W. Church Fenton !
A. On grass, Brayton; *H. F. Parsons.* Steeton. Saltaire. Shipley Glen !

524. Ptychogaster albus Corda.
N. Harrogate; *Exc.*, 1881.
A. On stumps, etc., Heaton Wood.

525. Didymium melanopus Fr.
T. Swinefleet; *H. F. Parsons.*

526. Didymium hemisphæricum Fr.
A. On dead twigs, Leeds district; *J. Abbott.*

527. Didymium squamulosum A. & G.
A. On leaves, grass, etc., in damp places, Ingrow.

528. Didymium nigripes Fr.
W. Near Church Fenton; *Excurs.*, 1880.
A. On rotten wood, Leeds district; *J. Abbott.*

529. Didymium serpula Fr.
A. On dead leaves, near Bingley.

530. Diderma vernicosum P.
On mosses, twigs, etc., in woods, common.

531. Diderma umbilicatum Pers. (*Chondrioderma radiatum* L.)
A. On an old stump, Hainworth Wood; *T. Hebden.*

532. Diderma nitens Klotsch. (*Chondrioderma difforme* Pers.)
On bark, twigs, dead leaves, etc.
N. Harrogate district; *T. Hick.*
A. Bingley, Hawksworth, etc. Fagley near Bradford.

533. Physarum Tussilaginis B. & Br. (*Chondrioderma Cookei* Rost.)
T. On butterbur near Worksop; *H. Friend.*

534. Physarum nutans Pers.
On decayed wood.
A. Selby; *W. N. Cheesman.* Meanwood ! Esholt, Saltaire, etc.

VAR. **aureum.**
N. Harrogate; *T. Hick.*
A. Saltaire. Esholt. Heaton.

535. Physarum leucophæum Fr.
A. On a stump, Newsholme Dean, 1884.

536. Physarum metallicum Berk.
A rare and extremely pretty species.
A. On a stump, Saltaire.

537. Physarum album Fr.
A. Shipley Glen. Nab Wood, Saltaire.

538. Angioridium sinuosum Grev.
A. Amongst dead leaves, Horsforth; *W. West.*

539. Craterium pedunculatum Trent.
A. On sticks, leaves, etc., Shipley Glen, Newsholme Dean.

540. Craterium minutum Fr.
A. On dead leaves, Nab Wood, Saltaire.

541. Craterium leucocephalum Ditm.
A. On wood, etc. Ingrow. Hainworth; *T. Hebden.* Adel; *W. West.*

542. Dictydium umbilicatum Schrad.
A. On a stump, Bingley, 1884; *W. West.* A most beautiful object.

543. Stemonitis fusca Roth.
N. On rotten wood, Harrogate district; *T. Hick.*
A. Newsholme Dean, Keighley. Selby; *W. N. Cheesman.*
C. Near Halifax: *W. West.*

544. Stemonitis typhoides DC.
A. On wood, Gateforth and Drax; *H. F. Parsons.*

545. Stemonitis ovata Fr. (*Comatricha Friesiana* De Bary. *Stemonitis obtusata* Fr.)
L. Helks Wood, Ingleton.
A. On rotten trunks, Esholt, Newsholme, Steeton.

546. Arcyria punicea Pers.
On rotten stumps, common.
A. Leeds; *J. Abbott.* Esholt, Saltaire, etc. Hensall and Bishops Wood; *H. F. Parsons.*

547. Arcyria incarnata Pers.
On rotten wood, general !
A. Church Fenton !

548. Arcyria ochroleuca Fr. (*Arcyria pomiformis* Roth.)
A. On a rotten stump, Shipley Glen.

549. Ophiotheca chrysosperma Curr.
A. Under bark. Steeton, Esholt, etc.

550. Trichia rubiformis Cooke. (*Trichia Ayresii* B. & Br. *Hemiarcyria* Pers.)
N. Harrogate; *T. Hick.*
A. On dead wood, Micklefield, 1881.

551. Trichia chrysosperma DC. (*Trichia turbinata* With.)
A. Leeds district; *J. Abbott.* Esholt. Barlow; *W. N. Cheesman.*
D. On decaying wood, Doncaster; *H. F. Parsons.*

552. Trichia fallax Pers.
A. On rotten wood, Esholt Spring Wood.

553. Trichia varia Pers.
A. On dead wood, Newsholme Dean, Shipley Glen.

VAR. **nigripes** Pers.
A. Esholt.

554. Perichæna abietina Fr.
A. On dead fir-trees, Esholt Woods.

555. Perichæna populina Fr.
On dead poplars, very common. Generally distributed.

556. Licea fragiformis Fr.
A. Camblesforth; *H. F. Parsons.* Carlton; *W. N. Cheesman.*

XI. NIDULARIACEI.

557. Cyathus vernicosus DC. *'Egg-cup' Fungus.*
A. On stubble. Leeds district; *J. Abbott.* Newland and Brayton; *H. F. Parsons.* Selby; *W. N. Cheesman.* In a cucumber-frame at Hainworth; *T. Hebden.* Undercliff near Bradford.

558. Cyathus striatus Hoff.
A. Bishops Wood; *W. N. Cheesman.*

559. Crucibulum vulgare Tul.
A. On dead twigs, etc. Leeds district; *J. Abbott.* Saltaire.

560. Sphærobolus stellatus Tode.
On twigs, rails, etc., common.
N. Harrogate; *T. Hick.*
A. Bradford, Esholt, Saltaire, etc.

CONIOMYCETES.

[SECTION A. GROWING ON DEAD OR DYING PLANTS.]

XII. SPHÆRONEMEI.

561. Leptostroma Spireæ Fr.
A. On dead stems of *Spiræa ulmaria*, Malham, etc.

562. Phoma asteriscus Berk.
A. On dead stems of *Heracleum sphondylium* near Ingrow.

563. Phoma samarorum Desm.
On the samari of the ash-tree, frequent.

564. Phoma nebulosum Berk.
On dead herbaceous stems, common.

565. Dothiora sphæroides Fr.
A. On ash-twigs, Nab Wood, Saltaire. Near Selby; *G. E. Massee.*

566. Acrospermum graminum Leb.
A. On dead grass, Saltaire.

567. Diplodia herbarum Lev.
On herbaceous stems. Common.

568. Vermicularia dematium Fr.
A. On dead stems of herbaceous plants, Saltaire. Leeds; *J. Abbott.*

569. Vermicularia trichella Grev.
A. On dead leaves of ivy, Bingley.

570. Vermicularia atramentaria B. & Br.
A. On dead potato-stems, Newsholme Dean.

571. Pilidium Acerinum Kze.
A. On dead leaves of sycamore, common. Saltaire.

572. Septoria Hippocastani B. & Br.
On horse-chestnut leaves, common.

573. Septoria graminum Desm.
A. On grass, Saltaire !

574. Septoria Ribis Desm.
W. On black-currant leaves, garden at Boston Spa 1880 !

575. Asteroma Rosæ DC.
On rose leaves, common.

576. Dinemasporium graminum Lev.
A. On dead leaves of grass, near Selby; *G. Massee.*

VAR. **herbarum** Cooke.
L. On dead stems of nettle, Clapham.
A. Calverley.

XIII. MELANCONIEI.

577. Stilbospora macrosperma Pers.
A. On oak-twigs, Nab Wood, Saltaire.

578. Coryneum disciforme Kze.
A. On birch-twigs, Saltaire.

579. Pestalozzia lignicola Cooke.
A. On wood, Saltaire.

580. Glæosporium Ficariæ Berk.
A. On leaves of *Ranunculus Ficaria*, Saltaire.

XIV. TORULACEI.

581. Torula ovalispora Berk.
A. On wood, frequent near Leeds; *J. Abbott.*

582. Torula pulveracea Corda.
A. On wood. Steeton. Micklefield, 1881.

583. Torula herbarum Lk.
A. On dead stems of thistles, Umbelliferæ, etc., common, Saltaire.

584. Torula graminis Desm.
A. On dead leaves of grass, Saltaire.

585. Torula sporendonema B. & Br.
A. Micklefield, 1881.
C. Liversedge, on cheese; *W. Fowler* and *J. E. Vize.*

586. Bispora monilioides Cd.
A. On wood, common. Malham, etc.

587. Septonema spilomeum Berk.
A. On fir-wood, Saltaire.

588. Sporidesmium alternariæ Cooke.
A. On moist paper, Bradford; *W. West.*

589. Sporidesmium pyriforme Corda.
A. On wood, Saltaire.

590. Sporidesmium lepraria B. & Br.
A. On wood in a timber-yard, Steeton.

591. Sporidesmium atrum Grev.
A. Steeton.

592. Sporidesmium abruptum B. & Br.
A. On dead wood, Steeton.

[SECTION B. PARASITIC ON LIVING PLANTS.]

XV. PUCCINIÆI.

593. Xenodochus carbonarius Schl.
On living leaves of Burnet (*Sanguisorba officinalis*), locally common.
W. Grass Woods, Grassington. Barden. Seavy Carr, Stutton (!) near Church Fenton.
A. Frizinghall.

594. Phragmidium mucronatum Link.
A. On leaves of roses, locally common. Malham, etc.

595. Phragmidium acuminatum Fr.
On living leaves of salad-burnet (*Poterium sanguisorba*).
L. Ingleton.
W. Arncliffe; *W. West.* Barden, etc.
A. Malham.
T. Between Thorpe Salvin and Harthill; *Hilderic Friend.*

596. Phragmidium bulbosum Schl. (*P. violaceum* Schultz.)
On bramble leaves, common in some parts of the Riding and apparently absent from others.
N. Beckwithshaw.
W. Near Ilkley.
T. Roche Abbey Valley; *J. H. Aveling* (1870). Lindrick Dale; *H. Friend.*

597. Phragmidium Rubi Pers.
On bramble-leaves, not so common as the last, and perhaps confounded with it.
N. Beckwithshaw.
Obs.—Common just outside the Riding, near Boroughbridge.

598. Phragmidium gracile Grev. (*Lecythea gyrosa* Lev.)
D. On raspberry leaves, Askern.

599. Phragmidium obtusum Link.
On leaves of Barren Strawberry (*Potentilla fragariastrum*).
W. Bolton Woods.
A. Frequent. Shipley Glen, Malham, etc.
T. Anston Crags; *H. Friend.*

600. Triphragmium Ulmariæ Link. (*Uromyces Ulmariæ* Lev.)
On leaves of meadow-sweet and rose, local.
R. Near Settle.
W. Arncliffe; *W. West.*
A. Malham.
T. Roche Abbey; and Anston Crags, very abundant, with its *Uredo* ('*Golden Rust*'), whole plants smothered in its yellow spores; *H. Friend.*

601. Puccinia graminis Pers. (*Uredo linearis* Eng. Fl.) '*Corn Mildew.*'
On culms and leaves of cereals and grass, locally abundant, and usually in proximity to barberry-bushes affected with *Æcidium berberidis* Pers.
W. Church Fenton.
A. Near Pontefract. Selby; *W. N. Cheesman.*
T. In this district common. Under Anston Crags; *H. Friend.*

602. Puccinia Phragmitis Schum. (*P. arundinacea* DC.)
D. On reeds, abundant about Goole; *T. Birks.*
T. Under Anston Crags; *H. Friend.*

603. Puccinia Magnusiana Köen.
On reeds.
A. Near Church Fenton.
D. Goole; *T. Birks.*

604. Puccinia rubigo-vera DC. (*Uredo rubigo-vera* DC.)
Very common on grasses, especially on *Holcus mollis*, *H. lanatus*, and *Avena elatior* (*W. West*).

605. Puccinia striola Lk. (*Uredo caricina* Eng. Fl.)
On sedges, frequent.
Obs.—We are inclined to think that two or three species of *Puccinia* affect our various sedges, but the matter requires investigation.

606. Puccinia coronata Cd.
R. On leaves of grass, not common, Giggleswick.

607. Puccinia sessilis Schr.
A. On *Phalaris arundinacea*, Airton near Malham, 1882.

608. Puccinia Baryi B. & Br. (*Lecythea Baryi*, Berk.)
W. On leaves of *Brachypodium sylvaticum*, not common, Bolton Woods, 1885.

609. Puccinia Anthoxanthi Fckl.
A. On leaves of *Anthoxanthum odoratum*. Not common, Nab Wood, Saltaire.

610. Puccinia Poarum Niel.
A. On *Poa annua*, not uncommon, and always near to coltsfoot which has been affected with *Æcidium tussilaginis*. Heaton, Saltaire, near Keighley, etc.

611. Puccinia obscura Schr.
On *Luzula campestris*, common in pastures.
W. Otley Chevin.
A. Saltaire. Keighley.

612. Puccinia vaginalium Lk.
Common on *Polygonum aviculare*.

613. Puccinia Primulæ Grev.
On leaves of *Primula vulgaris*.
L. Ingleton.
R. Giggleswick.

614. Puccinia Veronicarum DC.
On *Veronica chamædrys* and *V. montana*.
W. Bolton Woods.
A. Newsholme Dean.

615. Puccinia Glechomatis DC.
T. On ground-ivy, near Shireoaks, rare; *H. Friend.*

616. Puccinia Menthæ Pers.
Common in gardens and marshes on mints.
W. Seavy Carr, Stutton!
A. Frequent on *Mentha hirsuta*, Shipley Glen.
T. Shireoaks, Anston, etc.; *H. Friend.*

617. Puccinia Betonicæ DC.
A. On *Stachys Betonica*, locally common. Hawksworth; Baildon; *W. West.* Saltaire.

618. Puccinia Campanulæ Carm.
T. On harebell, near Worksop, very rare; *H. Friend.*

619. Puccinia Clinopodii DC.
L. On *Calamintha clinopodium*, rare. Helks Wood, Ingleton.

620. Puccinia Malvacearum Corda.
W. About Wighill and Walton!
D. On *Malva sylvestris*, locally common. Near Askern.

621. Puccinia sparsa Cooke.
A. On *Tragopogon pratensis*, not common, near Keighley; *T. Hebden.*

622. Puccinia syngenesiarum Link.
On leaves of thistles, common.

623. Puccinia suaveolens Pers. (*Trichobasis suaveolens* Lev.)
A. On *Carduus arvensis*, common. Saltaire, Frizinghall, etc.
D. Goole and Doncaster; *W. West.*

624. Puccinia Senecionis Lev.
A. On *Senecio aquatica*, Idle, 1886.

625. Puccinia glomerata Grev.
A. On *Senecio Jacobœa*, frequent. Saltaire, Bingley, etc.
T. Canal-side, Shireoaks; *H. Friend.*

626. Puccinia Cirsii Lasch.
W. On *Carduus heterophyllus*, rare. Grass Woods, Grassington, 1884.

627. Puccinia Chondrillæ Fckl.
On *Lactuca muralis*, locally common.
W. Barden. Bolton Woods.
A. Malham. Bingley.

628. Puccinia Andersoni B. & Br.
On *Carduus heterophyllus*, not common.
L. Helks Wood, Ingleton; *T. Hebden.*
W. Grassington.

629. Puccinia Lapsanæ Fckl.
On *Lapsana communis*, frequent.

630. Puccinia Centaureæ DC.
On *Centaurea nigra*, common.

631. Puccinia Hieracii Mart.
On various *Hieracia*, common.

632. Puccinia variabilis Grev.
W. On leaves of dandelion. Grassington. Bolton Woods.
T. In this district frequent; *H. Friend.*

633. Puccinia Valantiæ Pers.
A. On leaves of *Galium cruciatum*, rare. Near Malham, 1883.
T. Very fine on crosswort at 'Moses' Seat,' near Lindrick Dale; *H. Friend.*

634. Puccinia Galiorum Link.
A. On *Galium saxatile*, common. Baildon Moor, etc.

635. Puccinia Heraclei Grev.
A. On leaves of *Heracleum sphondylium*, not common. Church Fenton.

636. Puccinia umbelliferarum DC.
A. On *Myrrhis odorata*, *Bunium flexuosum*, etc., frequent. Bingley. Saltaire. Eldwick.

637. Puccinia Ægopodii Link.
W. On goutweed in shady hedges, Harewood!

638. Puccinia Saniculæ Grev.
W. On leaves of *Sanicula europæa*, not common. Bolton Woods.
T. Woods on county border, not common; *H. Friend.*

639. Puccinia Calthæ Link.
W. On leaves of *Caltha palustris*, rare. Church Fenton, 1881; *G. Massee.*

640. Puccinia Anemones Pers.
On leaves of wood-anemone, common. Generally distributed.

641. Puccinia Adoxæ DC. (*P. saxifragarum*; Cooke, Hbook.)
On leaves of *Adoxa moschatellina*, frequent.
W. Ilkley. Collingham Bank Wood; *J. Abbott!*
A. Bingley. Saltaire and Esholt; *W. West.*
T. Anston Crags, very abundant; *H. Friend.*

642. Puccinia Chrysosplenii Grev.
W. On leaves of *Chrysosplenium alternifolium*, rare. Bolton Woods.

643. Puccinia Violarum Lk.
On leaves of *Viola sylvatica*, etc., common. Generally distributed.

644. Puccinia Lychnidearum Lk.
W. On leaves of *Lychnis diurna*, locally common. Bolton Woods.
T. On campion leaves, Kiveton, plentiful; *H. Friend.*

645. Puccinia Epilobii DC.
A. On leaves of *Epilobium palustre*, rare. Hawksworth.

646. Puccinia pulverulenta Grev.
A. On *Epilobium hirsutum*, not common. Calverley.

647. Podisoma Juniperi Fr.
A. On branches of *Juniperus communis*, Wistow; *W. N. Cheesman.* Snaith; *H. F. Parsons.* A precedent state of *Ræstelia lacerata.*

XVI. CÆOMACEI.

648. Tilletia caries Tul. '*Bunt.*'
W. On wheat. Healaugh and Wetherby! 'Common.'
T. In corn, some years frequent; *H. Friend.*

649. Ustilago carbo Tul. (*Ustilago segetum* Ditm.) *Corn Smut.*
On the ears of cereals, common some years, others rarer, but generally distributed.

650. Ustilago longissima Tul.
A. On leaves of *Glyceria aquatica* and *G. fluitans*, common about Saltaire, Bingley, etc.
D. Burton Grange, Barnsley (!) *H. Johnson.*
T. Canal-side, Shireoaks to Kiveton; *H. Friend.*

651. Ustilago hypodytes Fr.
A. On leaves of grass? Saltaire.

652. Ustilago Salveii B. & Br.
A. On leaves of *Triticum repens*, Bolton near Bradford.

653. Ustilago antherarum Fr.
On the anthers of *Lychnis diurna*, common. Generally distributed.

654. Ustilago bromivora Waldh.
A. On flowers of *Bromus mollis*, locally common. Hainworth near Keighley; *T. Hebden.* Bradford. Below Shipley; *W. West.*

655. Ustilago Kuehniana Wolff.
A. On inflorescence of *Rumex acetosa*, common about Saltaire, Bradford, etc. Keighley; *T. Hebden.*

656. Urocystis Violæ B. & Br.
A. Not common, on cultivated violas. Hainworth; *T. Hebden.*

657. Urocystis occulta Preuss.
A. On leaves of *Triticum repens*, Saltaire.

658. Urocystis pompholygodes Schlecht.
On leaves of *Ranunculus repens* and *Anemone nemorosa*, common.
T. Frequent in this district; *H. Friend.*

659. Urocystis Fischeri Körn.
A. On leaves of *Carex glauca* in pastures, not common. Saltaire and Hawksworth.

660. Entyloma Ungerianum De Bary. (*Protomyces microsporus* Unger.)
T. Abundant on *Ranunculus Ficaria*; *H. Friend.*

661. Uromyces Poæ Rabenh.
A. On leaves of *Poa trivialis*, frequent, and always in proximity to *Ranunculus Ficaria* infested with *Æcidium ranunculacearum*, of which it is a condition. Saltaire, Esholt, Malham, etc.

662. Uromyces appendiculata Lev.
Frequent, on *Trifolium pratense.*
W. Ilkley. Collingham Bridge!
A. Saltaire.
T. Between Shireoaks and Anston; *H. Friend.*

663. Uromyces apiculosa Lev.
A. On *Rumex acetosa*, Saltaire.

664. Uromyces Ficariæ Lev.
On leaves of pilewort, in spring, common. Generally distributed. Usually with *Æcidium* on same plant.

665. Uromyces Junci Tul.
W. On rushes, Church Fenton, 1881; *W. West.*

666. Uromyces intrusa Lev.
M. On *Alchemilla vulgaris*, common. Saddleworth; *C. P. Hobkirk.*
T. Near the new parsonage, Anston; *H. Friend.*

667. Uromyces concentrica Lev.
A. On leaves of *Scilla nutans*, frequent. Saltaire, Bingley, etc.
T. Common in moist woods in this district; *H. Friend.*

668. Uromyces Gageæ Berk. (*U. ornithogali* Schr.)
D. On leaves of *Gagea lutea* at Sprotborough; *J. Abbott.*
T. Roche Abbey and Anston district; *H. Friend.*

669. Uromyces polygoni Fckl.
T. Lindrick Dale, Kiveton station, etc.; *H. Friend.*

670. Coleosporium Tussilaginis Lev.
On coltsfoot leaves, common.
T. Disused quarry, Anston Crags; *H. Friend.*

671. Coleosporium Petasites Lev.
On leaves of *Petasites vulgaris.*
W. Wetherby!
A. Camblesforth; *H. F. Parsons.* Bingley.
T. Near Worksop; *H. Friend.*

672. Coleosporium Campanulæ Lev.
On leaves of *Campanula rotundifolia*, not on other species hitherto.
W. Otley.
A. Malham.
T. Near Worksop; *H. Friend.*

673. Coleosporium Senecionis Fr.
A. On groundsel leaves, Saltaire.

674. Coleosporium Sonchi-arvensis Lev.
A. On leaves of *Sonchus* (? sp.). Church Fenton 1881; *G. Massee.*

675. Coleosporium Rhinanthacearum Lev.
W. On leaves of *Bartsia odontites*, Malham.

676. Melampsora salicina Lev.
On leaves of sallows, common.
T. Shireoaks, etc.; *H. Friend.*

677. Melampsora betulina Desm.
A. On leaves of *Betula alba*, Adel Bog! Bingley; *W. West.* Saltaire.
D. Thorne Waste; *H. F. Parsons.*

678. Melampsora Euphorbiæ Cast.
D. On leaves of *Euphorbia* (? sp.). Hook near Goole; *H. F. Parsons*
T. Very frequent near Worksop; *H. Friend.*

679. Melampsora populina Lev.
L. On leaves of *Populus nigra*, Clapham.
A. Bishops Wood; *W. West.*
D. Doncaster; *W. West.*

680. Cystopus cubicus Str.
On goatsbeard, rare.
A. Leeds; *J. Abbott.*
T. Near Worksop; *H. Friend.*

681. Cystopus candidus Lev.
On shepherd's purse and other cruciferous plants, very common. Generally distributed.

682. Uredo Potentillarum DC.
Common on *Rosaceæ.* Generally distributed.

683. Uredo filicum Desm.
R. On fronds of *Cystopteris fragilis*, Penyghent, 1885.
Obs.—Frequent on the same species in fern-cases.

684. Uredo confluens DC.
On leaves of *Mercurialis perennis.*
R. Gisburn.
W. Bolton Woods.
T. Anston Crags, not common; *H. Friend.*

685. Uredo Padi Kze.
Y. On leaves of *Prunus Padus*, rare. Hackfall!

686. Uredo Vacciniorum Pers.
A. On leaves of *Vaccinium Myrtillus.* Baildon Moor, Rombalds Moor, etc.

687. Uredo Caryophyllacearum Johnst.
R. On leaves of *Cerastium alpestre*, Penyghent.

688. Uredo bifrons Grev.
A. On leaves of *Rumex acetosa*, Wilsden.
T. Kiveton and Anston; *H. Friend.*

689. Trichobasis oblongata B.
A. On leaves of *Luzula pilosa* and *L. sylvatica*, common. Esholt, Holden Wood, Shipley Glen, Bingley, etc.

690. Trichobasis suaveolens Levy. (? = *Puccinia suaveolens*; *H.T.S.*)
T. On thistles, very plentiful in district; *H. Friend.*

691. Trichobasis Geranii B.
W. On a specimen of *Geranium columbinum* from Thorparch!
A. On *G. pyrenaicum* near Bingley!

692. Trichobasis fallens Cooke.
A. On leaves of *Vicia sepium*, not common. Near Malham.

693. Trichobasis Fabæ Lev.
A. On leaves of beans, frequent. Saltaire and Church Fenton.
T. Plentiful in this district; *H. Friend.*

694. Trichobasis Parnassiæ Cooke.
W. Kirk Deighton Quarry! Seavy Car, Stutton; *Excurs.*, 1880.

695. Lecythea saliceti Lev.
A. On leaves of sallows, common. Bishops Wood; *H. F. Parsons.*

696. Lecythea mixta Levy.
T. On willows, Shireoaks, common; *H. Friend.*

697. Lecythea Valerianæ Berk.
A. On valerian leaves, frequent. Nab Wood; Hawksworth; *W. West.*
T. On county border, near Worksop, rare; *H. Friend.*

698. Lecythea Lini Lev.
On leaves of *Linum catharticum*. Generally distributed.
L. Ingleton.
R. Near Settle.
A. Malham.
T. Shireoaks; *H. Friend.*

XVII. CHYTRIDIEI.

This order is absent from the Handbook, but appears in Dr. Cooke's more recent work on 'Microscopic Fungi.'

699. Protomyces Sagittariæ Fckl.
A. On leaves of *Sagittaria sagittifolia*, rare. Gilstead, in plenty, September 1879; *W. West.* Near Keighley, 1884.
T. Near Shireoaks, common; *H. Friend.*

700. Protomyces macrosporus Ung.
A. On petioles of *Ægopodium podagraria*. Skipton, 1886.
T. Near Shireoaks, common; *H. Friend.*

701. Protomyces Chrysosplenii B. & Br.
On leaves of *Chrysosplenium alternifolium*.
W. Bolton Woods.
A. Malham.

702. Synchytrium Taraxaci De Bary.
A. On leaves of dandelion. Wilsden.

703. Synchytrium Mercurialis Fckl.
On leaves of *Mercurialis perennis*. Common.
L. Clapham, etc.; *W. West.*
A. Shipley Glen. Bingley.
T. Anston Crags; *H. Friend.*

704. Synchytrium Anemones Wor.
A. On leaves of *Anemone nemorosa*. Saltaire.

XVIII. ÆCIDIACEI.

705. Rœstelia lacerata Tul.
On hawthorn leaves, etc., very rare.
R. Horton-in-Ribblesdale; *A. Crebbin.*
A. Hambleton; *H. F. Parsons.*

706. Rœstelia cancellata Reb.
T. On pear-trees at Dinnington, very plentiful, 1886; *H. Friend.*

707. Peridermium Pini Chev.
A. On firs, Bishop Wood; *J. Harrison* (*Nat.*, ix, 219).
T. On firs, Roche Abbey neighbourhood; *H. Friend.*

708. Peridermium acicolum Link.
T. On firs, Roche Abbey neighbourhood; *H. Friend.*

709. Æcidium leucospermum DC.
On leaves of *Anemone nemorosa*, not common.
W. Bolton Woods.
A. Newsholme Dean; *T. Hebden.*
T. Sparingly near Shireoaks; *H. Friend.*

710. Æcidium quadrifidum DC.
A. On leaves of cultivated anemones in gardens, Bradford.

711. Æcidium albescens Grev.
A. On leaves of *Adoxa moschatellina*, rare. Jennet's Cave, Malham.

712. Æcidium Epilobii DC.
On leaves of *Epilobium hirsutum*. Local.
A. Calverley.
D. Went-vale; *W. West.*
T. Plentiful in this district; *H. Friend.*

713. Æcidium Tragopogonis Pers.
A. On leaves, stems, and involucres of *Tragopogon pratensis*, not common. Near Ingrow and Keighley; *T. Hebden.*
T. Around Worksop, abundant; *H. Friend.*

714. Æcidium Berberidis Pers.
On leaves of the barberry.
W. Thorner; *W. West.*
A. Near Pontefract. Bradford, 1885, artificially produced from germinating spores of *Puccinia graminis.*
T. Roche Abbey Valley; *J. H. Aveling* in *Hist.* (1870).

715. Æcidium Thalictri Grev.
D. On leaves of *Thalictrum flavum*, Goole; *T. Birks.* Pollington; *P. F. Lee.*

716. Æcidium crassum Pers.
On leaves of *Rhamnus catharticus.*
R. Giggleswick.
A. Malham; *W. West.* Near Pontefract.
T. Roche Abbey Valley; *J. H. Aveling* in *Hist.* (1870). Thorpe, Shireoaks, Anston, etc.; *H. Friend.*

717. Æcidium Periclymeni DC.
A. On honeysuckle leaves, not common. Nab Wood, Saltaire, 1884.
T. Rare about Shireoaks, etc.; *H. Friend.*

718. Æcidium Ranunculacearum DC.
On leaves of *Ranunculus Ficaria* and *repens* in spring, common.
T. Very abundant; *H. Friend.*
Obs.—A condition of *Uromyces Poæ* Rbh., but not of *Uromyces Ficariæ.*

719. Æcidium Valerianacearum Duby.
On leaves of *Valeriana officinalis* and *V. dioica.*
W. Ilkley!
A. Hawksworth; *W. West.* Meanwood.
T. Rare near Worksop; *H. Friend.*

720. Æcidium Grossulariæ DC.
A. On gooseberry leaves and fruit, Malham.
T. Kiveton Park, 1885, rare; *H. Friend.*

721. Æcidium Urticæ DC.
On nettle leaves and stems, local.
W. Near Bolton Woods.
A. Cracoe; *J. Firth.* Bradford, artificially produced from germinating spores of *Puccinia Caricis* Schum.
D. Goole; *T. Birks.*
T. Common in this district; *H. Friend.*

722. Æcidium Convallariæ Schum.
Y. Fountains Abbey, on solomon's seal.
A. Garden, Meanwood, on lily.

723. Æcidium Prenanthis Pers.
A. On leaves of *Crepis paludosa.* Hawksworth. Newsholme Dean.

724. Æcidium Tussilaginis Pers.
On coltsfoot leaves, common.

725. Æcidium Lapsanæ Purt.
On leaves of *Lapsana communis*, frequent.
Y. Near Ripon.
A. Saltaire. Baildon; *W. West.*

726. Æcidium Taraxaci Grev.
W. On leaves of dandelion. Bolton Woods. Grassington.

727. Æcidium Bellidis DC.
A. On leaves of *Bellis perennis.* Frequent in pastures at Saltaire. Near Keighley; *T. Hebden.*
Obs.—This and the preceding four, + *Æcidium Jacobææ*, are varieties of *Æ. compositarum* according to Dr. Cooke in 1871 and 1878; *W. West.* The aggregate *Æ. compositarum* is recorded by Aveling for Roche Valley (T.), also from many other localities.

728. Æcidium Saniculæ Carm.
On leaves and petioles of *Sanicula europæa*, not common.
W. Bolton Woods.
T. Shireoaks; *H. Friend.*

729. Æcidium Violæ Schum.
On violets in spring, frequent.
L. Ingleton. Clapham; *W. West.*
R. On Penyghent.
A. Bishop Wood; *J. Harrison* (*Nat.*, 1884, ix, 219). Bingley.
D. Went-vale; *H. F. Parsons.* Womersley; *W. West.*
T. Roche Abbey! Roche Abbey Valley; *J. H. Aveling* in *Hist.* (1870). Anston, Shireoaks, etc.; *H. Friend.*

730. Æcidium Geranii DC.
R. On leaves of *Geranium pratense*, Horton-in-Ribblesdale.

731. Æcidium Menthæ DC.
W. Seavy Carr, Stutton!
A. On *Mentha viridis*, very common in gardens at Saltaire and on leaves of *Mentha hirsuta* in Shipley Glen. Manningham; *W. West.*
T. Old quarries, Shireoaks, very fine; *H. Friend.*

732. Æcidium Primulæ DC.
L. On primrose leaves in spring, not common. Ingleton; *T. Hebden.*
R. Giggleswick.

733. Æcidium rubellum Pers.
A. Bradford, artificially produced from germinating spores of *Puccinia Phragmitis* Schum.
D. On leaves of *Rumex*, Goole; *T. Birks.*

734. Æcidium Ari Berk.
W. On leaves of *Arum maculatum*, rare, Bolton Woods.

735. Æcidium Allii Grev.
On leaves of *Allium ursinum*.
W. Bolton Woods.
A. Airton near Malham; *W. West*.
T. Kiveton, abundant, 1885; *H. Friend*.

736. Æcidium aviculariæ Kze.
A. On leaves of *Polygonum aviculare*, Frizinghall.

HYPHOMYCETES.

XIX. ISARIACEI.

737. Isaria arachnophila Ditm.
On dead spiders, frequent.
W. Pool. Church Fenton; *W. West*.
A. Shipley Glen. Ingrow.

738. Isaria Friesii Mont.
T. Swinefleet; *H. F. Parsons*.

739. Anthina flammea Fr.
W. On fallen leaves, beech wood, Barwick-in-Elmet, 1876! Very pretty.

740. Ceratium Hydnoides A. & S.
W. On rotten wood. Church Fenton; *G. Massee*.

741. Pachnocybe subulata Berk.
A. On wood. Saltaire.

XX. STILBACEI.

742. Stilbum tomentosum Schrad.
On *Trichia*.
N. Harrogate; *T. Hick*.
A. Esholt, Ingrow, etc.

743. Stilbum fimetarium B. & Br.
A. On rabbit-dung. Heaton Wood near Bradford.

744. Stilbum pellucidum Schrad.
A. Horsforth; *W. West*.

745. Epicoccum neglectum Desm.
A. On decaying plants. Saltaire.

746. Epicoccum Equiseti Berk.
A. On dead *Equisetum hyemale*, near Ingrow, 1884.

XXI. DEMATIEI.

747. Arthrobotryum atrum B. & Br.
L. On dead stems of nettles, Ingleton, 1885.
A. Apperley, 1887.

748. Arthrobotryum stilboideum Ces.
A. On fungus-mycelium, Holden Wood near Keighley.

749. Dendryphium laxum B. & Br.
A. On wood, Steeton.

750. Helminthosporium folliculatum Corda.
A. On cabbage-stalks, etc., Saltaire.

751. Helminthosporium macrocarpum Grev.
A. On wood, Saltaire.

752. Helminthosporium reticulatum Cooke.
A. On dead ash-leaves, Cottingley.

753. Helminthosporium apicale B. & Br.
A. On rotten wood, Esholt.

754. Helicocoryne viridis Corda.
A. On dead wood, Steeton, 1884.

755. Cladosporium herbarum Lk.
Very common everywhere.

756. Cladosporium epiphyllum Nees.
A. On fallen leaves, Shipley Glen.

757. Cladosporium depressum B. & Br.
A. On leaves of *Angelica sylvestris*, Malham.

758. Cladosporium lignicolum Corda.
A. On dead wood, Newsholme Dean.

759. Gonatosporium puccinioides Corda.
A. On dead leaves of *Carex glauca*, Saltaire.

XXII. MUCEDINES.

760. Aspergillus glaucus Lk. '*Blue Mould.*'
Very common on cheese, bread, and all kinds of substances.

761. Aspergillus candidus Lk.
A. On decaying substances. On old leather, Bradford.

762. Aspergillus virens Lk.
A. On decaying Boleti, Bingley Woods.

763. Nematogonum aurantiacum Desm.
A. On wood in a timber-yard, Steeton, 1884. Very pretty.

764. Rhinotrichum lanosum Cooke.
A. On wall-paper, Saltaire.

765. Botrytis Tilletii Desm.
A. On leaves, etc., Leeds; *J. Abbott*.

766. Botrytis Jonesii B. & Br.
On dung, common. Is a condition of *Mucor mucedo*.

767. Peronospora infestans Mont. '*Potato-disease' fungus*.
Only too common some seasons. Has occurred in every district.

768. Peronospora nivea Ung.
A. On leaves of *Chærophyllum sylvestre* and other Umbelliferæ, Saltaire. Malham.

769. Peronospora pygmæa Ung.
A. On leaves of *Anemone nemorosa*, in woods, common. Bingley, Saltaire, etc.
T. Shireoaks; *H. Friend*.

770. Peronospora ganglioniformis Berk.
A. On leaves of groundsel and other plants, Cottingley.
T. On *Lapsana* near Worksop, frequent; *H. Friend*.

771. Peronospora parasitica Pers.
A. On cruciferous plants. On *Cardamine pratensis*, Newsholme Dean. On leaves of cabbages, Leeds; *J. Abbott*.
T. Near Worksop, common; *H. Friend*.

772. Peronospora Viciæ Berk.
T. On legumes, Shireoaks; *H. Friend*.

773. Peronospora Arenariæ Berk.
T. Roche Abbey, May 1887; *H. Friend*.

774. Peronospora Violæ De Bary.
T. Roche, Shireoaks, on violets; *H. Friend*.

775. Peronospora Urticæ Casp.
A. On nettle-leaves, not common, Malham, 1881.

776. Peronospora grisea Ung.
On leaves of *Veronica Beccabunga*, very common.
A. Saltaire, Bingley, etc.
T. Near Worksop; *H. Friend*.

777. Peronospora Ficariæ Tul.
On leaves of pilewort.
Y. Studley; *Exc.*, 1881.
A. Micklefield; *Exc.*, 1881.
T. In this district abundant; *H. Friend*.

778. Peronospora candida Fckl.
A. On primrose leaves, not common. Malham, 1881.

779. Peronospora Schleideniana De Bary.
Common on leaves of *Allium ursinum*, etc.
A. Airton, Malham, etc.
T. Shireoaks and Kiveton; *H. Friend*.

780. Peronospora obliqua Cooke.
Very common on *Rumex obtusifolius* and other docks.
T. About Worksop, common; *H. Friend*.

781. Peronospora alta Fckl.
A. On leaves of *Plantago major*, Saltaire, 1883.

782. Haplaria grisea Lk.
A. On decaying vegetables, Leeds; *J. Abbott*. Bolton near Bradford, 1887.

783. Polyactis vulgaris Lk.
On decaying cabbages, etc., common.

784. Polyactis cana Berk.
A. On lettuces, etc., Leeds; *J. Abbott*, 1872. Saltaire. Micklefield, 1881.

785. Polyactis cinerea Berk.
W. Church Fenton! *Exc.*, 1881.
A. On herbaceous stems, Leeds; *J. Abbott*. Bradford; *W. West*.

786. Penicillium crustaceum Fr.
Very common on all kinds of decaying substances.

787. Oidium chartarum Lk.
A. On damp paper. Bradford. Saltaire.

788. Oidium concentricum B. & Br.
A. On nettles, leaves, etc. Ingrow. Gilstead; *W. West*.

789. Stysanus stemonitis Corda.
A. On damp paper and other substances. Saltaire. Leeds; *J. Abbott*.

790. Monilia fasciculata Corda.
A. On decaying grass, Saltaire.

791. Dactylium roseum Berk.
A. On wood. Steeton, 1885.

792. Myxotrichum chartarum Kze.
A. On paper, straw, etc., in damp situations. Bradford; *W. West*. Saltaire.

793. Botryosporium pulchrum Corda.
A. On herbs. Leeds; *J. Abbott*.

794. Botryosporium diffusum Corda.
A. On decaying potato plants, Saltaire.

XXIII. SEPEDONIEI.

795. Sepedonium chrysospermum Lk.
On fungi in decay, especially *Boleti* and *Paxillus involutus*, very common.
A. Hambleton; *H. F. Parsons.*

796. Fusisporium mucophytum W.Sm.
C. Parasitic on *Volvaria gloiocephalus*, Huddersfield; *Hudd. Bot. Soc.*

PHYSOMYCETES.

XXIV. ANTENNARIEI.

797. Zasmidium cellare Fr.
A. On corks, bottles, etc., in cellars, Bradford.

XXV. MUCORINI.

798. Ascophora mucedo Tode.
On moist bread, etc., common.

799. Mucor mucedo L. '*Blue Mould.*'
On fruit, paste, surfaces of pots of jam, etc., common.

800. Mucor caninus P.
On cat's dung, etc., in cellars, etc., common.

801. Mucor fusiger Lk.
On decaying Agarics.
A. On *Mycena galericulatus*, Saltaire.

802. Mucor amethysteus Berk.
A. On decaying pears, Leeds; *J. Abbott.*

803. Mucor subtilissimus Berk.
A. Extremely minute, growing on onions affected with a *Sclerotium*, Saltaire.

804. Pilobolus crystallinus Tode.
On dung, very common in summer and autumn.

805. Hydrophora stercorea Tode.
A. On dung, Leeds; *J. Abbott.* Saltaire.

806. Acrostalagmus cinnabarinus Corda.
A. On decayed plants, Leeds; *J. Abbott.* On old leather, Bradford.

XXVI. SAPROLEGNIEI.

807. Saprolegnia ferax Kutz. '*Salmon disease.*'
A. Only too common in some seasons, on salmon and other fish in some of our rivers. Found amongst algæ at Hainworth; *T. Hebden.* *Empusa muscæ* Cohn is a terrestrial condition of this species, and is extremely common on house-flies in autumn.

ASCOMYCETES.

XXVII. PERISPORIACEI.

808. Lasiobotrys Loniceræ Kze.
A. On living leaves of honeysuckle, not common. Saltaire.

809. Sphærotheca pannosa Lev. *Rose Blight.*
Very common on leaves, branches, etc., of wild and cultivated roses.
T. Shireoaks; *H. Friend.*

810. Sphærotheca Castagnei Lev.
On leaves of meadow-sweet, burnet, etc., locally common.
W. Grassington.
A. Malham, Bingley (*W. West*), Saltaire, etc.

811. Phyllactinia guttata Lev. *Hazel Blight.*
On leaves of hawthorn, hazel, and forest trees, locally abundant.

812. Podosphæra clandestina Levy.
W. Thorner, May 1887; *W. West.*

813. Microsphæria Grossulariæ Levy.
T. Lindrick Dale, September 1886; *H. Friend.*

814. Erysiphe Linkii Lev.
A. On mugwort leaves; *J. Abbott.*

815. Erysiphe lamprocarpa Lev.
A. On leaves of *Plantago media*, Malham.

816. Erysiphe graminis DC.
Common on leaves of grasses. *Oidium monilioides* Link. is a condition of this species. Generally distributed.

817. Erysiphe tortilis Link.
T. Anston, September 1886, very abundant; *H. Friend.*

818. Erysiphe Martii Lk.
On several species of umbelliferous plants, locally common.
W. Church Fenton (!) *W. West.* Stutton!
A. Keighley; *W. West.* Saltaire.
D. Doncaster; *W. West.*

819. Erysiphe communis Schl.
T. Plentiful throughout this district; *H. Friend.*

820. Erysiphe Montagnei Lev.
On burdock leaves, not common.
W. Boston Spa; *W. West.* Church Fenton; *J. Abbott.*
A. Wilsden; *W. West.*

821. Erysiphe horridula Lev.
A. On leaves of *Stachys sylvatica.* Bingley; *W. West.* Common about Saltaire.

822. Chætomium elatum Kze.
Common on decayed leaves of grasses in damp places.

823. Ascotricha chartarum B.
A. On damp paper, Bradford.

824. Eurotium herbariorum Lk.
On dried plants in herbaria, very common; is a condition of *Aspergillus glaucus.*

XXVIII. ELVELLACEI.

825. Morchella esculenta Pers. '*The Morell.*'
In woods, etc., in spring and early summer, very local. Esculent, and really a *bonne bouche*, but chiefly used dried (like the Truffle) for giving a delicious mushroom flavour to soup, pies, etc.
W. Boston Spa and Collingham, near the Wharfe; *J. Abbott*! Bolton Woods; *W. West.* On banks of Wharfe at Barden, frequent.
A. On tan-pit heaps, Meanwood, shown me by *H. Barber*! Only found once in Airedale; several specimens, 1885; *W. H. Hind.* In a garden, Selby; *W. N. Cheesman.*
D. Went-vale; *H. F. Parsons.* Campsall Park, 1886.
T. Roche Abbey Valley; *J. H. Aveling* in *Hist.* (1870).

826. Morchella semilibera DC.
N. Kirk Deighton, 1887; *J. Beanland.*
D. Rare, Conisborough, 1883; *Hudd. Bot. Soc.*

827. Helvella crispa Fr.
In woods, etc., rare. Esculent.
N. Near Harrogate; *T. Hick.*
A. Nab Wood, Saltaire, 1886.
C. Bradley Wood, near Huddersfield, 1884; *Hudd. Bot. Soc.*

828. Helvella lacunosa Afz.
A. In woods, rare. Only known to occur in one locality, Heaton Wood near Bradford, where it is common every autumn.

829. Helvella elastica Bull.
A. Rare. Found once near Bingley; *T. Hebden.*

830. Mitrula cucullata Fr.
A. Amongst fir-needles, Adel Bog Wood!

831. Mitrula paludosa Fr.
A strikingly pretty species.
W. West Woods, Bramham!
A. On dead leaves and *Sphagnum* in the ditches of the fir-wood at the sluice-end of the 'Seven Arches' aqueduct, Adel, in plenty in 1877; *William Todd*! Cullingworth Bog; *T. Hebden.*

832. Spathularia flavida Pers.
R. In fir-woods, Slaidburn, 1876!
W. Otley Chevin Woods!

833. Leotia lubrica Pers.
Y. Studley; *Exc.*, 1881.
A. In woods, locally common, has only been found once in Airedale, at Cullingworth, 1884.

834. Vibrissea truncorum Fr.
A. Rare, only found once, at Cullingworth Bog, 1884; *T. Hebden.*

835. Geoglossum olivaceum Pers.
C. Occasionally in pastures in the Huddersfield district; *Hudd. Bot. Soc.*

836. Geoglossum viscosum Pers.
A. Hainworth Wood near Keighley; *T. Hebden.*

837. Geoglossum glabrum Pers.
A. On grassy road-sides, rare, near Haworth, 1884; *T. Hebden.*

838. Geoglossum hirsutum Pers.
Amongst grass, not common.
C. Huddersfield district; *Hudd. Bot. Soc.*
D. Askern; *Hudd. Bot. Soc.*

839. Geoglossum difforme Fr.
Y. Studley; *Exc.*, 1881.
A. Brayton; *W. N. Cheesman.* Saltaire.
D. Known since 1650. On grassy anthills in a close next Hampole Wood, as one goes from Hutton Pagnell; *Mr. Stonehouse*; *How, Phyt.*, 44; repeated by *Ray* in his *Catalogue* (1670), i, 125, with the additional information that he was present with Mr. Stonehouse at its discovery.

840. Peziza macropus Pers.
N. On the ground in woods, rather rare. Harrogate; *Exc.*, 1881.

841. Peziza tuberosa Bull.

In woods amongst *Anemone nemorosa*, usually attached to a *Sclerotium*.

W. Bolton Woods.

A. Shipley Glen.

842. Peziza venosa Pers.

On the ground amongst *Petasites vulgaris* and other plants in spring. Remarkable for its powerful odour of nitric acid, and one of the most delicious of edible fungi.

A. Cottingley. Camblesforth; *W. N. Cheesman.*

D. Conisborough; *Hudd. Bot. Soc.*

843. Peziza badia Pers.

N. Harrogate; *Exc.*, 1881.

A. Hainworth near Keighley; *T. Hebden.* Meanwood near Leeds.

844. Peziza cochleata Huds.

W. On moss on old walls, Lindley Valley!

845. Peziza leporina Batsch.

A. On the ground in Heaton Wood near Bradford, 1885.

846. Peziza aurantia Fr. '*Orange Cup Fungus.*'

On the ground in woods, etc., frequent.

A. Saltaire, Ingrow, Esholt, etc. Stainor Wood; *W. N. Cheesman.*

C. Huddersfield district; *Hudd. Bot. Soc.*

T. Roche Abbey Valley; *J. H. Aveling* in *Hist.* (1870).

847. Peziza repanda Wahl.

A. About the sewage works, Bradford, 1880.

848. Peziza cerea Sow.

A. On tan-beds at Smith's Tannery, Meanwood, and on the Gallop, Woodhouse Moor, Leeds; *J. Abbott.* Rare.

849. Peziza vesiculosa Bull.

On dung-heaps, in potato-fields, etc., common.

W. Tadcaster, in plenty, 1881, and hawked about for sale as edible; *B. Blaydes Thompson!*

A. Saltaire. Bingley. Esholt; *W. West.*

850. Peziza rutilans Fr.

A. On the ground in a plantation near Cottingley.

851. Peziza humosa Fr.

D. Goole; *T. Birks.*

852. Peziza granulata Bull.

Very common on cow-dung.

W. Sweep Wood, Wetherby; *G. E. Massee!*

A. Malham; Roundhay; *W. West.* Saltaire, Hawksworth, Keighley, etc.

C. Wibsey; *W. West.*

853. Peziza fusispora Berk.

A. On heathy ground in Shipley Glen, 1886.

854. Peziza coccinea Jacq. '*Crimson Cup Fungus.*'

On sticks in hedge-rows and woods, in spring, rare, but very beautiful.

Y. Studley Woods!

W. Bolton Woods; *W. West.* Bramhope; *J. Abbott.* Lane near Sweep Plantation, Wetherby!

A. Shipley Glen. Holden Wood near Keighley.

T. Roche Abbey Valley; *J. H. Aveling* in *Hist.* (1870).

855. Peziza hemisphærica Wigg.

A. On the ground in Nab Wood, Saltaire.

856. Peziza hirta Sch.

Y. On the ground, Hackfall, in shade!

857. Peziza trechispora B. & Br.

W. On the ground in Denton Wood near Ilkley.

858. Peziza umbrata Fr.

W. River-side, Ilkley.

A. On the ground amongst moss, Newsholme Dean.

859. Peziza scutellata L.

W. Dike by railway near Towton; *G. E. Massee!*

A. Not uncommon amongst moss in pastures at Marley and Cullingworth. Esholt; *W. West.* Micklefield; *W. N. Cheesman.*

860. Peziza stercorea Pers.

A. Common on cow-dung at Baildon; *W. West.* Saltaire, Hawksworth, etc.

861. Peziza ciliaris Schrad.

A. On dead oak-leaves in Nab Wood, Saltaire.

862. Peziza virginea Batsch.

N. Harrogate; *Exc.*, 1881.

W. Collingham Bank!

A. On stumps, etc., common. Saltaire, Bingley, Esholt, etc.

863. Peziza nivea Fr.

A. On old stumps, etc., Nab Wood, Saltaire.

864. Peziza calycina Schum.

On twigs of larch and Scotch fir.

Y. Studley; *Exc.*, 1881.

N. Harrogate; *Exc.*, 1881.

A. Steeton. Bingley Woods. Esholt; *W. West.*

865. Peziza bicolor Bull.

A. Selby and Rawcliffe; *H. F. Parsons.*

866. Peziza hyalina Pers.

N. Harrogate; *Exc.*, 1881.

A. On tree-stumps, Saltaire.

867. Peziza plano-umbilicata Grev.

W. Bolton Woods.

A. On dead stems of nettles, Apperley.

868. Peziza villosa Pers.

A. On dead stems about the Seven Arches, Saltaire.

869. Peziza apala B. & Br.

A. On dead rushes, common on Baildon Moor.

870. Peziza domestica Sow.

A. On white-washed walls in cellars, Saltaire, 1882.

871. Peziza mutabilis B. & Br.

A. On withered leaves of *Aira cæspitosa*, Shipley Glen, 1887.

872. Peziza fusca Pers.

A. On dead twigs, etc., in Nab Wood, Saltaire.

873. Peziza firma Pers.

A. On dead twigs, Nab Wood, Saltaire, common.

874. Peziza Curreyiana Tul.

A. Very common on the dead leaves of various species of *Juncus*, and always attached to a *Sclerotium*. Baildon Moor. Shipley Glen. Hawksworth.

875. Peziza coronata Bull.

A. On dead stems of thistles and nettles, Apperley and Calverley. Meanwood, exceptionally fine.

876. Peziza cyathoidea Bull.

A. On dead stems of *Chærophyllum sylvestre*, Fagley Woods, 1887.

877. Peziza vinosa A. & S.

A. On dead wood, Steeton.

878. Peziza cinerea Batsch.

N. Harrogate; *Exc.*, 1881.

W. Towton dike; *Exc.*, 1880!

A. On rotten wood, common, Saltaire, Calverley, etc.

879. Peziza hepatica Batsch.

M. Greenfield; *W. West.*

A. Amongst moss in Heaton Woods near Bradford, 1883.

C. Wibsey, etc.; *W. West.*

880. Peziza erumpens Grev.

A. On sycamore petioles, Bingley.

881. Peziza fusarioides Berk.

W. Bolton Woods.

A. On dead stems of nettles, Saltaire, Apperley.

882. Helotium aciculare Fr.

W. Bolton Woods.

A. About old stumps, Nab Wood, Saltaire. Horsforth; *W. West.*

C. Odsal Wood near Wibsey.

883. Helotium æruginosum Fr.

On dead oak and holly, local.

L. Helks Wood, Ingleton.

W. Bolton Woods; *W. West.*

A. Holden Wood near Keighley, very fine.

884. Helotium virgultorum Fr.

Y. Studley; *Exc.*, 1881.

885. Helotium lutescens Fr.

A. On dead twigs, Saltaire.

886. Helotium claro-flavum Berk.

W. On dead holly, Bolton Woods.

887. Helotium citrinum Fr.

A. On a dead stump, Nab Wood, Saltaire, 1887.

888. Helotium herbarum Fr.

A. On dead stems of *Stachys sylvatica*, Apperley, 1887.

889. Tympanis conspersa Fr.

A. On dead wood, Malham.

890. Ascobolus Crouani Cooke.

N. Harrogate.

A. Very common on the mortar of mossy walls, Horsforth, and at Horton and Allerton; *W. West.* Saltaire, Steeton, etc.

C. Wibsey; *W. West.*

891. Ascobolus Jungermaniæ B. & Br.

A. Amongst *Jungermaniæ*, Holden Wood near Keighley; *T. Hebden.*

892. Ascobolus furfuraceus Pers.

On cow-dung, extremely common.

893. Ascobolus ærugineus Fr.

A. On cow-dung, Saltaire.

894. Ascobolus immersus Pers.

A. On sheep-dung, Shipley Glen; *W. West.*

895. Ascobolus Kerverni Crouan.

A. On old cow-dung, Saltaire.

896. Ascobolus vinosus B.
A. On rabbit-dung, Leeds district; *J. Abbott.*

897. Ascobolus granuliformis Crouan.
A. On cow-dung, very common, Bradford Moor, Saltaire, etc.

898. Ascobolus saccharinus B. & Curr.
A. On an old leather boot at Greengates, near Apperley, December 1886.

899. Ascobolus ciliatus Schm.
W. Beamsley; *W. West.*
A. On cow-dung, common, Bradford, Saltaire, Bingley, Meanwood, etc.

900. Bulgaria inquinans Fr.
Probably generally distributed.
Y. Studley; *Exc.*, 1881.
N. Harrogate; *Exc.*, 1881.
W. Bolton Woods; *W. West.*
A. On dead tree-trunks, Bingley Woods. Steeton.

901. Bulgaria sarcoides Fr. '*Flesh Fungus.*'
Y. Studley; *Exc.*, 1881.
W. Horn Bank, Flint Mill, Wetherby!
A. On old stumps, etc., common, Esholt, Steeton, etc. Eccup
D. Goole; *H. F. Parsons.*
T. Roche Abbey Valley; *J. H. Aveling* in *Hist.* (1870).

XXIX. TUBERACEI.

902. Tuber æstivum L. '*Common Truffle.*'
In woods, subterranean, esculent and delicious.
W. Beech-woods, Thorparch, Wetherby, etc.!

903. Elaphomyces granulatus Fr.
A. One specimen that had been uprooted, Bingley Wood, 1885. Woods near Selby; *W. N. Cheesman.* (In West Riding?)

XXX. PHACIDIACEI.

904. Phacidium Vaccinii Fr.
W. On dead leaves of *Vaccinium Vitis-idæa*, Barden.

905. Phacidium coronatum Fr.
W. On dead oak-leaves, Harrogate; *Exc.*, 1881.

906. Phacidium Ilicis Fr.
L. Sedbergh; *W. West.*
W. Bolton Woods.
A. On dead holly-leaves, Nab Wood, Saltaire.

907. Heterosphæria patella Grev.
L. Ingleton.
A. On dead stems of umbelliferous plants, Ingrow.

908. Ectostroma Iridis Rabenh.
W. On *Iris* in Towton dike; *W. West* and party (1880)!

909. Rhytisma Andromedæ Fr.
D. & T. On living leaves of marsh-rosemary, Goole Moor, abundant; *W. West*; and Thorne Waste; *H. F. Parsons*!

910. Rhytisma punctatum Fr.
T. Roche Abbey Valley; *J. H. Aveling* in *Hist.* (1870).
Obs.—Doubtful if distinct from the succeeding species.

911. Rhytisma Acerinum Fr.
Y. Studley; *Exc.*, 1881.
W. Bolton Woods.
A. Common on fallen leaves of *Acer pseudo-platanus*, Shipley Glen, Nab Wood, etc. Meanwood; *J. Abbott.*
T. Roche Abbey Valley; *J. H. Aveling* in *Hist.* (1870).

912. Hysterium Fraxini Pers.
W. On an ash-twig, Wetherby!

913. Hysterium Juniperinum De Not.
R. On dead juniper leaves, Horton-in-Ribblesdale.

914. Colpoma Quercinum Wallr.
L. On oak-twigs, Helks Wood, Ingleton.

915. Stegia Ilicis Fr.
L. Dentdale; *W. West.*
N. Knaresborough; *W. West.*
W. Bolton Woods. Woodhall!
A. On holly-leaves, common, Saltaire, Horsforth, etc.; *W. West.*

916. Trochila Lauro-cerasi Fr.
Y. On leaves of cherry-laurel, Studley; *Exc.*, 1881.

XXXI. SPHÆRIACEI.

917. Torrubia entomorrhiza Fr.
A. On a dead pupa, Newsholme Dean; *T. Hebden.*

918. Torrubia militaris Fr.
A. On dead chrysalides of *Smerinthus populi* in a garden at Meanwood, 1876! Frequent on dead pupæ about Saltaire, Bingley, etc.

919. Claviceps purpurea Tul. '*Ergot*' (the stroma).
Very common some seasons on various grasses, chiefly *Glyceria fluitans* and *Lolium perenne.*
W. Wetherby!
A. Saltaire, Bingley, etc. Camblesforth; *H. F. Parsons.*
D. Askern; *P. F. Lee*!
T. Common on cocksfoot grass, Roche Abbey Valley; *J. H. Aveling* in *Hist.* (1870).

920. Epichloe Typhina Berk.
W. Aketon Bog, Spofforth; *T. Hick*!
A. On living grasses, Shipley Glen, Bingley, Hawksworth, etc.

921. Hypocrea rufa Fr. (*Trichoderma viride* Pers.)
A. On dead oak. Steeton. Bingley.

922. Hypomyces aurantius Tul.
W. On decaying *Polyporus*, Bolton Woods.

923. Hypomyces rosellus Tul.
N. Parasitic on *Corticium*, Harrogate; *Exc.*, 1881.

924. Hypomyces chrysospermus Tul.
Y. Studley; *Exc.*, 1881.

925. Nectria mammoidea Ph. & Pl.
A. On dead wood, Steeton, Bingley, and Esholt.

926. Nectria cinnabarina Fr.
On dead twigs, very common. General in distribution.
D. With *perithecia* at Doncaster; *H. F. Parsons.*

927. Nectria coccinea Fr.
A. On a decaying tree, Esholt.

928. Nectria aquifolia Berk.
W. On dead holly, Bolton Woods.

929. Nectria Peziza Fr.
A. On dead holly, Holden Wood near Keighley.

930. Nectria sanguinea Fr.
A. On dead twigs, Shipley Glen.

931. Xylaria polymorpha Grev.
L. On old stumps, Ingleton.
Y. Studley; *Exc.*, 1881.
W. Wetherby, Harewood, etc.

932. Xylaria hypoxylon Grev. '*Candle-snuff.*'
Common and generally distributed. The earliest published record is
T. Roche Abbey Valley; *J. H. Aveling* in *Hist.* (1870).

933. Xylaria carpophila Fr.
W. On beech-mast and leaves, Jackdaw Crag, Stutton, and Boston Spa!

934. Ustulina vulgaris Tul.
Y. Studley; *Exc.*, 1881.
N. On the Cowthorpe Oak stump, 1880!
W. On a tree-stump, Bolton Woods, 1880.

935. Hypoxylon fuscum Fr.
On dead branches, locally common.
L. Sedbergh; *W. West.*
R. Gisburn. Settle; *W. West.*
Y. Studley; *Exc.*, 1881.
W. Bolton Woods; *W. West.* Seavy Carr!
A. Steeton. Baildon; *W. West,*

936. Hypoxylon concentricum Grev.
W. On dead trees, Wetherby, 1881!
D. Goole (on ash) and Thorne Waste (on brick); *H. F. Parsons.*

937. Hypoxylon cohærens Fr.
W. Sweep Wood, Wetherby; *G. E. Massee*!

938. Hypoxylon multiforme Fr.
A. On dead birch-bark, Scotland Mills Wood, Seven Arches, Adel!

939. Eutypa lata Tul.
A. On dead twigs, Shipley Glen.

940. Dothidea Betulina Fr.
W. On birch-leaves near Church Fenton.
T. Roche Abbey Valley; *J. H. Aveling* in *Hist.* (1870).

941. Dothidea graminis Fr.
W. Near Ilkley.
A. On various grasses, Bingley.

942. Dothidea Junci Fr.
A. On stems of rushes, Baildon Moor.

943. Dothidea Filicina Fr.
A. Common on dead stems of bracken.

944. Dothidea Ulmi Fr.
T. Roche Abbey Valley; *J. H. Aveling* in *Hist.* (1870).

945. Diatrype verrucæformis Fr.
A. On dead branches, Micklefield; *Rev. J. E. Vize.*

946. Diatrype stigma Fr.
A. On dead hawthorn, Saltaire.

947. Diatrype disciformis Fr.
A. On dead branches, Bingley.

948. Diatrype ferruginea Fr.
Y. On hazel branches, Studley ; *Exc.*, 1881.

949. Diatrype Quercina Tul.
N. On oak-twigs, North Deighton near Wetherby !

950. Valsa leiphemia Fr.
A. On dead oak-twigs, Bingley.

951. Valsa pulchella Fr.
R. On branches of *Prunus Padus*, Linn Gill !

952. Valsa quaternata Fr. (*faginea*).
W. Crossley's wood-yard, Wetherby, on beech sticks !

953. Valsa Tiliæ Tul.
R. Horton-in-Ribblesdale ; *W. West*. But there are no lime-trees there !

954. Valsa Platanoides Berk.
A. On sycamore twigs, Saltaire.

955. Cucurbitaria Berberidis Gray.
T. Roche Abbey Valley ; *J. Bohler* in *Aveling's* list (1870).

956. Sphæria Brassicæ Klotsch.
A. On dead cabbage-stalks near Ingrow.

957. Sphæria aquila Fr.
A. On rotten wood, Holden Wood near Keighley.

958. Sphæria spermoides Hoffm.
A. On stumps, Horsforth ; *W. West*. Shipley Glen. Steeton.
C. North Dean ; *W. West* !

959. Sphæria bombarda Batsch.
T. Roche Abbey Valley.

960. Sphæria innumera B. & Br.
T. Roche Abbey Valley ; *J. H. Aveling* in *Hist.* (1870).

961. Sphæria pulvis-pyrius Pers.
A. On dead holly, Nab Wood, Saltaire. Micklefield.

962. Sphæria confluens Tode.
T. Roche Abbey ; *J. H. Aveling* in *Hist.* (1870).

963. Sphæria mammæformis Pers.
T. Roche Abbey ; *J. H. Aveling* in *Hist.* (1870).

964. Sphæria coprophila Fr.
A. On old cow-dung, Eldwick.

965. Sphæria moriformis Tode.
Y. On dead wood, Studley ; *Exc.*, 1881.

966. Sphæria celata Curr.
L. On wych-elm bark, Dent ; *W. West*.
W. Collingham near Wetherby !

967. Sphæria scirpicola DC.
W. On *Typha*, Towton Dike ; *Excurs.*, 1880 !

968. Sphæria millepunctata Grev.
On ash-twigs, common.

969. Sphæria ceuthosporoides Berk.
T. On cherry-laurels, Roche Abbey ; *J. Bohler* in *Aveling's* list (1870).

970. Sphæria herbarum Pers.
On herbaceous stems, common.

971. Sphæria rubella Pers.
W. Church Fenton ; *Excurs.*, 1880 !
A. On stems of thistles, etc., Heaton, Bingley, etc.

972. Sphæria Urticæ Rabh.
L. On nettle-stems, Clapham.

973. Sphæria acuta Moug.
On nettle-stems, common.

974. Sphæria acuminata Sow.
A. On stems of thistles, Heaton and Bingley.

975. Sphæria doliolum Pers.
W. On nettle-stems, Bolton Woods.

976. Sphæria planiuscula B. & Br.
W. On dead herbaceous stems near Ilkley.

977. Sphæria rostellata Fr.
A. On dead bramble, etc., Bingley.

978. Sphæria complanata Tode.
W. Towton dike, on *Typha* stems !
A. On dead herbaceous stems, Esholt.

979. Sphæria setacea Pers.
A. On petioles of sycamore, Saltaire, Bingley, Esholt.

980. Sphærella ostruthii Fr.
W. On leaves of *Angelica sylvestris*, Bolton Woods.

981. Sphærella Hederæ Sow.
A. On dead ivy-leaves, Cottingley.

982. Sphærella Brassicæcola De Not.
On decaying cabbage-leaves, common.

983. Sphærella Rumicis Desm.
A. On leaves of *Rumex*, Wilsden. Saltaire.

984. Pyrenophora phæocomes Fr.
A. On dead leaves of grasses, Rawdon.

985. Orbicula cyclospora Cooke.
A. On damp paper, Bradford.

986. Stigmatea Robertiani Fr.
A. Gordale ; *W. West*. On living leaves of herb-robert, Saltaire, Bingley, etc.

987. Capnodium sphæricum Cooke.
A. On dead leaves of *Veronica officinalis*, Hawksworth.

FRESH-WATER ALGÆ.

FRESH-WATER ALGÆ.

REVISED BY WILLIAM WEST, F.L.S.

[The arrangement of the following list, that of Cooke, in his recently-published 'British Fresh-water Algæ,' will probably be of most service to those who study this highly-interesting group of plants; we should, however, have preferred to classify them more in accordance with the requirements of the higher examinations of the various Universities. It is only our earnest desire after something like symmetry and completeness in the matter of a West Yorkshire Flora that impels us to offer a list so comparatively incomplete as that which follows—including in all little over three hundred and seventy forms of this very beautiful and sadly neglected class of plants.

For a large number of the records we are indebted to less than a dozen workers: the Rev. R. A. Gatty, Dr. H. F. Parsons, and Messrs. W. Barwell Turner, James Abbott, C. P. Hobkirk, W. H. Dikes, F. Emsley, C. B. Crawshaw, and E. Hunter. The information coming to us through these gentlemen, together with a few notices in Hooker's 'English Flora' (Vol. v, part 1), Hassall's 'Fresh-water Algæ' (2 vols., 1845), Norman's 'List of Hull Diatomaceæ' (which includes some West Riding stations), and papers in such journals as the *Naturalist*, forms the following florula—if such an incomplete enumeration merits even that diminutive term. Future observers may, however, prefer to have the few facts it contains—with the lesson it carries with it—rather than utter uncertainty as to what has and what has not been observed; and for such future Alexanders there is this encouragement—the results of their difficulties when conquered are more likely to be initial discoveries than mere confirmatory investigation upon a well-fought field; an incentive of no mean kind to a certain class of minds who do not care to be for ever following in the precise tracks beaten out for them by others. With respect to their distribution, and the climatic or lithologic conditions affecting them, we know little, so it will be as well to quote an excellent and suggestive summary by Mr. W. Barwell Turner. In the *Naturalist* for February 1886, with regard to one family, he wrote:—

'I would remark, with respect to the Desmids, that our knowledge of them is but slight; not a single "life-history" is perfectly known of a tribe numbering about 900 species, or with varieties 1,200. The altitude as affecting their presence has very rarely been noted, though we are aware that Didymoprium (*Bambusina*), Desmidium, Docidium, and Closterium *prefer* the low-lands. My own notion is that, while certain species are found everywhere, at moderately low levels, from the Pole to the Equator, being thoroughly cosmopolitan, geological influences have a decided 'something to do' with certain *local* species, and I have long held the opinion (as yet unshaken) (1) that *most* Desmids prefer the Igneous geo-

logical districts to others; (2) that *some* are but rarely found in districts newer than the Old Red Sandstone, e.g., *Staurastrum ophiura*, *S. cerastes*, *S. Brasiliense* (?), *S. arctiscon*, *Micrasterias brachyptera*, etc.; (3) that few are found in chalk districts, or on the greensand; and (4) that they are most abundant (at least the commoner species) when the water wherein they live contains a *small* quantity of dissolved iron, though they will not endure so much of this metal as will the Diatoms (with their yellow or brownish endochrome), the colour of which a somewhat large percentage of iron seems to enhance, while an excess kills them!'

Mr. Turner's observations are very valuable as well as interesting; but we may add that we have found *Didymoprium* in plenty at an elevation of 1,500 feet, even if it does generally 'prefer the lowlands.' Pre-carboniferous strata certainly yield the richest and most varied gatherings. Pools containing either *Utricularia*, *Sphagnum contortum*, *Hypnum scorpioides*, or *Hypnum revolvens* are particularly desmidiferous. With regard to the presence of iron in the water where these symmetric unicells love to dwell, we can always vouch for a deposit of oxide on the sides of the bottles within a few weeks after the collection of any desmid gathering. As far as our own experience goes Desmids seem to vary in their altitudinal distribution, for some species, as *Micrasterias Jenneri*, *M. crenata*, *M. mucronata*, and *Euastrum insigne* seem at home at about 1,500 feet, while some frequently occurring species like *Penium digitus* and *Tetmemorus granulatus* are frequent from this elevation down to the plains of the lowlands. We would fain have dwelt longer on this aspect of the subject, but refrain, feeling that there are not sufficient data for even a short disquisition on their distribution.

The records for which W. H. Dikes vouches are from MS. in a copy of Ralfs' Desmidiaceæ, formerly belonging to him, now in the Library of the Bradford Philosophical Society. Some have been confirmed by Mr. West; all bear internal evidence of reliability.

In the following list the initials A., E., and T. stand for Messrs. James Abbott, F. Emsley, and W. B. Turner.]

CLASS I.—CHLOROPHYLLOPHYCEÆ.

ORDER I.—COCCOPHYCEÆ.

I. PALMELLACEÆ.

1. Pleurococcus vulgaris Menegh. (*Chlorococcum*).
On bark, everywhere. Leeds, Wetherby, Goole, Dewsbury, etc.
It is probably only the gonidial state of Lichens.

2. Chlorococcum frustulosum Rabenh.
L. Cautley Spout; *W. West.*

3. Glæocystis ampla Kütz.
W. Bramhope; *J. Abbott.*
D. Goole Moor; *W. West.*

4. Glæocystis vesiculosa Näg.
A. Rawcliffe; *W. West.*

5. Glæocystis rupestris (Lyngb.) Rabh., Alg., iii, 30.
On rocks in glens.
L. Cautley Spout.
A. Baildon, Bingley, Keighley, Horton, etc.; *W. West.* Meanwood; *W. B. Turner.*

6. Urococcus Hookerianus Hass.
N. In the Knaresborough 'Dropping-well'; *Prof. G. J. Allman.*

7. Urococcus Allmanni Hass.
N. With the preceding; *Prof. G. J. Allman.*

8. Palmella hyalina Bréb.
A. Garforth; *H. F. Parsons.*

9. Palmella Mooreana Harv.
A. Leeds; *W. B. Turner.*
C. Soothill, Dewsbury; *C. B. Crawshaw.*

10. Palmella prodigiosa Mont.
A. Bradford; *W. West.*

11. Porphyridium cruentum Nägeli.
In patches on damp rocks or stones, resembling clotted blood.
W. Wetherby; *F. A. Lees.*
A. Leeds district; *J. Abbott.* Bradford, Bingley, etc.; *W. West.* Selby; *H. F. Parsons.*
D. Sheffield; *A. Carr.*

12. Botrydina vulgaris Bréb.
L. Ingleborough, Whernside, Howgill Fells, etc.; *W. West.*
R. Penyghent; *W. West.*
N. Harrogate; *A., E.,* and *T.*
A. Headingley; *W. B. Turner.*

13. Tetraspora gelatinosa (Vauch.).
A. Gordale; *C. B. Crawshaw.* Adel Bog; *A., E.,* and *T.*

14. Tetraspora bullosa Ag.
A. Roundhay; *A., E.,* and *T.* Scotland Mill Beck; *W. B. Turner.*
D. Goole; *H. F. Parsons.* Askern; *Rev. F. H. Allen.*

15. Botryococcus Braunii Kütz.
A. Bradford; *W. West.*

16. Apiocystis Brauniana Nägeli.
A. Chapeltown; Adel Bog; *A., E.,* and *T.*

17. Rhaphidium falcatum (Corda).
W. Ilkley; *W. H. Dikes.*
A. Chapel-Allerton; Adel; *A., E.,* and *T.* Bradford, Cullingworth, etc.; *W. West.*

18. Raphidium aciculare Braun.
A. Cullingworth; *T. Hebden.*

19. Rhaphidium duplex Kütz.
A. Moor-Allerton; Roundhay; *W. B. Turner.* Cullingworth, Baildon, etc.; *W. West.*

20. Dictyosphærium Ehrenbergianum Näg.
A. Bradford; *W. West.*

21. Hydrurus penicillatus Ag.
A. Newsholme Dean near Keighley; *T. Hebden.* Esholt; *W. West.*

II. PROTOCOCCACEÆ.

22. Protococcus viridis Cohn. (Ag. Rabh. Alg.)
A. Common about Leeds; *A., E.,* and *T.* Frequent about Bradford; *W. West.*

23. Scenedesmus quadricaudatus Bréb.
W. Ilkley; *W. West.* Bramhope; *A., E.,* and *T.*
A. Malham, etc.; *W. West.*

24. Scenedesmus acutus Meyen.
W. Arncliffe, etc.; *W. West.* Bramhope; *A., E.,* and *T.,* with var. *obliquus* Rabh.
A. Rawcliffe; *H. F. Parsons.*

25. Scenedesmus obtusus Meyen.
W. Bramhope; *A., E.,* and *T.*
A. Rawcliffe; *H. F. Parsons.* Bradford, Cullingworth, etc.; *W. West.*

26. Ophiocytium majus Näg.
W. Bramhope; *A., E.,* and *T.*
Obs.—This is not given in Cooke's Algæ as a British species.

27. Sciadium arbuscula A. Braun.
N. Harrogate; *A., E.,* and *T.*

28. Pediastrum angulosum Ehrenb.
C. Wakefield; *W. H. Dikes.*

29. Pediastrum Boryanum Turp.
Y. Markington; *Rev. J. S. Tute.*
A. Rawcliffe; Bingley; Manningham; Calverley; *W. West.*
C. Near Wakefield; *W. West.*
D. Askern; *Rev. F. H. Allen.*

VAR. **granulatum** Kütz.
A. Adel; *A., E.,* and *T.*

30. Pediastrum Ehrenbergii Br.
A. Rawcliffe; *H. F. Parsons.*

31. Pediastrum pertusum Kütz.
C. Wakefield; *W. H. Dikes.*

32. Characium ornithocephalum Br.
A. Baildon; *W. West.*

33. Characium Sieboldi Braun.
A. Keighley; *T. Hebden.* Bowling; *W. West.*

34. Hydrianum heteromorphum Reinsch.
C. Near Mirfield; *W. West.*

III. VOLVOCINEÆ.

35. Chlamydococcus pluvialis A. Braun.
In stagnant rain-water in the hollows of stones, everywhere.
N. Spout at Harrogate; *T. Hick.*
A. Leeds; *A., E.,* and *T.* Baildon; *W. West.*
Obs.—*C. nivalis*—the 'Red Snow' plant—is, according to Cooke, a form of this species.

36. Volvox globator L.
Pools, rare.
Y. Pool by Ure at Ripon; *J. Abbott.*
W. Pond at Bramhope; *J. Abbott.* Fish-pond farthest from the House, Bramham Park; *B. Saynor*!
A. 'Shepherd's Pond,' Roundhay Road; *Robert Duncan.* Hunslet old race-course; *J. Abbott.* Rabbit-hills, Rawcliffe; *T. Hick.*
D. Askern; *Rev. F. H. Allen.*

37. Pandorina morum Ehrenb.
W. Bramhope; *J. Abbott.* York; *A., E.,* and *T.*
A. Rawcliffe; *H. F. Parsons.* Very common about Bradford; *W. West.*

38. Gonium pectorale Müll.
A. Wortley near Leeds; *W. B. Turner.* Cullingworth, etc.; *W. West.*
D. Goole; *H. F. Parsons.*

ORDER II.—ZYGOPHYCEÆ.

IV. DESMIDIEÆ.

39. Gonatozygon Ralfsii De Bary.
A. Moor-Allerton! *W. B. Turner.*

40. Sphœrozosma vertebratum Ralfs.
A. Rawcliffe Common; *W. B. Turner.*

41. Sphœrozosma excavatum Ralfs.
A. Near Selby; *W. West.* Adel Bog; *W. B. Turner.* Hainworth near Keighley; *W. West.*

42. Sphœrozosma pulchellum Arch.
A. Adel; *W. B. Turner.*

43. Hyalotheca dissiliens Bréb.
W. & A. Common. Ilkley; Cullingworth; etc.; *W. West.*

44. Desmidium cylindricum Grev.
A. Adel Bog; *W. B. Turner.*

45. Desmidium Swartzii Kütz.
Not rare, but local.
W. Bramhope; *A., E.,* and *T.* Ilkley; *W. West.*

46. Docidium Ehrenbergii (Kütz.) Ralfs.
A. Moor-Allerton; Adel; *W. B. Turner.*

47. Docidium clavatum Kütz.
W. Ilkley; *W. H. Dikes.*
A. Meanwood; *W. B. Turner.*

48. Docidium nodulosum Bréb.
A. Adel Bog; *W. B. Turner.*

49. Docidium truncatum Bréb.
W. Ilkley; *W. H. Dikes.*

50. Docidium baculum Bréb.
W. Ilkley; *W. West.*
A. Cullingworth, etc.; *W. West.* Adel Bog; *W. B. Turner.*

51. Closterium lunula Müll.
W. & A. Common. Ilkley; Esholt; etc.; *W. West.*

52. Closterium acerosum (Schranck.) Ehrenb.
L. Ingleborough; *W. West.*
Y. Markington; *Rev. J. S. Tute.*
A. Cullingworth; *W. West.*

53. Closterium lanceolatum Kütz.
L. Ingleborough; *W. West.*
A. Roundhay; Adel Bog; *W. B. Turner.*

54. Closterium Ehrenbergii Menegh.
A. Hawksworth near Shipley; *H. F. Parsons.* Armley and Adel; *A., E.,* and *T.* Denholme, etc.; *W. West.*
C. Wakefield; *W. H. Dikes.*

VAR. **Malinvernianum** De Not.
A. Adel Bog; *W. B. Turner.*

55. Closterium moniliferum Bory.
A. Heaton; *W. West.* Adel Bog; *W. B. Turner.* Drax; *H. F. Parsons.*
D. Goole; *W. G. Tacey.*

56. Closterium Jenneri Ralfs.
C. Wakefield; *W. H. Dikes.*

57. Closterium Dianæ Ehrenb.
W. Askwith; *W. B. Turner.*
A. Bingley; Baildon; Cullingworth; *W. West.*

58. Closterium Venus Kütz.
W. Alwoodley; *W. B. Turner.*
A. Cullingworth; Eldwick; *W. West.*
Obs.—Cooke puts this under the last species, but it is only half the size and is more curved.

59. Closterium striolatum Ehrenb.
W. Ilkley; *W. H. Dikes.*
A. Baildon; Cullingworth; *W. West.*

VAR. **intermedium** Ralfs.
W. Ilkley; *W. H. Dikes.*
A. Baildon; Cullingworth; *W. West.*

60. Closterium costatum Corda.
Y. Markington; *Rev. J. S. Tute.*

61. Closterium juncidum Ralfs.
A. Moor-Allerton; *W. B. Turner.* Baildon; *W. West.*

62. Closterium lineatum Ehrenb.
A. Frizinghall; *W. West.*

63. Closterium rostratum Ehrenb.
W. Askwith Moor; *W. B. Turner.*
A. Bingley; *W. West.* Adel Bog; *A., E.,* and *T.*
D. Goole; *H. F. Parsons.*

64. Closterium setaceum Ehrenb.
A. Meanwood; *J. Abbott.* Roundhay and Adel Bog; *A., E.,* and *T.*
C. Wakefield; *W. H. Dikes.*

65. Closterium cornu Ehrenb.
W. Ilkley; *W. H. Dikes.*
A. Adel Bog; *A., E.,* and *T.*

66. Closterium acutum Ehrenb.
Y. Markington; *Rev. J. S. Tute.*
A. Adel Bog; *W. B. Turner.*

67. Closterium parvulum Näg.
A. Cullingworth; Eldwick; Bingley; *W. West.*
Obs.—This is not recorded in Cooke's Desmids.

68. Closterium eboracense Turn.
A. Chapel-Allerton; *W. B. Turner.*
D. Campsall; *W. B. Turner.*

69. Penium margaritaceum Bréb.
W. Ilkley; *W. H. Dikes.*

70. Penium Brebissonii Ralfs.
W. Ilkley ; *W. H. Dikes.*
A. Moor-Allerton ; *W. B. Turner.*
D. Goole Moor ; *H. F. Parsons.*

71. Penium digitus (Ehrenb.) Ralfs.
W. Ilkley ; *W. West.*
A. Cullingworth ; *W. West.* Adel Dam ; *W. B. Turner.*

72. Penium closterioides Ralfs.
W. Ilkley ; *W. H. Dikes.*

73. Penium oblongum De Bary.
A. Adel ; *W. B. Turner.*

74. Penium cylindrus Ehrenb.
A. Cullingworth ; *W. West.*

75. Penium Ralfsii De Bary.
A. Adel Bog ; *W. B. Turner.*

76. Cylindrocystis diplospora Lund.
L. Ingleborough ; *W. West.*
W. Blubberhouses ; *W. B. Turner.*
A. Adel Dam ; *W. B. Turner.* Cullingworth ; *W. West.*

77. Cylindrocystis crassa De Bary.
W. Blubberhouses ; *W. B. Turner.*

78. Mesotænium DeGreyii Turner (Naturalist, Feb. 1886, p. 34).
W. Blubberhouses ; *W. B. Turner.* 'May perhaps be deemed a *Penium.*'

79. Mesotænium Braunii De Bary. (*Palmoglœa macrococca* Kütz.)
Frequent in narrow glens on rocks and mosses ; *W. West.*
A. Adel Bog ; *W. B. Turner.*

80. Mesotænium chlamydosporum De Bary.
L. Cautley Spout ; *W. West.*

81. Tetmemorus granulatus Ralfs.
L. Austwick ; *W. West.*
W. Ilkley ; *W. H. Dikes.*
A. Cullingworth. Eldwick ; *W. West.* Roundhay and Adel Bog, near Leeds ; *A.*, *E.*, and *T.*

82. Tetmemorus Brebissonii Ralfs.
W. Blubberhouses ; *W. B. Turner.*
C. Halifax ; *W. West.*

83. Tetmemorus lævis Ralfs.
L. Ingleborough ; *W. West.*
W. Blubberhouses ; *W. B. Turner.* Ilkley ; *W. H. Dikes.*
A. Cullingworth ; *W. West.*

84. Spirotænia condensata Bréb.
W. Ilkley ; *W. H. Dikes.*
A. Adel Bog ; *W. B. Turner.* Cullingworth ; *W. West.*

85. Micrasterias rotata Ralfs.
Common ; *W. B. Turner.* In A. and W. districts, and probably in the others ; *W. West.*

86. Micrasterias denticulata Bréb.
Common, more so than last. Probably generally distributed where there is peat.
A. A verrucose form at Cullingworth and Eldwick ; *W. West.* Adel ; *W. B. Turner.*

87. Micrasterias fimbriata Ralfs.
A. Cullingworth ; Eldwick ; *W. West.* Adel Dam ; *W. B. Turner.*

88. Micrasterias Americana (Ehrenb.) Ralfs.
W. Askwith Moor ; *W. B. Russell* and *R. Hebblethwaite.*

89. Micrasterias truncata Corda.
L. Ingleborough ; *W. West.*
A. Cullingworth, Eldwick, etc. ; *W. West.*

VAR. **crenata** Bréb. (species).
W. Ilkley ; *W. H. Dikes.*
A. Cullingworth ; Eldwick ; *W. West.*

90. Micrasterias papillifera Bréb.
A. Adel Dam ; *W. B. Turner.*

91. Euastrum verrucosum Ehrenb.
W. Bramhope ; *W. B. Turner.* Ilkley ; *W. H. Dikes.*

92. Euastrum oblongum Ehrenb.
Common ; *W. B. Turner.*
A. Cullingworth, Adel, etc. ; *W. West.*

93. Euastrum crassum (Bréb.) Kütz.
A. Cullingworth ; *W. West.*

94. Euastrum ampullaceum Ralfs.
W. Ilkley ; *W. H. Dikes.*
A. Cullingworth ; *W. West.* Adel Dam ; *W. B. Turner.*

95. Euastrum insigne Hass.
W. Ilkley ; *W. H. Dikes.*

96. Euastrum didelta Turp.
W. Ilkley ; *W. H. Dikes.*
A. Cullingworth ; Eldwick ; *W. West.* Adel ; *W. B. Turner.*

97. Euastrum pectinatum Bréb.
A. Cullingworth ; *W. West.*

98. Euastrum ansatum Ehrenb.
A. Cullingworth ; Eldwick ; Baildon ; *W. West.*

99. Euastrum binale Ralfs.
L. Ingleborough ; *W. West.*
W. Blubberhouses ; *W. B. Turner.*
A. Adel ; *W. B. Turner.* Cullingworth ; Eldwick ; *W. West.*

100. Euastrum elegans Ehrenb.
A. Near Selby ; Cullingworth ; Eldwick ; *W. West.* Adel ; *W. B. Turner.*

101. Cosmarium plicatum Reinsch.
W. Blubberhouses ; *W. B. Turner.*

102. Cosmarium pyramidatum Bréb.
W. Bramhope ; *W. B. Turner.*
A. Cullingworth ; *W. West.*

103. Cosmarium quadratum Ralfs.
A. Eldwick ; *W. West.*

104. Cosmarium bioculatum Bréb.
A. Rawcliffe ; *H. F. Parsons.* Cullingworth ; *W. West.*

105. Cosmarium tinctum Ralfs.
A. Cullingworth ; *W. West.*

106. Cosmarium Meneghinii Bréb.
W. Blubberhouses ; *W. B. Turner.*
A. Cullingworth ; *W. West.*

107. Cosmarium crenatum Ralfs.
L. Cautley Spout ; *W. West.*
W. Arncliffe ; *W. West.*
A. Rawcliffe ; *H. F. Parsons.* Baildon, Cullingworth, etc. ; *W. West.*

108. Cosmarium Holmiense Lund.
A. Baildon ; *W. West.*

109. Cosmarium undulatum Cor.
W. Ilkley ; *W. H. Dikes.*
A. Cullingworth ; *W. West.*

110. Cosmarium tetraophthalmum Kütz.
W. Bramhope ; *W. B. Turner* and *J. Abbott.*
A. Eldwick ; *W. West.*

111. Cosmarium Brebissonii Meneg.
W. Blubberhouses ; *W. B. Turner.*
A. Cullingworth ; *W. West.*

112. Cosmarium margaritiferum Meneg.
Common.

113. Cosmarium botrytis Bory.
Y. Markington ; *Rev. J. S. Tute.*
A. Kildwick ; Cullingworth ; Frizinghall ; *W. West.*

114. Cosmarium biretum Bréb.
A. Cullingworth ; *W. West.*

115. Cosmarium Broomei Thur.
A. Cullingworth ; *W. West.*

116. Cosmarium ornatum Ralfs.
W. Blubberhouses ; *W. B. Turner.*

117. Cosmarium orbiculatum Ralfs.
A. Cullingworth ; *W. West.*

118. Cosmarium moniliforme (Turp.) Ralfs.
W. Ilkley ; *W. H. Dikes.*

119. Calocylindrus connatus Kirchn.
W. Ilkley ; *W. H. Dikes.*

120. Calocylindrus Cucurbita De Bary.
L. Ingleborough ; *W. West.*
A. Cullingworth ; *W. West.*

121. Xanthidium armatum Bréb.
W. Ilkley ; *W. H. Dikes.*
A. Adel ; *W. B. Turner.* Cullingworth ; *W. West.*

122. Xanthidium aculeatum Ehrenb.
W. Ilkley ; *W. H. Dikes.*

123. Arthrodesmus octocornis Ehrenb.
W. Bramhope ; *W. B. Turner.* Ilkley ; *W. H. Dikes.*

124. Arthrodesmus convergens Ehrenb.
W. Bramhope ; *J. Abbott.* Ilkley ; *W. H. Dikes.*

125. Arthrodesmus incus Bréb.
A. Adel Dam ; *W. B. Turner.*

126. Staurastrum dejectum Ralfs.
W. Bramhope ; *J. Abbott.*

VAR. **apiculatum** Bréb. (sp.)
A. Cullingworth ; *W. West.*

127. Staurastrum Dickiei Ralfs.
A. Cullingworth ; *W. West.*

128. Staurastrum cuspidatum Bréb.
W. Ilkley; *W. H. Dikes.*
A. Cullingworth; *W. West.*

129. Staurastrum furcigerum Ralfs.
W. Bramhope; *J. Abbott.*

130. Staurastrum polymorphum Bréb.
L. Ingleborough; *W. West.*

131. Staurastrum gracile Ralfs.
W. Bramhope; *J. Abbott.*
A. Rawcliffe; *H. F. Parsons.*
C. Wakefield; *W. H. Dikes.*

132. Staurastrum paradoxum Mey.
W. Blubberhouses; *W. B. Turner.*

133. Staurastrum alternans Bréb.
W. Blubberhouses; *W. B. Turner.*

134. Staurastrum teliferum Ralfs.
A. Adel Dam; *W. B. Turner.* Cullingworth; *W. West.*

135. Staurastrum muticum Bréb.
W. Ilkley; *W. H. Dikes.*

136. Staurastrum orbiculare Ralfs.
A. Cullingworth; *W. West.*

137. Staurastrum punctulatum Bréb.
A. Cullingworth, Kildwick, etc.; *W. West.*

138. Staurastrum dilatatum Ehrenb.
A. Cullingworth; *W. West.*

139. Staurastrum hirsutum Bréb.
A. Cullingworth; *W. West.*

140. Staurastrum orthostichum Lund.
A. Cullingworth; *W. West.*

141. Staurastrum pilosum Näg.
A. Cullingworth; *W. West.* Adel Dam; *W. B. Turner.*

142. Staurastrum Meriani Reinsch.
W. Blubberhouses; *W. B. Turner.*
A. Baildon; *W. West.*

143. Staurastrum cosmarioides Nordst.
W. Blubberhouses; *W. B. Turner.*

144. Staurastrum tumidum Bréb.
W. Ilkley; *W. H. Dikes.*

V. ZYGNEMACEÆ.

145. Zygnema pectinatum Ag.
C. Near Heckmondwike; *C. B. Crawshaw* and *C. P. Hobkirk.*

146. Zygnema Ralfsii Kütz.
W. Bramhope; *A., E.,* and *T.*
A. Adel; *A., E.,* and *T.*

147. Zygnema cruciatum Vauch.
L. Sedbergh; *W. West.*
W. Bramhope; *A., E.,* and *T.*
A. Baildon; Cullingworth; *W. West.*
C. Wibsey; *W. West.*

148. Zygnema anomalum Hass.
C. Sharleston; *H. F. Parsons.*

149. Zygnema pellucidum Hass. (*Spirogyra pellucida* Dillw.)
N. Harrogate; *W. B. Turner.*
A. Hunslet; *J. Abbott.*

150. Spirogyra crassa Kütz.
L. Sedbergh; *W. West.*
C. Near Heckmondwike; *C. B. Crawshaw* and *C. P. Hobkirk.*
D. Doncaster; *W. West.*

151. Spirogyra insignis Hass.
L. Near Sedbergh; *W. West.*
A. Near Bradford; and Wilsden; *W. West.*
C. Near Mirfield; *W. West.*

152. Spirogyra nitida Dillwyn.
N. Harrogate; *T. Hick.*
W. Bramhope; *J. Abbott.*
A. Adel Bog; *J. Abbott.* Common about Bradford *W. West.* Malham; and Wilsden; *W. West.*
C. Howley near Dewsbury; *C. B. Crawshaw* and *C. P. Hobkirk.*

153. Spirogyra condensata Vauch.
A. Baildon; *W. West.*
C. Howley near Dewsbury; *C. B. Crawshaw* and *C. P. Hobkirk.*

154. Spirogyra flavescens (Hass.) Cleve.
A. Cullingworth, etc.; *W. West.*
C. Soothill, Dewsbury; *C. B. Crawshaw.*

Vars. **gracilis** and **parva.**
D. In dam at Kexborough; *C. B. Crawshaw* and *C. P. Hobkirk.*

155. Spirogyra Grevilleana Hass.
A. Keighley, etc.; *W. West.*
C. Dewsbury; *C. B. Crawshaw.*

156. Spirogyra longata Vauch.
A. Baildon, etc.; *W. West.* Adel; *A., E.,* and *T.*

Var. **communis** Dillw.
A. Adel Bog; *W. B. Turner.* Frequent about Bradford; *W. West.*

157. Spirogyra porticalis Vauch.
A. Baildon, etc.; *W. West.* Roundhay; *A., E.,* and *T.*

Var. **quinina** Ag.
A. Frequent about Bradford; *W. West.*
C. Sharleston; *H. F. Parsons.*
D. Upper Bradfield; *R. A. Gatty.*

158. Spirogyra Weberi Kütz.
A. Cullingworth; *W. West.* Hainworth; *T. Hebden.*

159. Sirogonium sticticum Kütz.
A. 'Streamy' well-aerated pools, Baildon; Keighley; Hawksworth; *W. West.*

160. Zygogonium ericetorum De Bary, var. **aquaticum.**
Common on wet moors.
L. Ingleborough; *W. West.*
A. Baildon; Harden; *W. West.* Malham; *C. B. Crawshaw* and *C. P. Hobkirk.*
D. Goole; *W. West.*

Var. **terrestris.**
Common on damp heaths and on the ground in hilly woods.
W. Bramhope; *A., E.,* and *T.* Rombalds Moor; *W. West.*
A. Baildon; Thornton; *W. West.*

161. Mesocarpus nummuloides Hass.
A. Cullingworth; *W. West.*

162. Mesocarpus parvulus Hass.
W. Askwith; *W. B. Turner.*
A. Cullingworth; *W. West.* Newsholme Dean; *T. Hebden.*

163. Mesocarpus pleurocarpus De Bary.
W. Askham Bog; *T. Hick* and *W. West.*
A. Baildon; Rawcliffe near Selby, etc.; *W. West.*

164. Mesocarpus scalaris Hass.
A. Roundhay; *A., E.,* and *T.*
C. Dewsbury; *C. B. Crawshaw* and *C. P. Hobkirk.*

165. Staurospermum viride Kütz.
A. Horsforth; *J. Abbott.* Cullingworth; *T. Hebden* and *W. West.*
D. Near Rotherham; *W. West.*

166. Staurospermum gracillimum (Hass.).
L. Ingleborough; *W. West.*
W. Blubberhouses; *W. B. Turner.*

Order III.—SIPHOPHYCEÆ.

VI. BOTRYDIACEÆ.

167. Botrydium granulatum (L.) Greville.
A. On drying-up mud, Adel Dam, after it was drained; *James Abbott.*

VII. VAUCHERIACEÆ.

168. Vaucheria dichotoma Lyngb. var. **submarina** Ag.
D. Goole; *W. West.*

169. Vaucheria terrestris Lyngb.
A. Common about Bradford; *W. West.* Keighley; *T. Hebden.*
C. 'On damp earth, Grimescar'; '*Huddersfield: its Hist. and Nat. Hist.*'

170. Vaucheria sessilis Vauch.
A. Baildon, etc.; *W. West.* Adel Bog; *A., E.,* and *T.*

Var. **cæspitosa** Vauch.
L. Dent-in-Dentdale; *H. F. Parsons* and *W. West.*
Y. Markington near Ripley; *J. S. Tute.*
W. Arncliffe; *W. West.*

171. Vaucheria geminata Vauch.
N. Harrogate; *A., E.,* and *T.*
A. Old gardens, Headingley, near Leeds; *J. Abbott.* Common in Bradford district; *W. West.*

Var. **racemosa** Eng. Bot.
L. Ingleborough; *W. West.*
A. Meanwood; *J. Abbott.*

ORDER IV.—NEMATOPHYCEÆ.

VIII. ULVACEÆ.

172. Prasiola crispa Kütz.
Common; *W. West.* This urinophilous species can be found in any country-lane where there is a convenient corner for micturition.
L. Sedbergh; *W. West.*
W. Wetherby; *F. A. Lees.* Rombalds Moor; *H. F. Parsons.*
A. In Leeds town; *A., E.*, and *T.* Woodhouse Ridge; *J. Abbott.*
C. Copley; *H. F. Parsons.* Huddersfield; *C. P. Hobkirk.*
D. Upper Bradfield; *R. A. Gatty.*

173. Enteromorpha intestinalis Link. (*Solenia* Ag.)
W. Sewage drain, Sherburn!
A. Garforth near Leeds; *W. B. Turner.*
D. Goole; *H. F. Parsons.* Bradfield; *Rev. R. A. Gatty.*
T. In the Torne river or dike, Rossington!

IX. CONFERVACEÆ.

174. Microspora fugacissima Ag.
A. Frequent about Bradford; *W. West.* Leeds; *A., E.*, and *T.*
C. Dewsbury; *C. B. Crawshaw* and *C. P. Hobkirk.*

175. Microspora vulgaris Rabh.
A. Frequent and general; *W. West*; *A., E.*, and *T.*; etc.

176. Microspora floccosa Ag.
L. Sedbergh; *W. West.*
A. Frequent about Bradford; *W. West.* Chapel-Allerton; *A., E.*, and *T.*

177. Conferva tenerrima Kütz.
L. Ingleborough; *W. West.*
A. Very common about Bradford; *W. West.* Also about Adel and Leeds; *A., E.*, and *T.*

178. Conferva bombycina Ag.
A. Common about Bradford; *W. West.* Pontefract; *H. F. Parsons.*
C. Norland Moor; *H. F. Parsons.*
D. Goole; *H. F. Parsons.*

179. Conferva fontinalis Berk.
W. Blubberhouses; *W. B. Turner.*

180. Conferva Funkii Kütz.
A. Frizinghall; *W. West.*
Obs.—This is not in Cooke's Algæ.

181. Cladophora crispata Roth.
A. 'Common' about Leeds; *W. B. Turner.* Frequent about Bradford; *W. West.*
D. Bradfield; *Rev. R. A. Gatty.*

182. Cladophora glomerata L. (Dillw.).
W. Ilkley; *H. F. Parsons.* Boston Spa and Wetherby; *F. A. Lees.*
A. Common in every stream, river, canal, etc.; *W. West.* Moor-Allerton near Leeds; *W. B. Turner* and *J. Abbott.* Shipley Glen; Selby; *H. F. Parsons.* Sherburn, dyke; *F. A. Lees.* Kirkby Malham; *C. B. Crawshaw* and *C. P. Hobkirk.*
C. Coxley Valley, Horbury; *C. B. Crawshaw* and *C. P. Hobkirk.*
D. Upper Bradfield; *Rev. R. A. Gatty.* Conisborough; *H. F. Parsons.*

X. ŒDOGONIACEÆ.

183. Œdogonium cryptoporum Wittr.
A. Bradford district; *W. West.* This is var. β *vulgare* Wittr.

184. Œdogonium flavescens Wittr.
A. Bradford district; *W. West.*

185. Œdogonium Braunii Kütz.
A. Adel Bog; *A., E.*, and *T.* Baildon; *W. West.*

186. Œdogonium ciliatum Hass.
W. Bramhope; *W. B. Turner.*

187. Œdogonium tumidulum Kütz.
W. Bramhope; *J. Abbott.*
A. Adel; *A., E.*, and *T.*

188. Œdogonium Landsboroughii Hass.
C. Near Dewsbury; *W. West.*

189. Œdogonium rivulare (Le Clerc).
A. Baildon; *W. West.*
C. Dewsbury district; *C. B. Crawshaw* and *C. P. Hobkirk.*

190. Œdogonium princeps Wittr.
L. Near Sedbergh; *W. West.*

191. Œdogonium vesicatum (Lyngb.).
W. Bramhope; *J. Abbott.*

192. Œdogonium subsetaceum Kütz.
L. Near Sedbergh; *W. West.*

193. Bulbochœte setigera Ag.
A. Adel Bog, near Leeds; *J. Abbott.* Cullingworth; *W. West.*

XI. ULOTRICHEÆ.

194. Hormiscia moniliformis Kütz.
A. Cullingworth; *W. West.*

195. Hormiscia zonata W. & M.
A. Malham, Frizinghall, etc.; *W. West.* Gordale; *C. B. Crawshaw* and *C. P. Hobkirk.*

196. Hormiscia æqualis (Kütz.) VAR. **catæniformis** Kütz.
A. Malham Cove; *C. B. Crawshaw* and *C. P. Hobkirk.*

197. Hormiscia bicolor (Eng. Bot., I, t. 2288).
L. Near Sedbergh; *W. West.*

198. Ulothrix variabilis Kütz.
A. Malham; *C. B. Crawshaw* and *C. P. Hobkirk.*

199. Ulothrix tenerrima Kütz.
A. Baildon; *W. West.* Kirkby Malham; *C. B. Crawshaw* and *C. P. Hobkirk.*
D. Barnsley; *W. West.*

200. Ulothrix tenuis Kütz.
A. Marley; Eldwick; Frizinghall; *W. West.*

201. Ulothrix radicans Kütz.
A. Bradford, etc.; *W. West.* Roundhay; *A., E.*, and *T.* Gordale; *C. B. Crawshaw* and *C. P. Hobkirk.*

202. Ulothrix parietina (Vauch.) Kütz.
A. Bradford, etc.; *W. West.*
C. Dewsbury; *C. B. Crawshaw* and *C. P. Hobkirk.*

203. Schizogonium murale Kütz.
A. Bradford, etc.; *W. West.* Garforth; *A., E.*, and *T.*

XII. CHROOLEPIDEÆ.

204. Chroolepus aureus (L.) Kütz.
First record: Dr. Richardson in Ray, Syn., iii (1724), p. 57.
L. Dentdale! Ingleton; *W. West.*
R. Upper Ribblesdale, on lime-rock; *F. A. Lees.*
N. & W. Plentiful on the faces of the limestone rocks in the quarries about Kirk Deighton, Wetherby, Tadcaster, Knaresborough, and Sherburn; *F. A. Lees.*
W. Arncliffe, Bolton Woods, Ilkley, etc.; *W. West.*
A. Bingley; *W. West.* Snaith; *H. F. Parsons.*
D. Conisborough; *H. F. Parsons.*
T. Sheffield (South Anston); *A. Carr.* Roche Abbey!

XIII. CHÆTOPHORACEÆ.

205. Microthamnion vexator Cooke.
W. Blubberhouses; *W. B. Turner.*
A. Bingley, etc.; *W. West.*

206. Stigeoclonium protensum (Dillw.) Kütz.
W. Ilkley; *W. West.* Farnley; *W. B. Turner.* Burley; *W. Nuttall.*
A. Keighley, etc.; *W. West.*
C. Wibsey; *W. West.*
D. Well. Darton; *C. B. Crawshaw* and *C. P. Hobkirk.*

207. Stigeoclonium nanum (Dillw.) Kütz.
A. Near Bradford; *W. West.*
C. Near Dewsbury; *C. B. Crawshaw* and *C. P. Hobkirk.*

208. Stigeoclonium fastigiatum (Ralfs) Kütz.
C. Near Wakefield; *W. West.* Soothill, Dewsbury; *C. B. Crawshaw* and *C. P. Hobkirk.*

209. Draparnaldia glomerata Ag.
W. Ilkley; *W. West.*
A. Roundhay near Leeds; *W. B. Turner.* Eccleshill; *W. Nuttall.* Bingley, Keighley, etc.; *W. West.*
C. Dewsbury; *C. B. Crawshaw.*
Obs.—This is much the commoner species.

210. Draparnaldia plumosa (Vauch.) Ag.
W. Wetherby; *H. F. Parsons*!
A. Adel Bog; *A., E.*, and *T.* Clayton; *W. West.*
D. Upper Bradfield; *Rev. R. A. Gatty.*

211. Chætophora pisiformis Ag.
A. Near Selby; *W. West.*

212. Chætophora tuberculosa Ag.
W. York; *Thomas Hick.* Askham Bog; *W. West.*
A. Airton and Baildon; *W. West.* Keighley; *T. Hebden.*

213. Chætophora elegans Ag.
A. Adel, and Roundhay near Leeds; *W. B. Turner.* Selby; *H. F. Parsons.* Baildon, etc.; *W. West.*
C. Dewsbury; *C. B. Crawshaw* and *C. P. Hobkirk.*

214. Chætophora endivæfolia Ag.
R. Horton-in-Ribblesdale; *W. West.*
A. Bingley; Saltaire; *W. West.*

215. Coleochæte scutata Bréb.
A. Common about Bradford; *W. West.* Adel Bog; *James Abbott.*

CLASS II.—PHYCOCHROMOPHYCEÆ.

ORDER I.—CYSTIPHORÆ.

XIV. CHROOCOCCACEÆ.

216. Chroococcus turgidus Näg.
L. Cautley Spout; *W. West.*
A. Cullingworth; Steeton, etc.; *W. West.* Holden Gill; *T. Hebden.*

217. Glæocapsa polydermatica Kütz.
L. Cautley Spout; *W. West.*
N. Harrogate; *T. Hick* and *C. P. Hobkirk.*
W. Otley; *H. F. Parsons.*
A. Bingley; Baildon; Keighley; *W. West.*

218. Glæocapsa caldariorum Rabenh.
A. Greenhouse, Cullingworth; *T. Hebden.* Manningham Park greenhouses; *W. West.*

219. Aphanocapsa testacea Näg.
R. Settle; *H. F. Parsons.* 'Probably but a condition of *Collema.*' Not in Cooke's 'Algæ.'

220. Merismopedia glauca Näg.
A. Cullingworth; *T. Hebden* and *W. West.*

221. Tetrapedia setigera Archer.
A. Cullingworth; *T. Hebden.*

ORDER II.—NEMATOGENÆ.

XV. NOSTOCEÆ.

222. Nostoc muscorum Ag.
L. Near Ingleton; *W. West.*

223. Nostoc commune Vauch.
In water and damp places, frequent.
Y. Ripon; *F. A. Lees.*
N. Knaresborough; *W. B. Turner.*
W. Wetherby; *F. A. Lees.* Arncliffe; *W. West.*
A. Sherburn! Near Jennetts Cave, Malham; *C. P. Hobkirk.*
D. Wentbridge; *H. F. Parsons.*

224. Nostoc sphæricum Vauch.
L. Ingleton; *A., E.,* and *T.,* and *W. West.*
A. Near Bradford; *W. West.*

225. Nostoc rupestre Kütz.
L. Cautley Spout; Ingleborough; Clapham; *W. West.*
A. Malham, etc.; *W. West.*

226. Nostoc cæruleum Lyngb.
A. Baildon, etc.; *W. West.*

227. Nostoc pruniforme Ag.
W. Beamsley Beacon; *W. West.*

228. Nostoc verrucosum Vauch.
A. On rocks, Malham Cove; *H. F. Parsons.*

229. Cylindrospermum macrospermum Kütz.
A. Shady places in the gardens of Bradford; *W. West* and *J. E. Wilson.* Not in water.

230. Cylindrospermum catenatum Ralfs.
D. Goole; *W. G. Tacey.*

XVI. LYNGBYÆ.

231. Spirulina Jenneri Kütz.
D. Askern; *Rev. F. H. Allen.*

232. Spirulina oscillarioides Turp.
W. Bramhope; *A., E.,* and *T.*

233. Oscillaria tenerrima Kütz.
A. Bradford; *W. West.*

234. Oscillaria leptotricha Kütz.
A. Calverley; *W. West.*

235. Oscillaria subfusca Vauch.
A. Baildon; *W. West.*

236. Oscillaria tenuis Ag. (*C. contexta* Hass.)
L. Near Sedbergh; *W. West.*
A. In still water in spring, of a pale green colour. Potternewton near Leeds; *W. B. Turner.* Brick-ponds, Sherburn; *F. A. Lees.* Baildon; *W. West.*
C. Wakefield; *Rev. W. Fowler.*

237. Oscillaria muscorum Carm. MSS.
A. Allerton near Bradford; *W. West.*
C. Hebden Bridge; *W. West.*

238. Oscillaria limosa Ag.
On mud, pool-bottoms.
W. Quarry at Linton Springs near Wetherby; *F. A. Lees.*
A. Baildon; *W. West.*

239. Oscillaria irrigua Kütz.
L. Near Sedbergh; *W. West.*
W. Addingham; *W. West.*
A. Keighley; *W. West.*

240. Oscillaria nigra Vauch.
Very common in still water.
W. Wetherby; *F. A. Lees.*
A. Potternewton, Leeds; *W. B. Turner.* Bradford, etc.; *W. West.*
D. Goole; *H. F. Parsons.* Endcliffe, Sheffield; *Amos Carr.*

241. Oscillaria Frœlichii Kütz.
A. Common about Bradford; *W. West.*
C. Dewsbury; *C. B. Crawshaw* and *C. P. Hobkirk.*

242. Lyngbya ochracea Thur.
N. Harrogate; *W. West.*
W. Blubberhouses; *W. B. Turner.*
A. Pools, Hunslet; *W. B. Turner.* Keighley, etc.; *W. West.*

243. Lyngbya inundata Kütz. (*Oscillaria autumnalis* Carm. [partly].)
Common enough everywhere on damp earth, rocks, dripping walls, etc.
N. Harrogate; *A., E.,* and *T.*
W. Bramhope; *A., E.,* and *T.* Wetherby; *F. A. Lees.*
A. Chapel-Allerton; *A., E.,* and *T.* Bradford; *W. West.* Selby; *W. G. Tacey.*

244. Lyngbya cincinnata Kütz.
A. Potternewton; *A., E.,* and *T.*
Obs.—This is not recorded in Cooke's Algæ.

245. Lyngbya rupestris (Ag.). (*Oscillaria* Ag.)
L. Ingleton; *H. F. Parsons.*
W. Blubberhouses; *W. B. Turner.*

246. Plectonema mirabile Thur.
L. Ingleton; *W. West.*
A. Malham; *W. West.*

XVII. SCYTONEMEÆ.

247. Scytonema myochrous Ag.
L. Dripping limestone cliffs, Ingleton; *A., E., T.,* and *W. West.* Near Sedbergh; *W. West.*
W. Arncliffe; *W. West.*

248. Petalonema alatum Ehrenb.
A. Vertical cliffs of calcareous rock (dripping), Gordale; *W. West.*

249. Tolypothrix distorta (Müll.) Kütz.
A. Bradford; *W. West.*

250. Tolypothrix ægagropila Kütz. VAR. **muscicola** Kütz.
W. Dringhouses; *W. West.*

251. Stigonema ocellatum Thur.
L. Near Sedbergh; *W. West.*

252. Stigonema saxicolum Näg.
L. Ingleton; *W. West*; *T. Hebden.*

XVIII. CALOTRICHEÆ.

253. Rivularia calcarea Eng. Bot.
Stones in streams of limestone districts.
L. Ingleton; Austwick, etc.; *W. West.*
W. Arncliffe; *W. West.*
A. Malham Tarn; *H. F. Parsons*; *C. B. Crawshaw* and *C. P. Hobkirk.* Gordale; *W. West.*

254. Rivularia crustacea Carm.
L. Austwick; *W. West.*
A. Gordale, abundant; *W. West.*

255. Rivularia atra Agardh.
A. Roundhay; *W. B. Turner.*

256. Gloiotrichia natans Thur. (*Rivularia angulosa* Ag. *R. pruniformis* Purt.) '*Plum Laver.*'
W. On moss in ditches, Wetherby; *F. A. Lees.*
A. In Malham Tarn; *Gough's Camden* (1789).

CLASS III. RHODOPHYCEÆ.

XIX. CHANTRANSIACEÆ.

257. Chantransia violacea Kütz.
L. Ingleborough; *W. West.*
A. Eldwick; *W. West.*

258. Chantransia chalybea Lyngb.
A. Baildon; *W. West.* Hainworth; *T. Hebden.*
C. Near Halifax; *W. West.*

259. Chantransia pygmæa Kütz.
L. Ingleton; *W. West.*

XX. BATRACHOSPERMEÆ.

260. Batrachospermum moniliforme Roth.
W. Stream near Towton! Washburn Valley! Blubberhouses; *E. Howgate.*
A. Allerton, Baildon, etc.; *W. West.* Malham; *H. F. Parsons.* Leeds district; *W. B. Turner.* Meanwood; *J. Abbott.* Sherburn town drain; *F. A. Lees.* Malham Tarn; *C. B. Crawshaw* and *C. P. Hobkirk.*
C. Battye-ford near Mirfield; *H. F. Parsons.*
D. Upper Bradfield; *Rev. R. A. Gatty.*
T. Swinefleet warping-drain; *H. F. Parsons.*

Var. **stagnale** Ag.
A. Malham Cove; *C. B. Crawshaw* and *C. P. Hobkirk.*

261. Batrachospermum vagum Harv.
A. Kirkstall; *W. B. Turner* (List in *Nat.*, Oct. 1879). Bingley; *W. West.*

262. Batrachospermum atrum Harv.
W. Barkston; *H. F. Parsons.*
A. Meanwood; *J. Stubbins.*

XXI. HILDENBRANDTIACEÆ.

263. Hildenbrandtia rivularis (Liebm.) J. Ag. (*H. rosea*, var. β Kütz.)
T. On pebbles and stones (looking like gouts of blood) in the stream flowing through Anston Crags. First found, June 11th, 1887; *Hild. Friend.* The record in *Wesley Naturalist*, August 1886, was the first for Notts.; but not for England, since Kützing in his 'Species Algarum' (1849) records it as British; and a station for it is mentioned by Carter in *Journ. Bot.*, vol. i, p. 227 (1864). Cooke in his recent 'Brit. Fresh-water Algæ' does not mention it.

XXII. LEMANEACEÆ.

264. Lemanea fluviatilis Agardh.
On stones in rivers.
L. Ingleborough; Cautley Spout; *W. West.*
Y. Ripon; *W. Brunton.*
N. Near Harrogate; *T. Hick.* In the river Crimple, Spofforth!
W. Wharfe, Boston Spa, and Wetherby; *F. A. Lees.*
A. Kildwick; *W. West.* Shipley Glen; *H. F. Parsons* and *W. West.*
D. Upper Bradfield near Sheffield; *Rev. R. A. Gatty.*

XXIII. DELESSERIACEÆ.

265. Plocamium coccineum Lyngb.
D. Adventive, not native. A marine alga to be occasionally found in the river Ouse at Goole, washed up at high tides; *H. F. Parsons.*

XXIV. DIATOMACEÆ.

266. Diatoma vulgare Bory.
N. Harrogate, Ripley, and Knaresborough; *G. Norman.*
A. Adel, Headingley, and Roundhay; *W. B. Turner.* Hawksworth; *H. F. Parsons.* Extremely common in canals and rivers near Bradford, attached to *Cladophora glomerata*; *W. West.*
D. Goole; *E. Hunter.*

267. Diatoma elongatum Ag.
N. Ripley; *G. Norman.*
W. Blubberhouses; *W. B. Turner.*
A. Roundhay; *W. B. Turner.* Shipley; *W. West.*
C. Fixby; *W. West.*
D. Goole; *W. West.*

268. Diatoma Ehrenbergii Kütz.
Y. Markington; *Rev. J. S. Tute.*

269. Fragilaria virescens Ralfs.
A. Hawksworth Springs; *H. F. Parsons.* Common in Bradford district; *W. West.*

270. Fragilaria capucina Ralfs.
A. Chapeltown and Roundhay; *W. B. Turner.* Cullingworth, etc.; *W. West.*
D. Doncaster; *W. West.*

271. Meridion circulare Ag.
Common away from the lowlands.
A. Horsforth and Adel; *W. B. Turner.* Allerton and Esholt; *W. West.* Gordale Cove; *H. F. Parsons.*
C. Dewsbury; *C. B. Crawshaw* and *C. P. Hobkirk.*
D. Doncaster; *Rev. F. H. Allen.*

272. Himantidium pectinale Kütz.
W. Wetherby; *F. A. Lees.*
A. Bingley; Baildon; *W. West.*

273. Himantidium undulatum Sm.
W. Blubberhouses; *W. B. Turner.*
A. Baildon; *W. West.* Hawksworth; *E. Hunter.*

274. Himantidium gracile Ehrenb.
W. Blubberhouses, attenuate form; *W. B. Turner.*

275. Odontidium hyemale Kütz.
W. Blubberhouses; *W. B. Turner.*
A. Wilsden; *W. West.*

Forma **minima.**
W. Blubberhouses; *W. B. Turner.*
A. Roundhay; *W. B. Turner.* Shipley Glen; *W. West.*

276. Tabellaria ventricosa Kütz.
A. Baildon; and dam at Adel; *W. West.*

277. Tabellaria flocculosa Kütz.
L. Near Sedbergh; *W. West.*
W. Blubberhouses; *W. B. Turner.*
A. Horsforth; *W. B. Turner.* Common about Bradford; *W. West.*

278. Tabellaria fenestrata Kütz.
W. Blubberhouses; *W. B. Turner.*

279. Tetracyclus lacustris Kütz.
A. Roundhay; *W. B. Turner.*

280. Epithemia turgida Sm.
W. Bramhope; *J. Abbott.*
A. Baildon, Malham, Cullingworth, etc.; *W. West.*

281. Eunotia triodon Ehrenb.
A. Roundhay; *W. B. Turner.*

282. Ceratoneis Arcus Kütz. (*Eunotia* Sm.)
L. Calf Beck, Howgill (1879), and Cautley Spout (1887); *W. West.*

283. Cymbella ventricosa Ag.
L. Ingleton; *H. F. Parsons.*
A. Baildon; *W. West.* Roundhay; *W. B. Turner.*

284. Cymbella gastroides Kütz.
A. Baildon; *W. West.*
D. Goole; *E. Hunter.*
T. Roche Abbey Lake; *E. Hunter.*

285. Cymbella truncata Rabenh.
A. Esholt; *W. West.*

286. Cymbella cuspidata Kütz.
N. Harrogate; *G. Norman.*

287. Synedra ulna Ehrenb.
L. Near Cautley Spout; *W. West.*
Y. Markington; *Rev. J. S. Tute.*
N. Ripley and Harrogate; *G. Norman.*
W. Blubberhouses; *W. B. Turner.*
A. Baildon and Bingley; *W. West.*

288. Synedra longissima Sm.
A. Wilsden; *W. West.*

289. Synedra capitata Ehrenb.
A. Shepherd's Pond, Hunslet; *J. Abbott.* South Milford; *W. B. Turner.*
D. Goole; *E. Hunter.*
T. Roche Abbey; *E. Hunter.*

290. Synedra splendens Ehrenb.
A. Baildon and Bingley; *W. West.* Roundhay and South Milford; *W. B. Turner.*
D. Askern; *Rev. F. H. Allen.*

291. Synedra lanceolata Ehrenb.
A. Roundhay; *W. B. Turner.*

292. Pinnularia mesolepta Sm.
N. Harrogate; *G. Norman.*

293. Pinnularia interrupta Sm.
N. Harrogate; *G. Norman.*

294. Pinnularia major Ralfs.
N. Harrogate; *G. Norman.*
A. Baildon; *W. West.*

295. Pinnularia oblonga (Kütz.) Rabenh.
A. Baildon and Bingley; *W. West.*

296. Pinnularia cardinalis Ehrenb.
A. Baildon; *W. West.*

297. Pinnularia lata W. Sm.
W. Blubberhouses; *W. B. Turner.*

298. Pinnularia minor Ralfs.
A. Roundhay; *A.*, *E.*, and *T.*

299. Pinnularia viridis (Rabenh.) Sm.
N. Harrogate; *G. Norman.*
A. South Milford; *W. B. Turner.* Frequent about Bradford; *W. West.*
T. Roche Abbey; *E. Hunter.*

300. Pinnularia radiosa Rabenh.
N. Harrogate; *G. Norman.*
A. South Milford; *W. B. Turner.*
T. Roche Abbey; *E. Hunter.*

301. Pinnularia viridula Sm.
N. Killinghall; *G. Norman.*

302. Cyclotella operculata Kütz.
N. Ripley; *G. Norman.*
A. Roundhay; *J. Abbott.*
D. Goole; *E. Hunter.*

303. Melosira varians (Ag.) Greg.
Common in all the troughs in the millstone grit districts; *W. West.*
N. Harrogate; Ripley; *G. Norman.*
A. Adel Bog and Roundhay; *W. B. Turner.*
T. Roche Abbey; *E. Hunter.*

304. Melosira nummuloides Sm.
D. Goole; *E. Hunter.*

305. Melosira arenaria Moore.
A. Baildon, Kildwick, Gordale, etc.; *W. West.*

306. Melosira orichalcea Kütz.
T. Very rare in a boggy puddle, Thorne Moor; *G. Norman.*

307. Campylodiscus spiralis (Kütz.) Sm.
A. Baildon, Hawksworth, and Cullingworth; *W. West.*
T. Roche Abbey; *E. Hunter.*

308. Campylodiscus noricus Ehrenb.
A. Baildon and Cullingworth; *W. West.*

309. Stauroneis phænicenteron Ehrenb.
N. Harrogate; *G. Norman.*
A. South Milford; *W. B. Turner.* Frequent about Bradford; *W. West.*

310. Stauroneis linearis Sm.
N. Killinghall; *Geo. Norman.*

311. Amphora hyalina Kütz.
D. Goole; *E. Hunter.*

312. Amphora ovalis Kütz.
A. South Milford; *W. B. Turner.* Baildon; *W. West.*
T. Roche Abbey; *E. Hunter.*

313. Amphora littoralis Donkin.
T. Roche Abbey; *E. Hunter.*

314. Nitzschia sigmoidea Sm.
N. Harrogate; *G. Norman.*
A. Adel and South Milford; *W. B. Turner.* Baildon and Cullingworth; *W. West.*
D. Goole; *E. Hunter.* Askern; *F. H. Allen.*
T. Roche Abbey; *E. Hunter.*

315. Nitzschia sigma Sm.
D. Goole; *J. Harrison.*

316. Nitzschia tænia Sm.
A. Roundhay; *W. B. Turner.*

317. Nitzschia dubia Sm.
D. Goole; *J. Harrison.*

318. Nitzschia lanceolata Sm.
A. Roundhay; *W. B. Turner.* South Milford and Killingbeck; *W. B. Turner.*
T. Roche Abbey; *E. Hunter.*

319. Nitzschia minutissima Sm.
A. Roundhay; *W. B. Turner.*

320. Nitzschia Amphioxys Sm.
N. Killinghall and Harrogate; *G. Norman.*

321. Nitzschia linearis Sm.
A. Frizinghall; *W. West.*

322. Bacillaria paradoxa Gmel.
A. Goole, in warping-drain; *W. West.*

323. Gomphogramma rupestre Braun.
D. Goole; *J. Harrison.*

324. Gomphonema acuminatum Ehrenb.
A. Baildon; *W. West.* Roundhay; *W. B. Turner.*
D. Goole; *E. Hunter.*
T. Roche Abbey; *E. Hunter.*

325. Gomphonema geminatum Ehrenb.
L. Cautley Spout; *W. West.*
R. Near Ribblehead; *W. West.*
A. Roundhay; *W. B. Turner.*

326. Gomphonema olivaceum Ehrenb.
Very common on muddy stones in running brooks.
L. Near Sedbergh; *W. West.*
W. Kilnsey; *W. West.*
A. Roundhay; *W. B. Turner.* Bradford; *W. West.*

327. Gomphonema constrictum Ehrenb.
L. Sedbergh; *W. West.*
N. Harrogate; *G. Norman.*
C. Woolley; *W. West.*

328. Gomphonema curvatum Kütz.
N. Harrogate; *G. Norman.*
A. Canal at Holbeck; *W. West.*

329. Gomphonema tenellum Kütz.
A. Baildon; *W. West.*

330. Gomphonema capitatum Ehrenb.
A. Frequent about Bradford; *W. West.*

331. Gomphonema catena Ehrenb.
A. Roundhay; *W. B. Turner.*

332. Amphipleura pellucida Bréb.
A. Roundhay; *A.*, *E.*, and *T.*

333. Cymatopleura Solea Sm.
A. Roundhay; *W. B. Turner.* South Milford; *F. Emsley.*
D. Goole; *E. Hunter.* Askern; *F. H. Allen.*
T. Roche Abbey; *E. Hunter.*

VAR. **constricta** Grun.
A. Goole; *W. West.*

334. Cymatopleura elliptica (Bréb.).
A. Roundhay and South Milford; *W. B. Turner.* Horton; Cullingworth; *W. West.*
C. Fixby; *W. West.*
D. Goole; *E. Hunter.*
T. Roche Abbey; *E. Hunter.*

335. Surirella bifrons Turp.
A. Roundhay and South Milford; *W. B. Turner.*

336. Surirella biseriata Bréb.
W. Blubberhouses; *W. B. Turner.*
A. Cullingworth; Baildon; Eldwick; *W. West.*

337. Surirella ovata Turp.
A. Roundhay and South Milford; *W. B. Turner.*

338. Surirella striatula Turp.
D. Goole; *J. Harrison.*

339. Surirella elegans Ehrenb.
Y. Stones in river Skell, Fountains Abbey; *G. Sturt* (1884); *E. Grove.*

340. Surirella panduriformis Sm.
N. Harrogate; *G. Norman.*

341. Surirella Crumena Bréb.
N. Birk Crag, Harrogate; *G. Norman.*

342. Tryblionella gracilis Sm.
D. Goole; *J. Harrison.*

343. Tryblionella acuminata Sm.
D. Goole; *J. Harrison.*

344. Tryblionella minuta Kütz.
T. Roche Abbey; *E. Hunter.*

345. Tryblionella scutellum Sm. (*Cocconeis.*)
A. Roundhay; *W. B. Turner.*
D. Goole; *J. Harrison.*

346. Cocconeis pediculus Sm.
W. Wetherby.
A. Baildon and Esholt, on *Cladophora glomerata*; *W. West.* Roundhay; *W. B. Turner.*

347. Cocconeis placentula Ehrenb.
N. On *Cladophora*, Harrogate; *G. Norman.*

348. Cocconema lanceolatum Sm.
N. Pure, in a stream near Birk Crag, Harrogate; *G. Norman.*
A. Frequent about Bradford; *W. West.* Roundhay; *W. B. Turner.*
D. Askern; *F. H. Allen.*

349. Cocconema cymbiforme Sm.
A. Roundhay; *W. B. Turner.* Frequent about Bradford; *W. West.*

350. Cocconema Bœckii Sm.
T. Roche Abbey; *E. Hunter.*

351. Cocconema parvum Sm.
D. & T. Goole and Roche Abbey; *E. Hunter.*

352. Encyonema prostratum Ralfs.
N. Harrogate; *G. Norman.*
A. Malham Tarn; *H. F. Parsons.*

353. Encyonema cæspitosum Kütz.
L. Cautley Spout; *W. West.*
N. Birk Crag, Harrogate; *G. Norman.*

354. Achnanthidium microcephalum Kütz.
A. Chapel-Allerton; *A.*, *E.*, and *T.*

355. Achnanthidium flexellum Bréb.
A. Adel; *A.*, *E.*, and *T.*

356. Navicula affinis Ehrenb.
D. Goole; *J. Harrison.*

357. Navicula Reinhardti Grun.
Y. In river Skell, Fountains Abbey; *G. Sturt* (1884); *E. Grove.*

358. Navicula trochus Greg.
Y. In Skell, Fountains Abbey; *G. Sturt*; *E. Grove.*

359. Navicula amphigomphus Ehrenb.
N. Harrogate; *G. Norman.*

360. Navicula amphirhynchus Ehrenb.
N. Harrogate; *G. Norman.*
A. Roundhay; *A.*, *E.*, and *T.*
D. & T. Goole and Roche Abbey; *E. Hunter.*

361. Navicula cuspidata Kütz.
N. Puddle near Birk Crag, Harrogate; *G. Norman.*
W. Blubberhouses; *W. B. Turner.*
A. Baildon; *W. West.* Roundhay and South Milford; *W. B. Turner.*
T. Roche Abbey; *E. Hunter.*

362. Navicula rhomboides (Ehrenb.) Greg.
W. Blubberhouses; extends to 3,000 feet alt. in Britain; *W. B. Turner.*
A. South Milford; *W. B. Turner.*
T. Thorne Moor; *G. Norman.*

363. Navicula latissima Greg.
T. Roche Abbey; *E. Hunter.*

364. Navicula ovalis Sm.
T. Roche Abbey; *E. Hunter.*

365. Navicula inflata Kütz.
N. Killinghall; *G. Norman.*

366. Navicula lanceolata Sm.
A. Bradford; *W. West.*

367. Navicula Spencerii Quekett.
A. South Milford; *W. B. Turner.*
D. Goole; *E. Hunter.*

368. Navicula amphisbæna Bory.
N. Harrogate, copiously; *G. Norman.*
D. & T. Goole and Roche Abbey; *E. Hunter.*

369. Navicula rhynchocephala Kütz.
N. Harrogate; *G. Norman.*
D. & T. Goole and Roche Abbey; *E. Hunter.*

370. Navicula angustata Sm.
T. Thorne Moor; *G. Norman.*

371. Navicula attenuata Kütz. (*Gyrosigma* Auct.)
A. Baildon; *W. West.* South Milford; *W. B. Turner.*
D. & T. Goole and Roche Abbey; *E. Hunter.*

372. Navicula mutica Kütz.
T. Thorne Moor; *G. Norman.*

373. Gyrosigma (Pleurosigma) strigilis Sm.
D. & T. Goole and Roche Abbey; *E. Hunter.*

374. Gyrosigma lacustre Sm.
N. Harrogate; *G. Norman.*
A. Roundhay; *W. B. Turner.*
D. Askern; *F. H. Allen.*
T. Roche Abbey; *E. Hunter.*

375. Gyrosigma hippocampus Grim.
D. Goole; *E. Hunter.*

376. Gyrosigma acuminatum Sm.
A. Baildon; *W. West.* Roundhay; *W. B. Turner.*
D. Askern; *F. H. Allen.* Goole; *E. Hunter.*
T. Roche Abbey; *E. Hunter.*

377. Gyrosigma angulatum Sm. (and var. β).
A. South Milford; *W. B. Turner.*

378. Gyrosigma tenuissimum Sm.
A. South Milford; *W. B. Turner.*

379. Asterionella formosa Hass.
N. Not unfrequent in the wooden outlet to the pond at Ripley Castle; *G. Norman.* Near Cowthorpe; *W. West.*

APPENDICES.

APPENDICES.

ADDENDA: OMISSA.

[The following list of additional facts, and annotations upon those facts, turning up whilst the body of the work has been going through the press, is, from the nature of the case, probably in itself incomplete. There can be no finality in such matters, since fresh evidence is constantly being adduced to those on the look-out for it by the very fields themselves, the vegetation of every tract undergoing a natural and continual change, albeit in many cases one very slow and variable. We cannot leave out of the calculation the fact that from natural causes, without any interposition whatever of human agencies direct or indirect, a plant occurring in a given spot this year *may* not be there next year, may even be replaced by some other.]

2. Thalictrum minus L. (Cf. p. 112.)
A plant recorded under this aggregate name for Sedbergh (L. district) in Thompson's 'Florula Sedbergensis' (8vo, pp. 20. privately printed, 1886) is probably Var. *montanum* Wallr.; but may possibly be the small-leaved race of *T. flexuosum*. I have not been able to obtain sight of a specimen.

4. Thalictrum flavum L. (Cf. p. 113.)
N. Half a mile from Knaresborough, on the right-hand side of the York Road; *Archd. Pierson* in *Hargrove*, Ed. iii (1782).

22. Ranunculus hirsutus Curtis. (Cf. p. 122.)
D. Hooton Pagnell; *J. E. Kenyon* in list in *Bot. Chron.* (Aug. 1864). I should say it was little more than a Casual or corn-field weed here—the locality is on the Permian limestone, seven miles inland, to the north-west of Doncaster, and quite away from the alluvial levels whereon alone (in the West Riding) it has pretension to true indigenity.

[Epimedium alpinum L. (Cf. p. 128.)
L. Sedbergh; *J. Handley* in *Notes on Flora* (1885). Of course, Alien.]

36. Papaver Rhœas L.
37. Papaver dubium L. } (Cf. p. 129.)
Both of these are vulgarly styled *Cuproses* in the east of the Riding.

47. Raphanus Raphanistrum L. (Cf. p. 133.)
The local vulgar name for this, '*Runsh*,' was accidentally omitted; but the *Charlock* (also called *Catlock*) sometimes receives the same appellation.

[**Sisymbrium canescens** Nutt. Addition.
A pretty Alien Crucifer, glistening with pellucid-hoary setæ on stem and peduncles, native of North America.
A. On waste grassy ground by barns near the canal between Apperley and Calverley!]

61. Cardamine pratensis L. (Cf. p. 138.)
This also receives the rustic titles, *Mayflower* and *Bird's-eye*, in addition to those named.

[Barbarea præcox Br. (Cf. p. 141.)
C. Rail-bank near Thornhill Junction station, plentifully (1887)!]

73. Nasturtium palustre DC. (Cf. p. 142.)
C. Waste grounds by Spen beck, Spivey's Mill, Heckmondwike!

76. Cochlearia alpina Wats. (Cf. pp. 143-144.)
Apropos of the Obs., with regard to my reference of Merrett's plant 'on the *top* of Penigent Hill under the stones' to this species, Mr. Whitwell writes—'*Draba incana* is plentiful on the ledges of the uppermost precipice on the west side (of Penyghent) above Hunt Pot.' I do not see the connexion: the uppermost precipice is not the top—which is stony, barely-grassed gritstone, and *if* the *Draba* had 'foliis lunatis' still it does not grow actually on the summit.

[Neslia paniculata Desv. (Cf. p. 146.)
A. By barn, near canal, east of Apperley Bridge (1887), with other Casuals, brought with hen-corn!]

88. Lepidium campestre Br. (Cf. p. 149.)
C. Ryhill; *G. Roberts* (1886)!

[**Lepidium coronopifolium** Fisch. Alien. Addition.
A. Casually, on waste ground where manure had lain, in field by canal near Calverley! Doubtful how brought, or an allied prostrate-growing species with cordate-ovate silicles, pinnatifid lowest leaves, entire upper leaves, and very small densely racemed flowers with lilac-white petals. Perhaps *Lepidium Iberis*, but 'in too young a state to name with safety' (J.G.B.).]

95. Viola palustris L. (Cf. pp. 152-153.)
C. Soothill Wood (1887), dying out—very rare in this smoky colliery district!

108. Polygala depressa Wender. (Cf. pp. 158-159.)
C. Coxley Valley; *P. F. Lee* (1887)!

115. Silene anglica L. (Cf. p. 161.)
C. Spen Vale Mills, Heckmondwike,* 1887! A wool-waste Alien, here.

121. Lychnis Githago Lam. (Cf. p. 163.)
Locally termed *Popple* in the eastern part of the Riding.

129. Stellaria media With. (Cf. p. 166.)
Vulgarly and ubiquitously known as *Chickweed*, and also (because of its prevalence and uselessness?) as *Murrain* on the eastern sandstone tract.

130. Stellaria Holostea L. (Cf. p. 166.)
Its commonest vulgar name is *Stitchwort*, but *Cuckow-wort*, *Mayflower*, and *Milkmaid* are also occasionally heard on child and rustic lips within the Riding.

150. Hypericum dubium Leers. (Cf. p. 173.)
L. Howgill near Sedbergh! Additional for northern section of Lune tributaries' area.

[**Althæa hirsuta** L. Alien or Misnomer?
L. This is mentioned by Rev. W. Thompson, M.A., in *Florula Sedbergensis*, p. 17, as having occurred near Sedbergh as a 'Casual.']

[**Sida Abutilon** L. (*S. althæifolia.*) Addition.
An orange-flowered Indian Malvaceous Alien.
A. On drying-up mud-refuse taken from the Bradford beck by the Sewage Works at Frizinghall near Shipley; *W. West, junr.* (1887)! With this occurred many other Aliens, nearly all clearly brought in wool or hair from the east. Amongst them were the two next named *Erodiums*, *Carduus setosus*, *Chenopodium murale*, *C. glaucum*, *Sebtaria*, and a glabrous *Vicia* near *V. lævigata*, with a solitary semi-erect, stalked, smooth, six-seeded pod, and four pairs of linear-acute leaflets—perhaps a var. of *V. angustifolia.*]

[*Erodium moschatum L'Hérit. (Cf. pp. 182-183.)
A. Near Bradford—as a Casual; *H. T. Soppitt* (1886)!
C. Waste ground by Spen Vale Mills, Heckmondwike!
'Wool-waste' Aliens, both.]

[Erodium cygnorum Nees. (Cf. p. 183.)
C. Common End, Linthwaite; *Shaw* teste *Baker*; and under wall of field by 'Three Houses' Inn, Sandal, where shoddy manure or hen-corn had been out-thrown; *G. Roberts.*]

180. Impatiens parviflora DC. (Cf. p. 185.)
R. Ribble bank near Settle; *Miss R. F. Thompson* (1887)!

184. Rhamnus Frangula L. (Cf. pp. 186-187.)
A. Gascoigne Wood, Milford; *G. Roberts* (1887)! *R. catharticus* also sent from Kellington (A.)—a new station.

190. GENISTA ANGLICA L. (Cf. p. 188.)
Petty-whin is a local name, in actual use to-day and for long time by rustics, for this plant, and one never applied to the *Arrestharrow* in West Yorks. It cannot be a 'book-name,' as applied to *Genista*, although Britten and Holland state that it was invented by Turner for *Ononis arvensis*; if so, of that surely the spiny form, and even then singularly inapplicable, taking facies into account. But did not Turner make a mistake as to the species of 'fur' (i.e., the *Genista anglica* growing on Furze-hills Hildersham, Gamlinghay, Stow, and other places in Cambridgeshire), which he says was in that county, as well as in many places besides, called a 'Whine'? Babington in his Flora of Cambridgeshire correctly refers 'Petty-whin' to *Genista*, of which allocation, however, Britten and Holland take no note whatever in their Plant-name Dictionary. Yet it is the proper one at any rate for Yorkshire, in many moory districts of which *Genista* is abundantly prevalent, and *Ononis* scarce or absent.

191. GENISTA TINCTORIA L. (Cf. p. 188.)
The local name, derived from ancient use, in Ripon neighbourhood is '*Woad-wesh*,' i.e., Woad-wash.

194. ONONIS ARVENSIS Auct. (Cf. p. 190.)
Rustboirne is the rustic name on the dry calcareous fields in the east of the Riding.

197. MEDICAGO DENTICULATA Willd. (Cf. p. 191.)
A. Frizinghall, near the Sewage Works, in plenty (1887), with *Medicago maculata*, various Vetches, etc.
C. Heckmondwike in many places, impermanent, but regularly recurrent from fresh importations, about spinning mills, wool-waste-manured fields, etc. Other foreign medicks, such as *M. lappacea* Desr., *M. arabica* All., *M. Murex* Willd., and *M. marina* L., have also occurred, but rarelier.

[***Trigonella hamosa** Willd. Alien. Addition.
This, the Egyptian Fenugreek, has occurred as an alien at Sandal (*G. Roberts*!) and elsewhere, introduced with foreign 'forage' seed, or with cotton.]

[***Trigonella corniculata** L. Alien. Addition.
A. On waste ground where manure had lain, in field by canal at Calverley. Also at Sandal (C.); *G. Roberts*.]

213. LOTUS CORNICULATUS L. (Cf. p. 197.)
In the north-west of the Riding known to shepherds and children as *Butterjag*. A jag is a load (of fell-hay especially, 'sleds,' i.e., wooden sledges, being used for 'carrying' instead of wheeled waggons, and 'a jag' is a 'sled-full'), and the bushy bosses of the plant are usually covered with quite a load of butter-coloured blossoms. But *a jag* also means a slash or saw-tooth-like indentation in many northern dialects, and the word may have reference to the trifoliately-cut leaves, or even to the cut glove-finger like arrangement of the blooms in the cluster. Anyway, there hardly seems such obscurity as Prior implies. The suggested derivation from 'bottle-jacks' quoted in Britten and Holland (p. 79), is amended in Prior's 3rd Edition (1879) in favour of the much more likely corruption of the word from 'butter'd eggs.'

219. HIPPOCREPIS COMOSA L. (Cf. p. 199.)
D. Hooton Pagnell; *J. E. Kenyon* in *Bot. Chron.* (1864). The name only in the list, and the possibility of error must be borne in mind, since the plant is not known to occur on the Permian limestone elsewhere, in the North of England at any rate. It occurs on the mountain scar-lime, the oolite and the chalk.

[***Coronilla varia** L. Alien.
C. Hungerhills, Dewsbury (1886), 'gathered by a lady'; *P. F. Lee*, spn.! Doubtless a mere escape.]

[***Ornithopus scorpioides** Willd. (*Astrolobium* DC.).
C. Casual, Lofthouse; *Geo. Roberts* (1887)!
In connection with this, one may remark that it seems possible that Elihu Berry's Barnsley casual may have been this species, and not *O. ebracteatum*, as one more likely to be introduced with field-seeds. Cf. p. 199.]

[**Ononis mitissima** L. Addition.
A South European Casual.
A. Waste ground near barns, Apperley!]

232. PRUNUS SPINOSA L. (Cf. p. 204.)
Fruiting finely in the sheltered valley of the Rawthey near Cautley in 1887, but the season has perhaps been an exceptionally favourable one.

VAR. **fruticans** Weihe. (Var. *cöaetanea* Wimm.)
Not unfrequent in the lowland sandy districts, or on limestone, as about Ripon (Y.), Collingham (W.), Tickhill, Bawtry, and Sandbeck (T.), but seldom discriminated. This year, 1887, both the var. and the squat-growing early-flowering type—the true 'Blackthorn,' have fruited well in many places where they do not do so usually. The differences between the two are emphasised as the season advances. The type blooms whilst its patent, matted twigs are yet bare of leaves; the var. flowers nearly a month later, concurrently with the bursting of its leaf buds. By August, *P. fruticans* is, in facies, conspicuously unlike the type; it is a taller slenderer shrub, hardly so spinescent, more erect growing, almost a tree, its sloes less globular, more oval, and larger, drooping more on their curved stalks (in consequence of their size) than do the patent-erect little spheres of *eu-spinosa*, and they change colour, and ripen later. The leaves, too, after midsummer

has passed, differ; born later, they die later, and so whilst the leaves of the type are fading to a lighter, yellower virescence, and beginning to fall, those of var. *fruticans* are still of a dark full green. There are other minuter characters by which the two can be differentiated, such as the pubescence of pedicel, peduncle, leaf, and twig, but they are all comparative, though all tend to give *P. fruticans* a place midway between *spinosa* and *insititia*.

258. RUBUS SUBERECTUS Anders. (Cf. p. 213.)
CORRECTION.—The date of the earliest notice by Geo. Anderson (for Dallow Gill), is *not* 1824, as given on p. 213, line 24, but 1811; see *Linn. Trans.*, xi, 219.

259. RUBUS FISSUS Lindl. (Cf. p. 213.)
Usually in more open places than *R. suberectus*; *C. C. Babington*. By the light of this, it would seem probable that West's 'Bingley Wood' plant (see p. 214) may be the same as the Cullingworth one alluded to below. Mr. West, when he saw the latter, in situ, with me, was inclined to think so.
A. Fine and in plenty in boggy thicket on S.W. side of Bell-horse Lane, Cullingworth!
T. Drive-sides, Hatfield Moor; *P. F. Lee, spns.* (1887)!

[RUBUS HAMULOSUS Müll. (Cf. p. 214.) (**R. hamosus** Génév.)
Doreton, Yorks.; *J. G. Baker* (1865) in *Herb. Génév.*, given as *R. cordifolius* to Déséglise by Baker. It belongs to the first division of Génévier's *Discolores*, but seems to me more nearly allied to *rhamnifolius*; *C. C. Babington* in *Journ. Bot.*, vol. 24, p. 237 (August 1886).
N. Road-side, Poppleton, near York, Aug. 1886; *G. Webster* in *Rep. Bot. Exch. Club* for 1886, p. 147 (1887). In the same place Dr. Focke says of this, '*R. carpinifolius* Wh. et N.,' whilst Prof. C. C. Babington says, 'I have a series of *hamosus* in Herb. Génév.; they are like this (of Mr. Webster's), but do not seem to accord with his description. I call them *rhamnifolius*.'

261. RUBUS LINDLEIANUS Lees. (Cf. p. 214.)
N. About Brimham Rocks, common.
A. Quarry in Hainworth Wood, Keighley!
T. Hatfield Chase, frequent.

265. RUBUS PUBESCENS Wirtg. (Cf. p. 216.)
Prof. Babington wrote—too late for insertion in proof-slip—that this species is 'like the Surrey plant' which Watson and Newbould taught me to know as *R. thyrsoideus*, and he added that 'it is *Rubus nemophilus* Rip., according to Génévier.'

271. RUBUS MUCRONATUS Blox. (Cf. p. 217.)
A. Hainworth Wood, Keighley!

272. RUBUS FESTIVUS Müll. (Cf. p. 218.)
C. Soothill Wood; *P. F. Lee*!

276. RUBUS SCABER Angl. Auct. (Cf. p. 219.)
Prof. Babington writes, asking if this is 'the true *R. scaber*.' No; although marked '*scaber*' by Prof. Babington in 1879 on the sheet of Mr. Carr's Bradfield gatherings. Time will not allow of my submitting these again to Prof. Babington, but I have little doubt they are his *Babingtonii* (Bell-Salt.) and not the restricted *scaber* of Weihe & Nees, and Focke.

1043. Rubus podophyllus Müll. Addition to Flora.
C. Open copse, Soothill (1887!), coll. *P. F. Lee*; teste *C. C. Babington*. I do not know rightly to which species to refer this: from its habit, etc., as studied in situ I should say to *R. villicaulis*—it seems a glandulous *R. carpinifolius*, or between that and *R. villicaulis*. Its flat gray-green leaflets with patent teeth alternately big and little give it a peculiar facies.

278. RUBUS RADULA var. DENTICULATUS Bab. (Cf. p. 219.)
This is now named *Rubus Newbouldii* Bab. (cf. *Journ. Bot.*, 1886, p. 230).

282. RUBUS FLEXUOSUS M. & L. (Cf. p. 220.)
Prof. Babington writes that this is the right name (not *R. Guntheri*) of Edwin Lees's Hebden Bridge plant.

284. RUBUS SAXICOLUS Müll. (Cf. p. 220.)
Prof. Babington added, too late to alter in proof, that '*saxatilis* is the true name.'

291. RUBUS CHAMÆMORUS L. (Cf. p. 222.)
Add—'always with a north exposure' (*C. C. Babington*) to the stational distribution given.

296. ROSA SPINOSISSIMA L. '*Cat-whin*.' (Cf. p. 225.)
L. (upper division). Road-side bank, Howgill near Sedbergh! (alt. 600 ft.).

315. EPILOBIUM HIRSUTUM L. (Cf. p. 235.)
One of the most singular products of hybridisation between this and *E. parviflorum* is a remarkably sericeo-villose plant—the var. *villosissimum* of Koch, which is the *E. intermedium* Mérat. It has been referred to both *E. hirsutum* and *E. parviflorum*, but seems on the whole nearest the former, where Koch puts it. I have only one West Yorkshire record for it, however:
N. Harrogate; *W. R. Thorp*.

325. CIRCÆA ALPINA L. (Cf. p. 238.)
M. In a hollow called Wicken Hole, Seal Bark, Greenfield (*Mr. John Mellor*); *R. Buxton's Bot. Guide*, p. 4 (1849); rep. *Grindon* in his *Manchester Flora*, p. 282 (1859). The two earliest records for the locality reported by Whitehead with an implication of its extinction (in 1877).
L. Dentdale—perhaps var. *intermedia*; *M. B. Slater*, oral record only *ad auctor*, no specimen seen.

327. Myriophyllum spicatum L. (Cf. p. 238.)
A. Malham Tarn; *P. F. Lee*! The head station for Aire river-basin, and extending its upward range to 1,250 feet alt., 50 feet higher than Hooker gives as its altitudinal limit in the Highlands of Scotland.

329. Hippuris vulgaris L. (Cf. p. 239.)
A. Hambleton; *G. Roberts*!

348. Saxifraga stellaris L. (Cf. p. 247.)
The limit of descent of this Montane species is placed much too high at 1,500 feet. I saw it on August 1st, 1887, beside the stream below Cautley Spout (L.) at less than 600 feet above sea-level. The record in the table on page 55 requires altering, which will put the *Saxifraga* into the same category with *Meum* and *Polygonum viviparum*.

350. Saxifraga aizoides L. (Cf. p. 248.)
L. (upper division). Gill between Winder and Crook Fells, Sedbergh; *Rev. W. Thompson*; *J. Backhouse, junr.*; etc. (1887)

356. Parnassia palustris L. (Cf. pp. 250-251.)
L. (upper). In many spots in Dentdale and Rawtheydale about Sedbergh!

[Petroselinum segetum Koch. (Cf. p. 254.)
A record as of this, sent to me in 1887, upon investigation turned out to be *Silaus*!]

362. Sison Amomum L. (Cf. p. 254.)
D. Hooton Pagnell; *J. E. Kenyon* (*Bot. Chron.*, 1864). An overlooked record.

367. Pimpinella major Huds. (Cf. p. 255.)
Add an asterisk after the Low Moor (*Soppitt's*) record for this in C. river-basin. I have reason to believe it was only a Casual there.

[**Ammi majus** L. Casual. Additional to Flora.
A. On waste ground near barn, and in field where manure had lain, by canal between Apperley and Calverley (!) *H. T. Soppitt*, etc., 1886-7.]

376. Silaus pratensis Bess. (Cf. p. 259.)
A. Railway-bank, Apperley Bridge; *H. T. Soppitt.*

377. Meum Athamanticum Jacq. '*Balmony.*' (Cf. p. 259.)
L. Still (about Beck Houses and Howgill) in the Sedbergh localities, locally plentiful, in 1887! Smells like 'cattle-spice,' but any rustic use of it, in beast-pharmacy now-a-days, denied by local farmers, who regard it as a stinking pest.

380. Peucedanum Ostruthium Koch. (Cf. p. 260.)
L. Sedbergh; *J. Handley.* Has still a local use in felon and other affections of cattle. An addition to the flora of the river-basin.

385. Torilis infesta Spreng. (Cf. p. 262.)
L. ('Sedbergh, most northern habitat; *J. Handley* in *Notes on Flora*, p. 8 (1885).) Surely a Casual—'only one plant seen'—if name indeed be not an error, and No. 387 (or 388) the plant intended.

398. Viscum album L. (Cf. p. 266.)
D. Askern, common in 1840; *E. Lankester* in '*Account of Askern*' (1842). An overlooked record.

401. Sambucus Ebulus L. (Cf. p. 267.)
A. Bank by canal near Calverley station*, 1886; *H. T. Soppitt.* In plenty, but Alien, and recent too, there.

407. Galium boreale L. (Cf. p. 269.)
L. Rocks by the Lune in plenty, near Sedbergh; *J. Handley.* Additional river-basin area.

408. Galium cruciatum With. (Cf. p. 269.)
I have once heard this called '*Polygally*' in North Yorkshire, and singular to say in the very river valley (Yore) where the Giant Throat-wort is styled '*Polypoddy*'! Is this a misapplication of terms through ignorance, or a survival of a very ancient name, handed down—the district was formerly a stronghold of Romanism—from days when ceremonials were rifer than now? Can the *Galium* have ever had the honour, as a 'Cross-wort' (from the form of its tiny golden bloom), of being conveyed parishes round on withy-wands, etc., in the old solemn mockeries of Litany week, as well as the *Polygala*? Might it not have got the name *Polygally* from its use as a substitute for the Milkwort?

411. Galium Mollugo L. (Cf. p. 270.)
L. Lune-side, Sedbergh; *J. Handley*!

416. Galium Aparine L. (Cf. p. 271.)
An additional West Yorkshire local name for this plant is '*Sweet-hearts*'—in use in the Ainsty. It is also styled 'Snares' and 'Clegs,' but Hairough [Hair-rough], pronounced 'Hariff,' is its commonest rustic appellation.

[Centranthus ruber DC. (Cf. p. 273.)
'*Pretty Betty*' is the local name in South Yorkshire.]

432. Silybum Marianum Gaert. (Cf. p. 277.)
A. Casual on waste ground by Sewage Works, Frizinghall!

440. Carduus heterophyllus L. (Cf. p. 280.)
L. (upper). Lowgill, Sedbergh, Rawtheydale!

442. Carduus setosus Bess. (Cf. p. 281.)
(*C. arvensis* var. *integrifolium* Koch.)
A. Casually plentiful, in 1887, on refuse by road near Sewage Works, Frizinghall, Shipley!

447. Serratula tinctoria L. (Cf. p. 282.)
L. (upper). Lune-side, Sedbergh; *J. Handley.*

[Centaurea melitensis L. (Cf. p. 284.)
A. Waste ground, by barns, canal-side near Apperley; *H. T. Soppitt*; etc.! Alien; grain-brought.]

[Gnaphalium margaritaceum L. '*Pearly Everlasting.*' (Cf. p. 290.)
L. Sedbergh*; *J. Handley* in '*Notes.*' Alien; an Escape.]

468. Gnaphalium dioicum L. (Cf. p. 290.)
M. Greenfield; *W. Wilson, ms.* (1832). Older record.
L. (upper). Howgill Fells, Sedbergh! Addition.

472. Senecio erucifolius L. (Cf. p. 292.)
C. A very fine form of this, apparently a hybrid with *S. Jacobæa*, but not so nearly half-way between the two as Mr. Whitwell's Skipton plant, grows in great clumps, but very locally, in a strip of plantation off Soothill Wood, Howley!

475. Senecio saracenicus L. (Cf. p. 293.)
'Intermediate type'; *Baker, Flo. Lake Dist.*, 138—as if to be entered into calculation on the same terms as an indigenous plant, albeit admittedly 'a Denizen' only; yet 'Alien' in *Flo. Northumb. and Durham.* Once introduced it is very tenacious of life, as by the Rawthey river-side at Sedbergh, and may well grow, in the spots where planted, 'as plentifully' as *S. Jacobæa*, or more so, without that being any argument for its indigenity, as *Nicholson* in *Dillenius* (*Ray, Syn.*, iii, 177) seemed to think.

476. Doronicum Pardalianches L. (Cf. p. 293.)
L. Sedbergh*; *J. Handley* in '*Notes*' (1885). Escape.

480. Inula Conyza DC. (Cf. p. 295.)
D. Hooton Pagnell, N.W. of Doncaster; *J. E. Kenyon* (1864).

481. Inula dysenterica L. (Cf. p. 295.)
C. Field-side ditch-bank, Owlet Hurst Farm, near Heckmondwike! Sparingly; a rare plant on the Coal-measures.

[Hemizonia pungens Torr. & Gray. (Cf. p. 297.)
A. By barns near canal between Apperley and Calverley—a grain-brought Alien (1887)!]

487. Petasites officinalis Mœnch. Female. (Cf. p. 298.)
A. Aire-bank, near Apperley; *W. West.*

489. Eupatorium cannabinum L. (Cf. pp. 298-299.)
'*Hemp-wort*' is the local, rustic name in South Yorkshire.

499. Taraxacum palustre DC. (Cf. p. 302.)
C. Hunger-hills, Dewsbury; *P. F. Lee.*

500. Lactuca virosa L. (Cf. p. 302.)
A. Castle wall, Pontefract, 1887; *P. F. Lee*! New to river-basin.
C. I am sorry to say the Sandal Quarry plant can only have been a Casual in that station; there was not a trace of it in 1887! Few of the aliens of 1885-6 were surviving, but the *Lepidium Draba* was rather plentifully ripening its seed—the result of the exceptional summer?—which, like *Armoracia*, it rarely does with us, even in the stations where it most persists and spreads (by roots).

[Crepis biennis L. (Cf. p. 304.)
L. Sedbergh; *J. Handley* in '*Notes*' (1885), but inadvertently for *C. paludosa.*]

[512. Hieracium pallidum Fr. (Cf. p. 307.)
To be expected about Howgill. It grows on slate crags, higher up the same main river-basin, just outside the Riding; and should ultimately be found on the rocks by the Lune.
The recent receipt of a Hawkweed (which it was hoped would prove *H. Gibsoni*, but was only *vulgatum*) from the Attermire hills above Settle, suggests the remark—additional to those on p. 307—that a closer study of Mr. Backhouse's words (at pp. 44-45 and 53-54 of his Monograph of the *British Hieracia*) seems to show that (1) Mr. Backhouse only saw '*tracings*' of the Sherardian Herb. specimens, some of which were only 'forms of *H. pallidum*' lacking the hair-fringe to leaf of the type; whilst (2) other specimens from Richardson, in the same Herb., bearing the same number, were 'unquestionably forms of *Hieracium vulgatum.*' So that, in spite of the belief that the Gordale '*oreades*' was '*H. pallidum*' (p. 54), the evidence that true slate-rock *pallidum* ever occurred in Craven on lime-rock seems very shaky and unsubstantial.]

518. Hieracium tridentatum Fr. (Cf. p. 310.)
L. (upper). Lune-side, Sedbergh; *J. Handley*!

519. Hieracium prenanthoides Vill. (Cf. p. 310.)
R. Oustre-bank Wood, Settle; *Miss F. P. Thompson* (1887), spn.! A new locality.

520. Hieracium umbellatum L. (Cf. p. 310.)
L. Lune-banks, Sedbergh; *J. Handley, spn.*! An addition to river-basin, so far as Yorkshire is concerned, but it occurs higher up the same main stream in Westmorland; and no doubt search will reveal it a fact that all the forms of Hawk-weed already known there in the Lune basin grow likewise lower down, spate-brought, or because the physical conditions of the surface are similar.

521. HIERACIUM CROCATUM Fr. (Cf. p. 311.)
L. Lune-side, somewhere near Sedbergh; *J. Handley, spn.*! Here apparently much rarer, i.e., less numerous in individuals than the larger *H. 'corymbosum'* (with leaves broader and more amplexicaul), included with it by Baker in *Flo. Lake Dist.* I restrict the name *crocatum* to the rather wand-like narrowish-leaved plant with few heads.

1044. Hieracium Eupatorium Griseb. Addition to Flora.
H. corymbosum Fr., and Backh. Mon.
Area: No record in Top. Bot. Range: 600—450 feet?
Rocky banks, rocks by streams in the mountain slate area, locally plentiful. P. Aug.
First record: W. Thompson, 1886.
L. Sedbergh neighbourhood; *W. Thompson* in *Florula Sedbergensis*, p. 10 (*sub nom. nudus* corymbosum). Lune-side rocks (1887); *J. Handley, spns.*! Nearer *H. boreale* in facies than the preceding, with numerous corymbose heads and ovate leaves, glaucous and veined below, and semi-clasping. This species is, truly, new to the West Riding Flora; all the Malham and Attermire plants I have seen are the restricted *H. crocatum*, as I understand it. *H. Eupatorium* is less a crag than a river-bank form.

[XANTHIUM SPINOSUM L. 'Wool' Alien. (Cf. p. 312.)
A. Bradford district; *W. West*; etc.
C. Heckmondwike (1887)!]

526. CAMPANULA RAPUNCULOIDES L. (Cf. p. 314.)
L. By side of river Dee in Dentdale*; *J. Handley* in '*Notes.*' Product, doubtless, of a garden-grown root, out-thrown and carried down by stream until lodging in some creek. Certainly with no claim to being indigenous.
A. Seacroft; *W. Nelson*, but no specimen seen—probably a garden-escape, unless a misnomer and some depauperate form of *C. latifolia* meant.

532. VACCINIUM OXYCOCCOS L. (Cf. p. 316.)
A dale name is '*Mèa-wort*' or '*Mèa-berry*'; from A.S. *mer-isc*—through *mere*—originally 'a marsh,' and so, a swampy pasture on the moors; in which, in Yorkshire, the Cranberry frequently grows along with *Hydrocotyle, Anagallis, Sphagnum*, etc. The name *mea* is common in the higher dales for a pasture, e.g., 'Green Mèa Gill,' 'Mèa Fell,' etc.

539. CALLUNA VULGARIS Salisb. (Cf. p. 320.)
A curiously interesting local name for Heather in Langstrothdale and thereabout is '*Kid*,' from a Celtic word meaning something bound up—as a faggot—transferred in this case to the only thing, Ling, that was (or could well be) in that district gathered and tied up for

besoms, although bundles of it may have had other uses. The word is preserved on our maps, etc., in the name 'Kidstones'—the pass leading over from Buckden into Bishopdale.

546. VINCA MINOR L. (Cf. p. 323.)
L. Sedbergh*; *J. Handley*. Alien there.

[Gilia cotulæfolia Steud. Addition.
A Polemoniaceous, Californian Alien.
A. Waste-ground near barn by canal-side between Apperley and Calverley (!) 1887; *H. T. Soppitt*! This occurred along with many other fugitive Aliens—*Neslia paniculata* L., a pretty glandulous crucifer—*Sisymbrium canescens* Nutt., *Ononis mitissima* L., *Centaurea melitensis* L., a Californian composite, *Hemizonia pungens* T. & G., *Amsinckia lycopsoides* Lehm., *Phacelia congesta* Hook., and many other common Casuals—all, doubtless, brought in grain-sweepings, out-thrown for poultry, and all equally hardly worth the line of permanent record. For those who want a downright good botanical puzzle, and a rousing lesson in plant character by the way, these foreign weeds answer admirably, however. It is only necessary to purchase a stone or two of 'hen-corn,' at different mills, strew it at intervals, in spring, on waste garden ground, allowing fowls a few minutes to pick the wheat, and in a few short weeks, voila! a parterre more bewildering to the lover of the botanically curious in its variety than it is possible to attain otherways with so little trouble.

564. HYOSCYAMUS NIGER L. (Cf. p. 332.)
A. Among ordinary weeds on a waste fallow near Bierley Coal-pit, near Bradford, about a dozen plants with some *Datura Stramonium*, September 1887; *W. West, junr.*! Casuals both, of course, although new to the Aire river-basin, even for that category how introduced difficult to surmise.

[VERBASCUM VIRGATUM With. (Cf. p. 334.)
Near Wetherby, one large plant (September 1887) in a pit dug in a bed of river-gravel; *John Jackson*; *Nat.*, November 1887, p. 351. Certainly Alien, here, as everywhere else in North England; and, since not grown in gardens thereabout now, probably only another example of the enduring vitality seeds possess under favourable conditions. The plant that flowered in 1866 below Thorparch Bridge, it is not impossible may have been a seed of the same long dead individual which produced the Wetherby one, germinating and fructifying twenty-one years earlier because laid bare by some accident by so much sooner!]

575. LINARIA MINOR Desf. (Cf. p. 338.)
Linaria viscida Mœnch.—oldest name.
L. Ingleton and Sedbergh*, on rail-track cinders, in plenty, 1887! Addition for river-basin, but Casual therein. 'Colonist'; *J. G. Baker* in *Flora of Lake Dist.*
C. Haybeck, West Ardsley*; *P. F. Lee.* Hillam; *G. Roberts* (1887)!

576. MIMULUS LUTEUS L. (Cf. p. 339.)
L. Sedbergh, well-established for thirty years in one place; *J. Handley*. Additional district.

588. VERONICA SCUTELLATA L. (Cf. p. 342.)
A. Mr. Whitwell's Carlton 'Ing-fields' station is wrongly allotted to W. river-basin.
C. Stocks-moor, Dewsbury; *P. F. Lee*!

599. LATHRÆA SQUAMARIA L. (Cf. p. 347.)
A. Skipton 'Park' Woods; *T. W. Edmondson* (1886)!
C. Sycamore roots, Hopton; *C. P. Hobkirk.*

604. LYCOPUS EUROPÆUS L. (Cf. p. 349.)
First record: Hargrove, 1782, to be substituted for Salt, 1800.
N. Lane from Staveley into the Cars; *E. Hargrove, Hist.* cit. (1782).

633. STACHYS AMBIGUA Sm. (Cf. p. 360.)
C. Bullrush dam, Carlinghow near Batley; *P. F. Lee, spns.* (1887)! Not exactly Smith's plant; but clearly a hybrid, and retaining the strange peculiar sylvan *Wound-wort's* smell; although nearer *S. palustris* than *S. sylvatica* in the matter of breadth of leaf and length of stalk, etc. As the name '*ambigua*' is not—whatever it ought in strictness to be—restricted to Smith's half-way plant, but is given to a group of forms, varying in so far as they favour one or other parent, I consider interchangeable terms *sylvatica-palustris* or *palustris-sylvatica* much preferable, since if carefully applied they describe what is meant, and leave no one in that doubt which must exist, unless a comparatively long explanation is tacked on, as to how the name '*ambigua*' is to be understood.

641. LAMIUM INCISUM Willd. (Cf. p. 363.)
Lamium hybridum Vill.
C. Horbury, near corn-mill; *P. F. Lee* (1887)!

647. ECHIUM VULGARE L. (Cf. p. 365.)
C. [Gone from Sandal Castle ruins, in 1887 (!), and probably not there for thirty years, since Gissing did not repeat Dixon's station, though living nigh.] Overton, Thornhill Edge; *W. Rushforth* (1886-7). Casual, not a true Native, at Horbury?

652. MYOSOTIS REPENS Don. (Cf. p. 368.)
First record: William Wilson, apparently, in MS. dated 1832.
M. Greenfield; *W. Wilson*; *James Cash* in *Nat.*, June 1887, p. 182.
C. Bretton, Soothill; *P. F. Lee.*

657. ANCHUSA ARVENSIS Bieb. (Cf. p. 369.)
C. Horbury, near corn mill; *P. F. Lee*!

[ASPERUGO PROCUMBENS L. (Cf. p. 372.)
This plant, though in some plenty at Sandal in 1886, did not re-appear in 1887. This year, when I specially visited the quarry for the purpose of ascertaining, not a vestige of it was visible!]

[AMSINCKIA LYCOPSOIDES Lehm. Casual. (Cf. p. 372.)
A. By barns, at Calverley, with other Casuals, 1887!]

665. HOTTONIA PALUSTRIS L. (Cf. p. 374.)
C. Field-side, Horbury Bridge; *W. Rushforth*! Quite a new station, but the plant may have been there unreachable many years, as the ditch is (I am informed) a deep one, and only accessible in seasons of long drought, such as that of the summer of 1887!

670. LYSIMACHIA VULGARIS VAR. **angustifolia** Wats. (Cf. p. 377.)
Mr. Roberts's narrow-leaved 'Barnby Dun' Loosestrife is the above variety. The local name for the type in the vicinity of Doncaster (where, by the Carr dikes, it is abundant) is '*Yellow-saugh.*' A rustic near Askham Bog said *he* called it '*Yellow Rod*,' and 'had some in his garden.' I have never heard it styled 'Yellow Rocket,' nor even 'Loosestrife,' by the uneducated country-folk.

674. ANAGALLIS ARVENSIS L. (Cf. p. 379.)
Local names: *Shepherd's Weatherglass*; *Shepherd Red-eye*; and '*Peepers.*'

675. ANAGALLIS TENELLA L. (Cf. pp. 379-380.)
T. Hatfield Moor!

[AMARANTHUS RETROFLEXUS L. Alien. (Cf. p. 383.)
A. In plenty on refuse by Sewage Works, Frizinghall near Bradford *W. West, junr.* (1887)!

688. CHENOPODIUM MURALE L. (Cf. p. 384.)
A. On road-embankment made from Sewage Works refuse, Frizinghall, Bradford, with *C. glaucum* and *C. ambrosioides*: all Casuals in this situation.

687. CHENOPODIUM ALBUM L. (Cf. p. 384.)
Local name: '*Mixenweed*' (about Halifax).

[Atriplex hortensis L. VAR. **rubra** Loud. Encycl. Alien.
On waste ground by Spen Vale Mills, Heckmondwike (1887), with *Xanthium spinosum*, various Meliloti, etc.! Attracted much attention from working men, one of whom transplanted into a garden and tended it, with the view of growing it in quantity from seed as a 'bedding-out' plant, on account of the fine deep purple-red of its stems, leaves, and perianth-wings.]

[Rumex dentatus Camp. Alien.
A. In plenty where large heaps of old manure had lain, in a pasture by the canal off Calverley Wood; *H. T. Soppitt* (1886). Seen by myself in 1887. A curious plant named by Mr. J. G. Baker, and said by him to be 'allied to *pulcher*' and 'widely spread in the Mediterranean region,' with panicle leafy to the top, very dwarf and stunted in habit of growth, and differing from *conglomeratus* in

having strongly-toothed sepals, all tubercled, and with pedicels jointed near base. I thought at first it might be a hybrid, perhaps *R. Knafi* Celak—which I have not seen; but, from its situation amongst other aliens, and the fact of its abundance just where it grows without other of its supposed parents except *R. crispus* near, it must be an Alien. A curtain of mystery, not to be raised, often hangs over the exact how and whence these strangers come. In the case of this Dock, all its associates were species more or less natural to the south-east corner of Europe, just as those others from near the barn at Apperley (see p. 792) were all North American.]

[Polygonum arenarium W. et K. (*P. pulchellum* Lois.?)

A. Casual, on waste ground where manure had lain by canal near Calverley, 1887 !

A strikingly pretty, twiggy plant, procumbent when growing; characterised by its trigynous rosy flowers, on distinct pedicels, three in each leaf-axil, and by its large silvery, ciliated ocreæ. According to Nyman, a native of Hungary, South Russia, etc.]

716. Polygonum Bistorta L. (Cf. p. 391.)

Local name about Howgill, Sedbergh—'*Easter-edges.*'

[Asarum europæum L. (Cf. p. 393.)

C. 'Kirklees Park, probably introduced; *H. T. Soppitt*' in Lee's *Flora of Dewsbury*, p. 242. On inquiry, I find that a Mr. Spencer reported it to H.T.S., who has not himself gathered it. It was, of course, planted as a curiosity in the shrubbery borders where seen. I fear greatly that did we know—what we shall never learn now—the early history of the Asarabacca in its old Mytholmroyd, Settle and Pontefract stations, we should have to come to the conclusion it was equally a planted alien in those places. Were it turned up now-a-days in a Yorkshire wood for the first time, the 'bull's-eye' of critical inquiry turned upon it would soon light up what was dark about its origin; but not so with many of the old (even if not yet extinct) localities for suspicious natives, with real or supposed values to the simplist of bygone days—with the lives of the recorders, the time has passed in which we can with any hope of success trace out the causation of these occurrences.]

720. Empetrum nigrum L. (Cf. p. 394.)

Additional rustic names: '*Crow-ling*,' '*Cranes.*'

726. Ceratophyllum aquaticum Syme. (Cf. p. 396.)

T. In the Torne dikes.

739. Populus canescens Sm. (Cf. p. 401.)

A. Shrubberies by field-road, Eccleshill, Bradford ! Doubtless alien in this station; but, as a nurseryman's tree, much rarelier planted than the more handsome *P. alba*.

[Populus balsamifera L. Alien. (Cf. p. 402.)

A. Field corner, Wadsley Bridge, below Eccleshill, near Bradford ! Stray from garden near, but from its suckers, likely to be long persistent.]

741. Salix pentandra L. '*Black Withy.*' (Cf. p. 402.)

C. Coxley dam-side; *Mrs. P. F. Lee*. Perhaps planted there.

[Salix lanata L. Incognit; Error?

'In Seal Bark [Greenfield] I saw . . a *Salix* (male) like *caprea*, but more hoary (=*S. lanata*'; *W. Wilson*, MS. Note in Warrington Museum). Mr. Wilson was, I suppose, mistaken in referring his gathering to the Scottish *S. lanata*.]

762. Sparganium simplex Huds. (Cf. p. 411.)

A. Canal under Calverley Wood !

773. Potamogeton rufescens Schrad. (Cf. pp. 414-415.)

The localities given under C. and D. are both in South-west Yorkshire (vice-co. 63), for which Watsonian division the plant is an addition to *Topographical Botany*, ed. ii.

784. Potamogeton flabellatus Bab. (Cf. p. 416.)

A. Canal under Calverley Wood !

785. Zannichellia brachystemon Gay. (Cf. p. 41[illegible].)

D. Ditch near the river Went, near a cot at the foot of the wood, Brockerdale (1887); *P. F. Lee* !

792. Butomus umbellatus L. (Cf. p. 421.)

A. Margin of Leeds and Liverpool canal (south side) under Calverley Wood, 1887, in plenty and fine flower owing to the exceptionally favourable season ! Perhaps not an additional station, but in the manufacturing districts of West Yorkshire rarely seen in bloom, probably not since September 1874, when Mr. James Abbott, of Leeds, saw it in flower in the canal at Kirkstall, three miles further east than this Calverley locality (Cf. *Journ. Bot.*, N.S., vol. iii, p. 307).

798. Orchis Morio L. (Cf. p. 425.)

Mr. W. Whitwell has 'very much doubt as to the statement made (p. 425, line 18) that about York *O. Morio* is very much commoner than *O. mascula*.' I do not know; but in many *open* pasture-fields about Wetherby Cemetery I recollect distinctly a like preponderance of the green-winged species. *O. mascula* is, of course, much commoner in woods and broken ground, and in the pastures of the dales goes up much higher.

800. Orchis incarnata 'L.' (Cf. p. 426.)

A. Adel; *W. Kirkby* (sub nom. aggreg. *latifolia*).

808. Ophrys apifera Huds.

Page 430, lines 18, 19. Enclose 'Birtwhistle rocks' locality in square brackets. The 'rocks' are now the site of Botherby Wood, and the plant long extinct; teste *T. W. Edmondson*, 1887.

814. Epipactis latifolia Auct. (Cf. p. 433.)

C. Coxley Wood, and abundant in a copse near Bretton; *P. F. Lee*.

826. Tamus communis L. (Cf. p. 440.)

'*Lady-Seal*' is a local name of some interest, in use in the south of the Riding.

831. Fritillaria Meleagris L. (Cf. p. 443.)

Local name: '*Falfalaries.*'

838. Allium oleraceum L. (Cf. p. 446.)

C. Horbury; *W. Rushforth*; *P. F. Lee*, spn. !

882. Carex paniculata L. (Cf. p. 461.)

C. Near Dewsbury; *P. F. Lee* ! Rare in that district.

899. Carex ornithopoda Willd. (Cf. p. 466.)

Y. One or two plants, very dwarfed, in the vicinity of Mackershaw Wood, broken thickety ground, in 1887 !

900. Carex pilulifera L. (Cf. p. 466.)

C. Stocks-moor, Dewsbury, 1887; *P. F. Lee* !

902. Carex pallescens L. (Cf. p. 468.)

C. Hartley Bank, Horbury; *P. F. Lee* !

908. Carex lævigata Sm. (Cf. pp. 469-470.)

C. Coxley Wood; and Hartley Bank; *P. F. Lee*, *spns.* !

916. Carex Pseudo-cyperus L. (Cf. p. 473.)

T. Dike-sides, Hatfield Moor; *J. T. Hasselby* (1887) !

[Setaria viridis Beauv. (Cf. p. 475.)

A. With *Echinochloa* (one tuft) and *Setaria glauca* (?) on waste ground by the Bradford Sewage Works at Frizinghall (1887), in some plenty; *W. West, junr.* !]

923. Alopecurus agrestis L. (Cf. p. 476.)

C. Sandal; *P. F. Lee* (1886) !

933. Calamagrostis Epigejos Roth. (Cf. p. 479).

L. Sedbergh neighbourhood; *W. Thompson* in *Florula Sedbergensis*, p. 15 Possibly a little doubtful (I have not seen specimens), but the plant grows near Kirkby Lonsdale, in Furness, and as high as 750 feet in a hagg by Hawes in Wensleydale. It is a very local plant in North England, and one of those (like *Festuca sylvatica*) of which it is very difficult to rightly discern the rationale of distribution.

937. Aira cæspitosa L. (Cf. p. 480.)

A local name (near Bardsey, where it grows finely) is '*Hassock grass.*'

971. Festuca sciuroides Roth. (Cf. p. 491.)

C. Storrs Hill, Ossett; *P. F. Lee* ! Casual in road by Spen Vale Mills*, Heckmondwike, with other aliens !

CRYPTOGAMIA.

997. Hymenophyllum tunbridgense Sm. (Cf. p. 499.)

Addition to Obs.—The occurrence of this at Greenfield seems to rest on somewhat slender, unsafe evidence. A ms. note by William Wilson, dated June 15th, 1832, unearthed from among Wilson's papers by James Cash, and given in extenso in *Nat.*, 1887, p. 182, says 'In the vacancies underneath the lower blocks [Seal Bark rocks] was found *Hymenophyllum Wilsoni*. We *saw* H. tunbridgense in a *very small deep* hole, at a greater elevation under the first rocky knoll. This station has been long known to Jethro [Tinker].' Wilson's *Journal* (*Nat.*, loc. cit.) has a somewhat briefer entry for that same visit of June 10th, 1832. It is, simply, of his having seen 'two species of *Hymenophyllum*—chiefly *H. Wilsoni*, fr.' I consider the character of fimbriate involucre, or its absence, could not be certainly discerned at half-a-yard's distance or less, 'in a very small deep hole' especially, and for my own part I doubt if *H. tunbridgense* ever grew at Seal Bark. Ordinary *H. unilaterale*, grown in shade and drawn-up, begets quite the more gracile physiognomy of *H. tunbridgense*.

1006. Asplenium Adiantum-nigrum L. (Cf. p. 504.)

L. Sedbergh neighbourhood; *W. Thompson* in *Fl. Sedbergensis*. This species prefers grit or slate-rock soil—it does not like limestone.

1009. Ceterach officinarum Willd. (Cf. pp. 505-506.)

L. Sedbergh neighbourhood; *W. Thompson* in *Flor. Sedbergensis*, without exact station given; and the 'Middleton Head' locality is in Westmorland, as also was (by a few yards only) a spot near Lowgill, where it was gathered by *T. W. Edmondson* on Aug. 1st, 1887.

W. On Pool bridge yet, August 1887, in some plenty; *H. T. Soppitt*.

(W.? or Y.) On walls by the road near the summit of the pass from Coverdale Head leading to Kettlewell, but whether just within Coverdale (and consequently Yoredale) limits, or Wharfe (and so West Yorkshire) not now certainly recollected; *M. B. Slater*, orally, to author, 1887 ! The precise situation merits ascertainment, since Baker does not give *Ceterach* for Coverdale in his 'North Yorkshire.'

1031. Lycopodium alpinum L. (Cf. p. 517.

M. Greenfield Valley, 'lower down' than Seal Bark; *Jethro Tinker*, *W. Wilson, MS.* (1832), *Nat.*, 1887 182. Not actually seen by

Wilson. It was, doubtless, even then, dying out. Now (1887) those who know the district well regard it as having been extinct for many years. Even the vegetation of the high moors, within a few miles of the larger smoky village-towns of the Riding and adjacent county, are being surely if slowly affected detrimentally by the nameless mephitic influences of modern manufactures—the change, a lessening of vitality, being in its lines precisely parallel with the effect upon the blood, stature, nerves, and faces of the human inhabitants of those busy communities.

SUMMARY.—The two additional species in the foregoing amplificatory list bring up the total number of West Yorkshire Phanerogams, with Vascular Cryptogams (vide p. 522), to 1044.

CHARACEÆ.

1. CHARA FRAGILIS Desv. (Cf. p. 525.)
Y. Between Sharow and Bridge Hewick, 1881; *G. Nicholson*, comm. *H.* and *J. Groves*.

VAR. **Hedwigii** Ag.
Y. Near Ripon, 1881; *G. Nicholson*, comm. *H.* and *J. Groves*.

2. CHARA ASPERA Willd. (Cf. p. 525.)
The *lake* at Langould near Letwell, which is where Gerard Smith gathered this, is in part in Yorkshire, the south-east end being in Notts.

4. CHARA HISPIDA L. (Cf. p. 526.)
W. Pool in Scarthingwell Park, 1884; *H. F. Parsons*.

5. CHARA VULGARIS L. (Cf. p. 526.)
L. Dale beck, Ingleton, 1886; *C. Bailey*. Ingleborough, 1864; *W. P. Hiern*, comm. *H.* and *J. Groves*.
D. Goole, 1882; *T. Birks, junr.*—VAR. *longibracteata*, teste *H.* and *J. Groves*.

6. CHARA CONTRARIA Kütz. (Cf. p. 526.)
A. 'Kippax, 1834,' *Herb. Edinb.*; comm. et teste *H.* and *J. Groves*.

7. TOLYPELLA GLOMERATA Leonh. (Cf. p. 527.)
A. Armin; *W. P. Hiern*, comm. *H.* and *J. Groves*.

8. TOLYPELLA INTRICATA Leonh. (Cf. p. 527.)
D. Swinefleet Common, 1883; *T. Birks, junr.*

10. NITELLA FLEXILIS Ag. (Cf. p. 527.)
D. Near Sheffield, 1879; *F. T. Mott*, teste *H.* and *J. Groves*.

11. NITELLA OPACA Ag. (Cf. p. 527.)
T. Adlingfleet drain, 1882; *T. Birks, junr.*

BRYOPHYTA.

MUSCI.

5. SPHAGNUM INTERMEDIUM Hoffm. (Cf. p. 533.)
L. Cautley Spout, August 1887!

21. GYMNOSTOMUM MICROSTOMUM Hedw. (Cf. p. 536.)
(T. Roche Abbey Valley; *J. H. Aveling* in *Hist.* (1870). Accidentally omitted on p. 536, but possibly a case of misnomer.)

VAR. γ **elata** Schpr.
Mollia microstoma var. *elata* (Schpr.) Braith. Mon.
R. Settle (*Burgess*); *Braith., Brit. Moss Flora.*
A. Wet rocks, Janet's Cave, Malham (*Holt*, 1885); *Braith., Brit. Moss Flora*, part x, p. 301 (1887).
The above two records are inserted in *Braithwaite's Flora*, on faith of sterile gatherings 'apparently referable to this variety,' although the author adds 'it is difficult without fruit to distinguish it from *Mollia viridula* var. *densifolia*,' which is plentiful at Malham as recorded in the body of this work (p. 536).

27. RHABDOWEISIA FUGAX Hedw. (Cf. p. 537.)
Weisia striata Hook. & Tayl.
M. Seal Bark, Greenfield; *W. Wilson, MS.* (1832), *Nat.*, 1887, 182.
The first West Riding record; overlooked in the body of this work.

30. DICHODONTIUM FLAVESCENS Lindb. (Cf. p. 537.)
L. Howgill Fells; *M. B. Slater*!

31. DICRANELLA CRISPA Schpr. (Cf. p. 537.)
L. Cautley Spout, rocky bank of stream below the fall; *M. B. Slater*. Over-ripe, and not in good condition—after examination a little doubtful to *W. West* and myself!

33. DICRANELLA SQUARROSA Schpr. (Cf. p. 538.)
T. Roche Abbey Valley; *J. H. Aveling* in *Hist.* (1870).

40. DICRANUM SAXICOLA Ferguss. (Cf. p. 539.)
Perhaps this is the same thing as the var. *falcifolium* of *Braith., Moss Flora*, but I have had no opportunity of examining a specimen.

41. DICRANUM SCOPARIUM L. var. **turfosum** Milde. (Cf. p. 539.)
N. Moor near Blasham Beck, Lofthouse (*Wesley*), 1878; *Braith., Brit. Moss Flora*, p. 148. This is the same plant as is given in the body of this work (p. 539) for moors near Ramsgill. It was named '*paludosum*' by H. Boswell. The Aketon field plant is not like it, being much larger, stem longer and densely tomentose, with leaves broad-based, undulate, and patently squarrose. It *may* prove to be true *D. undulatum* Ehrh. This Aketon plant is very similar to a form named *Dicr. palustre* var. *rugifolium* by H. Boswell in 1878.

43. DICRANUM BONJEANII De Not. (Cf. p. 540.)
L. Cautley Spout; *M. B. Slater*!

[**Seligeria calcarea** (Dicks.) Schpr. Error.
Given for Roche Abbey Valley in the list of mosses at the end of *Aveling's Hist.* (1870); but doubtless some other small encrusted *Seligeria*, or, perhaps, *Eucladium* (which does grow on the moist Permian rock-faces there), was the species intended.

60. SELIGERIA RECURVATA Hedw. (Cf. p. 543.)
M. Seal Bark; *W. Wilson, MS.* (1832). The earliest notice for the same neighbourhood as Whitehead's 1867 record.

63. BLINDIA ACUTA B. & S. (Cf. p. 543.)
L. Cautley Spout; *M. B. Slater*!

Blindia trichodes Braith. (Cf. p. 543.)
Braithwaite having omitted mention of this in *Brit. Moss Flora* at page 124, gives it on pp. 297-8 of the concluding part of vol. i as a species, with the locality 'Marsden near Burnley; *Whitehead*, 1865!!' Marsden is not near Burnley, and the village quoted is the same as the more accurately localised 'Wessenden Valley' of this flora. Marsden lies at the embouchure of the valley of the Wessenden beck.

74. DIDYMODON RUBELLUS B. & S. VAR. **dentata** Schpr.
L. Ingleborough (*Cash*, 1880); *Braith., Brit. Moss Flora*, 261.
A. Malham; *W. West*!

75. DIDYMODON LURIDUS Hornsch. (Cf. p. 545.)
R. Banks of the Hodder (*Burgess* and *Holt*, 1886); *Braith., Brit. Moss Flora*, 262.

76. DIDYMODON FLEXIFOLIUS H. & T. (Cf. p. 545.)
M. Greenfield, slope towards the inn; *W. Wilson, MS.* (1832) in *Nat.*, 1887, 182.

77. DIDYMODON CYLINDRICUS B. & S. (Cf. p. 545.)
Mollia tenuirostris Lindb.
L. Dentdale, and Howgill (!) *M. B. Slater* (1887).
R. Ribble banks, Horton; *J. Shackleton* fide *M. B. Slater*.

78. DIDYMODON SINUOSUS Schpr. (Cf. p. 546.)
Dr. J. S. Wesley would appear to have gathered his specimens of this at Helmsley, in 1877, according to *Braith., Brit. Moss Flora*, 271.

82. TRICHOSTOMUM TOPHACEUM Brid. (Cf. p. 546.)
Barbula brevifolia (Dicks.) Lindb.
L. Clapham Springs (*Abbott*); *Braith., Brit. Moss Flora*, 264.
Braithwaite in his *Moss Flora* (loc. cit.) says that Dickson's *Bryum brevifolium* is certainly identical with the typical form of this plant—*T. tophaceum*, so that the later name must go.

84. TRICHOSTOMUM CRISPULUM Bruch. (Cf. p. 547.)
L. Sedbergh district; *M. B. Slater*. Dentdale, I believe.

91. BARBULA UNGUICULATA Hedw. VAR. **microcarpa** Schultz.
L. Baugh Fell; *W. West* (1879); *Braith., Brit. Moss Flora*, 276.

93. BARBULA RECURVIFOLIA Schpr. (Cf. p. 549.)
Barbula reflexa Brid.
R. Giggleswick Scar; *J. G. Baker* (1855); *Braith., Brit. Moss Flora*, 266.

[TORTULA GRACILIS Hook. & Grev. (*Barbula acuta* Brid.)
Dillenius's Ingleborough '*Tortula brevifolia*' was *Trichostomum tophaceum*, not *B. recurvifolia*, as suggested on p. 549 in the body of this work.]

94. BARBULA RIGIDULA Schpr. (Cf. p. 549.)
W. Wetherby and Boston Spa (*Wesley*, 1869); *Braith., Brit. Moss Flora*, 268.

95. BARBULA SPADICEA Mitt. (Cf. p. 549.)
L. Dent (*J. M. Barnes*, 1872); *Braith., Brit. Moss Flora*, 267.
W. Haselden [Heseltine] Gill (*J. Nowell*, 1866); *Id.*

96. BARBULA CYLINDRICA Schpr. (Cf. p. 549.)
L. Deepdale (*J. G. Baker*); *Braith., Brit. Moss Flora*, 270.
Y. Studley (*J. G. Baker*, 1856); *Id.*
W. Bolton Abbey (*G. E. Hunt*, 1867); *Id.*

101. BARBULA TORTUOSA W. & M. (Cf. p. 550.)
M. Greenfield, in some spots in fruit; *W. Wilson, MS.* (1832).

Vice 112. **Encalypta alpina** Sm. (*Leersia* Lindb., B.M.F.)
Hooker's Ingleborough moss (A.D. 1800), considered hitherto to belong to *E. rhabdocarpa*, is referred by Braithwaite (*Brit. Moss Flora*, 280)—without having examined the original specimens, since the '!!' is not appended to the citation of the Yorkshire locality—to this as a ver-species. It is the *Encalypta ciliata* β *alpina* of Hook. & Taylor. West's re-gathering of recent years is the same thing, by whatever name it has to go.

113. ENCALYPTA CILIATA Hedw. (1801); Hoffm. (1795).
Leersia laciniata Hedw. (1782); *Braith., Brit. Moss Flora.*
A. Malham Tarn (*Nowell*, 1854); *Braith., Brit. Moss Flora*, 283.

[**Grimmia commutata** Hübn. Error.
Reported, through some lapsus linguæ not clearly explained, for Howgill Fells. For history, and correction of error, vide *Naturalist*, Sept. and Oct. 1887, pp. 281 and 299.]

119. GRIMMIA TRICHOPHYLLA Grev. (Cf. p. 553.)
L. Wall, Low Haygarth, Rawtheydale!

143. ORTHOTRICHUM LYELLII H. & T. (Cf. p. 559.)
Before 'Ingleton' (line 17) *insert* 'On the stock of an ash-tree beside the river a quarter of a mile below Ingleton, May 1783'; *James Bolton*, spn.! Not then known, i.e., without a name, but recognised as 'a new species' by its first gatherer.

173. WEBERA ELONGATA Schpr. (Cf. p. 563.)
(M. Seal Bark, Greenfield (under-ripe); *W. Wilson* (1832); *James Cash*, *Nat.*, 1887, p. 182. No one seems to have confirmed this.)

204. MNIUM SERRATUM Brid. (*Bryum marginatum* Dicks.)
The statement (at p. 568, line 9 from foot) that this moss has been reported from every district except Mersey and Trent must be modified by the deletion of 'Mersey and,' since the species was reported, under Dickson's name, from Seal Bark, Greenfield, by William Wilson in *MS.* (1832); vide *James Cash* in *Nat.*, 1887, 182.

206. MNIUM STELLARE Hedw. (Cf. p. 569.)
L. Sedbergh; *M. B. Slater* (1887).

222. POLYTRICHUM FORMOSUM Hedw. (Cf. p. 571.)
L. Sedbergh, 1887!

227. DIPHYSCIUM FOLIOSUM (L.) Möhr.
M. Ravine near Seal Bark; *Ed. Hobson* (prior to 1832), *W. Wilson, MS.*; *Nat.*, 1887, 183.

348. Fissidens Curnowii Mitt. sp.
Fiss. bryoides VAR. *cæspitans* Schpr. (No. 396 *b*, Lond. Cat.).
W. Bolton Woods (*C. J. Wild*, 1876); *Braith.*, *Brit. Moss Flora*, 83. This moss, erected a species by Mitten, is adopted as such by Braithwaite, although admittedly 'very near' *F. bryoides*, and connected with that by intermediate forms. It was known to W. West as a variety.

233. FISSIDENS OSMUNDOIDES Hedw. (Cf. p. 573.)
L. Cautley Spout; *W. West*.

234. FISSIDENS CRASSIPES Wils. (Cf. p. 573.) *Fiss. incurvus* var. *β fontanus* Schpr.
This aquatic moss, adopted as a ver-species by Mitten and Braithwaite under Schimper's varietal name, has the additional station—
W. 'Ilkley (*Dr. Carrington*, 1855)' in *Brit. Moss Flora*, p. 84.

245. LEUCODON SCIUROIDES Schw. (Cf. p. 575.)
L. Beck Houses, Howgill!

291. HYOCOMIUM FLAGELLARE B. & S. (Cf. p. 584.)
M. Near Bills' o' Jack's; *W. Wilson, MS.* (1832); *James Cash*, *Nat.*, 1887, p. 183.

302. PLAGIOTHECIUM SYLVATICUM B. & S. (Cf. p. 586.
L. Sedbergh; *M. B. Slater*. New to drainage district.

342. HYPNUM SCORPIOIDES L. (Cf. p. 594.)
[W. Bramham Moor; *James Bolton*, spn.! Collected in 1783.] The earliest record for this fine moss, which is now probably extinct in the locality through drainage and cultivation. The only remaining bit of aboriginal bog now (1887) left at all likely to still nurture it is the 'White Car' field at the south-west corner of Bramham Park demesne, outside the wall, and search has been made there in vain, although it still produces *Narthecium*, *Genista anglica*, and other peat-loving species.

SUMMARY.—Total number of Mosses known for West Yorkshire—**348**, out of 580 British species, or about 60 per cent.

HEPATICÆ.

2. PREISSIA COMMUTATA Nees. (Cf. p. 602.)
L. (upper). Sedbergh; *M. B. Slater*.

15. LEJEUNIA ECHINATA Hook. (Cf. p. 605.)
L. (upper). Gill near Sedbergh, on road up Rawtheydale, 1887; *G. Stabler*!

22. PORELLA LÆVIGATA Lindb. (Cf. p. 606.)
L. (upper). Sedbergh district; *G. Stabler*! May be instantly distinguished from *P. platyphylla* by chewing a small portion, when *P. lævigata* at once produces a biting sensation on the tongue like pepper; an observation of Prof. Lindberg's, but which I have not seen in print.

30. ODONTOSCHISMA SPHAGNI Dum. (Cf. p. 608.)
A. By rill in boggy field—in process of reclamation in 1887!—adjoining Bell-horse Lane, Cullingworth; *Thos. Hebden*.

47. TRICHOCOLEA TOMENTELLA Nees. (Cf. p. 613.)
L. Cautley Spout (!) *W. West*.

64. MYLIA TAYLORI Gray. (Cf. p. 618.)
L. (upper). Howgill Fells; *Y. N. Excurs.*, 1887!

76. JUNGERMANIA BANTRIENSIS Hook. (Cf. p. 621.)
L. Sedbergh district; *M. B. Slater* (1887).

78. JUNGERMANIA SCHREBERI Nees. (Cf. p. 621.)
L. Sedbergh; *Y. N. Excurs.*, 1887!

92. NARDIA EMARGINATA Gray. (Cf. p. 624.)
L. Cautley Spout; *M. B. Slater* (1887).

SUMMARY.—The total number of Hepatics known for West Yorkshire remains at 108 species, as enumerated in the body of the work.

LICHENES.

235. Collema plicatile Ach. (**Collemodium** Nyl.)
Not Incognit, as supposed on page 632. Since it has turned up recently, the old Hailstonean record for Malham may be accepted. It is not so maritime as Leighton's limitation would imply.
L. On wet rocks in the bed of the river Rawthey towards Cautley, 1887; *J. A. Martindale*.
A. Malham; *S. Hailstone* (1812); rep. *Carrington*, *W. R. Flora*, 62.

9. COLLEMA FURVUM Ach. (Cf. p. 632.)
L. Clapham; *W. West*!

16. LEPTOGIUM LACERUM (Ach.). (Cf. p. 633.)
'In most parts of Craven'; *S. Hailstone* (1805). Assuredly rare now —can it be a decreasing species? If this and many other old records are to be accepted as trustworthy for their time (and no one has impugned their substantial accuracy), I believe close, special investigation would bring to light numberless instances all tending towards one conclusion, viz., to establish a lemma strongly favouring the view that the flora of wide areas (such as that of Craven) is undergoing even now a profound if gradual alteration, from some occult change in climatic conditions, mean humidity or what not, possibly in part the result of a century of cultivation, but certainly independent of any great increase in populosity, smokiness of air, etc. The thesis is a difficult one, needing time, residential leisure, and the drawing together of many threads of fact apparently unconnected, but the inquiry would be worth almost any pains called for to settle it.

26. SPHÆROPHORON FRAGILE Pers. (Cf. p. 634.)
A. Rocks in the woods near St. Ives, Bingley; *S. Hailstone* in *Whitaker's Craven* (1805).

35. CLADONIA PYXIDATA (L.) Fr. VAR. **fimbriata** Hoffm.
C. tubæformis Schær.
L. On walls near Sedbergh; *W. West*! Opposite Ingleborough Cave, very fine; *B. Carrington*.
W. Outershaw, Langstrothdale! Otley Chevin; *B. Carrington*.
One of the best marked of the varieties of a common polymorphic 'species,' ranked as a ver-species itself by Nylander and Crombie in their recent arrangements (vide *Grevillea*, Nos. 73, 74, 75, 77, 1886-7).

36. CLADONIA GRACILIS Hoffm. (Cf. p. 635.)
L. Near Cautley Spout; *J. A. Martindale*.

45. CLADONIA SYLVATICA Hoffm. (*Cladina* Nyl.) (Cf. p. 636.)
Very common from Cautley, Howgill, and Ingleborough down to Goole Moor and the mosses of the Vale of York; most, if not all of the so-called *Cladonia rangiferina*—the true 'Reindeer Moss,' being really

C. sylvatica, according to Mr. J. A. Martindale. The statements as to width of range, etc., given under *C. rangiferina* on page 636 must, therefore, be transferred from that species to the present one. Views as to the differentiation, or distinctness, of the two species have undergone a great change since Dr. Carrington's Lichen Florula for the Riding issued in 1862, since in that work *Cladonia sylvatica* finds no place at all! The result of this is that I am not prepared to cite more than one station in West Yorkshire—the south end of the Widdale Fell ridge—for the true *C. rangiferina*.

51. ALECTORIA JUBATA L. VAR. **chalybeiformis** Ach.
A. Rocks in woods near St. Ives, Bingley; *S. Hailstone* (1805).

60. PLATYSMA GLAUCUM (L.). (Cf. p. 639.)
A. [Beck-foot Lane near Bingley; *S. Hailstone* (1805)]. Gone?

66. PELTIGERA POLYDACTYLA Hoffm. (Cf. p. 640.)
A. Rombalds Moor near Helwick; *Richardson* and *Dillenius*; *Hailstone* in *Whitaker's Craven* (1805).

67. PELTIGERA HORIZONTALIS (L.). (Cf. p. 640.)
A. Bingley and Steeton Woods; *S. Hailstone* (1805).

77. PARMELIA OLIVACEA (L.). (Cf. p. 641.)
Mr. J. A. Martindale informs me that this species is now known to be very rare in Britain, and, as regards Yorkshire, all, or nearly all, the records for it on pages 641-2, are probably either *Parmelia fuliginosa* (on rocks) or *Parmelia subaurifera* (on trees). Two other closely allied 'species,' *P. exasperata* (Ach.) Nyl., and *P. prolixa* (Ach.) Nyl., will probably also be ascertained to be West Yorkshire species.

236. Parmelia subaurifera Nyl. Addition.
On trees, probably not so common as the next, but doubtless in other places besides those stated.
L. Sedbergh, here and there, but not well developed; *J. A. Martindale*.
W. Bramham Park!

Parmelia fuliginosa (Fr.) Nyl.
On rocks and walls mainly, but also on trees, probably common.
L. Sedbergh, on rocks and trees abundant, several plants in fruit; *J. A. Martindale*.
N. Pateley Bridge and Dacre Banks!

Obs.—This species is not numbered (as an addition), since it must be regarded as taking the place of No. 77, *P. olivacea*, although manifestly it would be unsafe to say that *all* the '*olivacea*' records belong to this—yet most of them certainly do; as stated above it is quite uncertain if we have the true *olivacea* at all.

82. PARMELIA SCORTEA Ach. (Cf. p. 642.)
L. Wall, Low Haygarth, Rawtheydale Austwick; *W. West*.

237. Parmelia Borreri (Turn.). Addition.
L. Tree near Sedbergh; *Jos. A. Martindale.* Dent; *W. West.* The first and only records for West Yorkshire.

86. Parmelia saxatilis (L.) var. **panniformis** (Ach.) Nyl. Addition.
On walls and rocks, pretty, and in facies distinct enough, but perhaps a form of *omphalodes.*
Y. and W. Walls, Countersett to Outershaw! By the Green Road—the old Settle pack-road. Some part in Y. and some in W. Beamsley; *W. West.*

238. Parmelia lævigata Sm. var. **revoluta** Flk. Addition.
L. Ingleton; *W. West,* teste *J. A. Martindale.*
At one time (Leighton, ed. ii) ranking as a var. of *P. tiliacea,* but both in facies and affinities much nearer some forms of *P. saxatilis.*

89. Physcia pityrea (Ach.) Nyl. (Cf. p. 644.)
R. Giggleswick; *W. West*!

239. Physcia aipolia (Ach.) Nyl. Addition.
L. Dent; *J. A. Martindale.*
R. & A. Trunks of old Ashes, Craven; *S. Hailstone* (1805).

240. Physcia tribacia (Ach.) Nyl. Addition.
Ph. erosa Borr., of Leight. ii, pro parte.
L. A few specimens on a wall, Sedbergh; *J. A. Martindale.*

98. Umbilicaria polyrhiza (L.). (Cf. p. 645.)
A. Harden Moor; *W. West*!

100. Pannaria rubiginosa (Thunb.). (Cf. p. 645.)
W. Bolton Woods; *W. West.*

103. Amphiloma lanuginosum Ach. (Cf. p. 646.)
R. Linn Gill near Ribblehead!
Not common, yet not to be called rare; it is very liable to be overpassed as an immature gonidiferous growth because of its powdery, granulose appearance.

104. Squamaria crassa (Huds.) (Cf. p. 646.)
R. Banks of rill, Thorns Gill, Ribblehead!
A. Malham; *S. Hailstone* (1805), confirmed by *W. West,* spns.!

241. Squamaria gelida (L.). Addition.
On sub-alpine rocks, two records only; but perhaps not really rare on the scar-limestone.
L. & A. Ingleton and Malham; *W. West,* spns.!

106. Placodium murorum Angl. Auct. (Cf. p. 646.)
A similar transference of statements made under one name to another is called for here, as in the case of the two Cladinæ. Hoffmann's *P. murorum* is anything but 'frequent' in Yorkshire (as stated by Carrington in West Riding Flora, 1862); indeed it is, probably, very rare, only the plants of Fountains Abbey, and perhaps Went-vale, being really referrible to it; most of this *P. 'murorum'* of Carrington, Windsor, and later collectors being *Plac. sympageum* Ach., but a small part may be *Plac. tegulare* Ehrenb.

107. Placodium callopismum (Ach.) var. **sympageum** (Ach.) Nyl. (Cf. p. 646.)
On scar-limestone rocks and walls, common in North-West Yorkshire.
L. Moughton Fell and Ingleton!
R. Ribblehead and Horton, common.
W. Outershaw!
A. Malham and Gordale (!) *W. West.* About Bank Newton, Skipton; *S. Hailstone* in *Whitaker's Craven* (1805) sub nom. *sympageus.*

242. Placodium tegulare Ehrenb. Addition.
L. Sedbergh, associated with *Physcia tribacia*; *J. A. Martindale.*

243. Placodium xantholytum Nyl. Addition.
Limestone rocks, rare.
L. Ingleton and Clapham; *W. West* teste *J. A. Martindale.*
R. Giggleswick; *W. West*!

110. Lecanora vitellina Ehrh. var. **coruscans** Ach.
L. Sedbergh, with the type; *J. A. Martindale.*

244. Lecanora glaucocarpa Wahl. Addition.
L. On limestone rock, Dent; *J. A. Martindale.*

113. Lecanora fuscata Schrad. (Cf. p. 647.)
L. Sedbergh; *J. A. Martindale.*

117. Lecanora varia Ehrh. var. **conizæa** Ach. (Cf. p. 648.)
L. A few poor specimens on larch trees, near Sedbergh; *J. A. Martindale.*

118. Lecanora atra (Huds.). (Cf. p. 648.)
L. Sedbergh, not very abundant; *J. A. Martindale.*

245. Lecanora intricata Schrad. Addition.
L. Sedbergh, here and there on rocks and stones; *J. A. Martindale.*

121. Lecanora sulphurea Ach. (Cf. p. 648.)
L. Sedbergh; *J. A. Martindale.*
A. Malham; *W. West*!

246. Lecanora galactina Ach. var. **dispersa** (Pers.).
L. Sedbergh, on limestone in a wall; *J. A. Martindale.*

122. Lecanora subfusca var. **campestris** Schær.
L. On limestones in walls, Sedbergh; *J. A. Martindale.*

126. Lecanora Dicksonii (Ach.). (Cf. p. 649.)
L. (upper). Howgill Fells, common (!) *J. A. Martindale.*

127. Lecanora badia Ach. (Cf. p. 649.)
L. Somewhat plentiful on rocks and walls, Howgill Fells (!) *J. A. Martindale.*

129. Lecanora aurantiaca (Lightf.). (Cf. p. 649.)
L. One specimen on a wall, Sedbergh; *J. A. Martindale.*

131. Lecanora ferruginea (Huds.). (Cf. p. 649.)
L. Parapets of a bridge, Sedbergh; *J. A. Martindale.*

133. Lecanora pyracea (Ach.). (Cf. p. 649.)
L. On rocks in the bed of the Rawthey; *J. A. Martindale.*

247. Pertusaria lactea L. (*P. velata* Turn.) Addition.
L. On rocks and walls, Sedbergh; *J. A. Martindale.*

142. Pertusaria amara (Ach.). (Sub *P. faginea.*) (Cf. p. 651.)
L. Sedbergh, on trees; *J. A. Martindale.*

147. Lecidea crustulata (Ach.).
L. Below the Spout at Cautley; *J. A. Martindale.*

248. Lecidea panæola (Ach.). Addition.
L. On rock at foot of Cautley Spout, one specimen, not characteristic; *J. A. Martindale.*

249. Lecidea lithophila Ach. Addition.
L. Dent; *J. A. Martindale.*

250. Lecidea lactea (Floerk.). Addition.
L. On a block of stone below Cautley Spout; *J. A. Martindale.*

158. Lecidea fusco-atra Ach. forma **grisella** Flk.
L. On walls of a bridge near Sedbergh; *J. A. Martindale.*

Var. fumosa Ach. (Cf. p. 652.)
L. Walls below Cautley Spout (!) *W. West,* etc.

160. Lecidea albocærulescens (Wulf.). (Cf. p. 653.).
L. On rocks near Cautley Spout; *J. A. Martindale.*

251. Lecidea speirea Ach. Addition.
L. On rocks at Cautley Spout; *J. A. Martindale.*

163. Lecidea canescens (Dicks.).
L. Sedbergh (!) *W. West.*

252. Lecidea coracina Ach. Addition.
L. On stones below Cautley Spout; *J. A. Martindale.*

167. Lecidea chalybeia Borr. (Cf. p. 653.)
L. Dent; *J. A. Martindale.*

168. Lecidea vesicularis Hoffm. (Cf. p. 653.
W. Hawkswick, Littondale; *W. West.*

170. Lecidea epipolia (Ach.). (Cf. p. 654.)
L. Sedbergh; *J. A. Martindale.*

172. Lecidea exanthematica (Sm.). (Cf. p. 654.)
L. Dent and Whernside, 1879; *W. West*!
A. Passing from Gordale to Malham Tarn; *S. Hailstone* (1805).

253. Lecidea citrinella Ach. Addition.
L. A few small specimens, on earth, in rock niches, at the foot of Cautley Spout; *J. A. Martindale.*

254. Lecidea chloroscotina Nyl. Addition.
L. On stones in the bed of the river Rawthey near Cautley; *J. A. Martindale.*

255. Lecidea lavata Ach. Addition.
L. Rawthey stream-side stones, Cautley; *J. A. Martindale.*

184. Lecidea concentrica Dav. var. **coarctata.** (Cf. p. 655.)
L. (lower). Clapham; *W. West*!
The type is plentiful on the plates of Yoredale flag- and lime-rock in Upper Garsdale and the west side of the Widdale Fell Ridge.

209. Endocarpon rufescens Ach. (Cf. p. 658.)
On sub-alpine earth, frequent, but often overlooked because microphylline, and so relatively inconspicuous compared with *Endocarpon miniatum* and *fluviatile.* It, too, prefers drier places among the niches of limestone rock.
R. Thorns and Linn Gills, Ribblehead!
W. Jackdaw Crag Quarry, Stutton, 1880, on the ground!

256. Verrucaria mauroides Schær. Addition.
L. Dent, April 1879; *W. West.*

223. Verrucaria epidermidis Ach. (Cf. p. 659.)
L. Dentdale; *W. West*!

257. Verrucaria subpyrenophora Leight. Addition.
L. Dent to Whernside, 1879; *W. West*!

258. Melanotheca gelatinosa (Chev.). Addition.
L. Dent, April 1881; *W. West*!

Summary.—The numerous additions to the list in the body of this work—the result of Mr. J. A. Martindale's valuable assistance, procured only at the eleventh hour, but none the less valuable, and gratefully acknowledged here—raise the sum total of West Yorkshire Lichens to 258. Compared with the known British species, the percentage of the whole is still far below that which the Mosses, Hepatics—and even the Fungi—show us: a safe indication that the West Riding, although probably not so rich as Westmorland or Scotia, is far from having been exhaustively worked. West Yorkshire should, surely, nourish three hundred forms?

FUNGI.

988. Agaricus (Amanita) virosus Fr. Addition.
A. Nab Wood, Saltaire, September 1887, not common; *H. T. Soppitt.*

989. Agaricus (Lepiota) mesomorphus Bull. Addition.
A. Fagley Woods near Calverley, rare, September 1887; *H. T. S.*

990. Agaricus (Tricholoma) ionides Bull. Addition.
W. Ilkley; *H. T. S.*
A. Pastures at Hawksworth; *H. T. S.*
D. Goole; *T. Bunker.*

991. Agaricus (Tricholoma) crassifolius Berk. Addition.
A. Hainworth Wood near Keighley (!); *H. T. S.*

992. Agaricus (Collybia) vertirugis Cooke. Addition.
W. Roadside near Pool, September 1887; *H. T. S.*

101. Agaricus (Mycena) leptocephalus Pers. (Cf. p. 672.)
T. Hatfield Woodhouse, 21st September, 1887; *H. T. S.*

993. Agaricus (Leptonia) chalybæus Pers. Addition.
A. Pastures near Bingley; *H. T. S.*

994. Agaricus (Naucoria) siparius Fr. Addition.
W. Wood near Otley Chevin; *H. T. S.* (September 1887).

195. Agaricus (Stropharia) squamosus Fr. (Cf. p. 680.)
T. Hatfield Woodhouse (September 21st, 1887); *H. T. S.*

995. Agaricus (Psilocybe) udus Pers. Addition.
A. Wet place, Baildon Moor, September 1887; *H. T. S.*!

996. Agaricus (Psathyra) conopileus Fr. Addition.
W. Pastures near Otley, September 1887; *H. T. S.*

997. Boletus striæpes Sec. Addition.
A. Nab Wood, Saltaire, October 1887; *H. T. S.*

998. Polyporus terrestris Fr. Addition.
A. On the ground in Shipley Glen, October 1887; *H. T. S.*

999. Cyphella capula Fr. Addition.
A. On nettle-stems, Apperley, October 1887; *H. T. S.*

258. Cortinarius cinnamomeus Fr. (Cf. p. 685.)
T. Lindholme, Hatfield Moor; *H. T. S.*

259. Cortinarius limonius Fr. (Cf. p. 685.)
T. Lindholme, Hatfield Moor, September 21st, 1887; *H. T. S.*

262. Cortinarius (Hygrocybe) castaneus Fr. (Cf. p. 685.)
T. Lindholme, Hatfield Moor; *H. T. S.*

302. Lactarius rufus Fr. (Cf. p. 689.)
W. Otley Chevin Wood; *H. T. Soppitt*!
T. Lindholme, Hatfield Moor; *H. T. S.*

307. Lactarius subdulcis Fr. (Cf. p. 689.)
T. Lindholme, Hatfield Moor; *H. T. S.*

331. Cantharellus aurantiacus Fries. (Cf. p. 691.)
W. Otley Chevin; *H. T. S.*—the *specimen* of an unusually deep, bright orange!

Obs.—The pater of the specific name—Fries—accidentally omitted on p. 691.

333. Cantharellus infundibuliformis Fr. (Cf. p. 692.)
A. Near Shipley, September 1887; *H. T. S.*!

340. Marasmius oreades Fr. (Cf. p. 692.)
T. Hatfield Chace, common; *Y. N. U. Excurs.* The rustic idea of causation locally was said to be that these coltstool' rings form where young horses have 'rolled' themselves on the turf!

597. Phragmidium Rubi Pers. (Cf. p. 713.)
T. Hatfield Chace; *H. T. S.* (September 1887).

1000. Trichosporium murinum Sacc. Addition.
A. On dead wood, Steeton; *H. T. S.*

1001. Peziza palearum Desm. Addition.
A. On *Dactylis*, Hainworth Wood Quarry; *H. T. S.*

1002. Peziza ciborioides Fr. Addition.
A. On dead oak leaves, Shipley Glen; *H. T. S.*

1003. Peziza fascicularis A. & S. Addition
L. On oak branches, Helks wood, Ingleton; *H. T. S.*

882. Peziza fusarioides Berk. (Cf. p. 735.)
L. Clapham.
N. Near Harrogate; *H. T. S.*

888. Helotium citrinum Fr. (Cf. p. 735.)
A. On dead cabbage-stalks, Frizinghall; *H. T. S.*

1004. Helotium serotinum Fr. Addition.
W. Church Fenton; *G. E. Massee.*
This and Nos. 1005 and 1007 are strictly omissions as well as Additions: they were known and should have appeared in the body of the work.

1005. Helotium calyculus Fr. Addition.
W. Church Fenton; *G. E. Massee.*

1006. Helotium lenticulare Fr. Addition.
W. On stumps, Bolton Woods; *H. T. Soppitt.*

1007. Rhytisma salicinum Fr. Addition.
W. On sallow leaves, Church Fenton; *G. E. Massee.*
An accidental omission from list in body of work.

909. Rhytisma Andromedæ Fr. (Cf. p. 737.)
T. Hatfield Moor, 21st September, 1887; *H. T. S.*!

1008. Trochila craterium Fr. Addition.
A. On dead ivy leaves, Cottingley; *H. T. S.*

936. Hypoxylon concentricum Grev. (Cf. p. 739.)
A. Stainor Wood; *W. N. Cheesman.*
T. Lindholme, Hatfield Moor, 21st September, 1887; *H. T. S.*

1009. Hysterium curvatum Fr. Addition.
A. On dead rose branches, Bingley; *H. T. S.*

Summary.—The additions made during the autumn of 1887 have brought the number of West Yorkshire species of Fungi up to 1,009! This is numerically creditable considering the few workers, and compares favourably with lists hitherto published for areas as large as our own, albeit the number is scarcely 25 per cent. of those now described as British; but the distribution through the various drainage districts of the known species is still, through lack of resident observers, very far from being well worked out.

ALGÆ.

8. Palmella hyalina Bréb. (Cf. p. 747.)
A. Cullingworth; *Thos. Hebden.*

36. Volvox globator L.
T. Hatfield Chase; *J. M. Kirk* (1887).

40. Sphærozosma vertebratum Ralfs. (Cf. p. 749.)
L. Ingleborough; *W. West.*

52. Closterium acerosum (Schrank) Ehrenb. (Cf. p. 750.)
T. Hatfield Chase; *J. M. Kirk* (1887).

56. Closterium Jenneri Ralfs. (Cf. p. 750.)
A. Cullingworth; *W. West.*

65. Closterium cornu Ehrenb. (Cf. p. 751.)
A. Cullingworth; *W. West.*

77. Cylindrocystis crassa De Bary. (Cf. p. 752.)
L. Cautley; *W. West.*

79. Mesotænium Braunii De Bary. (Cf. p. 752.)
L. Sedbergh; *W. West.*

201. Chroölepus aureus Kütz. (Cf. p. 762.)
A. Rylstone Road, Skipton; *W. Whitwell* in '*A Bachelor's Christmas Day*, 1883,' p. 24.

266. Diatoma vulgaris Bory. (Cf. p. 769.)
T. Hatfield; *J. M. Kirk* (1887).

270. Fragilaria capucina Ralfs. (Cf. p. 769.)
T. Hatfield; *J. M. Kirk.*

348. Cocconema lanceolatum Sm. (Cf. p. 775.)
T. Hatfield; *J. M. Kirk.*

380. Pinnularia nobilis Ehrenb. (*Navicula nobilis* Ktz.) Addition.
T. Hatfield Chase; *J. M. Kirk* (1887).

381. Frustulia olivacea Ktz. (*Gomphonema olivacea* (Lyngb.) Ktz.) Addition.
T. Hatfield; *J. M. Kirk* (1887).

382. Diadesmis confervacea Ktz. Addition.
T. Hatfield Chase; *J. M. Kirk* (1887).

Summary.—The total number of Fresh-Water Algæ in West Yorkshire is now 382.

BIBLIOGRAPHY II.

Additional Works, accidentally omitted or examined too late for inclusion in list at pp. 87—99.

[*The works here quoted are numbered consecutively onwards from the last of those on page 99.*]

1782.
186. Hargrove, E. Names of Plants growing in the Neighbourhood of Knaresborough. Hist. of Town, Castle, and Forest of Knaresborough. Third ed., 1782. 12mo. (58 entries only, in Linnean order, pp. 136-144. This list was drawn up by Archd. Pierson. Fifteen of the names are omitted from the 1832 edition.)

1788.
187. Marshall, Wm. The Rural Economy of Yorkshire. 2 vols., 8vo. London: T. Cadell. 1788. (Contains a good deal of information relating to the vulgar names of common plants.)

1836.

188. Higgins, Godfrey. Anacalypsis. Lond. 1836 (but preface dated Skellow Grange near Doncaster, May 1st, 1833). Vide p. 302, vol. i, for account of horizontally submerged Fir and Oak trees in soil six feet below tide mark, near the confluence of Ouse and Trent.

1837.

189. Francis, G. W. An analysis of the British Ferns and their allies. 8vo. Lond. Ed. ii in 1842.

1837—1848.

190. Deakin, Richd. Florigraphia Britannica. Vol. i, 1837; ii, 1845; iii, 1847; iv, 1848; and Ferns, 1848.

1849.

191. Buxton, Richard. A Botanical Guide to the Flowering Plants, Ferns, Mosses, and Algæ . . within eighteen miles of Manchester. Sm. 8vo. Lond. and Manr. 1849. A 2nd ed. in 1859. Contains some first West Riding records for Greenfield district. The Appendix, enumerating the Fresh-water Algæ, was by 'Mr.' (now Prof.) W. C. Williamson.

1851.

192. Leighton, Rev. W. A., B.A., F.L.S. The British Species of Angiocarpous Lichens. Lond.: Ray Society. 4to. (Vide p. 31 for the sole West Yorkshire Notice—that of *Thelotrema Lepadinum* in 'King's Wood, Roche Abbey! *Bohler's Lich. Brit.*')

1855.

193. Johnson, Charles. The Ferns of Great Britain. Illustrated by J. E. Sowerby. Roy. 8vo., Lond. (Part I bears on the cover, 1854; Part VIII, 1855.)

1859.

194. Grindon, Leo Hartley. The Manchester Flora. Lond. 8vo. W. White. 1859. (Contains several West Yorkshire records for Greenfield and Delph districts, they being 'within 18 miles of Manchester.')

1859-60.

195. Moore, Thomas, F.L.S. Nature Printed British Ferns. 2 vols., 8vo. Vol. i, 1859; Vol. ii, 1860.

1879.

196. Prior, R. C. A., M.D. On the Popular Names of British Plants. Third Edition. Lond. 1879.

1870.

197. Aveling, J. H., M.D. The History of Roche Abbey. Imp. 8vo. Worksop. 1870. Contains in an Addendum 'The Flora of Roche Abbey' (Contributed by J. Bohler?). Only received in time to be quoted in latter half of this work.

1882.

198. Lucas, Joseph. Studies in Nidderdale. Demy 8vo., pp. 292. Lond.: Elliot Stock. Contains 'The Modern Botany' at pp. 121-129, 243, and xxi, xxii.

1883.

199. Watson, H. C. (the late). Topographical Botany. Lond.: B. Quaritch. 1883. Demy 8vo., pp. xlviii, 612. Second Edition; corrected and enlarged by J. G. Baker and W. W. Newbould. (Contains some Newbouldian additions for West Yorkshire, previously unpublished, and the result of his own private investigation.)

1878-86.

200. Britten, James, F.L.S., and Holland, Robert. A Dictionary of English Plant Names. Three Parts. Demy 8vo. (English Dialect Society.) Part I, dated 1878. Part II, 1880. Part III, 1884; but combined Title-page says 1886. (Parts II and III received too late to be quoted.)

1885.

201. Handley, John. Sedbergh: A few Notes on its Flora. Sedbergh: Jackson and Son. 12mo.

202. W. W. [W. Whitwell]. A Bachelor's Christmas Day, 1883. 12mo. Pamphlet. Ashton-under-Lyne: Griffin and Sheard. Reprint from 'West Yorkshire Pioneer' of Dec. 24, 1885.

1886.

203. Anon. Contributions to a Fauna and Flora of West Yorkshire, Section B. Flora. Ranunculaceæ to Cruciferæ. Trans. Leeds Naturalists' Club, Vol. i, pp. 57-68. Leeds. 1886. Roy. 8vo. The numerous errors, duplicated specific names, localities outside limit, and misprints, render this work unquotable.

204. Anon [but W. B. Turner, by internal evidence]. West Yorkshire Algæ (exclusive of Diatomaceæ), pp. 69-78 of Trans. Leeds Naturalists' Club, vol. i, 1886. Clearly the result of knowledge and care.

205. Thompson, W., M.A., J.P. Florula Sedbergensis. Privately printed. 8vo, pp. 18, and 2 Addenda.

1887.

206. Braithwaite, R., M.D., F.L.S. The British Moss Flora. Vol. I. Acrocarpi (to Weberaceæ). Roy. 8vo. Lond.: L. Reeve and Co. 1887. (Was published in 10 parts—Part I, May 1880; Part X, June 1887).

207. Cash, James. The early botanical work of the late William Wilson. Naturalist, June 1887, and Reprinted.

208. Lee, P. Fox. The Flora of Dewsbury and Neighbourhood, in Trans. Yorkshire Naturalists' Union, Series E, pp. 225 to 248. Pub. 1887.

209. Watts, W. Marshall, D.Sc. A School Flora. Settle: Wildman. 8vo. 2nd Edition, enlarged and corrected. No date, but Preface dated April 1887.

Additional list of persons who have furnished information bearing on the Flora.

Backhouse, James, junr., York.
Edmondson, T. W., Skipton.
Foggitt, T. J. (the late), Thirsk.
Friend, Rev. Hilderic, Carlisle.
Gissing, A. F., Wakefield.
Handley, John, Sedbergh.
Hebden, Thomas, Hainworth.
Martindale, J. A., Staveley.
Stabler, George, Levens.
Thompson, Misses, Settle.
Thompson, Rev. W., M.A., Sedbergh.

INDEXES.

Compiled by W. Whitwell.

I. To Preliminary Chapters and Explanations.

II. To Families and Genera.

III. To Common and Rustic Names.

INDEX I.

Subjects in Preliminary Chapters and Sectional Explanations.

INDEX II.

FAMILIES AND GENERA.

Italics in names or page-figures indicate merely *synonymic* or *casual* mention. Names of Classes and Orders in small capitals.

INDEX III.

COMMON AND RUSTIC NAMES.